AF811957

Sustainability Sciences in Asia and Africa

Agroecosystem Dynamics and Sustainable Practices

Series Editors

Kamel A. Abd-Elsalam, Plant Pathology Research Institute
Agricultural Research Center
Cairo, Egypt

Olubukola O. Babalola, Faculty of Natural & Agricultural Sciences
North-West University
Mmabatho, South Africa

Editorial Board

Adel Abdel Azeem, President of Agricultural Research Center, Giza, Egypt
Salah Abdel-Momen, Former Minister of Agriculture, Giza, Egypt
Mohsen Abdel-Tawwab, Agriculture Research Center, Egypt
Aladdin Hamwieh, ICARDA Manager, Giza, Egypt
Hupenyu Allan Mupambwa, Head SANUMARC & Researcher, Sam Nujoma Campus, University of Namibia, Namibia
Heba I. Mohamed, Ain Shams University, Cairo, Egypt

Sustainability as a scientific endeavor requires broad understanding of the interconnections in our global environment. It is therefore positioned at the crossroads of research in natural sciences, technology and human sciences. This book series supports the global efforts by providing timely coverage of the progress, opportunities, and challenges towards achieving sustainable development goals (SDGs) in Asia and Africa. It fosters the research in transdisciplinary academic fields spanning across agriculture and global food security, health and diseases, access to clean water and affordable energy, climate resilience and effects of global climate change on biodiversity, water and land. This series is planned to narrate the stories of scientific development and progression towards SDGs from the regions of Asia and Africa. The overall aim of the series is to provide a comprehensive publication platform and act as a knowledge engine in the growth of sustainability sciences with a special focus on developing nations. The series constitute of four sub-series each focusing on a specific theme closely linked to a set of SDGs. Each sub-series is leaders and experts in agricultural sciences, environmental sciences, biomedicine and medical sciences respectively. Books in this series are useful for environmentalists, biotechnologists, plant scientists, medical practitioners, conservation biologists, policy makers, NGOs and scientific bodies working in the areas of sustainability sciences in the developing regions.

This book series publishes research associated with the prevailing opportunities and challenges faced by agroecosystems in Asia and Africa. It also explores the future scope of agroecological techniques and offers guidance on sustainable management of these systems. The series discusses important areas such as sustainable agricultural production, soil health, water resource management, biodiversity conservation, animal health and productivity, insect management strategies, genome editing and climate change mitigation. Further, it includes the challenges faced by African agriculture in arid and semi-arid regions, particularly in response to climate change. It highlights issues such as water shortages, soil degradation, desertification, and the need for sustainable farming techniques.

Sustainable agroecosystem maintenance entails strategies that improve overall environmental health and economic viability while limiting negative environmental impacts. The series takes a multidisciplinary approach to investigate creative ways for boosting sustainable agriculture, increasing food security, and improving lives for people in global south. This book series acts as a global publishing platform to collate and disseminate knowledge on fostering resilient and productive agroecosystems, thus enhancing accessibility, and fostering informed decision-making.

Arafat Abdel Hamed Abdel Latef
Editor

Oilseed Crops Under Abiotic Stress

Mitigation Strategies and Future Perspectives

Springer

Editor
Arafat Abdel Hamed Abdel Latef
Botany and Microbiology Department
South Valley University
Qena, Egypt

ISSN 2730-6771 ISSN 2730-678X (electronic)
Sustainability Sciences in Asia and Africa
ISSN 3059-247X ISSN 3059-2488 (electronic)
Agroecosystem Dynamics and Sustainable Practices
ISBN 978-981-96-8345-1 ISBN 978-981-96-8346-8 (eBook)
https://doi.org/10.1007/978-981-96-8346-8

© The Editor(s) (if applicable) and The Author(s), under exclusive license to Springer Nature Singapore Pte Ltd. 2025

This work is subject to copyright. All rights are solely and exclusively licensed by the Publisher, whether the whole or part of the material is concerned, specifically the rights of translation, reprinting, reuse of illustrations, recitation, broadcasting, reproduction on microfilms or in any other physical way, and transmission or information storage and retrieval, electronic adaptation, computer software, or by similar or dissimilar methodology now known or hereafter developed.
The use of general descriptive names, registered names, trademarks, service marks, etc. in this publication does not imply, even in the absence of a specific statement, that such names are exempt from the relevant protective laws and regulations and therefore free for general use.
The publisher, the authors and the editors are safe to assume that the advice and information in this book are believed to be true and accurate at the date of publication. Neither the publisher nor the authors or the editors give a warranty, expressed or implied, with respect to the material contained herein or for any errors or omissions that may have been made. The publisher remains neutral with regard to jurisdictional claims in published maps and institutional affiliations.

This Springer imprint is published by the registered company Springer Nature Singapore Pte Ltd.
The registered company address is: 152 Beach Road, #21-01/04 Gateway East, Singapore 189721, Singapore

If disposing of this product, please recycle the paper.

Preface

Oilseed crops play a crucial role in global agriculture, providing essential sources of edible oils, industrial materials, and bioenergy. However, like many other crop species, oilseeds are increasingly subjected to various abiotic stressors. These environmental challenges can severely limit growth, productivity, and quality, threatening food security and agricultural sustainability on a global scale.

Significant advancements have been made in recent years in the understanding and management of abiotic stress in crops. Among the various mitigation strategies, the use of biostimulants and phytohormones, the adoption of organic fertilizers, the exploration of advanced omics technologies, and the application of nanotechnology have emerged as promising methods to boost stress tolerance in oilseed crops. Despite these developments, there is still a critical need to consolidate this growing body of knowledge and provide clear, practical insights for researchers, agronomists, and policymakers.

The book, *Oilseed Crops Under Abiotic Stress: Mitigation Strategies and Future Perspectives*, aims to bridge this gap. It offers a comprehensive examination of innovative strategies for improving the resilience of oilseed crops under a variety of abiotic stresses. The chapters systematically discuss the role of biostimulants in alleviating diverse abiotic stresses, including salinity, drought, waterlogging, flooding, temperature extremes, nutrient deficiencies, heavy metal toxicity, and UV radiation. Furthermore, the book explores the interplay between phytohormones and stress responses, shedding light on hormonal regulation under various adverse conditions. Additional sections address the integration of organic fertilizers, advanced omics techniques, and nanotechnology in the development of stress-tolerant oilseed varieties. These chapters not only emphasize the complexity of abiotic stress interactions with oilseed crops but also highlight integrative, multidisciplinary strategies that could shape the future of sustainable oilseed production.

This book is a valuable resource for researchers, agronomists, plant biologists, and policymakers seeking sustainable solutions to improve oilseed crop productivity in challenging environmental conditions. By combining fundamental knowledge with emerging technologies, it aims to contribute to the development of climate-resilient agricultural practices, securing the future of oilseed production.

We wish to express our sincere gratitude to all the contributing authors for their expertise and dedication in shaping this volume. We also acknowledge the efforts of publishers and editors in bringing this compilation to fruition. We hope that this book will inspire further research and innovation in the field of abiotic stress mitigation, fostering a more sustainable and productive agricultural future.

Qena, Egypt Arafat Abdel Hamed Abdel Latef

Contents

Editor and Contributors

About the Editor

Arafat Abdel Hamed Abdel Latef is a Professor at the Department of Botany and Microbiology, Faculty of Science, South Valley University, Egypt. He has published more than 90 research articles in peer-reviewed journals and 27 book chapters in international books. He served as editor of six published international books. He serves as an editorial member of several reputed journals. Furthermore, he is an expert reviewer for over 150 international journals. He was one of the top 2% of scientists in the world, according to Stanford University, in 2020, 2021, 2022, 2023, and 2024. He has received several awards, such as Springer Nature Editor of Distinction Award 2025, the Egypt State Award of Excellence in Agricultural Sciences 2021, the South Valley University Award of Scientific Excellence 2019, the Top Peer Reviewer Award 2019 Powered by Web of Science and Publons, and the South Valley University Award of Scientific Publication for the years from 2010 to 2024.

Contributors

Arafat Abdel Hamed Abdel Latef Botany and Microbiology Department, Faculty of Science, South Valley University, Qena, Egypt

Muhammad Adnan Iqbal Ahmad Qarshi Laboratory, Qarshi Industries (Pvt.), Ltd, Hattar Haripur, Pakistan

Affan Afzal Department of Microbiology, Faculty of Veterinary and Animal Sciences, The Islamia University of Bahawalpur, Bahawalpur, Pakistan

Nabeel Ahmad School of Allied Sciences Dev Bhoomi Uttarakhand University, Dehradun, Uttarakhand, India

Nayab Ahmad Institute of Biotechnology and Genetic Engineering (IBGE), The University of Agriculture, Peshawar, Pakistan

Quaid Ahmad Biology Laboratory, University Public School, University of Peshawar, Peshawar, Pakistan
Department of Botany, University of Peshawar, Peshawar, Pakistan

Sidra Ahmad Institute of Biotechnology and Genetic Engineering (IBGE), The University of Agriculture, Peshawar, Pakistan

Rasheed Akbar School of Environment and Safety Engineering, Jiangsu University, Zhenjiang, China
Department of Entomology, Faculty of Physical and Applied Sciences, The University of Haripur, Haripur, Khyber Pakhtunkhwa, Pakistan

Sk Naim Aktar Department of Agronomy, Palli Siksha Bhavana (Institute of Agriculture), Birbhum, West Bengal, India

Amjad Ali Department of Sustainable Crop Production, Universita Cattolica del Sacro Cuore, Parmense, Italy

Baber Ali Department of Plant Sciences, Quaid-i-Azam University, Islamabad, Pakistan

Hailah M. Almohaimeed Department of Basic Science, College of Medicine, Princess Nourah bint Abdulrahman, University, Riyadh, Saudi Arabia

Taghreed S. Alnusaire Department of Biology, College of Science, Jouf University, Sakaka, Saudi Arabia

Ansa Institute of Chemistry, University of Sargodha, Sargodha, Pakistan

Saeed Ahmad Asad Department of Biosciences, COMSATS University Islamabad, Islamabad, Pakistan

Nasir Assad Institute of Chemistry, University of Sargodha, Sargodha, Pakistan

Rachid Azenzem Regional Center of Agricultural Research of Tanger, National Institute of Agricultural Research, Rabat, Morocco

Ali Bandehagh School of Molecular Sciences, Australian Plant Phenomics Network, ARC Training Centre in Predictive Breeding for Agricultural Futures and The UWA Institute of Agriculture, The University of Western Australia, Boorloo, WA, Australia

Soham Barik Department of Agronomy, Maharana Pratap University of Agriculture and Technology, Udaipur, Rajasthan, India

Maryam Bibi Department of Botany, University of Malakand, Chakdara, Lower Dir, Khyber Pakhtunkhwa, Pakistan

Soufiane Chakkour Faculty of Sciences, Ibn Tofail University, Kenitra, Morocco

Basma Chokrane Regional Center of Agricultural Research of Marrakech, National Institute of Agricultural Research, Rabat, Morocco

Anas Chraka Laboratory of Engineering Materials and Sustainable Energy, Faculty of Science, Abdelmalek Essaadi University, Tetouan, Morocco

Soumya Shephalika Dash Department of Agricultural Entomology, Palli Siksha Bhavana (Institute of Agriculture), Birbhum, West Bengal, India

Zahra Dehghanian Department of Biotechnology, Faculty of Agriculture, Azarbaijan Shahid Madani University, Tabriz, Iran

Hülya Doğan Yozgat Bozok University, Hemp Research Institute, Department of Agriculture and Food, Yozgat, Türkiye

Kaoutar El Issaoui Laboratory of Biology and Health, Department of Biology, Faculty of Sciences, Abdelmalek Essaadi University, Tetouan, Morocco

Sezai Ercisli Department of Horticulture, Faculty of Agriculture, Ataturk University, Erzurum, Türkiye

HGF Agro, Ata Teknokent, Erzurum, Türkiye

Hafsa Department of Chemistry, Bacha Khan University, Charsadda, Pakistan

Khizar Hayat Department of Botany, Islamia College Peshawar, Peshawar, Pakistan

Department of Biosciences, COMSATS University Islamabad, Islamabad, Pakistan

Sagun Hebmram Department of Agronomy, Palli Siksha Bhavana (Institute of Agriculture), Birbhum, West Bengal, India

Akbar Hossain Division of Soil Science, Bangladesh Wheat and Maize Research Institute, Dinajpur, Bangladesh

Babar Iqbal School of Environment and Safety Engineering, Jiangsu University, Zhenjiang, China

Majid Iqbal Institute of Geographic Sciences and Natural Resources Research, University of Chinese Academy of Sciences, Beijing, China

Rashid Iqbal Department of Agronomy, Faculty of Agriculture and Environment, The Islamia University of Bahawalpur, Bahawalpur, Pakistan

Department of Life Sciences, Western Caspian University, Baku, Azerbaijan

Sidra Iqbal Department of Plant Sciences, Quaid-i-Azam University Islamabad, Islamabad, Pakistan

Mazharul Islam Department of Agronomy, Kanas State University, Manhattan, KS, USA

Khalil Kadaoui Bio-Agrodiversity Team, Biology, Ecology, and Health Laboratory, Sciences Faculty, Abdelmalek Essaâdi University, Tétouan, Morocco

Kanwal Department of Plant Sciences, Quaid-i-Azam University, Islamabad, Pakistan

Alevcan Kaplan Department of Crop and Animal Production, Sason Vocational School, Batman University, Batman, Turkey

Karishma Department of Plant Biotechnology, Atta-Ur-Rahman School of Applied Biosciences (ASAB), National University of Sciences and Technology (NUST), Islamabad, Pakistan

Department of Agricultural Sciences and Technology, Atta-ur-Rahman School of Applied Biosciences (ASAB), National University of Sciences and Technology (NUST), Islamabad, Pakistan

Jalal Kassout Regional Center of Agricultural Research of Marrakech, National Institute of Agricultural Research, Rabat, Morocco

Amir Abdullah Khan School of Environment and Safety Engineering, Jiangsu University, Zhenjiang, China

Muhammad Nauman Khan Department of Botany, Islamia College Peshawar, Peshawar, Pakistan

University Public School, University of Peshawar, Peshawar, Pakistan

Biology Laboratory, University Public School, University of Peshawar, Peshawar, Pakistan

Santanu Kundu Division of Agronomy, ICAR-Indian Agricultural Research Institute, New Delhi, Delhi, India

Marzia Batool Laila Institute of Chemistry, University of Sargodha, Sargodha, Pakistan

Tabarak Malik Department of Biomedical Sciences, Institute of Health, Jimma University, Jimma, Ethiopia

Mouaad Amine Mazri Agro-Biotechnology Research Unit, Regional Center of Agricultural Research of Marrakech, National Institute of Agricultural Research, Rabat, Morocco

Muhammad Naeem-Ul-Hassan Institute of Chemistry, University of Sargodha, Sargodha, Pakistan

Sara Oulbi Regional Center of Agricultural Research of Marrakech, National Institute of Agricultural Research, Rabat, Morocco

Fethi Ahmet Ozdemir Department of Molecular Biology and Genetics, Faculty of Science and Art, Bingol University, Bingol, Turkey

Siddharth Shankar Patre Department of Agronomy, Palli Siksha Bhavana (Institute of Agriculture), Birbhum, West Bengal, India

Kalipada Pramanik Department of Agronomy, Palli Siksha Bhavana (Institute of Agriculture), Birbhum, West Bengal, India

Abdul Razzaq Qarshi Herb Research Center, Qarshi Industries (Pvt.) Ltd, Hattar, Haripur, Pakistan

Qarshi University, Lahore, Punjab, Pakistan

Abdelouahab Sahli Bio-Agrodiversity Team, Biology, Ecology, and Health Laboratory, Sciences Faculty, Abdelmalek Essaâdi University, Tétouan, Morocco

Rokayya Sami Department of Food Science and Nutrition, College of Sciences, Taif University, Taif, Saudi Arabia

Seyedeh-Somayyeh Shafiei-Masouleh Department of Genetics and Breeding, Ornamental Plants Research Center (OPRC), Horticultural Sciences Research Institute (HSRI), Agricultural Research, Education and Extension Organization (AREEO), Mahallat, Iran

Maksud Hasan Shah Department of Agronomy, Bidhan Chandra Krishi Viswavidyalaya, Nadia, West Bengal, India

Mona H. Soliman Botany and Microbiology Department, Faculty of Science, Cairo University, Giza, Egypt

Biology Department, Faculty of Science, Taibah University, Yanbu, Kingdom of Saudi Arabia

Nicolas L. Taylor School of Molecular Sciences, Australian Plant Phenomics Network, ARC Training Centre in Predictive Breeding for Agricultural Futures and The UWA Institute of Agriculture, The University of Western Australia, Boorloo, WA, Australia

Barkat Ullah Department of Botany, Islamia College Peshawar, Peshawar, Pakistan

Sher Ullah Institute of Chemistry, University of Sargodha, Sargodha, Pakistan

Sana Wahab Department of Plant Sciences, Quaid-i-Azam University, Islamabad, Pakistan

Nazima Wahid Department of Plant Sciences, Quaid-i-Azam University, Islamabad, Pakistan

Tabassum Yaseen Department of Botany, Bacha Khan University, Charsadda, Khyber Pakhtunkhwa, Pakistan

Aurang Zeb MOE Key Laboratory of Pollution Processes and Environmental Criteria, College of Environmental Science and Engineering, Nankai University, Tianjin, People's Republic of China

Chapter 1
Biostimulants as a Sustainable Strategy to Augment Salinity Tolerance in Oilseed Crops

Nazima Wahid, Sana Wahab, Sidra Iqbal, Kanwal, Muhammad Nauman Khan, Khizar Hayat, Karishma, Alevcan Kaplan, and Amjad Ali

Abstract Salinity stress is a significant impediment to the growth and production of oilseed crops worldwide, affecting plant physiology, nutrient uptake, and yield. As the demand for oilseed products increases, there is an urgent need for long-term solutions to improve crop resilience to salinity. Biostimulants, which include plant extracts, microbial inoculants, and organic compounds, have emerged as an effective and environmentally acceptable method to mitigate the adverse effects of salt stress. These chemicals enhance plant tolerance by promoting root development, improving nutrient absorption, regulating osmotic balance, and facilitating stress-related metabolic processes. This chapter investigates our current understanding of how biostimulants promote salinity tolerance in oilseed crops. It describes the methods by which biostimulants work, such as increased antioxidant activity, hormonal balance, and soil health. Furthermore, the possible integration of biostimulants

N. Wahid (✉) · S. Wahab · S. Iqbal · Kanwal
Department of Plant Sciences, Quaid-i-Azam University Islamabad, Islamabad, Pakistan

M. N. Khan
Department of Botany, Islamia College Peshawar, Peshawar, Pakistan

University Public School, University of Peshawar, Peshawar, Pakistan

K. Hayat
Department of Botany, Islamia College Peshawar, Peshawar, Pakistan

Department of Biosciences, COMSATS University Islamabad, Islamabad, Pakistan

Karishma
Department of Plant Biotechnology, Atta-Ur-Rahman School of Applied Biosciences (ASAB), National University of Sciences and Technology (NUST), Islamabad, Pakistan

A. Kaplan
Department of Crop and Animal Production, Sason Vocational School, Batman University, Batman, Turkey

A. Ali
Department of Sustainable Crop Production, Universita Cattolica del Sacro Cuore, Parmense, Italy

© The Author(s), under exclusive license to Springer Nature Singapore Pte Ltd. 2025
A. A. H. Abdel Latef (ed.), *Oilseed Crops Under Abiotic Stress*, Sustainability Sciences in Asia and Africa, https://doi.org/10.1007/978-981-96-8346-8_1

with other sustainable agricultural techniques is examined, emphasizing the importance of large-scale field trials to validate laboratory results. Future research should concentrate on optimizing biostimulant compositions, understanding their molecular mechanisms, and ensuring cost-effective manufacture to stimulate wider adoption. Finally, the application of biostimulants is a potential, long-term strategy for improving the growth and productivity of oilseed crops under saline circumstances. Advancing research and development in this area can help to achieve more resilient and climate-smart agriculture.

Keywords Biostimulants · Mitigation strategy · Salinity stress · Oilseed crops

1.1 Introduction

Increased crop productivity and production are required to meet the growing population's food demands. Oilseed crops are among the most significant crops known to man because of their significance as sources of edible oil for human use as well as nonedible oil for biofuels. With a growth rate of 4.1% annually over the last three decades, the oilseeds sector of global agriculture has been one of the most dynamic. It has surpassed the livestock product sector (Hongal et al. 2023). In the age of climate change, plants must withstand progressively harsher environments to survive. According to studies, giving plants biostimulants can increase their ability to withstand and survive abiotic challenges like salt. Concerns regarding whether or not the Earth can sustain everyone are raised by the fact that climate change is increasing the strain on the resources that provide food for humans and quickening the rate at which biotic and abiotic stresses affect crops. Abiotic stresses affect different plant species and often lead to crop quality degradation, which limits plant growth and survival. A few instances of these stressors are a lack of nutrients, high or low temperatures, drought, too much salt, and hypoxia or anoxia (Mariani and Ferrante 2017).

A broad variety of inputs can be categorized as biostimulants, as evidenced by the continual change in both the terms and the idea of plant biostimulants. Two projects by biostimulant business consortia, one in North America and the other in Europe, showcase the scope of the concept of biostimulants. Plant biostimulants are defined as follows by the European Biostimulant Committee (EBIC). Applying chemicals to plants or the rhizosphere to promote natural processes that enhance or benefit crop quality, abiotic stress tolerance, nutrient absorption, and nutrient efficiency is known as plant biostimulating. Therefore, by highlighting advantages associated with enhanced plant growth, stress tolerance, and quality, this definition differentiates among biological intervention and acquired defense against diseases and plant biostimulants. A number of demonstrated ways, including but not limited to; improving the efficiency of the plants metabolism to induce yield increases and enhanced crop quality; increasing plant tolerance to and recovery from abiotic stresses; facilitating nutrient assimilation, translocation, and use; enhancing quality attributes of produce, including sugar content, color, fruit seeding, etc.; rendering water use more efficient; enhancing certain physicochemical properties of the soil and fostering the development of complementary soil micro-organisms are only some of the ways that biostimulants encourage plant development and growth from seed germination to plant maturity in the agricultural life cycle (Calvo et al. 2014). Lastly, goods that contain some nutrients are included in the EBIC concept of biostimulants as long as they don't directly fertilize plants. "Biostimulants operate through different mechanisms than fertilizers, regardless of the presence of nutrients in the products" (Fan et al. 2013).

1.1.1 Overview of Salinity Stress in Agriculture

Salinity is one of the primary abiotic elements influencing agricultural productivity worldwide, particularly in arid and semiarid regions. Salinity affects more than 1000 million hectares of land globally, which drastically limits the total quantity of land that can be utilized for sustainable agriculture (Wicke et al. 2011; Dawood et al. 2024). It is anticipated that when drought and sea level rise are made worse by climate change, salinity stress in agriculture will become an increasingly pressing issue when the root zone of a soil has enough salt to prevent agricultural plants from growing, it is said to be saline (Ponnamperuma 1984). However, it is challenging to define saline soils precisely because salt damage varies depending on species, variety, development stage, ambient circumstances, and salt nature (Yadav et al. 2011). Mostly created by natural processes, salt-affected soils (SAS) are salty and sodic soils; nevertheless, over the past few decades, rapid human-induced secondary salinity has made large portions of the world's irrigated and dryland regions uncultivable (Munns et al. 2005).

The primary cause of rising water tables and salinity development in dryland areas is the removal of natural grasses and trees to make way for annual crop cultivation. In irrigated areas, salinization is caused by seepage from earthen canals, overwatering, clogged drainage systems, and irrigation with waters of questionable quality that are saline and sodic (Ambast 2006). Ion poisoning, nutritional imbalances, and limited plant water intake due to high soil salt concentrations all contribute to decreased growth and yield. Physical–chemical characteristics of acidic and saline soils vary. In salty soils, excess soluble salts, mainly in the form of sodium, calcium, and magnesium ion chlorides and sulfates, result in higher soil saturation extract electrical conductivity (ECe > 4 dS m^{-1}) (Sharma and Singh 2016). Salinity is the term used to describe high quantities of soluble salts, primarily NaCl, in soil or water. This disease, which affects more than 800 million hectares of land globally, is harmful to plant development and agricultural productivity. The issue is becoming worse due to a lack of water and a greater dependence on poor irrigation water.

(i) *Causes of Salinity Stress*

(a) *Natural Causes*: In areas where rainfall is inadequate to remove salts from the soil, high concentrations of salts can naturally build up in the soil. Soil salinity is also influenced by rock weathering.
(b) *Irrigation Practices*: Increased soil salinity is frequently the result of improper irrigation management, particularly in regions that use salt water. Long-term salt buildup in the root zone is caused by inadequate drainage.
(c) *Sea Level Rise*: The soil's salinity may increase due to salty water intrusion, particularly in coastal regions.

1.1.2 Importance of Oilseed Crops

Europe has a special opportunity for both traditional and novel oilseeds due to the growing demand for renewable feedstock to replace products made from petroleum. Because of their many applications and economic importance, oilseed crops are very important on a global scale (Zanetti et al. 2013). With the replacement of petroleum products with new, renewable industrial products, interest in the nonfood applications of oils derived from plants has grown over the past 20 years, opening up new markets. On the international market, *Brassica napus* and *Glycine max* are the two most popular oilseeds. While *Helianthus annuus* and other minor crops *Camelina sativa* and *Linum usitatissimum* lag far behind, *Brassica napus* is the main oilseed crop in Europe (Wittkop et al. 2009).

The industry can use oilseeds' innate ability to synthesize extremely complex molecular structures to replace a sizable portion of compounds derived from petroleum oil (Carlsson 2009). Plant-derived oils can have a significant impact when used as feedstock for various industries instead of petroleum. The highlights are enumerated as follows:

1. *Source of Edible Oils*

Edible oils used in baking, food processing, and cooking are mostly derived from oilseed crops like soybean, sunflower, rapeseed, and palm. All over the world, these oils are an essential part of many diets. The most significant seed legume in the world, soybean (*Glycine max*) accounts for 25% of all edible oil produced worldwide. Humans eat the seeds directly, and they may also be utilized as a source of oil and protein (Weiss 1950). Furthermore, in many parts of the world, plant leftovers are crucial as cattle feed. Groundnuts are interplanted in Asia and Africa between mature coconut trees or, in certain places, with maize, sorghum, and soybeans (Stalker et al. 1988). In 2010, 9.21 million hectares of soybeans were cultivated commercially in India. In a similar vein, in 2010, productivity and production levels rose to 1200 kg/ha and 10.4 million tons, respectively. India produces just 3% of the world's soybeans. In India, soybeans rank third in terms of oilseed crops, behind only mustard and groundnuts. According to Directorate of Oilseeds Development

(DOD), Government of India (GOI), Hyderabad (2014), soybean production fluctuated between 99.054 lakh tons in 2008–2009 and 127.364 lakh tons in 2010–2011 (Rai et al. 2016).

2. *Industrial Applications*

Oilseeds are essential for the production of nonfood products like paints, lubricants, soaps, cosmetics, and biodiesel. Oilseeds are a sustainable resource for bio-based products that are heavily relied upon by these industries.

3. *Nutritional Value*

Proteins, vitamins, and essential fatty acids are abundant in oilseeds. They are essential for human nutrition, help maintain cardiovascular health, and provide a source of protein for vegetarian and vegan diets. They also contribute to a balanced diet.

4. *Economic Importance*

Oilseeds are an important agricultural product that is vital to the economies of many nations. Farmers and nations alike earn a significant amount of money from the export and trade of oilseeds and the products they produce.

5. *Livestock Feed*

The livestock business and food production are supported by the by-products of oilseed processing, such as soybean meal, which is a vital source of protein for animal feed.

6. *Biodiesel Production*

Oilseeds, such as soybean and rapeseed, are important sources of raw materials for biodiesel, an environment-friendly fossil fuel substitute that lowers carbon emissions and supports renewable energy initiatives.

7. *Soil Health and Crop Rotation*

Oilseed crops are often utilized in crop rotation systems to improve soil health. Legumes, such as soybeans and peanuts, in particular, improve soil fertility through nitrogen fixation.

1.1.3 Biostimulants Function in Agriculture

The two main challenges facing the agricultural industry in recent years have been improving yields to feed the world's growing population and increasing the efficiency of resources, all while minimizing the sector's adverse impact on ecosystems and public health. Since fertilizers and pesticides are efficient instruments that assist farmers in maintaining a steady production across the growing seasons under both favorable and unfavorable circumstances, they are truly indispensable to agriculture. Over the past three decades, several technical innovations have been proposed to reduce the usage of synthetic agrochemicals such as pesticides and fertilizers, thereby improving the sustainability of agricultural production systems. Because they promote nutrient uptake, stress resilience, and plant growth, biostimulants are essential to modern agriculture. Biostimulants function by promoting the natural processes that occur in plants and soil, as opposed to fertilizers, which supply nutrients to plants directly. They consist of both synthetic and natural components, including seaweed extracts, plant extracts, humic compounds, and advantageous microbes. Natural plant biostimulants (PBs) are a potentially useful and environmentally beneficial innovation that enhance flowering in the establishment of plants, fruit development, crop output, nutrient utilization efficiency (NUE), and tolerance against a range of abiotic challenges (Colla et al. 2014). A plant biostimulant is an European Union (EU) fertilizing product whose purpose is to promote plant nutrition processes, regardless of the product's nutrient content, with the exclusive intent of enhancing one or more of the following qualities of the plant or the plant rhizosphere:

(i) Efficiency of nutrient use
(ii) Resilience to abiotic stress
(iii) Quality traits, or
(iv) Availability of restricted nutrients in the soil or rhizosphere (Rouphael and Colla 2020).

1.1.3.1 Key Roles of Biostimulants in Agriculture

1. *Improved Nutrient Uptake*: Plants that get biostimulants are better able to take in and utilize nutrients from the soil. This encourages improved plant development and lessens the need for excessive fertilizers.
2. *Stress Tolerance*: Cold, heat, drought and salinity are examples of abiotic stimuli that plants that get biostimulants are better equipped to withstand. They cause plants to activate defense mechanisms, which increases their resistance to adverse environmental conditions.
3. *Better Root Development*: A variety of biostimulants, particularly those that include microorganisms like mycorrhizae, encourage improved root development, which makes it easier for plants to utilize nutrients and water.
4. *Soil Health*: Biostimulants have the potential to cultivate a more wholesome and durable soil environment by augmenting microbial activity and improving soil structure. Long-term sustainability and increased soil fertility result from this.
5. *Yield and Quality Improvement*: Biostimulants have the potential to increase nutrient absorption and promote healthier plant development, leading to improved crop quality with reference to nutritional content, flavor and size alongside enhanced crop yield.
6. *Sustainability*: As the global agricultural industry shifts toward more sustainable practices, biostimulants assist in reducing the dependency on chemical inputs like pesticides and synthetic fertilizers, contributing to more environment-friendly farming practices (Fig. 1.1).

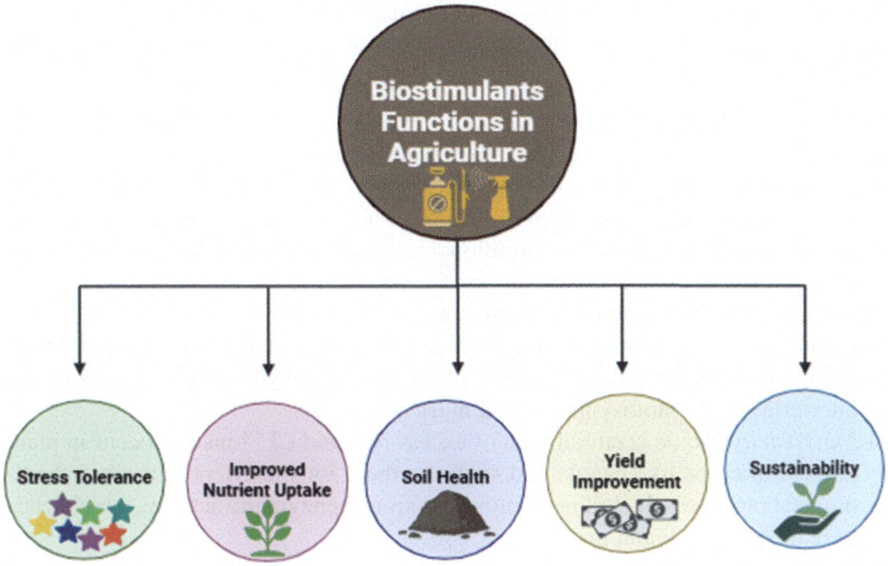

Fig. 1.1 Role of biostimulants and their application in agriculture

1.2 Salinity Stress and Its Impact on Oilseed Crops

Elevated soil salinity negatively impacts plant growth, development, and yield. We call this condition salinity stress. Abiotic stressors like heat, salt, drought, and cold seriously harm basic food crops. According to Food and Agriculture Organization (FAO) estimates, about 6% of the land on Earth is affected by salinity. Therefore, it would appear that plant and agricultural productivity are seriously threatened by salinity stress (Parihar et al. 2015). Twenty percent of irrigated land is affected by soil salinity, a serious worldwide issue that drastically lowers agricultural productivity (Qadir et al. 2014). The productivity and quality of oilseed crops, including mustard, soybean, canola, and sunflower, can be significantly impacted by salt stress. Around the world, oil seeds, pulses, and grains are important food crops that provide wholesome nutrients that are necessary for human health. However, salinity stress severely reduces their yield and quality, putting traditional agriculture in jeopardy by hindering its expansion and advancement. The severity of the problem is highlighted by the estimated $27.3 billion in yearly economic harm caused by saline soil on irrigated land. Significant alterations in physiological, molecular, and biochemical processes are brought about by salinity stress. Although diverse levels of adaptive mechanisms have been identified, a thorough comprehension of food crops' salinity tolerance still eludes us. The main human activities that cause soil salinity are deforestation and improper irrigation techniques. The socioeconomic consequences encompass decreased crop yields, lowered profit margins, joblessness, and a decrease in land value because of infertile soil. It is noteworthy that there isn't a review that focuses on mechanisms and responses to salinity tolerance in important food crops like oilseeds, barley, rye, oats, and pulses (Verma et al. 2024).

(a) *Impaired Nutrient Uptake*: The equilibrium of essential elements in the soil is disrupted by salt stress because it competes with nutrients, including magnesium (Mg), potassium (K), and calcium (Ca). This has an impact on how plants absorb nutrients, which results in a nutritional shortage and a decline in plant vigor. Osmotic stress, which is brought on by high soil salt concentrations, restricts how much water seedlings may take up.
(b) *Osmotic Stress*: Plants have a harder time absorbing water when the soil has high salinity content. This causes symptoms similar to dehydration in oilseed crops, interfering with photosynthesis and nutrient transfer.
(c) *Ion Toxicity*: Toxic accumulation of excess Na^+ and Cl^- ions can occur in plant tissues. In crops like canola and soybean, these ions interfere with vital physiological processes, including photosynthesis and enzymatic activity, which leads to reduced growth and production.
(d) *Oxidative Stress*: When plants are stressed by salt, they release reactive oxygen species (ROS). These reactive chemicals can damage proteins, lipids, and DNA, which further inhibits development. Oxidative damage is a common response of stressed oilseed crops to salt, which reduces the quality of the seeds.
(e) *Altered Metabolism and Lipid Composition*: Lipid metabolism is impacted by salinity stress, which is important for oilseed crops. It modifies the oil's fatty acid composition, which lowers the oil's quality and output from crops like sunflower and canola.
(f) *Reduced Flowering and Pod/Seed Formation*: Oilseed crops' reproductive organs frequently exhibit abnormalities as a result of salinity stress, which reduces seed set, reduces pod count, and impairs flowering. This has a direct effect on yield, especially for crops like soybean and mustard.
(g) *Reduced Germination and Growth of Seedlings*: Seedlings' ability to absorb water is limited by osmotic stress, which is caused by elevated levels of salt in the soil. Oilseed crops experience reduced seedling growth due to low germination rates, such as those found in sunflower and mustard (Fig. 1.2).

1.2.1 Mechanisms of Salinity Stress

Globally, there is a problem with salt stress, and in the future, more arable land will unavoidably be damaged by salinity, especially due to climate change. Since salt stress is a complex stress that plants undergo, building a plant's resistance to it is difficult. Efforts to boost plants' resistance to salt stress have hitherto mostly focused on redox regulation, enhanced antioxidant defense mechanisms, Na^+ exclusion or compartmentalization, and other tactics. Additionally, a plethora of studies have

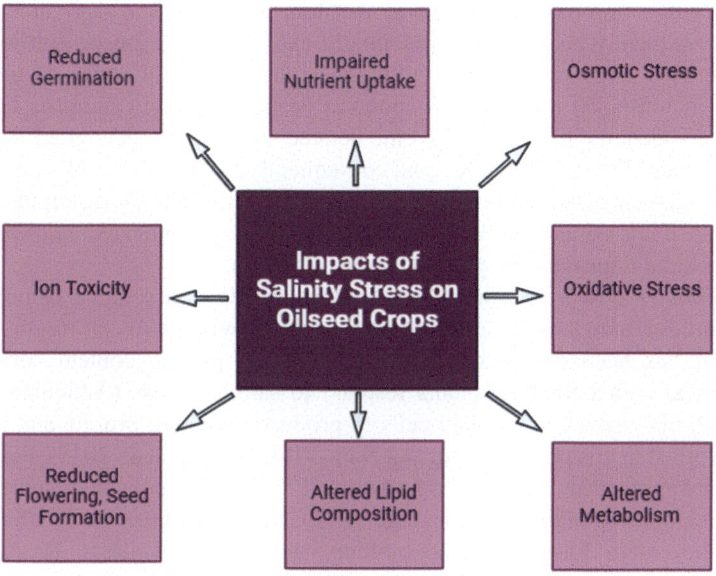

Fig. 1.2 Salinity stress impacts on oilseed crops

demonstrated the advantageous effects of various substances, including nutrients, hormones, amino acids, and organic osmolytes; nonetheless, our knowledge of plants' ability to tolerate salinity stress is continually growing (Khan et al. 2020).

1. *Osmotic Adjustment Substances Accumulation*

When exposed to salt stress, plants frequently react by storing organic solutes that are compatible with their environment, including soluble glycine betaine, sugars, as well as proline. Without affecting the cell metabolism, these soluble, low-molecular-weight substances protect the membrane structure. Additionally, they serve as osmoprotectants, supporting the detoxification of ROS and intracellular osmotic correction (Sharma et al. 2019). Plants that undergo stress caused by salt and dry out release large amounts of the quaternary ammonium compound glycine betaine (Slama et al. 2015; Zhu et al. 2022). The chloroplast produces glycine betaine, which the thylakoid membrane accumulates to maintain the effectiveness of photosynthetic processes (Alasvandyari and Mahdavi 2018). In response to salinity stress, many plants synthesize glycine betaine to lessen the negative effects of saline surroundings and maintain the cell's osmotic state. By employing transgenic techniques to alter its metabolic pathways, glycine betaine has been comprehensively investigated. The halophyte plant *Suaeda liaotungensis*'s betaine aldehyde dehydrogenase gene, for example, is overexpressed in transgenic tobacco plants. It has been demonstrated that this gene creates an enzyme with a markedly increased tolerance to salt, which is responsible for converting betaine aldehyde to betaine (Li et al. 2003). When the *Arthrobacter globiformis* choline oxidase gene was inserted into Indica rice, it was able to withstand salt stress at concentrations as high as 150 mM. Choline oxidase catalyzes a two-step oxidation pathway that transforms choline to glycine betaine, which subsequently accumulates, potentially explaining the greater salt tolerance. Other methods were employed in addition to the successful usage of transgenic plants to increase salt tolerance. In common beans (*Phaseolus vulgaris* L.), as an example, exogenous injection of glycine betaine significantly decreased the uptake of Na^+, increased the uptake of K^+, and consequently maintained an elevated ratio of K^+/Na^+, enhancing the common beans' tolerance to salt. A different team of researchers demonstrated how exogenously supplying *Dalbergia odorifera* with glycine betaine improved the plant (Cisse et al. 2021).

Arthrobacter globiformis introduced choline oxidase, a suitable osmolyte and antioxidant that increases under salt stress and helps with plant cell turgor, into the Proline. It has been discovered that the increased proline content serves as a physiological signal for how plants respond to salinity stress (Abdelhamid et al. 2013). Salinity stress causes plant cells to produce and store proline and activates genes involved in this process (Kim and Nam 2013; Nguyen et al. 2013). Pyrroline-5 carboxylate synthetase, more often known as P5CS, is the enzyme responsible for starting the first step of proline biosynthesis. P5CS1 is one of the two P5CS isoforms that are necessary for the proline buildup caused by salt stress. Research has indicated that the absence of P5CS1 enhances *Arabidopsis thaliana*'s susceptibility to salt stress. Proline administered exogenously dramatically raised (Jannin et al. 2012). By generating more dehydrin proteins, which protect against stress, and by preserving the activity of antioxidant enzymes when exposed to salinity, the desert plant *Pancratium maritimum* L. boosts its resistance to salt (Khedr et al. 2003).

The three primary soluble sugar types that affect a plant's ability to withstand salt stress are glucose, sucrose, and trehalose. Under salt stress, soluble enzymes are shielded from the toxicity of elevated intracellular inorganic ion concentrations by a larger accumulation of soluble carbohydrates. For instance, molecules may be shielded against osmotic damage by trehalose's unique capacity to absorb water reversibly (Delorge et al. 2014). When transgenic Arabidopsis plants overexpressing the salt-related wheat gene TaSST were exposed to NaCl, their levels of soluble sugar were higher than those of wild-type plants. Soluble sugars do not hinder polyols since they are compatible solutes and ROS scavengers. Mannitol is an acyclic polyol that is essential for osmotic control and enhances salt tolerance in higher plants (Pujni et al. 2007). When Indica rice was genetically modified to produce the E. coli mannitol-1-phosphate dehydrogenase gene (mt/D), a gene implicated in mannitol production, transgenic rice plants demonstrated a link between improved salt tolerance and mannitol accumulation (Pujni et al. 2007). Therefore, developing crops that can withstand salt may be aided by understanding the pertinent mechanisms of osmolytes.

2. *The Balance of Ions and Segmentation*

Salt stress results in an ionic imbalance that impacts various facets of plant development as well as growth. For example, reduced absorption of mineral nutrients under salt stress causes disruptions to root architecture, cellular metabolism, and photosynthesis. Plants have developed defense mechanisms against salinity stress, though, by managing ion influx and compartmentalization to preserve intracellular ion homeostasis. The term "ion homeostasis" describes a fundamental dynamic process in plants that involves taking up necessary ions and eliminating harmful ions via an energetically expensive gradient. Glycophytes and halophytes alike are susceptible to higher cytoplasmic Na^+ concentrations. Primary plant sodium absorption at the root-soil interface is most likely mediated by two nonselective cation channels, glutamate receptors (GLRs) and cyclic nucleotide-gated channels (CNGCs). Furthermore, plants use aquaporins and high-affinity potassium transporters (HKTs) for Na^+ uptake (Demidchik and Maathuis 2007). The apoplastic route facilitates the translocation of Na^+ via roots to shoots. It first passes through the root epidermis symplast before entering the xylem's tracheids. Eventually, it becomes most noticeable on the shoots, especially the leaf blades (Munns et al. 2005). To protect the cytoplasm from the detrimental effects of Na^+, plants have evolved a variety of defense mechanisms. These defense mechanisms include optimizing the quantity of Na^+ that is compartmentalized into the vacuole, reducing the amount that enters the cell, and raising the amount that is rejected from the cell.

3. *High-Affinity Potassium Transporters (HKTs)*

The Trk/Ktr/HKT superfamily includes HKT transporters, often known as monovalent cation transporters. By keeping Na^+ ions out of the plants' fragile shoot tissues, they have been thoroughly investigated in a range of plants and are crucial for salt tolerance (Jaime-Perez et al. 2017; Wang et al. 2014). HKT*s* are categorized as Na^+/K^+ symporters because of their ability to moderate the import of Na^+ due to their transport selectivity. Moreover, specific HKT members may mediate the porousness of Mg^{2+}/Ca^{2+} across plant cell plasma membranes. Four MPM repeats determine the structural characteristic of HKT*s*, where "M" stands for the transmembrane segment and "P" for the "pore-loop domain." The structural determinant found in the first P site, PA, along with the repeat M1A-PA-M2A-M1D-PD-M2D combines to figure out the HKT1-type and HKT-2 type. HKT2-type proteins function as Na^+/K^+ symporters because they contain a Gly residue at the MPAM motif, which controls the permeability of either Na^+ or K^+. HKT transporters belong to the HKT1-type group and only show a Na^+ permeability gradient when Ser is replaced with Gly. They impede the migration of Na^+ ions and their accumulation in shoots by extracting excess sodium from xylem sap and encapsulating it in xylem parenchyma cells, as shown in Fig. 1.3.

4. *Oxidative Stress and Antioxidant Protection Under Conditions of Salinity Stress*

The powerful oxidative properties of ROS generated by salt lead to oxidative destruction of membrane lipids, nucleic acids and proteins, ultimately resulting in irreversible metabolic failure. ROS generated by salt are probably formed via electron transport pathways in mitochondria and chloroplasts, although plants have antioxidant enzymes and nonenzymatic compounds to detoxify them. Antioxidant enzymes are catalase (CAT), ascorbate peroxidase (APX), superoxide dismutase (SOD), and peroxidase (POX), while nonenzymatic antioxidant compounds include glutathione (GSH), Ascorbic Acid (ASC), and carotenoids (Azeem et al. 2023;

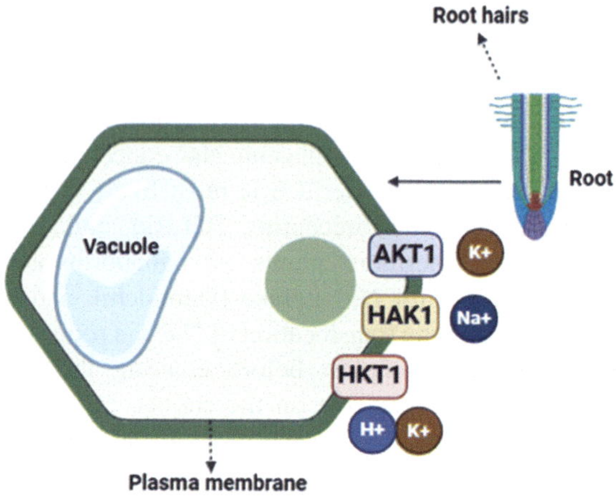

Fig. 1.3 Mechanism of high-affinity potassium transporters in plants

Gupta and Huang 2014). These antioxidant chemicals and enzymes quench or scavenge the elevated levels of ROS in conditions of salt stress. These ROS include O_2^-, OH^-, singlet oxygen ($^1O_2$), and H_2O_2. Plant cells start the process of gradual detoxification by producing SOD. As the body's initial line of defense, SOD works by converting superoxide radicals to oxygen and hydrogen peroxide, therefore lowering hydroxyl radicals. Because superoxide radicals decrease metal ions (Fe^{3+} and Cu^{2+}) and produce hydroxyl radicals as a byproduct, they can damage lipids and cellular membranes.

Subsequently, CAT and POX degrade the H_2O_2, and their metabolites produced under salt stress. Numerous studies have demonstrated the connection between salt tolerance and elevated antioxidant activity. The salt-tolerant variety of *A. tricolor*, VA14, for instance, has higher levels of ascorbate, APX, and SOD, which help with ROS detoxification, according to a recent study on the foliage of the plant (Sarker and Oba 2020). Increased malondialdehyde (MDA) production is an indicator of membrane damage brought on by salt stress (Hussain and Boney 1973). A comparison of the salt tolerance of multiple wheat genotypes revealed that the salt-tolerant varieties produced less MDA, which was connected to a reduced degree of membrane lipid peroxidation. A different study found that overexpressing the wild-type soybean peroxidase gene GsPRX9 increased the antioxidant response and improved salt tolerance (Jin et al. 2019). Plants under salt stress are known to produce antioxidants called anthocyanins. Because the Arabidopsis mutant gene anthocyanin-impaired-response-1 (air1) is unable to accumulate anthocyanins under salt stress, it is implicated in salt tolerance by modulating different phases of the flavonoid and anthocyanin synthesis pathways (Van Oosten et al. 2013). These findings highlight the preventive functions of several compounds, pigments, and enzymes that enhance plant tolerance to salt and lessen oxidative damage.

1.2.2 Salinity Affects Oilseed Crop Growth and Production

One important environmental stressor that has an impact on oilseed crop development and productivity is salinity. Salinity has physiological and biochemical effects on these crops, which could eventually lead to decreased yields (Fig. 1.4). The primary effects are as follows:

1. *Reduced Germination Rate*: Inadequate germination happens when seedlings don't have enough water because the soil has a lot of salt in it.
2. *Ion Toxicity*: Toxic ions like salt and chloride can accumulate around the seeds and disrupt the regular metabolic processes required for seed germination.

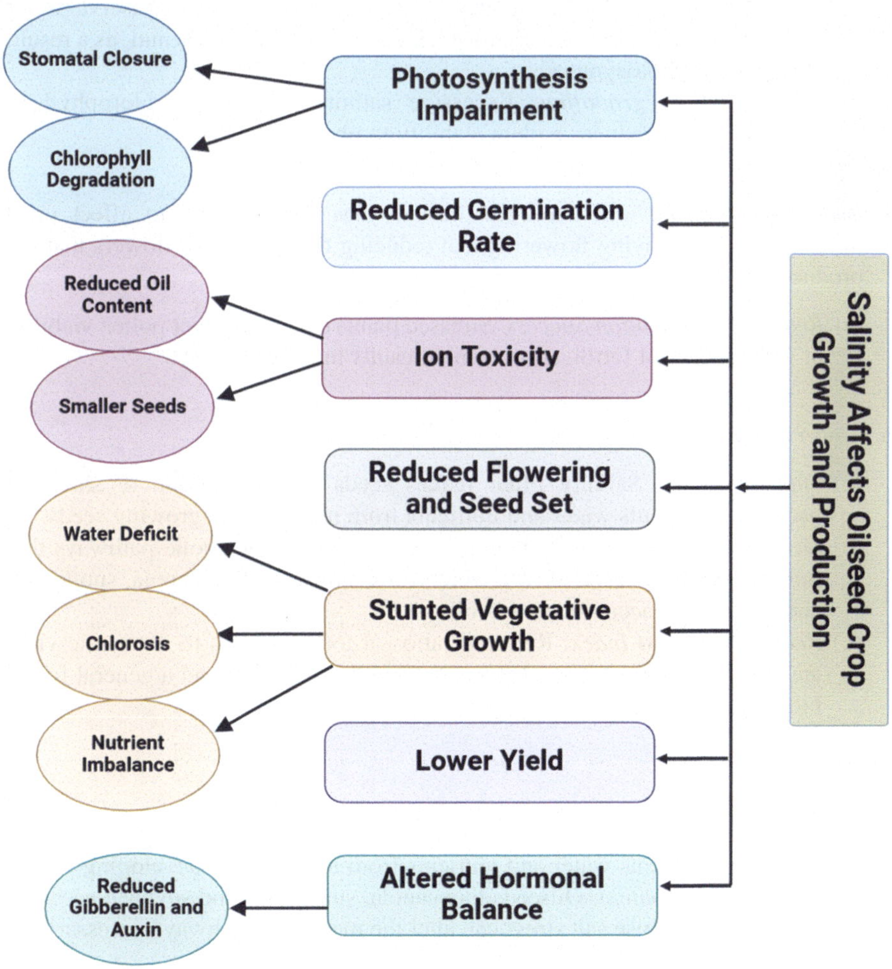

Fig. 1.4 Salinity effects on the growth and production of oilseed crops

3. *Stunted Vegetative Growth*:
   (a) *Water Deficit*: Salt causes a plant's capacity to absorb water to be diminished, which results in smaller leaves and lessened cell elongation.
   (b) *Nutrient Imbalance*: The absorption of essential minerals, such as potassium, calcium, and magnesium, which are all required for healthy growth, might be hampered by an excess of sodium.
   (c) *Chlorosis*: Salinity frequently results in nutrient deficiencies, particularly in potassium, which impairs the production of chlorophyll and causes leaves to turn yellow.

4. *Photosynthesis Impairment*:
   (a) *Stomatal Closure*: Plants frequently shut their stomata to conserve water while under salt stress, which lowers their absorption of CO_2 and, as a result, slows down photosynthesis.
   (b) *Chlorophyll Degradation*: Excessive salinity can cause chlorophyll to degrade, which reduces a plant's ability to photosynthesize.

5. *Reduced Flowering and Seed Set*: Salinity has the potential to affect yield immediately by delaying flowering and reducing the number of flowers that are produced.
   (a) *Reduced Fertilization Success*: Stressed plants can have lower pollen viability or less successful fertilization, which results in fewer seeds.

6. *Lower Yield*:
   (a) *Smaller Seeds*: Salinity often makes seeds smaller and more shriveled because it prevents water and nutrients from reaching the growing seeds.
   (b) *Decreased Oil Content*: Salt stress can change the metabolic pathways that lipid biosynthesis depends on, which causes oilseeds like canola, sunflower, and soybean to accumulate less oil.
   (c) *Reduced Harvest Index*: Reduced ratios of total biomass to economic yield are the result of the two factors, poor seed development and a general fall in biomass.

7. *Ion Toxicity*:
   (a) *Smaller Seeds*: Salinity often makes seeds smaller and more shriveled because it prevents water and nutrients from reaching the developing seeds.
   (b) *Reduced Oil Content*: Oilseeds like canola, sunflower, and soybean accumulate less oil because salt stress can alter the metabolic pathways necessary for lipid biosynthesis.

(c) *Reduced Harvest Index*: Both poor seed development and an overall decline in biomass lead to a reduction in the ratio of economic yield to total biomass.

8. *Altered Hormonal Balance*: Abscisic acid (ABA) production increases: In reaction to salt, plants usually create more ABA. ABA is a stress hormone that controls stomatal closure and can cause a reduction in growth rates.

   (a) *Reduced Gibberellins and Auxins*: In saline environments, these growth-promoting hormones are frequently reduced, which further impedes development and growth.

9. *Yield Variability Among Oilseeds*:

   (a) *Sunflower*: Has a moderate tolerance for salinity, but as salinity levels rise, yields decrease.
   (b) *Canola*: Significant yield loss occurs when salinity levels are high, particularly during the reproductive stage.
   (c) *Soybean*: Salinity reduces pod set and seed development; moderate sensitivity.
   (d) *Sesame*: Compared to other oilseed crops, it is relatively more tolerant of salinity, but it still loses yields in situations with high salinity.

1.2.3 Current Challenges in Managing Salinity in Oilseed Cultivation

Degraded soils and marginally quality water resources offer a growing opportunity for agricultural output as a result of the world's population growth. This is particularly true given the extensive environmental deterioration, water shortage, and loss of arable land that come with climate change. Proposals of various agronomic and biological strategies to enhance the agricultural efficiency of degraded, salt-affected fields have been made because of the belief that even small gains in productivity could have a major positive influence on farmers' incomes and the quality of the environment. However, an examination of the literature shows that many of these conventional methods have drawbacks that restrict their ability to fully address these problems. Because sodic and salty soils cannot be restored without fresh water, acute freshwater shortage has forced the updating of existing practices to ensure that water of poor quality may be used for soil restoration. It has been determined that climate change poses a severe danger to water and land resources, with agriculture in salt-affected areas most likely to suffer from the predicted adjustments to crop growth circumstances. Taking this into consideration, this study evaluates the weaknesses of the present salinity management systems and proposes the research goals for the future to preserve and boost agricultural output in salt-affected soils (Sharma and Singh 2017).

1. *Soil and Water Salinization*:
   (a) *Natural Salinity*: Due to the weathering of parent rock material or the deposition of salts from ancient seabeds, salinity is often inherent in the soil.
   (b) *Irrigation-Induced Salinity*: Inadequate irrigation practices can lead to a buildup of salt in the soil, which aggravates salinity issues, especially in areas with poor drainage. Saline irrigation water is another component in this.
   (c) *Groundwater Salinity*: Due to capillary action, high-salinity shallow groundwater in many agricultural areas can rise and cause secondary salinization in the root zone.
2. *Limited Availability of Freshwater for Irrigation*:
   (a) *Water Scarcity*: Due to climate change, excessive resource extraction, and competition from other industries, freshwater availability for irrigation is becoming more and scarcer.
   (b) *Dependency on Saline Water*: Farmers in some regions irrigate their crops using brackish or saline water, which has the effect of hastening soil salinization and decreasing crop yields.
3. *Varietal Limitations*:
   (a) *Limited Salt-Tolerant Varieties*: Although salt-tolerant oilseed crop development has progressed, the availability of salt-tolerant sunflower, canola, and soybean cultivars remains limited.
   (b) *Lower Genetic Diversity*: Breeding programs that prioritize high-yielding cultivars may result in a lower genetic diversity, which makes it more difficult to create crops that can survive in salty environments.
4. *Soil Structure Degradation*:
   (a) *Inadequate Soil Drainage*: In salty soils, insufficient drainage and sodium buildup can cause compaction, deterioration of the soil's structure, and decreased aeration. Water infiltration and root growth are hampered by this.
   (b) *Loss of Soil Fertility*: Excessive salinity can cause hazardous levels of sodium and chloride while upsetting the soil's nutrient balance. This reduces the availability of vital nutrients like potassium and phosphorus.
5. *Declining Soil Organic Matter*:
   (a) *Decreased Resilience*: Reduced soil organic matter affects water retention, soil structure, and the soil's ability to tolerate salt. This is often the outcome of continuous monocropping or excessive use of chemical inputs.
   (b) *Soil Compaction*: Decreases in salt leaching and water penetration can also result from low levels of organic matter in the soil.

6. *Policy and Institutional Barriers*:
   (a) *Weak Regulatory Frameworks*: Policies for conserving water resources and halting salinization are either nonexistent or inadequately implemented in many areas. This results in unsustainable agricultural practices, inadequate water management, and excessive irrigation.
   (b) *Fragmented Approaches*: Progress in addressing salinity challenges is hampered by a lack of coordinated efforts among research institutions, agricultural policy, and water management.
7. *Biotic Stress Synergies*:
   (a) *Combined Stress*: Pests, illnesses, and nutrient shortages are some examples of stressors that frequently coexist with salinity. These complex interactions have a synergistic effect that lowers oilseed crop yields even more.
   (b) *Weakened Plant Defenses*: The natural defenses of plants are weakened by salt stress, which increases their susceptibility to pathogen attacks and complicates management.
8. *Salinity in Marginal Lands*:
   (a) *Extension of Areas Prone to Salinity*: It becomes more difficult to sustain oilseed cultivation as agriculture spreads into marginal, semiarid, and coastal areas because of the increased salinity exposure.
9. *Lack of Salt-Tolerant Cover Crops*:
   (a) *Inadequate Crop Rotation Options*: One of the most crucial practices for preserving soil health and controlling salinity is crop rotation, but it is hampered by the absence of salt-tolerant cover crops.
   (b) *Effect on Recovery of Soil*: In the absence of appropriate cover crops, the soil's diminished ability to recover from salt stress exacerbates long-term salinization issues.
10. *Breeding and Biotechnology*: More genetic engineering and biotechnology research on salt-tolerant crops may result in the creation of cultivars that are more resistant to salinity.
11. *Sustainable Water Management*: It's critical to employ saline-resistant crops in areas where salinity problems exist, as well as to make improvements in irrigation techniques like the use of precision irrigation (like drip irrigation).
12. *Integrated Soil Management*: To preserve soil health and lessen the effects of salinity, it is essential to use soil conditioners, organic amendments, and enhanced drainage systems.
13. *Awareness and Training*: To effectively manage salinity, farmers require improved access to information about sustainable practices as well as assistance from governmental or nongovernmental organizations.

1.3 Biostimulants

1.3.1 Overview of Biostimulants

Plants benefit greatly from plant biostimulants because they increase their production and ability to withstand stress. This is closely related to their capacity to affect plant metabolism, cellular structure, and physiological functions. Because plant biostimulants are complex and diverse in their chemical composition, they affect plants through a range of metabolic pathways and regulatory networks. This study covers the mechanisms of action of biostimulants produced from smoke, seaweed extracts, protein hydrolysates, biopolymers, humic chemicals, and beneficial microbes (Table 1.1). It highlights how these compounds can influence transcriptional regulatory networks, promote or control protein and enzyme function, and imitate the effects of hormones. The present knowledge of the physiological, genetic, and molecular processes influenced by plant biostimulants and how these result in enhanced plant development, growth, productivity, and stress tolerance is discussed in this chapter. It also identifies promising areas of research that center on the mechanisms underlying plant biostimulant action (Omoarelojie et al. 2021).

Any substance, regardless of nutritional value, which is introduced into plants to improve crop quality traits, resistance to abiotic stress, and/or nutritional efficiency, is referred to as a plant biostimulant. Plant biostimulants also include manufactured goods that contain combinations of these materials and/or microbes. Arguments based on scientific understanding of the traits, modes of action, and variety of impacts of biostimulants on horticultural and agricultural plants lend credence to this theory. The proposed definition also aims to make it easier to employ biostimulants in future regulations, especially in the EU, by clarifying the distinctions between them and fertilizers, insecticides, and biocontrol agents. Plant biostimulants, therefore, also include commercial goods that contain combinations of these materials and/or microbes. To facilitate future regulation, especially in the EU, that encompasses biostimulants, the suggested definition also seeks to define the distinctions between them and pesticides, fertilizers, and biocontrol agents (Du Jardin 2015).

1.3.2 Types of Biostimulants

Plant biostimulants are regarded as one of the most well-liked sustainable alternatives due to its inherent qualities and capacity to supplant conventional farming methods. Biostimulants (source-based rather than constitution-based) include seaweed extracts (SWE), humic substances, protein hydrolysates, and microbes. These substances have demonstrated potential in raising plant development, reducing the harmful effects of stress, as well as enhancing agricultural output and quality. However, because of their diverse content and multimolecular structure,

Table 1.1 Different marketed biostimulants and their active ingredients

S. No	Biostimulant	Active ingredients
1	Radifarm	Amino acids, glycosides, saponins, betaines, polysaccharides, organic acids, vitamins, and trace elements
2	Megafol	Amino acids (glycine, glutamic acid), betaines, proteins, vitamins (B5, PP, B1, B6), auxins, gibberellins, and cytokinins
3	Viva	Folic acid, vitamins (B6 and PP), polysaccharides, humic acids, proteins, peptides, and amino acids
4	Benefit	Amino acids, nucleotides, enzymatic proteins, and vitamins
5	Kendal	Urea, oligosaccharides, glutathione, protein hydrolysates, and saponins
6	Bio-algeen S-92	Seaweed (*Ascophyllum nodosum*) extract
7	Ruter AA	Free amino acids and both macro- and micronutrients
8	Ergonfill	Animal protein hydrolysates, cysteine, folic acid, and keratin derivatives
9	Asahi SL (Atonik)	Sodium para-nitrophenolate (0.3%), sodium ortho-nitrophenolate (0.2%), and sodium 5-nitroguaiacolate (0.1%)
10	Biozyme	Seaweed (*Ascophyllum nodosum*) extract, plant hormones (GA3, IAA, zeatin), and chelated micronutrients
11	Goëmar BM 86	Extract of seaweed (*Ascophyllum nodosum*)
12	Tablet	Spores of *Rhizophagus intraradices* and *Trichoderma atroviride*
13	KE-Plantasalva	Biomolasses and herb extracts, with or without sea salt
14	Fermented Plant Extract	Sugar cane molasses, lactic acid and photosynthetic bacteria, yeasts, extracts from grasses, weeds, garlic, and chili peppers
15	Biplantol Universal	Macro- and micronutrients, germanium, and uronic acids
16	Equisetum Extract	Flavonoids, plant acids, glycosides, and silicon
17	Algreen	Seaweed extract (*Sargassum* sp., *Ascophyllum nodosum*, *Laminaria* sp.), plant hormones, vitamins, free amino acids, and alginic acid
18	Grow-plex SP	Liquid humate
19	Roots 2	Seaweed, humic acid, and vitamins
20	Root and Shoot Builder	Seaweed extract (*Ascophyllum nodosum*), micronutrients, amino acids, and natural chelating agents
21	Bio Rhizotonic	Algae extract and vitamins
22	Root Juice	Seaweed extract, humic acids, and fulvic acids
23	Bio Root	Plant and mineral-derived organic acids, humates, alfalfa meal, soybean meal, brewer's yeast, K-sulfate, rock phosphate, and sea kelp
24	Slavol	Nitrogen-fixing and phosphate-mineralizing bacteria, and auxins
25	Tytanit	Titanium ascorbate

Parađiković et al. (2019)

commercially available biostimulants are challenging to comprehend in terms of their basic mechanisms (Baltazar et al. 2021).

(a) *Seaweed Extract*

A class of biostimulants known as "seaweed extracts" is generally used to describe a group of algae species that are processed, most frequently macroalgae (seaweeds). The species used have different biostimulant chemical compositions and levels of utility (Ugarte et al. 2010; Abdel Latef et al. 2017, Dawood et al. 2025). According to Goni et al. (2016), commercial commodities' constituents differ greatly based on the species selected, the collecting stage, and the extraction technique employed by each company. The most popular extraction method is alkaline hydrolysis, which is followed by super-critical fluid, microwave-assisted, water-based, pressured liquid extraction, enzyme-assisted, and ultrasound-assisted (Shukla et al. 2019). Macroscopic algae are distinguished from other products formed from organic matter by the significant number of particular carbohydrates found in the ulvan, fucoidan, laminarin, and ulvan of brown and red algae, as well as in the carrageenan of the latter (Goñi et al. 2020). Carefully remove highly sulfated components such as fucoidan, carrageenan, and ulvan to minimize the generation of sulfhydryl compounds, which could hinder crop growth (Craigie 2011).

When seaweed extract components are applied to crops, the primary effect is stress reduction since they contain antioxidant qualities that may reduce ROS-induced cell damage during abiotic or biotic stress (Sible et al. 2021). New research indicates that a particular polysaccharide-rich commercial extract of *Ascophyllum nodosum,* called Super Fifty, alters several transcriptomic, metabolic, and lipid processes (Omidbakhshfard et al. 2020; Rasul et al. 2021; Shekhar et al. 2012; Staykov et al. 2020). The authors use a variety of pathways to demonstrate how these adjustments lead to significant phenotypic changes, including decreased ROS, decreased electrolyte leakage, enhanced plant development, and resistance to oxidative and abiotic stressors. Accordingly, recent studies are beginning to alter our understanding of how SWE works by suggesting that a few polysaccharides, along with various nongrowth hormone components, are the most likely sources of the effects seen in plants (Baltazar et al. 2021).

(b) *Microbial Inoculants*

Organic matter is broken down by microorganisms in a complicated process that results in a wide range of byproducts throughout the degradation route (Nardi et al. 2007). Conventional wisdom states that soil organic matter (SOM) at some point consists of humus, a tough and recalcitrant mixture of chemical compounds. The total amount of organic matter in soils can be as much as 60%, composed of these components. That group is categorized as fulvic acids (FA), humic acids (HA), and humin (alkali and acid-insoluble). Although HA is acid-insoluble, it is soluble in alkali (Lamar 2020). Such substances are resistant to degradation, according to traditional viewpoints; however, more recent viewpoints consider the idea that biological matter breaks down continuously and that materials that were once thought to be stable go through reversible processes. The soil microbiome may therefore be activated by these substances (Lehmann and Kleber 2015). Levite, deteriorated organic waste, and peat soils are a few of the materials that can be used to make humic acids (Niemeyer et al. 1992). The structure, content, and impact on crop and soil systems are also influenced by the source and stage of HA or FA decomposition. Although the alkali/acid extraction methods of HA and FA are different, their elemental makeup and molecular sizes vary as well, which may provide light on any potential discrepancies in their biostimulant activity (Sible et al. 2021). This category includes micro-algae, filamentous fungi, yeast, and bacteria. They are kept apart from other organic materials such as composted manure, water, plants, and soil. They are added to the soil to stimulate metabolic processes that raise crop production. Their production of volatile organic compounds (VOCs) and increased resilience to abiotic stressors may also directly affect plants. Additionally, by solubilizing minerals and fixing nitrogen, they increase the availability of nutrients. They also alter the hormonal balance of a plant by stimulating the synthesis of plant hormones such as auxins and cytokinins. According to Ruzzi and Aroca (2015) and Turan et al. (2017), plant growth-promoting rhizobacteria (PGPR) induce physical, chemical, and biological activities in plants that improve their responses to abiotic stressors. There are advantages to microorganisms that increase nutrient and water intake and form a protective biofilm layer on the root surface (Franzoni 2020).

(c) *Humic Substances (HS)*

With a variety of established biological actions, HS, also known as HA and FA, have long been recognized to function as biostimulants. HS are made up of organic compounds that are left over after dead biota in soils decompose. They are largely immune to microbial degradation and have a very diverse molecular makeup. These chemicals have been assigned several responsibilities concerning the activities of soil and plants. Plant growth may be impacted either directly or indirectly by heavy metals or HS. HS has indirect benefits on plant development through improved microbial populations, cation exchange capacity, soil structure, and pH buffering qualities (Garcia-Mina et al. 2004; Gerke 2021; Dawood et al. 2019; Nardi et al. 2017, 2018). Even though further investigation is required to fully comprehend the outcomes and mode of action of HS, these biostimulants are intriguing (Baltazar et al. 2021).

Humic substances include humins, fulvic acid, and humic acid. A naturally occurring component of organic matter, HS are created by the metabolism of soil microbes and the decomposition of plant, animal, and microbiological wastes (Franzoni 2020). Applying humic compounds has been shown to accelerate plant root development and maturation. This is demonstrated by improved water and nutrient absorption as well as a greater ability to withstand environmental challenges (Canellas et al. 2015; Nardi et al. 2016). It is uncertain how the HS impact plant physiology. This is due to the chemical complexity as well as the range and quantity of plant responses they alter when applied. Furthermore, a robust correlation has been documented between the medium parameters and the bioactivity of HS (Canellas et al. 2012). These complex aggregates' beneficial effects may be attributed to IAA-independent processes as well as the hormone-like action of certain of their constituents. For instance, HS can stimulate plant development in a manner akin to auxins by triggering the plasma membrane's H^+ ATPase (Dobbss et al. 2010; Nardi et al. 2000, 2002).

1.3.3 Mechanisms of Action of Biostimulants

Plant fundamental metabolism, which is responsible for many of the physiological activities the plant conducts to meet its basic needs, includes the effective use and absorption of nutrients alongside the metabolism of carbon (C) energy. Therefore, for appropriate plant development, nutrients from the soil must be ingested and integrated. Nutrients not only function as cofactors but also as vital components for the manufacture of enzymes and metabolites. In spinach, tomato, maize, oyster tree, kinnow mandarin, rose-scented geranium, and perennial wall rocket, it has been demonstrated that a variety of biostimulants increase the concentrations of P, S, Ca, Mg, K, Fe, and Zn (Nasir et al. 2016; Regni et al. 2021; Rouphael et al. 2017; Rouphael et al. 2018; Ali et al. 2021). Alfalfa-based PH treatment of tomato plants resulted in increased transcription of ion-uptake-related transporters, such as sulfate transporters, transporter protein, phosphate transporter PT2, and Cu transporters (Ertani et al. 2017).

Thus, biostimulants may be able to lower the dose of fertilizer used by increasing its efficiency. Additionally, by changing the Pb, Ni, Cu, and Cd concentrations, higher plant-derived biostimulants (hPDBs) offer protection against the excessive buildup of *potentially harmful elements (PHEs)* in plants grown in contaminated soils (Howladar 2014; Rouphael et al. 2023; Desoky et al. 2019). Additionally, it has been demonstrated that the negative effects of salt stress can be mitigated by the use of biostimulants. An imbalance in the K/Na ratio leads to ionic and osmotic stress and hinders the appropriate absorption of other ions. The biostimulants found in plants helped to partially restore this imbalance (Desoky et al. 2019; Abbas et al. 2010; Abdel Latef et al. 2019; Abdou et al. 2018; Huang et al. 2021). Additionally, this method proved successful in balancing the concentrations of other necessary ions, including S, K, Mn, P, and Ca (Desoky et al. 2020; Rady and Mohamed 2015).

Nitrogen is one of the most important nutrients that plants ingest because it is fixed as organic N from nitrate and then used to produce amino acids. Higher N concentrations have been seen in the leaves of several crops after hPDB spraying. This effect has also been shown to occur when there is salt stress (Rady et al. 2019; Rady and Mohamed 2015; Abdalla 2013; Cirillo et al. 2017; Di Mola et al. 2020). Because of this rise in N concentration, fertilizer dose may be decreased while still keeping nitrate levels within the legally mandated ranges established by the European Commission (Di Mola et al. 2019). Higher transcription of genes encoding the ammonium (AMT1.1 and AMT1.2) and nitrate (NTR2) transporters was shown to be linked to higher N assimilation. This increase in movement from the soil to the root was attributed to these genes. Higher levels of N assimilation were also associated with increased expression of genes related to N metabolism and amino acid synthesis, such as ferredoxin-dependent glutamate synthase, glutamate dehydrogenase, glutamine synthetase (GS), nitrate reductase (NR), nitrite reductase, and aspartate aminotransferase (GS1 and GS2) (Ertani et al. 2017).

NR, GS, as well as glutamate synthase (GOGAT) activities have also been demonstrated to be elevated in maize (Baglieri et al. 2014; Ertani et al. 2009). Furthermore, it was demonstrated that treating salinity-exposed maize plants with a PH produced from alfalfa boosted their GS and GOGAT activities. Therefore, increased N consumption and subsequent incorporation into amino acids boost N's bioavailability in plant tissues (Carillo et al. 2019). Furthermore, it has been discovered that tomato plants treated with hPDB*s* have higher transcript levels of nodulin MtN21, glutathione-conjugate transporter MRP4, amino acid transporter AAT1, organic cation/carnitine transporters, and other N-associated enzymes (aspartyl protease, glutamate dehydrogenase, and serine decarboxylase) in addition to genes encoding proteins related to protein synthesis and modification (aminoacyl-tRNA synthetases, elongation factors Tu and 1-alpha, translation initiation factors, and ubiquitin-conjugating enzymes) (Ertani et al. 2017).

Maize, rice, wheat, mung beans, and common bean plants exhibit an improvement in some parameters as a result of this spike in protein synthesis that occurs after the injection of biostimulants, or total proteins. It has been shown that when plants are stressed, treatment increases their content of protein and amino acids in contrast to when they are not treated. Proline is an amino acid that has been connected to stress reactions; treated plants also exhibited higher levels of this amino acid (Bajwa et al. 2018; Baglieri et al. 2014; Abdel Latef et al. 2019; Lorenzo et al. 2019). Bifunctional aspartokinase/homoserine dehydrogenase 2, which is crucial for protein metabolism, was produced in greater amounts in grapevines cultivated in water-scarce conditions when commercial biostimulant formulations like Trainer® and Stimtide® were used (Bavaresco et al. 2020). Plant growth and development depend on sugars and carbohydrates since they are necessary for the production of structural molecules and energy. It has been demonstrated that applying biostimulants to plants increases the amount of soluble sugars in a variety of plant species, such as habanero pepper, rocket, tomato, kinnow mandarin, and maize. The transcription of genes encoding important enzymes for sugar metabolism (polygalacturonase, pectinesterase, starch synthase, sucrose synthase, cellulose synthase, and inositol oxygenase) and C metabolism (such as fumarate dehydrogenase, malate dehydrogenase, phosphoenolpyruvate carboxylase, and phosphoenolpyruvate carboxylase kinase 2) was also increased by biostimulants (Rouphael et al. 2017; Ertani et al. 2017; Nasir et al. 2016).

Isocitrate dehydrogenase, citrate synthase, malate dehydrogenase transcription and citrate synthase and activity were all boosted in monocotyledons, such as maize, with administration of hPDBs (Schiavon et al. 2008). A dry blueberry cold extract or a cellulosolitic dry apple hydrolysate treatment of maize plants produced an interesting reaction. Fructose content increased while glucose and sucrose concentration decreased. The treatment increased glucose consumption by promoting plant respiration, which limited glucose availability for sucrose synthesis and caused fructose to accumulate instead (Ertani et al. 2011). Other fundamental metabolites, such as nucleosides, which are composed of ribose or deoxyribose pentose linked to form their chemical structure, are synthesized with the help of carbohydrates and sugars. It has been demonstrated that treating habanero pepper fruits with various hPDB*s* causes an increase in thymidine buildup (Ertani et al. 2014).

When hPDB*s* are applied to stressed crops, the concentration of soluble sugar often rises, which benefits both monocotyledons and dicotyledons in terms of osmoprotection (Desoky et al. 2019; Latif and Mohamed 2016). According to Huang et al. (2021), because α-amylase hydrolyzes starch into maltose units, it may be the cause of the increased maltose levels in camelina and wheat treated with sorghum water extract in salinized circumstances. Additionally, as demonstrated by the concentration of formazan that was collected throughout the viability test, salinized pea plants treated with narrow-leaf cattail extracts had higher respiration rates than untreated plants and maintained metabolic activity (Ghezal et al. 2016). The transcription of genes encoding lipases has risen in tomato plants fed with an alfalfa-based PH, according to studies on lipid metabolism. In contrast, some investigations have revealed an accumulation of structural lipids such as hydroxystearate and hexadecanediol (Ceccarelli et al. 2021). Under conditions of decreased water availability, tomato plants treated with Trainer® showed the accumulation of sterol and both positive and negative alterations in membrane lipids (Paul et al. 2019).

Sadly, a lot of biostimulant medications lack a recognized mechanism of action or a well-defined mode of action. Because certain fraudulent items undermine the market for all participants, many people believe that biostimulants are all just "snake oils," a derogatory phrase that implies the product is worthless. It can be particularly difficult to reconcile multicomponent biostimulants since some of their components may have documented mechanisms of action while others lack demonstrable functional effects. Furthermore, physiologically insignificant quantities of known required elements, amino acids, plant hormones, etc., are frequently found in multicomponent biostimulants; the way of action of these substances is known, but the amounts, when taken at prescribed rates, are meaningless. We thus suggest that, rather than pursuing unambiguous proof of the "mode of action," a more practical and attainable regulatory aim for many of the multicomponent biostimulants already on the market would be the demonstration of a clear "mechanism of action." We suggest that the current focus of biostimulant research and validation be on characterizing the mechanism of action rather than the need to identify a mode of action, given the difficulties of defining a "mode of action" for a multicomponent product with the complexity of a biostimulant and the need for the biostimulant market to attain credibility. This is the authorized technique for a number of medicinal medications (Fig. 1.5).

Future advancements in analytical technology, bioinformatics, systems biology, and other essentially novel approaches may lead to a deeper comprehension of the workings and potential processes of these substances. Although this proposal implies that a mode of action demonstration may not be necessary for the development and marketing of a biostimulant, it is still in the best interests of the companies

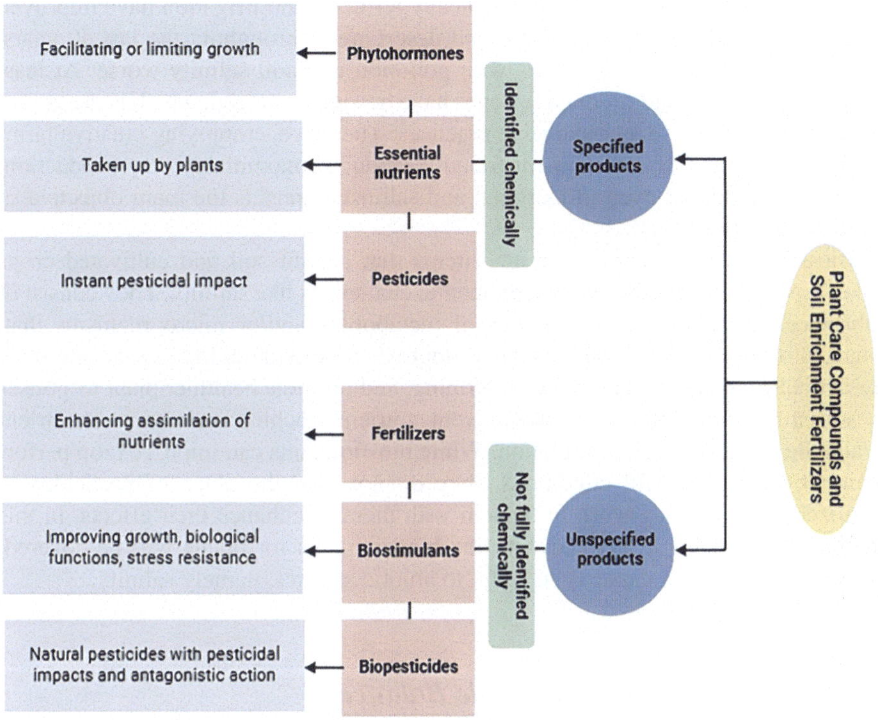

Fig. 1.5 Different plant compounds and their mechanisms of action

producing these products to investigate and comprehend the mode of action in order to enhance the product and optimize its application for a range of cropping systems and environments (Yakhin et al. 2017).

1.4 Mechanisms of Salinity Tolerance Induced by Biostimulants

The population of the planet has surpassed 7.63 billion, more than quadrupling from 1.8 billion in 1915. The United Nations has released its most recent projections, which suggest that by 2050, we will have 9.7 billion people on Earth. The population's exponential growth, the shift from rural to urban areas, the increase in per capita food consumption, and dietary changes brought on by the rising affluence of emerging nations are some of the factors that are expected to cause the global food demand to increase from 59% to 98% over the next 30 years. However, achieving a balance between agricultural output maximization and environmental sustainability would be challenging. Mechanization and innovative management techniques that prioritize heavy fertilizer and irrigation have improved agricultural productivity in semiarid and desert areas throughout the last 50 years. They have, however, made issues with pollution and soil salinity worse. At least 20% of all irrigated and productive zones have become salinized, which is the largest adverse effect of these unsustainable practices. Therefore, employing creative farming techniques and/or environment-friendly products, maximizing crop production, output, and potential even in marginal and salinized areas is the main objective of modern agriculture.

Biostimulants are among the new items that benefit soil and cultivated crops alike, especially in the face of environmental challenges like salinity. They consist of substances, metabolites, or mixtures of metabolites and/or microorganisms that, when introduced to the soil or plants, improve food availability, lessen the need for agricultural chemical products in farming, and promote healthier plant responses to stressful conditions. They also prevent nutrient leaching and increase nutrient availability, uptake, and assimilation. While biostimulants can improve crop performance even in stressful conditions, they do not take the place of pesticides or fertilizers. Rather, they work in tandem with them to enhance their effects. In this session, the benefits associated with the biostimulants for plants will be covered, especially about their capacity to adapt to abiotic stresses, namely salinity.

1.4.1 Enhancement of Osmotic Balance

Adaptive processes are utilized by plants to withstand diverse abiotic stressors. Through the employment of several low-molecular-weight compounds known as osmoprotectants, the osmoprotection system can lessen the negative consequences of abiotic stresses. Osmoprotectants include sugars, polyamines, amino acids, and substances containing quaternary ammonium. As safe and efficient metabolic signals, these compounds also scavenge ROS produced under stressful conditions and stabilize cellular structures and enzymes. Plants that are better suited to abiotic stressors must be produced, considering the recent and significant changes to the global environment. By passing laborious traditional and classical breeding methods, genetic engineering is a revolutionary way of transferring genes involved in the production of osmoprotectants from one plant to another. Here, we examine new research on osmoprotectants and transgenic plants that have been modified to have certain osmoprotectant qualities (Zulfiqar et al. 2020).

1.4.2 Antioxidant Activity and Stress Enzyme Modulation

In reaction to a range of abiotic stimuli, plant cells frequently create reactive oxygen species. These compounds can be hazardous or serve as signaling molecules, depending on the concentration at which they are present. Based on the plants' natural antioxidant defense mechanism, crop health may be improved by carefully controlling the production of ROS during stress. When applied exogenously, biostimulants are naturally occurring, environmentally acceptable compounds that improve resistance to abiotic stress. The possible effects of certain biostimulants on the antioxidant defense system activity in important agricultural plants under stress, as well as their additional functions in the management of abiotic stressors. Biostimulants can reduce oxidative stress, although their wider applicability is severely limited by their crop growth stage, dose, kind, and variety (Hasanuzzaman et al. 2021).

Enzymatic and nonenzymatic plant antioxidant systems are essential for maintaining redox and osmotic balance, which supports the natural defenses of plants against biotic and abiotic stressors. Reactive species generation and stress-response elicitation are closely related, which leads to the buildup of hyperreactive, lethal chemicals over physiological thresholds. Reactive oxygen species are removed by plant antioxidant systems, which also maintain tissue and cell ROS concentrations within acceptable bounds. They accomplish this by maintaining redox equilibrium through physiological and metabolic processes. "Plant biostimulants" are compounds that can both exacerbate and ameliorate the consequences of stress on plants. These substances can be inorganic or organic. Through the regulation of plant antioxidant systems, they attenuate stress-induced damage and limit rapid growth and development. The present knowledge of how biostimulants alter and stimulate plant antioxidant systems in relation to enhancing stress tolerance is positioned in this part. Additionally, we point out areas that merit further investigation and may yield useful information for crop improvement initiatives (Omoarelojie et al. 2021).

Because ROS have two roles, cellular ROS balance is necessary for plants. They are signaling molecules that help plants adapt to harsh environments and control their growth and development. But when their concentration rises during stress, they have negative effects on cellular metabolism (Halliwell 2006). The complex antioxidant system found in plants aids in the maintenance of the right ratio of their cellular redox potential. However, species and genotype have a substantial impact on the strength and capacity of antioxidant defense (Montoro et al. 2005). In antioxidant-based plant defense mechanisms, cellular ROS equilibrium is required because ROS have two functions in plants. As signaling molecules, they not only regulate the growth and development of plants but also aid in their adaptation to challenging conditions. However, when their concentration rises under stress, they have negative effects on cellular metabolism (Halliwell 2006). Plants possess an intricate antioxidant system that helps them maintain the proper balance of their cellular redox potential. However, there is a significant species- and genotype-dependent variation in the potency and potential of antioxidant defense (Duan et al. 2021). In order to prevent or postpone cellular damage, plants use both enzymatic and nonenzymatic defense mechanisms that eliminate ROS or reduce their oxidative activity (Dumont and Rivoal 2019).

1.4.3 Improvement of Nutrient Uptake and Utilization

Modern agriculture uses a great deal of ineffective fertilizer, much of which is discharged into the environment and degrades the ecology. Increasing crop nutrient absorption by using biostimulants is one method to minimize fertilizer usage without impairing plant nutrition. Plant biostimulants are defined broadly in this study to cover substances that are sometimes categorized as biopesticides or biofertilizers. When applied to plants, seeds, or growing substrates in certain formulations, plant biostimulants, as opposed to fertilizers and pesticides, can change a plant's physiological processes in a way that may be beneficial to the plant's growth, development, or stress response. This description contains a lot of components, four of which will be covered in this article: amino acids, seaweed extract, humic substances, and microorganisms that aid in plant growth. The advantages of using biostimulants on plants and the foundational processes that make them possible will be the main topics of this section. These techniques include improvements in symbiotic relationships, plant physiology, root morphology, and the structure or solubility of nutrients in the soil. Future research paths should focus on identifying the most promising compounds, isolating the active constituents, and clarifying the mechanisms through which they influence dietary consumption. Experiments carried out in fields and greenhouses must continually highlight the benefits and mechanisms (Halpern et al. 2015).

Both microbial and nonmicrobial plant biostimulants can improve soil exploration, root system design, macro- and micronutrient solubilization, and nutrient usage efficiency (NUE), particularly nitrogen (De Pascale et al. 2017). Research has indicated that the topical application of biostimulants derived from seaweed and vegetables, such as protein hydrolysates or extracts from tropical plants or the brown macroalgae (*Ecklonia maxima*), is a sustainable way to boost greenhouse baby lettuce yield and NUE in low-input cropping systems (Di Mola et al. 2019).

When legume-derived protein hydrolysates, namely seaweed extract, were applied to plants that were infused with tropical plant extract or that were growing under both ideal and suboptimal N fertilization conditions (0 and 10 kg ha^{-1}), the authors of the study discovered that fresh yield increased dramatically. Nevertheless, when administered to plants that were fertilized with abundant N (20 and 30 kg ha^{-1}), the positive effects of plant biostimulants were less noticeable. The same study team saw comparable outcomes when two additional noteworthy greenhouses leafy crops, lamb's lettuce and baby spinach, were grown under both optimal and suboptimal N regimes and treated with hydrolysates of proteins obtained from legumes. Applying vegetal-based foliar biostimulants significantly improved the N absorption and utilization efficiency of leafy crops (19% and 18%, respectively, for baby spinach and 50% and 73%, respectively, for lamb's lettuce). In lamb's lettuce, baby spinach, and baby lettuce, higher agronomic performance and consumption efficiency were found to be associated with enhanced photosynthetic activity and biochemical status (higher levels of carotenoids and total, a, and b chlorophyll) (Di Mola et al. 2019). In greenhouse lettuce grown under three distinct N conditions

(0, 70, and 140 kg ha^{-1}), the treatment of both nonmicrobial (vegetal biopolymer containing amino acids, peptides, and vitamins) and microbial (*Trichoderma virens*) biostimulants showed synergistic biostimulant action (Rouphael and Colla 2020).

1.4.4 Stimulation of Hormonal Balance and Growth Regulators

Plant growth regulators are used extensively in modern agriculture, winemaking, and horticulture. The hormonal state of the plant is actively impacted by "true" plant growth regulators. Hormone receptor blockers, hormone synthesis or translocation inhibitors, and synthetic analogues of plant hormones are among them. A phytotoxic effect that is either transient or confined is how "atypical" plant growth regulators work. There are about forty active substances that are used, either separately or in mixtures. Many processes related to plant development can be actively controlled in cultivated plants. These include inducing flowering and fruiting, accelerating or delaying seed germination, breaking dormancy in woody perennials, stimulating or reducing shoot elongation, reducing or increasing fruit set, and accelerating or delaying senescence processes like fruit ripening and defoliation. The benefits that are experienced include improved crop management, increased yield and quality of harvested products, and improved storage and shelf life. Plant growth regulators were first systematically used in the 1930s. Presently, yearly worldwide sales hover around US$ 1.2 billion (Rademacher 2015).

Plants have a special ability to adapt their growth to constantly shifting environmental circumstances. Plants are remarkably versatile because they can grow new organs and maintain populations of stem cells throughout their entire life cycle. Signaling chemicals called plant hormones, including gibberellin, brassinosteroid, ethylene, auxin, cytokinin, jasmonic acid, strigolactone, and abscisic acid, regulate and synchronize these developmental processes. The molecular elements of each hormone pathway's signal transduction and perception have been broken down by physiological and genetic research. However, it's now evident that hormones work through several linear channels. A complex network of interactions and feedback loops between the various hormone pathways determines the outcome of the activity of the individual hormones (Vanstraelen and Benková 2012).

Plant hormones are necessary for plants to be able to adjust to their ever-changing environment because they control development, growth, source/sink transitions, nutrition, and distribution. Auxin, cytokinin, as well as brassinosteroids, are important under environmental stress, even though ABA is the hormone that reacts to stress the most. This is becoming increasingly evident. Plant hormones are involved in several processes in recent studies. Numerous plant hormones, which can interact antagonistically or cooperatively, influence how plants react to abiotic stress. This knowledge of the molecular mechanisms behind hormone production, signaling, and function enables the development of transgenic agricultural plants with enhanced tolerance to abiotic stress by altering the hormone biosynthesis pathways (Peleg and Blumwald 2011).

Inqlab-91, a salt-tolerant cultivar of *Triticum aestivum* L., and MH-97, a salt-intolerant cultivar, were the two cultivars whose primed and unprimed seeds were sown in a field salinized with 15 dS m^{-1} NaCl in order to clarify the physicochemical changes resulting from GA_3-priming that induce salt tolerance in wheat. When both cultivars were subjected to salt stress, they reacted differentially to three different GA_3 dosages (100, 150, and 200 mg L^{-1}); the most noticeable reaction was shown with the 150 mg L^{-1} GA_3 dose. In mature wheat plants cultivated in salt, seed priming with GA_3 altered the order of ion accumulation between shoots and roots. After applying 150 mg L^{-1} of GA_3, the concentrations of K^+ and Ca^{2+} in the roots increased in both wheat cultivars, while the concentrations of Na^+ in the roots and shoots decreased. There were mixed results from GA_3-priming regarding the gaseous exchange parameters and auxin levels in the salt-stressed plants of both wheat varieties. The opposite was true in saline conditions for the salt-tolerant cultivar; however, all GA_3 doses decreased the levels of leaf-free ABA in the salt-intolerant cultivar. In either wheat cultivar under salt stress, priming with GA_3 (150 mg L^{-1}) markedly increased the salicylic acid (SA) level. After being treated with GA_3 (100–150 mg L^{-1}), the plants of both wheat cultivars showed lower amounts of putrescine (Put) and spermidine (Spd). Given that the salt-stressed plants of the salt-intolerant cultivar treated with GA_3 had lower concentrations of ABA and polyamines (Put and Spd), they were probably less stressed than the control group. Thus, it was shown that the physiological changes brought about by GA_3-priming included the absorption and partitioning of ions (within shoots and roots) as well as the modification of hormone homeostasis under saline conditions (Iqbal and Ashraf 2013).

Finding the physiologically active chemicals and their molecular mechanism of action, as well as analyzing the effects on plant development, would ultimately be necessary to fully comprehend the biological activity of complex biostimulant formulations (Henda and Bordenave-Juchereau 2013). Many molecular methods, including proteomics, microarrays, metabolomics, and transcriptomics, have been used to determine the active components of biostimulants. These techniques have been applied in combination with biostimulants to study changes in gene expression after the administration of biostimulants (Jannin et al. 2012). Much more has to be discovered regarding the functions underlying growth responses and stress alleviation, as well as the impacts of complex biostimulants and their constituents on the transcriptome and genome of plants (Khan et al. 2011).

Numerous biostimulants have been shown to solely activate genes and improve plant responses to biotic and abiotic stress, which complicates the search for their exact mechanisms of action. Hence, experimental procedures must be developed to produce pertinent and reproducible stress circumstances in order to guarantee that the findings of any molecular instrument used to study gene activity are pertinent to the stated impacts on plant productivity. The role of signaling molecules in plants' reactions to environmental stimuli has lately been thoroughly studied in plant biology. There are various stages involved in signal transmission: signaling molecules (ligands) are synthesized, translocated, bound to receptors, followed by cellular reactions, and eventually removed from the system. Secondary messengers, also known as intracellular signaling mediators, are activated as soon as a signaling molecule attaches to its receptor. This is the first physiological activity that occurs. This sets off a sequence of subsequent biological processes. A variety of substances, including as gases, ions, nitric oxide, lipids, cAMP, ADP-ribose, sugars, inositol-1,4,5-triphosphate, phosphoinosides, cGMP, nucleotides, Ca_2^+, cyclic small GTPase, and 1,2-diacylglycerol, can function as secondary messengers (Zhao et al. 2005).

While lipophilic substances are the main source of cytosol-mediated activity, water-soluble molecules generally show membrane-mediated action. Plant growth and metabolic processes are controlled by a complex network of interdependent hormones and cofactors whose actions are closely related. Hormones play a major part in this process. Hormonal states in plants are known to be influenced by biostimulants, which include complex organic materials, seaweed, humic chemicals, free amino acids, antitranspirants, and crude extracts of both lower and higher plants (Yakhin et al. 2012; Rathore et al. 2009). Even though biostimulants may also contain molecules that mimic hormones, it is possible that they will cause treated plants to produce hormones on their own. Numerous biostimulants contain organic acids, glycosides, amino acids, and polysaccharides, which can stimulate or function as precursors of endogenous plant hormones (Jannin et al. 2012). It is therefore possible to link the activity of natural biostimulants derived from higher plants, animals, microbes, algae, and humate-based raw materials to hormones or traits similar to hormones (Yakhin et al. 2017).

1.5 Application of Biostimulants in Oilseed Crops

Biostimulants are applied to soil as solutions, liquids, powders, or granules to prepare it (Kocira et al. 2018). Applications of nitrogen and humic acid compounds to the soil are common (Drobek et al. 2019). Furthermore, biostimulants can be administered topically and absorbed by plants that get water through irrigation systems. Sprayed on *Phaseolus vulgaris* L., *Kelpak SL* (sea weed extract) is one example of this (Kocira et al. 2018). Biostimulants can be dipped in seed or applied as a seed treatment prior to planting. The application of biostimulants produced from sewage sludge topically increased the levels of micro and macronutrients in maize leaves (Tejada et al. 2016). Higher quantities of carotene and chlorophyll are seen when *Ecklonia maxima* seaweed extract is applied topically to *Brassica oleracea* (Rengasamy et al. 2016). Increase in the volume of biostimulants decreased seed vigor in the case of soybean (Brzezinski et al. 2017). The use of biostimulants on crops has the potential to influence growth and production characteristics, resulting in higher yields and better quality. Meanwhile, biostimulants are cheaper than fertilizers and are eco-friendly in nature; the use of biostimulants could minimize hazardous effects caused by fertilizers and pesticides, like water pollution, contamination of food, and danger to birds and other animals. Crops treated with biostimulants may experience changes in growth and production characteristics, resulting in higher yields and better-quality products. Meanwhile, biostimulants are cheaper than fertilizers and are eco-friendly in nature; use of biostimulants could minimize hazardous effects caused by fertilizers and pesticides, like water pollution, contamination of food, and danger to birds and other animals.

1.5.1 Methods of Application

Plant biostimulants influence growth, yield as well and quality of oil seeds. Biostimulants are naturally available or synthetic. They influence the production and productivity of crop plants. These are substances or microorganisms or a combination of both with influence on nutrient uptake, growth, yield and quality of seeds, they also protect seeds and seedlings from abiotic stress, nutrient deficiencies and enhance the seed and seedling vigor. Biostimulants include things like fulvic acids, seaweed extract, polyphenols, helpful fungus, inorganic salts, amino acids, humic acids, plant derivatives, beneficial bacteria, beneficial symbiosis and waste products from industry and agriculture. In India, biostimulants are regulated under Fertilizer Control Order. Bio stimulants are applied as soil or foliar application, seed treatment or seedling dip.

1.5.2 Timing and Dosage of Biostimulant Application

Aschophyllum nodusum extract influences the growth and development of soybeans by acting as a biostimulant. At 60 DAS (days after sowing), plants that received 500 ml/ha of biozyme crop with half of the suggested nitrogen, phosphorus and potassium (NPK) continued to have 30% more trifoliate, 14.20% more nodules per plant, 25.12% more fresh weight nodules, and 17.51% more dry weight nodules than the control (recommended fertilizer dose) (Tandon and Dubey 2015). Because auxins, a plant hormone, are present in seaweed extract, *A. nodosum* extract contains cytokinin, IAA, amino acids, and vitamins. These substances stimulate the formation of axillary buds through cell division and elongation, which may have an impact on plant height (Stirk et al. 2004; Sanderson et al. 1987).

The effects of growth regulators, including auxin, gibberellin, and cytokinin, on soybean crops are found to have a beneficial effect up to a specific concentration. Hormonal imbalance caused by large doses results in detrimental physiological effects. Two applications of the terra sorb complex (0.5%) enhanced pod count per plant by 32% and seed count per plant by 38% in soybeans as compared to the control (Kocira et al. 2018). It is due to the application of the terra sorb complex, which contain free amino acid, which works like gibberellic acid (Colla et al. 2014). The quantity of pods and seeds on soybeans rose when *Achophyllum nodosum* extract, which contains free amino acids, was applied (Kocira et al. 2018). A similar effect was seen in the quantity of pods and seeds produced by applying red algae (*Kppaphycus alvarezii*) extract topically (Rathore et al. 2009). When polyphenols are applied, identical outcomes to the first pod's height rise can be observed (Kocira et al. 2018). Improved plant height (118.07 cm), branch count (6.67), and leaf count per plant (37.73) were noted in fodder cowpea plants treated with humic acid (8 ml/L). This forage cowpea yielded higher yield metrics than the control, such as forage yield of 13.91%, pod length of 43.4%, number of pods per plant of 33.9%, and 100-seed weight of 10.2% (Hongal et al. 2023). It was linked to how foliar humic acid treatment improved photosynthetic assimilation in plants (Delfine et al. 2005; El-Hak et al. 2012). Humic acid, a commercial chemical made up of 44–58% carbon, 42–46% oxygen, 6–8% hydrogen, and 0.5–4% nitrogen, along with other distinct elements, improved soil fertility and increased the availability of nutrients (Larcher 2003). This could be the reason for the promotion of plant development and growth.

(a) *Influence of Biostimulants on Yield of Oilseed Crops*

When soybeans were double-sprayed with biostimulants (amino acids) at greater concentrations, the yield rose by 25% above the control (Kocira et al. 2018). Increased yield can result from the terra sorb complex's free amino acid content (Colla et al. 2014). The highest yield in soybeans was obtained when nozzle API 2003 was employed, which was 11.35% greater than nozzle 6MS03C, when biostimulants were sprayed foliarly using two different nozzles, according to Krawczuk et al. (2021). The effects of applying foliar fertilizer to maize crops using different nozzles were studied by Aloni et al. (2006). Compared to the control, the yield rose (3.277 t/ha) when the biozyme *Aschophyllum nodusum* was treated together with half of the necessary NPK (Tandon and Dubey 2015). It is a result of the action of the cytokinin, IAA, amino acid, and vitamin-containing *A. nodosum* extract (Taylor and Wilkinson 1977) seaweed extract's auxins and cytokinins, which encourage the growth and development of additional reproductive structures like flowers. While cytokinins engage in cell division, initiate the transfer of photoassimilates, and mobilize nutrients to sinks, auxin promotes the formation of pollen tubes and reproductive structures like fruits, as well as flowers (Aloni et al. 2006; Buchanan et al. 2015).

1.5.3 Case Studies: Biostimulant Use in Specific Oilseed Crops

(a) *Canola (Brassica napus, Brassica rapa)*

A common oilseed crop, canola, also referred to as rapeseed (*Brassica napus* L.), is the third-largest source of vegetable oil in the world. Seeds contain about 40–45% oil and 18–25% protein. Linoleic acid accounts for 18% and oleic acid for 60% of edible oil. A major issue in dry and semiarid regions of the world, where canola is commonly farmed, is salinity (Hu et al. 2006). Fresh weight/plant, dry weight/plant, and canola plant height tended to increase gradually as a result of the several treatments that were studied (mycorrhizae, humic acid, algal extract, and a combination of all). When contrasted to the control treatment, every administered therapy led to a statistically significant rise in every character under investigation.

(b) *Sunflower (Helianthus annuus)*

The performance of sunflower seedlings can impact the quality of oil and seed output. Biostimulators enhance the circumstances for seedlings and can boost plant immunity while reducing the effects of stress. This study aimed to determine which inbred line responded best to biostimulator therapy and to assess the impact of biostimulators on sunflower seedlings. Responses to several biostimulants have shown that the native Dilbar variety of sunflowers shows positive growth, development, and yield output. Depending on the kind, applying biostimulants, immunostimulants, or microbial fertilizers increased the number of seeds in the basket. In one basket, there were between 14.9 and 303 more seeds than in the control group. These seeds were big and had a full core. As a result, 1000 seeds turned out to weigh more than 102.7–107.8 g.

(c) *Soybean (Glycine max)*

Advances in plant breeding technologies aim to increase yields and improve product quality by minimizing threats to the natural environment. Application of biostimulants is a practice that increases product yield and quality, especially after plants have been exposed to stress conditions (Kocira 2019). Kocira (2019) applied Terra Sorb Complex biostimulant in two concentrations (0.3% and 0.5%) as single or double spraying during the growing season and noted that the number and concentration of biostimulant applications changed the yield and quality of the product as well as the nutraceutical and antioxidant potential of soybean. Tandon and Dubey (2015) conducted a field experiment to investigate the effects of Biozyme (*Ascophyllum nodosum*) on soybean. They reported that Biozyme granule and plant plus spray application with half of the recommended NPK showed significant effects on all growth and yield parameters of soybean. Melo et al. (2024) studied the agronomic performance of soybeans with seeds treated with algae-based biostimulants, observing greater grain yield and grain mass, as well as greater profitability at both harvests. In summary, different types of studies using biostimulants in soybean crops can be observed, reinforcing the importance of verticalizing the field production of this crop. The use of biostimulants in soybeans is a potential alternative input to enhance crop growth and development. Studies show that the application of biostimulants (via foliar or seed) in the soybean crop can provide increases in production variables. It is understood that further research is needed to provide better information on agronomic interests (Da Silva et al. 2023).

(d) *Peanut/Groundnut (Arachis hypogaea)*

The use of biostimulants in agriculture is suitable for promoting the initial development of various crops; however, reports on their impact on peanut cultivation are scarce. Aguilera et al. (2024) conducted a study to investigate the effect of application doses of Biostimulant Seed+ on the initial development of two pistachio cultivars under controlled greenhouse conditions. Their results showed that the use of two improved contrasting cultivars with upright and horizontal growth habits gave different responses to the applied biostimulant doses along with crop development of pistachio cultivars, especially for the cultivar IAC Tatu ST. Application of 9 mL kg^{-1} biostimulant Seed+ in seed treatment could be an effective alternative to promote the initial development of pistachio cultivars cultivated in sandy soil of Cerrado Sul-Mato-Grossense under controlled greenhouse conditions. Eswaran et al. (2024), in their study on the growth regulation of *Arachis hypogaea* L under drought stress by osmolyte-producing microbial biostimulants, indicated that *Acinetobacter* sp. AC06 and *Bacillus* sp. BA01 could be considered as osmolyte-producing microbial biostimulants to simultaneously induce osmotic tolerance and metabolic changes in peanuts under drought stress. Many biostimulant trials continue to be conducted by various researchers.

(e) *Cottonseed (Gossypium subsp.)*

It is known that biostimulants are used to balance vegetative and reproductive growth in cotton (*Gossypium hirsutum* L.) and to increase cotton seed yield and fiber quality. Therefore, Silva et al. (2016) established a field experiment to investigate the effectiveness, nutrition, production and technological quality of seed treatment with biostimulants for cotton fiber. The experiment was arranged in a completely randomized block design with four replicates and five treatments (control group, untreated group, Booster®, Stimulate®, Improver® and Biozyme®). Nutrient contents in leaves, yield and technological quality of fiber were evaluated. The results showed that biostimulant application to seeds increased N, K, S and Fe contents in cotton leaves, but had no effect on product yield. However, these products caused changes in fiber properties related to length uniformity, micronaire, length, and strength of fiber. Mantzoukas et al. (2023) reported that increasing commercial demands from textile and food industries have created a strong pressure on the cultivation of cotton and its derivatives to obtain high yields, at the same time, cotton has high nutrient and irrigation requirements and is highly susceptible to insect pests. Therefore, adhering to the theory that coating cotton seeds with beneficial fungi could solve these problems, they investigated the biostimulant and bioinsecticidal effects of coating cotton seeds with endophytic *Beauveria bassiana* under semifield conditions and studied the growth of cotton using (A) conventional seeds and (B) seeds coated with the entomopathogenic fungus *Beauveria bassiana* (Hypocreales: Cordycipitaceae). Besides plant growth, a significantly lower aphid population was recorded in inoculated plants after the fifth week compared to the control in *Aphis gossypii* (Hemiptera: Aphididae) infestation monitored for 6 weeks. They demonstrated that *B. bassiana* coating processes have a significant biostimulatory and bioinsecticidal effect. Biostimulant studies will continue, as cotton is an industrially important plant.

(f) *Castor Bean (Ricinus communis)*

In drylands, low soil fertility, imbalance of nutrients and inability of crops to absorb some nutrients are among the main constraints in crop production. The use of biofertilizers and biostimulants can be a suitable solution (Esmaeilian et al. 2023). Esmaeilian et al. (2023) conducted a 2-year field experiment as a factorial experiment based on randomized complete block design with three replications to study the response of castor oil plant to arbuscular mycorrhizal fungi (AMF) and amino acid biostimulants (AAB) (foliar, soil and combined application methods). The findings showed that plants inoculated with AMF showed better performance: biological and seed yields increased by an average of 20.9% and 26.4% during the 2 years of the experiment. Yield components increased by 14.0% to 18.6% and water efficiency (WP) improved by 13.9%. Seed oil content increased by 5.1%: Linoleic acid, a fatty acid, showed the highest response with an increase of 10.0%. Among the AAB applications, the combined application method (CAM) brought the best results. The highest increases in biological and seed yields (68.4% and 63.2%) were obtained from CAM application. WP boosted by 67.8% and seed oil content increased by 9.4%. Among fatty acids, the highest increase (24.4%) was noted for linoleic acid. The results indicated that the castor oil plant showed a positive and significant response to biostimulants, indicating that replacing chemicals with biofertilizers could be a promising approach for the cultivation of the castor oil plant in arid climates.

(g) *Safflower (Carthamus tinctorius)*

Application of growth regulators (GR) as an effective technique to reduce the harmful effects of environmental stresses on plants has received much attention (Davari et al. 2022). Davari et al. (2022) studied the possibility of modulating drought stress damage in safflower using exogenous paclobutrazol (PBZ) and an amino acid-based biostimulant (BS) as GR. A 2-year experiment was conducted on two safflower genotypes, Sina (a thorny cultivar) and Faraman (a thornless cultivar), with different levels of GR (control, 50 mg PBZ L^{-1}, 100 mg PBZ L^{-1}, and 500 mg BS L^{-1}) under two irrigation regimes: water deficit (WD) and well-watered (WW). It was reported that there was a significant decrease in photosynthetic pigment content, relative water content (RWC), seed number, 1000 seed weight, biomass, seed yield and oil content compared to well-watered conditions due to water deficit stress, while there was a significant increase in proline content, electrolyte leaching and antioxidant enzyme activities. Under stress conditions, foliar application of PBZ and BS improved photosynthetic pigments and proline content, relative water content (RWC), cell membrane stability, the enzymatic defense system and dry matter accumulation, compensated for stress damage in Sina cultivar and subsequently increased seed yield. On the other hand, GR applications did not have any ameliorative effect on seed yield and biomass of the Faraman cultivar under drought stress. In general, it was shown that PBZ and BS can be used to alleviate the harmful effects of drought stress on safflower, depending on the genotype.

In addition to the biostimulant applications applied to the important industrial plants mentioned above under stress conditions, studies on other side industrial plants such as *Elaeis guineensis*, *Cocos nucifera*, *Olea europaea*, *Sesamum indicum*, *Linum usitatissimum*, *Brassica juncea*, *Zea mays*, *Oryza sativa*, *Jatropha curcas*, *Camelina sativa*, *Nigella sativa*, *Cucurbita pepo*, *Cannabis sativa*, *Oenothera biennis*, *Perilla frutescens*, *Vernicia fordii*, *Moringa oleifera*, *Guizotia abyssinica*, *Azadirachta indica*, *Pongamia pinnata*, *Attalea speciose*, *Madhuca longifolia*, *Shorea robusta*, *Persea americana*, *Vitis vinifera*, etc. continue at various levels.

1.6 Benefits and Limitations of Biostimulant Use

The biostimulants encompass a broad spectrum of goods, such as microbial inoculant formulations, derivatives of amino acids, humic acid, and seaweed extracts. With dwindling natural resources, the oilseeds industry in Asian and African nations is finding it difficult to feed a population that is expanding quickly. For improved yields, quality, and nutrient management in these areas, the application of biostimulants has become essential. In addition to stimulating the plasma membrane and increasing the soil's ability for cation exchange, humic material also improves the amount of nutrients that roots can absorb. The H^+ ATPases and antioxidants such as glycine and L-proline help scavenge-free radicals and control ion fluxes across membranes, while melatonins help seedlings resist stress more effectively. Due to their hormonal activity, gibberellic acids, brassinosteroids, abscisic acid, salicylic acid, and jasmonic acid are biostimulants that serve a variety of purposes. While plant extracts function as stress relievers, seaweeds and other plant materials containing cytokinins and auxins promote growth.

1.6.1 Advantages of Biostimulant Application in Saline Conditions

Natural compounds called biostimulants can promote plant development even in the face of abiotic and hydric stressors, potentially lowering the demand for fertilizers. Their minimal requirements support the efficient operation of the plant's essential operations, which produce high-yield, high-quality goods; algae, seaweed extracts, bacteria, protein hydrolysates, and beneficial fungi are among the active ingredients utilized as biostimulants. Humic and fulvic acids are also included. Although it can deliver the beneficial effects of several components, it can be formulated with one or more components. One can use biostimulants as soil or foliar treatments. These might be results of hydrolysis, fermentation, or extraction. They can lessen air pollution and soil deterioration while being environmentally safe. When used with horticultural crops, particularly fruits and vegetables, they work better. They could be used in conjunction with sustainable agricultural production methods.

1.6.2 Limitations and Challenges in the Adaptation of Biostimulant

Addition of chemicals, either organic or inorganic, and/or microbes to plants can stimulate several processes that boost growth, productivity, and resilience to stress. Leaves or the soil near the root zone can be sprayed with them. They come in soluble powder, liquid, and granular forms. With a potential market size of 5.6 billion dollars by 2026, these compounds are growing at a rate of 12% annually. Eco-friendly agronomic instruments are considered to be biostimulants (Lau et al. 2022; Bulgari et al. 2019). Numerous factors, such as the desire for more environment-friendly alternatives to synthetic pesticides and modifications to agricultural and environmental regulations in light of climate change, have contributed to the growth of the biostimulant industry (Lau et al. 2022). Biostimulants have little to no toxicity and don't accumulate over time (Sangiorgio et al. 2020). Although there are numerous varieties of biostimulants, they can be divided into six main classes according to the source of the raw material: Seaweed, humic substances, protein hydrolysates, plant extracts, inorganic chemicals, and microbes (Franzoni et al. 2021). Extraction of different plant components, including stems, leaves, flowers, and fruits, can provide some biostimulants (Franzoni 2020). Numerous bioactive compounds found in these extracts can activate specific physiological processes in plants, improving their general health (Zulfiqar et al. 2020).

1.6.3 Economic Consideration and Cost–Benefit Analysis

Biostimulants are becoming more and more popular in modern agriculture. These substances and microbes are applied to soils or plants in order to boost agricultural output, growth, and vigor. Their ability to increase plant stress tolerance and nutrient usage efficiency makes them particularly desirable from an environmental and financial standpoint. When evaluating the economic viability of biostimulants, a cost–benefit analysis (CBA) is essential, considering both direct and indirect economic impacts.

(i) *Cost of Biostimulant Use*

The cost of biostimulants varies significantly depending on their type, formulation, and the application rates required for different crops. Products like amino acids, microbial inoculants, humic acids, seaweed extracts, and other naturally occurring substances can all be considered biostimulants. Generally, these products are priced higher than traditional fertilizers or pesticides, primarily due to the research, development, and manufacturing processes involved in producing them (Yakhin et al. 2017). Despite the upfront costs, biostimulants often provide long-term savings through reduced input costs. For example, when biostimulants improve the efficiency of nutrient uptake or increase the tolerance of plants to biotic and abiotic challenges, farmers can use less synthetic fertilizers, water or pest control methods (Calvo et al. 2014). Additionally, reduced dependency on chemical inputs can lower the costs associated with environmental compliance or damage mitigation in the long run (Rouphael and Colla 2020).

(ii) *Benefits of Biostimulant Use*

The economic benefits of using biostimulants are primarily derived from increased crop yield, improved crop quality, and reduced input costs. According to studies, biostimulants can increase the production and development of crops, including fruits, vegetables, and grains. According to Bulgari et al. (2019), biostimulants administered in both controlled and field settings resulted in a considerable boost in tomato and pepper yield. According to Van Oosten et al. (2017), using biostimulants could improve crop quality as well, resulting in higher pricing on the market by enhancing fruit and vegetable size, nutritional content, and shelf life. Additional benefits include increased drought tolerance and resilience to pests and diseases, which lessen the need for treatments and increase farmers' income.

(iii) *Cost–Benefit Analysis (CBA)*

Conducting a cost–benefit analysis (CBA) involves comparing the costs associated with biostimulant use against the financial gains from increased crop yield, improved quality, and reduced input costs. Research indicates that while the initial investment in biostimulants may be higher, the long-term economic returns are favorable, particularly when the environmental and agronomic benefits are factored in. According to Colla et al. (2014), growers could anticipate a return of $4 to $7 for every dollar invested in biostimulants because of increased yields and decreased input requirements. This indicates that horticultural crops treated with biostimulants produced a 4:1 to 7:1 return on investment (ROI). Moreover, farmers using biostimulants report lower costs related to water usage and pest management, as these products often improve plant resilience to environmental stressors.

(iv) *Environmental and Economic Synergies*

The environmental benefits of biostimulants, such as improved soil health, reduced nutrient leaching, and enhanced carbon sequestration, also contribute indirectly to economic advantages. Biostimulants can assist in mitigating the environmental costs of intensive farming, such as soil deterioration and water pollution, by encouraging more sustainable agricultural practices (Du Jardin 2015). These benefits can result in long-term economic savings by maintaining or enhancing the productivity of agricultural land and reducing regulatory costs associated with environmental damage.

(v) *Challenges in Economic Adoption*

Despite these advantages, the widespread adoption of biostimulants faces certain economic and practical challenges. The variability in the effectiveness of biostimulants across different crops, soil types, and climatic conditions makes it difficult for farmers to predict consistent returns on their investment (Rouphael and

Colla 2020). Adoption may also be restricted by the exorbitant expense of biostimulant chemicals and the absence of uniform laws controlling their application in various areas.

1.7 Future Perspectives and Research Directions

1. *Emerging Trends in Biostimulant Research*

Future research should concentrate on the development of novel biostimulant compounds that are especially designed to improve salt tolerance in diverse oilseed crops. These formulations may include natural plant extracts, microbial inoculants, and organic chemicals that work together to increase crop performance under salty circumstances.

2. *Understanding Mechanisms of Action*

Although biostimulants have shown potential in reducing salt stress, the underlying processes are not well understood. Future research should seek to understand the physiological, pharmacological, and molecular routes by which biostimulants improve salt tolerance. This insight will assist in creating more successful goods and tactics.

3. *Integration with Crop Management Practices*

To optimize the advantages of biostimulants, future studies should look at how these products might be used with other sustainable agriculture techniques. For example, integrating biostimulants with improved irrigation, soil management, and fertilizer management may result in more robust oilseed crop production systems.

4. *Field Trials and Large-Scale Studies*

While several studies have shown that biostimulants are beneficial in controlled situations, there is a need for large field trials in various geographies and soil types. These studies will help establish biostimulants' efficacy in real-world settings, paving the way for wider deployment.

5. *Development of Stress-Specific Biostimulants*

Given the variation in salinity stress between places, it is critical to create stress-specific biostimulants capable of addressing regional difficulties. Research should concentrate on developing products that meet the unique demands of oilseed crops under various levels of saline stress.

6. *Omics Approaches*

Advanced tools such as genomics, transcriptomics, proteomics, and metabolomics can be employed to study the impact of biostimulants on oilseed crops at a molecular level. These approaches can provide insights into how biostimulants alter gene expression, protein synthesis, and metabolic pathways to enhance salinity tolerance.

7. *Exploration of Indigenous Microorganisms*

There is a need to investigate the potential of indigenous microbes as biostimulants, particularly those that are naturally adapted to salty environments. These microorganisms can be used to improve crop salinity tolerance through activities such as nitrogen fixation, phosphorus solubilization, and the synthesis of stress-related metabolites.

8. *Cost-Effective and Sustainable Production*

The widespread use of biostimulants will be determined by their cost-effectiveness. To make biostimulants more accessible to farmers, researchers should look for sustainable and low-cost sources, such as agricultural waste, byproducts, or renewable resources.

9. *Climate Change Adaptation*

As climate change exacerbates soil salinity issues, biostimulants present a possible alternative for dealing with these challenges. Future studies should look into how biostimulants might be used in climate-smart agriculture practices to help oilseed crops maintain productivity in increasingly demanding environmental circumstances. By addressing these research topics, biostimulants' promise to reduce salt stress in oilseed crops can be fully realized, resulting in more sustainable and resilient agriculture systems.

1.8 Conclusion

Salinity stress has a substantial impact on the productivity of oilseed crops, affecting their growth, yield, and quality. The use of biostimulants has emerged as a potential, environmentally benign approach to improving salt tolerance in these crops. Biostimulants, which include natural plant extracts, microbial inoculants, and organic substances, improve physiological and biochemical processes, promote root growth, increase nutrient uptake, and activate stress-related pathways. While significant progress has been made in understanding the benefits of biostimulants, more study is needed to completely understand the mechanisms by which they reduce salt stress and optimize their formulations for specific crop requirements. Large-scale field experiments and integration with other sustainable farming methods are required to assure their efficacy and widespread adoption. Furthermore, the development of low-cost, high-quality, and regionally adapted biostimulant products can increase their use, allowing farmers to control salinity stress more effectively. In conclusion, biostimulants represent a sustainable and innovative approach to enhancing the resilience of oilseed crops against salinity stress. Continued research and development in this field will not only improve crop performance but also contribute to more sustainable and climate-resilient agricultural practices. To summarize, biostimulants are a sustainable and unique strategy to increasing the resilience of oilseed crops to salinity stress. Continued research and development in this discipline will help to increase crop performance while also contributing to more sustainable and climate-resilient agriculture methods.

References

Abbas W, Ashraf M, Akram NA (2010) Alleviation of salt-induced adverse effects in eggplant (*Solanum melongena* L.) by glycinebetaine and sugarbeet extracts. Sci Hortic 125(3):188–195

Abdalla MM (2013) The potential of *Moringa oleifera* extract as a biostimulant in enhancing the growth, biochemical and hormonal contents in rocket (*Eruca vesicaria* subsp. *sativa*) plants. Int J Plant Physiol Biochem 5(3):42–49

Abdel Latef AA, Srivastava AK, Saber H, Alwaleed EA, Tran L-SP (2017) *Sargassum muticum* and *Jania rubens* regulate amino acid metabolism to improve growth and alleviate salinity in chickpea. Sci Rep 7:10537

Abdel Latef AAH, Mostofa MG, Rahman MM, Abdel-Farid IB, Tran L-SP (2019) Extracts from yeast and carrot roots enhance maize performance under seawater-induced salt stress by altering physio-biochemical characteristics of stressed plants. J Plant Growth Regul 38(3):966–979

Abdelhamid MT, Rady MM, Osman AS, Abdalla MA (2013) Exogenous application of proline alleviates salt-induced oxidative stress in *Phaseolus vulgaris* L. plants. J Hortic Sci Biotechnol 88(4):439–446

Abdou M, Badran F, Ahmed E, Taha R, Abdel-Mola M (2018) Effect of compost and some natural stimulant treatments on: II. Corms production and chemical constituents of (Gladiolus grandiflorus cv. Peter Pears) plants. Sci J Flowers Ornamental Plants 5(2):115–126

Aguilera J, do Nascimento ACV, Rocha AS, de Oliveira IS, Guedes JVFL, Steiner F, Zuffo AM (2024) Impact of applying the biostimulant Seed+ on initial growth of peanut cultivars. Trends Agric Environ Sci:e240001. https://doi.org/10.46420/TAES.e240001

Alasvandyari F, Mahdavi B (2018) Effect of glycine betaine and salinity on photosynthetic pigments and ion concentration of safflower. Desert 23(2):265–271

Ali O, Ramsubhag A, Jayaraman J (2021) Biostimulant properties of seaweed extracts in plants: implications towards sustainable crop production. Plan Theory 10(3):531

Aloni R, Aloni E, Langhans M, Ullrich CI (2006) Role of auxin in regulating Arabidopsis flower development. Planta 223:315–328

Ambast S (2006) Land levelling, shaping, planning and design in irrigation command areas. Sustainable Irrigated Agriculture

Azeem M, Pirjan K, Qasim M, Mahmood A, Javed T, Muhammad H, Yang S, Dong R, Ali B, Rahimi M (2023) Salinity stress improves antioxidant potential by modulating physio-biochemical responses in *Moringa oleifera* Lam. Sci Rep 13(1):2895

Baglieri A, Cadili V, Monterumici CM, Gennari M, Tabasso S, Montoneri E, Nardi S, Negre M (2014) Fertilization of bean plants with tomato plants hydrolysates. Effect on biomass production, chlorophyll content and N assimilation. Sci Hortic 176:194–199

Bajwa AA, Farooq M, Nawaz A (2018) Seed priming with sorghum extracts and benzyl aminopurine improves the tolerance against salt stress in wheat (*Triticum aestivum* L.). Physiol Mol Biol Plants 24:239–249

Baltazar M, Correia S, Guinan KJ, Sujeeth N, Bragança R, Gonçalves B (2021) Recent advances in the molecular effects of biostimulants in plants: An overview. Biomol Ther 11(8):1096

Bavaresco L, Lucini L, Squeri C, Zamboni M, Frioni T (2020) Protein hydrolysates modulate leaf proteome and metabolome in water-stressed grapevines. Sci Hortic 270:109413

Brzezinski CR, Abati J, Henning FA, Henning AA, França JB, Krzyzanowski FC, Zucareli C (2017) Spray volumes in the industrial treatment on the physiological quality of soybean seeds with different levels of vigor. J Seed Sci 39:174–181

Buchanan BB, Gruissem W, Jones RL (2015) Biochemistry and molecular biology of plants. Wiley

Bulgari R, Franzoni G, Ferrante A (2019) Biostimulants application in horticultural crops under abiotic stress conditions. Agronomy 9(6):306

Calvo P, Nelson L, Kloepper JW (2014) Agricultural uses of plant biostimulants. Plant Soil 383:3–41

Canellas L, Dobbss L, Oliveira A, Chagas J, Aguiar N, Rumjanek V, Novotny E, Olivares F, Spaccini R, Piccolo A (2012) Chemical properties of humic matter as related to induction of plant lateral roots. Eur J Soil Sci 63(3):315–324

Canellas LP, Olivares FL, Aguiar NO, Jones DL, Nebbioso A, Mazzei P, Piccolo A (2015) Humic and fulvic acids as biostimulants in horticulture. Sci Hortic 196:15–27

Carillo P, Colla G, Fusco GM, Dell'Aversana E, El-Nakhel C, Giordano M, Pannico A, Cozzolino E, Mori M, Reynaud H (2019) Morphological and physiological responses induced by protein hydrolysate-based biostimulant and nitrogen rates in greenhouse spinach. Agronomy 9(8):450

Carlsson AS (2009) Plant oils as feedstock alternatives to petroleum-A short survey of potential oil crop platforms. Biochimie 91(6):665–670

Ceccarelli AV, Miras-Moreno B, Buffagni V, Senizza B, Pii Y, Cardarelli M, Rouphael Y, Colla G, Lucini L (2021) Foliar application of different vegetal-derived protein hydrolysates distinctively modulates tomato root development and metabolism. Plan Theory 10(2):326

Cirillo C, Rouphael Y, Pannico A, El-Nakhel C, Colla G, De Pascale S (2017) Application of protein hydrolysate-based biostimulant as new approach to improve performance of bedding plants. In: International symposium on greener cities for more efficient ecosystem services in a climate changing world 1215, pp 443–448

Cisse EHM, Miao LF, Yang F, Huang JF, Li DD, Zhang J (2021) Gly betaine surpasses melatonin to improve salt tolerance in *Dalbergia odorifera*. Front Plant Sci 12:588847

Colla G, Rouphael Y, Canaguier R, Svecova E, Cardarelli M (2014) Biostimulant action of a plant-derived protein hydrolysate produced through enzymatic hydrolysis. Front Plant Sci 5:448

Craigie JS (2011) Seaweed extract stimuli in plant science and agriculture. J Appl Phycol 23:371–393

Da Silva JHB, da Silva AV, da Silva CM, Amorim PMD, do Carmo EVB, Da Sliva Carvalho KD, da Cunha Martins AHP, de Barros Franca KC, Custódio BA, da Silva EM, dos Santos Filho F, da Silva ODV, da Mata DA (2023) Soybean response (*Glycine max* L.) under the use of biostimulants: a literature review. Sci Electron Arch 16(10):8–13

Davari K, Rokhzadi A, Mohammadi K, Pasari B (2022) Paclobutrazol and amino acid-based biostimulant as beneficial compounds in alleviating the drought stress effects on safflower (*Carthamus tinctorius* L.). J Soil Sci Plant Nutr 22:674–690

Dawood MG, Abdel-Baky YR, El-Awadi MES, Bakhoum GS (2019) Enhancement quality and quantity of faba bean plants grown under sandy soil conditions by nicotinamide and/or humic acid application. Bull Natl Res Cent 43:1–8

Dawood MFA, Tahjib-Ul-Arif M, Sohag AAM, Abdel Latef AA (2024) Role of acetic acid and nitric oxide against salinity and lithium stress in canola (*Brassica napus* L.). Plan Theory 13(1):51

Dawood MFA, Tahjib-Ul-Arif Md, Shirazy BJ, Abdel Latef AAH (2025) Unraveling the role of κ-carrageenan on the combined effect of drought and chromium stress in wheat (Triticum aestivium L.). Plant Physiology and Biochemistry. 223, J109781

De Pascale S, Rouphael Y, Colla G (2017) Plant biostimulants: innovative tool for enhancing plant nutrition in organic farming. Eur J Hortic Sci 82(6):277–285

Delfine S, Tognetti R, Desiderio E, Alvino A (2005) Effect of foliar application of N and humic acids on growth and yield of durum wheat. Agron Sustain Dev 25(2):183–191

Delorge I, Janiak M, Carpentier S, Van Dijck P (2014) Fine tuning of trehalose biosynthesis and hydrolysis as novel tools for the generation of abiotic stress tolerant plants. Front Plant Sci 5:147

Demidchik V, Maathuis FJ (2007) Physiological roles of nonselective cation channels in plants: from salt stress to signalling and development. New Phytol 175(3):387–404

Desoky E-SM, Elrys AS, Rady MM (2019) Integrative moringa and licorice extracts application improves *Capsicum annuum* fruit yield and declines its contaminant contents on a heavy metals-contaminated saline soil. Ecotoxicol Environ Saf 169:50–60

Desoky E-SM, El-Maghraby LM, Awad AE, Abdo AI, Rady MM, Semida WM (2020) Fennel and ammi seed extracts modulate antioxidant defence system and alleviate salinity stress in cowpea (Vigna unguiculata). Sci Hortic 272:109576

Di Mola I, Ottaiano L, Cozzolino E, Senatore M, Giordano M, El-Nakhel C, Sacco A, Rouphael Y, Colla G, Mori M (2019) Plant-based biostimulants influence the agronomical, physiological, and qualitative responses of baby rocket leaves under diverse nitrogen conditions. Plan Theory 8(11):522

Di Mola I, Cozzolino E, Ottaiano L, Nocerino S, Rouphael Y, Colla G, El-Nakhel C, Mori M (2020) Nitrogen use and uptake efficiency and crop performance of baby spinach (*Spinacia oleracea* L.) and Lamb's Lettuce (*Valerianella locusta* L.) grown under variable sub-optimal N regimes combined with plant-based biostimulant application. Agronomy 10(2):278

Dobbss LB, Pasqualoto Canellas L, Lopes Olivares FB, Oliveira Aguiar NL, Peres LEP, Azevedo M, Spaccini R, Piccolo A, Facanha AR (2010) Bioactivity of chemically transformed humic matter from vermicompost on plant root growth. J Agric Food Chem 58(6):3681–3688

Drobek M, Frąc M, Cybulska J (2019) Plant biostimulants: importance of the quality and yield of horticultural crops and the improvement of plant tolerance to abiotic stress-A review. Agronomy 9(6):335

Du Jardin P (2015) Plant biostimulants: definition, concept, main categories and regulation. Sci Hortic 196:3–14

Duan Y, Santiago FEM, Dos Reis AR, de Figueiredo MA, Zhou S, Thannhauser TW, Li L (2021) Genotypic variation of flavonols and antioxidant capacity in broccoli. Food Chem 338:127997

Dumont S, Rivoal J (2019) Consequences of oxidative stress on plant glycolytic and respiratory metabolism. Front Plant Sci 10:166

El-Hak S, Ahmed A, Moustafa Y (2012) Effect of foliar application with two antioxidants and humic acid on growth, yield and yield components of peas (*Pisum sativum* L.). J Hortic Sci Ornamental Plants 4(3):318–328

Ertani A, Cavani L, Pizzeghello D, Brandellero E, Altissimo A, Ciavatta C, Nardi S (2009) Biostimulant activity of two protein hydrolyzates in the growth and nitrogen metabolism of maize seedlings. J Plant Nutr Soil Sci 172(2):237–244

Ertani A, Schiavon M, Altissimo A, Franceschi C, Nardi S (2011) Phenol-containing organic substances stimulate phenylpropanoid metabolism in *Zea mays*. J Plant Nutr Soil Sci 174(3): 496–503

Ertani A, Pizzeghello D, Francioso O, Sambo P, Sanchez-Cortes S, Nardi S (2014) *Capsicum chinensis* L. growth and nutraceutical properties are enhanced by biostimulants in a long-term period: chemical and metabolomic approaches. Front Plant Sci 5:375

Ertani A, Schiavon M, Nardi S (2017) Transcriptome-wide identification of differentially expressed genes in *Solanum lycopersicon* L. in response to an alfalfa-protein hydrolysate using microarrays. Front Plant Sci 8:1159

Esmaeilian Y, Babaeian M, Caballero-Calvo A (2023) Optimization of castor bean (*Ricinus communis* L.) cultivation methods using biostimulants in an arid climate. Euro-Mediterr J Environ Integr 8:823–834

Eswaran SUD, Sundaram L, Perveen K, Bukhari NA, Sayyed RZ (2024) Osmolyte-producing microbial biostimulants regulate the growth of *Arachis hypogaea* L. under drought stress. BMC Microbiol 24(1):165

Fan D, Hodges DM, Critchley AT, Prithiviraj B (2013) A commercial extract of brown macroalga (*Ascophyllum nodosum*) affects yield and the nutritional quality of spinach in vitro. Commun Soil Sci Plant Anal 44(12):1873–1884

Franzoni G (2020) Mechanisms of action of biostimulants in crops. P.hD dessertation submmited to Università degli Studi di Milano. 1–242
Franzoni G, Bulgari R, Ferrante A (2021) Maceration time affects the efficacy of borage extracts as potential biostimulant on rocket salad. Agronomy 11(11):2182
Garcia-Mina J, Antolin M, Sanchez-Diaz M (2004) Metal-humic complexes and plant micronutrient uptake: a study based on different plant species cultivated in diverse soil types. Plant Soil 258:57–68
Gerke J (2021) The effect of humic substances on phosphate and iron acquisition by higher plants: qualitative and quantitative aspects. J Plant Nutr Soil Sci 184(3):329–338
Ghezal N, Rinez I, Sbai H, Saad I, Farooq M, Rinez A, Zribi I, Haouala R (2016) Improvement of Pisum sativum salt stress tolerance by bio-priming their seeds using Typha angustifolia leaves aqueous extract. S Afr J Bot 105:240–250
Goni O, Fort A, Quille P, McKeown PC, Spillane C, O'Connell S (2016) Comparative transcriptome analysis of two *Ascophyllum nodosum* extract biostimulants: same seaweed but different. J Agric Food Chem 64(14):2980–2989
Goñi O, Quille P, O'Connell S (2020) Seaweed carbohydrates. In: The chemical biology of plant biostimulants, pp 57–95
Gupta B, Huang B (2014) Mechanism of salinity tolerance in plants: physiological, biochemical, and molecular characterization. Int J Genomics 2014(1):701596
Halliwell B (2006) Reactive species and antioxidants. Redox biology is a fundamental theme of aerobic life. Plant Physiol 141(2):312–322
Halpern M, Bar-Tal A, Ofek M, Minz D, Muller T, Yermiyahu U (2015) The use of biostimulants for enhancing nutrient uptake. Adv Agron 130:141–174
Henda YB, Bordenave-Juchereau S (2013) Using marine cryptides against metabolic syndrome. In: Bioactive compounds from marine foods: plant and animal sources, pp 95–112
Hongal G, Siddaraju R, Bai SK, Vishwanath K, Ramanappa T, Pavan A (2023) Bio-stimulants in oil seed crops-its influence on growth, yield and quality. Mysore J Agric Sci 57(4):11–24
Howladar SM (2014) A novel Moringa oleifera leaf extract can mitigate the stress effects of salinity and cadmium in bean (*Phaseolus vulgaris* L.) plants. Ecotoxicol Environ Saf 100:69–75
Hu X, Sullivan-Gilbert M, Gupta M, Thompson SA (2006) Mapping of the loci controlling oleic and linolenic acid contents and development of fad2 and fad3 allele-specific markers in canola (*Brassica napus* L.). Theor Appl Genet 113:497–507
Huang P, He L, Abbas A, Hussain S, Hussain S, Du D, Hafeez MB, Balooch S, Zahra N, Ren X (2021) Seed priming with sorghum water extract improves the performance of camelina (*Camelina sativa* (L.) crantz.) under salt stress. Plan Theory 10(4):749
Hussain A, Boney A (1973) Hydrophilic growth inhibitors from *Laminaria* and *Ascophyllum*. New Phytol 72(2):403–410
Iqbal M, Ashraf M (2013) Gibberellic acid mediated induction of salt tolerance in wheat plants: growth, ionic partitioning, photosynthesis, yield and hormonal homeostasis. Environ Exp Bot 86:76–85
Jaime-Perez N, Chaza P, José B, García Sogo B, Atarés Huerta A, Athman A, Byrt CS, Olias R, Asins MJ, Gilliham M (2017) The sodium transporter encoded by the HKT1;2 gene modulates sodium/potassium homeostasis in tomato shoots under salinity. Plant Cell Environ 40(5): 658–671
Jannin L, Arkoun M, Ourry A, Laîné P, Goux D, Garnica M, Fuentes M, Francisco SS, Baigorri R, Cruz F (2012) Microarray analysis of humic acid effects on *Brassica napus* growth: involvement of N, C and S metabolisms. Plant Soil 359:297–319
Jin T, Sun Y, Zhao R, Shan Z, Gai J, Li Y (2019) Overexpression of peroxidase gene GsPRX9 confers salt tolerance in soybean. Int J Mol Sci 20(15):3745
Khan W, Hiltz D, Critchley AT, Prithiviraj B (2011) Bioassay to detect Ascophyllum nodosum extract-induced cytokinin-like activity in *Arabidopsis thaliana*. J Appl Phycol 23:409–414

Khan W-u-D, Tanveer M, Shaukat R, Ali M, Pirdad F (2020) An overview of salinity tolerance mechanism in plants. In: Salt and drought stress tolerance in plants: signaling networks and adaptive mechanisms, pp 1–16

Khedr AHA, Abbas MA, Wahid AAA, Quick WP, Abogadallah GM (2003) Proline induces the expression of salt-stress-responsive proteins and may improve the adaptation of Pancratium maritimum L. to salt-stress. J Exp Bot 54(392):2553–2562

Kim GB, Nam YW (2013) A novel Δ1-pyrroline-5-carboxylate synthetase gene of *Medicago truncatula* plays a predominant role in stress-induced proline accumulation during symbiotic nitrogen fixation. J Plant Physiol 170(3):291–302

Kocira S (2019) Effect of amino acid biostimulant on the yield and nutraceutical potential of soybean. Chilean J Agric Res 79(1):17–25

Kocira A, Świeca M, Kocira S, Złotek U, Jakubczyk A (2018) Enhancement of yield, nutritional and nutraceutical properties of two common bean cultivars following the application of seaweed extract (*Ecklonia maxima*). Saudi J Biol Sci 25(3):563–571

Krawczuk A, Parafiniuk S, Przywara A, Huyghebaert B, Rabier F, Limbourg Q, Mostade O, Kocira S (2021) Technical parameters of spraying with a biostimulant as a determinant of biometrical properties and yield of soybean seeds. Agric Eng 25(1):171–179

Lamar RT (2020) Possible role for electron shuttling capacity in elicitation of PB activity of humic substances on plant growth enhancement. In: The chemical biology of plant biostimulants, pp 97–121

Larcher W (2003) Physiological plant ecology: ecophysiology and stress physiology of functional groups. Springer Science & Business Media

Latif H, Mohamed H (2016) Exogenous applications of moringa leaf extract effect on retrotransposon, ultrastructural and biochemical contents of common bean plants under environmental stresses. S Afr J Bot 106:221–231

Lau S-E, Teo WFA, Teoh EY, Tan BC (2022) Microbiome engineering and plant biostimulants for sustainable crop improvement and mitigation of biotic and abiotic stresses. Discov Food 2(1):9

Lehmann J, Kleber M (2015) The contentious nature of soil organic matter. Nature 528(7580): 60–68

Li QL, Gao XR, Yu XH, Wang XZ, An LJ (2003) Molecular cloning and characterization of betaine aldehyde dehydrogenase gene from *Suaeda liaotungensis* and its use in improved tolerance to salinity in transgenic tobacco. Biotechnol Lett 25:1431–1436

Lorenzo P, Souza-Alonso P, Guisande-Collazo A, Freitas H (2019) Influence of Acacia dealbata Link bark extracts on the growth of *Allium cepa* L. plants under high salinity conditions. J Sci Food Agric 99(8):4072–4081

Mantzoukas S, Papantzikos V, Katsogiannou S, Papanikou A, Koukidis C, Servis D, Eliopoulus P, Patakioutas G (2023) Biostimulant and bioinsecticidal effect of coating cotton seeds with endophytic *Beauveria bassiana* in semi-field conditions. Microorganisms 11(8):2050

Mariani L, Ferrante A (2017) Agronomic management for enhancing plant tolerance to abiotic stresses-drought, salinity, hypoxia, and lodging. Horticulturae 3(4):52

Melo GB, da Silva AG, da Costa AC, Alves da Silva A, Rosa M, Bessa LA, ... & Vitorino LC (2024) Foliar application of biostimulant mitigates water stress effects on soybean. Agronomy 14(3):414

Montoro P, Braca A, Pizza C, De Tommasi N (2005) Structure-antioxidant activity relationships of flavonoids isolated from different plant species. Food Chem 92(2):349–355

Munns R, Goyal S, Passioura J (2005) Salinity stress and its mitigation. University of California, Davis

Nardi S, Pizzeghello D, Reniero F, Rascio N (2000) Chemical and biochemical properties of humic substances isolated from forest soils and plant growth. Soil Sci Soc Am J 64(2):639–645

Nardi S, Pizzeghello D, Muscolo A, Vianello A (2002) Physiological effects of humic substances on higher plants. Soil Biol Biochem 34(11):1527–1536

Nardi S, Muscolo A, Vaccaro S, Baiano S, Spaccini R, Piccolo A (2007) Relationship between molecular characteristics of soil humic fractions and glycolytic pathway and krebs cycle in maize seedlings. Soil Biol Biochem 39(12):3138–3146

Nardi S, Pizzeghello D, Schiavon M, Ertani A (2016) Plant biostimulants: physiological responses induced by protein hydrolyzed-based products and humic substances in plant metabolism. Sci Agric 73(1):18–23

Nardi S, Ertani A, Francioso O (2017) Soil-root cross-talking: the role of humic substances. J Plant Nutr Soil Sci 180(1):5–13

Nardi S, Pizzeghello D, Ertani A (2018) Hormone-like activity of the soil organic matter. Appl Soil Ecol 123:517–520

Nasir M, Khan AS, Basra SA, Malik AU (2016) Foliar application of moringa leaf extract, potassium and zinc influence yield and fruit quality of 'Kinnow'mandarin. Sci Hortic 210: 227–235

Nguyen ML, Kim GB, Hyun SH, Lee SY, Lee CY, Choi HK, Choi HK, Nam YW (2013) Physiological and metabolomic analysis of a knockout mutant suggests a critical role of MtP5CS3 gene in osmotic stress tolerance of *Medicago truncatula*. Euphytica 193:101–120

Niemeyer J, Chen Y, Bollag JM (1992) Characterization of humic acids, composts, and peat by diffuse reflectance Fourier-transform infrared spectroscopy. Soil Sci Soc Am J 56(1):135–140

Omidbakhshfard MA, Sujeeth N, Gupta S, Omranian N, Guinan KJ, Brotman Y, Nikoloski Z, Fernie AR, Mueller-Roeber B, Gechev TS (2020) A biostimulant obtained from the seaweed *Ascophyllum nodosum* protects *Arabidopsis thaliana* from severe oxidative stress. Int J Mol Sci 21(2):474

Omoarelojie LO, Kulkarni MG, Finnie JF, Van Staden J (2021) Modes of action of biostimulants in plants. In: Biostimulants for crops from seed germination to plant development. Elsevier, pp 445–459

Parađiković N, Teklić T, Zeljković S, Lisjak M, Špoljarević M (2019) Biostimulants research in some horticultural plant species-A review. Food Energy Secur 8(2):e00162

Parihar P, Singh S, Singh R, Singh VP, Prasad SM (2015) Effect of salinity stress on plants and its tolerance strategies: a review. Environ Sci Pollut Res 22:4056–4075

Paul K, Sorrentino M, Lucini L, Rouphael Y, Cardarelli M, Bonini P, Miras Moreno MB, Reynaud H, Canaguier R, Trtílek M (2019) A combined phenotypic and metabolomic approach for elucidating the biostimulant action of a plant-derived protein hydrolysate on tomato grown under limited water availability. Front Plant Sci 10:493

Peleg Z, Blumwald E (2011) Hormone balance and abiotic stress tolerance in crop plants. Curr Opin Plant Biol 14(3):290–295

Ponnamperuma F (1984) Effects of flooding on soils. Flooding Plant Growth 10(45):1454–1458

Pujni D, Chaudhary A, Rajam M (2007) Increased tolerance to salinity and drought in transgenic indica rice by mannitol accumulation. J Plant Biochem Biotechnol 16:1–7

Qadir M, Quillérou E, Nangia V, Murtaza G, Singh M, Thomas RJ, Drechsel P, Noble AD (2014) Economics of salt-induced land degradation and restoration. In: Natural resources forum, vol 4. Wiley Online Library, pp 282–295

Rademacher W (2015) Plant growth regulators: backgrounds and uses in plant production. J Plant Growth Regul 34:845–872

Rady MM, Mohamed GF (2015) Modulation of salt stress effects on the growth, physio-chemical attributes and yields of Phaseolus vulgaris L. plants by the combined application of salicylic acid and *Moringa oleifera* leaf extract. Sci Hortic 193:105–113

Rady MM, Talaat NB, Abdelhamid MT, Shawky BT, Desoky E-SM (2019) Maize (*Zea mays* L.) grains extract mitigates the deleterious effects of salt stress on common bean (*Phaseolus vulgaris* L.) growth and physiology. J Hortic Sci Biotechnol 94(6):777–789

Rai S, Deeksha Charak DC, Rajeev Bharat RB (2016) Scenario of oilseed crops across the globe. Plant Arch 16(1):125–132

Rasul F, Gupta S, Olas JJ, Gechev T, Sujeeth N, Mueller-Roeber B (2021) Priming with a seaweed extract strongly improves drought tolerance in Arabidopsis. Int J Mol Sci 22(3):1469

Rathore S, Chaudhary D, Boricha G, Ghosh A, Bhatt B, Zodape S, Patolia J (2009) Effect of seaweed extract on the growth, yield and nutrient uptake of soybean (*Glycine max*) under rainfed conditions. S Afr J Bot 75(2):351–355

Regni L, Del Buono D, Miras-Moreno B, Senizza B, Lucini L, Trevisan M, Morelli Venturi D, Costantino F, Proietti P (2021) Biostimulant effects of an aqueous extract of duckweed (*Lemna minor* L.) on physiological and biochemical traits in the olive tree. Agriculture 11(12):1299

Rengasamy KR, Kulkarni MG, Pendota SC, Van Staden J (2016) Enhancing growth, phytochemical constituents and aphid resistance capacity in cabbage with foliar application of eckol-a biologically active phenolic molecule from brown seaweed. New Biotechnol 33(2):273–279

Rouphael Y, Colla G (2020) Biostimulants in agriculture, vol 11. Frontiers Media SA

Rouphael Y, Cardarelli M, Bonini P, Colla G (2017) Synergistic action of a microbial-based biostimulant and a plant derived-protein hydrolysate enhances lettuce tolerance to alkalinity and salinity. Front Plant Sci 8:131

Rouphael Y, Giordano M, Cardarelli M, Cozzolino E, Mori M, Kyriacou MC, Bonini P, Colla G (2018) Plant-and seaweed-based extracts increase yield but differentially modulate nutritional quality of greenhouse spinach through biostimulant action. Agronomy 8(7):126

Rouphael Y, Carillo P, Ciriello M, Formisano L, El-Nakhel C, Ganugi P, Fiorini A, Miras Moreno B, Zhang L, Cardarelli M (2023) Copper boosts the biostimulant activity of a vegetal-derived protein hydrolysate in basil: morpho-physiological and metabolomics insights. Front Plant Sci 14:1235686

Ruzzi M, Aroca R (2015) Plant growth-promoting rhizobacteria act as biostimulants in horticulture. Sci Hortic 196:124–134

Sanderson KJ, Jameson PE, Zabkiewicz JA (1987) Auxin in a seaweed extract: identification and quantitation of indole-3-acetic acid by gas chromatography-mass spectrometry. J Plant Physiol 129(3–4):363–367

Sangiorgio D, Cellini A, Donati I, Pastore C, Onofrietti C, Spinelli F (2020) Facing climate change: application of microbial biostimulants to mitigate stress in horticultural crops. Agronomy 10(6): 794

Sarker U, Oba S (2020) The response of salinity stress-induced A. tricolor to growth, anatomy, physiology, non-enzymatic and enzymatic antioxidants. Front Plant Sci 11:559876

Schiavon M, Ertani A, Nardi S (2008) Effects of an alfalfa protein hydrolysate on the gene expression and activity of enzymes of the tricarboxylic acid (TCA) cycle and nitrogen metabolism in *Zea mays* L. J Agric Food Chem 56(24):11800–11808

Sharma P, Singh A (2016) Overview of salinity management in agriculture with emphasis on India. Quality seed production, processing and certification of selected field and vegetable crops in salt affected areas Training Manual. ICAR-CSSRI, Karnal, pp 1–7

Sharma DK, Singh A (2017) Current trends and emerging challenges in sustainable management of salt-affected soils: a critical appraisal. In: Bioremediation of salt affected soils: an Indian perspective, pp 1–40

Sharma A, Shahzad B, Kumar V, Kohli SK, Sidhu GPS, Bali AS, Handa N, Kapoor D, Bhardwaj R, Zheng B (2019) Phytohormones regulate accumulation of osmolytes under abiotic stress. Biomol Ther 9(7):285

Shekhar SH, Lyons G, McRoberts C, McCall D, Carmichael E, Andrews F, McCormack R (2012) Brown seaweed species from Strangford Lough: compositional analyses of seaweed species and biostimulant formulations by rapid instrumental methods. J Appl Phycol 24:1141–1157

Shukla PS, Mantin EG, Adil M, Bajpai S, Critchley AT, Prithiviraj B (2019) Ascophyllum nodosum-based biostimulants: sustainable applications in agriculture for the stimulation of plant growth, stress tolerance, and disease management. Front Plant Sci 10:462648

Sible CN, Seebauer JR, Below FE (2021) Plant biostimulants: a categorical review, their implications for row crop production, and relation to soil health indicators. Agronomy 11(7):1297

Silva RDA, Santos JL, Oliveira LS, Soares MR, Santos S (2016) Biostimulants on mineral nutrition and fiber quality of cotton crop. Rev Bras Eng Agric Ambiental 20:1062–1066

Slama I, Abdelly C, Bouchereau A, Flowers T, Savouré A (2015) Diversity, distribution and roles of osmoprotective compounds accumulated in halophytes under abiotic stress. Ann Bot 115(3): 433–447

Stalker DM, McBride KE, Malyj LD (1988) Herbicide resistance in transgenic plants expressing a bacterial detoxification gene. Science 242(4877):419–423

Staykov NS, Angelov M, Petrov V, Minkov P, Kanojia A, Guinan KJ, Alseekh S, Fernie AR, Sujeeth N, Gechev TS (2020) An *Ascophyllum nodosum*-derived biostimulant protects model and crop plants from oxidative stress. Metabolites 11(1):24

Stirk W, Arthur G, Lourens A, Novak O, Strnad M, Van Staden J (2004) Changes in cytokinin and auxin concentrations in seaweed concentrates when stored at an elevated temperature. J Appl Phycol 16:31–39

Tandon S, Dubey A (2015) Effects of Biozyme (*Ascophyllum nodosum*) biostimulant on growth and development of soybean [*Glycine max* (L.) Merill]. Commun Soil Sci Plant Anal 46(7): 845–858

Taylor IE, Wilkinson AJ (1977) The occurrence of gibberellins and gibberellin-like substances in algae. Phycologia 16(1):37–42

Tejada M, Rodríguez-Morgado B, Gómez I, Franco-Andreu L, Benítez C, Parrado J (2016) Use of biofertilizers obtained from sewage sludges on maize yield. Eur J Agron 78:13–19

Turan M, Yildirim E, Kitir N, Unek C, Nikerel E, Ozdemir BS, Güneş A, NEP M (2017) Beneficial role of plant growth-promoting bacteria in vegetable production under abiotic stress. In: Microbial strategies for vegetable production, pp 151–166

Ugarte RA, Craigie JS, Critchley AT (2010) Fucoid flora of the rocky intertidal of the Canadian Maritimes: implications for the future with rapid climate change. In: Seaweeds and their role in globally changing environments, pp 69–90

Van Oosten MJ, Sharkhuu A, Batelli G, Bressan RA, Maggio A (2013) The Arabidopsis thaliana mutant air1 implicates SOS3 in the regulation of anthocyanins under salt stress. Plant Mol Biol 83:405–415

Van Oosten MJ, Pepe O, De Pascale S, Silletti S, Maggio A (2017) The role of biostimulants and bioeffectors as alleviators of abiotic stress in crop plants. Chem Biol Technol Agric 4:1–12

Vanstraelen M, Benková E (2012) Hormonal interactions in the regulation of plant development. Annu Rev Cell Dev Biol 28(1):463–487

Verma O, Sharma S, Kumar V, Singh T, Kumar R, Auji R (2024) Salinity stress effect on staple food crops and novel mitigation strategies. Biologia 79:1–16

Wang TT, Ren ZJ, Liu ZQ, Feng X, Guo RQ, Li BG, Li LG, Jing HC (2014) SbHKT1; 4, a member of the high-affinity potassium transporter gene family from *Sorghum bicolor*, functions to maintain optimal Na^+/K^+ balance under Na^+ stress. J Integr Plant Biol 56(3):315–332

Weiss MG (1950) Variabilty of agronomic and seed compositional characters in soybeans as influenced by variety and time of planting, vol 1017. Department of Agriculture

Wicke B, Smeets E, Dornburg V, Vashev B, Gaiser T, Turkenburg W, Faaij A (2011) The global technical and economic potential of bioenergy from salt-affected soils. Energy Environ Sci 4(8): 2669–2681

Wittkop B, Snowdon R, Friedt W (2009) Status and perspectives of breeding for enhanced yield and quality of oilseed crops for Europe. Euphytica 170:131–140

Yadav S, Irfan M, Ahmad A, Hayat S (2011) Causes of salinity and plant manifestations to salt stress: a review. J Environ Biol 32(5):667

Yakhin O, Lubyanov A, Yakhin I (2012) Changes in cytokinin, auxin, and abscisic acid contents in wheat seedlings treated with the growth regulator Stifun. Russ J Plant Physiol 59:398–405

Yakhin OI, Lubyanov AA, Yakhin IA, Brown PH (2017) Biostimulants in plant science: a global perspective. Front Plant Sci 7:2049

Zanetti F, Monti A, Berti MT (2013) Challenges and opportunities for new industrial oilseed crops in EU-27: a review. Ind Crop Prod 50:580–595

Zhao J, Davis LC, Verpoorte R (2005) Elicitor signal transduction leading to production of plant secondary metabolites. Biotechnol Adv 23(4):283–333

Zhu M, Li Q, Zhang Y, Zhang M, Li Z (2022) Glycine betaine increases salt tolerance in maize (*Zea mays* L.) by regulating Na^+ homeostasis. Frontiers. Plant Sci 13:978304

Zulfiqar F, Casadesús A, Brockman H, Munné-Bosch S (2020) An overview of plant-based natural biostimulants for sustainable horticulture with a particular focus on moringa leaf extracts. Plant Sci 295:110194

Chapter 2
Use of Biostimulants to Improve Drought Tolerance in Oilseed Crops

Sidra Ahmad, Nayab Ahmad, Muhammad Nauman Khan, Sezai Ercisli, and Rashid Iqbal

Abstract The productivity of essential oilseed crops, including soybean (*Glycine max* L.), sunflower (*Helianthus annuus* L.) and rapeseed, is increasingly threatened by the increasing rate of drought stress, aggravated by climate change. In this situation, biostimulants present a viable natural approach for improving crop resilience. This chapter highlights the regulatory roles of biostimulants at various growth stages and molecular levels, emphasizing their impact on seed germination, physiological responses and transcriptional alterations induced by drought stress. Biostimulants boost drought tolerance by stimulating primary and secondary metabolism, promoting vigorous root development for improved water and nutrient absorption, regulating stomatal function and activating stress-responsive signaling pathways. The effectiveness of biostimulants is dependent upon strategic aspects including application time, dose and mechanism of action. Moreover, the incorporation of biostimulants with traditional agricultural inputs and precision agriculture technologies can markedly enhance crop output. Novel formulations, comprising encapsulated biostimulant chemicals in conjunction with elicitors and developments in genetic changes or transgenes for drought tolerance, present new synergistic prospects for increasing plant resilience. This chapter emphasizes the potential of

---

S. Ahmad · N. Ahmad
Institute of Biotechnology and Genetic Engineering (IBGE), The University of Agriculture, Peshawar, Pakistan

M. N. Khan (✉)
Department of Botany, Islamia College Peshawar, Peshawar, Pakistan

University Public School, University of Peshawar, Peshawar, Pakistan
e-mail: nomiflora@uop.edu.pk

S. Ercisli
Department of Horticulture, Faculty of Agriculture, Ataturk University, Erzurum, Türkiye

HGF Agro, Ata Teknokent, Erzurum, Türkiye

R. Iqbal
Department of Agronomy, Faculty of Agriculture and Environment, The Islamia University of Bahawalpur, Bahawalpur, Pakistan

Department of Life Sciences, Western Caspian University, Baku, Azerbaijan

© The Author(s), under exclusive license to Springer Nature Singapore Pte Ltd. 2025

A. A. H. Abdel Latef (ed.), *Oilseed Crops Under Abiotic Stress*, Sustainability Sciences in Asia and Africa, https://doi.org/10.1007/978-981-96-8346-8_2

biostimulants as a crucial tool for sustainable agriculture in oilseed crops against problems posed by climate change, namely drought.

Keywords Biostimulants · Drought stress tolerance · Oilseed crops · Sustainable agriculture

2.1 Introduction

Climate change is resulting in more severe drought, which is of course a disaster for the agricultural industry (Nazari et al. 2023). Drought has a substantial impact on water availability (Akram et al. 2024). It can affect vegetable oil, biofuels and livestock feed from these crops in the global food supply (Hanjra et al. 2010). Oilseed crops have been more at risk when compared to other cereals because of high water requirement during the vegetative stage of crop growth and hence drought specifically results causing yield losses, which has economic value (Ahmadi et al. 2009). In soybeans, the utilization of biofertilizers such as *Bacillus amyloliquefaciens* and mycorrhiza could amplify drought tolerance, which enhances yield under drought stress (Sheteiwy et al. 2021). Biostimulants have been proposed as an innovation to tackle some of these challenges that can maximize the natural physiological and biochemical processes in plants, which in turn might be able to make them more resilient against abiotic stress. Biostimulants, unlike fertilizers or pesticides, work indirectly by improving the plant's capacity to deal with environmental stressors such as droughts (Hasanuzzaman et al. 2021). One of the reasons for which they are popular in agriculture is due to their positive effects on water nutrient uptake, root development and overall health of plants without causing any harm to the surroundings or organisms (Halpern et al. 2015).

Oilseed crops are an essential part of worldwide agriculture and industry, providing edible oils, animal feed components or biofuels. Worldwide, the demand for higher-quality cooking oils (soybean oil, sunflower oil and rapeseed) is growing as the population is increasing day by day (Khanal et al. 2021). Oilseed crops also contribute by providing protein-rich by-products used as feed ingredients for animal production, which is critical to meet the increasing demand of animal products (Wadhwa et al. 2016). However, the yield potential of these crops depends on highly sensitive water availability, especially during critical growth stages including germination, flowering and pod formation (Anjum et al. 2017).

Drought stress is responsible for significant reductions in oilseed crop yields, resulting in severe economic consequences on food production and also representing a threat to food security (Kambiranda et al. 2011; Khayatnezhad et al. 2022). In soybeans, drought stress during the reproductive phase can lead to decreased seed set and ultimately low yields (Poudel et al. 2023). Likewise, drought stress during these stages in sunflower and rapeseed compromises reproductive success that influences oil-related traits, leading to a reduction of yield quantity and quality (Altaf et al. 2023). Ever since the problem of water stress has been recognized, researchers have tried to cope with it using traditional strategies for drought management, such as irrigation or the development of new varieties specific to drought tolerance. Nevertheless, there are significant drawbacks to these systems. Irrigating crops in water-scarce areas is frequently impractical, and overusing groundwater resources can result in long-term environmental damage. On the other hand, breeding for genetic improvement in drought tolerance is a time-consuming process. Biostimulants contribute as an effective and green strategy to improve drought tolerance of oilseed crops by means of the enhancement of innate defense mechanisms (Mansour et al. 2023; Mikiciuk et al. 2024). Many microbial and nonmicrobial biostimulants increase plant growth, development and response to stress. Biostimulants can be differentiated based on types of microbial, including plant growth-promoting rhizobacteria (PGPR) and arbuscular mycorrhizal fungi (AMF). Biostimulants of both types enhance root development, increase nutrient absorption and alleviate stress-signaling pathways to augment resistance against drought in several ways (Sun et al. 2024). Symbiotic properties of PGPR and AMF microbial biostimulants with plant roots improve their capacity for water uptake as well as nutrient absorption from the soil. PGPR produce phytohormones such as indole-3-acetic acid (IAA). PGPR also belongs to genera that are capable in the production of producing IAA, like *Azospirillum, Bacillus* or *Pseudomonas*. These bacteria help in the better development of roots by elongation and branching. With a more optimized root architecture, those crops can extract water from deeper soil layers (Grover et al. 2021).

AMF help expand the plant's root system by connecting them through hyphae to increase the area over which water and nutrients are absorbed (Abdel Latef and Chaoxing 2011a, b, 2024). Under drought conditions, the nutrient phosphorus is often limited in the soil. AMF are especially efficient at increasing phosphorus uptake (Yooyongwech et al. 2023). In addition, the mycorrhizal association of AMF on plant roots improves soil structure. Water can also be held in place within aggregates with a high porosity to aid conservation and hydration for plants during drought (Das et al. 2022). Biostimulants are nonmicrobial and include seaweed extracts, humic substances or amino acids that act through different mechanisms to confer drought tolerance in oilseed crops. Seaweed extracts such as *Ascophyllum nodosum* and *Ecklonia maxima*, have bioactive compounds like polysaccharides, phytohormones and antioxidants that boost root development and improve moisture holding capacity in the soil profile, which reduces water stress. These extracts induce root growth, increase water-use efficiency and enhance turgidity of cells in plants under stress (Han et al. 2024; Sabatino et al. 2023).

Humic substances are natural organic constituents in soil, including humin, hemicellulose and fulvic acid, which can improve the structure of soils to retain water. Humic substances can also increase the availability of nutrients and improve root growth, thus enabling plants to resist drought stress. Proline and glutamine are amino acids that function as osmoprotectants and adjust the osmotic potential during drought. They provide structural rigidity in plants by stabilizing proteins and membranes during osmotic dehydration, aiding the plant to maintain cellular integrity (Anderback et al. 2021). Several factors such as the application timing, dosage and delivery method, can affect how biostimulants contribute to strengthening drought tolerance. Different application methods have advantages and limitations. Some biostimulants can be applied through foliar sprays or soil drenches. This allows for rapid uptake through the leaves and is ideal when dealing with short-term drought stress while soil applications place biostimulants at or near the root zone, leading to long-term improvements in root development, allowing roots, as well as younger seedlings better water uptake (Bulgari et al. 2015; Matthews et al. 2024).

Besides response variability, the uptake of biostimulants on an industrial scale and in agriculture is influenced by economic constraints. Although biostimulants potentially provide long-term advantages in terms of higher resistance to drought and reduced input costs. Furthermore, the absence of harmonized regulations on the production and use of biostimulants implies failure to guarantee both quality standards and performance levels. To address these challenges, focused research on biostimulants should be pursued along with the establishment of regulatory frameworks and financial incentives to make their adoption more widespread in agriculture (Bashir et al. 2024). Ongoing research to develop better formulations, delivery systems and integration with other agricultural technologies can enhance the role of biostimulants in agriculture. The proliferation of precision agriculture based on data-driven tools for improved crop management opens a new avenue in solutions to enhance efficacy and application strategies by making them more targeted and efficient. Moreover, the scope of combining biostimulants with genetically improved drought-tolerant crops is very promising in coping with climate change (Mikiciuk et al. 2024).

2.2 Overview of Drought Stress in Oilseed Crops

Drought stress is the main environmental challenge seriously affecting crop productivity all over the world. Oilseed crops, such as soybean (*Glycine max*), sunflower (*Helianthus annuus*) and rapeseed are highly susceptible to drought stress, which can significantly impair their growth, reduce seed yield, and affect oil quality by disrupting photosynthesis, nutrient uptake, and reproductive development. *Brassica napus* is more sensitive to drought. With the increased frequency and severity of global climate change-driven drought events, research into the effects that this type of stress has on oilseed crops is an urgent priority for researchers and agriculturalists. In this section, we summarize the impacts of drought on oilseed yield and irrigation usage practices that have been predicted to be affected by such a change (Elferjani et al. 2018; Raza et al. 2017). Drought stresses reduce oilseed crop yield significantly due to the reduced water availability during critical growth stages like germination, flowering and pod filling. In soybeans, water deficit imposed during the reproductive stage may reduce seed set markedly and lead to as high as 50% yield loss in extreme drought stress. Likewise, in sunflowers, drought stress during the flowering phase decreases the pollen viability and seed filling. Rapeseed is extremely sensitive to water deficits, especially during flowering and silique-filling stages, which lead to both quantitative as well qualitative yield losses, such as reductions in seed oil content (Pokharel, 2020).

Numerous studies have reported the magnitude of drought-induced yield damage in oilseed crops. In the United States, severe drought in 2012 reduced soybean yields by an estimated 20% over the nondrought years' average (Boyer et al. 2013). In another study, it was observed that the severe drought conditions in Spain caused sunflower yields to drop by a huge amount under stress (García-López et al. 2014). The yield of rapeseed also faces consistent reductions because of frequent droughts in significant rapeseed growing regions around India (Raza, 2021). However, drought conditions reduce oilseed crops' yield and have a great influence on soil health as well as water availability. When this condition persists over an extended period, soil moisture content decreases and crops may face water stress. Thus, it results in decreased uptake of nutrients since most of the essential nutrient elements (like N or K) are dissolved into soil water and are less accessible to plants under drought conditions. Poor soil structure can also be caused by the reduced activity of important components like density, water and nutrient-holding abilities if the microbiological community is dormant under drought stress (Bhattacharya et al. 2021; Singh et al. 2022a, b).

Some crops may access water stored in deeper soil layers through their root networks, but shallow-rooted oilseed crops like sunflower and rapeseed have a low capacity to utilize deep-soil moisture reserves and are therefore prone to drought. This can cause compaction of the earth in dry conditions and decreased infiltration, which aggravates water scarcity problems by increasing runoff with more erosion that seriously deteriorates soil health (Debaeke et al. 2021). It is extremely difficult to manage water effectively in drought-stricken areas. Moreover, traditional irrigation cannot satisfy crop water requirements during an extreme drought event, especially in arid and semiarid regions. In addition, the rapid economic development in many areas with agriculture and a falling rain pattern above most agricultural zones has resulted in persistent aquifer depletion coupled with long-term reduced water availability (Sharma, Smakhtin, 2006). Therefore, farmers in drought-prone regions have been pushed to take new pathways of water and improve the resilience of oilseed crops under various levels of water stress through efficient use, like drip irrigation & mulching (Tyagi et al. 2020).

In South Asia, India and Pakistan are major producers of the rapeseed-mustard species that often fail in drought-affected years. These drought events have not only resulted in lower yields per unit land area but have also increased the competition from different sectors for water resources, thereby further affecting agricultural productivity (Bhushan et al. 2015; Kumar, 2019). In Sub-Saharan Africa, where soybean serves as both a staple for food security and a cash crop for some export markets, drought stress also has long been identified as one of the primary factors limiting production. In these districts, which are prone to recurrent droughts and where there is a lack of proper infrastructure for water storage, the challenge has become severe, with farmers finding it difficult even to maintain their crop yields under moisture-scarce conditions (Payne, 2010).

In many of these vulnerable regions, drought is anticipated to become more frequent and severe as climate change unfolds. The Intergovernmental Panel on Climate Change (IPCC) has projected an increase in droughts, which are likely to be more frequent and severe over the future years at several prime oilseed regions due to a rise in global temperatures (Jaime et al. 2018). Over the longer term, this could shrink oilseed crop yields and imperil global food security and agricultural sustainability.

Drought stress presents a serious threat to the productivity of oilseed crops because these agricultural systems strongly need sustainable methods to strengthen plant resistance. Previous research findings demonstrated that different biostimulants, including silicon, seaweed extracts, microbial inoculants and calcium improve drought resistance by modifying physiological processes and biochemical functions in plants. Brassica crops had better drought resistance with respect to photosynthetic activities and stomatal conductance as well as by maintaining osmotic system balance under foliar application of orthosilicic acid (Sujata et al. 2023). Integrated treatment decreased the values of hydrogen peroxide (H_2O_2) and malondialdehyde (MDA) contents, also simultaneously increased the values of the antioxidant enzyme capacity and soluble sugar.

Scientists observed synergistic effects of microbial biostimulant ProbioHumus with $CaCO_3$ soil application through their study of *Brassica napus* under drought stress (Gavelienė et al. 2025). Through this treatment, plants obtained superior water retention capabilities as relative water content (RWC) was highly increased while proline levels increased up to fourfold. The photosynthetic pigments increased together with improved stomatal control, which allowed the plants to tolerate drought conditions. The research data indicate that bio-stimulated microbial applications and calcium administration together boosted the drought tolerance of oilseed crops.

The research evaluated plant PGPR to determine their effect on *Arachis hypogaea* drought tolerance under different field capacity levels (Eswaran et al. 2024). The growth-promoting bacteria of *Acinetobacter calcoaceticus* and *Bacillus amyloliquefaciens* enhanced the production of phytohormones and osmolytes alongside phosphate solubilization ability. The drought stress condition led to increased levels of proline along with salicylic acid, trehalose and glycine betaine in both strains, which improved water retention responses while boosting biomass production. The plant's antioxidant enzymes, such as ascorbate peroxidase and superoxide dismutase and catalase, showed increased activity to reinforce stress tolerance mechanisms.

Research on drought tolerance using biostimulants for canola (*Brassica napus*) was conducted (Passandideh et al. 2022). The effects of amino acid foliar application, humic acid and fulvic acid foliar application were evaluated. The yield performance of biostimulants under nonstress circumstances was better than the control group and demonstrated their strongest effect in the combination use. The yield improvement under drought stress with humic acid and through the combined application of nutrients showed a high increase and seaweed extract also contributed to an increase in yield.

Cumin (*Cuminum cyminum*) seed yield, along with essential oil yield and physiological characteristics under drought stress was investigated when treated with seaweed extract combined with humic acid (Rahgoshahi et al. 2023). The combination of severe drought reduced both seed yield and essential oil yield. The combined treatment of seaweed extract alongside humic acid produced good results since it minimized drought-related stress effects through enhanced total chlorophyll levels in addition to increased proline accumulation and better antioxidant levels (peroxidase and catalase). The observed results demonstrated that humic acid, alongside seaweed extract, shows great promise for improving drought tolerance in medicinal crops subjected to water scarcity.

The effects of stress modulation approaches through intercropping under rainfed environments were assessed for safflower (*Carthamus tinctorius*) and chickpea (*Cicer arietinum*) in a recent study conducted by Mosalman et al. (2024). This research examined how different combinations of salicylic acid and selenium affected intercropped systems. Single safflower cultivation using selenium treatment resulted in maximum safflower seed yield. The combination of intercropping with selenium led to increased safflower oil content and elevated unsaturated fatty acids, together with better chlorophyll content along with higher proline accumulation and antioxidant enzyme activities (catalase, ascorbate peroxidase and superoxide dismutase). This study demonstrated the value of combining stress management substances such as selenium with salicylic acid when using intercrop systems to increase oilseed and legume crop resistance to drought conditions.

Scientific research conducted by Cirillo et al. (2022) examined biostimulant effects on oil yield and nutritional value of olive (*Olea europaea* cv. Racioppella) trees in southern Italy. The researchers conducted a study on a tropical plant extract and a glycine betaine-based product and a *Trichoderma* spp.-based biostimulant during May to October 2021. Treated olive drupes and olives with Trichoderma showed a substantial polyphenol increase compared to the control group and more polyphenols in the Trichoderma-treated olive oil. The two other treatments using glycine betaine and tropical plant extract enhanced polyphenol elevation. The beneficial bioactive substances demonstrated improvement of drupe and olive oil antioxidant capacity as well as total anthocyanin content and fatty acid characteristics, which indicated their effectiveness toward yield and nutritional value improvements in olive farming.

2.3 Importance of Enhancing Drought Tolerance

Drought stress remains an important yield constraint on global agriculture, and this is particularly true for oilseed crops. In addition to soybeans and sunflowers, these crops are a key part of the global food system. Improving drought tolerance in these crops is not only a scientific challenge but also an economic and environmental urgency. It is important to develop crop varieties that can withstand these conditions so as not only to maintain agricultural productivity but also limit effects on ecosystems.

Oilseed crops are among the most important in global agriculture, with many uses including fuel and high-quality protein. These include vegetable oils derived from crops like soybean, sunflower and rapeseed that are used extensively in food products, cooking oils and industrial applications. Global demand for these oils is increasing due to population growth, changes in dietary patterns and increased use of vegetable oils in food and nonfood industries (Vinnichek et al. 2019). They are not only used by the food industry, but oilseed crops also play a role in renewable energy supply through biofuel feedstock. Being the major oilseed crop in the world, soybeans account for more than 50% of global oilseed crops production. This is followed by Sunflower and Rapeseed, which are both major contributors to oil production in general as well as for biofuels. Moreover, these crops are a major source of by-products like protein meals for livestock feeding that add to their economic value (Adeleke et al. 2020).

The economic importance of oilseed crops causes them to be sensitive to drought stress. Drought-induced yield losses can lead to severe economic costs both for farmers and industries based on these crops. For instance, in the United States in 2012, a severe drought impacted soybean yields by 20% relative to nondrought years, with subsequent large negative implications for the profit margins of producers and an abrupt escalation of global prices (Boyer et al. 2013). Likewise, severe drought in Europe resulted in major losses of grain yield and quality for sunflower seed and rapeseed due to the decrease in average water availability by over 40%, which affected up to 25% of the total harvested area. Limited water availability due to recurrent droughts in rapeseed growing areas of India has resulted in yield losses above 30%. However, the economic impacts are so significant that they emphasize an increasing need for drought-resistant oilseed varieties in order to help mitigate financial risks associated with seasonal climate variability and protect the longer-term sustainability of oilseed industries at a large scale (Nadeem et al. 2024).

A global food security concern is the improvement of drought tolerance in oilseed crops as well. The global population is still rapidly increasing, so demand for both food and oilseeds is high. It means that agricultural systems face a greater strain to develop their production with less input. Because oilseed crops are very sensitive to water deficits at flowering and especially during the seed filling phase. Accordingly, increasing the drought resistance of oilseed crops is a key strategy to stabilize crop production in these areas and achieve sustainable production levels for vegetable oils, bioenergy feeds and animal protein (Attia et al. 2021).

Among the increasing challenges of climate change, one of the greatest opportunities is to maintain productivity through drought-tolerant crops. Over 80% of all agricultural losses worldwide are due to droughts. The development of such drought-tolerant oilseed crops, which could yield viable amounts even when grown in dry areas or during dry years, is critical for adaptive agriculture methods that would alleviate the substantial losses and maintain a functional supply chain to ensure continuous availability of nonedible oils (Attia et al. 2021). It should also be noted that increasing the drought tolerance of these crops could stabilize fluctuations in global food markets. Drought relief leads to large and erratic price changes, which again distort food security among producers as well as consumers (Mishra et al. 2021).

Apart from their effects on economics and food security, drought-tolerant oilseed crops play an important role in enhancing the climate resilience of agriculture. Climate models indicate that droughts occur more frequently and with greater intensity in many parts of the world, especially those located on arid or semiarid soils where oilseed production is significant. The development of drought-tolerant oilseed crops not only increases productivity under these conditions but also reduces the dependency on water-intensive methods in agriculture (Srivastav et al. 2021).

Similarly, there are also significant environmental and ecological gains associated with the enhancement of drought tolerance in oilseed crops. The most significant benefit of drought-tolerant crops is their reduced irrigation requirement, leading to direct water conservation. In water-scarce regions, a continuous struggle lies in the supply of water resources (Levy et al. 2013). This means that for countries like Australia, where irrigation is often required to sustain crop production in arid and semiarid areas, the development and deployment of drought-tolerant oilseed varieties give rise to lower requirements. Therefore, this is important in order to ensure the long-term sustainability of agriculture practices in these areas (Nadeem et al. 2024; Rao et al. 2023). To make up for the loss of water, users also tend to engage in high-water-consuming practices like extensive irrigation that can lead to soil degradation, salinization and over-extraction of groundwater tables. These environmental problems can be potentially mitigated by drought-tolerant oilseed crops (Tarolli et al. 2024).

Drought-tolerant oilseed crops can help keep crop productivity in drought-prone areas, thereby supporting the functionality of agricultural landscapes and preventing land from being converted to other uses. These are often areas where ecosystems are exceptionally fragile, and they show great sensitivity to variations in land use and water availability (Azadi et al. 2018). These ecosystems can be preserved with the development and broad selection of drought-resistant oilseed crops, which leads to less demand for open irrigation installations necessary as well as reducing the ecological footprint in agricultural activities. In addition, maintaining agricultural production in water-limiting regions can serve to preserve the natural flora and fauna that would otherwise be at risk due to desertification (Ameen et al. 2024).

2.4 Understanding Biostimulants

Biostimulants are attracting unprecedented interest in contemporary agriculture because they can boost plant growth and resilience to abiotic stresses while facilitating sustainable farming. Biostimulants, unlike their traditional fertilizer and pesticide counterparts, work through multiple different mechanisms to benefit plant health without directly offering those nutrients or fighting pests. Their capacity to ameliorate plant tolerance toward environmental stresses, such as drought stress is of major importance with ongoing consideration for climate change events and water resources (Mandal et al. 2023).

2.4.1 Definition and Classification of Biostimulants

Biostimulants are defined as unique substances or microorganisms that, when applied to plants, are intended to enhance plant growth and nutrient use efficiency, from all abiotic stress and quality traits of a crop without giving either direct nutrition or pesticidal activity. Biostimulants act differently than fertilizers, which contain vital nutrients such as nitrogen, phosphorus and potassium that encourage plant growth; or pesticides to repel pests and diseases. By activating the natural defenses and growth mechanisms of plants, they enhance plant vitality, making it more resistant to various environmental challenges. The global climate crisis has driven an increasing demand for the development of sustainable agricultural solutions and today, biostimulants are considered very important in crop management practices (Malik et al. 2020; Rady et al. 2025).

Biostimulants also contribute to healthier soil, facilitating nutrient uptake and improving crop yields at the same time. Such multispectral action helps biostimulants promote overall farming system resilience, thereby positioning them

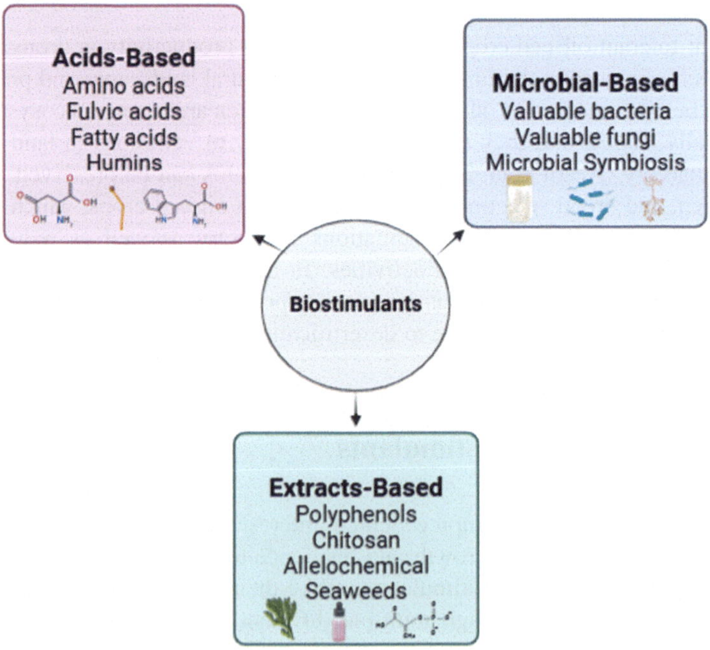

Fig. 2.1 Biostimulants and their categories

as a pivotal component of integrated crop management. This is a unique function of biostimulants, unlike traditional agricultural inputs that are largely geared toward nutrients or managing biotic threats (Rajesaheb et al. 2024). Biostimulants can be grouped according to their origin and composition (Fig. 2.1). They are classified into two main types, microbial biostimulants and nonmicrobial biostimulants. This classification serves to understand the mode of action between different biostimulants that are dedicated to certain agronomic goals (Papa et al. 2022).

2.4.1.1 Microbial Biostimulants

Microbial biostimulants include a beneficial microorganism that includes PGPR, mycorrhiza and certain Rhizobia. These microorganisms contribute to plant growth and stress tolerance in a number of ways. One of the major functions of microbial-based biostimulants is to increase nutrient uptake due to intensive root development and modification by plant–microbe interaction in the rhizosphere. It includes the production of phytohormones such as auxins by PGPR which stimulates elongation and branching roots for water uptake efficiency and nutrient availability from soil solution (Sun et al. 2023). Moreover, some rhizobacteria are responsible for nitrogen fixation which facilitates the conversion of atmospheric N_2 into a form accessible to plants effectively benefiting their growth in potentially nitrogen-limited soils (Kumari et al. 2022).

One group of such beneficial soil fungi, the AMF, specifically AMF establish symbiotic relationships with plant roots. By creating wider hyphal networks in surrounding soils, AMF enhance the nutrient uptake of plants, especially phosphorus. Phosphorus is a critical nutrient as it is immobilized in the soil and mycorrhizal fungi might enhance its availability to plants under drought stress. Microbial biostimulants help protect the plant in addition to improving nutrient uptake. Some representatives of PGPR species can elicit systemic resistance in plants which results in plant protection against pathogen attack by stimulating a plant's innate immunity (Olowe et al. 2020). Utilization of microbial biostimulants in sustainable agriculture has gained popularity by supporting soil biodiversity and function. Help with long-term soil health and fertility by encouraging good microbial activity in the soil which can reduce the requirement for chemical fertilizers (Kumar et al. 2021).

2.4.1.2 Nonmicrobial Biostimulants

Nonmicrobial biostimulants are a wide range of biological and synthetic bioactive compounds affecting plant growth, nutrition efficiency and tolerance to abiotic stress by direct biochemical interactions with cells or tissues in plants. While microbial biostimulants are based on living organisms such as bacteria and fungi, nonmicrobial biostimulants are a collection of compounds that regulate the physiological processes of plants. In recent years, these biostimulants have emerged as attractive options in agricultural systems because of their ability to improve plant tolerance against various environmental stresses like drought or salinity and underutilized nutrient sources (Ishrat et al. 2022; Nephali et al. 2020). The varieties of the nonmicrobial biostimulants that are most commonly used, such as seaweed extracts, humic substances, amino acids and protein hydrolysates, each work on unique mechanisms to supplement plant growth and protective action against stress conditions (Ciriello et al. 2024).

2.4.1.3 Seaweed Extracts

Algae extracts are nonmicrobial biostimulants because of their high content of bioactive compounds that affect physiological processes within plants. Seaweed extracts are usually prepared from species such as *Ascophyllum nodosum* and *Ecklonia maxima*, which in turn contain a huge number of biologically active molecules like phytohormones (e. g., cytokinins, auxins or abscisic acid), polysaccharides, vitamins along with essential minerals. These metabolites are involved in many physiological processes of plants, especially under abiotic stress conditions, including drought and salinity that contribute to modulating plant growth and promoting defense responses (Hariharan et al. 2024; Kumar et al. 2024). Seaweed extracts contain large amounts of phytohormones, which determine how plants grow and respond to environmental stress. For example, cytokinins stimulate cellular division and growth that inhibits chlorophyll degradation, delaying leaf senescence to help plants sustain their photosynthetic capacity under stress conditions. This hormone can help prevent the leaves from premature aging. The preservation of leaf function by cytokinins is likely to be fundamental for continued growth and productivity under drought conditions. By contrast, abscisic acid (ABA) is important for the control of stomatal closure to decrease water loss through transpiration. Under drought stress, ABA accumulation induces stomatal movement that ensures minimum water loss from the plant to balance the little water available in the root area (Trivedi et al. 2021). This mechanism is essential for the survival of plants within water-limiting environments, as it allows the plant to limit transpirational loss while still engaging in basal levels of gas exchange required for maintaining metabolic function (Jeyakumar, 2005). Furthermore, seaweed extract polysaccharides also improve soil structure by inducing the formation of soil aggregates to enhance soil water retention and reduce evaporative losses. Its effects are naturally positive in sandy and degraded soils which usually have poor water retention. Polysaccharides improve the physical structure of soil and make water available for plant uptake characterized by low moisture and drought conditions. In addition, seaweed extracts are sources of micronutrients and vitamins that aid overall plant health and stress tolerance for managing crop responses to the environment (Papnai et al. 2022).

2.4.1.4 Humic and Fulvic Acids

Humic substances within the class of nonmicrobial biostimulants include humic acids and fulvic acids. The products are extracted from the biological decomposition of plant and animal residues, which help to promote soil health by improving key traits including soil fertility, structure and nutrient access. They are biochemically complex molecules that help promote the structural integrity of soil by binding together individual particles and increasing water-holding capacity. These large aggregates reduce water infiltration and make available the stored soil moisture for plants during drought conditions (Vikram et al. 2022). The humic acids also aid the physical quality of soils, allowing more roots to penetrate deeper layers optimizing access to available water and nutrients that are crucial in plant survival during stress. Moreover, humic acids promote rhizosphere microbiota and necessitate nutrient cycling and plant-available nutrients (Cozzolino et al. 2021).

Fulvic acids are highly soluble in water due to their smaller molecular size which is easily taken up by roots. Upon entering the plant, fulvic acids function as chelating agents that bind to important minerals like iron, zinc and manganese and drive them into the cells of plants. This is particularly important in soils with low levels of nutrients and limits the level to which micronutrients become available (Barzgar et al. 2024). Fulvic acids increase the availability of all nutrients that are bound in soil and aid microorganisms to help plants uptake and add growth-promoting elements from soils even during low temperatures or drought conditions. Natural products can be a safe and effective resource to manage plant growth which is based on reducing the synthetic NPK fertilizer input (Liu, 2024). Furthermore, they provide guidance for agro-ecological sustainability under low nutrient conditions as fulvic acids have the capability of amelioration in soil about the bioavailability of nutrients that leads toward improved plant stress tolerance and subsequently elevate crop production (Alsudays et al. 2024).

2.4.1.5 Amino Acids and Protein Hydrolysates

Another important family of nonmicrobial biostimulants is amino acids and protein hydrolysates which are crucial for plant metabolism, especially under stress conditions such as drought. Osmotic adjustment is a plant process that enables plants to sustain cell turgor and water balance during the stages of drought conditions, where amino acids such as proline, glutamine and glycine play an important role. Through their role in controlling the osmotic potential inside plant cells, these systems enable plants to maintain a level of hydration that stabilizes cellular structures and allows enzymatic reactions under drought conditions (Johnson et al. 2024). Among osmoprotectants, proline is one of the most vital components being transported into plant tissues under water stress conditions. Consequently, its function in osmoregulation is important for the plant to adapt the plant under water stress conditions mitigated by drought and it stabilizes proteins, membranes and enzymes in addition to scavenging reactive oxygen species (ROS) that are created during oxidative stress. Proline helps plant tissues fill with the osmolyte, resulting in decreased cellular dehydration and preventing cell collapse under extremely dehydrating conditions (Ozturk et al. 2021). The protein hydrolysate, a mixture of peptides and free amino acids available for plants, is an easy source to supply nitrogen and carbon. They are derived from the enzymatic hydrolysis of proteins to produce compounds that function as precursors for the synthesis of stress-proteins and enzymes. The use of protein hydrolysates aids in the recovery of plants by stimulating cell division, growth and tissue repair after stress events. Protein hydrolysates not only provide building blocks for protein biosynthesis but are also activators of metabolic pathways that help plants efficiently cope with environmental stresses such as drought (Vanani et al. 2020). For example, these compounds may induce the synthesis of antioxidant enzymes from catalase and superoxide dismutase systems that participate in ROS neutralization to avoid oxidative injury during stress events (Limón-Pacheco et al. 2009). Amino acids also influence hormonal homeostasis in the plant, most probably under stress conditions. They regulate the production of plant hormones like auxins and cytokinins, which are essential to root elongation as well as nutrient acquisition or stress resistance. These resources are used in the best way, supporting the survival and growth of plants during drought stress. By regulating hormone levels, amino acids can enhance a plant's response to water deficit conditions (EL Sabagh et al. 2022). Biostimulants are also classified as organic and synthetic biostimulants based on their origin. Biostimulants of organic origin are obtained from plant and animal byproducts, as well as microorganisms. Because of their sustainability and eco-friendly nature, these biostimulants are commonly used in organic farming systems. In contrast, synthetic biostimulants are not natural but chemically derived and function similarly to natural biostimulants (Tolisano et al. 2023). This heightened interest in biostimulants is due to their ability to decrease the use of synthetic fertilizers and pesticides, which are known for causing harmful effects on the environment. Biostimulants play a role in creating productive and less harmful agricultural systems by inducing natural biological processes that promote the growth of plants under stress conditions (Hamid et al. 2021) as given in Table 2.1.

2.5 Biostimulants and Drought Tolerance

Drought is a major adverse abiotic factor that affects crop productivity worldwide. Biostimulants offer a potential tool for the realization of resilient agricultural systems that can enhance drought stress tolerance in plants by regulating physiological and biochemical pathways. Biostimulants, as opposed to traditional inputs, act on the metabolism of plants by enhancing their natural processes for coping with drought such as water uptake and root development improved stress signaling (Nephali et al. 2020) (Table 2.2).

Table 2.1 Biostimulants for enhancing drought tolerance in oilseed crops: application methods, dosage, mode of action and impact on yield

Oilseed crop	Biostimulant used	Application method	Dosage	Timing of application	Mode of action	Drought tolerance improvement	References
Canola	Tecamin (flowering fruit set biostimulant)	Foliar spray	Silicon: 2 kg/ha, sulfur: 1–2 L/ha, Tecamin: 1–2 L/ha	During flowering	Silicon: Enhances structural integrity, sulfur: Improves grain weight and biomass, Tecamin: Improves pod numbers, grain yield and biomass	Increased grain yield by 14.12%, improved biomass, enhanced grain yield and number of grains per pod	Payman et al. (2022)
Rapeseed (autumnal)	None (irrigation interval study)	Irrigation	I1: 7 days, I2: 10 days, I3: 14 days (intervals)	After Rosette stage (March 2007)	N/A (effect of irrigation interval)	Higher seed yield with 7-day and 10-day irrigation intervals (37.58% and 38.38% seed oil content)	Zarei et al. (2010)
Camelina, canola	Selenium (Na_2SeO_3)	Seed priming, foliar spray	Seed priming: 75 μM, foliar spray: 7.06 μM	Vegetative and reproductive stages	Enhances antioxidant enzymes, osmoprotectants (e.g., proline, GB), chlorophyll content	Improved yield, better physiological response (e.g., osmotic potential, water potential), increased yield components (branches, seed weight), better drought stress tolerance	Ahmad et al. (2021)
Winter rapeseed (*Brassica napus*)	Nitroxin (plant growth-promoting rhizobacteria)	Soil application	1 L ha^{-1} (liquid biofertilizer with 108 CFU per mL)	Applied during drought stress conditions (after 80, 130 and 180 mm evaporation)	Increases soil nutrients (K, P, Fe, Cu), decreases pH and EC, influences seed yield components	Nitroxin alleviates drought stress effects under mild stress conditions, improving seed yield, biological yield and oil yield. Effects were less pronounced	Safari et al. (2023)

(continued)

Table 2.1 (continued)

Oilseed crop	Biostimulant used	Application method	Dosage	Timing of application	Mode of action	Drought tolerance improvement	References
						under severe drought stress	
Brassica (RH-725)	Seaweed extract (SWE)	Foliar application	OSA: 2 mL/L and 4 ml/L; SWE: 3 mL/L and 4 ml/L	Vegetative stage	Improves photosynthesis, water retention and antioxidant activity while reducing stress markers	OSA and SWE enhance drought tolerance, improving yields under drought conditions	Sujata et al. (2023)
Cumin (*Cuminum cyminum* L.)	Humic acid and seaweed extract	Foliar application	Humic acid: 300 mg/L; seaweed extract: 2%	During drought stress	Enhances chlorophyll content, antioxidant activity and proline accumulation while improving physiological traits	Combined application of humic acid and seaweed extract effectively mitigates drought stress, improving seed and oil yields	Rahgoshahi et al. (2023)
Winter oilseed rape	Lumibio Kelta (biostimulant)	Seed treatment	General treatment at two levels: With and without the biostimulant	At sowing	Increases root length and seed yield by improving plant growth parameters	Biostimulant treatment increased root length and seed yield, especially under drought stress conditions, improving plant stand and yield	Bečka et al. (2024)
Soybean (*Glycine max* L.)	Calcium chloride ($CaCl_2$), seaweed extract, humic acid	Seed priming	$CaCl_2$: 6 g/L (others unspecified)	Presowing (seed priming)	Enhances seed growth and resilience to water stress	Improved growth parameters under 80% water requirement with $CaCl_2$ priming and biofertilization (*rhizobium japonicum* + mycorrhiza)	Seadh et al. (2024)
Safflower	Salicylic acid (SA)	Foliar application,	SA: 1 mM, Se: 10 mM	Presowing/vegetative stage		Increased oil yield, higher unsaturated fatty	Mosalman et al. (2024)

		intercropping systems			Modifies stress response, enhances antioxidant activity	acids, enhanced chlorophyll and antioxidant activity under rainfed conditions; highest seed and oil yields with Se treatment in sole cropping	
Sesame (Shandawil 3)	Spirulina + chlorella, chlorella, spirulina, vinasse	Foliar spray	Spirulina + chlorella: Mixed; chlorella: 2%, spirulina: 2%	Preflowering, vegetative stage	Enhanced plant growth (height, branches, capsules), improved seed yield, seed quality, germination and protein content	Improved seed yield, seed oil content and biological yield. Best results when spirulina + chlorella were applied at the optimal harvest date (120 days)	Kishk (2024)
Sesame	Salicylic acid (SA)	Foliar spray	SA: 1.5 mM; BWE: 2000 ppm; LE: 5000 ppm	Preflowering, vegetative stage	Mitigates drought-induced oxidative stress, regulates osmoprotectants, enhances antioxidant defense system, increases mineral nutrients and photosynthesis systems	BWE most effective in alleviating drought stress. Enhanced plant growth, yield, photosynthetic activity and osmoprotectants	Pourghasemian et al. (2020)
Soybean	*Gracilaria tenuistipitata* (red seaweed)	Foliar spray	0.0%, 5.0% and 10.0% v/v	Well-watered (80% FC) and drought (40% FC)	Contains bioactive compounds, including antioxidants, which help reduce stress	Increased grain yield by 54.87% under drought, 23.97% under well-watered conditions	Mannan et al. (2023)

Table 2.2 Classification and mechanisms of biostimulants

Category	Subtype	Mechanisms	Targeted stress responses	Examples	References
Microbial biostimulants	Rhizobacteria, mycorrhizae	Enhances root architecture, promotes nutrient uptake	Improves water absorption and root growth	*Bacillus subtilis*, *Rhizophagus irregularis*	Rouphael et al. (2015)
Organic acids	Humic and fulvic acids	Modulates ion transport, enhances soil nutrient uptake	Improves water retention	Liqhumus, Fulvagra	Mackiewicz-Walec, Olszewska (2023)
Plant extracts	Seaweed extracts	Provides phytohormones, increases antioxidant activities	Reduces oxidative stress	*Ascophyllum nodosum*	Khan et al. (2009); Kumari et al. (2023)
Hydrolysates	Protein and amino acids	Supports osmoregulation, stabilizes proteins	Maintains cellular homeostasis	Animal/plant hydrolysates	González-Morales et al. (2021)
Hydrolysates	Protein hydrolysates	Enhances antioxidant activity, regulates metabolism	Mitigates drought and heat stress	Fish protein hydrolysates	Leporino et al. (2024)
Inorganic compounds	Organosilicone	Increased the wheat yield by optimizing the availability of soil nutrients	Improves soil microbial community structure	Bacillus, Massilia, Pseudomonas and Talaromyces	Wang et al. (2025)
Beneficial fungi	*Trichoderma* spp., endophytes	Improves nutrient cycling, induces stress tolerance	Enhances root resilience and defense response	*Trichoderma harzianum*, fungal endophytes	Gowtham et al. (2024)
Bacterial metabolites	Siderophores, exopolysaccharides	Chelates nutrients, promotes soil aggregation	Enhances nutrient availability and water retention	*Pseudomonas fluorescens*	Morcillo et al. (2021)
Biochar	Carbon-based biochar	Improves soil structure, enhances microbial activity	Increases water-holding capacity	Agricultural waste biochar	Adhikari et al. (2022)
Antioxidants	Vitamin C, polyphenols	Scavenges reactive oxygen species (ROS)	Reduces oxidative stress and lipid peroxidation	Ascorbic acid, plant-derived polyphenols	Gill et al. (2010)
Vermicompost	Organic vermicast	Enhances nutrient bioavailability and soil health	Increases soil moisture and fertility	Earthworm-processed organic matter	Toor et al. (2024)

2.5.1 Role of Biostimulants in Stress Mitigation

Biostimulants are one of the most essential factors in assisting plants to avoid harmful impacts from drought stress by improving their efficiency to sustain growth, yield and physiological traits under water deficit conditions. It provides an overview of the main mechanisms operating in biostimulants, whether through water uptake, plant root system architecture or stress-signaling pathway modulation and outlines their interconnection (Franzoni et al. 2022). The most important role of biostimulants in the optimization of drought is related to their function regarding root development. Biostimulants, including those combining microbial such as AMF and PGPR, are an essential part of root architecture improvement (Sun et al. 2024). AMF are beneficial microbes that form symbiotic relationships with plant roots which facilitate the expansion of root networks, increasing surface area for water and nutrient uptake. AMF form a huge web of mycelium extending well beyond the root zone, allowing plants to tap into water and phosphorus sources that would otherwise be unreachable. This more extensive, deeper root system allows plants to keep pulling in water even when things get dry and the plant stays hydrated even in drought conditions (Püschel et al. 2020). PGPR, in contrast to pathogens, produce phytohormones like IAA leading to more root branching and elongation which is essential for the rapid microbe multiplying phase of mutualistic existence. This increases root development, allowing the plant to absorb more soil moisture, especially in drought. This role is particularly important in drought-prone areas, where the persistence of a deep root system can be critical to ensure plant survival (Etesami et al. 2024). Biostimulants also increase water uptake and retention, which enhances the plant's WUE (water use efficiency). Some biostimulants, such as many of those derived from (e.g., seaweed extract) which contain polysaccharides and bioactive compounds including phytohormones, also positively influence for antitranspiration effect on leaf or enhance the ability of plants to uptake water. These compounds increase the radicular hydraulic conductivity or in simple terms ability of plant roots to transport water through their tissues (Devin et al. 2023). Finally, biostimulants can also activate aquaporin proteins which are membrane-bound water channels integral to the facilitation of water movement across cell membranes. Plant water uptake and transport from the roots to leaves are facilitated by aquaporin activity. By this mechanism, biostimulants can help plants maintain cellular hydration under conditions of soil dryness (Brunetti et al. 2023).

The modulation of stress signaling pathways is another critical role of biostimulants in drought stress mitigation. Drought initiates various signaling pathways in plants via changes in phytohormone levels and expression of stress-responsive genes (Lephatsi et al. 2021). Further, ABA is a key phytohormone in the response to drought experienced by plants and it acts prominently through modulation of stomatal closure wherein which aims to reduce transpiration losses. Biostimulants that are based on seaweed extracts and amino acids stimulate the synthesis of ABA or simulate its effects, increasing plant capacity to control stomatal closure (Hariharan et al. 2024). Additionally, biostimulants can also be prime plants for drought subsequent events by encouraging systemic acquired resistance (SAR) and induced systemic resistance (ISR). These defense mechanisms are a "preconditioning" of the plant stress responses, ultimately constituted by an activation of many genes and proteins involved with preserving or increasing resilience to abiotic stresses. Biostimulants prime the plant to defend itself, leading in turn allows plants to better endure drought conditions, accelerating their recovery from stress events (Olowe et al. 2020).

2.5.2 *Physiological and Biochemical Responses to Biostimulants*

Biostimulants applications modulate various physiological and biochemical aspects in plants which enable them to improve drought resistance. These responses are crucial for maintaining plant function and productivity under water-limited conditions, which take place on cellular to molecular scales (Ali et al. 2024a, b). One of the key indirect responses elicited by biostimulants is improved regulation of stomata to regulate water loss under drought conditions. Small openings on the surface of leaves responsible for gas exchange and water vapor release are called stomata. Plants close their stomata in times of drought stress to conserve water from being lost via transpiration. However, excessive stomatal closure which results from prolonged exposure can restrict CO_2 uptake and thus decrease photosynthesis rates leading to stunted growth (Ramzan et al. 2022).

Biostimulants, such as seaweed extracts are essential to help modulate the dynamic balance between stomatal closure and gas exchange for plants to maintain metabolism with less water stress. Selection for natural variation increased the range of responses in stomatal function, thereby regulating to minimize loss from water stress on growth and yield (Santaniello et al. 2017). Osmotic adjustment which is the accumulation of compatible solutes such as proline, glycine betaine and soluble sugars provides cell osmolarity allowing production to maintain turgor pressure in cells which protects cellular structures under reduced water potential is another critical physiological response involved. While amino acids and protein hydrolysates can promote the synthesis of these osmolytes which helps in plant adaptation leading to a better performance under water deficit conditions (Sanders et al. 2012).

Osmotic adjustment is an essential mechanism for maintaining cell activities in dehydration conditions because such a process can protect cells against wilting damage and maintain the integrity of cellular membranes, proteins, etc. (Fig. 2.2). By maintaining the osmotic balance, plants can continue to grow even when water becomes scarce by biostimulants. Besides these physiological changes, biostimulants also activate significant biochemical responses contributing to an increased plant tolerance to drought which is one of the most important biochemical responses is promoting the antioxidant defense system. Drought stress usually induces an overproduction of ROS and by-products from the oxidative damage to

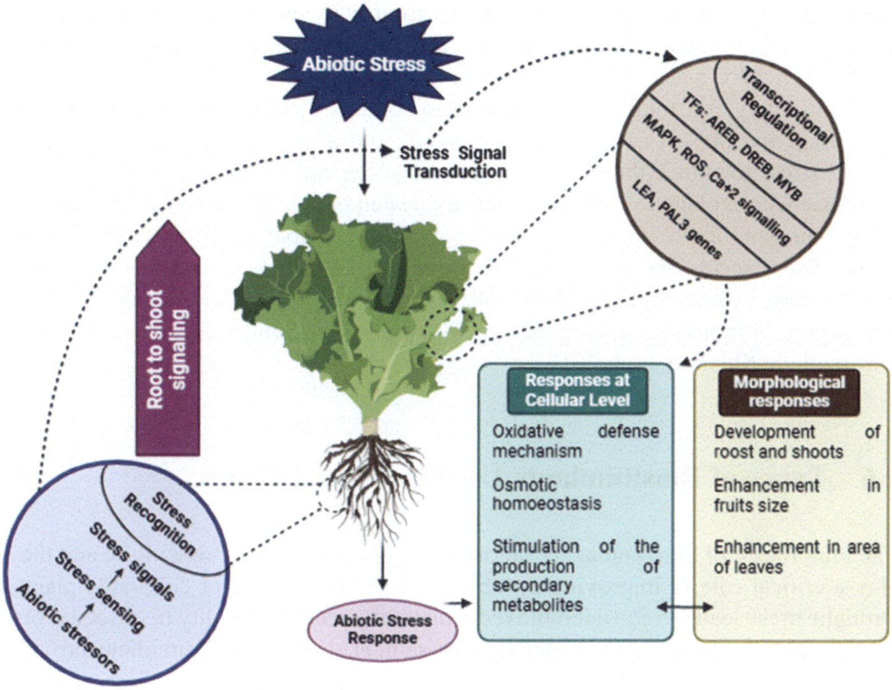

Fig. 2.2 Biostimulants and their abiotic stress responses in oilseed crops

various cellular molecules (lipids, proteins or nucleic acids). Plants predominantly rely on the activity of antioxidant enzymes, such as superoxide dismutase, catalase and peroxidases to reduce ROS levels and counteract oxidative stress in cells under drought conditions (Hasanuzzaman et al. 2021).

A study suggested that phosphate-solubilizing bacteria (PSB) application may directly stimulate the activities of antioxidative enzymes, thereby increasing the capability of plants to scavenge ROS and protect cells from oxidative damage (Azizi et al. 2024). Another study found increased activity under salt stress with biostimulants containing humic substances or seaweed extracts (Gürsoy, 2024). However, further research has shown that extracts of seaweed and Aloe Vera can enhance antioxidant enzymes catalase and peroxidase, resulting in better mitigation against oxidative stress during drought. This greater antioxidant activity helps cells survive and be productive even under water-limited conditions so, higher plant photosynthetic rates are maintained (Naeem et al. 2020).

Furthermore, the increased production of osmolytes including proline and soluble sugars due to biostimulants is another significant biochemical response. Their synthesis is crucial for plant stress tolerance, as these compounds function as compatible solutes in supporting plants to maintain osmotic balance and preserving the structural integrity of cellular components (Eswaran et al. 2024). Especially, proline is involved in oxidative stress and dehydration tolerance which helps to protect proteins, including enzymes and membranes by the elimination of damaging reactive oxygen species. They activate proline accumulation to allow the plant to prepare better for water stress (Shafi et al. 2019).

Additionally, biostimulants can provoke secondary metabolite production such as terpenoids and phenol compounds, flavonoids and alkaloids. This may contribute toward stress tolerance through enhancement of the cell wall, induction of defense strategies within plants, such as drought-associated oxidative effects and deleterious effects on gene expression induced during these possible challenges (Mrid et al. 2021). These secondary metabolites are involved in the protection against environmental stress by scavenging free radicals, stabilizing cell walls and modulating gene expression responsible for response to desiccation under severe conditions (Neeragunda Shivaraj et al. 2018).

2.6 Types of Biostimulants Used in Oilseed Crops

The importance of biostimulants cannot be neglected in today's agriculture and they play a critical role in improving the drought tolerance ability of cultivated plants. Drought stress leads to considerable reductions of yields and quality in oilseed crops such as soybean, sunflower or rapeseed (Hosseini et al. 2023). Biostimulants provide a sustainable and efficient solution for developing the resistance of these crops to water stress conditions. Different types of biostimulants are regularly applied, all with unique action mechanisms. Here we have discussed the role of microbial biostimulants, seaweed extracts, humic substances, amino acids and protein hydrolysates in detail, which reveals their mechanism during drought stress conditions (Sun et al. 2024).

2.6.1 Microbial Biostimulants (e.g., Rhizobacteria, Mycorrhizae)

Microbial biostimulants such as PGPR and AMF have been widely studied for improving drought tolerance in oilseed crops through interaction with the root system, involved in better uptake of water and nutrients (Darakeh et al. 2022). PGPR is a group of bacteria that colonize plant roots and induce plant growth by diverse mechanisms. Some common PGPR bacteria belong to genera as *Azospirillum, Bacillus* and *Pseudomonas*. The produced phytohormones and IAA by these bacteria can enhance root elongation and branching which leads to a larger surface for absorbing water from the soil thereby increasing drought tolerance of oilseed crops (Grover et al. 2021). Besides, PGPR can increase the availability of crucial nutrient elements predominantly nitrogen and phosphorus. Azospirillum species promote total nitrogen fixation in the form of atmospheric nitrogen into a usable soil available nitrogen making it significant to nutrient uptake, especially during water stress when the process is limited (Fukami et al. 2018).

PGPR are also able to produce exopolysaccharides, essential for the alteration of soil and saving water. These compounds increase soil particle aggregation and lead to an improvement of moisture-retention capacity in the rhizosphere which is beneficial for root systems in the time of water shortage (Naseem et al. 2018). PGPR use is associated with improving root development and drought tolerance, which result in increased yields under water-limited conditions, like oilseed crops such as sunflower (Saeed et al. 2016). AMF form symbiotic relationships with the roots of most plant species, including oilseed crops. Fungal hyphae can grow much further from the root than unicellular bacteria and thus are a crucial means by which plants gain access to water. AMF enhance root hydraulic conductivity and an increase in the osmoprotectants occurs. Proline within plant tissues leads to drought stress alleviation. These osmoprotectants cause plants to maintain their cell turgor during water-deficit conditions, prevent dehydration and tissue damage (Jajoo et al. 2021).

AMF, on the other hand, creates soil structure by binding up to aggregates with glomalin, which is a glycoprotein that improves porosity and water retention in soils. Mycorrhizal inoculation has been demonstrated to decrease the risk of prolonged drought stress in oilseed crops such as soybean and rapeseed by increasing root biomass and improving water and nutrient use efficiency (Agnihotri et al. 2024). In general, bioinoculants in terms of microbial biostimulants, especially PGPR and AMF, are important to increase the drought tolerance ability of oilseed crops by root stimulation causing better roots development, plant water and nutrient accumulations (Sarkar et al. 2021).

2.6.2 Seaweed Extracts and Their Benefits

Biostimulants based on seaweed extracts, containing diverse bioactive molecules beneficial in plant development and stress alleviation are commonly utilized (Ali et al. 2021). These extracts are derived primarily from brown seaweeds, such as *Ascophyllum nodosum* and *Ecklonia maxima* and contain many beneficial compounds such as polysaccharides, phytohormones and antioxidants which may contribute to enhanced drought tolerance in oilseed crops (El Boukhari et al. 2020).

Seaweed extracts contain polysaccharides like alginates, laminarin and fucoidan which can improve the soil structure by facilitating soil aggregation, therefore enhancing and stabilizing the water-holding capacity of soils. These substances also boost beneficial soil microorganisms and thus enhance the health of soils which directly fuels crop productivity (Akbar et al. 2024). These seaweed extracts contain a range of essential phytohormones such as cytokinins, auxins and ABA along with polysaccharides. These hormones regulate plant growth and development, specifically when the plants are under stress conditions like drought.

Cytokinins cause cell division and limit the promotion of leaf senescence, while ABA induces stomatal closure to limit water loss from transpiration (Dong et al. 2009).

The application of seaweed extracts has been reported to noticeably enhance drought tolerance in oilseed crops (Kasim et al. 2015). Seaweed extracts improve drought tolerance by priming the plants to grow roots, as one of their key components enables gene expression for lateral root growth. This effect is highly associated with the presence of auxins in extracts, which enhance root length and lateral root formation (Ghaderiardakani et al. 2019).

Seaweed extracts have been shown to improve root biomass and water uptake efficiency under environmental stresses (Dawood et al. 2025; Rahgoshahi et al. 2023; Shukla et al. 2018). A study reported the positive role of seaweed in the enhancement of root dry matter in sunflower and rapeseed during drought stress conditions (Sujata et al. 2023). Seaweed extracts additionally help to balance stomatal closure which leads to less water loss during times of low moisture availability (Ahmed et al. 2024). The presence of ABA in seaweed extracts improves drought tolerance through stomatal closure by maintaining plant turgor pressure, whereas cytokinins prevent the onset of leaf senescence under water stress conditions for an elongated period and allow sustained photosynthesis (Luo et al. 2021).

Seaweed extracts are a rich source of antioxidants including ascorbic acid and tocopherols for ROS scavenging in drought stress. This decrease in oxidative damage preserves cellular functionality and supports plant sustainability under water-deficient conditions. These characteristics together with the root formation, water holding capacity and antioxidant properties indicate that seaweed extracts could serve as biostimulants offering a tool to enhance drought tolerance of oilseed crops (Ramzan et al. 2022).

2.6.3 Humic Substances and Their Role in Drought Tolerance

Humic substances, namely, humic and fulvic acids, are the organic compounds produced during the breakdown of plant and animal sources. They are known to be good at improving soil structure, nutrient availability and water-holding capacity. Humic substances are heterogeneous complex molecular structures, which bind to soil minerals and roots resulting in physical, chemical and biological improvement of the soils (Guo et al. 2019). The hydrophilic nature of these larger and more complex humic acids enhances soil aggregation by improving soil porosity, infiltration and organic matter content along with increasing cation exchange capacity. It favors water retention in the root zone, which is fundamental during dry periods and has a much higher porosity. Because their molecules are smaller, fulvic acids tend to be more readily absorbed by plant roots and can increase the uptake of vital nutrients including many micronutrients like iron, zinc or manganese (Turan et al. 2022).

Humic substances improved drought tolerance of oilseed crops by promoting root growth and enhancing water and nutrient uptake (Hayati et al. 2022). The addition of humic acid leads to improved soil structure which increases water storage capacity within the root zone so that plants have a source of water available during drought (Ennab et al. 2023). This is especially useful in sandy soils with low water retention. Moreover, humic substances stimulate the activity of soil microorganisms which are involved in nutrient cycling and maintenance of high levels of fertility (Jin et al. 2024).

Humic substances when applied to oilseed crops such as soybean and sunflower can improve root growth as well as water use efficiency in relation to drought tolerance. Humic substances can assist in water stress by improving soil moisture holding capacity. These compounds promote growth, development and productivity of oilseed crops under limited moisture conditions (de Moura et al. 2023). In addition, humic substances have been shown to induce many mechanisms of defense in the plant such as the production of endogenous antioxidants which are implicated against oxidative stress induced by water deficit (Badr et al. 2024).

2.6.4 Amino Acids and Protein Hydrolysates in Drought Stress Management

Amino acids and protein hydrolysates are commonly used as biostimulants that can improve plant metabolism and stress tolerance. Stress tolerance mediates several physiological processes that are important for drought response, as well as their requirements for protein and enzyme synthesis that build up plant resistance (Sun et al. 2024). Plants need amino acids like proline, glutamine and glycine to grow properly. Protein Hydrolysates are the combination of peptides and free amino acids derived from the enzymatic hydrolysis of proteins. Hydrolysates are a good source of amino acids directly absorbable and usable by plants for growth promotion and stress adaptation (Pasković et al. 2024).

One of the main mechanisms for enhancing drought tolerance is due to better osmotic adjustment. In general, proline also acts as an osmoprotectant and plays a key role in cell turgor maintenance by stabilizing cellular structures and protecting genes from denaturation. Under extreme drought conditions, the plants accumulate proline in order to retain water within their cells and prevent dehydration (Singh et al. 2015).

Amino acids, namely glutamine and glycine are also implicated in nitrogen assimilation pathways of stress-responsive proteins which can enhance plant rigidity (Ganie, 2021). In addition, protein hydrolysates could trigger antioxidant biosynthesis including superoxide dismutase and catalyze to protect plants against oxidative damage resulting from ROS accumulated during drought stress period. The use of amino acids and protein hydrolysates has been reported to increase the biomass production, water-use efficiency and growth characteristics of plants in oilseed crops under drought conditions (Sun et al. 2024). These biostimulants have significant roles in drought-stress management and oilseed crop productivity at suboptimal water conditions by providing essential metabolites, enhancing the stress response systems of plants (Bhupenchandra et al. 2022).

2.7 Application Strategies of Biostimulants in Oilseed Crops

Based on previous findings, the application of biostimulants in oilseed crops (soybean, sunflower and especially rapeseed) has shown potential to enhance plant resistance to some abiotic stresses that are relevant for promoting resilience against drought (Sujata et al., 2023). Nevertheless, the performance of biostimulants relies heavily on timing, dosage and application mode. Furthermore, biostimulants are typically used in conjunction with conventional agricultural inputs such as fertilizers and irrigation to maximize crop growth capacity (Calvo et al. 2014). Both the timing and dosage of biostimulant application are important variables that can affect their success in enhancing drought tolerance as well as overall plant growth. At the time of sowing, it is therefore necessary to know how long and precisely one should apply biostimulants to obtain an advantage over enhancing stress resilience. The correct timing of biostimulant application is closely related to the growth stage and environmental conditions under which plants grow. Normally, biostimulants are applied at specific plant stages for drought stress management when the plants are more susceptible to water deficiency including the germination phase, early vegetative growth or reproductive stage (Dalal et al. 2019).

The application before stress creates some damage to establish biostimulants alerting plant defense mechanisms against future water scarcity. These applications at an early stage, such as during seed germination or seedling establishment are an effective way to improve root development along with access to water and ultimately contribute toward drought tolerance (Romero-Munar et al. 2023). For example, the application of microbial biostimulants like PGPR in early stages has been shown to help improve root architecture and enable enhanced water and nutrient access from deeper soil layers. Another example is the application of seaweed extracts on growth responses and drought tolerance by enhancing the accumulation of osmolytes and other stress-related compounds (Christou et al. 2022). However, when applied mid-season or late in the season on a crop that is already suffering from drought stress, biostimulants are less likely to have an effect. Improve nutrient use efficiency and reduce the influence of water stress on seed formation, simply by applying biostimulants during reproductive stages (Jiménez-Arias et al. 2022). The recommended dosage of biostimulants varies from type to type, crop species and local environmental conditions (Fig. 2.3).

Overuse of biostimulants can also result in nutrient imbalances, phytotoxicity or wasteful resources (Bartucca et al. 2022). However, for microbial biostimulants, the

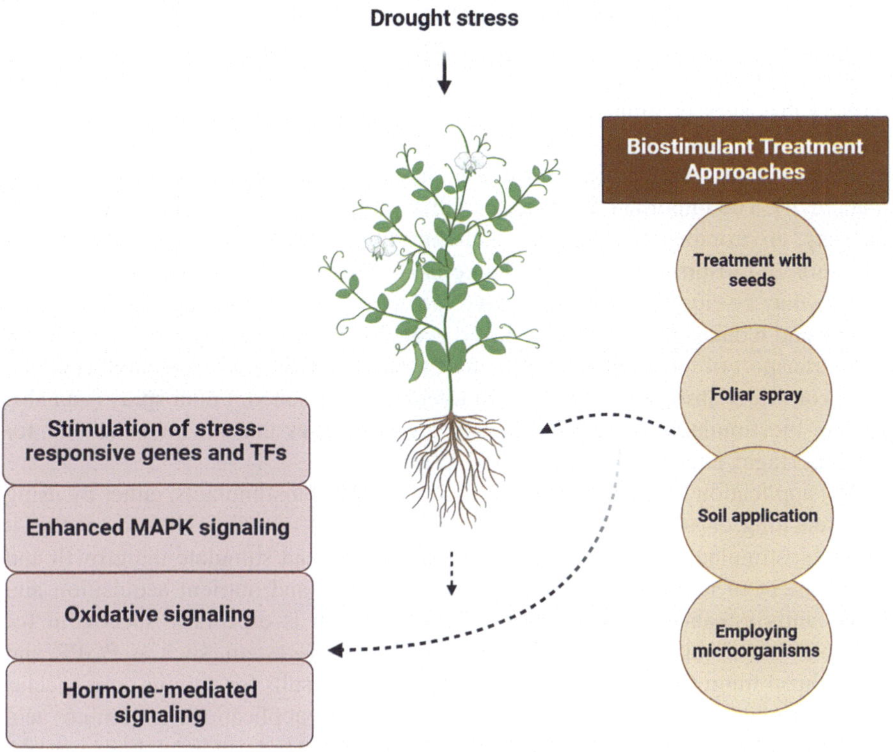

Fig. 2.3 Different applications of biostimulants under drought stress

inoculum concentration should be tightly controlled to achieve effective colonization of roots and stems without outcompeting existing plant-associated microbes in the soil. For instance, a study reported that soybean and sunflower growth and drought resilience were greatly improved by an application of PGPR (Afroz et al. 2024). Dosage recommendations for nonmicrobial biostimulants like seaweed extracts, humic substances and amino acids are usually made by the degree of active ingredient to be used. Seaweed extracts are generally used at rates of 1 to 2 L/ha depending on the crop species and growth stage (Tarafdar, 2022).

2.7.1 Foliar Versus Soil Application Methods

Biostimulants can be applied through foliar spray, which means sprayed directly on leaves or soil application, either to the surface of a planting medium in pots and greenhouse systems or by incorporation into planting media implanted with irrigation systems. There are several application methods, each with its advantages and limitations (Abd-El Hameed et al. 2023). Foliar application is one popular method to apply biostimulants, along with nonmicrobial biostimulants such as humic substances or amino acids. A key benefit of foliar application is the fast uptake by the plant, making it well-suited to stress conditions. This is why foliar application works best for short-term drought stress or nutrient deficiencies (Ciriello et al. 2024). The important attributes of foliar spray efficiency are related to the leaf anatomy, environmental conditions including humidity rate, temperature and concentration of biostimulant solution. Therefore, under drought conditions when plant stomata may be closed to conserve water, the uptake of biostimulants into leaves can decline and reduce their efficacy (Elsayed et al. 2023). Foliar sprays are less efficient in the transport of biostimulants to the root system, which is where many beneficial effects on water absorption and drought tolerance are elicited. Foliar sprays can also provide biostimulants to be used at key growth stages that are most critical for different stages under stress (Sattar et al. 2024).

Soil application is the method often used to induce biostimulants, either by using soil drenching or any other method. The main benefit of soil application is that it brings biostimulants to the root zone directly, which can stimulate the growth and development of roots so as to improve water uptake and nutrient acquisition and significantly enhance soil physiology. This approach is especially successful for microbial biostimulants that need to colonize the root system, such as PGPR and mycorrhizal fungi (Halpern et al. 2015). Differential results for various crop species and biostimulant types were observed when foliar applications of amino acid formulations were compared to the same ones applied at the soil level. For example, foliar application of seaweed extract at the reproductive stage in rapeseed resulted in increased leaf water content and decreased oxidative damage which contributed to higher drought tolerance. In this sense, there were positive results when using the soil application of humic substances in sunflower due to improved root growth and water use efficiency which enabled better drought tolerance and yield stability. In conclusion, the choice of application methods should be adjusted to both the objectives pursued with biostimulant treatment and crop particularities (Hariharan et al. 2024; Sujata et al. 2023).

2.7.2 Integrating Biostimulants with Other Agricultural Practices

Biostimulants are frequently used in combination with conventional agricultural practices such as fertilizers, irrigation systems and pest management to deliver maximum benefits. Employing biostimulants in practice is often combined with existing fertilizers. By promoting root growth and maintaining soil structure, they can help plants take up nutrients more easily from any applied fertilizers (Ali et al. 2024a, b). For example, the coapplication of microbial biostimulants such as PGPR with nitrogen and phosphorus fertilizers increased nutrient use efficiency and drought tolerance in oilseed crops. In the same way, seaweed extracts and humic substances, when sprayed alongside fertilizers on a crop, can cause increased nutrient absorption, taking crop nutrition to another level in terms of productivity (Singh et al. 2022a, b).

Additionally, integrated biostimulant-fertilizer applications help reduce the over-exploitation of fertilizer and have a positive effect on sustainable agriculture. Biostimulants increase the efficiency of nutrient use by crops and mitigate against nutrient leaching and soil degradation, in addition to promoting more sustainable agricultural practices (Shahrajabian et al. 2021). Biostimulants are commonly used in combination with irrigation systems to enhance water use efficiency and mitigate the effects of drought (Chen et al. 2023). Research indicates the use of biostimulants with deficit irrigation strategies which may boost drought tolerance in oilseed crops and similarly reduce water usage. For instance, the application of humic substances and microbial biostimulants via drip irrigation has been demonstrated to increase water retention in soil at the whole root zone depth conferring drought tolerance improvement in sunflower (*Helianthus annuus*) and soybean (*Glycine max*) (Sun et al. 2024). Biostimulants can also help to strengthen plant resilience against biotic and abiotic stressors in integrated pest nutritional management systems (Asif et al. 2023). For example, some microbial biostimulants such as mycorrhizal fungi and PGPR, not only enhance plant tolerance to drought but also increase their defense mechanism against pests and diseases by the induction of systemic resistance (Rai et al. 2021). Likewise, using amino acids and seaweed extracts could activate the natural antidefense resistance in the plant to a minimum dependence on chemical pesticides (Ali et al. 2021).

2.8 Conclusion

Importantly, the role of biostimulants in drought tolerance can help to contribute to sustainable and innovative options for future challenges, mainly due to climate change and resource limitations faced by oilseed crops. Biostimulants such as microbial inocula, seaweed extracts, humic substances and amino acids are biologically based products that work through the natural physiological and biochemical processes of plants to improve root architecture, increase water and nutrient uptake, regulate stomatal function and help mitigate drought-induced oxidative stress. Biostimulants incorporated into agricultural practices have been proven to increase water use efficiency and boost crop yields in conditions of limited supply, thus contributing to long-term sustainability, particularly for oilseed crops. The optimization of the use of biostimulants would be greatly influenced by precision agriculture technologies which make applications more precise and efficient. Even with these advancements, there are still obstacles to overcome in terms of economic viability, registration and regulatory requirements as well as standardization of biostimulant products. The key to ensuring the uptake of biostimulants in agriculture is overcoming these hurdles through effective government measures, subsidies and research programs. In the future, biostimulants could become an integral part of sustainable farming by reducing chemical inputs and meeting the worldwide need to produce more food with less land amid increasing environmental stress. However, overcoming these challenges and ensuring biostimulants penetrate agriculture require significant government actions, such as incentive programs for farmers to produce the costly, convoluted molecules themselves in addition to research and development efforts. Biostimulants may therefore play a key role in sustainable agriculture to reduce chemical inputs and satisfy the global imperative of producing more food with less land, under heightened environmental stress.

References

Abd-El Hameed SM, Elsayed YA, Shoeip A (2023) Influence of organic-biofertilizers and foliar spray with different biostimulants on increasing yield and essential oil of sweet marjoram (Marjorana hortensis L.) plant. Hortic Res J 1(4):60–76

Abdel Latef AA, Chaoxing H (2011a) Effect of arbuscular mycorrhizal fungi on growth, mineral nutrition, antioxidant enzymes activity and fruit yield of tomato grown under salinity stress. Sci Hortic 127:228–233

Abdel Latef AA, Chaoxing H (2011b) Arbuscular mycorrhizal influence on growth, photosynthetic pigments, osmotic adjustment and oxidative stress in tomato plants subjected to low temperature stress. Acta Physiol Plant 33:1217–1225

Abdel Latef AA, Chaoxing H (2014) Does the inoculation with *Glomus mosseae* improve salt tolerance in pepper plants? J Plant Growth Regul 33:644–653

Adeleke BS, Babalola OO (2020) Oilseed crop sunflower (Helianthus annuus) as a source of food: nutritional and health benefits. Food Sci Nutr 8(9):4666–4684

Adhikari S, Timms W, Mahmud MAP (2022) Optimising water holding capacity and hydrophobicity of biochar for soil amendment—a review. Sci Total Environ 851:158043

Afroz M, Muzahid-E-Rahman M, Akhter MS, Bhor SA, Islam S (2024) Root-colonizing endophytes as biostimulants: context, mechanisms of actions, and their potential use for ensuring agricultural sustainability. In: Plant endophytes and secondary metabolites. Elsevier, pp 331–374

Agnihotri R, Pandey A, Sharma MP, Prakash A, Ramesh A, Maheshwari HS, Verma RK, Nargund R, Billore SD (2024) Enhanced soil carbon storage and arbuscular mycorrhizal fungal biomass in a long-term nutrient management under soybean-based cropping system. Environ Sci Pollut Res 1:1–17

Ahmad Z, Anjum S, Skalický M, Waraich EA, Tariq RMS, Ayub MA, Hossain A, Hassan MM, Brestič M, Islam MS, Habib-ur-Rahman M, Wasaya A, Iqbal MA, Sabagh AE (2021) Selenium alleviates the adverse effect of drought in oilseed crops Camelina (Camelina Sativa L.) and Canola (Brassica Napus L.). Molecules 26 (6):1699. https://doi.org/10.3390/molecules26061699

Ahmadi M, Bahrani M (2009) Yield and yield components of rapeseed as influenced by water stress at different growth stages and nitrogen levels. Am.-Eurasian J Agric. Environ Sci 5(6):755–761

Ahmed M, Ullah H, Himanshu SK, García-Caparrós P, Tisarum R, Cha-um S, Datta A (2024) Ascophyllum nodosum seaweed extract and potassium alleviate drought damage in tomato by improving plant water relations, photosynthetic performance, and stomatal function. J Appl Phycol 1:1–14

Akbar WA, Rahim HU, Rutigliano FA (2024) Microbial-and seaweed-based biopolymers: sources, extractions and implications for soil quality improvement and environmental sustainability—a review. J Environ Manag 359:120964

Akram NA, Fatima K, Kong H, Zafar N, Mahmood S, Ashraf M, Abdel Latef AA (2024) Interactive effect of drought stress and l-methionine on the growth and physio-biochemical changes in broccoli (*Brassica oleracea* L. var. italica): leaf and head. J Plant Growth Regul 43:1954–1966

Ali O, Ramsubhag A, Jayaraman J (2021) Biostimulant properties of seaweed extracts in plants: implications towards sustainable crop production. Plan Theory 10(3):531

Ali A, Niu G, Masabni J, Ferrante A, Cocetta G (2024a) Integrated nutrient management of fruits, vegetables, and crops through the use of biostimulants, soilless cultivation, and traditional and modern approaches—a mini review. Agriculture 14(8):1330

Ali S, Akhtar MS, Siraj M, Zaman W (2024b) Molecular communication of microbial plant biostimulants in the rhizosphere under abiotic stress conditions. Int J Mol Sci 25(22):12424

Alsudays IM, Alshammary FH, Alabdallah NM, Alatawi A, Alotaibi MM, Alwutayd KM, Alharbi MM, Alghanem SM, Alzuaibr FM, Gharib HS (2024) Applications of humic and fulvic acid under saline soil conditions to improve growth and yield in barley. BMC Plant Biol 24(1):191

Altaf M, Sarwar S, Iqbal J, Ur Rahman S, Mustafa HSB, Nazir S, Habib I, Nawaz M (2023) Revealing the yield and quality responses of soybean advanced lines under semi-arid conditions. Adv Life Sci 10(1):54–60

Ameen M, Zia MA, Najeeb Alawadi HF, Naqve M, Mahmood A, Shahzad AN, Khan BA, Alhammad BA, Aljabri M, Seleiman MF (2024) Exogenous application of selenium on sunflower (Helianthus annuus L.) to enhance drought stress tolerance by morpho-physiological and biochemical adaptations. Front Plant Sci 15:1427420

Anderback T-M (2021) Biowastes for plant production

Anjum SA, Ashraf U, Zohaib A, Tanveer M, Naeem M, Ali I, Tabassum T, Nazir U (2017) Growth and developmental responses of crop plants under drought stress: a review. Zemdirbyste-Agric 104(3):267

Asif A, Ali M, Qadir M, Karthikeyan R, Singh Z, Khangura R, Di Gioia F, Ahmed ZF (2023) Enhancing crop resilience by harnessing the synergistic effects of biostimulants against abiotic stress. Front Plant Sci 14:1276117

Attia Z, Pogoda CS, Reinert S, Kane NC, Hulke BS (2021) Breeding for sustainable oilseed crop yield and quality in a changing climate. Theor Appl Genet 134(6):1817–1827

Azadi H, Keramati P, Taheri F, Rafiaani P, Teklemariam D, Gebrehiwot K, Hosseininia G, Van Passel S, Lebailly P, Witlox F (2018) Agricultural land conversion: reviewing drought impacts and coping strategies. Int J Disaster Risk Reduct 31:184–195

Azizi A, Bagnazari M, Mohammadi M (2024) Seaweed and phosphate-solubilizing bacteria biofertilizers ameliorate physiochemical traits and essential oil content of Calendula officinalis L. under drought stress. Sci Hortic 328:112653

Badr EA, Bakhoum GS, Al-Ashkar I, Islam MS, Sabagh AE, Abdelhamid MT (2024) Enhancing canola yield and photosynthesis under water stress with hydrogel polymers

Bartucca ML, Cerri M, Del Buono D, Forni C (2022) Use of biostimulants as a new approach for the improvement of phytoremediation performance—a review. Plan Theory 11(15):1946

Barzgar Z, Towfighi H, Fathi Gerdelidani A, Shahbazi K, Heidari A (2024) Effect of different kinds of humic and fulvic acids on the release of manganese from calcareous soils. Iran J Soil Water Res 55(7):1151–1165

Bashir S, Kadir NHA, Naji YAM, Ramesh S (2024) Commercialization and market potential of Nanobiostimulants. In: Nanobiostimulants: emerging strategies for agricultural sustainability. Springer, pp 399–417

Bečka D, Bečková L, Satranský M, Pazderů K (2024) Effect of seed weight and biostimulant seed treatment on establishment, growth and yield parameters of winter oilseed rape. Plant Soil Environ 70(5):296

Bhattacharya A, Bhattacharya A (2021) Dry matter production, partitioning, and seed yield under soil water deficit: a review. Soil Water Deficit and Physiological Issues in Plants, pp 585–702

Bhupenchandra I, Chongtham SK, Devi EL, Choudhary AK, Salam MD, Sahoo MR, Bhutia TL, Devi SH, Thounaojam AS, Behera C (2022) Role of biostimulants in mitigating the effects of climate change on crop performance. Front Plant Sci 13:967665

Bhushan C, Srinidhi A, Kumar V, Singh G (2015) Lived anomaly: how to enable farmers in India cope with extreme weather events. Centre for Science and Environment, New Delhi

Boyer J, Byrne P, Cassman K, Cooper M, Delmer D, Greene T, Gruis F, Habben J, Hausmann N, Kenny N (2013) The US drought of 2012 in perspective: a call to action. Glob Food Sec 2(3): 139–143

Brunetti C, Alderotti F, Sillo F, Magno R, Brilli L, Gori A, Centritto M, Ferrini F, Pasquini D, Saurer M (2023) InvesBgaBng holm oak forest dieback through a mulBscale approach: physiological measurements and visual assessment data with remote sensing. In: XII SIBV (Italian Society of Plant Biology) Congress. Società Italiana di Biologia Vegetale, pp 1–153

Bulgari R, Cocetta G, Trivellini A, Vernieri P, Ferrante A (2015) Biostimulants and crop responses: a review. Biol Agric Hortic 31(1):1–17

Calvo P, Nelson L, Kloepper JW (2014) Agricultural uses of plant biostimulants. Plant Soil 383:3–41

Chen C-H, Lin K-H, Chang Y-S, Chang Y-J (2023) Application of water-saving irrigation and biostimulants on the agronomic performance of maize (Zea mays). Process Saf Environ Prot 177:1377–1386

Christou A, Agathokleous E, Fotopoulos V (2022) Safeguarding food security: hormesis-based plant priming to the rescue. Curr Opin Environ Sci Health 28:100374

Ciriello M, Fusco GM, Woodrow P, Carillo P, Rouphael Y (2024) Unravelling the nexus of plant response to non-microbial biostimulants under stress conditions. Plant Stress 1:100421

Cirillo A, De Luca L, Graziani G, Cepparulo M, El-Nakhel C, Giordano M, Rouphael Y, Ritieni A, Romano R, Di Vaio C (2022) Biostimulants application on Olea europaea L. in Mediterranean conditions increase the production and bioactive compounds of drupes and oil. Agriculture 12 (12):2173

Cozzolino V, Monda H, Savy D, Di Meo V, Vinci G, Smalla K (2021) Cooperation among phosphate-solubilizing bacteria, humic acids and arbuscular mycorrhizal fungi induces soil microbiome shifts and enhances plant nutrient uptake. Chem Biol Technol Agric 8:1–18

Dalal A, Bourstein R, Haish N, Shenhar I, Wallach R, Moshelion M (2019) Dynamic physiological phenotyping of drought-stressed pepper plants treated with "productivity-enhancing" and "survivability-enhancing" biostimulants. Front Plant Sci 10:905

Darakeh SASS, Weisany W, Tahir NA-R, Schenk PM (2022) Physiological and biochemical responses of black cumin to vermicompost and plant biostimulants: Arbuscular mycorrhizal and plant growth-promoting rhizobacteria. Ind Crop Prod 188:115557

Das D, Ullah H, Himanshu SK, Tisarum R, Cha-Um S, Datta A (2022) Arbuscular mycorrhizal fungi inoculation and phosphorus application improve growth, physiological traits, and grain yield of rice under alternate wetting and drying irrigation. J Plant Physiol 278:153829

Dawood MFA, Tahjib-Ul-Arif Md, Shirazy BJ, Abdel Latef AAH (2025) Unraveling the role of κ-carrageenan on the combined effect of drought and chromium stress in wheat (Triticum aestivium L.). Plant Physiology and Biochemistry. 223, J109781.

de Moura OVT, Berbara RLL, de Oliveira Torchia DF, Da Silva HFO, de Castro TAVT, Tavares OCH, Rodrigues NF, Zonta E, Santos LA, García AC (2023) Humic foliar application as sustainable technology for improving the growth, yield, and abiotic stress protection of agricultural crops. A review. J Saudi Soc Agric Sci 22(8):493

Debaeke P, Casadebaig P, Langlade NB (2021) New challenges for sunflower ideotyping in changing environments and more ecological cropping systems. OCL 28:29

Devin SR, Prudencio ÁS, Mahdavi SME, Rubio M, Martínez-García PJ, Martínez-Gómez P (2023) Orchard management and incorporation of biochemical and molecular strategies for improving drought tolerance in fruit tree crops. Plan Theory 12(4):773

Dong H, Niu Y, Kong X, Luo Z (2009) Effects of early-fruit removal on endogenous cytokinins and abscisic acid in relation to leaf senescence in cotton. Plant Growth Regul 59:93–101

El Boukhari MEM, Barakate M, Bouhia Y, Lyamlouli K (2020) Trends in seaweed extract based biostimulants: manufacturing process and beneficial effect on soil-plant systems. Plan Theory 9(3):359

El Sabagh A, Islam MS, Hossain A, Iqbal MA, Mubeen M, Waleed M, Reginato M, Battaglia M, Ahmed S, Rehman A (2022) Phytohormones as growth regulators during abiotic stress tolerance in plants. Front Agron 4:765068

Elferjani R, Soolanayakanahally R (2018) Canola responses to drought, heat, and combined stress: shared and specific effects on carbon assimilation, seed yield, and oil composition. Front Plant Sci 9. https://doi.org/10.3389/fpls.2018.01224

Elsayed SI, Sabra AS, Omer EA (2023) Role of plant extracts and biostimulant in mitigating plant drought and salinity stress. In: Climate-resilient agriculture, vol 2. Agro-Biotechnological Advancement for Crop Production. Springer, pp 571–602

Ennab HA, Mohamed AH, El-Hoseiny HM, Omar AA, Hassan IF, Gaballah MS, Khalil SE, Mira AM, Abd El-Khalek AF, Alam-Eldein SM (2023) Humic acid improves the resilience to salinity stress of drip-irrigated Mexican lime trees in saline clay soils. Agronomy 13(7):1680

Eswaran SUD, Sundaram L, Perveen K, Bukhari NA, Sayyed R (2024) Osmolyte-producing microbial biostimulants regulate the growth of Arachis hypogaea L. under drought stress. BMC Microbiol 24(1):165

Etesami H, Glick BR (2024) Bacterial indole-3-acetic acid: a key regulator for plant growth, plant-microbe interactions, and agricultural adaptive resilience. Microbiol Res 1:127602

Franzoni G, Cocetta G, Prinsi B, Ferrante A, Espen L (2022) Biostimulants on crops: their impact under abiotic stress conditions. Horticulturae 8(3):189

Fukami J, Cerezini P, Hungria M (2018) Azospirillum: benefits that go far beyond biological nitrogen fixation. AMB Express 8(1):73

Ganie SA (2021) Amino acids other than proline and their participation in abiotic stress tolerance. In: Compatible solutes engineering for crop plants facing climate change. Springer, pp 47–96

García-López J, Lorite IJ, García-Ruiz R, Domínguez J (2014) Evaluation of three simulation approaches for assessing yield of rainfed sunflower in a Mediterranean environment for climate change impact modelling. Clim Chang 124:147–162

Gavelienė V, Mockevičiūtė R, Jankovska-Bortkevič E, Šveikauskas V, Zareyan M, Žalnierius T, Jankauskienė J, Jurkonienė S (2025) Synergistic effects of microbial biostimulants and calcium in alleviating drought stress in oilseed rape. Microorganisms 13(3):530

Ghaderiardakani F, Collas E, Damiano DK, Tagg K, Graham NS, Coates JC (2019) Effects of green seaweed extract on Arabidopsis early development suggest roles for hormone signalling in plant responses to algal fertilisers. Sci Rep 9(1):1983

Gill SS, Tuteja N (2010) Reactive oxygen species and antioxidant machinery in abiotic stress tolerance in crop plants. Plant Physiol Biochem 48(12):909–930. https://doi.org/10.1016/j.plaphy.2010.08.016

González-Morales S, Solís-Gaona S, Valdés-Caballero MV, Juárez-Maldonado A, Loredo-Treviño A, Benavides-Mendoza A (2021) Transcriptomics of biostimulation of plants under abiotic stress. Front Genet 12:583888

Gowtham H, Hema P, Murali M, Shilpa N, Nataraj K, Basavaraj G, Singh SB, Aiyaz M, Udayashankar A, Amruthesh KN (2024) Fungal endophytes as mitigators against biotic and abiotic stresses in crop plants. J Fungi 10(2):116

Grover M, Bodhankar S, Sharma A, Sharma P, Singh J, Nain L (2021) PGPR mediated alterations in root traits: way toward sustainable crop production. Front Sustain Food Syst 4:618230

Guo X-x, Liu H-t, Wu S-b (2019) Humic substances developed during organic waste composting: formation mechanisms, structural properties, and agronomic functions. Sci Total Environ 662: 501–510

Gürsoy M (2024) Alone or combined effect of seaweed and humic acid applications on rapeseed (Brassica napus L.) under salinity stress. J Soil Sci Plant Nutr:1–13

Halpern M, Bar-Tal A, Ofek M, Minz D, Muller T, Yermiyahu U (2015) The use of biostimulants for enhancing nutrient uptake. Adv Agron 130:141–174
Hamid B, Zaman M, Farooq S, Fatima S, Sayyed RZ, Baba ZA, Sheikh TA, Reddy MS, El Enshasy H, Gafur A (2021) Bacterial plant biostimulants: a sustainable way towards improving growth, productivity, and health of crops. Sustain For 13(5):2856
Han M, Kasim S, Yang Z, Deng X, Saidi NB, Uddin MK, Shuib EM (2024) Plant extracts as biostimulant agents: a promising strategy for managing environmental stress in sustainable agriculture. Phyton 93(9):0031–9457
Hanjra MA, Qureshi ME (2010) Global water crisis and future food security in an era of climate change. Food Policy 35(5):365–377
Hariharan G, Vathshalyan N, Galahitigama H, Wimalasiri U, Kumara GDK (2024) Potential of foliar application of seaweed extracts as a biostimulant for abiotic stress alleviation on crop production. Rev Agric Sci 12:295–312
Hasanuzzaman M, Parvin K, Bardhan K, Nahar K, Anee TI, Masud AAC, Fotopoulos V (2021) Biostimulants for the regulation of reactive oxygen species metabolism in plants under abiotic stress. Cells 10(10):2537
Hayati A, Hosseini SM, Rahimi MM, Kelidari A (2022) The effect of humic acid and nano-chelate iron on the absorption of some nutrients, grain yield, oil, and essential oil of (Nigella sativa) under drought stress. Commun Soil Sci Plant Anal 53(14):1744–1755
Hosseini P, Mohsenifar K, Rajaie M, Babaeinejad T (2023) Plant growth regulators affecting canola (Brasica Napus L.) biochemistry including oil yield under drought stress. Physiol Mol Biol Plants 29(11):1663–1674
Ishrat F, Afrasiab H, Chaudhury FA (2022) Effects of salinity stress on growth and physio-biochemical parameters of three pea (Pisum sativum l.) cultivars of different maturity duration. Adv Life Sci 9(3):380–391
Jaime R, Alcántara JM, Manzaneda AJ, Rey PJ (2018) Climate change decreases suitable areas for rapeseed cultivation in Europe but provides new opportunities for white mustard as an alternative oilseed for biofuel production. PLoS One 13(11):e0207124. https://doi.org/10.1371/journal.pone.0207124
Jajoo A, Mathur S (2021) Role of arbuscular mycorrhizal fungi as an underground saviuor for protecting plants from abiotic stresses. Physiol Mol Biol Plants 27(11):2589–2603
Jeyakumar P (2005) Role of growth substances in conservation agriculture. Department of Crop Physiology, Tamil Nadu Univ—641003
Jiménez-Arias D, Hernándiz AE, Morales-Sierra S, García-García AL, García-Machado FJ, Luis JC, Borges AA (2022) Applying biostimulants to combat water deficit in crop plants: research and debate. Agronomy 12(3):571
Jin Y, Yuan Y, Liu Z, Gai S, Cheng K, Yang F (2024) Effect of humic substances on nitrogen cycling in soil-plant ecosystems: advances, issues, and future perspectives. J Environ Manag 351:119738
Johnson R, Joel JM, Puthur JT (2024) Biostimulants: the futuristic sustainable approach for alleviating crop productivity and abiotic stress tolerance. J Plant Growth Regul 43(3):659–674
Kambiranda DM, Vasanthaiah HK, Katam R, Ananga A, Basha SM, Naik K (2011) Impact of drought stress on peanut (Arachis hypogaea L.) productivity and food safety. Plants Environ 1: 249–272
Kasim WA, Hamada EA, El-Din NS, Eskander S (2015) Influence of seaweed extracts on the growth, some metabolic activities and yield of wheat grown under drought stress. Int J Agron Agric Res 7(2):173–189
Khan W, Rayirath UP, Subramanian S, Jithesh MN, Rayorath P, Hodges DM, Critchley AT, Craigie JS, Norrie J, Prithiviraj B (2009) Seaweed extracts as biostimulants of plant growth and development. J Plant Growth Regul 28:386–399
Khanal A, Shah A (2021) Oilseeds to biodiesel and renewable jet fuel: an overview of feedstock production, logistics, and conversion. Biofuels Bioprod Biorefin 15(3):913–930

Khayatnezhad M, Gholamin R, Keynoos H (2022) Investigation of the effect of end season drought stress on morphological on Durum wheat genotypes. Adv Life Sci 9(3):284–288

Kishk AM (2024) Effect of plant biostimulants, harvesting dates, and storage periods on Sesame (Sesamum indicum L.) seed quality in newly reclaimed soils. Egypt. J Agron 46 (2):329–341

Kumar R (2019) Emerging challenges of water scarcity in India: the way ahead. Int J Innov Stud Sociol Humanit 4(4):6–28

Kumar P, Singh RP (2021) Microbial diversity and multifunctional microbial biostimulants for agricultural sustainability. Clim Resil Environ Sustain Approaches: Global Lessons and Local Challenges, pp 141–184

Kumar G, Nanda S, Singh SK, Kumar S, Singh D, Singh BN, Mukherjee A (2024) Seaweed extracts: enhancing plant resilience to biotic and abiotic stresses. Front Mar Sci 11:1457500

Kumari M, Swarupa P, Kesari KK, Kumar A (2022) Microbial inoculants as plant biostimulants: a review on risk status. Life 13(1):12

Kumari S, Sehrawat KD, Phogat D, Sehrawat AR, Chaudhary R, Sushkova SN, Voloshina MS, Rajput VD, Shmaraeva AN, Marc RA (2023) Ascophyllum nodosum (L.) Le Jolis, a pivotal biostimulant toward sustainable agriculture: a comprehensive review. Agriculture 13 (6):1179

Lephatsi MM, Meyer V, Piater LA, Dubery IA, Tugizimana F (2021) Plant responses to abiotic stresses and rhizobacterial biostimulants: metabolomics and epigenetics perspectives. Meta 11(7):457

Leporino M, Rouphael Y, Bonini P, Colla G, Cardarelli M (2024) Protein hydrolysates enhance recovery from drought stress in tomato plants: phenomic and metabolomic insights. Front Plant Sci 15:1357316

Levy D, Coleman WK, Veilleux RE (2013) Adaptation of potato to water shortage: irrigation management and enhancement of tolerance to drought and salinity. Am J Potato Res 90:186–206

Li J, Van Gerrewey T, Geelen D (2022) A meta-analysis of biostimulant yield effectiveness in field trials. Front Plant Sci 13:836702

Limón-Pacheco J, Gonsebatt ME (2009) The role of antioxidants and antioxidant-related enzymes in protective responses to environmentally induced oxidative stress. Mutat Res Genet Toxicol Environ Mutagen 674(1–2):137–147

Liu Y (2024) Multifaceted ability of organic fertilizers to improve crop productivity and abiotic stress tolerance: review and perspectives. Agronomy 14(6):1141. https://doi.org/10.3390/agronomy14061141

Luo Y, Li W, Huang C, Yang J, Jin M, Chen J, Pang D, Chang Y, Li Y, Wang Z (2021) Exogenous abscisic acid coordinating leaf senescence and transport of assimilates into wheat grains under drought stress by regulating hormones homeostasis. Crop J 9(4):901–914

Mackiewicz-Walec E, Olszewska M (2023) Biostimulants in the production of forage grasses and turfgrasses. Agriculture 13(9):1796

Malik A, Mor VS, Tokas J, Punia H, Malik S, Malik K, Sangwan S, Tomar S, Singh P, Singh N (2020) Biostimulant-treated seedlings under sustainable agriculture: a global perspective facing climate change. Agronomy 11(1):14

Mandal S, Anand U, López-Bucio J, Kumar M, Lal MK, Tiwari RK, Dey A (2023) Biostimulants and environmental stress mitigation in crops: a novel and emerging approach for agricultural sustainability under climate change. Environ Res 233:116357

Mannan MA, Yasmin A, Sarker U, Bari N, Dola DB, Higuchi H, Ercisli S, Ali D, Alarifi S (2023) Biostimulant red seaweed (Gracilaria tenuistipitata var. liui) extracts spray improves yield and drought tolerance in soybean. PeerJ 11:e15588

Mansour E, El-Sobky E-SE, Abdul-Hamid MI, Abdallah E, Zedan AM, Serag AM, Silvar C, El-Hendawy S, Desoky E-SM (2023) Enhancing drought tolerance and water productivity of diverse maize hybrids (Zea mays) using exogenously applied biostimulants under varying irrigation levels. Agronomy 13(5):1320

Matthews S, Siddiqui Y, Ali A (2024) Unleashing the power of bio-stimulants for enhanced crop growth, productivity, and quality: a comprehensive review. J Plant Nutr:1–23

Mikiciuk G, Miller T, Kisiel A, Cembrowska-Lech D, Mikiciuk M, Łobodzińska A, Bokszczanin K (2024) Harnessing beneficial microbes for drought tolerance: a review of ecological and agricultural innovations. Agriculture 14(12):2228

Mishra A, Bruno E, Zilberman D (2021) Compound natural and human disasters: managing drought and COVID-19 to sustain global agriculture and food sectors. Sci Total Environ 754: 142210

Morcillo RJL, Manzanera M (2021) The effects of plant-associated bacterial exopolysaccharides on plant abiotic stress tolerance. Meta 11(6). https://doi.org/10.3390/metabo11060337

Mosalman S, Rezaei-Chiyaneh E, Mahdavikia H, Dolatabadian A, Siddique KH (2024) Enhancing rainfed safflower yield, oil content, and fatty acid composition through intercropping with chickpea and stress-modifier biostimulants. Front Agron 6:1389045

Mrid RB, Benmrid B, Hafsa J, Boukcim H, Sobeh M, Yasri A (2021) Secondary metabolites as biostimulant and bioprotectant agents: a review. Sci Total Environ 777:146204

Nadeem F, Rehman A, Ullah A, Farooq M, Siddique KH (2024) 7 managing drought in semi-arid regions through improved varieties and choice of species. In: Managing soil drought. CRC Press, pp 212–234

Naeem M, Shabbir A, Ansari AA, Aftab T, Khan MMA, Uddin M (2020) Hyacinth bean (Lablab purpureus L.)—an underutilised crop with future potential. Sci Hortic 272:109551

Naseem H, Ahsan M, Shahid MA, Khan N (2018) Exopolysaccharides producing rhizobacteria and their role in plant growth and drought tolerance. J Basic Microbiol 58(12):1009–1022

Nazari M, Ghasemi-Soloklui AA, Kordrostami M, Abdel Latef AA (2023) Deciphering the response of medicinal plants to abiotic stressors: a focus on drought and salinity. Plant Stress 10:100255

Neeragunda Shivaraj Y, Barbara P, Gugi B, Vicré-Gibouin M, Driouich A, Ramasandra Govind S, Devaraja A, Kambalagere Y (2018) Perspectives on structural, physiological, cellular, and molecular responses to desiccation in resurrection plants. Scientifica 1:9464592

Nephali L, Piater LA, Dubery IA, Patterson V, Huyser J, Burgess K, Tugizimana F (2020) Biostimulants for plant growth and mitigation of abiotic stresses: a metabolomics perspective. Meta 10(12):505

Olowe OM, Akanmu AO, Asemoloye MD (2020) Exploration of microbial stimulants for induction of systemic resistance in plant disease management. Ann Appl Biol 177(3):282–293

Ozturk M, Turkyilmaz Unal B, García-Caparrós P, Khursheed A, Gul A, Hasanuzzaman M (2021) Osmoregulation and its actions during the drought stress in plants. Physiol Plant 172(2): 1321–1335

Papa S, Fusco GM, Ciriello M, Formisano L, Woo SL, De Pascale S, Rouphael Y, Carillo P (2022) Microbial and non-microbial biostimulants as innovative tools to increase macro and trace element mineral composition of tomato and spinach. Horticulturae 8(12):1157

Papnai N, Chaurasiya DK, Sahni S (2022) Biostimulants: concept, types and way to enhance soil health. Int J Plant Soil Sci 34(20):24–40

Pasković I, Popović L, Pongrac P, Polić Pasković M, Kos T, Jovanov P, Franić M (2024) Protein hydrolysates—production, effects on plant metabolism, and use in agriculture. Horticulturae 10(10):1041

Passandideh M, Rajaie M, Zeinalzadeh-Tabrizi H (2022) Effect of some plant growth biostimulants on increasing canola (Brassica napus L.) tolerance to drought stress. Env. Stresses Crop Sci 15(4):1023–1035

Payman Y, Babak P, Asad R, Khosro M (2022) Foliar application of silicon, sulfur, and flowering fruit set biostimulant on canola. Gesunde Pflanzen 74(1):193–203

Payne W (2010) Farming systems and food security in sub-Saharan Africa. Food security and soil quality:23–56

Pokharel M (2020) Physiological characterization of winter canola under heat and drought stress during flowering and pod-filling stages. Kansas State University

Poudel S, Vennam RR, Shrestha A, Reddy KR, Wijewardane NK, Reddy KN, Bheemanahalli R (2023) Resilience of soybean cultivars to drought stress during flowering and early-seed setting stages. Sci Rep 13(1):1277

Pourghasemian N, Moradi R, Naghizadeh M, Landberg T (2020) Mitigating drought stress in sesame by foliar application of salicylic acid, beeswax waste and licorice extract. Agric Water Manag 231:105997

Püschel D, Bitterlich M, Rydlová J, Jansa J (2020) Facilitation of plant water uptake by an arbuscular mycorrhizal fungus: a Gordian knot of roots and hyphae. Mycorrhiza 30(2):299–313

Rahgoshahi M, Laghari KPK, Rahimi M, Kelidari A, Keshavarzi K (2023) Physiological enhancement of seed yield and essential oil yield in cumin under drought stress through humic acid and seaweed extract. Russ J Plant Physiol 70(6):147

Rai N, Rai SP, Sarma BK (2021) Prospects for abiotic stress tolerance in crops utilizing phyto-and bio-stimulants. Front Sustain Food Syst 5:754853

Rajesaheb KS, Subramanian S, Boominathan P, Thenmozhi S, Gnanachitra M (2024) Bio-stimulant in improving crop yield and soil health. Commun Soil Sci Plant Anal:1–36

Ramzan F, Younis A (2022) Use of biostimulants in tolerance of drought stress in agricultural crops. Emerging Plant Growth Regulators in Agriculture. Elsevier, pp 429–446

Rao S, Silva-Perez V, Armstrong R, Tefera AT, Brand J, Riffkin P, Rosewarne G (2023) Pulse ideotypes for abiotic constraint alleviation in Australia. Plant Soil 492(1):1–30

Rady MM, Abdelfattah M. A., Desoky EM, Azzam CR, Abdel Latef AA, Mahmoud AEM, Mohamed IAA (2025) A novel group of bio-stimulators: their attenuation of abiotic stress impacts in crop plants. J Plant Nutr 48(13):2201–2216.

Raza A (2021) Eco-physiological and biochemical responses of rapeseed (Brassica napus L.) to abiotic stresses: consequences and mitigation strategies. J. Plant Growth Regul 40(4): 1368–1388

Raza MAS, Shahid AM, Saleem MF, Khan I, Ahmad S, Ali MU, Iqbal R (2017) Effects and management strategies to mitigate drought stress in oilseed rape (Brassica Napus L.): a review. Zemdirb-Agric 104(1):85–94. https://doi.org/10.13080/z-a.2017.104.012

Romero-Munar A, Aroca R, Zamarreño AM, García-Mina JM, Perez-Hernández N, Ruiz-Lozano JM (2023) Dual inoculation with Rhizophagus irregularis and Bacillus megaterium improves maize tolerance to combined drought and high temperature stress by enhancing root hydraulics, photosynthesis and hormonal responses. Int J Mol Sci 24(6):5193

Rouphael Y, Franken P, Schneider C, Schwarz D, Giovannetti M, Agnolucci M, De Pascale S, Bonini P, Colla G (2015) Arbuscular mycorrhizal fungi act as biostimulants in horticultural crops. Sci Hortic 196:91–108

Sabatino L, Consentino BB, Rouphael Y, Baldassano S, De Pasquale C, Ntatsi G (2023) Ecklonia maxima-derivate seaweed extract supply as mitigation strategy to alleviate drought stress in chicory plants. Sci Hortic 312:111856

Saeed M, Noshin I, Abida A, Naveed R, Roomina M, Fatima B, Nazima B (2016) Effect of drought stress on Brassica crops and its mitigation by inoculation of PGPR. Int J Biosci 9:282–291

Safari MR, Dadashi MR, Faraji A, Armin M (2023) Effect of biofertilizer and drought stress on quantitative and qualitative traits in some winter rapeseed (Brassica napus L.) cultivars. Rom Agric Res 40:403

Sanders GJ, Arndt SK (2012) Osmotic adjustment under drought conditions. In: Plant responses to drought stress: from morphological to molecular features. Springer, pp 199–229

Santaniello A, Scartazza A, Gresta F, Loreti E, Biasone A, Di Tommaso D, Piaggesi A, Perata P (2017) Ascophyllum nodosum seaweed extract alleviates drought stress in Arabidopsis by affecting photosynthetic performance and related gene expression. Front Plant Sci 8:1362

Sarkar D, Singh S, Parihar M, Rakshit A (2021) Seed bio-priming with microbial inoculants: a tailored approach towards improved crop performance, nutritional security, and agricultural sustainability for smallholder farmers. Curr Res Environ Sustain 3:100093

Sattar A, El-Yazied AA, Alharbi BM, El-Gawad HGA, Abbas ZK, El-Absy KM, Mahmoud SF, Althaqafi MM, Darwish DBE, Al-Harbi NA (2024) Application of biostimulants alleviated drought stress in sugar beet (Beta vulgaris L.) by improving oxidative defense system, osmolytes accumulation and root yield. J Soil Sci. Plant Nutr: 1–17

Seadh SE, El-Moneam A, El-Saidy AE, Attia A, Ghozzy A (2024) Effect of seed priming and biofertilization treatments on vegetative growth characters of soybean under water stress conditions. J Plant Prod 15(11):749–760

Shafi A, Zahoor I, Mushtaq U (2019) Proline accumulation and oxidative stress: diverse roles and mechanism of tolerance and adaptation under salinity stress. Salt Stress, Microbes, and Plant Interactions: Mechanisms and Molecular Approaches 2:269–300

Shahrajabian MH, Chaski C, Polyzos N, Petropoulos SA (2021) Biostimulants application: a low input cropping management tool for sustainable farming of vegetables. Biomol Ther 11(5):698

Sharma BR, Smakhtin VU (2006) Potential of water harvesting as a strategic tool for drought mitigation. Int Water Manag Inst (IWMI)

Sheteiwy MS, Abd Elgawad H, Xiong YC, Macovei A, Brestic M, Skalicky M, Shaghaleh H, Alhaj Hamoud Y, El-Sawah AM (2021) Inoculation with Bacillus amyloliquefaciens and mycorrhiza confers tolerance to drought stress and improve seed yield and quality of soybean plant. Physiol Plant 172(4):2153–2169

Shukla PS, Shotton K, Norman E, Neily W, Critchley AT, Prithiviraj B (2018) Seaweed extract improve drought tolerance of soybean by regulating stress-response genes. AoB Plants 10(1): plx051

Singh M, Kumar J, Singh S, Singh VP, Prasad SM (2015) Roles of osmoprotectants in improving salinity and drought tolerance in plants: a review. Rev Environ Sci Biotechnol 14:407–426

Singh A, Singh S, Kurella A, Verma A, Mahatama M, Venkatesh I (2022a) Plant bio-stimulants, their functions and use in enhancing stress tolerance in oilseeds. In: New and future developments in microbial biotechnology and bioengineering. Elsevier, pp 239–259

Singh AL, Rani K, Jena R, Kona P, Reddy KK, Gangadhara K (2022b) Chapter 8—microbes-based bio-stimulants towards sustainable oilseeds production: nutrient recycling and genetics involved. In: Singh H, Vaishnav A (eds) New and future developments in microbial biotechnology and bioengineering. Elsevier, pp 111–130. https://doi.org/10.1016/B978-0-323-85163-3.00013-2

Srivastav AL, Dhyani R, Ranjan M, Madhav S, Sillanpää M (2021) Climate-resilient strategies for sustainable management of water resources and agriculture. Environ Sci Pollut Res 28(31): 41576–41595

Sujata GV, Baliyan V, Avtar R, Mehrotra S (2023) Alleviating drought stress in Brassica juncea (L.) Czern & Coss. by foliar application of biostimulants—orthosilicic acid and seaweed extract. Appl Biochem Biotechnol 195(1):693–721

Sun W, Shahrajabian MH (2023) The application of arbuscular mycorrhizal fungi as microbial biostimulant, sustainable approaches in modern agriculture. Plan Theory 12(17):3101

Sun W, Shahrajabian MH, Kuang Y, Wang N (2024) Amino acids biostimulants and protein hydrolysates in agricultural sciences. Plan Theory 13(2):210

Tarafdar J (2022) Biostimulants for sustainable crop production. New and Future Developments in Microbial Biotechnology and Bioengineering. Elsevier, In, pp 299–313

Tarolli P, Luo J, Park E, Barcaccia G, Masin R (2024) Soil salinization in agriculture: mitigation and adaptation strategies combining nature-based solutions and bioengineering. iScience 27(2)

Tolisano C, Del Buono D (2023) Biobased: biostimulants and biogenic nanoparticles enter the scene. Sci Total Environ 885:163912

Toor MD, Ay A, Ullah I, Demirkaya S, Kızılkaya R, Mihoub A, Zia A, Jamal A, Ghfar AA, Di Serio A (2024) Vermicompost rate effects on soil fertility and morpho-physio-biochemical traits of lettuce. Horticulturae 10(4):418

Trivedi K, Gopalakrishnan VAK, Kumar R, Ghosh A (2021) Transcriptional analysis of maize leaf tissue treated with seaweed extract under drought stress. Front Sustain Food Syst 5:774978

Turan M, Ekinci M, Kul R, Kocaman A, Argin S, Zhirkova AM, Perminova IV, Yildirim E (2022) Foliar applications of humic substances together with Fe/nano Fe to increase the iron content and growth parameters of spinach (*Spinacia oleracea* L.). Agronomy 12(9):2044

Tyagi V, Nagargade M, Singh RK (2020) Agronomic interventions for drought management in crops. In: New frontiers in stress management for durable agriculture. Springer, pp 461–476

Vanani FR, Shabani L, Sabzalian MR, Dehghanian F, Winner L (2020) Comparative physiological and proteomic analysis indicates lower shock response to drought stress conditions in a self-pollinating perennial ryegrass. PLoS One 15(6):e0234317

Vikram N, Sagar A, Gangwar C, Husain R, Kewat RN (2022) Properties of humic acid substances and their effect in soil quality and plant health. In: Humus and Humic Substances-Recent Advances

Vinnichek L, Pogorelova E, Dergunov (2019) A oilseed market: global trends. IOP Conf Ser Earth Environ Sci 1:012030

Wadhwa M, Bakshi M (2016) Application of waste-derived proteins in the animal feed industry. Protein byproducts. Elsevier, pp 161–192

Wang Z, Tu S, Shehzad K, Hou J, Xiong S, Cao M (2025) Comparative study of organosilicon and inorganic silicon in reducing cadmium accumulation in wheat: Insights into rhizosphere microbial communities and molecular regulation mechanisms. J Hazard Mater 492:138061

Yooyongwech S, Tisarum R, Samphumphuang T, Phisalaphong M, Cha-um S (2023) Integrated strength of osmotic potential and phosphorus to achieve grain yield of rice under water deficit by arbuscular mycorrhiza fungi. Sci Rep 13, 5999

Zarei G, Shamsi, H & Dehghani SM (2010). The effect of drought stress on yield, yield components and seed oil content of three autumnal rapeseed cultivars (Brassica napus L.). Journal of Research in Agricultural Science 06:29–36

Chapter 3
Use of Biostimulants to Enhance Waterlogging and Flooding Tolerance in Oilseed Crops

Majid Iqbal, Affan Afzal, Muhammad Nauman Khan, Quaid Ahmad, Tabassum Yaseen, Alevcan Kaplan, Nazima Wahid, and Rokayya Sami

Abstract The adverse susceptibility of oilseed cultivars to prolonged edaphic saturation, inundation events, and persistent waterlogging and flooding significantly hinders progress toward global agricultural pillars. These abiotic stressors produce significant perturbations in pathways for nutrient acquisition, metabolic homeostasis, and physiological functionality, leading to substantial yield decrements. The development of agricultural biotechnology stemming from the use of biostimulatory compounds on contemporary crops, such as humic derivatives, proteolytic

M. Iqbal (✉)
Institute of Geographic Sciences and Natural Resources Research, University of Chinese Academy of Sciences, Beijing, China
e-mail: miqbal@igsnrr.ac.cn

A. Afzal
Department of Microbiology, Faculty of Veterinary and Animal Sciences, The Islamia University of Bahawalpur, Bahawalpur, Pakistan

M. N. Khan
Department of Botany, Islamia College Peshawar, Peshawar, Pakistan

Biology Laboratory, University Public School, University of Peshawar, Peshawar, Pakistan

Q. Ahmad
Biology Laboratory, University Public School, University of Peshawar, Peshawar, Pakistan

Department of Botany, University of Peshawar, Peshawar, Pakistan

T. Yaseen
Department of Botany, Bacha Khan University Charsadda, Khyber Pakhtunkhwa, Pakistan

A. Kaplan
Department of Crop and Animal Production, Sason Vocational School, Batman University, Batman, Turkey

N. Wahid
Department of Plant Sciences, Quaid-i-Azam University Islamabad, Islamabad, Pakistan

R. Sami
Department of Food Science and Nutrition, College of Sciences, Taif University, Taif, Saudi Arabia

© The Author(s), under exclusive license to Springer Nature Singapore Pte Ltd. 2025
A. A. H. Abdel Latef (ed.), *Oilseed Crops Under Abiotic Stress*, Sustainability Sciences in Asia and Africa, https://doi.org/10.1007/978-981-96-8346-8_3

hydrolysates, beneficial microbial consortia, and algal extracts, has already shown efficacy in counteracting the devastating consequences of excessive hydrological conditions. This chapter aims to quantitatively explain the resilience threshold of oilseed crops to the effects of waterlogging and flooding stress regimes. These bioactive compounds are supported by empirical evidence that they promote rhizospheric oxygenation processes, facilitate nutrient assimilation kinetics, and augment antioxidative defense mechanisms to improve the growth of plants in harsh conditions. Basic biochemical modulation of this kind substantially improves crop adaptability to such hydrological stresses. Remarkably, the morphophysiological development indices, agronomic yield components' parameters, and stress tolerance parameters are well augmented by amino acid constituents, rhizospheric microbial inoculants, and algal derivatives in affected plant systems. The existing comprehensive analysis integrates extant scientific paradigms harmoniously to identify the important lacunae of knowledge and a means to successfully counter waterlogging-induced physiological perturbation in a complex agronomic and biostimulant management situation. Despite the limitations, the research outcomes emphasized the importance of continuing with investigative efforts to ascertain the best strategies for executing application methods, dosage regimens, and time frames in biostimulant application for the sustainability of agricultural practice for an improvement in the food security maturities in hydrologically vulnerable agroecosystems.

Keywords Biostimulants impacts · Waterlogging tolerance · Oilseed crops · Flooding

3.1 Introduction

3.1.1 Overview of Oilseed Crops and Their Importance

Oilseed crops are essential in different industries as oil origin and as unprocessed elements. These crops are associated with many plant families. Over 40 oilseed crops are cultivated, but few are used as significant commercial material (Tokel and Erkencioglu 2021). Large areas for cultivating and sustaining sectors are available in Iran for oilseed crop growth. Statisticians reported that from abroad, they are still importing about 85–90% of the required oil (Rostami Ahmadvandi and Faghihi 2021). Soybean, sunflower, groundnut, castor, and most importantly, sesame oil seed crops are the sustainable source of energy, these most chosen crops are also associated with power generation. The water consumption and production costs used for the growth of sunflower, soybean, and canola crops are lower (Chauhan et al. 2021). Soybean and sunflower crops are the signal growing season crops. Although the oilseed crops occupy a small portion of the trade market, but still used to complete the nutritional requirements of both humans and animals (Tokel and Erkencioglu 2021).

When it comes to riveting the oilseed crop's nutritional values, the valuable nutritional product for the animal is the protein component after the removal of oil (Bobo et al. 2021). Polyunsaturated fatty acids in the oil extracted from the oilseed crops and their beneficial impact on the health of humans (Schulte et al. 2013). The demand for the production of these crops is increasing because of their antioxidant properties (Subrahmaniyan et al. 2024). In human food, the most important fatty acids present in vegetable oils are polyunsaturated, especially linoleic acid, omega-3, and 6 (Djuricic and Calder 2021). The first and the oldest (about 3000 years ago) oilseed crop used by man was sesame (*Sesamum indicum* L., *Pedaliaceae*), for various purposes. Although researchers do not give much attention to its study and it remains neglected at the agriculture level, it is still known as the "Queen of oilseed" (Yadav et al. 2022). The two previous decades drastically enhanced the requirements for sesame seeds because these seeds can survive in intense ground conditions, harsh climates, and environments. These seeds also contain a content of antioxidant oil and protein portion, which makes these seeds more beneficial for human use (Hussain et al. 2023a, b). The primary aim is to enhance the quantity of seeds and the quality of oil obtained from oilseed crops, ensuring that the oil is suitable for both food and industrial purposes. The bred oilseed plants can also be formed to achieve such objectives, and these crops should have the property to tolerate the biotic and abiotic conditions in their environment. However, this breeding mechanism is still under discussion related to the crops that are annual oil crops (Rathnakumar and Sujatha 2022).

3.1.2 Impact of Waterlogging and Flooding on Oilseed Crops

Waterlogging and flooding are hazards to the sustainability of agriculture, and crops are damaged because of these harsh weather patterns. As there is an increase worldwide in the graph of heavy rainfall and affecting the production of crops, waterlogging is contributing too in this tense weather issue by decreasing the development of the oilseed crops and the growth of these crops (Lindsey et al. 2024). Waterlogging not only influences the metabolic reaction and the production of hormones in these plants but also disturbs the aerobic respiration in the rhizosphere and the hormone signaling pathways in oilseed crops (Yang et al. 2023). Due to the waterlogging stress, plants develop various mechanisms that help them to survive as adventitious roots and aerenchyma. The development of these adaptations is necessary because the physiological changes not only decrease the production of the oilseed crops but also disturb the reproductive cycle (Srivastava et al. 2024).

The rhizosphere faces difficulty in the availability of normal air movement because waterlogging causes the saturation of the soil pores. When there is a deficiency in the oxygen (O_2) required by the plant, the respiration process shifts from aerobic to anaerobic, leading to a decrease in energy production (Pais et al. 2022). Because of this conversion, all the metabolic reaction happening in the plant also changes and a spontaneous decrease in the availability of energy in plants occurs. The suppression in the process, which needs O_2, such as carbon assimilation and photosynthate utilization, is because of the short supply of O_2 to oilseed crops (Bhaduri et al. 2017). If oxygen is in deficit, the root growth is disturbed, and the functional relationship of roots with shoots is disrupted, due to a rise in the quantity of CO_2. The soil that undergoes waterlogging stress suffers from the accumulation of chronic and anaerobic microbial metabolite toxicity, as well as the rise in the concentration of ethylene, manganese, and carbon dioxide. Therefore, to manage such a situation, there must be some tolerance genes and profitable processes. To make the oilseed crops such as sesame crops tolerant to waterlogging stresses, a thorough understanding of this mechanism is required (Habibullah et al. 2021).

About 10% of the world's productive farming land faces severe impacts from flooding which destroys crops during their development stage. The rising global temperatures according to hydro-climatology simulation models are expected to enhance flooding events and their occurrence duration which will potentially generate large-scale crop production losses worldwide (Wu et al. 2025). Flooding occurs under two distinct forms which are known as submergence and waterlogging. When a plant entirely sinks beneath water, we call it submergence, while waterlogging describes the flooding occurrence that stops at the plant root area. The plant loses its oxygen absorption capacity when submerged in water because respiration breaks down and water fills air spaces creating oxygen-deficient conditions (hypoxia) that force metabolic changes for alternative ATP generation leading to impaired plant development (Li et al. 2021). The combination of flooding damage leads plants to experience major yield reduction because they suffer from diminished stomatal conductance and photosynthesis together with nitrogen fixation as well as diminishes both shoot growth and root development and nutrient absorption (Aslam et al. 2023). The essential oilseed crop Sesame (*Sesamum indicum* L.) from India, China and Burma ranks first among cultivated plants due to its membership of the Pedaliaceae family. The plant demonstrates good tolerance to drought conditions, yet it cannot stand waterlogging since it loses half to three-quarters of its potential yield when grown in waterlogged soils. The intersection of waterlogged soil and associated stress factors first causes plant premature aging along with leaf discoloration that eventually leads to leaf drop while affecting nitrogen fixation processes and damaging plant development and production. The study shows waterlogging tolerance is present genetically in plants because they exhibit unique morphological and anatomical features as well as biochemical and physiological responses (Habibullah et al. 2021; Pan et al. 2021).

3.1.3 Biostimulants: Definition and Importance in Modern Agriculture

Biostimulants represent modern agronomic instruments that occupy a middle spot between fertilizer and plant growth regulator usage according to their substance proportions (Miranda et al. 2024). According to current regulations on fertilizers, biostimulants include such compounds and microbes that enhance plant growth when applied to plants or root zones by optimizing nutrient uptake while providing better stress tolerance and enhancing product quality without altering its nutrient value (Franzoni et al. 2022). Different extraction techniques help transform assorted organic substances into nonmicrobial biostimulant products that enhance crop growth and tolerance through bioactive compounds that enhance survival under adverse environmental conditions (Ciriello et al. 2025). The intricate character of raw organic substances affects the composition of product end-products, thus making it unclear which specific compounds display active effects on biological responses. The biostimulant action stems from specific compounds including nonnutritional minerals along with vitamins amino acids and chitin-chitosan as well as poly- and oligosaccharides and hormones naturally present in the raw materials (Shukla 2022).

Biostimulants establish a leading position in circular agricultural systems because they derive from waste and agro-industrial by-products. Strategy matches circular economy principles through its implementation of resource-efficient sustainability in current agricultural operations (Puglia et al. 2021). The double metabolic mechanism lies behind the biostimulant operating principles. Biostimulants have either a dual function by stimulating photosynthetic activity and accompanying metabolic product formation in primary metabolism or they trigger particular biosynthetic pathways in secondary metabolism. Such dual action allows for the broad application of these materials in agricultural production systems. Scientific analysis of biostimulant efficacy becomes complicated because their effects stem from compound synergies instead of singular molecular activities (Baltazar et al. 2021; Shiade et al. 2024). Multiple interacting factors in biostimulants lead to challenging conditions for identifying all active ingredients. Variant biostimulant compositions affect their productivity directly making it a fundamental obstacle during their research and usage sphere. Scientists have introduced measures to decrease source material variation and develop standardized extraction procedures because of compositional inconsistency issues. Biostimulant production processes seek to generate products with anticipated performance metrics and dependable agricultural effectiveness across different applications (Chabili et al. 2024).

Biostimulants represent a subcategory of agricultural inputs, which differs from fertilizers in their effect to enhance plant growth through their application. Increasing global adoption by them is also a strategic response to agricultural challenges induced by the expanding world population's needs. These substances, native to nature and compatible with the environment, benefit the plant because they enhance vegetative development, increase the uptake of mineral nutrients, and raise levels of tolerance to abiotic stress conditions (Bahuguna et al. 2022). The biostimulants derived from diverse natural microbial communities and their biochemical products, such as organic acids, proteins, enzymatic compounds, and phytohormones are currently being used in contemporary agricultural systems (Castiglione et al. 2021). The bioactive substances optimize plant–soil interactions by boosting nutrient availability, and nutrient uptake efficiency from the added fertilizers as well as from the base soil reserve and crop residue decomposition (Friedel and Ardakani 2021). Leading manufacturers of sophisticated biostimulant technologies have developed specified physiological mechanisms and agronomic benefits based on nanosurfactant additives, protein hydrolysates, humic substances, and seaweed extracts. These advanced biostimulant technologies have integrated well into organic farming systems and have great potential to increase crop productivity and soil fertility parameters, enhance pathogen resistance, and increase the agricultural self-sufficiency of organic producers in terms of profitability (Ali et al. 2024).

Because of this, contemporary agricultural systems confront increasing productivity challenges with the sustainability of ecological integrity. With the intensification of the requirement for environmentally friendly farming practices, it has become imperative to protect the fertile soil and prevent increasing erosion of soil biodiversity. Such context has led to the development of microbial biostimulants as profound technological innovations that would facilitate securing yield with enhanced nutritional potentials while at the same time alleviating the effects of environmental fluctuations (Kumar and Singh 2021). Individual microbial strains, such as synergistic microbial consortia, or with organic matrices acting as plant biostimulants (PBs), in which plant growth-promoting microorganisms (PGPM) are used through a variety of different application strategies. This includes the evaluation of the effectiveness as well as the multifunctional role of biological tools that improve fruit quality parameters at the cost of lowering soil degradation processes (Tabacchioni et al. 2021). The suitability of identifying and selecting beneficial microorganisms and microbial combinations, and the development of sophisticated methodologies for standing microbial biostimulants will increase with the expanding adoption in the agricultural sectors. This process for selection of microbial biostimulants must include forward-looking approaches capable of dealing with future agricultural challenges and microbial biostimulants must be recognized as essential counterparts in building resilient and sustainable food production systems for the unknown future (Castiglione et al. 2021).

Natural sources, including biostimulants, have recently gained use to enhance growth or defense. Biostimulant development is a new solution to agricultural product safety and it is widely used in agricultural technology (Jiang et al. 2024).

3.1.4 The Role of Biostimulants in Enhancing Stress Tolerance

Exposure to abiotic conditions leads to significant improvement in plant performance and stress resilience when using biostimulant applications. The abilities of these biological compounds to equip plants with resistance mechanisms against unfavorable environments such as drought, elevated temperatures, and salinity allow them to self-recover from damage induced by the stress (Bahuguna et al. 2022). The combination of better nutrient acquisition, higher quality of products, and a higher yield together permits agricultural producers to gain substantial economic benefits (Gedeon et al. 2022).

The category of biostimulants encompasses different components such as microbial inoculants, amino acid derivatives, humic substances (HS), and seaweed extracts, and some companies are selling them. Since biostimulant application is an essential strategy in optimizing nutrient management for oilseed production sectors in Asian and African countries characterized by the dual challenge of supporting rapidly growing populations in a limited natural resource environment, their use contributes to enhancing both yield potential and crop quality (Singh et al. 2022a, b).

Biostimulants have shown their efficacy as the correct way of boosting plant defense against strains like biotic and abiotic. The application of naturally derived substances from plant extracts, microbial inoculants, and humic compounds particularly under adverse environmental conditions increases plant growth and developmental capacities (da Silva et al. 2021).

The biostimulants-induced stress tolerance is a result of complex changes in morphological and molecular processes. Mainly acting through stimulating the production of antioxidant enzyme systems, such as superoxide dismutase (SOD), catalase (CAT), and peroxidase (POD), their capacity to protect plants from oxidative stress conditions (e.g., drought, salinity, and temperature extremes) is mostly achieved (Fujita and Hasanuzzaman 2022).

The uptake and assimilation of nutrients improve when plants receive treatment with either humic or fulvic acids because these substances improve root growth and enhance soil nutrient accessibility during nutrient shortage (Alsudays et al. 2024). The osmotic balance of cells becomes stable when biostimulants promote osmoprotectant accumulation including proline glycine betaine and sugars which enhances drought tolerance through improved water retention abilities of plants (Jiménez-Arias et al. 2021). Through biostimulant applications the phytohormone composition changes because auxins and gibberellins, cytokinins and abscisic acid (ABA) levels adjust to control the stress response mechanisms and stimulate adaptive growth processes (EL Sabagh et al. 2022). Biostimulant microbes promote plant disease protection by activating systemic acquired resistance and induced systemic resistance which lowers disease emergence (Khalid et al. 2025).

Types of Biostimulants for Stress Mitigation

- Seaweed extracts, which contain bioactive polysaccharides together with mineral components, promote root development while improving tolerance to stress and providing yield stability functions (Deolu-Ajayi et al. 2022).
- *Bacillus* spp. *and Pseudomonas* spp. together with mycorrhizal fungi function as microbial inoculants that create healthy soil that boosts nutrient intake capabilities against disease threats (Samantaray et al. 2024).
- Stress recovery happens faster when stress-related molecules are formed from amino acids and peptides while performing their role as protein-building substances (Ghosh and Roychoudhury 2024).

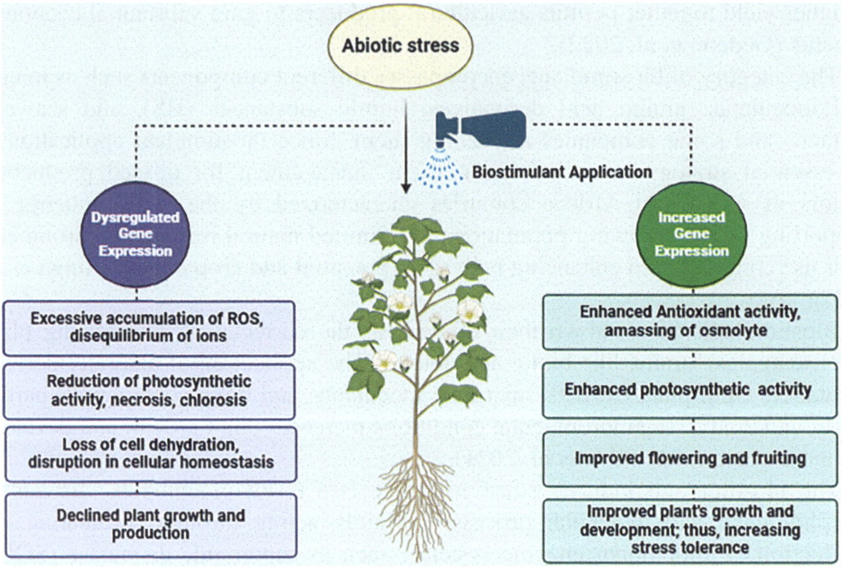

Fig. 3.1 Application of biostimulants in oilseed crops under abiotic stress

Farmers acquire earth-friendly plant stress protection technologies through the implementation of biostimulants. Biostimulants improve antioxidant mechanisms while simultaneously boosting nutrient intake and maintaining hormone balance and osmotic regulation systems; thus, they demonstrate unique agricultural value (Kajrolkar 2025; Omoarelojie et al. 2021) (Fig. 3.1).

3.1.5 Waterlogging and Flooding Stress in Oilseed Crops

The oilseed crop grown in China is *Brassica napus* L., in the middle and lower reaches of the Yangtze River region but is susceptible to waterlogging (Ke et al. 2023; Li et al. 2023). The present study focused on physiological parameters, cellular architecture modifications, and expression patterns of differential genes in rapeseed induced by waterlogging stress to the development of a rapeseed waterlogging-tolerant cultivar (Hong et al. 2024). Results showed that rapeseed genotypes tolerant to waterlogging had high soluble sugar concentrations and elevated activities of antioxidant enzymes, especially in the roots (Hussain et al. 2023a, b). Conversely, waterlogging sensitivity genotypes changed to a greater degree in malondialdehyde, proline, and hydrogen peroxide levels after the stress was applied (Hasanuzzaman et al. 2022). The analysis of leaf tissue ultrastructure showed that waterlogging stress led to cellular space, and cell wall changes and caused the loss of integrity of mitochondrial and chloroplast structures (Shi et al. 2023). Increased xylem area and frequency of vessels, mitochondrial degradation, and vacuolar contraction were the root anatomical modifications. In waterlogging-sensitive genotypes, these cellular perturbations were more severe (Xu et al. 2022). Gene transcription involved with anaerobic respiratory pathways and flavonoid biosynthesis networks were significantly different from the baseline (Jiang et al. 2023; Luan et al. 2025). Waterlogging stress conditions resulted in different transcriptional responses of different rapeseed varieties. It is concluded that rape seed genotypes exhibit different waterlogging stress responses at the morphological, physiological, cellular and transcriptional levels. The waterlogging-tolerant genotypes seem to limit stress impact through modulation of the antioxidant defense system. These findings should be informative to breeding programs aimed at developing rapeseed varieties that are resistant to waterlogging (Hong et al. 2024). Waterlogging of soybean (*Glycine max*) soil creates severe hypoxic or anoxic conditions which result in severe morphophysiological decline throughout the entire plant structure. The lack of oxygen leads to elevated reactive oxygen species (ROS) production together with impaired antioxidant defenses, particularly during extended periods of flooding (Adegoye et al. 2023). This oxidative imbalance within the plant system disrupts multiple life-sustaining processes because it affects hydraulic conductivity as well as nutrient acquisition along with stomatal regulation while impairing photosynthetic capacity, enzymatic functionality and phytohormonal homeostasis (Sharmin et al. 2024). Soybean shows the exceptional capacity to adapt its phenotypic characteristics when it faces waterlogging stress. The development of adventitious roots stands out as the most essential among all morphological adaptations shown by plants. The supplementary root structure of *Glycine max* serves as a vital mechanism to acquire oxygen while sustaining basic metabolic processes despite unfavorable waterlogging effects which helps the plant overcome soil saturation problems (Sathi et al. 2022).

3.1.5.1 Mechanisms of Waterlogging and Flooding Stress

The germination, viability of the plant, leaf damage, the extension of stem rate, the change in the color of the leaf, the rate coefficient due to waterlogging stress, and the productivity rate of the crop, all are assessed numerically and show the indication of the tolerance which showed after waterlogging (Jiang et al. 2022; Li and Rao 2024). Reduced photosynthetic activity is caused by a result of deterioration of chlorophyll and degeneration of the leaf followed by waterlogging, which leads to a reduction in the opening of stomata (Ren et al. 2023). Limited conductance of stomata impairs the processes of transport of oxygen in the tissue of the plant directly. As a result, restrictions between two cells, to the transport of oxygen and respiration in the mitochondria and those that have physiological and biochemical activities in normal situations. In addition, the plant waterlogging tolerance can be restricted by gaseous exchange because plant waterlogging can cause rapid accumulation or degradation of plant hormones (Manghwar et al. 2024).

The air spaces within waterlogged soil diminish substantially because gas exchange between the atmosphere and soil is reduced seriously. Soil water creates a situation of decreased oxygen diffusion that reaches only one ten-thousandth of normal airflow oxygen registry levels (Pais et al. 2022). Oxygen deprivation caused by the event further impairs root respiration functions and results in damaged root capabilities as well as severe metabolic deficiencies throughout plant tissues (Zhang et al. 2025). Plants under hypoxic conditions first use glycolysis and ethanol fermentation pathways for energy production before anoxic conditions arise. Waterlogged conditions prolonging more than usual and anaerobic respiration create environments where toxic compounds such as lactic acid along with ethanol and aldehydes can build up (Benkeblia 2021). The increase of hydrogen peroxide among ROS reaches significant levels which results in cellular damage leading to plant death before its normal lifespan. The change in gaseous exchange under waterlogged conditions triggers fast hormone production along with hormonal breakdown (Shields et al. 2021). The hormonal changes created by waterlogged conditions damage the waterlogging tolerance systems found in affected plants as reported. Most plant species show deteriorated performance under waterlogged conditions but various species have developed specialized adaptations that help them protect against waterlogging damage. Each species employs different waterlogging adaptation techniques which determine its ability to adapt to flooding conditions (Pan et al. 2021).

The metabolic processes of trees undergo fundamental changes due to waterlogging which requires higher carbohydrate utilization for each period to produce the energy required for reduced fermentation pathway effectiveness (Zhang et al. 2024). The changes in metabolism lead to insufficient compensatory capacity when faced with hostile environmental conditions. Evidence indicates plants fail to overcome energy deficit problems even with increased carbohydrate consumption because severe flooding conditions lead to plant death or growth restriction (Nasrullah et al. 2022). Waterlogging induces systemic impacts on plant functioning which extend from altered metabolism to modifications in morphological development and growth characteristics alongside different metabolic activities (Jia et al. 2021; Tyagi et al. 2023). Waterlogging activates ethylene production at elevated levels and creates a signaling molecule yet prevents critical physiological functions due to excessive ethylene concentrations. The biological reaction of plants to waterlogging stress uses multiple parallel mechanisms that work together by adjusting responses across various levels of biological organization. Morphological structural adaptations, energy metabolism pathway reconfiguration hormonal synthesis and regulation adjustments specialized signal transduction pathway activation. The series of complete adaptive mechanisms together form the foundation for how tree populations develop their tolerance to waterlogging environments (Zhang et al. 2025).

3.1.5.2 Physiological and Biochemical Impact on Oilseed Crops

Physiological and Biochemical Impacts of Waterlogging

- Limitations of oxygen flow to the rhizosphere from waterlogged conditions lead to hypoxic or anoxic conditions that enhance root decomposition while also causing air sufficiency along with root damage (Ejiri et al. 2021).

- Waterlogging produces changes in nutrient solubility that lead to essential nutrient deficiency of nitrogen and phosphorus with simultaneous accumulation of dangerous substances including iron and manganese (Radulov and Berbecea 2024; Singh et al. 2022a, b).
- The combination of waterlogged environments reduces chlorophyll content and lowers stomatal conductance to block photosynthetic operations, thus reducing the growth and productivity (Manghwar et al. 2024; Topali et al. 2024).
- ROS aggregation in cells causes cellular destruction as well as the breakdown of enzymes due to excessive ROS production (Juan et al. 2021).
- Root growth reductions occur due to hormonal changes which simultaneously lead to premature senescence because of elevated ethylene levels impaired auxin transport and improper abscisic acid (ABA) regulation (Bhoi et al. 2024).

The stomatal conductance levels significantly decrease among waterlogged plants because plants show reduced CO_2 assimilation capacity as chlorosis affects their leaves (Toral-Juárez et al. 2021). A diminished net CO_2 accumulation occurs because waterlogging blocks the plant water absorption capacity alongside vital substances which include phosphorus, calcium, magnesium, iron, manganese, and molybdenum restricting all developmental stages and organic matter generation (Pandey et al. 2021). Waterlogging stress causes destructive effects on various components of the photosynthetic mechanism through the following points: reduced activity of key photosynthetic enzymes (Ateeq et al. 2023), alterations in chloroplast structural organization (De Castro et al. 2022), and damage to photosynthetic reaction centers (Li and Rao 2024). Plants growing in waterlogged conditions show reduced chlorophyll content, especially in chlorophyll a and b pigments (Orsák et al. 2023; Sharma et al. 2022). The group of physiological disorders manifests as reduced photosynthetic rate which in turn impacts crop production yield negatively. Plants undergo oxidative damage through waterlogging or flooding incidents which reduces their electron transport chain to such an extent that ROS production increases and destroys both cellular function and integrity (Sharma et al. 2021).

Water logging leads to damage to plant root systems and triggers several biochemical changes such as decreased enzyme activity of critical enzymes including glycolate oxidase, phosphoglycolate, and ribulose bisphosphate carboxylase (RuBPC) (Pan et al. 2021). Furthermore, waterlogging stress is responsible for damage to the chloroplast membrane structure, leading to a reduction in the efficiency of electrons in photosynthesis and to decrease in the functionality of photosystem II (Sharma et al. 2021). This compromised photosynthetic capacity when waterlogged results in increased production of ROS including Superoxide ($O_2•$), Singlet oxygen ($^1O_2$), Hydrogen peroxide (H_2O_2), and Hydroxyl radicals ($OH•$) (Wang et al. 2022). The reactive molecules have detrimental effects on many plant tissues and functions. ROS generation has also been identified in multiple cellular organelles such as mitochondria, chloroplasts, and peroxisomes. Elevated intracellular ROS accumulation is a hallmark of oxygen deficiency in waterlogged environments, leading to lipid peroxidation, protein degradation, and enzyme inactivation.

In addition to being destructive, excess ROS molecules can also serve as important signaling compounds inside stressed plant cells and to trigger several different stress response mechanisms and adaptive pathways (Srivastava et al. 2024).

After the treatment with ethephon, the soybean plants under waterlogging conditions have better efficiency in eliminating the toxic effect of ROS by neutralizing and elevating the metabolism of glutathione activity and glutathione S-transferase DHAR2 upregulation (Srivastava et al. 2024). Whereas the enzymatic activity profiles of the youngest fully developed green leaves from control and waterlogged plants are significantly different in senescent genotypes, they are similar in younger genotypes. A particularly noteworthy point is that waterlogged plants always have greater levels of malondialdehyde (MDA), a biological sensor that senses the damage of lipid proliferation induced by ROS (Firdous et al. 2025). Rapeseed is a widely cultivated crop and contributes greatly to the world of edible oil (Banaś et al. 2023). During growth stages, abiotic stress constraints impact plant metabolism and several inconveniences related to soil water storage capacity and physiology (Rahim et al. 2024). The demand for more and more productivity in the plants is increasing day by day to ensure the stability of food globally (Hemathilake and Gunathilake 2022). One of the most severe problems faced by waterlogged plants is energy insufficiency due to hypoxic conditions and mainly in the suppression of root respiration function (Gu et al. 2021). Mitochondrial respiration processes are disrupted by a poor electron transport chain, to avoid adenosine triphosphate (ATP) synthesis as a result of limited oxygen availability (Jardim-Messeder et al. 2022). Metabolic adaptations plants have to compensate for the energy deficit resulting from waterlogging stress by activating alternative pathways for energy generation, particularly by glycolysis and ethanol fermentation. These metabolic adaptations are important critical survival mechanisms that will continue to allow metabolic function when aerobic respiration is impaired (Raza 2021).

Adaptive Mechanisms in Oilseed Crops

- The development of aerenchyma tissue occurs within specified oilseed crops to enable oxygen transport between shoots and roots (Manghwar et al. 2024).
- Waterlogged plants receive a better oxygen supply by creating new roots from any part of their body (Jing et al. 2022).
- Survival under anaerobic situations involves activating different biochemical pathways of anaerobic respiration (Gu et al. 2021).

3.1.6 Biostimulants: Mechanism of Action

Biostimulants are substances that, when applied are intended to enhance the growth of plants and stimulate various natural processes occurring in plants, by the root-zone treatment or by the foliar treatment (leaf spraying). This helps effectively to the absorption of nutrients available to plants, stimulates the physiological development processes boosts the stress adaptivity, even without considering the value of content in nutrients (Begum et al. 2022). Unlike conventional fertilizer, algal extract contains compounds that serve as biostimulants, by which they are applied, don't contain nutrients, and stimulate defense and growth response (Kapoore et al. 2021). Seaweed extracts treated plants generally acquire nutrients better and grow more, and better, too (Hussain et al. 2021). Rapeseed plants which are exposed and treated with the extract of *Ascophyllum nodosum,* increased their abilities to acquire nitrogen and sulfur. This was due to the overexpression of root transporters associated with the absorption of beneficial nutrients like nitrogen (N), iron (Fe), and sulfur (S) that code for BnNRT1.1/BnNRT2.1 and BnSultr4.1/BnSultr4.2 genes, as shown by transcription studies. It is explained to be due to the transcriptome upregulation of the genes coding for sulfate, iron, and nitrate transporters. In addition, evidence was given for increased transcription or activity of membrane nutrient transporters on plant nutrient acquisition in seaweed extract-treated plants (Ali et al. 2021) as well as increased cytokinins involved in the proliferation of cells, in elongation of root and shoot of a plant, in morphogenesis of flower part of plant in the transportation, regulation of the dormant and germination stage of seed, retardation in the aging process of plant and the supremacy in the tip of root and shoot (Stirk et al. 2020).

3.1.6.1 Categories of Biostimulants

A more sustainable way to increase agricultural productivity has been achieved through the usage of microbial and nonmicrobial plant biostimulants (Papa et al. 2022).

Microbial Biostimulants Continued demand exists to utilize such technological innovations that are environmentally sustainable tools during the production period and these tools will ensure the long-term viability of agricultural production systems (Kaushal et al. 2023). Nevertheless, adverse changes in the global climate can jeopardize crop yield. Biostimulants can be integrated with chemical elements to increase agricultural productivity, through the use of plant growth-promoting rhizobacteria (PGPR) and advantageous fungi (Hasan et al. 2024). A subgroup of the diverse class of biostimulants, defined as based on microorganisms is commonly referred to as biostimulants based on microbes that activate the biochemical pathway that increases the uptake and potency of nutrients, and endure the abiotic stress as well as increase crop production, which can alleviate the detrimental consequences of agrotechnology (Castiglione et al. 2021). Beneficial microbes used in the application can alter the biological composition of soil and enhance the development of the microbes, which not only improves productivity but also boosts the yield and contributes to the replenishment of the soil's microbial community (Ortiz and Sansinenea 2022). They may also regulate shoot targets by stomata and xylem hydraulic conductance, root targets by root zone water availability, root ethylene, and auxin levels as well as whole plant responses by ROS scavenging, osmo-protection, and membrane stability (Sun and Shahrajabian 2023).

While microbial biostimulants represent an emerging and effective input of agrotech, they are still quite unexplored from the perspective of most farmers. However, to put forward this technique as sustainable green-labeled technology, we require substantial attempts, to amplify crop production, which would significantly mark the era of biotechnology to the twenty-first century (Cellini et al. 2021).

Direct and indirect mechanisms are the two mechanisms of action that are used by microbial plant growth stimulators (Saeed et al. 2021). The production of those substances that increase the absorption of nutrients is demonstrated by the direct method while processes like the solubilization of zinc and phosphorus (Rawat et al. 2021). The production of siderophore (Khasheii et al. 2021), ammonia, and hydrogen cyanide, the biological formation of indole acetic acid (Vaishnav et al. 2022), and antioxidant enzymes and the processes of biological nitrogen fixation are all encompassed by the indirect mechanism of microbial biostimulants (Rohela and Saini 2022). Applying the hormone-like stimulant molecules produced by fungi and bacteria not only alleviates pressure conditions for plant productivity but additionally promotes growth and development (Baltazar et al. 2021). Regulating the molecular processes, the microbial biostimulants protect the plants from stress conditions such as endogenous metabolite synthesis and the interaction of the plant microbes involved in this regulation (Lephatsi et al. 2022). These protective stimulant molecules are produced by a pathway known as the shikimate pathway. In this pathway, after the elicitation of microbes, the production of phenylpropanoids falls out the help by phenylpropanoids, this is known as induced systemic resistance because it helps the plant to manage the external factors that hinder its growth (Jain et al. 2024). Microbial biostimulants include PGPR and arbuscular mycorrhiza fungi (AMF). These biostimulants develop plant–root relationships and enhance productivity (Darakeh et al. 2022).

Biostimulants can be derived naturally or synthetically, in addition, to escalating the growth of the plant also compelling them to become resistant to environmental stresses. These biostimulants work without involving the microorganisms, far apart from fertilizers and pesticides which provide nourishment and protect against pests, activating the internal natural mechanism of plants to bear the harsh environment and cope with it. The natural extracts and substances from which nonmicrobial biostimulants are derived include humic substances, complex organic substances, amino acids and N-containing substances, chitosan, and seaweed extracts (Benito et al. 2024).

The European Commission defined organic nonmicrobial plant biostimulants through three essential components including humic acids (HA), protein hydrolysates (PH), and seaweed extracts (SWE) in its 2016 draft framework. Statistical analysis shows that both HA and PH control half of the market whereas SWE maintains a significant 37% share (Rouphael and Colla 2018). Nonmicrobial biostimulants that use vegetal protein hydrolysates (V-PHs) and plant extracts (PEs) show enhanced crop productivity because their diverse applications activate several physiological processes in various growing environments. Multiple synergistic pathways allow these biostimulants to work including stimulating enzymatic activity and implementing phytohormone signaling together with root system modifications. The simultaneous action of these mechanisms enhances plant water and nutrient uptake capacity so the plants achieve optimal growth under difficult environmental conditions (Vultaggio et al. 2024).

3.1.6.2 Effects of Biostimulant on Root Growth, Nutrient Uptake, and Stress Tolerance

With complex biochemical mechanisms, biostimulants improve plant performance by root growth, nutrient uptake enhancement, as well as stress tolerance in the plants (Baltazar et al. 2021). Different biostimulants, i.e., microbial and nonmicrobial, have different effects on the development of plants, especially oilseed crops (Ancín et al. 2024).

a) Humic Substances

-Origin and Structural Characteristics

The decomposition of biological residues and soil microorganism metabolism together yield essential humic substances which make up the fundamental parts of soil organic matter. The heterogeneous organic compounds are grouped into three classification categories distribution patterns: humins, fulvic acids, and humic acids. Under plant root influence, HS behaves through a complex association/dissociation process that creates supramolecular colloids which are affected significantly by rhizosphere proton efflux and exudate release. The exact characteristics of humic substances existing in soils emerge from highly complex organic matter-soil microbe-plant relationships (Arumugam and Rabert 2022).

The biological mechanisms of plant development benefit greatly from the application of both HA and fulvic acid (FA). The combination of these compounds activates seed germination factors along with accelerating vegetative growth mechanisms aiding plant biomass growth supplying essential minerals and strengthening plants against environmental threats (Nardi et al. 2021). Different horticultural plant species exhibit notable yield enhancements when treated with Applications of HA and FA per research findings. Complex organic materials retain water effectively yet play substantial roles in plant physiological responses which create stress resistance together with better water retention in soil. Scientists research to explore how fulvic acid combined with seaweed extract works as an antistress agent in soybean plant production. These biostimulants allowed plants to keep high photosynthetic rates in unfavorable environmental situations, thus showing their effectiveness for stress minimization in agricultural production (Hongal et al. 2023).

Research has proven that applying exogenous HA leads to physiological function improvements along with biochemical benefits in different agricultural plant species

(Rathor et al. 2023). The positive effects of exogenous organic acid treatments consist of higher antioxidant enzyme activities together with better chlorophyll synthesis improved photosynthetic performance and greater nucleic acid content. FA possesses the ability to improve secondary metabolic processes within particular crop types. Treatment of pepper crops with FA as an application to foliage enhanced the levels of total phenolic compounds along with both carbohydrates and carotenoids (Kanabar et al. 2024). Applications of FA improve saffron secondary metabolites because they increase nutrient chelation which leads to improved nutrient uptake mechanisms (Aboueshaghi et al. 2023).

b) Physiological Mechanisms in Plant Biostimulation

The favorable biological effects of humic substances emerge through enhanced root nutrition levels which result from grouped interrelated mechanisms (Maffia et al. 2025). The main process includes improved acquisition of essential nutrients by polyanionic humic substances which boost soil cation exchange capacity (Ampong et al. 2022). The bioavailability of phosphorus receives an enhancement from HS because these substances inhibit the precipitation of calcium phosphate leading to higher phosphorus solubility in the rhizosphere (Yuan et al. 2023). Plasma membrane H^+-ATPases become activated through a second critical mechanism that uses released energy from ATP hydrolysis to create transmembrane electrochemical potential gradients. The established electrochemical gradients allow active nutrient transmembrane transport of nitrate and other vital elements through root cell membranes. This proton-pumping activity of plasma membrane enzymes plays a substantial role in both plant wall loosening and cellular expansion which leads to organ development (Du Jardin 2015).

c) Seaweed Extracts

- *Compositional Basis and Physiological Effects*

Bioactive compounds found in seaweed are diverse unbound amino acids, minerals like zinc and sulfur, and a variety of organic constituents that aid in plant abiotic stress reducibility and photosynthetic efficiency (Ali et al. 2021). The use of Ascophyllum nodosum extract during the asymptomatic developmental stages of *Brassica* cultivar RH 725 enhanced several physiological parameters when applied to either irrigated or water-limited plants. Improvements in photosynthetic capacity, stomatal conductance, transpiration rates, hydration level, osmotic potential osmoregulation level, photosynthetic light emission, soluble solute content, and accumulation of proteins were made. The physiological enhancement described earlier is because of cytokinin, and indole-3-acetic acid (IAA) in the bioactive constituents of seaweed extracts (Hongal et al. 2023; Sible et al. 2021).

d) Extraction Methodologies

Two classes of methodologies are used to extract bioactive compounds from the seaweed matrices, namely physical methods (thermal processing, pressure treatments, and microwave-assisted extractions), and chemical methods (solvent-based, acid-mediated, and alkaline extractions) (Matos et al. 2021; Perez-Vazquez et al. 2023). Extraction protocols have to select appropriate protocols to address the inherent compositional complexity of seaweed materials and their biologically active molecules that are responsible for biostimulant properties while preserving their structural and functional integrity (Quitério et al. 2022). Alkaline extraction with high pressure is the predominant industrial-scale processing and exhibits maximal effectiveness despite the possibility of degradation of certain hormonal constellations. Advantages of this method include better extraction efficiency and controlled degradation of polysaccharides into oligomeric derivatives (appearing as predominant components of the biological activity of seaweed extracts (Hines et al. 2021). The plant uptake efficiency as well as the subsequent physiological responses is dependent on the method of seaweed extract application employed. However, most of the contemporary agricultural practices apply foliar or rhizosphere methods of delivering biostimulants to achieve better biostimulant efficacy and crop response outcomes (Rahgoshahi et al. 2023).

e) Effect of SWE on Plant Growth

SWE can reduce or cure severe pest invasion and deterioration of crops by nematode parasites, but studies have also shown that it can reduce damage inflicted by nematodes to sunflowers. *A. nodosum* extract-treated rapeseed plants had higher amounts of nitrogen and sulfur. That was caused by the overexpression of the coding genes for root transporters taking up nitrogen, iron, and sulfur, namely, BnNRT1.1/BnNRT2.1 and BnSultr4.1/BnSultr4.2 (Ali et al. 2021).

f) Protein-Based Products

As protein-based products, PH contains mixtures of peptides with amino acids of organic matter and amino acids such as glutamate, glutamine, proline, and glycine betaine (Monterisi et al. 2024). The activity of PH may coordinate the modulation of carbon and nitrogen pathways to promote nitrogen assimilation in plants. Biomolecules sourced from soybean, rice, wheat, and maize, have been described as PPHs (plant protein hydrolysates), and are represented as refined protein forms with peptides, oligopeptides, and amino acids. CSL (corn steep liquor) protein hydrolysate contributes to augmenting wheat germination (Wazeer et al. 2024). Protein hydrolysates have an important capacity to enhance crop performance, particularly under environmental stress conditions (Shahrajabian et al. 2022).

After solid-state fermentation of soybean meal (SBM), water-soluble soy protein hydrolysates (SPHs) were studied for biostimulant activity on rapeseed growth. As compared to algae fertilizers found in the market with the conventional algae fertilizer, the growth of the root and nutrient uptake increases, and the biomass of plants, and leaf area also enhanced, it was reported that SPHs in water-soluble form increase shoot dry weight and chlorophyll content by 50–200% in rapeseed (Jain and Badve 2022). PH application increased the concentration of nitrate, amino acids, and ureides in soybean leaves. The expression of genes such as nitrate reductase, urease, and asparagine are also promoted. Furthermore, it improved productivity with an increased number and weight of pods and grains. Thus, it can be argued that PH can be considered to increase soybean productivity under nonstressed conditions (Engel et al. 2024).

g) Plant Growth-Promoting Rhizobacteria and Arbuscular Mycorrhizal Fungi

Biostimulant applications to agro-production systems provide such benefits as strategic supplementation with beneficial rhizosphere microorganisms including specialized fungal symbionts and PGPR to conventional chemical inputs (Parmar et al. 2023; Prasad 2023). Biostimulants exert diverse agronomic benefits across multiple crop parameters: enhancement of product visual quality characteristics; activation of plant immune response pathways; stimulation of defensive biomolecule synthesis; facilitation of heavy metal remediation in contaminated soils; optimization of overall crop performance metrics; reduction of nutrient leaching processes; enhancement of root architectural development; acceleration of seed germination dynamics; improvement of tolerance mechanisms against various biological and environmental factors; expedition of crop establishment phases; and optimization of nutrient acquisition and utilization efficiency (Sun et al. 2024).

h) Functional Mechanisms of PGPR

PGPR are specialized microbial communities associated with roots and whose activity promotes plant development by multiple physiological mechanisms (Kong and Liu 2022). The benefits of such beneficial organisms are that they stimulate plant growth by producing phytohormones and bioactive compounds, at the same time facilitating enhanced mineral nutrition acquisition. Like most other microorganisms, PGPR can effectively colonize diverse ecological niches within the rhizosphere throughout all developmental stages of crop growth. The influence of PGPR in plant development works through both direct and indirect pathways. There are direct mechanisms such as modulation of endogenous phytohormone profiles and improvement of acquisition processes of essential nutrients including potassium, phosphorus, nitrogen, and several micronutrients (Nagrale et al. 2023). Suppression of pathogen-induced growth inhibition is the main plant pathogen-negative indirect mechanism, which is accomplished through various biocontrol activities. Many

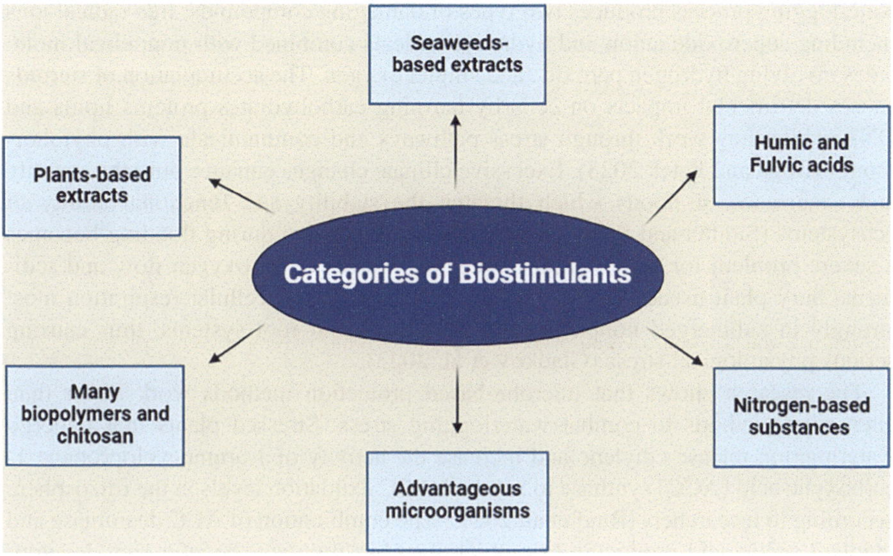

Fig. 3.2 Various categories of biostimulants

adaptive mechanisms to salinity stress in plants are relatively facilitated by PGPR, such as the facilitation of osmolyte accumulation, enhancement of nutrient uptake processes, promotion of biological nitrogen fixation, and solubilization of phosphorus and mineral elements (Vocciante et al. 2022). Synergistic application of biochar combined with PGPR improves the chemical composition and biological properties of plants, promoting plant growth and biomass. Upgraded Agricultural tools of plant photosynthesis, transpiration, and stomatal conductance using biochar and PGPR. The joint application of PGPR strain SIR03 and biochar helps to improve the productivity and growth of sesame plants under different environmental conditions and field capacity (Kang et al. 2022). AMF symbiosis with plant root systems expands the plant root functional boundary and optimally absorbs gravitating nutrients such as phosphorus and nitrogen, in the appropriate soil conditions (El Malahi et al. 2025). AMF protective benefits result from their ability to decrease oxidative stress, speed up nutrient and water absorption, and alter gene expression levels in stress response and signaling pathways. AMF efficacy is notably very sensitive to soil conditions and environmental factors (Sun and Shahrajabian 2023) (Fig. 3.2).

3.1.7 Application of Biostimulants in Flooding and Waterlogging Conditions

Waterlogging leads to both hypoxic and anoxic environments that damage plant metabolism because ROS substances build up and initiate oxidative stress. The waterlogging process produces two types of damaging compounds: free radical ions including superoxide anion and hydroxyl radicals combined with nonradical molecules involving hydrogen peroxide and singlet oxygen. The accumulation of steroids causes detrimental impacts on cells by harming carbohydrates proteins lipids and DNA while they work through stress pathways and communicate with phytohormones (Buga and Petek 2023). Excessive climate changes enhance both the severity and occurrence of floods which threaten the stability and functional ability of ecosystems (Siddha and Sahu 2022). The lack of oxygen during flooding becomes a severe problem for plants because stagnant water blocks oxygen flow and sediments bury plant tissues. Cutting off oxygen flow affects cellular respiration most strongly in submerged nonphotosynthetic tissues and root systems, thus causing serious physiological stress (Gladkov et al. 2023).

The research shows that microbe-based protection methods work better than alternative methods to combat waterlogging stress. Stressed plants that undergo waterlogging release ethylene and increase the activity of 1-aminocyclopropane-1-carboxylic acid (ACC) synthase to increase ACC exudation levels in the rhizosphere according to researchers (Rauf et al. 2021). The combination of ACC deaminase and indole-3-acetic acid-producing Microbial plant biostimulants benefits root development while strengthening the entire plant growth. The beneficial microorganisms use the ACC released from plants to gain an advantage in occupying the rhizosphere area. *Pseudomonas veronii* treatment demonstrated to decrease sesame plant waterlogging stress by protecting leaf chlorophyll along with fluorescence indicators and promoting both root and shoot growth and biomass formation according to Khan et al. (Ali et al. 2022). The seedlings of rapeseed experience growth hindrance due to waterlogging stress in addition to reduced root function and changes in bacterial communities within the rhizosphere. Research indicates that organic and microbial fertilizers when applied in combination improve seedling development and recovery of rapeseed crops after waterlogging incidents (Wu et al. 2024).

HS help the plant to overcome stressful conditions (flooding, salinity, and drought) and to grow in a healthy environment. To attain this goal, humic acids help with nutrient uptake, availability of micronutrients, and the transportation of nutrients (Chandio et al. 2021).

During the waterlogging condition, the concentration of zinc decreased, to elevate the concentration plants are treated with humic acid. However, during drought or other environmental stresses, such effect of foliar applied HA is not seen. The research was done on maize plant that was under waterlogging stress and reported that humic acid substances increase the Zinc concentration potentially (Pitann et al. 2024). FA-based biostimulants help soybean plants tolerate the water deficit conditions because of waterlogging (do Rosário Rosa et al. 2021).

Soluble salt accumulation in soil enhances the rate of evaporation in environments in arid and humid regions caused by soil flooding (Li et al. 2022; Naorem et al. 2023). In developing countries, the arable land faces a lot of deficiency in the production of plants because of this. Different experiments were done in waterlogging and salinization stresses, the result was a decreased productivity rate of crops. Waterlogging or salinization in plant roots is more drastic than any other stress and reduces the output of the crop (Tahjib-Ul-Arif et al. 2023). Biostimulant seaweed extract is applied to the different parts of plants especially in the root area by a spraying method, or added to the soil directly. This application has a positive influence on plant growth. Various bioactive compounds are present in SWEs, these compounds, and the metabolic pathways of these biostimulants indicate that it is an elicitor-like agent (Ali et al. 2021). Enhanced growth in plants and tolerance response against stress because of the presence of growth-promoting phytohormones as well as inorganic elements. SWEs show mechanisms like natural regulators that affect the entire growth of the plant to deal with environmental stress like flooding and salinity (Chanthini et al. 2022).

IAA and gibberellins are plant-based compounds produced from PGPR. PGPR generates enzymes. These bioactive enzymes help the plants in waterlogging. The use of PGPR methodology is an eco-friendly environmental methodology, which not only increases the growth but also increases the harvest of crops (Kumar et al. 2024).

Complex biological matrices contain various amino acids, peptides, and polypeptides, which are derived from chemical, enzymatic, and thermal hydrolysis of protein content. The mixture of these molecules is known as protein hydrolysates (Mohammad et al. 2023). The PH obtained from plants and animals, reported in research done on horticulture plants demonstrated that these biostimulants increase the stress tolerance ability of plants. This ability enables them to survive in this environmental stress, and the proper mechanism at the genetic and metabolite levels is not demonstrated yet, due to a lack of information regarding the composition of protein hydrolysates (Franzoni et al. 2022).

3.1.8 Future Perspectives

While biostimulants have great potential to help oilseed crops cope with flooding and waterlogging stress, current difficulties prevent them from being adopted more widely. Consistent formulations fail in uncertain fields with unreasonable variability in source materials (SWE, PH, and HS) in their raw form. Thus, they point to a critical need for standardized production protocols. Research has shown that these compounds can increase the oxygenation of roots, increase the acquisition of nutrients, and augment antioxidant defense systems, but the specifics of their mechanisms in terms of waterlogging stress have not been elaborated.

It is recommended to employ novel omics tools like transcriptomics, proteomics, and metabolomics to solve intricate biostimulant pairs and stress response pathways in oilseed crops for future investigations. By improving their ability to recover from oilseed crops these advanced formulations would greatly enhance the recovery ability and maintain yield stability to flooding events to ultimately leading to more resilient and sustainable production systems that respond to a diversity of environmental conditions. In addition, field-scale validation studies in various agroecosystems are needed for the optimization of application timing, dosage recommendations, and the economic reality of oilseed varieties under different waterlogging conditions.

There is an underutilized opportunity for integrating biostimulants with additional agricultural practices to improve waterlogging tolerance in oilseed crops. The potential collaboration of biostimulants and management strategies (precision irrigation, biochar amendments, and microbe applications) should be investigated by the researchers. Simultaneously improving flood resilience plant vigor and nutrient efficiency may be supported by the transition toward more sustainable farming systems for the transition with fewer chemical inputs through this comprehensive approach. A promising circular economy solution for the biostimulant production methodologies is to explore novel ways of deriving these from agricultural byproducts and food industry waste streams. Turning these underutilized resources into useful plant stress mitigation compounds would make economically viable alternatives from an environmental point of view while enhancing crop performance under water-logged conditions. Potential waste materials would be transformed into valuable agricultural inputs to maximize resource efficiency within the production chain. There is still room for extensive multiyear field trials in many true growing environments to validate biostimulant effectiveness. These studies must attempt to optimize application methods, prescript timing protocol, and dosage requirements while considering soil interaction of oilseed types with regional characteristics of soil, oilseed varieties, and farming practices. Involving scientists, industry partners, and producers, this type of research approach—collaborative, in which scientists work with those close to the ground, in this case, scientists, industry partners, and producers—can lead to targeted, location-specific adaptation strategies that enhance the stability of food and resilience in oilseed production systems for when waterlogging is increasing all the time, are more frequent than ever.

Future research will target closing these gaps in knowledge and addressing these challenges regarding developing advanced biostimulant technologies engineered specifically to improve oilseed crop resilience to waterlogging events. These innovations promise to significantly enhance the agricultural sustainability of foods through the provision of farmers with tools that can dependably protect yield potential under rather problematic soil moisture conditions, thereby providing stability in agriculture's economic and social realms as climate scenarios become even more unpredictable.

3.1.9 Conclusion

Oilseed crops contribute to global agriculture, significantly to edible oil production, animal feed, and biofuel industries. However, these crops face increasing challenges from climate change and unpredictable weather patterns. One of the most pressing threats is waterlogging and flooding, which can severely damage root systems, reduce nutrient uptake, and ultimately decrease harvest quality. Biostimulants appeared as a sustainable approach to crop resilience promotion and productivity under such adverse environmental conditions. Biostimulants have been demonstrated to reduce the effects of waterlogging. These compounds improve physiological processes such as root aeration, antioxidant activity, osmotic balance, and nutrient absorption. As a result, plants treated with biostimulants often exhibit enhanced stress tolerance, quicker recovery from water damage, and better overall yield performance. Have ability to boost plant development in stress makes them valuable tools for sustainable oilseed production. Incorporating biostimulants into agricultural systems not only strengthens crop tolerance to waterlogging but also reduces the dependency on synthetic fertilizers and other chemical inputs, contributing to more environmentally friendly farming practices. However, the impact of biostimulants varies with crop type depending on crop type, soil condition, application timing, and environmental context. Therefore, to access the optimal quantity, methods, and crop-specific responses an ongoing investigation is required. To fully harness the benefits of biostimulants in oilseed farming, collaboration among researchers, agricultural industries, and farmers is crucial. By advancing biostimulant science and promoting its practical application, we can help fortify oilseed crops against climate-induced stresses and contribute to global food security and sustainable agriculture.

References

Aboueshaghi RS, Omidi H, Bostani A (2023) Assessment of changes in secondary metabolites and growth of saffron under organic fertilizers and drought. J Plant Nutr 46(3):386–400

Adegoye GA, Olorunwa OJ, Alsajri FA, Walne CH, Wijewandana C, Kethireddy SR, Reddy KN, Reddy KR (2023) Waterlogging effects on soybean physiology and hyperspectral reflectance during the reproductive stage. Agriculture 13(4):844

Ali O, Ramsubhag A, Jayaraman J (2021) Biostimulant properties of seaweed extracts in plants: implications towards sustainable crop production. Plan Theory 10(3):531

Ali S, Moon Y-S, Hamayun M, Khan MA, Bibi K, Lee I-J (2022) Pragmatic role of microbial plant biostimulants in abiotic stress relief in crop plants. J Plant Interact 17(1):705–718

Ali A, Niu G, Masabni J, Ferrante A, Cocetta G (2024) Integrated nutrient management of fruits, vegetables, and crops through the use of biostimulants, soilless cultivation, and traditional and modern approaches—a mini review. Agriculture 14(8):1330

Alsudays IM, Alshammary FH, Alabdallah NM, Alatawi A, Alotaibi MM, Alwutayd KM, Alharbi MM, Alghanem SM, Alzuaibr FM, Gharib HS (2024) Applications of humic and fulvic acid under saline soil conditions to improve growth and yield in barley. BMC Plant Biol 24(1):191

Ampong K, Thilakaranthna MS, Gorim LY (2022) Understanding the role of humic acids on crop performance and soil health. Front Agronomy 4:848621

Ancín M, Soba D, Picazo PJ, Gámez AL, Le Page JF, Houdusse D, Aranjuelo I (2024) Optimizing oilseed rape growth: exploring the effect of foliar biostimulants on the interplay among metabolism, phenology, and yield. Physiol Plant 176(5):e14561

Arumugam R, Rabert GA (2022) Plant biostimulants: overview of categories and effects. Biostimulants: Exploring sources and applications, pp 1–29

Aslam A, Mahmood A, Ur-Rehman H, Li C, Liang X, Shao J, Negm S, Moustafa M, Aamer M, Hassan MU (2023) Plant adaptation to flooding stress under changing climate conditions: ongoing breakthroughs and future challenges. Plan Theory 12(22):3824

Ateeq M, Khan AH, Zhang D, Alam SM, Shen W, Wei M, Meng J, Shen X, Pan J, Zhu K (2023) Comprehensive physio-biochemical and transcriptomic characterization to decipher the network of key genes under waterlogging stress and its recuperation in Prunus persica. Tree Physiol 43(7):1265–1283

Bahuguna A, Sharma S, Rai A, Bhardwaj R, Sahoo SK, Pandey A, Yadav B (2022) Advance technology for biostimulants in agriculture. In: New and future developments in microbial biotechnology and bioengineering. Elsevier, pp 393–412

Baltazar M, Correia S, Guinan KJ, Sujeeth N, Bragança R, Gonçalves B (2021) Recent advances in the molecular effects of biostimulants in plants: an overview. Biomol Ther 11(8):1096

Banaś K, Piwowar A, Harasym J (2023) The potential of rapeseed (canola) oil nutritional benefits wide spreading via oleogelation. Food Biosci 56:103162

Begum M, Sarmah B, Kandali GG, Baruah VJ, Borkotoki B, Talukdar L, Barua HJ (2022) Foliar application of microbial and plant-based biostimulants on plant nutrition. In: Biostimulants: exploring sources and applications. Springer, pp 193–220

Benito P, Celdrán M, Bellón J, Arbona V, González-Guzmán M, Porcel R, Yenush L, Mulet JM (2024) The combination of a microbial and a non-microbial biostimulant increases yield in lettuce (Lactuca sativa) under salt stress conditions by up-regulating cytokinin biosynthesis. J Integr Plant Biol 66(10):2140–2157

Benkeblia N (2021) Physiological and biochemical response of tropical fruits to hypoxia/anoxia. Front Plant Sci 12:670803

Bhaduri D, Mandal A, Chakraborty K, Chatterjee D, Dey R (2017) Interlinked chemical-biological processes in anoxic waterlogged soil—a review. Indian J Agric Sci 87(12):1587–1599

Bhoi A, Yadu B, Chandra J, Keshavkant S (2024) Cross-talk of strigolactones with abscisic acid, gibberellins, ethylene, and other hormones. In: Strigolactones. Elsevier, pp 103–126

Bobo G, Nicolau-Lapeña I, Aguiló-Aguayo I (2021) Conventional oils and oilseeds: composition and nutritional importance. Oil and oilseed processing: opportunities and challenges:23–40

Buga N, Petek M (2023) Use of biostimulants to alleviate anoxic stress in waterlogged cabbage (Brassica oleracea var. capitata)—a review. Agriculture 13(12):2223

Castiglione AM, Mannino G, Contartese V, Bertea CM, Ertani A (2021) Microbial biostimulants as response to modern agriculture needs: composition, role and application of these innovative products. Plan Theory 10(8):1533

Cellini A, Spinelli F, Donati I, Ryu C-M, Kloepper JW (2021) Bacterial volatile compound-based tools for crop management and quality. Trends Plant Sci 26(9):968–983

Chabili A, Minaoui F, Hakkoum Z, Douma M, Meddich A, Loudiki M (2024) A comprehensive review of microalgae and cyanobacteria-based biostimulants for agriculture uses. Plan Theory 13(2):159

Chandio W, Pirzada T, Majid A, Rashid F (2021) Extraction and characterization of humic acid from agriculture soil and its effect on wheat (Triticum Indicum) seed growth. J Innov Sci 7:205

Chanthini KM-P, Senthil-Nathan S, Pavithra G-S, Malarvizhi P, Murugan P, Deva-Andrews A, Janaki M, Sivanesh H, Ramasubramanian R, Stanley-Raja V (2022) Aqueous seaweed extract alleviates salinity-induced toxicities in rice plants (Oryza sativa L.) by modulating their physiology and biochemistry. Agriculture 12(12):2049

Chauhan J, Choudhury P, Singh K (2021) Production, varietal improvement programme and seed availability of annual oilseeds in India: current scenario and future prospects. J Oilseed Res 38(1):1–18

Ciriello M, Pannico A, Rouphael Y, Basile B (2025) Enhancing yield, physiological, and quality traits of strawberry cultivated under organic management by applying different non-microbial biostimulants. Plan Theory 14(5):712

Darakeh SASS, Weisany W, Tahir NA-R, Schenk PM (2022) Physiological and biochemical responses of black cumin to vermicompost and plant biostimulants: arbuscular mycorrhizal and plant growth-promoting rhizobacteria. Ind Crop Prod 188:115557

De Castro J, Hill RD, Stasolla C, Badea A (2022) Waterlogging stress physiology in barley. Agronomy 12(4):780

Deolu-Ajayi AO, van der Meer IM, Van der Werf A, Karlova R (2022) The power of seaweeds as plant biostimulants to boost crop production under abiotic stress. Plant Cell Environ 45(9): 2537–2553

Djuricic I, Calder PC (2021) Beneficial outcomes of omega-6 and omega-3 polyunsaturated fatty acids on human health: an update for 2021. Nutrients 13(7):2421

do Rosário Rosa V, dos Santos ALF, da Silva AA, Sab MPV, Germino GH, Cardoso FB, de Almeida Silva M (2021) Increased soybean tolerance to water deficiency through biostimulant based on fulvic acids and Ascophyllum nodosum (L.) seaweed extract. Plant Physiol Biochem 158:228–243

de Andrade da Silva MSR, dos Santos BMS, da Silva CSRA, da Silva CSRA, Antunes LFS, dos Santos RM, Santos CHB, Rigobelo EC (2021) Humic substances in combination with plant growth-promoting bacteria as an alternative for sustainable agriculture. Front Microbiol 12: 719653

Du Jardin P (2015) Plant biostimulants: definition, concept, main categories and regulation. Sci Hortic 196:3–14

Ejiri M, Fukao T, Miyashita T, Shiono K (2021) A barrier to radial oxygen loss helps the root system cope with waterlogging-induced hypoxia. Breed Sci 71(1):40–50

El Malahi S, Touhami D, Ganoudi M, Ennami M, Maach M, Idrissi Hassani LM, Jauregui I (2025) Harnessing the power of microbial allies: AMF and PGPR as biostimulants for sustainable bioeconomy development in the global south. In: Sustainable bioeconomy development in the global south: volume I status and perspectives. Springer, pp 261–291

El Sabagh A, Islam MS, Hossain A, Iqbal MA, Mubeen M, Waleed M, Reginato M, Battaglia M, Ahmed S, Rehman A (2022) Phytohormones as growth regulators during abiotic stress tolerance in plants. Front Agronomy 4:765068

Engel, D. C. H., Feltrim, D., Rodrigues, M., Baptistella, J. L. C., & Mazzafera, P. (2024). Soybean productivity is increased with protein hydrolysate seed and foliar application

Firdous SM, Mallik S, Paria B (2025) Antioxidant effects of medicinal plants for the treatment of epilepsy. Antioxidants: Nature's Defense Against Disease, pp 441–489

Franzoni G, Cocetta G, Prinsi B, Ferrante A, Espen L (2022) Biostimulants on crops: their impact under abiotic stress conditions. Horticulturae 8(3):189

Friedel JK, Ardakani MR (2021) Soil nutrient dynamics and plant-induced nutrient mobilisation in organic and low-input farming systems: conceptual framework and relevance. Biol Agric Horticult 37(1):1–24

Fujita, M., & Hasanuzzaman, M. (2022). Approaches to enhancing antioxidant defense in plants. 11, 925): MDPI

Gedeon S, Ioannou A, Balestrini R, Fotopoulos V, Antoniou C (2022) Application of biostimulants in tomato plants (Solanum lycopersicum) to enhance plant growth and salt stress tolerance. Plan Theory 11(22):3082

Ghosh P, Roychoudhury A (2024) Plant peptides involved in abiotic and biotic stress responses and reactive oxygen species (ROS) signaling. J Plant Growth Regul 43(5):1410–1427

Gladkov EA, Tereshonok DV, Stepanova AY, Gladkova OV (2023) Plant–microbe interactions under the action of heavy metals and under the conditions of flooding. Diversity 15(2):175

Gu X, Xue L, Lu L, Xiao J, Song G, Xie M, Zhang H (2021) Melatonin enhances the waterlogging tolerance of Prunus persica by modulating antioxidant metabolism and anaerobic respiration. J Plant Growth Regul 40:2178–2190

Habibullah M, Sarkar S, Islam MM, Ahmed KU, Rahman MZ, Awad MF, ElSayed AI, Mansour E, Hossain MS (2021) Assessing the response of diverse sesame genotypes to waterlogging durations at different plant growth stages. Plan Theory 10(11):2294

Hasan A, Tabassum B, Hashim M, Khan N (2024) Role of plant growth promoting rhizobacteria (PGPR) as a plant growth enhancer for sustainable agriculture: A review. Bacteria 3(2):59–75

Hasanuzzaman M, Ahmed N, Saha T, Rahman M, Rahman K, Alam MM, Rohman MM, Nahar K (2022) Exogenous salicylic acid and kinetin modulate reactive oxygen species metabolism and

glyoxalase system to confer waterlogging stress tolerance in soybean (Glycine max L.). Plant Stress 3:100057
Hemathilake D, Gunathilake D (2022) Agricultural productivity and food supply to meet increased demands. In: Future foods. Elsevier, pp 539–553
Hines S, van der Zwan T, Shiell K, Shotton K, Prithiviraj B (2021) Alkaline extract of the seaweed Ascophyllum nodosum stimulates arbuscular mycorrhizal fungi and their endomycorrhization of plant roots. Sci Rep 11(1):13491
Hong B, Zhou B, Zhao D, Liao L, Chang T, Wu X, Wu J, Yao M, Chen H, Mao J (2024) Yield, cell structure and physiological and biochemical characteristics of rapeseed under waterlogging stress. BMC Plant Biol 24(1):941
Hongal G, Siddaraju R, Bai SK, Vishwanath K, Ramanappa T, Pavan A (2023) Bio-stimulants in oil seed crops-its influence on growth, yield and quality. Mysore J Agric Sci 57(4)
Hussain HI, Kasinadhuni N, Arioli T (2021) The effect of seaweed extract on tomato plant growth, productivity and soil. J Appl Phycol 33(2):1305–1314
Hussain MA, Naeem A, Pitann B, Mühling KH (2023a) High sulfur (S) supplementation imparts waterlogging tolerance to oilseed rape (Brassica napus L.) through upregulating s metabolism and antioxidant pathways. J Plant Growth Regul 42(12):7591–7605
Hussain M, Ul-Allah S, Farooq S (2023b) Sesame (Sesamum indicum L.). In: Neglected and underutilized crops. Elsevier, pp 733–755
Jain BM, Badve MP (2022) A novel process for synthesis of soybean protein hydrolysates and study of its effectiveness as a biostimulant and emulsifier. Chem Eng Proc-Process Intensification 174:108880
Jain D, Bisht S, Parvez A, Singh K, Bhaskar P, Koubouris G (2024) Effective biotic elicitors for augmentation of secondary metabolite production in medicinal plants. Agriculture 14(6):796
Jardim-Messeder D, Margis-Pinheiro M, Sachetto-Martins G (2022) Salicylic acid and adenine nucleotides regulate the electron transport system and ROS production in plant mitochondria. Biochimica et Biophysica Acta (BBA)-Bioenergetics 1863(6):148559
Jia W, Ma M, Chen J, Wu S (2021) Plant morphological, physiological and anatomical adaption to flooding stress and the underlying molecular mechanisms. Int J Mol Sci 22(3):1088
Jiang X, Mao D, Zhu M, Wang X, Li C, Zhu X, Guo W, Ding J (2022) Evaluating the waterlogging tolerance of wheat cultivars during the early growth stage using the comprehensive evaluation value and digital image analysis. Agriculture 12(3):384
Jiang J, Huang H, Gao Q, Li Y, Xiang H, Zeng W, Xu L, Liu X, Li J, Mi Q (2023) Effects of editing DFR genes on flowers, leaves, and roots of tobacco. BMC Plant Biol 23(1):349
Jiang Y, Yue Y, Wang Z, Lu C, Yin Z, Li Y, Ding X (2024) Plant biostimulant as an environmentally friendly alternative to modern agriculture. J Agric Food Chem 72(10):5107–5121
Jiménez-Arias D, García-Machado FJ, Morales-Sierra S, García-García AL, Herrera AJ, Valdés F, Luis JC, Borges AA (2021) A beginner's guide to osmoprotection by biostimulants. Plan Theory 10(2):363
Jing X, Su W, Fan S, Luo H, Chu H (2022) Ecological strategy of phyllostachys heteroclada oliver in the riparian zone based on ecological stoichiometry. Front Plant Sci 13:974124
Juan CA, Pérez de la Lastra JM, Plou FJ, Pérez-Lebeña E (2021) The chemistry of reactive oxygen species (ROS) revisited: outlining their role in biological macromolecules (DNA, lipids and proteins) and induced pathologies. Int J Mol Sci 22(9):4642
Kajrolkar A (2025) Advancements in plant nutrition: enhancing crop yield and quality. Premier J Plant Biol 3:100013
Kanabar P, Wu Y, Nandwani D (2024) Enhancing sustainable cultivation of organic bell pepper through fulvic acid (FA) application: impact on phytochemicals and antioxidant capacity under open-field conditions. Sustain For 16(16):6745
Kang S-M, Imran M, Shaffique S, Kwon E-H, Park Y-S, Lee I-J (2022) Growth and photosynthetic characteristics of sesame seedlings with gibberellin-producing Rhodobacter sphaeroides SIR03 and biochar. Int J Plant Biol 13(3):257–269

Kapoore RV, Wood EE, Llewellyn CA (2021) Algae biostimulants: a critical look at microalgal biostimulants for sustainable agricultural practices. Biotechnol Adv 49:107754
Kaushal P, Ali N, Saini S, Pati PK, Pati AM (2023) Physiological and molecular insight of microbial biostimulants for sustainable agriculture. Front Plant Sci 14:1041413
Ke H, Peng H, Wenbo G (2023) The distribution characteristics of waterlogging disaster risk of winter rape (Brassica napus L.) at seedling stage in Southwest China. J Agric 13(10):62
Khalid B, Javed M, Ashraf M, Saeed H, Shaheen M, Riaz T, Riaz R, Nawaz S (2025) Microbial bio stimulants as sustainable strategies for enhancing plant resistance to viral diseases: mechanisms and applications. Hosts Viruses 12:93–110
Khasheii B, Mahmoodi P, Mohammadzadeh A (2021) Siderophores: importance in bacterial pathogenesis and applications in medicine and industry. Microbiol Res 250:126790
Kong Z, Liu H (2022) Modification of rhizosphere microbial communities: a possible mechanism of plant growth promoting rhizobacteria enhancing plant growth and fitness. Front Plant Sci 13: 920813
Kumar P, Singh RP (2021) Microbial diversity and multifunctional microbial biostimulants for agricultural sustainability. Climate Resilience and Environmental Sustainability Approaches: Global Lessons and Local Challenges, pp 141–184
Kumar M, Gupta A, Vandana P, Tiwari LD, Patel MK, Siddique KH (2024) Recent advances of plant growth-promoting rhizobacteria (PGPR)-mediated drought and waterlogging stress tolerance in plants for sustainable agriculture. Microbial Biostimulants for Plant Growth and Abiotic Stress Amelioration, pp 315–344
Lephatsi M, Nephali L, Meyer V, Piater LA, Buthelezi N, Dubery IA, Opperman H, Brand M, Huyser J, Tugizimana F (2022) Molecular mechanisms associated with microbial biostimulant-mediated growth enhancement, priming and drought stress tolerance in maize plants. Sci Rep 12(1):10450
Li S, Rao L (2024) Response of growth and chlorophyll fluorescence parameters of mulberry seedlings to waterlogging stress. Sci Rep 14(1):25078
Li J, Iqbal S, Zhang Y, Chen Y, Tan Z, Ali U, Guo L (2021) Transcriptome analysis reveals genes of flooding-tolerant and flooding-sensitive rapeseeds differentially respond to flooding at the germination stage. Plan Theory 10(4):693
Li X, Guo M, Wang H (2022) Impact of soil texture and salt type on salt precipitation and evaporation under different hydraulic conditions. Hydrol Process 36(11):e14763
Li L, Zhang L, Tang J, Xing H, Zhao L, Jie H, Jie Y (2023) Waterlogging increases greenhouse gas release and decreases yield in winter rapeseed (Brassica napus L.) seedlings. Sci Rep 13(1): 18673
Lindsey AJ, Ortez OA, Thomison PR, Carter PR, Coulter JA, Roth GW, Carrijo DR, Quinn DJ, Licht MA (2024) Severe storm damage and short-term weather stresses on corn: A review. Crop Sci 64(3):1129–1166
Luan H, Gao J, Li Y, Qu X, Yang J, Qian X, Xu M, Sun M, Xu X, Shen H (2025) Integrated metabolomic and transcriptomic strategies to reveal adaptive mechanisms in barley plant during germination stage under waterlogging stress. J Plant Res:1–14
Maffia A, Oliva M, Marra F, Mallamaci C, Nardi S, Muscolo A (2025) Humic substances: bridging ecology and agriculture for a greener future. Agronomy 15(2):410
Manghwar H, Hussain A, Alam I, Khoso MA, Ali Q, Liu F (2024) Waterlogging stress in plants: unraveling the mechanisms and impacts on growth, development, and productivity. Environ Exp Bot 105824
Matos GS, Pereira SG, Genisheva ZA, Gomes AM, Teixeira JA, Rocha CM (2021) Advances in extraction methods to recover added-value compounds from seaweeds: sustainability and functionality. Food Secur 10(3):516
Miranda AM, Hernandez-Tenorio F, Villalta F, Vargas GJ, Sáez AA (2024) Advances in the development of biofertilizers and biostimulants from microalgae. Biology 13(3):199

Mohammad SS, da Silva Ferreira MV, Jacintho Barbosa MI, Barbosa Junior JL (2023) Characteristics of enzymatic hydrolysis of protein from different food sources and potential separation techniques. Curr Nutr Food Sci 19(6):590–601

Monterisi S, Garcia-Perez P, Buffagni V, Zuluaga MYA, Ciriello M, Formisano L, El-Nakhel C, Cardarelli M, Colla G, Rouphael Y (2024) Unravelling the biostimulant activity of a protein hydrolysate in lettuce plants under optimal and low N availability: a multi-omics approach. Physiol Plant 176(3):e14357

Nagrale D, Chaurasia A, Kumar S, Gawande S, Hiremani N, Shankar R, Gokte-Narkhedkar N, Renu, & Prasad, Y. (2023) PGPR: the treasure of multifarious beneficial microorganisms for nutrient mobilization, pest biocontrol and plant growth promotion in field crops. World J Microbiol Biotechnol 39(4):100

Naorem A, Jayaraman S, Dang YP, Dalal RC, Sinha NK, Rao CS, Patra AK (2023) Soil constraints in an arid environment—challenges, prospects, and implications. Agronomy 13(1):220

Nardi S, Schiavon M, Francioso O (2021) Chemical structure and biological activity of humic substances define their role as plant growth promoters. Molecules 26(8):2256

Nasrullah, Ali S, Umar M, Sun L, Naeem M, Yasmin H, Khan N (2022) Flooding tolerance in plants: from physiological and molecular perspectives. Rev Bras Bot 45(4):1161–1176

Omoarelojie LO, Kulkarni MG, Finnie JF, Van Staden J (2021) Biostimulants and the modulation of plant antioxidant systems and properties. In: Biostimulants for crops from seed germination to plant development. Elsevier, pp 333–363

Orsák M, Kotíková Z, Hnilička F, Lachman J (2023) Effect of long-term drought and waterlogging stress on photosynthetic pigments in potato. Plant, Soil Environ 69(4)

Ortiz A, Sansinenea E (2022) The role of beneficial microorganisms in soil quality and plant health. Sustain For 14(9):5358

Pais IP, Moreira R, Semedo JN, Ramalho JC, Lidon FC, Coutinho J, Maçãs B, Scotti-Campos P (2022) Wheat crop under waterlogging: potential soil and plant effects. Plan Theory 12(1):149

Pan J, Sharif R, Xu X, Chen X (2021) Mechanisms of waterlogging tolerance in plants: research progress and prospects. Front Plant Sci 11:627331

Pandey AK, Singh AG, Gadhiya AR, Kumar S, Singh D, Mehta R (2021) Current approaches in horticultural crops to mitigate waterlogging stress. Stress Tolerance in Horticultural Crops:289–299

Papa S, Fusco GM, Ciriello M, Formisano L, Woo SL, De Pascale S, Rouphael Y, Carillo P (2022) Microbial and non-microbial biostimulants as innovative tools to increase macro and trace element mineral composition of tomato and spinach. Horticulturae 8(12):1157

Parmar P, Kumar R, Neha Y, Srivatsan V (2023) Microalgae as next generation plant growth additives: functions, applications, challenges and circular bioeconomy based solutions. Front Plant Sci 14:1073546

Perez-Vazquez A, Carpena M, Barciela P, Cassani L, Simal-Gandara J, Prieto MA (2023) Pressurized liquid extraction for the recovery of bioactive compounds from seaweeds for food industry application: a review. Antioxidants 12(3):612

Pitann B, Khan K, Mühling KH (2024) Does humic acid foliar application affect growth and nutrient status of water-stressed maize? Plant-Environ Interact 5(3):e10156

Prasad K (2023) Symbiotic endophytes of glomalin AM fungi, rhizobium, and PGPR potential bio stimulants to intensive global food production for sustainable agriculture system. J Microb Res 2(2):2836–2187

Puglia D, Pezzolla D, Gigliotti G, Torre L, Bartucca ML, Del Buono D (2021) The opportunity of valorizing agricultural waste, through its conversion into biostimulants, biofertilizers, and biopolymers. Sustain For 13(5):2710

Quitério E, Grosso C, Ferraz R, Delerue-Matos C, Soares C (2022) A critical comparison of the advanced extraction techniques applied to obtain health-promoting compounds from seaweeds. Mar Drugs 20(11):677

Radulov I, Berbecea A (2024) Nutrient management for sustainable soil fertility

Rahgoshahi M, Laghari KPK, Rahimi M, Kelidari A, Keshavarzi K (2023) Physiological enhancement of seed yield and essential oil yield in cumin under drought stress through humic acid and seaweed extract. Russ J Plant Physiol 70(6):147

Rahim HU, Ali W, Uddin M, Ahmad S, Khan K, Bibi H, Alatalo JM (2024) Abiotic stresses in soils, their effects on plants, and mitigation strategies: a literature review. Chem Ecol:1–34

Rathnakumar A, Sujatha M (2022) Breeding major oilseed crops: prospects and future research needs. Accelerated plant breeding. Oil Crops 4:1–40

Rathor P, Gorim LY, Thilakarathna MS (2023) Plant physiological and molecular responses triggered by humic based biostimulants-a way forward to sustainable agriculture. Plant Soil 492(1):31–60

Rauf M, Awais M, Ud-Din A, Ali K, Gul H, Rahman MM, Hamayun M, Arif M (2021) Molecular mechanisms of the 1-aminocyclopropane-1-carboxylic acid (ACC) deaminase producing Trichoderma asperellum MAP 1 in enhancing wheat tolerance to waterlogging stress. Front Plant Sci 11:614971

Rawat P, Das S, Shankhdhar D, Shankhdhar S (2021) Phosphate-solubilizing microorganisms: mechanism and their role in phosphate solubilization and uptake. J Soil Sci Plant Nutr 21(1): 49–68

Raza A (2021) Eco-physiological and biochemical responses of rapeseed (Brassica napus L.) to abiotic stresses: consequences and mitigation strategies. J Plant Growth Regul 40(4):1368–1388

Ren B, Yu W, Liu P, Zhao B, Zhang J (2023) Responses of photosynthetic characteristics and leaf senescence in summer maize to simultaneous stresses of waterlogging and shading. Crop J 11(1):269–277

Rohela GK, Saini P (2022) Nitrogen-fixing biofertilizers and biostimulants. In: Microbial biostimulants for sustainable agriculture and environmental bioremediation. CRC Press, pp 83–100

Rostami Ahmadvandi H, Faghihi A (2021) Adapted oilseed crops with the ability to grow economically in dryland conditions in Iran. Agrotech Ind Crop 1(3):122–128

Rouphael Y, Colla G (2018) Synergistic biostimulatory action: designing the next generation of plant biostimulants for sustainable agriculture. Front Plant Sci 9:1655

Saeed Q, Xiukang W, Haider FU, Kučerik J, Mumtaz MZ, Holatko J, Naseem M, Kintl A, Ejaz M, Naveed M (2021) Rhizosphere bacteria in plant growth promotion, biocontrol, and bioremediation of contaminated sites: a comprehensive review of effects and mechanisms. Int J Mol Sci 22(19):10529

Samantaray A, Chattaraj S, Mitra D, Ganguly A, Kumar R, Gaur A, Mohapatra PKD, de Los Santos-Villalobos S, Rani A, Thatoi H (2024) Advances in microbial based bio-inoculum for amelioration of soil health and sustainable crop production. Curr Res Microb Sci 100251

Sathi KS, Masud AAC, Anee TI, Rahman K, Ahmed N, Hasanuzzaman M (2022) Soybean plants under waterlogging stress: responses and adaptation mechanisms. In: Managing plant production under changing environment. Springer, pp 103–134

Schulte LR, Ballard T, Samarakoon T, Yao L, Vadlani P, Staggenborg S, Rezac M (2013) Increased growing temperature reduces content of polyunsaturated fatty acids in four oilseed crops. Ind Crop Prod 51:212–219

Shahrajabian MH, Cheng Q, Sun W (2022) The effects of amino acids, phenols and protein hydrolysates as biostimulants on sustainable crop production and alleviated stress. Recent Pat Biotechnol 16(4):319–328

Sharma S, Sharma J, Soni V, Kalaji HM, Elsheery NI (2021) Waterlogging tolerance: a review on regulative morpho-physiological homeostasis of crop plants. J Water Land Dev:16–28

Sharma S, Bhatt U, Sharma J, Darkalt A, Mojski J, Soni V (2022) Effect of different waterlogging periods on biochemistry, growth, and chlorophyll a fluorescence of Arachis hypogaea L. Front Plant Sci 13:1006258

Sharmin RA, Karikari B, Bhuiyan MR, Kong K, Yu Z, Zhang C, Zhao T (2024) Comparative morpho-physiological, biochemical, and gene expressional analyses uncover mechanisms of waterlogging tolerance in two soybean introgression lines. Plan Theory 13(7):1011

Shi F, Pan Z, Dai P, Shen Y, Lu Y, Han B (2023) Effect of waterlogging stress on leaf anatomical structure and ultrastructure of Phoebe sheareri seedlings. Forests 14(7):1294

Shiade SRG, Zand-Silakhoor A, Fathi A, Rahimi R, Minkina T, Rajput VD, Zulfiqar U, Chaudhary T (2024) Plant metabolites and signaling pathways in response to biotic and abiotic stresses: exploring bio stimulant applications. Plant Stress 1:100454

Shields HJ, Traa A, Van Raamsdonk JM (2021) Beneficial and detrimental effects of reactive oxygen species on lifespan: a comprehensive review of comparative and experimental studies. Front Cell Develop Biol 9:628157

Shukla U (2022) Effect of different bio-stimulantss on growth, flowering, yield and post-harvest quality of Dahlia (Dahlia variabilis L.) under Western Uttar Pradesh conditions. Sardar Vallabhbhai Patel University of Agriculture and Technology

Sible CN, Seebauer JR, Below FE (2021) Plant biostimulants: a categorical review, their implications for row crop production, and relation to soil health indicators. Agronomy 11(7):1297

Siddha S, Sahu P (2022) Impact of climate change on the river ecosystem. In: Ecological significance of river ecosystems. Elsevier, pp 79–104

Singh A, Singh S, Kurella A, Verma A, Mahatama M, Venkatesh I (2022a) Plant bio-stimulants, their functions and use in enhancing stress tolerance in oilseeds. In: New and future developments in microbial biotechnology and bioengineering. Elsevier, pp 239–259

Singh S, Singh S, Singh G, Kaur G (2022b) Soil waterlogging: cause, impact, and management. Soil Constraints on Crop Production 62

Srivastava AK, Shukla SK, Singh A, Singh A, Jindal S, Malik P (2024) Biochemical, physiological and molecular aspects of waterlogging tolerance in economically important oilseed crops rapeseed, sesame and soybean. Genetika 56(1):221–241

Stirk WA, Rengasamy KR, Kulkarni MG, van Staden J (2020) Plant biostimulants from seaweed: an overview. The Chemical Biology of Plant Biostimulants, pp 31–55

Subrahmaniyan K, Veeramani P, Mahalingam A, Ranjith R, Harishma PH, Abarajitha P (2024) A comprehensive review on nutritional and antioxidant properties of Sesame (Sesamum indicum L.) seed oil with its therapeutic utilization as phytomedicine. Ann Phytomedicine 13(2):1–9

Sun W, Shahrajabian MH (2023) The application of arbuscular mycorrhizal fungi as microbial biostimulant, sustainable approaches in modern agriculture. Plan Theory 12(17):3101

Sun W, Shahrajabian MH, Soleymani A (2024) The roles of plant-growth-promoting rhizobacteria (PGPR)-based biostimulants for agricultural production systems. Plan Theory 13(5):613

Tabacchioni S, Passato S, Ambrosino P, Huang L, Caldara M, Cantale C, Hett J, Del Fiore A, Fiore A, Schlüter A (2021) Identification of beneficial microbial consortia and bioactive compounds with potential as plant biostimulants for a sustainable agriculture. Microorganisms 9(2):426

Tahjib-Ul-Arif M, Hasan MT, Rahman MA, Nuruzzaman M, Rahman AS, Hasanuzzaman M, Haque MR, Hossain MA, Latef AAHA, Murata Y (2023) Plant response to combined salinity and waterlogging stress: current research progress and future prospects. Plant Stress 7:100137

Tokel D, Erkencioglu BN (2021) Production and trade of oil crops, and their contribution to the world economy. Oil crop genomics:415–427

Topali C, Antonopoulou C, Chatzissavvidis C (2024) Effect of waterlogging on growth and productivity of fruit crops. Horticulturae 10(6):623

Toral-Juárez MA, Avila RT, Cardoso AA, Brito FA, Machado KL, Almeida WL, Souza RP, Martins SC, DaMatta FM (2021) Drought-tolerant coffee plants display increased tolerance to waterlogging and post-waterlogging reoxygenation. Environ Exp Bot 182:104311

Tyagi A, Ali S, Park S, Bae H (2023) Exploring the potential of multiomics and other integrative approaches for improving waterlogging tolerance in plants. Plan Theory 12(7):1544

Vaishnav A, Kumar R, Singh HB, Sarma BK (2022) Extending the benefits of PGPR to bioremediation of nitrile pollution in crop lands for enhancing crop productivity. Sci Total Environ 826: 154170

Vocciante M, Grifoni M, Fusini D, Petruzzelli G, Franchi E (2022) The role of plant growth-promoting rhizobacteria (PGPR) in mitigating plant's environmental stresses. Appl Sci 12(3):1231

Vultaggio L, Ciriello M, Campana E, Bellitto P, Consentino BB, Rouphael Y, Colla G, Mancuso F, La Bella S, Napoli S (2024) Single or blended application of non-microbial plant-based biostimulants and trichoderma atroviride as a new strategy to enhance greenhouse cherry tomato performance. Plan Theory 13(21):3048

Wang S, Hu J, Ren B, Liu P, Zhao B, Zhang J (2022) Effects of hydrogen peroxide priming on yield, photosynthetic capacity and chlorophyll fluorescence of waterlogged summer maize. Front Plant Sci 13:1042920

Wazeer H, Shridhar Gaonkar S, Doria E, Pagano A, Balestrazzi A, Macovei A (2024) Plant-based biostimulants for seeds in the context of circular economy and sustainability. Plan Theory 13(7):1004

Wu X, Wu J, Zhou B, Hong B, Zhao D, Guan M (2024) Effects of fertilization patterns on the growth of rapeseed seedlings and rhizosphere microorganisms under flooding stress. Agronomy 14(3):525

Wu C, Florez-Palacios L, Acuna A, Harrison D, Rogers D, Carlin J, Mozzoni L, Nguyen HT, Shannon G, Vieira CC (2025) Impact of flooding at the early reproductive growth stage on soybean yield and seed composition. Crop Sci 65(1):e21397

Xu F, Cai H, Zhang X, Su M, Zhou H, Li X, Zhang M, Hu Y, Gu C, Du J (2022) Comparison of waterlogging tolerance of three peach rootstock seedlings based on physiological, anatomical and ultra-structural changes. Horticulturae 8(8):720

Yadav R, Kalia S, Rangan P, Pradheep K, Rao GP, Kaur V, Pandey R, Rai V, Vasimalla CC, Langyan S (2022) Current research trends and prospects for yield and quality improvement in sesame, an important oilseed crop. Front Plant Sci 13:863521

Yang L, Li N, Liu Y, Miao P, Liu J, Wang Z (2023) Updates and prospects: morphological, physiological, and molecular regulation in crop response to waterlogging stress. Agronomy 13(10):2599

Yuan Y, Tang C, Jin Y, Cheng K, Yang F (2023) Contribution of exogenous humic substances to phosphorus availability in soil-plant ecosystem: a review. Crit Rev Environ Sci Technol 53(10):1085–1102

Zhang P, He J, Lyu D, Qin S (2024) Physiological and biochemical mechanisms of Prunus sachalinensis root for coping with short-term waterlogging and subsequent recovery. J Soil Sci Plant Nutr 24(2):2997–3010

Zhang Y, Chen X, Geng S, Zhang X (2025) A review of soil waterlogging impacts, mechanisms, and adaptive strategies. Front Plant Sci 16:1545912

Chapter 4
Use of Biostimulants to Enhance Temperature Tolerance in Oilseed Crops

Amir Abdullah Khan, Taghreed S. Alnusaire, Rasheed Akbar, Babar Iqbal, Aurang Zeb, and Mona H. Soliman

Abstract Abiotic stresses, particularly temperature extremes, pose significant challenges to the productivity and sustainability of oilseed crops. In recent years, biostimulants have emerged as promising tools for mitigating the adverse effects of temperature stress, leading to improved growth, development, and yield stability in these crops. This chapter provides a comprehensive overview of the role of biostimulants, such as plant growth-promoting rhizobacteria (PGPR), seaweed extracts, humic substances, and protein hydrolysates, in modulating physiological and molecular responses of oilseed crops under temperature stress. A thorough discussion is held regarding the mechanisms that underlie biostimulant-mediated elevated temperature tolerance, such as upregulated stress-responsive genes, osmolyte buildup, hormonal adjustment, and improved oxidative systems for protection. Furthermore, the chapter stresses how crucial biostimulants serve the purpose of preserving transmembrane stability, boosting absorbance of nutrients, and

Authors Amir Abdullah Khan and Taghreed S. Alnusaire have equally contributed to this chapter.

A. A. Khan (✉) · B. Iqbal
School of Environment and Safety Engineering, Jiangsu University, Zhenjiang, China

T. S. Alnusaire
Department of Biology, College of Science, Jouf University, Sakaka, Saudi Arabia

R. Akbar
School of Environment and Safety Engineering, Jiangsu University, Zhenjiang, China

Department of Entomology, Faculty of Physical and Applied Sciences, The University of Haripur, Haripur, Khyber Pakhtunkhwa, Pakistan

A. Zeb
MOE Key Laboratory of Pollution Processes and Environmental Criteria, College of Environmental Science and Engineering, Nankai University, Tianjin, People's Republic of China

M. H. Soliman
Botany and Microbiology Department, Faculty of Science, Cairo University, Giza, Egypt

Biology Department, Faculty of Science, Taibah University, Yanbu, Kingdom of Saudi Arabia
e-mail: hmona@sci.cu.edu.eg

© The Author(s), under exclusive license to Springer Nature Singapore Pte Ltd. 2025

A. A. H. Abdel Latef (ed.), *Oilseed Crops Under Abiotic Stress*, Sustainability Sciences in Asia and Africa, https://doi.org/10.1007/978-981-96-8346-8_4

promoting a plant's ultimate resistance to shifts in temperature. Amid changing climates, the possible application of biostimulants as environmentally acceptable, sustainable ways to increase oilseed crops' tolerance for heat is also investigated. In order to secure supplies of food in an ever-changing environment, this chapter culminates by identifying prospective study possibilities, with an especially strong focus on the inclusion of biostimulants throughout crop management approaches.

Keywords Biostimulants · Extreme temperatures · Oilseed crops · Plant growth-promoting rhizobacteria (PGPR) · Antioxidant defense mechanisms · Osmolytes · Heat resilience

4.1 Introduction

4.1.1 Impact of Temperature Stress on Oil Seed Crops

As disrupting important physiological, biochemical, and metabolic pathways, temperature stress, which covers the two extremes of heat and cold, significantly impacts the establishment, growth, and yield of the oilseed crops (Khan et al. 2025a; Meena et al. 2025). Due to its detrimental effects on pollen survival, fertilization being successful, and seed set along exposure to heat substantially decreases yields, in particular during critical phenological periods such as sprouting, blossoming and kernel filling. Abundant reactive oxygen species (ROS) from elevated temperatures promote membrane deterioration, worsen cellular stress, and decrease the rate of photosynthesis by lowering the activity of enzymes and chlorophyll concentration (Kourani et al. 2022; Khan et al. 2025b). The subsequent events include a spike in oxidative damage to lipids, denaturing of proteins, and enzymatic failure, which eventually contributes to a decline in both the amount and the quality of oil buildup (Medina et al. 2021). By lowering the production of enzymes, altering the hormonal equilibrium of hormonal substances, as well as hindering water and nutritional intake, cold stress, on the other hand, minimizes the emergence of seeds and the growth of seedlings (Soualiou et al. 2022). Low temperatures over a prolonged period build freezing complications, influence respiration within the mitochondria, result in membrane rigidity, and trigger metabolic discrepancies limiting growth and oil biosynthesis (Zhang et al. 2022). Moreover, the method by which high temperatures change the regulation of stress-responsive genes, such as those related to antioxidant defense, osmolyte buildup up, and heat-shock protein manufacturing, highlights the complex molecular mechanisms that regulate temperature tolerance in oilseed crops (Akbar et al. 2024; Perveen et al. 2025). Considering the growing severity of temperature variations caused by changing climates, we must cultivate thermo-tolerant varieties, incorporate stimulants of growth, as well as apply climate-smart agronomic approaches to safeguard oilseed crop output and assure global food security (Malik et al. 2020; Sangiorgio et al. 2020; Zulfiqar et al. 2024).

Since it disturbs fundamental physiological, biochemical, and metabolic processes, temperature stress, which spans both extremes of heat and cold, severely compromises the development, yield, and growth of oilseed crops (Meena et al. 2025). Particularly in important phenological periods, including germination, blooming, and seed filling, heat stress's negative impacts on pollen viability, fertilization success, and seed set cause notable yield declines (Farooq et al. 2017). Extreme temperatures develop a high amount of ROS, and that worsens the effects of oxidative stress, quickens membrane breakdown and minimizes the efficiency of photosynthesis by altering the amount of chlorophyll and enzyme activity (Muhammad et al. 2021; Khan et al. 2024). That follows higher levels of peroxidation of oils, protein degradation, and enzyme malfunction, which finally lowers the quantity and quality of oil accumulation (Clodoveo et al. 2014). Conversely, cold exposure reduces the activity of enzymes, alters hormonal balance, and hinders the absorption of nutrients as well as water, therefore lowering the chance of seed germination and the development of seedlings. The low temperatures create chilling injuries that inhibit respiration in mitochondria, induce membrane rigidification, and cause metabolic imbalances, restricting growth and oil synthesis (Lozano-Castellón et al. 2022). Breed for thermo-tolerant cultivars, include biostimulants, and apply climate-smart agronomic practices given the increasing frequency of temperature changes brought on by climate change. This will help to protect oilseed crop yield and guarantee global food security by highlighting the complex molecular control of temperature tolerance in oilseed crops via the way temperature extremes alter the expression of stress-responsive genes, including those involved in osmolyte accumulation, antioxidant defense, and heat-shock protein synthesis (Dias et al. 2022; Graziani et al. 2022).

4.1.2 Biostimulants' Function in Temperature Tolerance

Temperature stress, that is, both extremes of heat and cold, affects plant development, metabolism, and productivity. A broad category of organic substances and beneficial microbes called biostimulants has emerged as a practical strategy to raise tolerance of crops to temperature rise (Awasthi et al. 2015). These molecules boost the resistance of plants to damage from oxidation, membrane instabilities, and metabolic processes disparities that are brought on by heat by governing biological and molecular responses (Balbino 2017). One of the key ways biostimulants give temperature tolerance is by altering antioxidant defense mechanisms (Hasanuzzaman et al. 2021). Extremes of temperature generate excess ROS, which induces oxidative stress and cellular damage (Mrid et al. 2021). By raising the activity of antioxidant enzymes such as peroxidase (POD), catalase (CAT), and superoxide dismutase (SOD), biostimulants like humic substances, seaweed extracts, and microbial inoculants help to lower ROS buildup. Studies show that by reducing lipid peroxidation and boosting antioxidant pathways, the bioactive compounds found in polysaccharides produced from seaweed improve thermotolerance (Bhupenchandra et al. 2022; Alharbi et al. 2023). Moreover, crucial for maintaining the osmotic balance among cells under extreme temperatures are biostimulants. Amino acid-based biostimulants, such as proline and glycine betaine, support proteins and the membranes while minimizing the consequences of heat-induced water retention by osmoprotective action (Bulgari et al. 2019; Franzoni et al. 2022). Moreover, these chemical compounds enhance their ability to tolerate high temperatures via altering the regulation of genes connected to stress-responsive channels (Niu et al. 2022). Additionally, a couple of microbial biostimulants which improve water and nutrient absorption, which are critical for maintaining the equilibrium of cells over temperature fluctuations, are mycorrhizal fungi and rhizobacteria that promote plant growth (PGPR) (Backer et al. 2018; Kumari et al. 2022). To generate systemic tolerance, some PGPR bacteria—including *Bacillus* and *Pseudomonas*, change the amounts of phytohormones, particularly salicylic acid (SA) and abscisic acid (ABA), which are crucial controllers of temperature stress responses (Khan et al. 2020, 2025a; Ranjan et al. 2024). Also affected by biostimulants are the hormonal signaling systems regulating plant response to temperature stress. In response to heat and cold stress, protein hydrolysates and seaweed extracts stimulate cytokinin and auxin synthesis, therefore promoting root elongation, cell division, and photosynthetic efficiency (Askari-Khorasgani et al. 2019; Franzoni et al. 2022). Additionally, it is also discovered that brassinosteroids, a type of plant hormone commonly present in biostimulant formulations, also result in heat tolerance in plants which stabilizes cell membranes and increases stress-related gene expression (Mandal et al. 2023; Johnson et al. 2024). Biostimulants molecularly regulate the synthesis of cold-responsive genes and heat-shock proteins (HSPs), which are vital for the folding of proteins, integrity of membranes, and metabolic responses to temperature stress (González-Morales et al. 2021). Studies demonstrating avoidance of protein denaturation reveal higher expression of HSPs in plants treated with biostimulants, therefore improving their tolerance to high temperatures (Zahra et al. 2021; Jahed et al. 2023; Lee et al. 2024). Cold-responsive transcription factors (CBFs) are likewise raised in response to biostimulant therapies via regulating downstream stress-responsive genes, hence boosting cold tolerance (Ahmad et al. 2024). Combining biostimulants with traditional agronomic practices provides a long-term fix to lower the detrimental effects of crop temperature stress (Asif et al. 2023). Particularly for horticultural plants, oilseed crops, and grain crops, research in the field has demonstrated that applied topically biostimulants increase both yield and quality under temperature-stressed situations. Still, more study is required to maximize biostimulant compositions and understand their interactions with plant metabolism under different environmental conditions (Bulgari et al. 2019; Mandal et al. 2023).

4.2 Types of Biostimulants and Their Mechanisms

A vast class of natural substances and bacteria, biostimulants enhance plant growth, stress tolerance, and production through physiological and biochemical mechanisms (Mrid et al. 2021; Rady et al. 2025). Though they can be generally separated into microbial and nonmicrobial (organic and inorganic) forms, each of these biostimulants has a different way of action to lower temperature stress. Their effectiveness in lowering heat stress can be mostly attributed to their capacity to enhance water and nutrient absorption, control stress-responsive genes, strengthen antioxidant defenses, and alter plant metabolism (Ansari et al. 2023).

4.2.1 Microbial Biostimulants

Microbial biostimulants including endophytic bacteria, mycorrhizal fungi, and PGPR increase tolerance to stress in plants by means of several mechanisms as shown in Fig. 4.1 (Ansari et al. 2023). Particularly auxins, SA, and ABA, which are vital for plants to adapt with temperature, PGPR strains including *Bacillus*,

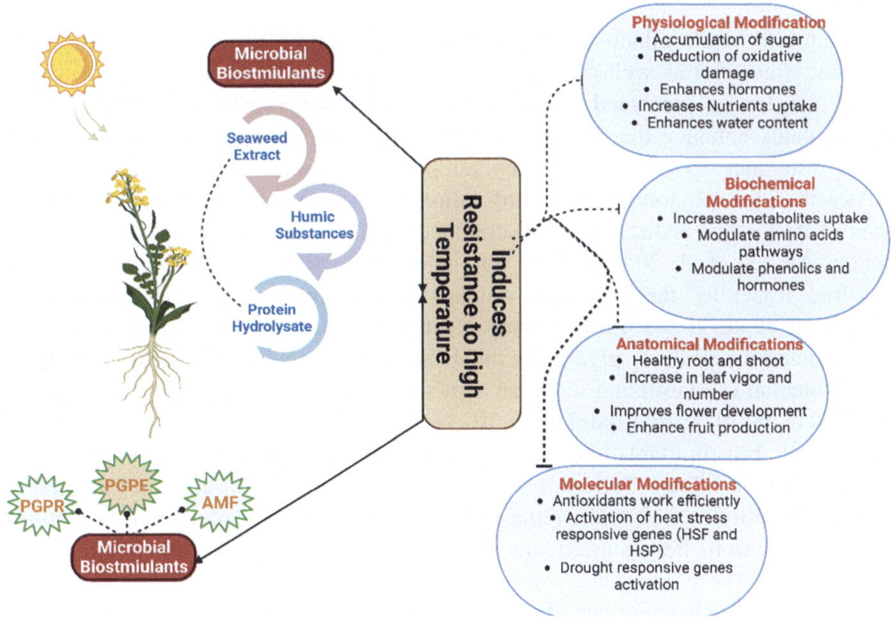

Fig. 4.1 Role of microbial and nonmicrobial biostimulants in enhancing plant thermotolerance through multifaceted modifications

Pseudomonas, and *Azospirillum* influence the synthesis of phytohormones. Furthermore, these bacteria enhance root structure, therefore enhancing the absorption of nutrients and water under either heat or cold stress (Backer et al. 2018; Soualiou et al. 2022; Ranjan et al. 2024).

Particularly arbuscular mycorrhizal fungi (AMF), which at first enhance osmotic equilibrium under stress, along with raising the absorption of nutrients, predominantly of phosphate and nitrogen, and root systems develop symbiotic relations with mycorrhizal fungi (Bano and Uzair 2021; Wahab et al. 2023). AMF also generates systemic tolerance by means of activation of stress-related genes and antioxidant enzymes. Studies on plants colonized by AMF demonstrate improved photosynthetic efficiency and membrane stability under heat stress, therefore mitigating the deleterious effects of high temperatures on metabolic activities (Bhupenchandra et al. 2024). Produced by endophytic bacteria, exopolysaccharides (EPS) help to retain soil moisture and reduce desiccation stress, hence enhancing thermotolerance. Moreover, by encouraging the synthesis of cold-responsive transcription factors and HSPs, microbial biostimulants defend cellular structures from temperature-induced damage (Askari-Khorasgani et al. 2019).

4.2.2 Organic and Inorganic Biostimulants

Nonmicrobial biostimulants include inorganic elements like silicon (Si), selenium (Se), and zinc (Zn) as well as organic compounds including protein hydrolysates, humic substances, seaweed extracts, and amino acids (Sharma 2022). These biostimulants enhance the defenses of cells and shifts chemical reactions to raise plants' tolerance to stress caused by temperatures (Jmaili et al. 2025). Rich in polysaccharides, phytohormones, and antioxidants, seaweed extracts under conditions of heat stress reduce stomatal conductance and improve photosynthetic efficiency (Ahmed et al. 2024a). They also boost the action of antioxidant enzymes, therefore lessening the oxidative damage caused by extremes in temperature (Hariharan et al. 2024). Protein hydrolysates and biostimulants based on amino acids, including proline, glycine betaine, and glutamate, under stress, acting as osmoprotectants, stabilizing cellular structures and sustaining enzyme activity (Elsayed et al. 2023; Er-Rqaibi et al. 2025).

Inorganic biostimulants like silicon and selenium are necessary to enhance plant defense systems (Trejo-Téllez et al. 2023). Enhancing structural integrity helps silicon deposition in cell walls reduce heat-induced cellular damage. As cofactors, zinc and selenium help antioxidant enzymes to scavenge ROS and inhibit lipid peroxidation. These inorganic compounds improve plant thermotolerance much more since they also regulate the expression of genes reacting to stress (Sarkar et al. 2022).

4.2.3 *Mode of Action in Heat Stress Mitigation*

Biostimulants lower heat stress using a combination of physiological, pharmacological, and molecular mechanisms (Bhupenchandra et al. 2022). Among the key routes is the improvement of antioxidant defense systems. Stress caused by heat generates too many ROS, and that, by oxidation breaks down proteins, DNA, and other building blocks and membranes, as shown in Fig. 4.1. Biostimulants effectively reduce ROS accumulation and maintain cell homostasis by triggering both nonenzymatic (ascorbic acid, glutathione, and flavonoids) and enzyme-mediated (SOD, CAT, and POD) antioxidant substances (Awasthi et al. 2015; Bhupenchandra et al. 2022, 2024). One essential method of action is osmotic regulation. Heat stress-induced excessive evaporation promotes metabolic changes and dehydration in cells (Medina et al. 2021; Niu et al. 2022). Among the biostimulants that boost osmolyte accumulation and hence enhance water retention and macromolecule stability are seaweed extracts, microbial exopolysaccharides, and amino acids. Moreover, humic and silicon compounds regulate ion homeostasis, hence reducing heat-induced electrolyte loss (Backer et al. 2018). Hormonal signaling networks under control by biostimulants also control plant responses to heat stress. Control of cytokinin, ABA, SA, and jasmonic acid (JA) enhances stress-responsive gene activation, improves stomatal control, and stimulates root development (Nephali et al. 2020). Particularly affecting the hormonal interaction needed for thermotolerance are seaweed extracts and microbial inoculants (Sangiorgio et al. 2020).

4.3 Biostimulant-Induced Physiological and Biochemical Responses

Biostimulants reduce heat stress by means of a mix of physiological, pharmacological, and molecular processes (Johnson et al. 2024). Protective mechanisms, including osmotic correction, membrane stability, HSP synthesis, osmotic correction, and improved photosynthetic efficiency are set off when utilized. These processes taken together reduce the detrimental consequences of heat stress and ensure continuous development and metabolic stability against very high temperatures (Lee et al. 2024).

4.3.1 *Proteins Under Heat Stress and the Activation of Antioxidant Enzymes*

Heat stress affects plant metabolism mostly by causing cellular dysfunction, protein denaturation, and aggregation (Rai et al. 2021). Protein folding, stability, and refolding depend on HSPs, particularly HSP70, HSP90, and tiny HSPs, under high temperatures (Timperio et al. 2008). When biostimulants including seaweed extracts, protein hydrolysates, and microbial inoculates raise the expression of genes associated with heat stress, plants are better equipped to survive it (Vaseva et al. 2022). Plants treated with biostimulants have shown increasing accumulation of HSPs, which protect proteins from heat breakdown and sustain enzymatic activity under stress (Campobenedetto et al. 2020a, b). Apart from HSPs, biostimulants also stimulate antioxidant defense systems to lower oxidative stress resulting from ROS accumulation (Hasanuzzaman et al. 2021). Among other biological components, heat stress generates too many ROS including superoxide radicals (O_2^-) and hydrogen peroxide (H_2O_2), which damage lipids, proteins, and nucleic acids (Desoky et al. 2019). Biostimulants—particularly those containing humic chemicals, antioxidants derived from seaweed, and microbial bioeffectors—activate important antioxidant enzymes like SOD, CAT, and POD. By means of their capacity to scavenge ROS, these enzymes reduce oxidative damage and preserve cell integrity. Reduced lipid peroxidation and enhanced thermotolerance have been associated with the higher antioxidant enzyme activity of plants treated with biostimulant (Campobenedetto et al. 2020a).

4.3.2 Osmoprotectants and Membrane Stability

In plants, too much water loss under heat stress results in cellular dehydration and unstable membranes (Awasthi et al. 2015). Examples of osmoprotectants that assist in osmotic adjustment and macromolecule stabilization are proline, glycine betaine, and polyamines; biostimulants help offset these implications by motivating their buildup (Bulgari et al. 2019). Promoting the production of these osmoprotectants, amino acid-based biostimulants, and microbial inoculants such as PGPR and AMF helps cells maintain turgor to be better and prevents heat-induced wilting (Bano and Uzair 2021). Membrane stability is an essential component in deciding a plant's capacity to resist heat stress (Niu and Xiang 2018). Heat stress reduces the integrity of the lipid bilayer, hence membrane permeability and electrolyte leakage rise (Ristic et al. 2007). Biostimulants comprising Si, Se, and humic substances increase membrane structure by changing the lipid composition and increasing the quantities of stabilizing proteins (Asthir 2015). Studies reveal that plants treated with biostimulants have higher membrane stability indices (MSI) and less electrolyte leakage when subjected to heat stress, therefore enhancing cellular function and stress tolerance (Lakhdar et al. 2023). Furthermore, EPS produced by some bacteria enhances water retention and reduces desiccation stress, therefore improving root–soil interactions and indirectly supporting membrane integrity (Matthews et al. 2025).

4.3.3 Photosynthetic Efficiency and Growth Regulation

Heat stress has a major impact on photosynthesis because of its negative consequences on chlorophyll biosynthesis, reduction in CO_2 fixation, and disturbance of electron transport chains (Mathur et al. 2014). Biostimulants improve photosynthetic performance by raising chlorophyll content, protecting photosystem II (PSII), and preserving Rubisco activity at high-temperature efficiency (Francesca et al. 2020). Rich in phytohormones and antioxidants, seaweed extracts modulate stomatal conductance and improve the stability of the light-harvesting complex, hence increasing the photosynthetic activity under heat stress (Muhammad et al. 2021). Furthermore changing the hormonal balance to influence plant development are biostimulants. Essential for cell division and elongation, cytokinins and auxins are regulated by biostimulants, which also sustain growth even under temperature stress (Nephali et al. 2020). Protein hydrolysates and microbial inoculants boost cytokinin production, hence fostering root and shoot development. Moreover, brassinosteroids—often included in biostimulant formulations—have been shown to boost photosynthetic efficiency by enhancing carbon absorption under heat stress and stabilizing chloroplast membranes (Johnson et al. 2024).

4.4 Enhancement of Yield Under Temperature Stress Caused by Biostimulants

Temperature stress—especially heat stress—significantly reduces crop output because it influences key physiological processes like root development, pollen viability, and grain filling (Farooq et al. 2017). To raise plant tolerance to temperature extremes, biostimulants boost water retention, strengthen reproductive systems, and alter crop-specific stress reactions (Shrestha et al. 2022). Their use promotes sustainable crop production by lowering the detrimental effects of high temperatures on plant development and output (Ristic et al. 2007).

4.4.1 Root Development and Water Retention

Root architecture is crucial for plants' response to temperature stress since it facilitates their absorption of water and nutrients (El Amrani 2023). Heat stress often reduces root biomass and hence influences water absorption, aggravating plant dryness (Wahid et al. 2007). By encouraging root extension, lateral root development, and root hair density, biostimulants include microbial inoculates (*Azospirillum*, *Pseudomonas*, *Bacillus*, and AMF) that raise the plant's capacity to absorb soil moisture at high temperatures (Grover et al. 2021; Mikiciuk et al. 2024). EPS produced by bacteria improve water and soil structure, therefore reducing the effects of heat-induced drought stress by raising the water retention capacity of the soil (Zainab et al. 2020). Moreover, humic and silicon ingredient in biostimulants help to promote ion homeostasis and root membrane stability, so lowering too high transpiration losses. The benefits of biostimulant-treated plants' rose root development are improved drought resistance and higher yield stability under temperature stress conditions (Khan et al. 2024, 2025b).

4.4.2 Prevention of Pollen Viability and Seed Production

Heat stress shows how sensitive reproductive development is to temperature variations by producing pollen sterility, lower fertilization rates, and reduced seed set (Arshad et al. 2017). Biostimulants reduce these effects by maintaining floral structures, improving pollen viability, and controlling hormone balance over the reproductive phases (Gupta et al. 2024). High in cytokinins and polyamines, seaweed extracts and protein hydrolysates help to prevent heat-induced pollen abortion by enhancing sugar metabolism and ROS scavenging in growing anthers (Zulfiqar et al. 2024). Moreover, silicon-based biostimulants enhance pollen tube growth and stigma sensitivity, therefore ensuring efficient fertilization. Microbial biostimulants, particularly PGPR strains, which help to synthesize protective osmolytes such as proline and glycine betaine, shield pollen grains from oxidative stress and desiccation (Nephali et al. 2020; Niu et al. 2022; Sharma 2022). Moreover, biostimulants guarantee suitable seed generation and reduce the possibility of pollen sterility by regulating heat-responsive genes engaged in reproductive development. The preservation of reproductive structures by biostimulants is a significant component in yield stability during temperature stress (Sharma 2022).

4.4.3 Effects of Biostimulants on Particular Crops

Different crop species have different metabolic pathways, growth patterns, and stress tolerance systems that influence their thermotolerance and the stress-reducing effect of biostimulants (Hasanuzzaman et al. 2021). At high temperatures, for instance, biostimulants boost antioxidant activity and photosynthetic efficiency, therefore enhancing grain filling and spikelet fertility in cereal crops such as rice and wheat (Albahri et al. 2023). Conversely, in oilseed crops like soybean and canola, biostimulants mostly increase lipid synthesis and membrane stability, therefore ensuring the optimal level of seed oil even in the presence of heat stress (Shah 2020; Nazari et al. 2022). Legumes, which are naturally linked with nitrogen-fixing bacteria, react powerfully to microbial biostimulants. Legume inoculations containing PGPR improve nitrogen absorption, chlorophyll stability, and reproductive success under high temperatures (Khan et al. 2021). Seaweed extracts and silicon treatments have been shown to boost fruit set and reduce flower drop in fruiting crops like tomatoes and peppers by changing hormonal balance and osmotic adjustment systems (Farouk 2023). Including biostimulants into crop management strategies is generally a good way to keep yield stability under temperature pressure. More research is required to maximize the efficacy of crop-specific biostimulant formulations and application timing and guarantee resistance in many different agricultural environments (Trejo-Téllez et al. 2023).

4.5 Progress in the Use of Biostimulants to Address Heat Stress in Oilseed Crops

Recent advancements in biostimulant applications aim to maximize formulations, integrate with modern agricultural technologies, and apply focused treatment strategies to raise oilseed crops' tolerance to heat stress (Bhupenchandra et al. 2022; Ancín et al. 2024). Research on how to preserve the yield of oilseed crops including soybean, canola, sunflower, and sesame is vital given their great economic value as global temperatures rise. Developments in biostimulant formulations, their synergistic uses with fertilizers and stress-resilient breeding, and the inclusion of biotechnological tools in precision agriculture are helping to build the next generation of stress-mitigating technologies for oilseed crops (Timperio et al. 2008).

4.5.1 Formulation and Application Strategies

The way biostimulants reduce heat stress depends much on their formulation, method of action, and timing of application (Bulgari et al. 2019; Campobenedetto et al. 2020a). Slow-release formulations created by developments in formulation technology provide liquid concentrates, extended stress protection, and microencapsulated biostimulants. For example, microbial biostimulants—such as strains of *Bacillus*, *Pseudomonas*, and *Azospirillum*—are being introduced to seed coatings more often to ensure early root colonization and stress adaptability during germination (Mandal et al. 2023). For oilseed crops, foliar treatments of humic chemicals, silicon-based biostimulants, and seaweed extracts have been refined for rapid stress release (Sun et al. 2024). Studies have shown that using biostimulants during pivotal growth phases, such as blooming and seed filling, considerably improve pollen viability, pod development, and oil quality under heat stress (Voko 2020; Memela 2022; Gupta et al. 2024). Moreover, the exact distribution of bioactive compounds made feasible by nanotechnology is enhancing the bioavailability of stress-relieving compounds in oilseed crops.

4.5.2 Synergistic Fertilizer–Stress-Resilient Breeding

Using biostimulants with nutritional management strategies increases the general tolerance of oilseed crops to heat stress (Ahmad et al. 2021). Under stress, biostimulants maximize plant metabolic processes by working in tandem with macronutrient fertilizers (such as potassium, phosphorus, and nitrogen) or micronutrient supplements (such as silicon, boron, and zinc) (Singhal et al. 2023; Khan et al. 2024, 2025b). Using silicon and amino acid-based biostimulants alongside nitrogen fertilizers improves osmotic balance, cell wall stability, and antioxidant activity, hence reducing heat-induced oxidative damage (Khan et al. 2023, 2025b). Including biostimulants into stress-resilient breeding programs has also attracted interest as a long-term approach to improve the heat tolerance of oilseed crops (Malik et al. 2020). Biostimulants can be used to induce epigenetic modifications, activate genes sensitive to stress, and boost the synthesis of secondary metabolites, therefore providing possible routes toward producing heat-tolerant crop varieties. Combining biostimulants with genetically engineered soybean cultivars with superior root architecture, efficient water-use strategies, and enhanced reproductive tolerance to heat stress would help to maximize output potential (Sharma 2022).

4.5.3 Integration of Biotechnology and Smart Agriculture

Combining contemporary biotechnology techniques with biostimulants and precision agricultural technologies is transforming heat stress control in oilseed crops (Oyebamiji et al. 2024). Utilizing transcriptomics, metabolomics, and high-throughput phenotyping, researchers can identify stress-responsive channels targeted by biostimulant treatments (Akbar et al. 2024). Researchers are also looking at how biostimulants alter heat-shock proteins, enzymes that scavenge ROS, and osmolyte generation utilizing synthetic biology methods including CRISPR-based gene editing (Arshad et al. 2017).

Smart agriculture technologies including automated fertigation systems, AI-driven predictive models, and remote sensing based on current environmental stress levels are optimizing the usage of biostimulants (Ahmed et al. 2024b). Sensor-based monitoring systems have enabled the exact application of biostimulants by spotting early indicators of heat stress in oilseed crops, therefore reducing resource waste and enhancing stress resilience (Trejo-Téllez et al. 2023; Zulfiqar et al. 2024).

4.5.4 The Role of Biostimulants in Increasing Oilseed Crops' Resistance to Heat Stress

Abiotic stresses, particularly heat and drought, severely impair the physiological homeostasis and yield of oilseed crops. These stresses change hormone balance, interfere with photosynthetic efficiency, and disturb cellular metabolism, therefore lowering production. By enhancing stress-responsive pathways, maximizing physiological and biochemical processes, and improving the efficacy of food absorption, applying biostimulants has become a reasonable approach to minimize these negative consequences. Several kinds of biostimulants, including silicon-based chemicals, microbial inoculants, protein hydrolysates, humic substances, and seaweed extracts, have shown great efficacy in providing oilseed crops thermotolerance using a range of processes (Table 4.1).

Table 4.1 Different biostimulant and their effect on oilseed crops under heat stress

Oilseed crop	Biostimulant type	Specific biostimulant	Effect on plant under temperature stress	Reference
Brassica napus	Orthosilicic acid and seaweed extract	Orthosilicic acid, seaweed extract	Enhanced plant development and yield while mitigating the adverse effects of high temperatures and water stress	Goyal et al. (2023)
Brassica napus	Microbial biostimulant	ProbioHumus	Augmented drought resilience, optimized biochemical mechanisms inside plant tissues	Gavelienė et al. (2025)
General oilseed crops	Various biostimulants	Not specified	Enhanced resistance of plants to high temperatures through a variety of processes, such as the modification of metabolic pathways and gene expression	Bulgari et al. (2019)
Glycine max	Seaweed extract	*Kappaphycus alvarezii*	Increased resilience to stress and increased yield with less fertilizer input	Trivedi et al. (2023)
Helianthus annuus	Humic and fulvic acids	Humic acid, fulvic acid	Enhanced output and better absorption of nutrients in stressful situations	Akladious and Mohamed (2018)
Arachis hypogaea	Protein hydrolysates	Amino acid-based biostimulants	Enhanced capacity for drought, yield, and growth	Sun et al. (2024)
Gossypium hirsutum	Silicon-based biostimulants	Potassium silicate	Increased capacity to grow and adapt to abiotic stressors	Tripathi et al. (2016) and Nazim et al. (2024)
Sesamum indicum	Plant growth-promoting rhizobacteria (PGPR)	*Bacillus subtilis*	Increased resilience to stress, growth, and yield in drought-prone environments	Lima et al. (2023)
Linum usitatissimum	Chitosan-based biostimulants	Chitosan	Enhanced capacity for stress tolerance, growth, and yield in salty environments	Ahmad et al. (2019)
Gossypium hirsutum	Seaweed extracts	*Ascophyllum nodosum*	Improved growth, yield, and resilience to drought stress	Shukla et al. (2019)

4.5.4.1 Organic and Silicon-Based Biostimulants for Reducing Temperature Stress

Because they enhance food solubility, increase root surface area, and alter hormone signaling, organic biostimulants like fulvic and humic acids have also been linked to stress reduction. In *Helianthus annuus*, treatments with fulvic acid and humic acid improved biomass accumulation and nutrient uptake efficiency in adverse environmental circumstances (Akladious and Mohamed 2018). Similarly, it has been demonstrated that protein hydrolysates, especially those based on amino acids, control nitrogen uptake, strengthen enzymatic defenses, and increase drought tolerance in *Arachis hypogaea*, all of which support long-term growth and yield (Sun et al. 2024).

Organic biostimulants such as fulvic and humic acids have also been connected to stress lowering since they improve food solubility, increase root surface area, and change hormone signaling. Treatments combining fulvic acid and humic acid in *Helianthus annuus* enhanced biomass accumulation and nutrient absorption efficiency in unfavorable environmental conditions (Akladious and Mohamed 2018). All of which boost long-term development and yield, protein hydrolysates—especially those based on amino acids—have been shown to limit nitrogen intake, strengthen enzymatic defenses, and raise drought tolerance in *Arachis hypogaea* (Table 4.1).

4.5.4.2 Seaweed and Microbial Biostimulants: Molecular and Physiological Interactions

Microbial biostimulants—especially plant growth-promoting rhizobacteria, or PGPR—change how plants respond to heat and drought stress by stimulating phytohormone control, increasing antioxidant enzyme activity, and improving water absorption. Improved physiological performance, higher biomass accumulation, and higher seed yield under water-deficit conditions show *Bacillus subtilis* inoculation in *Sesamum indicum* significantly increases drought resistance (Lima et al. 2023). Likewise, the use of microbial biostimulants. Humus in *Brassica napus* resulted in enhanced drought resistance, most likely as a result of its impact on osmolyte accumulation and antioxidant enzyme systems, which maximize intracellular homeostasis in stressful circumstances (Gavelienė et al. 2025).

Because of their richness of bioactive compounds, including phytohormones, polysaccharides, and polyphenols, biostimulants produced from seaweed are well known for their potential to raise resistance to abiotic stress. Suggesting a prospective utility in nutrient-use efficiency, *Kappaphycus alvarezii* extract boosted yield and stress tolerance when applied to Glycine max, hence lowering fertilizer reliance. By perhaps enhancing osmotic balance and upregulating genes that react to stress, using extract from *Ascophyllum nodosum* to *Gossypium hirsutum* also boosted drought stress tolerance (Shukla et al. 2019).

4.5.4.3 Modification of Gene Expression and Metabolic Pathways Induced by Biostimulants

Apart from their physiological advantages, biostimulants have been demonstrated to affect metabolic pathways linked to thermotolerance and gene expression. A meta-analysis of biostimulant applications in oilseed crops (Bulgari et al. 2019) revealed that their capacity to change transcriptional networks linked with HSPs, antioxidant defense systems, and osmolyte biosynthesis pathways resulted in enhanced adaptation to high temperatures. The emphasis of the metabolic reprogramming shown to boost stress tolerance on the necessity of more research on the molecular mechanisms behind the biostimulant efficacy.

4.5.5 Stress-Induced Physiological and Biochemical Changes in Oilseed Crops by Biostimulants

Applying biostimulants to oilseed crops is a realistic technique for boosting plant tolerance to abiotic stresses, notably heat and drought, and is becoming increasingly widely regarded. Biostimulants improve food intake, stress tolerance, and yield stability by means of many physiological and biochemical routes. Table 4.2 lists various types of biostimulants with considerable potential in changing metabolic paths and improving antioxidant defense mechanisms: humic materials, seaweed extracts, protein hydrolysates, silicon-based compounds, chitosan, and plant growth-promoting rhizobacteria.

4.5.5.1 Biostimulants Based on Silicon and Humic Substances in Stress Tolerance and Nutrient Uptake

Humic compounds (HS) greatly increase nutrient uptake by enhancing root architecture and nutrient transporter genes. Under stress, humic substance treatment in *Brassica napus* was found to maximize nutrient absorption by raising root nitrate absorption, which was associated to the upregulation of nitrate and sulfur transporter genes (Jannin et al. 2012). Silicon-based biostimulants have been studied closely for their ability to raise abiotic stress tolerance since they enhance cellular integrity and lower oxidative damage (Table 4.2). Most likely as a result of a significant decrease in lipid peroxidation and an increase in antioxidant enzyme activity, silicon application in *Brassica juncea* resulted in enhanced plant growth and heat tolerance, thus lowering cellular damage caused by oxidative stress (Parveen et al. 2022).

Table 4.2 Physiological and biochemical impact of organic and inorganic biostimulants on oilseed crops

Oilseed crop	Biostimulant type	Physiological effect	Biochemical effect	Reference
Brassica napus	Humic substances (HS)	Higher root nitrate and nutrition absorption	Nitrate/sulfur transporter gene upregulation	Jannin et al. (2012)
Glycine max	Seaweed extracts	Improved drought-related photosynthesis and growth	Increased antioxidant enzymes and osmolytes	Sharma et al. (2014)
Helianthus annuus	Seaweed extracts	Enhanced growth and yield under stressful	An increase in antioxidant enzymes and proline	Gomaa and Aldaby (2023)
Arachis hypogaea	Protein hydrolysates	Increase in growth and yield	Hype in antioxidants accumulation	Colla et al. (2015)
Brassica juncea	Silicon-based biostimulants	Enhanced growth and tolerance to heat	Decrease in lipid peroxidation and hype in antioxidants	Parveen et al. (2022)
Sesamum indicum	Plant growth-promoting rhizobacteria (PGPR)	Enhanced growth and yield	Improved antioxidants and proline production	Kafi et al. (2022)
Linum usitatissimum	Chitosan-based biostimulants	Enhanced growth and yield	Improved osmolytes production and antioxidants	Ahmad et al. (2019)
Gossypium hirsutum	Seaweed extracts	Enhanced growth and yield	Improved antioxidants and proline production	Zhang and Ervin (2004)

4.5.5.2 Enhancing Photosynthesis and Antioxidant Defense with Seaweed Extracts and Protein Hydrolysates

The bioactive compounds present in seaweed extracts increase stress tolerance by raising photosynthetic efficiency, upregulating antioxidant defense mechanisms, and changing osmolyte accumulation. Under drought conditions, seaweed extracts were reported to stimulate photosynthetic activity in *Glycine max* by accumulating antioxidant enzymes and osmolytes. This enhanced cellular homeostasis and water retention, which in turn improved water retention and water retention and Similarly, *Helianthus annuus* treated with seaweed extracts demonstrated enhanced growth and yield under stress as well as considerable increases in proline accumulation and antioxidant activity, therefore indicating a vital role in ROS detoxification and osmotic adjustment (Gomaa and Aldaby 2023). *Gossypium hirsutum* also responded well to seaweed extract treatments, indicating higher growth and yield coupled with increased proline production and antioxidant activity. This supports its drought adaptation mechanisms even more (Table 4.2).

Made primarily of amino acids and peptides, protein hydrolysates have been found to enhance plant growth and production by altering nitrogen metabolism and stress-responsive pathways. The considerable rise in *Arachis hypogaea* resulting from protein hydrolysate treatments served to further establish its ability to lower oxidative stress and improve metabolic efficiency in demanding conditions. Rising antioxidant buildup connected these changes (Colla et al. 2015).

4.5.5.3 Biostimulants Based on Microbes and Chitosan in Osmotic Regulation and Growth Promotion

Because they boost biochemical responses and drive plant growth, microbial inoculants—PGPR in particular—are indispensable for stress relief. In *Sesamum indicum*, *Bacillus subtilis* inoculation raised plant growth and seed yield as well as proline biosynthesis and antioxidant enzyme activity, all of which improve stress tolerance by scavenging reactive oxygen species and so regulating osmotic pressure (Kafi et al. 2022).

Chitosan-based biostimulants have attracted interest because they can induce stress tolerance by altering osmoprotectants and antioxidant defense systems. Applying chitosan to *Linum usitatissimum* produced better growth and yield as well as osmolytes and accumulations of antioxidants, implying a strong resistance against abiotic stress (Ahmad et al. 2019).

4.6 Challenges and Future Perspectives

Though their considerable potential to lower heat stress in oilseed crops, several factors prohibit biostimulants from being extensively employed and efficient. Among these challenges are variances in field performance, regulatory constraints, and ignorance of their molecular process under stress. Combining biostimulants with conventional agronomic techniques while preserving affordability and environmental sustainability presents yet another significant challenge. Policy support, technical advancements, and targeted research will help to overcome these limitations so that biostimulants may play a greater part in climate-resilient agriculture.

4.6.1 Limitations in Commercial Adoption

Variability in the environment, changes in soil microbiomes, and crop-specific sensitivities limit the commercial use of biostimulants for oilseed crop heat stress mitigation. Unlike synthetic fertilizers and pesticides, the efficacy of biostimulants can be influenced by microbial competition, application time, and climate, resulting in varying effects.

Absence of consistent formulations and regulatory norms further impedes commercialization. Before they are launched onto the market, there is little scientific evidence of the exact effects of many biostimulants on heat stress tolerance. Furthermore, since biostimulants are initially more costly and take more time to demonstrate effects than conventional inputs, farmers could be reluctant to employ them. Improving farmer acceptance will entail establishing long-term financial gains and increasing awareness via extension programs.

4.6.2 Research Gaps and Potential Areas of Improvement

Though biostimulants show promise in enhancing resistance to heat stress, much more study is needed in this area. Exactly how different biostimulants provide thermotolerance at the molecular and physiological levels for oilseed crops is yet unknown. For example, even while seaweed extracts and microbial inoculants are known to boost antioxidant defenses, more study is required to find how precisely they interact with plant hormone signaling systems under heat stress.

Another crucial area of research is optimizing biostimulant formulations to boost their stability and efficacy in demanding environments. With the development of multicomponent biostimulants, which combine microbial inoculants with bioactive compounds such as polyamines, osmoprotectants, and secondary metabolites, more consistent stress protection could be achievable. Moreover, the use of omics-based approaches—more especially, genomics, proteomics, and metabolomics—can help to identify new biomarkers for biostimulant-induced stress tolerance, thereby enabling the development of tailored formulations.

4.7 Conclusion

Using biostimulants on oilseed crops to lower temperature stress is a logical, environmentally friendly approach to boost plant resistance against always changing climate patterns. Important physiological and biochemical processes are significantly impacted by heat stress, which at last lowers yield and oil quality by diminishing photosynthetic efficiency, compromising membrane stability, so upsetting reproductive development. Working in a variety of ways, organic extracts, inorganic compounds, and microbial inoculants—all of which are biostimulants—increase the activities of antioxidant enzymes, activate heat-shock proteins, build up osmoprotectants, and so affect the expression of genes reacting to stress. These advantages taken together improve root structure, water retention capacity, and reproductive success, therefore ensuring continuous output even in hot climates. Although biostimulants have great potential, formulation problems, variable field performance, and lack of regulatory criteria are hindering their commercial usage. More study is needed to define the precise molecular pathways affected by various biostimulants, ideal formulations for various agroecosystems, and create synergistic approaches combining biostimulants with precision agriculture technology and stress-resilient breeding programs. Fresh techniques driven by CRISpen-driven genetic changes and omics-based profiling enable new technologies powered by them to improve the efficacy of biostimulants and adapt them to particular crop requirements. Maximizing the advantages of biostimulants for oilseed crops will call for a whole strategy incorporating smart agricultural technologies, environmentally friendly agronomic methods, and data-driven application techniques ahead. Policy assistance and farmer education will also inspire invention in this field by way of multidisciplinary interactions among plant physiologists, agronomists, and biotechnologists. Eventually, including biostimulants into a whole stress-reducing strategy will assist to create climate-resilient oilseed-producing systems, therefore ensuring food security and agricultural sustainability as global temperatures rise.

References

Ahmad W, Zahir A, Nadeem M, Garros L, Drouet S, Renouard S, Doussot J, Giglioli-Guivarc'h N, Hano C, Abbasi BH (2019) Enhanced production of lignans and neolignans in chitosan-treated flax (Linum usitatissimum L.) cell cultures. Process Biochem 79:155–165

Ahmad M, Waraich EA, Skalicky M, Hussain S, Zulfiqar U, Anjum MZ, Habib Ur Rahman M, Brestic M, Ratnasekera D, Lamilla-Tamayo L (2021) Adaptation strategies to improve the resistance of oilseed crops to heat stress under a changing climate: an overview. Front Plant Sci 12:767150

Ahmad Y, Haakim Z, Iqbal J, Abbasi BA, Mahmood T, Kazi M (2024) Technological innovations for abiotic stress resistance in horticultural crops. In: OMICs-based techniques for global food security. Wiley, Hoboken, pp 233–244

Ahmed M, Ullah H, Himanshu SK, García-Caparrós P, Tisarum R, Cha-Um S, Datta A (2024a) Ascophyllum nodosum seaweed extract and potassium alleviate drought damage in tomato by improving plant water relations, photosynthetic performance, and stomatal function. J Appl Phycol 36:2255–2268

Ahmed N, Zhang B, Deng L, Bozdar B, Li J, Chachar S, Chachar Z, Jahan I, Talpur A, Gishkori MS (2024b) Advancing horizons in vegetable cultivation: a journey from ageold practices to high-tech greenhouse cultivation—a review. Front Plant Sci 15:1357153

Akbar R, Sun J, Bo Y, Khattak WA, Khan AA, Jin C, Zeb U, Ullah N, Abbas A, Liu W (2024) Understanding the influence of secondary metabolites in plant invasion strategies: A comprehensive review. Plan Theory 13:3162

Akladious SA, Mohamed HI (2018) Ameliorative effects of calcium nitrate and humic acid on the growth, yield component and biochemical attribute of pepper (Capsicum annuum) plants grown under salt stress. Sci Hortic 236:244–250

Albahri G, Alyamani AA, Badran A, Hijazi A, Nasser M, Maresca M, Baydoun E (2023) Enhancing essential grains yield for sustainable food security and bio-safe agriculture through latest innovative approaches. Agronomy 13:1709

Alharbi K, Khan AA, Alhaithloul HAS, Al-Harbi NA, Al-Qahtani SM, Aloufi SS, Abdulmajeed AM, Muneer MA, Alghanem SM, Zia-Ur-Rehman M (2023) Synergistic effect of β-sitosterol and biochar application for improving plant growth of Thymus vulgaris under heat stress. Chemosphere 340:139832

Ancín M, Soba D, Picazo PJ, Gámez AL, Le Page JF, Houdusse D, Aranjuelo I (2024) Optimizing oilseed rape growth: exploring the effect of foliar biostimulants on the interplay among metabolism, phenology, and yield. Physiol Plant 176:e14561

Ansari M, Devi BM, Sarkar A, Chattopadhyay A, Satnami L, Balu P, Choudhary M, Shahid MA, Jailani AAK (2023) Microbial exudates as biostimulants: role in plant growth promotion and stress mitigation. J Xenobiot 13:572–603

Arshad MS, Farooq M, Asch F, Krishna JS, Prasad PV, Siddique KH (2017) Thermal stress impacts reproductive development and grain yield in rice. Plant Physiol Biochem 115:57–72

Asif A, Ali M, Qadir M, Karthikeyan R, Singh Z, Khangura R, Di Gioia F, Ahmed ZF (2023) Enhancing crop resilience by harnessing the synergistic effects of biostimulants against abiotic stress. Front Plant Sci 14:1276117

Askari-Khorasgani O, Hatterman-Valenti H, Flores Pardo FB, Pessarakli M (2019) Plant and symbiont metabolic regulation and biostimulants application improve symbiotic performance and cold acclimation. J Plant Nutr 42:2151–2163

Asthir B (2015) Protective mechanisms of heat tolerance in crop plants. J Plant Interact 10:202–210

Awasthi R, Bhandari K, Nayyar H (2015) Temperature stress and redox homeostasis in agricultural crops. Front Environ Sci 3:11

Backer R, Rokem JS, Ilangumaran G, Lamont J, Praslickova D, Ricci E, Subramanian S, Smith DL (2018) Plant growth-promoting rhizobacteria: context, mechanisms of action, and roadmap to commercialization of biostimulants for sustainable agriculture. Front Plant Sci 9:1473

Balbino S (2017) Vegetable oil yield and composition influenced by environmental stress factors. In: Oilseed crops: yield and adaptations under environmental stress. Wiley-Blackwell, Chichester, pp 80–101

Bano SA, Uzair B (2021) Arbuscular mycorrhizal fungi (AMF) for improved plant health and production. In: Microbial biotechnology in crop protection. Springer, pp 147–169

Bhupenchandra I, Chongtham SK, Devi EL, Choudhary AK, Salam MD, Sahoo MR, Bhutia TL, Devi SH, Thounaojam AS, Behera C (2022) Role of biostimulants in mitigating the effects of climate change on crop performance. Front Plant Sci 13:967665

Bhupenchandra I, Chongtham SK, Devi AG, Dutta P, Sahoo MR, Mohanty S, Kumar S, Choudhary AK, Devi EL, Sinyorita S (2024) Unlocking the potential of arbuscular mycorrhizal fungi: exploring role in plant growth promotion, nutrient uptake mechanisms, biotic stress alleviation, and sustaining agricultural production systems. J Plant Growth Regul:1–39

Bulgari R, Franzoni G, Ferrante A (2019) Biostimulants application in horticultural crops under abiotic stress conditions. Agronomy 9:306

Campobenedetto C, Grange E, Mannino G, Van Arkel J, Beekwilder J, Karlova R, Garabello C, Contartese V, Bertea CM (2020a) A biostimulant seed treatment improved heat stress tolerance during cucumber seed germination by acting on the antioxidant system and glyoxylate cycle. Front Plant Sci 11:836

Campobenedetto C, Mannino G, Agliassa C, Acquadro A, Contartese V, Garabello C, Bertea CM (2020b) Transcriptome analyses and antioxidant activity profiling reveal the role of a lignin-derived biostimulant seed treatment in enhancing heat stress tolerance in soybean. Plan Theory 9:1308

Clodoveo ML, Hbaieb RH, Kotti F, Mugnozza GS, Gargouri M (2014) Mechanical strategies to increase nutritional and sensory quality of virgin olive oil by modulating the endogenous enzyme activities. Compr Rev Food Sci Food Saf 13:135–154

Colla G, Nardi S, Cardarelli M, Ertani A, Lucini L, Canaguier R, Rouphael Y (2015) Protein hydrolysates as biostimulants in horticulture. Sci Hortic 196:28–38

Desoky E-SM, Elsayed AI, Merwad A-RM, Rady MM (2019) Stimulating antioxidant defenses, antioxidant gene expression, and salt tolerance in Pisum sativum seedling by pretreatment using licorice root extract (LRE) as an organic biostimulant. Plant Physiol Biochem 142:292–302

Dias MC, Araújo M, Silva S, Santos C (2022) Sustainable olive culture under climate change: the potential of biostimulants. Horticulturae 8:1048

El Amrani B (2023) Exploring the importance of root architecture plasticity in plant adaptation to environmental constraints. Plant Species Biol 38:234–244
Elsayed SI, Sabra AS, Omer EA (2023) Role of plant extracts and biostimulant in mitigating plant drought and salinity stress. In: Climate-resilient agriculture, vol 2: agro-biotechnological advancement for crop production. Springer, pp 571–602
Er-Rqaibi S, Lyamlouli K, El Yacoubi H, El Boukhari MEM (2025) Effect of crude extract and polysaccharides derived from Fucus spiralis on radish plants Raphanus sativus L. agrophysiological traits under drought stress. BMC Plant Biol 25:46
Farooq M, Nadeem F, Gogoi N, Ullah A, Alghamdi SS, Nayyar H, Siddique KH (2017) Heat stress in grain legumes during reproductive and grain-filling phases. Crop Pasture Sci 68:985–1005
Farouk S (2023) Role of biostimulants in plant's life cycle. In: Biostimulants in alleviation of metal toxicity in plants. Elsevier, pp 75–106
Francesca S, Arena C, Hay Mele B, Schettini C, Ambrosino P, Barone A, Rigano MM (2020) The use of a plant-based biostimulant improves plant performances and fruit quality in tomato plants grown at elevated temperatures. Agronomy 10:363
Franzoni G, Cocetta G, Prinsi B, Ferrante A, Espen L (2022) Biostimulants on crops: their impact under abiotic stress conditions. Horticulturae 8:189
Gavelienė V, Mockevičiūtė R, Jankovska-Bortkevič E, Šveikauskas V, Zareyan M, Žalnierius T, Jankauskienė J, Jurkonienė S (2025) Synergistic effects of microbial biostimulants and calcium in alleviating drought stress in oilseed rape. Microorganisms 13:530
Gomaa M, Aldaby ES (2023) Macroalgal-derived alginate/wastepaper hydrogel to alleviate sunflower drought stress. Planta 257(6):112
González-Morales S, Solís-Gaona S, Valdés-Caballero MV, Juárez-Maldonado A, Loredo-Treviño A, Benavides-Mendoza A (2021) Transcriptomics of biostimulation of plants under abiotic stress. Front Genet 12:583888
Goyal V, Kumari A, Avtar R, Baliyan V, Mehrotra S (2023) Orthosilicic acid and seaweed extract alleviate the deteriorative effects of high temperature stress in Brassica juncea (L.) Czern & Coss. SILICON 15:4909–4919
Graziani G, Cirillo A, Giannini P, Conti S, El-Nakhel C, Rouphael Y, Ritieni A, Di Vaio C (2022) Biostimulants improve plant growth and bioactive compounds of young olive trees under abiotic stress conditions. Agriculture 12:227
Grover M, Bodhankar S, Sharma A, Sharma P, Singh J, Nain L (2021) PGPR mediated alterations in root traits: way toward sustainable crop production. Front Sustain Food Syst 4:618230
Gupta S, Novák O, Kulkarni MG, Doležalova I, Van Staden J, Doležal K (2024) Unleashing the potential of biostimulants in stimulating pollen germination and tube growth. J Plant Growth Regul 43:1–32
Hariharan G, Vathshalyan N, Galahitigama H, Wimalasiri U, Kumara GDK (2024) Potential of foliar application of seaweed extracts as a biostimulant for abiotic stress alleviation on crop production. Rev Agric Sci 12:295–312
Hasanuzzaman M, Parvin K, Bardhan K, Nahar K, Anee TI, Masud AAC, Fotopoulos V (2021) Biostimulants for the regulation of reactive oxygen species metabolism in plants under abiotic stress. Cells 10:2537
Jahed KR, Saini AK, Sherif SM (2023) Coping with the cold: unveiling cryoprotectants, molecular signaling pathways, and strategies for cold stress resilience. Front Plant Sci 14:1246093
Jannin L, Arkoun M, Ourry A, Laîné P, Goux D, Garnica M, Etienne P (2012) Microarray analysis of humic acid effects on brassica napus growth: involvement of N, C and S metabolisms. Plant and Soil 359:297–319
Jmaili K, Asbai Z, Waddi K, Bahlaouan B, Silkina A, Boutaleb N (2025) Non-microbial biostimulants for plant growth and abiotic stress mitigation: a review of recent scientific innovations. Int J Environ Stud 82:1–31
Johnson R, Joel JM, Puthur JT (2024) Biostimulants: the futuristic sustainable approach for alleviating crop productivity and abiotic stress tolerance. J Plant Growth Regul 43:659–674
Kafi M, Nabati J, Rezazadeh EB, Oskoueian A, Soureshjani HK (2022) Single and poly capsule sesame (*Sesamum indicum L.*) productivity in response to plant growth-promoting rhizobacteria and foliar application of silicon, potassium, and calcium. Acta Physiol Plant 44(10):103

Khan N, Bano A, Ali S, Babar MA (2020) Crosstalk amongst phytohormones from planta and PGPR under biotic and abiotic stresses. Plant Growth Regul 90:189–203
Khan AA, Wang T, Hussain T, Amna, Ali F, Shi F, Latef AAHA, Ali OM, Hayat K, Mehmood S (2021) Halotolerant-Koccuria rhizophila (14asp)-induced amendment of salt stress in pea plants by limiting Na+ uptake and elevating production of antioxidants. Agronomy 11:1907
Khan AA, Alhaithloul HAS, Alghanem SM, Senousy HH, Alhoqail WA, Shi F, Ali I, Soliman MH (2023) Petrosolinum crispum under stressful conditions. In: Medicinal plant responses to stressful conditions. CRC Press, pp 299–310
Khan AA, Hayat K, Wang Y-F, Alhaithloul HAS, Alghanem SMS, Alhoqail WA, Soliman MH (2024) Biochar versus heavy metals under changing environmental conditions. In: Biochar-assisted remediation of contaminated soils under changing climate. Elsevier, pp 167–200
Khan AA, Wang Y-F, Akbar R, Alhoqail WA (2025a) Mechanistic insights and future perspectives of drought stress management in staple crops. Front Plant Sci 16:1547452
Khan AA, Wang Y-F, Zeb A, Hayat K, Alhoqail WA, Soliman MH (2025b) Beneficial elements and their roles against soil pollution. In: Beneficial elements for remediation of heavy metals in polluted soil. Elsevier, pp 1–32
Kourani M, Mohareb F, Rezwan FI, Anastasiadi M, Hammond JP (2022) Genetic and physiological responses to heat stress in Brassica napus. Front Plant Sci 13:832147
Kumari M, Swarupa P, Kesari KK, Kumar A (2022) Microbial inoculants as plant biostimulants: a review on risk status. Life 13:12
Lakhdar A, Trigui M, Montemurro F (2023) An overview of biostimulants' effects in saline soils. Agronomy 13:2092
Lee Z, Lim JA, Harikrishna JA, Islam T, Abd Rahim MH, Yaacob JS (2024) Regulation of plant responses to temperature stress: a key factor in food security and for mitigating effects of climate change. Int J Plant Prod 18:141–159
Lima GBPD, Gomes EF, Rocha GMGD, Silva FDA, Fernandes PD, Machado AP, Fernandes-Junior PI, Melo ASD, Arriel NHC, Gondim TMDS (2023) Bacilli rhizobacteria as biostimulants of growth and production of sesame cultivars under water deficit. Plan Theory 12:1337
Lozano-Castellón J, López-Yerena A, Domínguez-López I, Siscart-Serra A, Fraga N, Sámano S, López-Sabater C, Lamuela-Raventós RM, Vallverdú-Queralt A, Pérez M (2022) Extra virgin olive oil: a comprehensive review of efforts to ensure its authenticity, traceability, and safety. Compr Rev Food Sci Food Saf 21:2639–2664
Malik A, Mor VS, Tokas J, Punia H, Malik S, Malik K, Sangwan S, Tomar S, Singh P, Singh N (2020) Biostimulant-treated seedlings under sustainable agriculture: a global perspective facing climate change. Agronomy 11:14
Mandal S, Anand U, López-Bucio J, Kumar M, Lal MK, Tiwari RK, Dey A (2023) Biostimulants and environmental stress mitigation in crops: a novel and emerging approach for agricultural sustainability under climate change. Environ Res 233:116357
Mathur S, Agrawal D, Jajoo A (2014) Photosynthesis: response to high temperature stress. J Photochem Photobiol B Biol 137:116–126
Matthews S, Siddiqui Y, Ali A (2025) Unleashing the power of bio-stimulants for enhanced crop growth, productivity, and quality: a comprehensive review. J Plant Nutr 48:703–725
Medina E, Kim S-H, Yun M, Choi W-G (2021) Recapitulation of the function and role of ROS generated in response to heat stress in plants. Plan Theory 10:371
Meena H, Kiran BU, Bindu H (2025) Genetic enhancement of abiotic stress tolerance in oilseeds through contemporary breeding approaches. In: Breeding climate resilient and future ready oilseed crops. Springer, pp 43–99
Memela U (2022) The physiological effects of heat stress on anthesis and pollination in domesticated sunflowers (Helianthus annuus L.). University of Pretoria
Mikiciuk G, Miller T, Kisiel A, Cembrowska-Lech D, Mikiciuk M, Łobodzińska A, Bokszczanin K (2024) Harnessing beneficial microbes for drought tolerance: a review of ecological and agricultural innovations. Agriculture 14:2228
Mrid RB, Benmrid B, Hafsa J, Boukcim H, Sobeh M, Yasri A (2021) Secondary metabolites as biostimulant and bioprotectant agents: a review. Sci Total Environ 777:146204

Muhammad I, Shalmani A, Ali M, Yang Q-H, Ahmad H, Li FB (2021) Mechanisms regulating the dynamics of photosynthesis under abiotic stresses. Front Plant Sci 11:615942

Nazari M, Yaghoubian I, Smith DL (2022) The stimulatory effect of Thuricin 17, a PGPR-produced bacteriocin, on canola (Brassica, napus L.) germination and vegetative growth under stressful temperatures. Front Plant Sci 13:1079180

Nazim M, Li X, Tariq A, Shahzad K, Muhammad M, Dawei Z, Zeng F (2024) Exogenous potassium silicate improves drought tolerance in cotton genotypes by modulating growth, gas exchange and antioxidant metabolism. J Crop Health 76:883–901

Nephali L, Piater LA, Dubery IA, Patterson V, Huyser J, Burgess K, Tugizimana F (2020) Biostimulants for plant growth and mitigation of abiotic stresses: a metabolomics perspective. Meta 10:505

Niu Y, Xiang Y (2018) An overview of biomembrane functions in plant responses to high-temperature stress. Front Plant Sci 9:915

Niu C, Wang G, Sui J, Liu G, Ma F, Bao Z (2022) Biostimulants alleviate temperature stress in tomato seedlings. Sci Hortic 293:110712

Oyebamiji YO, Adigun BA, Shamsudin NAA, Ikmal AM, Salisu MA, Malike FA, Lateef AA (2024) Recent advancements in mitigating abiotic stresses in crops. Horticulturae 10:156

Parveen A, Mumtaz S, Saleem MH, Hussain I, Perveen S, Thind S (2022) Silicon and nanosilicon mediated heat stress tolerance in plants. In: Silicon and nano-silicon in environmental stress management and crop quality improvement. Academic Press, pp 153–159

Perveen A, Sheheryar S, Ahmad F, Mustafa G, Moura AA, Campos FA, Domont GB, Nishan U, Ullah R, Ibrahim MA (2025) Integrative physiological, biochemical, and proteomic analysis of the leaves of two cotton genotypes under heat stress. PLoS One 20:e0316630

Rady MM, Abdelfattah MA, Desoky EM, Azzam CR, Abdel Latef AA, Mahmoud AEM, Mohamed IAA (2025) A novel group of bio-stimulators: their attenuation of abiotic stress impacts in crop plants. J Plant Nutr:1–16. https://doi.org/10.1080/01904167.2025.2471895

Rai A, Rai GK, Dubey R (2021) Heat stress and its effects on plant growth and metabolism. In: Abiotic stress tolerance mechanisms in plants. CRC Press, pp 203–265

Ranjan A, Rajput VD, Prazdnova EV, Gurnani M, Sharma S, Bhardwaj P, Shende SS, Mandzhieva SS, Sushkova S, Minkina T (2024) Augmenting abiotic stress tolerance and root architecture: the function of phytohormone-producing PGPR and their interaction with nanoparticles. S Afr J Bot 167:612–629

Ristic Z, Bukovnik U, Prasad PV (2007) Correlation between heat stability of thylakoid membranes and loss of chlorophyll in winter wheat under heat stress. Crop Sci 47:2067–2073

Sangiorgio D, Cellini A, Donati I, Pastore C, Onofrietti C, Spinelli F (2020) Facing climate change: application of microbial biostimulants to mitigate stress in horticultural crops. Agronomy 10: 794

Sarkar MM, Mathur P, Roy S (2022) Silicon and nano-silicon: new frontiers of biostimulants for plant growth and stress amelioration. In: Silicon and nano-silicon in environmental stress management and crop quality improvement. Elsevier, pp 17–36

Shah A (2020) Efficacy of citrus bioflavonoid, and cell free supernatant from the novel bacterial strain (SL43) on seed germination and growth of Canola and soybean under salt stress and optimal growth conditions. McGill University

Sharma HS, Fleming C, Selby C, Rao JR, Martin T (2014) Plant biostimulants: a review on the processing of macroalgae and use of extracts for crop management to reduce abiotic and biotic stresses. J Appl Phycol 26:465–490

Sharma S (2022) Biostimulants: an eco-friendly approach in reducing heavy metal toxicity. NewBioWorld 4:1–7

Shrestha S, Mahat J, Shrestha J, Madhav K, Paudel K (2022) Influence of high-temperature stress on rice growth and development. A review. Heliyon:8

Shukla PS, Mantin EG, Adil M, Bajpai S, Critchley AT, Prithiviraj B (2019) Ascophyllum nodosum-based biostimulants: sustainable applications in agriculture for the stimulation of plant growth, stress tolerance, and disease management. Front Plant Sci 10:462648

Singhal RK, Fahad S, Kumar P, Choyal P, Javed T, Jinger D, Singh P, Saha D, Md P, Bose B (2023) Beneficial elements: new players in improving nutrient use efficiency and abiotic stress tolerance. Plant Growth Regul 100:237–265

Soualiou S, Duan F, Li X, Zhou W (2022) Crop production under cold stress: an understanding of plant responses, acclimation processes, and management strategies. Plant Physiol Biochem 190: 47–61

Sun W, Shahrajabian MH, Kuang Y, Wang N (2024) Amino acids biostimulants and protein hydrolysates in agricultural sciences. Plan Theory 13:210

Timperio AM, Egidi MG, Zolla L (2008) Proteomics applied on plant abiotic stresses: role of heat shock proteins (HSP). J Proteome 71:391–411

Trejo-Téllez LI, Gómez-Trejo LF, Gómez-Merino FC (2023) Biostimulant effects and concentration patterns of beneficial elements in plants. In: Beneficial chemical elements of plants: recent developments and future prospects. Wiley, Hoboken, pp 58–102

Tripathi DK, Singh S, Singh VP, Prasad SM, Chauhan DK, Dubey NK (2016) Silicon nanoparticles more efficiently alleviate arsenate toxicity than silicon in maize cultiver and hybrid differing in arsenate tolerance. Front Environ Sci 4:46

Trivedi K, Anand KV, Vaghela P, Critchley AT, Shukla PS, Ghosh A (2023) A review of the current status of Kappaphycus alvarezii-based biostimulants in sustainable agriculture. J Appl Phycol 35:3087–3111

Vaseva II, Simova-Stoilova L, Kostadinova A, Yuperlieva-Mateeva B, Karakicheva T, Vassileva V (2022) Heat-stress-mitigating effects of a protein-hydrolysate-based biostimulant are linked to changes in protease, DHN, and HSP gene expression in maize. Agronomy 12:1127

Voko MP (2020) Impacts of climate change on cowpea (Vigna unguiculata L. Walp) treated with biostimulants. Master's thesis, University of KwaZulu-Natal, South Africa, pp 1–154

Wahab A, Muhammad M, Munir A, Abdi G, Zaman W, Ayaz A, Khizar C, Reddy SPP (2023) Role of arbuscular mycorrhizal fungi in regulating growth, enhancing productivity, and potentially influencing ecosystems under abiotic and biotic stresses. Plan Theory 12:3102

Wahid A, Gelani S, Ashraf M, Foolad MR (2007) Heat tolerance in plants: an overview. Environ Exp Bot 61:199–223

Zahra N, Shaukat K, Hafeez MB, Raza A, Hussain S, Chaudhary MT, Akram MZ, Kakavand SN, Saddiq MS, Wahid A (2021) Physiological and molecular responses to high, chilling, and freezing temperature in plant growth and production: consequences and mitigation possibilities. In: Harsh environment and plant resilience: molecular and functional aspects. Springer, Cham, pp 235–290

Zainab N, Din BU, Javed MT, Afridi MS, Mukhtar T, Kamran MA, Khan AA, Ali J, Jatoi WN, Munis MFH (2020) Deciphering metal toxicity responses of flax (Linum usitatissimum L.) with exopolysaccharide and ACC-deaminase producing bacteria in industrially contaminated soils. Plant Physiol Biochem 152:90–99

Zhang Y, Dai T, Liu Y, Wang J, Wang Q, Zhu W (2022) Effect of exogenous glycine betaine on the germination of tomato seeds under cold stress. Int J Mol Sci 23:10474

Zhang X, Ervin EH (2004) Cytokinin-containing seaweed and humic acid extracts associated with creeping bentgrass leaf cytokinins and drought resistance. Crop Sci 44(5):1737–1745

Zulfiqar F, Moosa A, Ali HM, Bermejo NF, Munné-Bosch S (2024) Biostimulants: a sufficiently effective tool for sustainable agriculture in the era of climate change? Plant Physiol Biochem:108699

Chapter 5
Mitigating Nutrient Deficiencies in Oilseeds Through Biostimulant Applications

Amir Abdullah Khan, Hailah M. Almohaimeed, Rasheed Akbar, Babar Iqbal, Aurang Zeb, and Mona H. Soliman

Abstract Nutrient deficiency is a major limitation to the growth, development, and yield of oilseed crops, often resulting in reduced productivity and compromised crop quality. Biostimulants, including plant growth-promoting rhizobacteria (PGPR), humic substances, and microbial inoculants, offer an innovative approach to enhancing nutrient uptake and mitigating the effects of nutrient deficiencies in oilseed crops. This chapter examines the role of biostimulants in improving nutrient acquisition in oilseed crops, with a focus on their ability to mobilize essential nutrients such as nitrogen, phosphorus, and potassium. The underlying mechanisms, including root architecture modification, enhanced enzymatic activity, and improved soil microbial interactions, are discussed in detail. Additionally, the chapter explores how biostimulants modulate hormonal balance and stress signaling pathways to

Authors Amir Abdullah Khan and Hailah M. Almohaimeed have equally contributed to this chapter.

A. A. Khan (✉) · B. Iqbal
School of Environment and Safety Engineering, Jiangsu University, Zhenjiang, China

H. M. Almohaimeed (✉)
Department of Basic Science, College of Medicine, Princess Nourah bint Abdulrahman, University, Riyadh, Saudi Arabia
e-mail: hmalmohaimeed@pnu.edu.sa

R. Akbar
School of Environment and Safety Engineering, Jiangsu University, Zhenjiang, China

Department of Entomology, Faculty of Physical and Applied Sciences, The University of Haripur, Haripur, Khyber Pakhtunkhwa, Pakistan

A. Zeb
MOE Key Laboratory of Pollution Processes and Environmental Criteria, College of Environmental Science and Engineering, Nankai University, Tianjin, People's Republic of China

M. H. Soliman
Botany and Microbiology Department, Faculty of Science, Cairo University, Giza, Egypt

Biology Department, Faculty of Science, Taibah University, Yanbu, Kingdom of Saudi Arabia
e-mail: hmona@sci.cu.edu.eg

© The Author(s), under exclusive license to Springer Nature Singapore Pte Ltd. 2025

A. A. H. Abdel Latef (ed.), *Oilseed Crops Under Abiotic Stress*, Sustainability Sciences in Asia and Africa, https://doi.org/10.1007/978-981-96-8346-8_5

overcome nutrient limitations. The potential of biostimulants as sustainable, eco-friendly tools for optimizing nutrient use efficiency, improving oilseed crops' resilience, and maintaining productivity under nutrient-limited conditions is also highlighted. Finally, future perspectives on the integration of biostimulants into nutrient management strategies for oilseed crops are proposed.

Keywords Biostimulants · Nutrient deficiency · Oilseed crops · PGPR · Nutrient uptake · Root architecture · Nutrient use efficiency

5.1 Introduction

5.1.1 Oilseed Crops' Worldwide Significance in Agriculture and Nutrition

A vital component of world agriculture, oilseed crops like mustard (*Brassica juncea*), sunflower (*Helianthus annuus*), canola (*Brassica napus*), and soybean (*Glycine max*) play a wide range of roles in industrial uses, nutritional health, and food security (Abiodun 2017). About 348.8 million metric tons of soybeans were produced globally in 2022, making up about 60% of all oilseed production, according to the Food and Agriculture Organization (Olszak-Przybyś and Korbecka-Glinka 2024). With a combined yearly production of over 90 million metric tons, canola, sunflower, and mustard account for a sizeable portion of the remaining worldwide oilseed production (Zandberg et al. 2022).

Both people and animals rely on these crops for their protein-rich meals and edible vegetable oils. According to Hartman et al. (2016), soybeans are an essential part of animal diets because they provide around 70% of the world's protein sources. Because of its advantageous fatty acid composition, especially its high oleic acid and low saturated fat content, canola oil is prized (Sudarić et al. 2019). Owing to their unique nutritional and culinary qualities, sunflower and mustard oils are extensively used in Europe and South Asia, respectively (Kaur et al. 2019). In addition to their usage in food, oilseeds support a variety of industrial sectors such as lubricants, biofuel, and bioplastics, which increases their strategic and financial significance in both developed and developing nations (Rahman et al. 2018; Kaur et al. 2019).

5.1.2 The Frequency and Effects of Nutrient Deficits in Oilseed Plants

Despite their importance, oilseed crops are especially susceptible to nutritional shortages that limit their quality and productivity, as shown in Fig. 5.1. Various

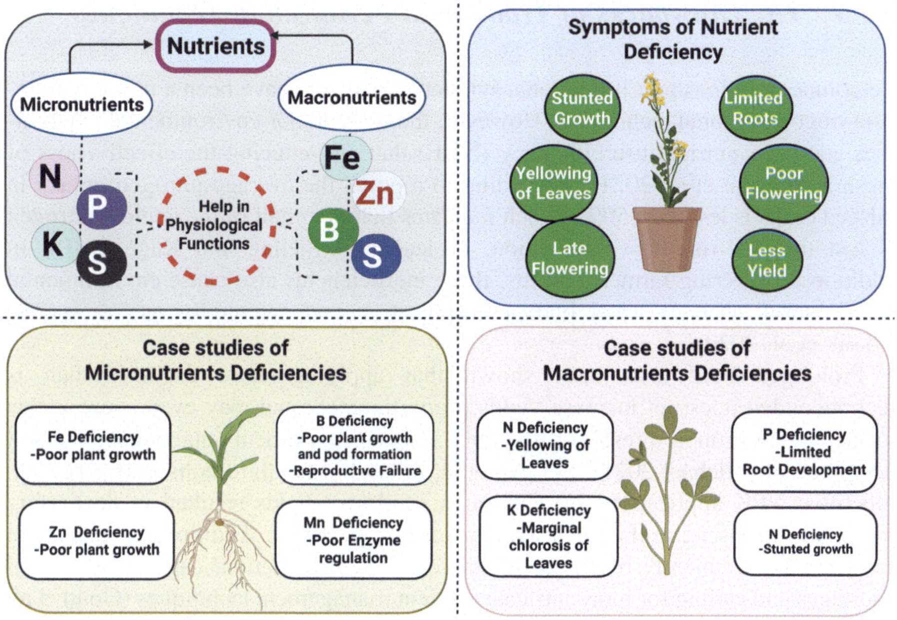

Fig. 5.1 Different types of nutrients, their deficiency symptoms, and case studies' report of both micronutrients and macronutrients

agroecological zones have been shown to have nutrient imbalances, particularly nitrogen (N), phosphorus (P), potassium (K), sulfur (S), iron (Fe), zinc (Zn), and boron (B) (Behera et al. 2021). For instance, approximately 65% of mustard-growing regions in India are lacking in S and B, according to Singh et al. (2022), which also found that this results in decreased seed yield and oil content (Akhatar et al. 2025). Likewise, in Brazilian soybean systems, deficits in micronutrients, specifically zinc and iron, have been linked to poor nodulation and chlorosis, resulting in yield losses of more than 15% in certain areas (Bagale 2021).

Photosynthesis, protein synthesis, and enzyme activation are crucial metabolic activities that are physiologically impacted by dietary shortages. For example, a phosphorus deficit is known to limit ATP synthesis and root growth, which harms biomass accumulation and reproductive development (Malhotra et al. 2018; Akbar et al. 2024; Khan et al. 2025a, b). Oilseed rape's sulfur shortage decreases chlorophyll production and oil accumulation in seeds, leading to a notable change in fatty acid profiles (Zhu et al. 2023). Further aggravating yield losses include deficits in zinc and boron, which have been connected to floral abortion and decreased pollen viability (Rerkasem et al. 2020; Thakur et al. 2023).

5.1.3 The Drawbacks of Traditional Fertilization Techniques

To compensate for these limitations, synthetic fertilizers have been a major component of conventional techniques. However, there are major environmental externalities and poor nutrient use efficiency (NUE) that are reducing the effectiveness of these inputs (Jat et al. 2021). According to reports, the average nitrogen uptake in oilseed crops is less than 50%, which indicates that over half of the applied nitrogen is lost through runoff, volatilization, or leaching (Sieling and Kage 2010). In addition to lowering farmers' profits, these inefficiencies also cause environmental deterioration, such as water body eutrophication and greenhouse gas emissions (Malik et al. 2020).

Prolonged field tests have shown that applying more fertilizer than is recommended does not increase yields proportionately and may even cause yields to stagnate or drop as a result of microbial imbalance, nutrient antagonism, and soil acidification (Michael 2021). For example, following the third consecutive season, repetitive NPK applications on Ukrainian sunflower fields resulted in decreasing marginal increases in oil content (Kachanova et al. 2023). Additionally, persistent fertilizer use frequently results in nontarget nutrient depletion, causing additional shortages and calling for more intricate nutrient management techniques (Goud et al. 2022).

5.1.4 Biostimulants in the Face of an Urgent Need for Long-Term Solutions

Biostimulants are a new generation of fertilizers that have the potential to overcome the drawbacks of traditional fertilizers by increasing nutrient uptake and crop resistance in nutrient-poor environments (Khan et al. 2023; Soliman et al. 2023; Khan et al. 2024). Plants and soils can benefit from biostimulants, which are chemicals or microbes that boost nutrient intake, efficiency, and stress tolerance through stimulating natural processes (Khan et al. 2021). Among the many items included in this category are microbial inoculants (such as *Azospirillum*, *Bacillus*, and *Trichoderma*), humic substances, protein hydrolysates, extracts from seaweed, and chemicals derived from silicon (dos Reis et al. 2024).

According to mechanistic research, microbial biostimulants can fix atmospheric nitrogen, mobilize micronutrients through the synthesis of siderophores, and solubilize phosphorus in forms that would otherwise be unavailable (Parihar et al. 2022). Under low P circumstances, *Bacillus subtilis* inoculation increased phosphorus uptake and increased seed output in canola (Iqbal et al. 2023). Similarly, it has been demonstrated that seaweed extracts high in phytohormones and polysaccharides enhance root growth in sunflower plants grown in nutrient-stressed conditions by upregulating nutrient transporter genes (Ali et al. 2021).

An increasing amount of research backs the use of biostimulants in nutrient management programs to supplement or even partially replace synthetic fertilizers, particularly in low-input and sustainable systems (Ali et al. 2024). The categorization of biostimulants has been codified by the implementation of international legislation, such as the Fertilizing Products Regulation (EU 2019/1009) of the European Union, which has also supported the greater adoption of biostimulants. Along with this, oilseeds and other important crops are the focus of public-private partnerships and international research initiatives looking for individualized biostimulant solutions (Liebig et al. 2019; Barros-Rodríguez et al. 2020). The importance of thoroughly investigating how biostimulants can help oilseed crops overcome nutritional deficiencies is shown by these recent developments taken together. This chapter aims to provide a comprehensive overview of the current state of scientific knowledge, emphasizing case studies, mechanistic insights, and prospective directions for future research.

5.2 Nutrient Deficiency in Oilseed Crops: An Overview

5.2.1 Important Nutrients Limiting Oilseed Productivity

The formation of seeds, reproduction, and vegetative growth in oilseed crops all depend on an adequate supply of macro- and micronutrients. Micronutrients such as sulfur, iron, zinc, and boron sometimes restrict oilseed system productivity in various agroecological zones, but macronutrients like as nitrogen, phosphorus, and potassium are of paramount importance (Patra 2023; Staniak et al. 2024; Welch 2024). To synthesize nucleic acids, chlorophyll, and amino acids, nitrogen is essential. N supply has a direct impact on biomass buildup, nodulation (in legumes), and oil content in oilseed crops like mustard and soybean (Zhu et al. 2023; Staniak et al. 2024). Root growth, seed generation, and energy transmission (ATP) are all processes that phosphorus is relevant to. According to Zhao et al., its absence has been demonstrated to minimize the oil content of soybeans and impede the development of reproductive organs in canola (Raboanatahiry et al. 2021). Sunflowers often exhibit potassium deficiency, which is linked to decreased seed weight and oil quality (Dar et al. 2021). Potassium controls osmotic equilibrium, enzyme activation, and transport mechanisms (Sardans and Peñuelas 2021). Certain micronutrients have specific functions. According to Siwik-Ziomek and Szczepanek (2019), sulfur is essential for the manufacture of cysteine, methionine, and oil, especially in Brassica species that have a high requirement for sulfur (Zenda et al. 2021). According to one study, electron transport and chlorophyll production are both impacted by iron deficiency in soybeans, which in turn hinders photosynthesis and N-fixation (Staniak et al. 2024). Zinc modulates auxin production, protein activity, and antioxidant protection. Sunflowers deficient in zinc demonstrate stunted development and inadequate seed filling (Kabir et al. 2021; Hamzah Saleem et al. 2022). Boron is essential for pollen viability, cell wall formation, and membrane stability. Mustard and canola exhibit heightened sensitivity to boron shortage, frequently leading to flower abortion and diminished pod formation (Kohli et al. 2023). These deficits are ubiquitous, as confirmed by agronomic surveys. In South Asia, for instance, 70–80% of soils used to cultivate mustard lack sufficient amounts of S and (Kumar et al. 2021; Rahman et al. 2024). Similarly, soybean fields grown on heavily worn Oxisols in Brazil frequently exhibit shortages in zinc and iron. Sunflower K and P shortages are commonly reported in Central and Eastern Europe as a result of uneven fertilization and soil exhaustion (de Melo Benites et al. 2023; Coka 2024).

5.2.2 Visible and Invisible Signs of Nutrient Deficit

Different physiological and morphological symptoms indicate the disruption of particular metabolic processes and show up in oilseed crops when nutrients aren't present. As a result of its great mobility, nitrogen deficit causes stunted development, decreased leaf area, and uniform chlorosis of older leaves (Wani et al. 2022). For plants growing in soils with lower phosphorus levels, symptoms include stunted root development and dark green, purplish foliage (Ghimirey et al. 2024). Older leaves with marginal chlorosis and necrosis, wilting, and weak stems are symptoms of potassium shortage (Hashimi et al. 2023; Ghimirey et al. 2024). There are distinct signs of micronutrient deficiency. Because iron is immobile in calcareous soils, a lack of iron causes interveinal chlorosis in early leaves (Yadav et al. 2024). In sunflowers, a zinc deficit causes the internodes to become shorter and the midribs of the leaves to develop chlorotic bands (Saha et al. 2023). In Brassica species in particular, sulfur shortage causes younger leaves to turn yellow and maturity to be delayed, a symptom that is commonly mistaken for nitrogen deficiency ((Zenda et al. 2021; Tsytsiura 2023). In mustard and canola, a lack of boron leads to brittle stems, sterile flowers, and hollow stems that are hollow (Dhaliwal et al. 2022).

Changes in anatomy and ultrastructure support these morphological indicators. As an example, a lack of B vitamins might cause problems with cell division and pollen tube development by interfering with pectin cross-linking in the cell wall (Balk et al. 2023). Less lateral root development and lower root hair density are two morphological changes that occur in roots when P and Zn stress are applied (Khan et al. 2025b). These changes further restrict the uptake of nutrients.

5.2.3 Exposure to Nutrient Deficiencies by Crop

5.2.3.1 Soybean

Symbiotic nitrogen fixation helps soybeans meet some of their nitrogen needs. Having insufficient amounts of molybdenum (Mo) and iron (Fe) hampers nitrogenase activity and the growth of nodules (Kafeel et al. 2023; Li et al. 2023b). A significant limitation in calcareous soils found in the Midwest of the United States and certain regions of China is iron deficiency chlorosis (IDC), which causes leaf chlorosis, decreased biomass, and a decrease in seed output (Merry et al. 2022; Sun et al. 2022). It has been found that a molybdenum shortage hinders the activity of nitrate reductase and biological N-fixation (Elseedy et al. 2023; Li et al. 2023b). Root surface area is reduced in soybeans with P and Zn shortages, which significantly impacts water and nutrient uptake (Müller et al. 2021).

5.2.3.2 Canola

Due to its high S requirement for glucosinolate and oil production, canola is particularly vulnerable to B and S shortages (Raza 2021; Stansly et al. 2021). Deficiency of S leads to slowed growth, chlorosis of younger leaves, and substantial decreases in seed oil content. Follicle abnormalities, impaired pod development, and a rise in pod abortion are all symptoms of vitamin B insufficiency (Waraich et al. 2022). Reduced seed set occurs as a result of changes in flowering intensity and leaf area index caused by nitrogen constraint in canola (Batool et al. 2022).

5.2.3.3 Sunflower

A lack of potassium, phosphorus, and zinc can have a devastating effect on sunflowers. According to Stoicea et al. (2022), photosynthate translocation is hindered by K deficit, which in turn lowers seed weight and oil yield. Reduced head diameter and seed number are symptoms of phosphorus shortage, which also slows reproductive development (Abbaszadeh-Dahaji et al. 2021). Several field investigations conducted in Turkey and India have demonstrated that zinc shortage hinders auxin metabolism and enzyme activation, leading to a delay in flowering and seed sterility (Nisar et al. 2023).

5.2.3.4 Mustard

Inadequate levels of vitamin B and S are highly noticeable in Indian mustard. Significantly less oil buildup is observed when S shortage changes the chloroplast structure and lipid metabolism (Shah et al. 2022). Studies conducted across the Indo-Gangetic Plains have linked B insufficiency to problems with pollen germination and reduced anther dehiscence, which in turn can lead to yield penalties of up to 40% (Sharma and Kumar 2024). Furthermore, different fatty acid compositions and reduced oil content were observed in mustard seeds that were B-deficient.

5.2.4 *Effects on Composition of Seeds, Oil Quality, and Yield*

Inadequate nutrition has a major impact on the quality and market value of oilseeds, in addition to reducing output. As an example, canola seeds with a sulfur deficiency have a 10–15% lower oil content, different fatty acid composition (such as more linolenic acid), and much lower protein content (Mohammadi and Mosleh Ghahfarokhi 2023). In soybeans, a lack of P and Mo hinders seed development, which in turn causes the seed size, protein content, and oil concentration to decrease (Rasheed et al. 2025). There is a decrease in oil content and an increase in free fatty acid levels in sunflower seeds that are zinc deficient, which in turn reduces their industrial value and shelf life (Szatkowski et al. 2024). When crops are under stress from nutrients, yields drop significantly. Yield reductions of up to 30% in iron-deficient soybean, 40% in sulfur-deficient mustard, and 25% in B-deficient canola have been documented in various studies (He et al. 2021; Yan et al. 2021; Safdar et al. 2023). Nutrient stress may influence the expression of genes involved in lipid metabolism, antioxidant defense, and hormone control, according to biochemical investigations (Raza et al. 2022). One example is the molecular explanation for lower oil accumulation in Brassica species, which is explained by the downregulation of fatty acid synthase and oleosin gene expression caused by S and B deficits (Jia et al. 2022).

5.3 Biostimulants: Classification and Role in Plant Nutrition

5.3.1 *An Overview of Biostimulants and Their Regulation*

"Biostimulants" are compounds or microbes that, regardless of their nutrient content, are administered to plants in order to improve their nutritional efficiency, abiotic stress tolerance, and crop quality attributes (Franzoni et al. 2022). Incorporating this concept into EU Regulation (EU) 2019/1009, which establishes a legal framework differentiating biostimulants from fertilizers and pesticides, the European Union has shown its support for the European Biostimulant Industry Council (EBIC). Biostimulants, in contrast to pesticides and fertilizers, mostly work by improving nutrient absorption and utilization through physiological and biochemical processes (Chernikova et al. 2021; Makhaye et al. 2021). Rather than classifying biostimulants according to their chemical make-up, the regulatory framework established by Regulation (EU) 2019/1009 groups them according to their claims and action mechanism. This allows for a wide variety of products to be grouped under a single functional category. The global community is currently working to standardize several aspects of product characterization, safety evaluations, and efficacy testing to bring regulations into harmony (Johnson et al. 2025). The biostimulant business relies on these endeavors to maintain product uniformity and encourage innovation.

5.3.2 Categorization of Biostimulants Used to Treat Nutrient Deficiencies

It is possible to categorize biostimulants according to their place of origin and the way they work. Biostimulants' capacity to improve nutrient availability, uptake, and absorption makes them an important tool in the fight against nutrient deficiencies. Some of the most important categories are:

5.3.2.1 Beneficial Microbes (PGPR, Mycorrhizae, N-fixers)

Numerous fungal symbionts, including arbuscular mycorrhizal fungi (AMF) such as *Glomus* spp., and plant growth-promoting rhizobacteria (PGPR) like *Azospirillum*, *Pseudomonas*, *Bacillus*, and *Rhizobium*, have demonstrated prominent functions in the solubilization and acquisition of nutrients. By producing organic acids, bacteria like *Bacillus megaterium* and *Pseudomonas fluorescens* improve phosphorus solubilization; meanwhile, *Rhizobium* and *Azotobacter* allow for biological nitrogen fixation (Zainab et al. 2020; Khan et al. 2021; Younes et al. 2023).

5.3.2.2 Humic and Fulvic Acids

By changing the shape of the roots and making the cell membranes more permeable, humic substances—and especially fulvic and humic acids—improve the uptake of nutrients. Further, they can increase micronutrient availability and solubility in the rhizosphere by forming chelates with them (Canellas et al. 2015). Humic acids facilitate ion absorption by upregulating root plasma membrane H^+ -ATPase activity (Pettit 2006). In oilseed crops, applications of humic substances have been shown to enhance uptake of potassium and calcium, especially under nutrient-deprived conditions (Siwik-Ziomek and Szczepanek 2019).

5.3.2.3 Hydrolyzed Proteins and Extracts

Root architecture and nutrient transporter expression can be influenced by protein hydrolysates sourced from either plants or animals, which include free amino acids and short peptides. In response to nutritional stress, amino acids such as glutamate and glycine activate pathways that take nutrients into the plant (Santi et al. 2017; Sun et al. 2024). By increasing root hair density and metabolic activity, amino acid-based biostimulants boosted nitrogen and phosphorus uptake in nutrient-deficient canola and sunflower plants, according to field studies (Colla et al. 2015).

5.3.2.4 Seaweed Extracts

A variety of polysaccharides (alginates), phenolics, cytokinins, auxins, and betaines are abundant in seaweed extracts, especially those from *Ecklonia maxima* and *Ascophyllum nodosum* (Stirk et al. 2020). By stimulating root development and amplifying the expression of genes that transport nutrients, these chemicals increase the efficiency with which plants consume these nutrients (Mughunth et al. 2024). Under nutrient-limited circumstances, biostimulants derived from seaweed have enhanced the uptake of nitrogen, potassium, and micronutrients in mustard and soybean, increasing oil content and seed yield (Kocira et al. 2019).

5.3.3 *Method of Use and Technology Development*

There are a number of ways to apply biostimulants, such as coating seeds, foliar spraying, soil soaking, and priming seeds. Both the biostimulant and the target nutrient dictate the optimal application method. To promote efficient rhizosphere colonization, microbial inoculants are typically sprayed onto seeds or mixed into soil. On the other hand, foliar sprays containing amino acid products and seaweed extracts are used to quickly ease nutrient stress (Shehata et al. 2011). Technology advancements in formulation, like nanoencapsulation and carriers based on polymers, have improved the biostimulants' stability, shelf-life, and ability to be delivered to specific areas (Mughunth et al. 2024). Researchers found that under stressful situations, nanoformulations of humic compounds and amino acid complexes improved nutrition absorption more effectively than standard formulations (Ali et al. 2021). Moreover, oilseed crops such as sunflower and mustard have demonstrated synergistic effects in nutrient-deficient soils when coformulated with humic acids or protein hydrolysates (Nikoogoftar-Sedghi et al. 2023).

5.4 Biostimulants' Metabolism: How They Relieve Malnutrition

Protein hydrolysates, microbial inoculants, humic acids, seaweed extracts, and other biostimulants have been the subject of extensive research on their ability to improve plant development and reduce nutrient deficits, with a focus on oilseed crops. These compounds enhance plant health by mobilizing nutrients and facilitating their uptake through a number of different pathways. With an emphasis on the effects of biostimulants under nutrient-stress situations, this section delves deeply into these mechanisms.

5.4.1 Improving Root Structure and Rhizosphere Function

One of the main ways biostimulants help with nutritional deficits is by making the roots stronger. Research has demonstrated that biostimulants such as seaweed extracts, humic acids, and PGPR can improve nutrient uptake by increasing root surface area, lateral root formation, and root hair production. This is especially true in soils that are poor in nutrients (Rajesaheb et al. 2025).

5.4.1.1 Root Architecture Modification

Research shows that biostimulants can lengthen roots and stimulate root hair development by affecting the expression of auxin production and transport genes. In sunflower and canola crops, for instance, PGPR strains like *Pseudomonas fluorescens* have been demonstrated to enhance root surface area and lateral root development (Li et al. 2022; Johnson et al. 2024). This adaptation aids plant nutrient exploration, especially in areas with few resources.

5.4.1.2 Rhizosphere Activity

In addition to improving nutrient uptake, biostimulants boost microbial activity in the rhizosphere. When humic compounds and seaweed extracts are present, helpful microbes like phosphorus-solubilizing and nitrogen-fixing bacteria can multiply and liberate soil nutrients that have been bound (Sible et al. 2021; Rohela and Saini 2022). These biostimulants not only boost rhizodeposition, but they also encourage beneficial microbial populations that aid in nutrient uptake and nutrient cycling.

5.4.2 Improving Nutrient Solubilization and Mobilization

The availability of nutrients, such as P, K, Fe and Zn, can be constrained by factors such as soil pH, organic matter concentration, and mineralization methods. Biostimulants increase the production of organic acids, siderophores, and enzymes, which aid in nutrient solubilization and mobilization, thereby helping to overcome these restrictions (Halpern et al. 2015; Chernikova et al. 2021).

5.4.2.1 Phosphorus Solubilization

The production of organic acids by numerous PGPR strains, including Bacillus and Pseudomonas species, facilitates the solubilization of inorganic phosphate and increases its availability to plants (Akhatar et al. 2025). Low accessible P content in soils is a common constraint in oilseed production; however, this helps a lot.

5.4.2.2 Siderophore Production and Micronutrient Mobilization

Biostimulants like PGPR and mycorrhizal fungi release siderophores, which help plants absorb micronutrients like iron by chelating them. Soils deficient in micronutrients can have their iron and zinc uptake hindered, but biostimulants such as AMF can enhance it, according to studies conducted on sunflowers and soybeans (Sakthieaswari et al. 2022; Grewal et al. 2023).

5.4.3 Enhancement of Nutrient Transporters and Efflux Routes

In situations of nutritional stress, biostimulants can improve nutrient absorption by increasing the expression and activity of nutrient transporters. Inorganic nutrients like Fe, N, and P are actively taken up by these transporters (Dhaliwal et al. 2022; Franzoni et al. 2022; Grewal et al. 2023). Biostimulants can upregulate the expression of genes that code for nutrient transporters, including nitrate transporters (NRTs), phosphate transporters (PHTs), and iron-regulated transporters (IRTs). One example is that in nitrogen-limited conditions, seaweed extracts can improve soybean nitrogen uptake by increasing the expression of the AtNRT2 gene, which encodes a high-affinity nitrate transporter (Sakthieaswari et al. 2022). A more efficient intake of nutrients and their subsequent transfer to other plant sections are both achieved by increasing the activity of nutrient transporters.

5.4.4 Regulation of Hormonal Signaling and Crosstalk in Response to Nutrient Stress

Plants' hormonal networks play a significant role in their responses to nutritional stress, and these networks can be influenced by biostimulants. Auxins, cytokinins, ethylene, and abscisic acid (ABA) are phytohormones that regulate nutrient intake and how plants respond to stress (EL Sabagh et al. 2022). Nutrient uptake and root development are enhanced by biostimulants such as humic acids and protein hydrolysates, which disrupt the hormonal balance (Sun et al. 2024). Auxins regulate root growth and help plants absorb nutrients in ecosystems where these nutrients are scarce. Enzyme activity known as aminocyclopropane-1-carboxylate (ACC) deaminase: this enzyme is produced by many PGPR bacteria, including Pseudomonas spp., and it reduces the amounts of ethylene in plants (Gamalero et al. 2023). A decrease in ethylene levels, which is linked to reduced root development in plants that are nutrient-stressed, is one mechanism by which biostimulants promote root growth and enhance nutrient absorption (Glick and Nascimento 2021).

5.4.5 Activation of Antioxidant and Stress-Responsive Mechanisms

When plants don't get enough nutrients, they experience oxidative stress, which harms their cells by letting reactive oxygen species (ROS) build up. By bolstering antioxidant defense systems, biostimulants aid in reducing oxidative damage. The activation of antioxidant enzymes, including peroxidase (POD), superoxide dismutase (SOD), and catalase (CAT), has been demonstrated by biostimulants including PGPR and seaweed extracts (Hasanuzzaman et al. 2021). When nutrients are scarce, these enzymes protect cells from harm by ROS (Mosa et al. 2021). Research on sunflowers and mustards, for example, has demonstrated that applying biostimulants improves plant survival and growth in the face of nutritional deficits by reducing oxidative stress markers. In addition to regulating genes that respond to stress and those that ROS, biostimulants activate stress-related transcription factors, including members of the AP2/ERF family. The plant's tolerance for nutritional deficits and resilience to abiotic stress are both enhanced by these molecular pathways (Hariharan et al. 2024).

5.4.6 Maximizing Nutrient Use Efficiency via Fertilizer Synergy

When used in conjunction with chemical fertilizers, biostimulants can boost NUE by increasing nutrient uptake, decreasing losses, and encouraging long-term nutrient cycling (Li et al. 2023a). Applying biostimulants like PGPR or seaweed extracts in conjunction with fertilizers improves fertilizer efficiency, according to field trials conducted on oilseed crops such as soybean and canola. This is accomplished by increasing the intake of nutrients, decreasing their leaching, and encouraging their remobilization from older tissues to growing seeds (Castiglione et al. 2021). NUE Index and Fertilizer Decrease: Research has shown that biostimulant-fertilizer mixes can lower the demand for heavy fertilizer applications while preserving or enhancing crop production. For instance, Singh et al. (2024) found that sunflowers and other crops with less nitrogen fertilizer treatment and the usage of humic acids had an improved NUE. Sustainable and ecologically friendly farming methods are advanced by this complementary strategy.

5.5 Research on Oilseed Crops Based on Experiment and Case Studies

To investigate the effects of biostimulants on oilseed crop performance, nutrient uptake, and stress tolerance, this section compiles data from case studies, greenhouse trials, and experimental data collected in the field (Halpern et al. 2015; Singh et al. 2022). In particular, it examines the use of biostimulants in soybean, canola, sunflower, and mustard crops to show how these products may help remedy nutrient deficits and improve crop output and quality when nutrients are scarce (Halpern et al. 2015).

5.5.1 Soybean: Field and Greenhouse Studies

To address nutritional inadequacies, soybeans have been extensively studied as a biostimulant (Tables 5.1 and 5.2). Several biostimulants have demonstrated potential in improving soybean yields, such as microbial inoculants (such as PGPR), seaweed extracts, and humic acids (Sankhala et al. 2024). Soybean NUE can be enhanced with the use of microbial inoculants, specifically PGPRs like Bacillus and Pseudomonas strains (Quadriya et al. 2024). As an example, Zahir et al. showed that in nitrogen-deficient conditions, soybean output could be increased by 15–25% through inoculation with Pseudomonas species (Quadriya et al. 2024). Hinsinger et al. (2015) found that microbial inoculants boost root development and nutrient uptake, particularly of K and P, by increasing rhizosphere activity and releasing soil-bound nutrients. Improved root morphology, including more surface area and density of root hairs, was seen in field trials that investigated the use of seaweed extracts. In a study conducted by Rathore et al. (2009), it was found that seaweed extract treatments improved P uptake in soybean plants. This was achieved by enhancing the solubilization of inorganic P in the soil and encouraging root development. Furthermore, research has demonstrated that seaweed extracts mitigate the harmful consequences of nutrition and water stress by enhancing osmotic adjustment (Kocira et al. 2019). The possibility of biostimulants and chemical fertilizers working in tandem has been the subject of multiple investigations. While reducing fertilizer inputs by 25% and boosting yield by 20%, a field study conducted by Kang et al. found that soybean NUE was greatly enhanced when PGPR inoculants were combined with reduced nitrogen fertilization (Arif et al. 2017). These findings highlight the promise of biostimulants to lessen reliance on synthetic fertilizers and improve nutrient uptake efficiency.

Table 5.1 Experimental and case study evidence on the use of biostimulants in oilseed crops

Plant	Biostimulant source	Effects	Outcome	Reference
Soybean (*Glycine max*)	PGPR (*Pseudomonas, Bacillus*)	Enhanced NUE, increased N uptake, improved rhizosphere activity	Yield increase (15–25%), improved K and P uptake	Hinsinger et al. (2015), and Quadriya et al. (2024)
Soybean (*Glycine max*)	Seaweed extracts	Improved root morphology, enhanced P solubilization	Better P uptake and osmotic adjustment under stress	Rathore et al. (2009) and Kocira et al. (2019)
Soybean (*Glycine max*)	PGPR + reduced N fertilization	Synergistic effect on NUE	25% less N input, 20% yield increase	Arif et al. (2017)
Canola (*Brassica napus*)	Protein hydrolysates, amino acids	Enhanced N and S uptake, improved root development	15–20% seed yield increase	Abd ElAzeiz et al. (2024)
Sunflower (*Helianthus annuus*)	Seaweed extracts	Improved K and P uptake, chlorophyll content, root architecture	Improved photosynthesis and plant growth	Petropoulos et al. (2022) and Osman and Salem (2011)
Sunflower (*Helianthus annuus*)	Humic acids + seaweed	Enhanced nutrient solubilization, microbial activity	18% increase in seed output, better antioxidant status	Osman and Salem (2011)
Mustard (*Brassica juncea*)	PGPR (Azotobacter, Bacillus) + compost extracts	Improved N fixation, P mobilization, soil microbial activity	20–30% N uptake, 25% increase in seed yield, 15% rise in oil content	Arif et al. (2017), Singh et al. (2023), and Vikram et al. (2022)

5.5.2 *Canola: Microbial and Nonmicrobial Biostimulant Applications*

One such oilseed crop that reaps the rewards of biostimulant treatments, especially when nutrients are few, is canola. Microbiological and nonmicrobiological biostimulants are equally important in canola for increasing production, stress resilience, and nutrient uptake, according to the research (Passandideh et al. 2022; Abd ElAzeiz et al. 2024). The effects of nonmicrobial biostimulants on canola's nutrient uptake have been studied in field conditions. These include humic compounds, protein hydrolysates, and biostimulants based on amino acids. Applying protein hydrolysates enhanced nitrogen and sulfur uptake, leading to a 15–20% increase in seed yield, as shown in a field trial by Abd ElAzeiz et al. (2024). Amino acids included in protein hydrolysates promote root growth and enhance nutrient uptake efficiency in settings with low nutrients by increasing the expression of nutrient transporters (Tables 5.1 and 5.2).

Table 5.2 Oilseed crop nutrient deficiency mitigation and the role of different biostimulant classes

Biostimulant type	Active component(s)	Target nutrient deficiency	Oilseed crop	Mode of application	Observed effect	Reference
PGPR (*Bacillus subtilis*)	ACC deaminase, IAA	Nitrogen (N)	Soybean	Seed treatment	↑ Root biomass, ↑ N uptake, ↑ yield	Dubey et al. (2024)
Seaweed extract (*Ascophyllum nodosum*)	Cytokinins, polysaccharides	Phosphorus (P)	Canola	Foliar spray	↑ P mobilization, ↑ chlorophyll content	Hariharan et al. (2024)
Humic substances	Fulvic acids, humic acids	Zinc (Zn), Iron (Fe)	Sunflower	Soil drench	↑ Micronutrient availability, ↑ shoot length	Grammenou et al. (2023)
Amino acid hydrolysate	Glutamic acid, proline	Nitrogen (N)	Mustard	Foliar spray	↑ N assimilation, ↑ leaf area index	Sun et al. (2024)
Mycorrhizal fungi (*Glomus intraradices*)	Arbuscular mycorrhizae	Phosphorus (P)	Soybean	Soil inoculation	↑ P uptake, ↑ pod number	Thioub et al. (2019)
PGPR (*Azospirillum brasilense*)	Nitrogenase enzymes	Nitrogen (N)	Sunflower	Seed coating	↑ N fixation, ↑ biomass accumulation	dos Reis et al. (2024)
Seaweed extract (*Ecklonia maxima*)	Auxins, betaines	Potassium (K)	Canola	Foliar spray	↑ K uptake, ↑ drought tolerance	Singh et al. (2022)
Humic substances + PGPR	Humic acids, beneficial microbes	Multiple nutrients	Mustard	Soil drench	↑ Nutrient availability, ↑ yield	Rigobelo (2024)
Amino acid hydrolysate + Mycorrhizae	Amino acids, mycorrhizal spores	Phosphorus (P)	Soybean	Soil application	↑ P uptake, ↑ root length	Rocha et al. (2024)
Seaweed extract (*Ascophyllum nodosum*)	Cytokinins, polysaccharides	Nitrogen (N)	Canola	Foliar spray	↑ N assimilation, ↑ seed weight	Jannin et al. (2013)

5.5.3 Sunflower: Response to Seaweed Extracts and Humic Substances

Applying biostimulants, especially extracts from seaweed and humic compounds to sunflowers (*Helianthus annuus*) greatly enhances their nutrient intake and stress tolerance (ur Rehman et al. 2018; de Moura et al. 2023). Research conducted by Petropoulos et al. (2022) demonstrated that seaweed extracts, when applied to sunflower plants, enhanced their uptake of potassium and phosphorus, leading to notable improvements in root length, surface area, and density. Improved access to soil-bound nutrients was seen as a result of the extracts' ability to promote nutrient solubilization and root growth. Furthermore, sunflowers that were exposed to seaweed extracts showed an increase in chlorophyll content, leading to improved photosynthetic efficiency and general plant development (Osman and Salem 2011). In the case of nutritional deficits, humic compounds have been demonstrated to enhance sunflower development. Humic acid and seaweed extract, when applied to sunflower plants in a greenhouse setting, resulted in an 18% increase in seed output compared to plants that were not treated (Osman and Salem 2011). Essential minerals like phosphate and micronutrients were more easily soluble when the humic compounds prompted microbial activity in the rhizosphere (Tables 5.1 and 5.2). Sunflowers are incredibly vulnerable to growth- and yield-limiting phosphate and potassium deficits. By increasing nutrient mobilization and stress resilience, biostimulants such as seaweed extracts can ameliorate these deficits. Reduced oxidative stress and greater levels of antioxidants (e.g., SOD, CAT) were seen in sunflower plants treated with seaweed extracts in a field trial that was conducted under nutritional stress conditions (Osman and Salem 2011).

5.5.4 Mustard: Integrated Use of PGPR and Organic Stimulants

In nutrient-limited environments, mustard (*Brassica juncea*) is best grown with a mix of PGPR and organic biostimulants such as compost extracts and protein hydrolysates (Caruso et al. 2019; Haider et al. 2023). In a field trial, mustard plants that were inoculated with PGPR and then treated with compost extracts had their nitrogen fixation and phosphorus mobilization improved (Arif et al. 2017). According to Singh et al. (2023), nitrogen uptake was increased by 20–30% when Azotobacter and Bacillus strains were applied. Compost extracts, on the other hand, increased soil nutrient availability and rhizosphere microbial activity (Vikram et al. 2022). For soils that are low in nutrients, a combination of PGPR and organic stimulants can increase nutrient availability. Mustard grown with PGPR and organic biostimulants had a seed production increase of 25% and an oil content improvement of 15% even in low-nutrient environments (Vikram et al. 2022).

5.5.5 *Analyzing the Differences: Increased Yield, Nutrient Content, and NUE*

The effects of biostimulants on mustard, canola, sunflower, and soybean are compared to show how diverse they are in boosting yield, nutrient content, and NUE as shown in Fig. 5.1 (Stirk et al. 2020; Sun et al. 2024). When applied to crops of all kinds, biostimulants greatly enhance root growth, nutrient absorption, and stress resilience, especially in nutrient-poor environments. Soybean and canola, for instance, reduced fertilizer inputs and increased yield significantly, demonstrating the greatest NUE benefits. Using seaweed extracts and PGPR, sunflower and mustard plants showed significant improvements in oil content and seed composition (Sankhala et al. 2024). In terms of improving nutrient uptake and NUE, a meta-analysis of research across different crops shows that microbial biostimulants, especially PGPR, provide the most consistent outcomes. Significantly beneficial effects were also observed using nonmicrobial biostimulants, including humic compounds and seaweed extracts, particularly with regard to enhancing root architecture and nutrient solubilization (Parihar et al. 2022).

5.6 Future Directions

5.6.1 *Omics-Guided Biostimulant Research*

New opportunities for molecular level investigation of plant responses to biostimulants have emerged with the integration of omics technologies, which include genomics, transcriptomics, proteomics, and metabolomics. For instance, in a study conducted by Sun et al. (2024), transcriptome analyses were used to find that protein hydrolysate-treated maize exhibited activation of genes related to nutrient transport and stress tolerance. In order to design products with a higher degree of specificity, these methods make it easier to find biomarkers for biostimulant response. The use of omics-driven research in conjunction with ML and big data analytics has the potential to completely transform the way biostimulants are designed and used.

5.6.2 *Customized Solutions for Targeted Nutrient Deficiencies*

Meticulously tailored biostimulant compositions to remedy individual nutritional deficits are the wave of the future. Stirk et al. (2020) state that oilseed crops can now be more effectively treated for iron and zinc shortages thanks to new multifunctional formulations that combine humic compounds, microbial consortia, and micronutrients. It is also possible to enhance the efficacy of biostimulants by developing them specifically for use during certain growth stages or stress periods. There is hope for precision agriculture in the ongoing research into tailored delivery methods, including inoculants that target specific roots or time-release capsules.

5.6.3 Harnessing Biostimulants for Digital and Precision Agriculture

A new era in site-specific biostimulant application has dawned thanks to digital agriculture instruments such as real-time soil sensors, GPS-guided equipment, and remote sensing. Biostimulant efficiency and input waste can be improved with precision agricultural platforms by optimizing application timing, dosage, and method.

The use of biostimulants in conjunction with drone-based monitoring greatly enhanced nutrient management in sunflower fields, according to research by Choudhary et al. (2021). Converging digital farming with biostimulant technology has the potential to make oilseed crop systems more resilient and productive.

5.7 Conclusion

Through enhanced nutrient uptake, soil health enhancement, and abiotic stress reduction, this chapter has examined the varied impact of biostimulants in reducing nutritional shortages in oilseed crops. The effectiveness of many classes of biostimulants in enhancing crop performance under nutrient-limited circumstances has been supported by evidence from field experiments, laboratory studies, and real-world applications. An environmentally friendly technique to improve oilseed crop yields, nutrient use efficiency, and oil quality without using synthetic fertilizers is by using biostimulants. They help make farming more financially viable by optimizing input costs, and they also help make agriculture more environmentally sustainable when used in nutrient management strategies. In the future, scientists should use omics technologies to better understand how biostimulants work, create individualized formulations to address certain nutritional challenges, and find ways to incorporate biostimulants into precision farming systems. Biostimulants have the potential to revolutionize sustainable oilseed crop production, but only if researchers, legislators, and industry stakeholders work together, conduct long-term field studies, and harmonize regulations.

References

Abbaszadeh-Dahaji P, Atajan FA, Omidvari M, Tahan V, Kariman K (2021) Mitigation of copper stress in maize (Zea mays) and sunflower (Helianthus annuus) plants by copper-resistant Pseudomonas strains. Curr Microbiol 78:1335–1343

Abd Elazeiz EH, Yousif EE, Elbatrawy WS (2024) Enhancement the quality and productivity of canola (Brassica napus L.) using some bio stimulant applications under saline soil conditions. J Glob Agric Ecol 16:8

Abiodun OA (2017) The role of oilseed crops in human diet and industrial use. In: Oilseed crops: yield and adaptations under environmental stress. Wiley-Blackwell, Chichester, pp 249–263

Akbar R, Sun J, Bo Y, Khattak WA, Khan AA, Jin C, Zeb U, Ullah N, Abbas A, Liu W (2024) Understanding the influence of secondary metabolites in plant invasion strategies: a comprehensive review. Plan Theory 13:3162

Akhatar J, Upadhyay P, Kumar H (2025) Crop cultivation and hybrid seed production strategies in rapeseed-mustard. In: Hybrid seed production for boosting crop yields: applications, challenges and opportunities. Springer, pp 177–224

Ali O, Ramsubhag A, Jayaraman J (2021) Biostimulant properties of seaweed extracts in plants: implications towards sustainable crop production. Plan Theory 10:531

Ali A, Niu G, Masabni J, Ferrante A, Cocetta G (2024) Integrated nutrient management of fruits, vegetables, and crops through the use of biostimulants, soilless cultivation, and traditional and modern approaches—a mini review. Agriculture 14:1330

Arif MS, Shahzad SM, Riaz M, Yasmeen T, Shahzad T, Akhtar MJ, Bragazza L, Buttler A (2017) Nitrogen-enriched compost application combined with plant growth-promoting rhizobacteria (PGPR) improves seed quality and nutrient use efficiency of sunflower. J Plant Nutr Soil Sci 180:464–473

Bagale S (2021) Nutrient management for soybean crops. Int J Agron 2021:3304634

Balk M, Sofia P, Neffe AT, Tirelli N (2023) Lignin, the lignification process, and advanced, lignin-based materials. Int J Mol Sci 24:11668

Barros-Rodríguez A, Rangseekaew P, Lasudee K, Pathom-Aree W, Manzanera M (2020) Regulatory risks associated with bacteria as biostimulants and biofertilizers in the frame of the European Regulation (EU) 2019/1009. Sci Total Environ 740:140239

Batool M, El-Badri AM, Hassan MU, Haiyun Y, Chunyun W, Zhenkun Y, Jie K, Wang B, Zhou G (2022) Drought stress in Brassica napus: effects, tolerance mechanisms, and management strategies. J Plant Growth Regul:1–25

Behera SK, Shukla AK, Prakash C, Tripathi A, Kumar A, Trivedi V (2021) Establishing management zones of soil sulfur and micronutrients for sustainable crop production. Land Degrad Dev 32:3614–3625

Canellas LP, Olivares FL, Aguiar NO, Jones DL, Nebbioso A, Mazzei P, Piccolo A (2015) Humic and fulvic acids as biostimulants in horticulture. Sci Hortic 196:15–27

Caruso G, De Pascale S, Cozzolino E, Giordano M, El-Nakhel C, Cuciniello A, Cenvinzo V, Colla G, Rouphael Y (2019) Protein hydrolysate or plant extract-based biostimulants enhanced yield and quality performances of greenhouse perennial wall rocket grown in different seasons. Plan Theory 8:208

Castiglione AM, Mannino G, Contartese V, Bertea CM, Ertani A (2021) Microbial biostimulants as response to modern agriculture needs: composition, role and application of these innovative products. Plan Theory 10:1533

Chernikova, O., Mazhaysky, Y., Buryak, S., Seregina, T., and Ampleeva, L. (2021). Comparative analysis of the use of biostimulants on the main types of soil

Choudhary V, Kumar KK, Sahu B, Muthappa SK, Das ADAA (2021) Agronomic innovations in biotic stress management and its combined effect with abiotic stresses in crop production. Indian J Agron 66:237–257

Coka SZ (2024) Response of cowpea to integrated mycoroot™ inoculation, biochar-compost mixture and soil variation: drought tolerance and performance

Colla G, Nardi S, Cardarelli M, Ertani A, Lucini L, Canaguier R, Rouphael Y (2015) Protein hydrolysates as biostimulants in horticulture. Sci Hortic 196:28–38
Dar JS, Cheema MA, Rehmani MIA, Khuhro S, Rajput S, Virk AL, Hussain S, Bashir MA, Alghanem SM, Al-Zuaibr FM (2021) Potassium fertilization improves growth, yield and seed quality of sunflower (Helianthus annuus L.) under drought stress at different growth stages. PLoS One 16:e0256075
De Melo Benites V, Schaefer CE, Machado PLO, Polidoro JC, Da Silva Teixeira R (2023) Insights into Brazilian soils and sustainable agriculture scenarios. In: The soils of brazil. Springer, Cham, pp 471–486
De Moura OVT, Berbara RLL, De Oliveira Torchia DF, Da Silva HFO, De Castro TAVT, Tavares OCH, Rodrigues NF, Zonta E, Santos LA, García AC (2023) Humic foliar application as sustainable technology for improving the growth, yield, and abiotic stress protection of agricultural crops. a review. J Saudi Soc Agric Sci 22:493–513
Dhaliwal SS, Sharma V, Shukla AK (2022) Impact of micronutrients in mitigation of abiotic stresses in soils and plants—a progressive step toward crop security and nutritional quality. Adv Agron 173:1–78
Dos Reis GA, Martínez-Burgos WJ, Pozzan R, Pastrana Puche Y, Ocán-Torres D, De Queiroz Fonseca Mota P, Rodrigues C, Lima Serra J, Scapini T, Karp SG (2024) Comprehensive review of microbial inoculants: agricultural applications, technology trends in patents, and regulatory frameworks. Sustain For 16:8720
Dubey A, Malla MA, Kumar A, Khan ML, Kumari S (2024) Seed bio-priming with ACC deaminase-producing bacterial strains alleviates impact of drought stress in soybean (Glycine max (L.) Merr.). Rhizosphere 30:100873
El Sabagh A, Islam MS, Hossain A, Iqbal MA, Mubeen M, Waleed M, Reginato M, Battaglia M, Ahmed S, Rehman A (2022) Phytohormones as growth regulators during abiotic stress tolerance in plants. Front Agron 4:765068
Elseedy M, Helaly DS, El-Sherpiny M, Taha A (2023) Drought stress effects on vegetable crop productivity: melatonin's mitigating role and rhizobium inoculants for enhanced nitrogen fixation under water scarcity. J Soil Sci Agric Eng 14:255–265
Franzoni G, Cocetta G, Prinsi B, Ferrante A, Espen L (2022) Biostimulants on crops: their impact under abiotic stress conditions. Horticulturae 8:189
Gamalero E, Lingua G, Glick BR (2023) Ethylene, ACC, and the plant growth-promoting enzyme ACC deaminase. Biology 12:1043
Ghimirey V, Chaurasia J, Marahatta S (2024) Plant nutrition disorders: insights from clinic analyses and their impact on plant health. Agric Ext Dev Ctries 2:09–17
Glick BR, Nascimento FX (2021) Pseudomonas 1-aminocyclopropane-1-carboxylate (ACC) deaminase and its role in beneficial plant-microbe interactions. Microorganisms 9:2467
Goud BR, Raghavendra M, Prasad PS, Hatti V, Halli HM, Nayaka GV, Suresh G, Maheshwari KS, Adilakshmi G, Reddy GP (2022) Sustainable management and restoration of the fertility of damaged soils. Agric Issues Polic 113
Grammenou A, Petropoulos SA, Thalassinos G, Rinklebe J, Shaheen SM, Antoniadis V (2023) Biostimulants in the soil–plant interface: agro-environmental implications—a review. Earth Syst Environ 7:583–600
Grewal S, Goel S, Kumari S (2023) Biostimulants and regulation of Cr, Pb, Fe, Zn, Hg, Cu, Mn, and Ni stress in plants. In: Biostimulants in alleviation of metal toxicity in plants. Elsevier, pp 149–176
Haider RZ, Jaffar MT, Khan MN, Asghar HN (2023) Interactive effect of PGPR and organic amendments on the growth and yield attributes of Brassica juncea. Plant Environ 4:29–36
Halpern M, Bar-Tal A, Ofek M, Minz D, Muller T, Yermiyahu U (2015) The use of biostimulants for enhancing nutrient uptake. Adv Agron 130:141–174
Hamzah Saleem M, Usman K, Rizwan M, Al Jabri H, Alsafran M (2022) Functions and strategies for enhancing zinc availability in plants for sustainable agriculture. Front Plant Sci 13:1033092

Hariharan G, Vathshalyan N, Galahitigama H, Wimalasiri U, Kumara GDK (2024) Potential of foliar application of seaweed extracts as a biostimulant for abiotic stress alleviation on crop production. Rev Agric Sci 12:295–312

Hartman GL, Pawlowski ML, Herman TK, Eastburn D (2016) Organically grown soybean production in the USA: constraints and management of pathogens and insect pests. Agronomy 6:16

Hasanuzzaman M, Parvin K, Bardhan K, Nahar K, Anee TI, Masud AAC, Fotopoulos V (2021) Biostimulants for the regulation of reactive oxygen species metabolism in plants under abiotic stress. Cells 10:2537

Hashimi MH, Abad Q, Shafiqi SG (2023) A review of diagnostic techniques of visual symptoms of nutrients deficiencies in plant. Int J Agric Res. (ISSN 2814-189X) 6:01–09

He M, Wang S, Zhang C, Liu L, Zhang J, Qiu S, Wang H, Yang G, Xue S, Shi L (2021) Genetic variation of BnaA3. NIP5; 1 expressing in the lateral root cap contributes to boron deficiency tolerance in Brassica napus. PLoS Genet 17:e1009661

Hinsinger P, Herrmann L, Lesueur D, Robin A, Trap J, Waithaisong K, Plassard C (2015) Impact of roots, microorganisms and microfauna on the fate of soil phosphorus in the rhizosphere. In: Annual plant reviews volume 48: phosphorus metabolism in plants. Wiley Blackwell, Chichester, pp 375–407

Iqbal M, Naveed M, Sanaullah M, Brtnicky M, Hussain MI, Kucerik J, Holatko J, Mustafa A (2023) Plant microbe mediated enhancement in growth and yield of canola (Brassica napus L.) plant through auxin production and increased nutrient acquisition. J Soils Sediments 23:1233–1249

Jannin L, Arkoun M, Etienne P, Laîné P, Goux D, Garnica M, Fuentes M, Francisco SS, Baigorri R, Cruz F (2013) Brassica napus growth is promoted by Ascophyllum nodosum (L.) Le Jol. seaweed extract: microarray analysis and physiological characterization of N, C, and S metabolisms. J Plant Growth Regul 32:31–52

Jat RK, Bijarniya D, Kakraliya SK, Sapkota TB, Kakraliya M, Jat ML (2021) Precision nutrient rates and placement in conservation maize-wheat system: effects on crop productivity, profitability, nutrient-use efficiency, and environmental footprints. Agronomy 11:2320

Jia Y, Yao M, He X, Xiong X, Guan M, Liu Z, Guan C, Qian L (2022) Transcriptome and regional association analyses reveal the effects of oleosin genes on the accumulation of oil content in Brassica napus. Plan Theory 11:3140

Johnson R, Joel JM, Puthur JT (2024) Biostimulants: the futuristic sustainable approach for alleviating crop productivity and abiotic stress tolerance. J Plant Growth Regul 43:659–674

Johnson R, Joel JM, Puthur JT (2025) Effect of biostimulants on yield and fruit quality of different crops. In: Biostimulants for improving reproductive growth and crop yield. Elsevier, pp 289–304

Kabir AH, Tahura S, Elseehy MM, El-Shehawi AM (2021) Molecular characterization of Fe-acquisition genes causing decreased Fe uptake and photosynthetic inefficiency in Fe-deficient sunflower. Sci Rep 11:5537

Kachanova T, Manushkina T, Petrova O, Shevchuk N (2023) Productivity of high-oleic sunflower when grown in the conditions of the Southern Steppe of Ukraine. Ukr Black Sea Region Agrar Sci:27

Kafeel U, Jahan U, Khan FA (2023) Role of mineral nutrients in biological nitrogen fixation. In: Sustainable plant nutrition. Elsevier, pp 87–106

Kaur R, Sharma AK, Rani R, Mawlong I, Rai P (2019) Medicinal qualities of mustard oil and its role in human health against chronic diseases: a review. Asian J Dairy Food Res 38:98–104

Khan AA, Wang T, Hussain T, Amna, Ali F, Shi F, Latef AAHA, Ali OM, Hayat K, Mehmood S (2021) Halotolerant-Koccuria rhizophila (14asp)-induced amendment of salt stress in pea plants by limiting Na+ uptake and elevating production of antioxidants. Agronomy 11:1907

Khan AA, Alhaithloul HAS, Alghanem SM, Senousy HH, Alhoqail WA, Shi F, Ali I, Soliman MH (2023) Petrosolinum crispum under Stressful Conditions. In: Medicinal plant responses to stressful conditions. CRC Press, pp 299–310

Khan AA, Hayat K, Wang Y-F, Alhaithloul HAS, Alghanem SMS, Alhoqail WA, Soliman MH (2024) Biochar versus heavy metals under changing environmental conditions. In: Biochar-assisted remediation of contaminated soils under changing climate. Elsevier, pp 167–200
Khan AA, Wang Y-F, Akbar R, Alhoqail WA (2025a) Mechanistic insights and future perspectives of drought stress management in staple crops. Front Plant Sci 16:1547452
Khan AA, Wang Y-F, Zeb A, Hayat K, Alhoqail WA, Soliman MH (2025b) Beneficial elements and their roles against soil pollution. In: Beneficial elements for remediation of heavy metals in polluted soil. Elsevier, pp 1–32
Kocira S, Szparaga A, Kuboń M, Czerwińska E, Piskier T (2019) Morphological and biochemical responses of Glycine max (L.) Merr. to the use of seaweed extract. Agronomy 9:93
Kohli SK, Kaur H, Khanna K, Handa N, Bhardwaj R, Rinklebe J, Ahmad P (2023) Boron in plants: uptake, deficiency and biological potential. Plant Growth Regul 100:267–282
Kumar S, Meena RS, Singh RK, Munir TM, Datta R, Danish S, Yadav GS, Kumar S (2021) Soil microbial and nutrient dynamics under different sowings environment of Indian mustard (Brassica juncea L.) in rice based cropping system. Sci Rep 11:5289
Li M, Zhu Y, Li S, Zhang W, Yin C, Lin Y (2022) Regulation of phytohormones on the growth and development of plant root hair. Front Plant Sci 13:865302
Li J, Ma H, Ma H, Lei F, He D, Huang X, Yang H, Fan G (2023a) Comprehensive effects of N reduction combined with biostimulants on N use efficiency and yield of the winter wheat–summer maize rotation system. Agronomy 13:2319
Li M, Zhang P, Guo Z, Cao W, Gao L, Li Y, Tian CF, Chen Q, Shen Y, Ren F (2023b) Molybdenum nanofertilizer boosts biological nitrogen fixation and yield of soybean through delaying nodule senescence and nutrition enhancement. ACS Nano 17:14761–14774
Liebig N, Salaun M, Monnier C (2019) Development of standards and guidance documents for biostimulants approval under European Fertilizer Regulation (EU) 2019/1009 qrcode. EAS Expert Artic. Eur. Biostimulants Approv. https://www.eurofins.com/agroscience-services/about-us/latest-news/eas-exp e rt-article-on-european-biostimulants-approval/. Accessed 11.9. 22
Makhaye G, Mofokeng MM, Tesfay S, Aremu AO, Van Staden J, Amoo SO (2021) Influence of plant biostimulant application on seed germination. In: Biostimulants for crops from seed germination to plant development. Academic, Amsterdam, pp 109–135
Malhotra H, Vandana, Sharma S, Pandey R (2018) Phosphorus nutrition: plant growth in response to deficiency and excess. In: Plant nutrients and abiotic stress tolerance. Wiley, Hoboken, pp 171–190
Malik A, Mor VS, Tokas J, Punia H, Malik S, Malik K, Sangwan S, Tomar S, Singh P, Singh N (2020) Biostimulant-treated seedlings under sustainable agriculture: a global perspective facing climate change. Agronomy 11:14
Merry R, Dobbels AA, Sadok W, Naeve S, Stupar RM, Lorenz AJ (2022) Iron deficiency in soybean. Crop Sci 62:36–52
Michael PS (2021) Role of organic fertilizers in the management of nutrient deficiency, acidity, and toxicity in acid soils–a review. J Glob Agric Ecol 12:19–30
Mohammadi M, Mosleh Ghahfarokhi Z (2023) Impact of nitrogen and sulfur applications on seed yield, protein, oil content, and fatty acid composition of Canola (Brassica napus L.). ACS Agric Sci Technol 3:906–910
Mosa WF, Sas-Paszt L, Gornik K, Ali HM, Salem MZ (2021) Vegetative growth, yield, and fruit quality of guava (*Psidium guajava* L.) cv. maamoura as affected by some biostimulants. Bioresources 16(4):7379
Mughunth R, Velmurugan S, Mohanalakshmi M, Vanitha K (2024) A review of seaweed extract's potential as a biostimulant to enhance growth and mitigate stress in horticulture crops. Sci Hortic 334:113312
Müller M, Schneider JR, Klein VA, Da Silva E, Da Silva Júnior JP, Souza AM, Chavarria G (2021) Soybean root growth in response to chemical, physical, and biological soil variations. Front Plant Sci 12:602569

Nikoogoftar-Sedghi M, Rabiei V, Razavi F, Molaei S, Khadivi A (2023) The effect of foliar application of Ascophyllum nodosum (L.) Le Jol. seaweed extract on biochemical traits related to abiotic stresses in pistachio (Pistacia vera L. cv. Kaleh-Ghoochi). BMC Plant Biol 23:635

Nisar M, Aqeel M, Sattar A, Sher A, Ijaz M, Ul-Allah S, Rasheed U, Al-Qahtani SM, Al-Harbi NA, Alzuaibr FM (2023) Exogenous application of silicon and sulfate improved drought tolerance in sunflowers through modulation of morpho-physiological and antioxidant defense mechanisms. J Soil Sci Plant Nutr 23:5060–5069

Olszak-Przybyś H, Korbecka-Glinka G (2024) The diversity of seed-borne fungi associated with soybean grown in Southern Poland. Pathogens 13:769

Osman HE, Salem O (2011) Effect of seaweed extracts as foliar spray on sunflower yield and oil content. Egypt J Phycol 12:57–70

Parihar P, Singh P, Patidar JK (2022) Biostimulants for improving nutritional quality in legumes. In: New and future developments in microbial biotechnology and bioengineering. Elsevier, pp 261–275

Passandideh M, Rajaie M, Zeinalzadeh-Tabrizi H (2022) Effect of some plant growth biostimulants on increasing canola (Brassica napus L.) tolerance to drought stress. Environ Stresses Crop Sci 15:1023–1035

Patra S (2023) Review an investigation of the effects of macro and micro nutrients on the production of high quality seed. Int J Theor Appl Sci 15(2):16–21

Petropoulos SA, Sami R, Benajiba N, Zewail RM, Mohamed MH (2022) The response of globe artichoke plants to potassium fertilization combined with the foliar spraying of seaweed extract. Agronomy 12:490

Pettit RE (2006) Organic matter, humus, humate, humic acid, fulvic acid and humin. The Wonderful World of Humus and Carbon

Quadriya H, Rajendran G, Mir MI, Surekha K, Hameeda B (2024) Role of Pseudomonas lini and Brevundimonas nasdae to enhance nitrogen use efficiency (NUE) and yield of Oryza sativa L. Int J Plant Prod 18:271–287

Raboanatahiry N, Li H, Yu L, Li M (2021) Rapeseed (Brassica napus): processing, utilization, and genetic improvement. Agronomy 11:1776

Rahman M, Khatun A, Liu L, Barkla BJ (2018) Brassicaceae mustards: traditional and agronomic uses in Australia and New Zealand. Molecules 23:231

Rahman M, Khatun A, Liu L, Barkla BJ (2024) Brassicaceae mustards: phytochemical constituents, pharmacological effects, and mechanisms of action against human disease. Int J Mol Sci 25: 9039

Rajesaheb KS, Subramanian S, Boominathan P, Thenmozhi S, Gnanachitra M (2025) Bio-stimulant in improving crop yield and soil health. Commun Soil Sci Plant Anal 56:458–493

Rasheed MU, Lihong W, Mahmood A, Rehman HU, Javaid MM, Ali B, Ishfaq F, Ameen M, Hassan MU, Hashem A (2025) Impact of nitrogen, Sulphur, and foliar applied Thiourea on growth, oil yield, and fatty acid profile of Canola. Pol J Environ Stud 34:2337–2347

Rathore S, Chaudhary D, Boricha G, Ghosh A, Bhatt B, Zodape S, Patolia J (2009) Effect of seaweed extract on the growth, yield and nutrient uptake of soybean (Glycine max) under rainfed conditions. S Afr J Bot 75:351–355

Raza A (2021) Eco-physiological and biochemical responses of rapeseed (Brassica napus L.) to abiotic stresses: consequences and mitigation strategies. J Plant Growth Regul 40:1368–1388

Raza A, Salehi H, Rahman MA, Zahid Z, Madadkar Haghjou M, Najafi-Kakavand S, Charagh S, Osman HS, Albaqami M, Zhuang Y (2022) Plant hormones and neurotransmitter interactions mediate antioxidant defenses under induced oxidative stress in plants. Front Plant Sci 13: 961872

Rerkasem B, Jamjod S, Pusadee T (2020) Productivity limiting impacts of boron deficiency, a review. Plant Soil 455:23–40

Rigobelo EC (2024) Humic substances in combination with PGPR. In: Microbial services for cereal crops: reducing costs and environmental impact. Springer, pp 153–169

Rocha TM, Marcelino PRF, Da Costa RAM, Rubio-Ribeaux D, Barbosa FG, Da Silva SS (2024) Agricultural bioinputs obtained by solid-state fermentation: from production in biorefineries to sustainable agriculture. Sustain For 16:1076

Rohela GK, Saini P (2022) Nitrogen-fixing biofertilizers and biostimulants. In: Microbial biostimulants for sustainable agriculture and environmental bioremediation. CRC Press, pp 83–100

Safdar ME, Qamar R, Javed A, Nadeem MA, Javeed HMR, Farooq S, Głowacka A, Michałek S, Alwahibi MS, Elshikh MS (2023) Combined application of boron and zinc improves seed and oil yields and oil quality of oilseed rape (Brassica napus L.). Agronomy 13:2020

Saha B, Reddy KM, Saha S, Fatima A, Solankey SS (2023) Nutritional stress management in vegetable crops under changing climate scenario. In: Advances in research on vegetable production under a changing climate vol 2. Springer, pp 85–108

Sakthieaswari P, Kannan A, Baby S (2022) Role of mycorrhizosphere as a biostimulant and its impact on plant growth, nutrient uptake and stress management. In: New and future developments in microbial biotechnology and bioengineering. Elsevier, pp 319–336

Sankhala J, Waktaliya A, Raghuvanshi R, Shende R, Maske G, Barman D (2024) Harnessing biostimulants to mitigate abiotic stress in soybean production: an overview. J Sci Res Rep 30: 53–65

Santi C, Zamboni A, Varanini Z, Pandolfini T (2017) Growth stimulatory effects and genome-wide transcriptional changes produced by protein hydrolysates in maize seedlings. Front Plant Sci 8: 433

Sardans J, Peñuelas J (2021) Potassium control of plant functions: ecological and agricultural implications. Plan Theory 10:419

Shah SH, Parrey ZA, Islam S, Tyagi A, Ahmad A, Mohammad F (2022) Exogenously applied Sulphur improves growth, photosynthetic efficiency, enzymatic activities, mineral nutrient contents, yield and quality of Brassica juncea L. Sustain For 14:14441

Sharma M, Kumar P (2024) Biofortification of Indian mustard (Brassica juncea L.) by foliar application of agrochemicals and growth regulators for improved nutrient assimilation, oil quality, and crop yield. Biopesticides Int:20

Shehata S, Abdel-Azem HS, Abou El-Yazied A, El-Gizawy A (2011) Effect of foliar spraying with amino acids and seaweed extract on growth chemical constitutes, yield and its quality of celeriac plant. Eur J Sci Res 58:257–265

Sible CN, Seebauer JR, Below FE (2021) Plant biostimulants: a categorical review, their implications for row crop production, and relation to soil health indicators. Agronomy 11:1297

Sieling K, Kage H (2010) Efficient N management using winter oilseed rape. A review. Agron Sustain Dev 30:271–279

Singh A, Singh S, Kurella A, Verma A, Mahatama M, Venkatesh I (2022) Plant bio-stimulants, their functions and use in enhancing stress tolerance in oilseeds. In: New and future developments in microbial biotechnology and bioengineering. Elsevier, pp 239–259

Singh S, Sarkar D, Rakesh S, Singh RK, Rakshit A (2023) Biopriming with bacillus subtilis enhanced the Sulphur use efficiency of indian mustard under graded levels of Sulphur fertilization. Agronomy 13:974

Singh A, Sharma A, Singh O, Rajput VD, Movsesyan HS, Minkina T, Alexiou A, Papadakis M, Singh RK, Singh S (2024) In-depth exploration of nanoparticles for enhanced nutrient use efficiency and abiotic stresses management: present insights and future horizons. Plant Stress:100576

Siwik-Ziomek A, Szczepanek M (2019) Soil extracellular enzyme activities and uptake of N by oilseed rape depending on fertilization and seaweed biostimulant application. Agronomy 9:480

Soliman MH, Alnusairi GS, Khan AA, Alnusaire TS, Fakhr MA, Abdulmajeed AM, Aldesuquy HS, Yahya M, Najeeb U (2023) Biochar and selenium nanoparticles induce water transporter genes for sustaining carbon assimilation and grain production in salt-stressed wheat. J Plant Growth Regul 42:1522–1543

Staniak M, Szpunar-Krok E, Wilczewski E, Kocira A, Podleśny J (2024) The function of macronutrients in helping soybeans to overcome the negative effects of drought stress. Agronomy 14: 1744

Stansly TL, Seepaul R, George S, Small I, Wright D (2021) The effects of sulfur fertility on yield and seed components in oilseed Brassica carinata. Crop Sci 61:4229–4238

Stirk WA, Rengasamy KR, Kulkarni MG, Van Staden J (2020) Plant biostimulants from seaweed: an overview. In: The chemical biology of plant biostimulants. Wiley, Hoboken, pp 31–55

Stoicea P, Chiurciu IA, Soare E, Iorga AM, Dinu TA, Tudor VC, Gîdea M, David L (2022) Impact of reducing fertilizers and pesticides on sunflower production in Romania versus EU countries. Sustain For 14:8334

Sudarić A, Kočar MM, Duvnjak T, Zdunić Z, Kulundžić AM (2019) Improving seed quality of soybean suitable for growing in Europe. In: Soybean for human consumption and animal feed. IntechOpen

Sun M, Chen S, Kurle JE (2022) Interactive effects of soybean cyst nematode, arbuscular-mycorrhizal fungi, and soil pH on chlorophyll content and plant growth of soybean. Phytobiomes J 6:95–105

Sun W, Shahrajabian MH, Kuang Y, Wang N (2024) Amino acids biostimulants and protein hydrolysates in agricultural sciences. Plan Theory 13:210

Szatkowski A, Antoszkiewicz Z, Purwin C, Jankowski KJ (2024) Oilseed radish: nitrogen and sulfur management strategies for seed yield and quality—a case study in Poland. Agriculture 14: 755

Thakur S, Sinha A, Ghosh Bag A (2023) Boron-A critical element for fruit nutrition. Commun Soil Sci Plant Anal 54:2899–2914

Thioub M, Ewusi-Mensah N, Sarkodie-Addo J, Adjei-Gyapong T (2019) Arbuscular mycorrhizal fungi inoculation enhances phosphorus use efficiency and soybean productivity on a Haplic Acrisol. Soil Tillage Res 192:174–186

Tsytsiura Y (2023) Viral diseases of cruciferous plants: prevalence, biology and development cycle, effective control tactics. Publishing House "Baltija Publishing"

Ur Rehman H, Alharby HF, Alzahrani Y, Rady MM (2018) Magnesium and organic biostimulant integrative application induces physiological and biochemical changes in sunflower plants and its harvested progeny on sandy soil. Plant Physiol Biochem 126:97–105

Vikram K, Meena S, Kumar S, Ranjan R, Nivetha N, Paul S (2022) Influence of medium-term application of rhizobacteria on mustard yield and soil properties under different irrigation systems. Rhizosphere 24:100608

Wani BA, Rashid S, Rashid K, Javid H, Magray JA, Ul Qadir R, Islam T (2022) Nitrogen deficiency in plants. In: Advances in plant nitrogen metabolism. CRC Press, pp 28–37

Waraich EA, Hussain A, Ahmad Z, Ahmad M, Barutçular C (2022) Foliar application of sulfur improved growth, yield and physiological attributes of canola (Brassica napus L.) under heat stress conditions. J Plant Nutr 45:369–379

Welch RM (2024) Importance of seed mineral nutrient reserves in crop growth and development. In: Mineral nutrition of crops. CRC Press, pp 205–226

Yadav N, Dhakad A, Saini S (2024) Chapter-7 symptoms of nutrient deficiencies and excesses. Comprehensive Guide to Horticultural Practices and Management, 75

Yan L, Li S, Riaz M, Jiang C (2021) Proline metabolism and biosynthesis behave differently in response to boron-deficiency and toxicity in Brassica napus. Plant Physiol Biochem 167:529–540

Younes NA, Anik TR, Rahman MM, Wardany AA, Dawood MF, Tran L-SP, Latef AA, Mostofa MG (2023) Effects of microbial biostimulants (Trichoderma album and Bacillus megaterium) on growth, quality attributes, and yield of onion under field conditions. Heliyon 9:e14203

Zainab N, Din BU, Javed MT, Afridi MS, Mukhtar T, Kamran MA, Khan AA, Ali J, Jatoi WN, Munis MFH (2020) Deciphering metal toxicity responses of flax (Linum usitatissimum L.) with exopolysaccharide and ACC-deaminase producing bacteria in industrially contaminated soils. Plant Physiol Biochem 152:90–99

Zandberg JD, Fernandez CT, Danilevicz MF, Thomas WJ, Edwards D, Batley J (2022) The global assessment of oilseed brassica crop species yield, yield stability and the underlying genetics. Plan Theory 11:2740

Zenda T, Liu S, Dong A, Duan H (2021) Revisiting Sulphur—the once neglected nutrient: it's roles in plant growth, metabolism, stress tolerance and crop production. Agriculture 11:626

Zhu J, Dai W, Chen B, Cai G, Wu X, Yan G (2023) Research progress on the effect of nitrogen on rapeseed between seed yield and oil content and its regulation mechanism. Int J Mol Sci 24: 14504

Chapter 6
Use of Biostimulants to Increase Heavy Metal Tolerance in Oilseed Crops

Muhammad Adnan, Tabassum Yaseen, Hafsa, Muhammad Nauman Khan, Khizar Hayat, Sana Wahab, Nazima Wahid, and Alevcan Kaplan

Abstract Heavy metal contamination poses a significant barrier to oilseed crop cultivation, leading to reduced growth and yield losses, as well as potential food safety concerns. Biostimulants have garnered interest as a long-term strategy for enhancing plant tolerance to heavy metal stress through physiological, biochemical, and molecular pathways. This chapter examines numerous forms of biostimulants, such as humic compounds, protein hydrolysates, seaweed extracts, and microbial inoculants, emphasizing their function in enhancing nutrient absorption, activating antioxidant defense mechanisms, and lowering metal toxicity of oilseed crops. Using biostimulants improves crop resilience while promoting soil health and sustainable agriculture practices. While their potential is great, further study is needed to optimize application tactics and better understand how they interact with plant metabolism under heavy metal stress.

Keywords Biostimulants · Crop resilience · Heavy metals · Sustainable agriculture

---

M. Adnan (✉)
Iqbal Ahmad Qarshi Laboratory, Qarshi Industries (Pvt.), Ltd, Haripur, Pakistan

T. Yaseen
Department of Botany, Bacha Khan University, Charsadda, Pakistan

Hafsa
Department of Chemistry, Bacha Khan University, Charsadda, Pakistan

M. N. Khan · K. Hayat
Department of Botany, Islamia College Peshawar, Peshawar, Pakistan

S. Wahab · N. Wahid
Department of Plant Sciences, Quaid-i-Azam University, Islamabad, Pakistan

A. Kaplan
Department of Crop and Animal Production, Sason Vocational School, Batman University, Batman, Turkey

© The Author(s), under exclusive license to Springer Nature Singapore Pte Ltd. 2025

A. A. H. Abdel Latef (ed.), *Oilseed Crops Under Abiotic Stress*, Sustainability Sciences in Asia and Africa, https://doi.org/10.1007/978-981-96-8346-8_6

6.1 Introduction

The growing industrialization and urbanization have led to widespread heavy metal contamination in agricultural soils, posing a significant threat to global food security. Heavy metals (HMs) such as cadmium (Cd), lead (Pb), arsenic (As), and mercury (Hg) accumulate in soil and are absorbed by crops, including oilseeds, at toxic levels. These metals interfere with physiological and biochemical processes in plants, such as photosynthesis, nutrient uptake, and enzymatic activity, ultimately reducing crop growth and productivity (El-Shahir et al. 2021). Oilseed crops, including sunflower (*Helianthus annuus*), canola (*Brassica napus*), and soybean (*Glycine max*), are particularly vulnerable to heavy metal stress due to their significant cultivation in contaminated soils. Therefore, developing strategies to mitigate heavy metal toxicity is a pressing need. Biostimulants have emerged as an innovative and sustainable solution for enhancing the tolerance of crops to abiotic stresses, including heavy metal contamination. These substances, derived from natural or synthetic origins, promote plant growth and improve stress resilience through various mechanisms. Biostimulants include categories such as amino acids, seaweed extracts, humic substances, microbial inoculants, and protein hydrolysates, each with unique properties that mitigate the harmful effects of heavy metals on plants (Calvo et al. 2014).

1. *Heavy Metal Stress in Oilseed Crops*

Heavy metal contamination disrupts plant metabolism by inducing oxidative stress, reducing chlorophyll content, and impairing nutrient uptake (Kumar et al. 2022). In oilseed crops, heavy metals reduce seed quality and oil yield, affecting their economic and nutritional value (Nazari et al. 2023). For instance, cadmium toxicity in sunflower leads to stunted growth and reduced oil accumulation (Shahid et al. 2016), while lead exposure in soybean negatively affects seed protein and oil content (Ali and Gill 2022). These challenges necessitate strategies that enhance the resilience of oilseed crops to heavy metal stress. Further, the negative impacts of HMs on the physiological, ecological aspects, as well as growth and development of oilseed crops are illustrated in Fig. 6.1.

2. *Role of Biostimulants in Enhancing Heavy Metal Tolerance*

   Biostimulants improve heavy metal tolerance in plants through diverse mechanisms:

   (i) *Chelation and Sequestration of Heavy Metals*: Humic and fulvic acids bind to heavy metals, reducing their bioavailability and toxicity in soil and plant tissues (Fakhar et al. 2022).
   (ii) *Improvement of Antioxidant Defense*: Biostimulants stimulate the production of enzymatic and nonenzymatic antioxidants, such as superoxide dismutase and ascorbate, which scavenge reactive oxygen species generated by heavy metal stress (Bartucca et al. 2022).

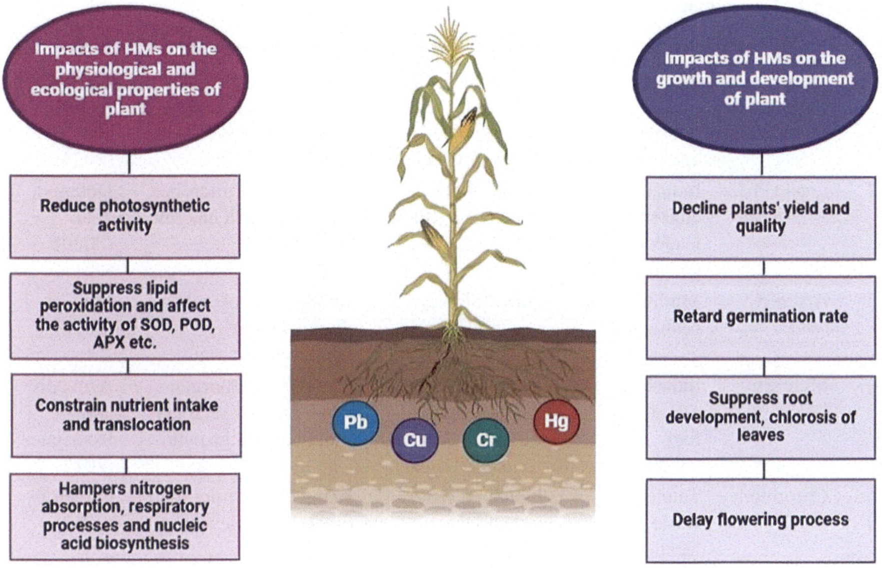

Fig. 6.1 Negative impacts of heavy metals on the different aspects of oilseed crops

(iii) *Enhanced Root Architecture*: Certain biostimulants improve root growth and activity, facilitating better nutrient uptake and mitigating the toxic effects of heavy metals (Halpern et al. 2015).

(iv) *Regulation of Stress-Responsive Genes*: Biostimulants influence the expression of genes involved in stress tolerance, aiding in the detoxification of heavy metals (Nephali et al. 2020).

6.2 Heavy Metal Contamination in Soil

Heavy metal contamination in soil is a global environmental concern, primarily driven by industrial activities, mining, agricultural practices, and urbanization (Sherameti and Varma 2015, Zainab et al. 2021). Heavy metals persist in the environment due to their nonbiodegradable nature, leading to long-term ecological and agricultural challenges (Sherameti and Varma 2015). Their accumulation in soil poses a significant threat to plant health, food safety, and human health (Alloway 2012), as given in Table 6.1.

1. *Sources of Heavy Metal Contamination*

   (i) *Industrial Activities*: Mining, smelting, and manufacturing processes release heavy metals into the environment as waste, leading to soil contamination (Tóth et al. 2016).

Table 6.1 Heavy metal sources, contamination mechanisms, and effects

S#	Heavy metal	Primary sources	Mechanisms of contamination	Effects on soil and plants	References
1	Cadmium (Cd)	Fertilizers, mining, sewage sludge	Adsorption onto clay particles, water infiltration	Inhibits nutrient uptake, causes oxidative stress in plants	Kumar et al. (2024)
2	Lead (Pb)	Industrial emissions, leaded gasoline	Atmospheric deposition, runoff	Toxic to soil microbes, interferes with enzyme activities	Dotaniya et al. (2020)
3	Arsenic (As)	Mining, pesticides	Adsorption and accumulation in topsoil	Inhibits photosynthesis, disrupts plant metabolism	Zhao et al. (2010)
4	Mercury (Hg)	Industrial waste, fossil fuel combustion	Deposition from the atmosphere, runoff	Toxic to microorganisms, impairs seed germination and growth	Azevedo and Rodriguez (2012)
5	Chromium (Cr)	Tannery waste, electroplating	Leaching from industrial effluents	Reduces soil microbial diversity, toxic to root development	Oliveira (2012)
6	Nickel (Ni)	Mining, industrial waste	Accumulates in soil organic matter	Alters enzymatic activities, reduces chlorophyll content	Ahmad and Ashraf (2012)

(ii) *Agricultural Practices*: The extensive use of pesticides, fertilizers, and sewage sludge introduces heavy metals like Cd, Pb, and As into the soil (Li et al. 2019).

(iii) *Urbanization*: Urban runoff and improper waste disposal contribute to heavy metal deposition in soils around cities (Li et al. 2019).

(iv) *Atmospheric Deposition*: Emissions from vehicles and industries deposit heavy metals onto the soil through precipitation (He et al. 2015).

2. *Impact of Heavy Metals on Soil Health*

Heavy metals disrupt soil microbial activity and nutrient cycling, leading to reduced soil fertility. They also form complexes with organic matter, making essential nutrients unavailable to plants (Mani and Kumar 2014). The toxic effects of heavy metals impair plant growth by disrupting physiological processes such as photosynthesis, respiration, and enzymatic activities.

3. *Challenges of Heavy Metal Contamination*

Heavy metal contamination is persistent due to the nonbiodegradability of these elements. Their bioaccumulation in plants leads to trophic transfer in food chains, raising concerns about food safety and human health (Nagajyoti et al. 2010).

Furthermore, contaminated soils pose difficulties in agricultural productivity, requiring remediation strategies for sustainable crop cultivation.

4. *Remediation Strategies*

   (i) *Phytoremediation*: The use of plants like *Brassica juncea* and *Helianthus annuus* to extract or stabilize heavy metals in contaminated soils (Ali et al. 2013).
   (ii) *Chemical Amendments*: Adding materials such as lime or biochar to immobilize heavy metals and reduce their bioavailability (Hasan et al. 2019).
   (iii) *Bioremediation*: Employing soil microorganisms to degrade or transform heavy metals into less toxic forms (Mani and Kumar 2014).

6.3 Importance of Oilseed Crops

Oilseed crops are essential components of global agriculture and economies, serving as vital sources of edible oils, animal feed, biofuels, and raw materials for industrial applications. Their adaptability to diverse agroclimatic conditions and significant contribution to nutritional security, sustainable agriculture, and economic development make them indispensable in modern agricultural systems (Table 6.2).

(i) *Source of Edible Oils*: Oilseed crops such as soybean (*Glycine max*), sunflower (*Helianthus annuus*), canola (*Brassica napus*), and groundnut (*Arachis hypogaea*) are primary sources of cooking oil worldwide. They supply essential fatty acids, including omega-3 and omega-6, which are crucial for human health (Oilseeds U 2020).
(ii) *Nutritional Contribution*: These crops are rich in proteins, vitamins, and minerals. For example, soybean provides high-quality protein, while canola oil contains low levels of saturated fats and high monounsaturated fats, promoting cardiovascular health (Durrett et al. 2008).
(iii) *Biofuel Production*: With increasing energy demands, oilseeds such as jatropha (*Jatropha curcas*) and palm (*Elaeis guineensis*) have gained prominence as feedstocks for biodiesel production (Embaye et al. 2018). These renewable energy sources reduce greenhouse gas emissions and dependency on fossil fuels (Demirbas 2009).
(iv) *Economic Significance*: Oilseed crops contribute substantially to global trade. Countries like India, the USA, and Brazil rely heavily on oilseed exports for economic growth (Chekhova 2021). The oilseed industry also supports millions of livelihoods in farming and processing sectors (Lafarga 2021).
(v) *Soil Health and Crop Rotation*: Oilseeds such as mustard and sunflower are often included in crop rotations to improve soil fertility, prevent pest buildup, and enhance farm sustainability (Eynck et al. 2013).

Table 6.2 Contributions and uses of oilseed crops

S#	Oilseed crop	Primary products	Nutritional/industrial importance	Economic significance	References
1	Soybean	Edible oil, soy protein	Rich in protein and omega-3 fatty acids	Major export crop, widely grown in the USA and Brazil	Goldsmith (2008)
2	Sunflower	Edible oil, sunflower meal	High in vitamin E, poly-unsaturated fats	Key crop in Europe and Asia	Tursunov et al. (2023)
3	Canola (rapeseed)	Canola oil, animal feed	Low saturated fat, high monounsaturated fats	Significant for bio-diesel production	Friedt et al. (2018)
4	Groundnut (peanut)	Edible oil, peanut butter	Rich in protein, vitamin B, and minerals	Major crop in Asia and Africa	Neves et al. (2023)
5	Palm	Palm oil, palm kernel oil	Used in cooking, biofuels, and cosmetics	Top export crop for Indonesia and Malaysia	Soomro et al. (2023)
6	Castor	Castor oil	Industrial applications in lubricants, cosmetics, and biopolymers	Cultivated in arid regions	Pari et al. (2020)
7	Jatropha	Biofuel	Nonedible oil used in biodiesel production	Emerging crop for renewable energy markets	Islam et al. (2011)
8	Mustard	Mustard oil, animal feed	Rich in omega-3 fatty acids; enhances soil health in crop rotations	Popular in South Asian cuisines	Rakow (2004)

(vi) *Industrial Uses*: Nonedible oils from seeds like castor (*Ricinus communis*) are used in the production of cosmetics, lubricants, paints, and bio-polymers, showcasing the versatility of oilseed crops (Gunstone 2009).

6.4 Role of Biostimulants in Agriculture

Biostimulants are natural or synthetic substances applied to plants or the soil to enhance crop growth, nutrient uptake, stress resistance, and overall yield as depicted in Fig. 6.2. Unlike fertilizers, biostimulants do not provide direct nutrients but influence the plant's physiological and biochemical processes, making them vital for sustainable agricultural practices (Table 6.3).

1. *Key Roles of Biostimulants in Agriculture*

   (i) *Enhancing Nutrient Uptake*: Biostimulants such as humic acids and fulvic acids improve nutrient availability in the soil by chelating nutrients and

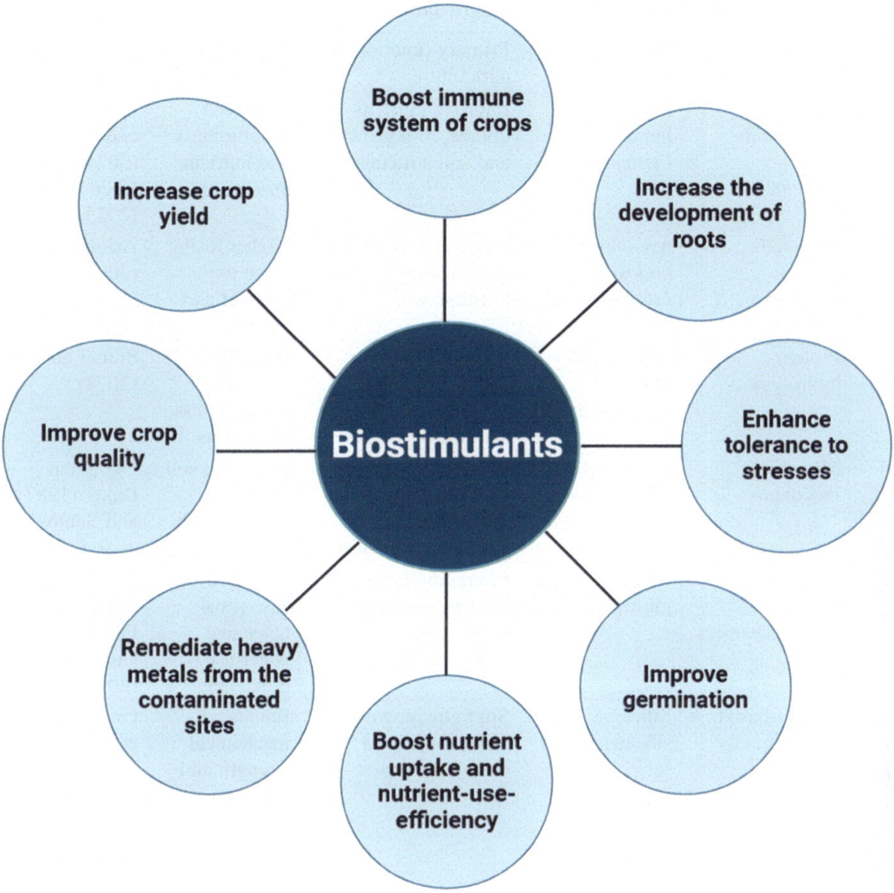

Fig. 6.2 Positive impacts of biostimulants on the different aspects of plants

enhancing root growth. This leads to better nutrient acquisition, particularly in nutrient-deficient soils (Du Jardin 2015).

(ii) *Improving Plant Stress Tolerance*: Under abiotic stresses like drought, salinity, and extreme temperatures, biostimulants such as seaweed extracts and protein hydrolysates activate stress-related pathways in plants. They enhance antioxidant activities, osmotic adjustment, and hormonal balance (Rouphael and Colla 2018).

(iii) *Stimulating Plant Growth*: Plant growth regulators promote cell division, elongation, and differentiation. This leads to improved plant height, leaf area, and overall biomass (Yakhin et al. 2017).

(iv) *Enhancing Soil Microbial Activity*: Biostimulants such as microbial inoculants stimulate beneficial soil microorganisms, enhancing soil fertility,

Table 6.3 Types of biostimulants and their agricultural benefits

S#	Types of biostimulants	Examples	Primary functions in agriculture	Impact on crops	References
1	Humic and fulvic acids	Potassium humate, Leonardite	Improve nutrient uptake, root growth, and soil structure	Enhanced root biomass and nutrient absorption	Shahrajabian and Sun (2024) and Chen et al. (2025)
2	Seaweed extracts	*Ascophyllum nodosum*, *Laminaria* sp.	Increase stress tolerance, stimulate plant hormones	Higher resistance to drought and salinity	Arioli et al. (2024)
3	Protein hydrolysates	Animal and plant hydrolysates	Enhance nitrogen assimilation and stress-related enzyme activities	Increased yield under abiotic stress conditions	Sun et al. (2024)
4	Microbial inoculants	Rhizobacteria, mycorrhiza	Promote nitrogen fixation, phosphorus solubilization, and pathogen resistance	Improved soil health and plant growth	Okon and Baker (1987) and Santos et al. (2019)
5	Chitosan and other polysaccharides	Chitosan, beta-glucans	Elicit defense mechanisms, improve water retention	Increased disease resistance and water use efficiency	Orzali et al. (2017) and Picos-Corrales et al. (2023)
6	Silicon-based biostimulants	Silicic acid, silicates	Strengthening cell walls, improved drought and pest resistance	Enhanced mechanical strength and stress resilience	Grankina et al. (2022)

organic matter decomposition, and nutrient cycling (Wadduwage et al. 2023).

(v) *Boosting Crop Yield and Quality*: The application of biostimulants improves crop yield and enhances the quality of produce, such as higher oil content in oilseeds, increased sugar levels in fruits, and improved shelf life of vegetables (Calvo et al. 2014).

6.5 Types of Biostimulants Used in Oilseed Crops under Heavy Metal Stress

Biostimulants are synthetic or natural substances that boost stress tolerance, nutrient absorption and plant growth. Biostimulants mitigate toxicity and increase the productivity of the oilseed crops, including sunflower, soybean, rapeseed, and flax under heavy metal stress conditions.

6.5.1 Use of Plant-Derived Extracts to Boost HMs Stress Tolerance in Oilseed Crops

The study of Bashir et al. (2021) explored the effects of various types of plant straw biochar on sunflower. Results indicated that the application of biochar significantly reduced the cadmium mobility in soil. The study showed that rice husk induced biochar significantly minimized oxidative stress and reduced catalase, peroxidase and superoxide dismutase activities by 49%, 40.5%, and 46.5%, respectively as compared to the control. Houben et al. (2013) stated that the addition of miscanthus-derived biochar at a rate of 5% and 10% minimized the metal contents in shoot of *Brassica napus* by 47% and 75%, for cadmium, 65% and 91% for zinc, and 33% and 59% for lead. Biochar applied at 2% significantly increased the fresh biomass of root and shoot, total photosynthetic pigments and lycopene contents, respectively in rapeseed (Kamran et al. 2020). Furthermore, biochar also enhanced the activities of antioxidant enzymes through reducing the oxidative stress in rapeseed leaves (Kamran et al. 2020). Spermidine (Spd), a phytohormone involved in various abiotic stresses. The results of Gu et al. (2022) revealed that exogenous application of Spd improved the rice seedling tolerance to cadmium and lead stress by increasing plant height, dry and fresh weight of roots and shoots as well as root length. In addition, Spd decreased the contents of H_2O_2, O_2^-, MDA, Cd, and Pb and increased photosynthetic contents including chlorophyll and carotenoids in roots and shoots of rice seedlings under heavy metal stress (Gu et al. 2022).

6.5.2 Use of Fresh and Marine (Seaweed) Algae Extract to Boost HMs Stress Tolerance in Oilseed Crops

Seaweeds are nutrient-rich and can improve soil fertility by adding essential organic contents and nutrients to the soil. *Sargassum ilicifolium* used both alone and in combination with plant growth-promoting rhizobacteria (PGPR), revealed a high level of biodegradation capacity of heavy metals such as lead (99.06% degradation) and nickel (98.8% degradation) (Rahman et al. 2024). The morphological features such as shoot and root length and fresh weight of sunflower were increased under lead and nickel contaminations (Rahman et al. 2024). Furthermore, the biochemical analysis including proline, total phenols, peroxidase (POD), superoxide dismutase (SOD), and 2,2-Diphenyl-1-picrylhydrazyl)-radical (DPPH) scavenging activity was improved under heavy metal stress (Rahman et al. 2024). Copper plays a significant role in influencing the photosynthesis, plant growth, biomass and cadmium accumulation characteristics of hemp plant (uz Zaman et al. 2025). Biostimulants significantly reduce the cadmium accumulation in the hemp. However, the seaweed extract (SWE) mitigated cadmium accumulation by improving antioxidant enzyme activity, plant growth as well as osmoprotectants levels in hemp under cadmium stress (uz Zaman et al. 2025).

6.5.3 Use of Humic Substances to Boost HMs Stress Tolerance in Oilseed Crops

Humic acid alleviates heavy metal toxicity in rapeseed by enhancing plant growth, decreasing metal accumulation, and minimizing oxidative stress, particularly the most pronounced effect observed at 1000 mg/kg (HA2) (Boysan Canal et al. 2022). The application of HA in heavy metal-contaminated soil significantly reduced the metal accumulation in roots and shoots of the rapeseed. Humic acid applied in heavy metal-contaminated soil has a pronounced effect in improving rapeseed growth through minimizing heavy metal and oxidative stress (Boysan Canal et al. 2022). The study conducted by Asadi Aghbolaghi et al. (2022) revealed that Humic acid reduced the adverse effects of Cd stress by enhancing seed vigor and germination percentage in pumpkin seeds. Humic acid also improved the activity antioxidant enzymes such as SOD, CAT, and POD, as well as glyoxylate cycle enzymes (including isocitrate lyase and malate synthase) (Asadi Aghbolaghi et al. 2022). These biochemical changes indicated improved tolerance to oxidative stress, enabling the seeds to better withstand cadmium stress.

6.5.4 Use of Microbial Biostimulants to Boost HMs Stress Tolerance in Oilseed Crops

Inoculation with arbuscular mycorrhizal fungi (AMF) plays a vital role in improving date palm growth under heavy metals stress (Abdel Latef 2013; Ghadbane et al. 2021). The study conducted by Ghadbane et al. (2021) showed the potential of isolates from AMF, *Glomus* spp. (A01), and PGPR, *Rhizobium leguminosarum* (GM01) for phytoremediation and managing lead absorption by date palm. The result demonstrated that date palm plants treated with A01 and GM01 improved growth and biomass accumulation through increasing root and shoot fresh weight, root length as well as germination index. PGPR significantly mitigated the adverse effects of cadmium in date palm plants. The authors proposed that PGPR contributed to improving nutrient availability, suppression of pathogens, and increasing root branching in date palm. A study conducted by Elsadig et al. (2017) showed that a substantially improved the nutritional content including nitrogen (N), phosphorus (P), potassium (K), calcium (Ca), magnesium (Mg), iron (Fe), zinc (Zn), manganese (Mn), and copper (Cu) in the date palm fruits of the "Khenazi" cultivar. Therefore, organic manure serves as an organic matter enhancer and nutrient reservoir, which contributes to reducing heavy metal contamination.

6.6 Impact of Heavy Metals on Oilseed Crops

Heavy metal contamination in agricultural systems is a pressing issue affecting crop growth, yield, and quality. Metals can accumulate in soil due to industrial waste, pesticide use, and contaminated irrigation water, directly impacting oilseed crops. Oilseed crops, such as sunflower (Abeed et al. 2022; Gopal and Khurana 2011), canola (Ramazani and Kargar 2023), soybean (Miransari 2016), and sesame (Chandrasekaran et al. 2014), are particularly susceptible to heavy metal stress due to their physiological and biochemical sensitivities.

1. *Physiological and Biochemical Effects*

   (i) *Disruption of Germination and Growth*: Heavy metals inhibit seed germination, root elongation, and shoot growth by interfering with water uptake and enzyme activity. Cadmium, for instance, reduces germination rates and seedling vigor in sunflower and mustard (Ul Hassan et al. 2017).

   (ii) *Photosynthesis Inhibition*: Heavy metals like Pb and Cd disrupt photosynthesis by damaging chloroplast structures, reducing chlorophyll synthesis, and impairing light reactions. In rapeseed, Pb toxicity decreases photosynthetic efficiency, reducing biomass production (Mearaji et al. 2021).

   (iii) *Oxidative Stress*: Heavy metals induce the overproduction of reactive oxygen species (ROS), leading to oxidative damage in lipids, proteins, and DNA. This imbalance negatively affects metabolic processes in oilseed crops, particularly in soybean and sesame (Abirami 2023) (Miransari 2016).

   (iv) *Nutrient Imbalance*: Metals like Cd and Cr compete with essential nutrients (e.g., Zn and Fe) for uptake, causing deficiencies and metabolic dysfunction. For example, cadmium reduces zinc uptake, impairing enzyme activities in soybean (Shaffique et al. 2023).

   (v) *Yield Reduction*: Heavy metal stress directly reduces flowering, fruit set, and seed development in oilseed crops. Yield losses in sunflower and canola are notable under cadmium and arsenic contamination (Azeem et al. 2017; BILAL et al. 2024).

2. *Heavy Metal Stress Mechanisms in Oilseed Crops*

   (i) *Metal Uptake and Transport*: Heavy metals are absorbed by roots and transported to shoots and seeds. This accumulation disrupts cellular functions and reduces crop productivity. In sunflower, cadmium is preferentially transported to aerial parts, impacting growth and oil quality (Ali and Gill 2022).

   (ii) *Enzymatic and Hormonal Disruption*: Metals interfere with enzymatic pathways and disrupt hormonal balance, leading to abnormal growth patterns (Ul Hassan et al. 2017).

   (iii) *Antioxidant Defense Systems*: Oilseed crops activate antioxidant enzymes such as superoxide dismutase (SOD) and catalase (CAT) to mitigate ROS damage (Ul Hassan et al. 2017). However, excessive metal exposure overwhelms these defenses, causing cellular damage.

3. *Management Strategies*

   (i) *Biostimulants*: Application of biostimulants can enhance heavy metal tolerance by improving nutrient uptake, chelating metals, and activating stress response pathways.
   (ii) *Phytoremediation*: Oilseed crops like sunflower are used for phytoextraction, a strategy to remediate metal-contaminated soils (Rizwan et al. 2016).
   (iii) *Soil Amendments*: Organic matter and biochar reduce metal bioavailability, improving soil conditions for crop growth (de Abreu et al. 2012).
   (iv) *Breeding and Biotechnology*: Advances in genomics and marker-assisted selection (MAS) aim to develop heavy metal-tolerant oilseed varieties (Elango et al. 2022).

6.7 Mechanisms of Biostimulant Action in Heavy Metal Stress Reduction

Biostimulants, composed of various organic and natural compounds, play a crucial role in mitigating the adverse effects of heavy metal stress in plants, including oilseed crops. They enhance plant growth, nutrient uptake, and stress tolerance by modulating physiological, biochemical, and molecular mechanisms. Below is a detailed description of the mechanisms by which biostimulants help reduce heavy metal stress:

1. *Chelation of Heavy Metals*

Certain biostimulants contain compounds like organic acids, amino acids, and polysaccharides that can chelate heavy metals, reducing their bioavailability and toxicity (Sears 2013). These chelators bind to metal ions in the soil, forming complexes that are less likely to be absorbed by plant roots, thereby protecting plants from metal toxicity (Sears 2013). For example, humic and fulvic acids in biostimulants act as chelating agents, immobilizing heavy metals like cadmium (Cd) and lead (Pb) (Rathor et al. 2023).

2. *Enhanced Antioxidant Defense*

Heavy metals induce oxidative stress by generating ROS such as superoxide radicals and hydrogen peroxide (Mansoor et al. 2023). Biostimulants enhance the activity of antioxidant enzymes like SOD, CAT, and POD, which neutralize ROS and prevent cellular damage (Akensous et al. 2022). Seaweed extracts, for instance, stimulate antioxidant enzyme activities, reducing oxidative damage in crops exposed to metal stress (Khan et al. 2009).

3. *Improved Nutrient Uptake and Utilization*

Heavy metal stress disrupts nutrient uptake by competing with essential nutrients (e.g., Zn, Fe, and Mn) (Kaur et al. 2021). Biostimulants facilitate the uptake and translocation of essential nutrients, restoring metabolic balance (Kaur et al. 2021). Amino acid-based biostimulants, such as those containing glycine and glutamic acid, enhance nutrient absorption and mitigate competition from heavy metals like cadmium and chromium (Rouphael and Colla 2020a).

4. *Activation of Stress-Responsive Pathways*

Biostimulants modulate stress-related signaling pathways, activating genes involved in detoxification and stress tolerance (Franzoni et al. 2022). They upregulate the synthesis of stress-related proteins such as heat shock proteins (HSPs) and metallothioneins, which bind heavy metals and reduce their toxicity (Ali et al. 2020). Hormone-like compounds in biostimulants, including auxins, cytokinins, and abscisic acid analogs, help plants adapt to stress conditions (Nephali et al. 2020).

5. *Reduction of Metal Uptake*

Certain biostimulants, particularly those containing seaweed extracts or microbial inoculants, alter root cell permeability, limiting heavy metal uptake (Deolu-Ajayi et al. 2022). They also promote root exudation of organic compounds, which immobilize heavy metals in the rhizosphere, reducing their availability to plants.

6. *Improvement in Soil Health*

Biostimulants improve soil structure and microbial activity, enhancing the overall soil environment and reducing heavy metal toxicity. Microbial biostimulants, such as those containing nitrogen-fixing or phosphate-solubilizing bacteria, promote the biodegradation of toxic compounds and improve plant resilience to metal stress (Ali et al. 2022). For example, AMF in biostimulants enhance the sequestration of metals in soil, preventing their uptake by plants (Yadav et al. 2024).

7. *Promotion of Phytochelatin and Metallothionein Production*

Biostimulants stimulate the production of phytochelatins and metallothioneins, which are metal-binding peptides that detoxify heavy metals within plant cells (Carillo et al. 2023). These peptides sequester heavy metals into vacuoles, preventing their interference with metabolic processes (Gupta et al. 2013).

8. *Osmotic Regulation*

Heavy metal stress can disrupt water uptake and cause osmotic stress in plants. Biostimulants containing betaines, proline, and other osmolytes help maintain osmotic balance, protecting plants from dehydration and ion toxicity (Nephali et al. 2020).

6.8 Applications of Biostimulants in Oilseed Crops

Biostimulants have emerged as a promising solution to enhance the growth, yield, and stress resilience of oilseed crops such as sunflower, canola, soybean, and mustard (Singh et al. 2022). These natural or synthetic substances improve plant vigor by modulating physiological and biochemical pathways, improving nutrient availability, and boosting resistance to abiotic and biotic stresses (Singh et al. 2022). Below is a detailed discussion of their applications in oilseed crops (Table 6.4)

1. *Enhancing Seed Germination and Early Growth*

Biostimulants improve seed germination rates and early seedling vigor, which are critical for the successful establishment of oilseed crops (Makhaye et al. 2021). For example, amino acid-based biostimulants have been shown to enhance seedling growth in sunflower by stimulating cell division and elongation (Rouphael and Colla 2020a).

2. *Improving Nutrient Uptake*

Biostimulants facilitate the uptake of essential macro- and micronutrients in oilseed crops by enhancing root architecture and nutrient solubility (Siwik-Ziomek and Szczepanek 2018). Humic and fulvic acids improve nutrient absorption and translocation, leading to better plant growth and higher oil content in crops like canola (Calvo et al. 2014).

3. *Alleviating Abiotic Stresses*

Oilseed crops often face abiotic stresses such as drought, salinity, and heat. Biostimulants help mitigate these stresses by enhancing antioxidant enzyme activity,

Table 6.4 Applications and mechanisms of oilseed crops

S#	Application	Mechanism	Example in oilseed crops	References
1	Seed germination and early growth	Enhances seedling vigor through hormone-like activity	Improved germination in sunflower	Rouphael and Colla (2020a)
2	Nutrient uptake	Enhances root architecture and nutrient solubility	Increased N and P uptake in canola	Calvo et al. (2014)
3	Abiotic stress tolerance	Boosts antioxidant activity and osmotic regulation	Drought tolerance in soybean	Khan et al. (2009)
4	Heavy metal stress reduction	Chelates metals and enhances antioxidant defenses	Reduced cadmium toxicity in sunflower	Rathor et al. (2023)
5	Enhanced photosynthesis	Increases chlorophyll synthesis and photosynthetic efficiency	Higher photosynthetic rates in canola	Colla et al. (2015)
6	Improved oil yield and quality	Modulates lipid metabolism and fatty acid biosynthesis	Enhanced oil yield in soybean	Mannan et al. (2023)
7	Soil health and microbial activity	Promotes beneficial microbes and nutrient cycling	Better phosphorus uptake in mustard	Yadav et al. (2024)

reducing oxidative damage and modulating osmotic balance through the accumulation of osmolytes like proline and betaines (Ma et al. 2022). Seaweed extracts, for instance, improve drought tolerance in soybean by upregulating stress-related genes and enhancing water use efficiency (Khan et al. 2009).

4. *Boosting Resistance to Heavy Metal Stress*

Biostimulants play a vital role in reducing heavy metal toxicity by chelating metals, enhancing phytoremediation, and improving antioxidant defenses (Akensous et al. 2022). Applications of seaweed extracts and microbial inoculants in sunflower have been reported to mitigate cadmium and lead toxicity (Rathor et al. 2023).

5. *Enhancing Photosynthesis and Biomass Production*

Biostimulants stimulate chlorophyll synthesis and improve photosynthetic efficiency, leading to higher biomass and oil yield (Akensous et al. 2022). Cytokinin-containing biostimulants increase photosynthetic rates in canola under normal and stress conditions (Colla et al. 2015).

6. *Promoting Yield and Oil Quality*

Biostimulants enhance the quantity and quality of oil in oilseed crops by influencing lipid metabolism and fatty acid biosynthesis (Zouari et al. 2020). For example, the application of algae-based biostimulants in soybean increased oil yield and improved the concentration of unsaturated fatty acids (Mannan et al. 2023).

7. *Stimulating Soil Health and Microbial Activity*

Biostimulants improve soil health by promoting beneficial microbial activity, enhancing organic matter decomposition, and improving nutrient cycling (Wadduwage et al. 2023). AMF-based biostimulants improve phosphorus availability in mustard, boosting overall plant growth and productivity (Yadav et al. 2024).

8. *Reducing Chemical Input Requirements*

Biostimulants can reduce dependency on chemical fertilizers and pesticides by enhancing natural plant defenses and nutrient use efficiency (Rouphael and Colla 2020a). This sustainable approach lowers production costs and minimizes environmental impact, making it especially beneficial for oilseed crop production.

6.8.1 Methods of Application

The application methods for biostimulants in agriculture play a critical role in determining their efficacy. These methods are designed to ensure optimal delivery of biostimulant compounds to plants, allowing for improved uptake, utilization, and benefits (Rouphael and Colla 2020b). The key methods include foliar spraying, soil application, seed treatment, fertigation, and root dipping, each with its specific advantages and applications.

1. *Foliar Spraying*

Foliar application involves spraying biostimulant solutions directly onto plant leaves, allowing for rapid absorption through stomata and cuticles. This method is widely used for providing quick nutritional and stress relief benefits, especially under abiotic stress conditions like drought or salinity. Foliar spraying delivers biostimulants such as amino acids, seaweed extracts, and micronutrients efficiently to plant tissues bypassing soil-related limitations like nutrient immobilization (Ibraheim 2014). For example, the foliar application of seaweed extracts has been shown to improve photosynthetic efficiency and antioxidant enzyme activity in soybean (Khan et al. 2009). This method is particularly useful in oilseed crops such as sunflower, where enhanced chlorophyll synthesis and nutrient uptake have been observed following foliar treatment (Csótó et al. 2024).

2. *Soil Application*

Biostimulants can be applied directly to the soil as granules, powders, or liquid solutions. This method improves soil fertility and enhances root–soil interactions by promoting beneficial microbial activity and nutrient solubility (Chernikova et al. 2021). Humic acids and microbial-based biostimulants are commonly applied through this method to enhance root growth and nutrient availability (Savarese et al. 2022). In canola, soil-applied humic substances have been found to improve root development and water retention, enabling better plant performance under water stress (Calvo et al. 2014). This approach is also effective for reducing heavy metal toxicity, as certain biostimulants can immobilize metals in the rhizosphere and prevent their uptake by plants (Rathor et al. 2023).

3. *Seed Treatment*

Seed treatment involves coating or soaking seeds with biostimulant solutions before planting. This method is designed to enhance germination, seedling vigor, and resistance to early-stage stress (Chen et al. 2024). Biostimulants like amino acids, protein hydrolysates, and microbial inoculants have been used to treat oilseed crop seeds, resulting in better germination rates and initial growth. For instance, the application of protein hydrolysates to sunflower seeds has been reported to improve seedling emergence and stress tolerance (Rouphael and Colla 2020a). Seed treatment is also a sustainable and cost-effective method, as it requires smaller quantities of biostimulants compared to other application methods (Chen et al. 2024).

4. *Fertigation*

Fertigation involves the application of biostimulants through irrigation systems, allowing the simultaneous delivery of water and nutrients (Agliassa et al. 2021). This method is efficient for large-scale applications and ensures uniform distribution of biostimulants across fields. Biostimulants like seaweed extracts, humic acids, and microbial consortia are often applied via fertigation to improve nutrient use efficiency and plant resilience (Agliassa et al. 2021). In soybean, fertigation with seaweed-based biostimulants has been found to enhance root growth and nodulation, contributing to higher yields under drought conditions (Mannan et al. 2023).

5. *Root Dipping*

Root dipping involves immersing the roots of seedlings or transplants in a biostimulant solution before planting (Dutta et al. 2023). This method is particularly effective for improving root establishment and early growth. It is commonly used for transplantable oilseed crops like mustard, where root-dipping in microbial-based biostimulants has been shown to promote root elongation and improve resistance to soil-borne pathogens (Yadav et al. 2024). Root dipping also helps in reducing transplant shock, ensuring better survival rates and faster establishment in field conditions.

6.9 Future Perspectives

The use of biostimulants to improve heavy metal tolerance in oilseed crops is a possible path toward sustainable agriculture. However, numerous areas need more investigation in order to maximize their efficiency and assure large-scale application.

- *Mechanistic Understanding*: Future study should concentrate on determining the specific biochemical and molecular pathways via which biostimulants reduce heavy metal stress. This involves looking at their effects on antioxidant pathways, metal chelation, and root exudation processes.
- *Optimization of Biostimulant Formulations*: It is critical to produce customized biostimulant formulations that are particular to various oilseed crops and degrees of soil pollution. To find the most effective combinations of bioactive chemicals, comprehensive field experiments and laboratory investigations are necessary.
- *Synergistic Approaches*: Combining biostimulants with other sustainable agriculture methods, such as phytoremediation, soil amendments, and microbial inoculants, might increase their effectiveness in alleviating heavy metal stress.
- *Genetic Interactions and Breeding Programs*: Understanding how biostimulants interact with plant genetics might help produce oilseed crops that are more resistant to heavy metals. Combining biostimulant treatments with genetically enhanced types may provide better outcomes.
- *Field-Scale Applications and Economic Viability*: Large-scale field experiments are required to determine the long-term advantages and cost-effectiveness of biostimulant usage under various agroclimatic conditions. Economic feasibility studies will aid in their implementation by farmers and policymakers.
- *Regulatory Framework and Environmental Safety*: Standardizing the usage of biostimulants via regulatory frameworks is critical for ensuring their safety and efficacy. Additionally, examining their environmental effect will help to minimize any unanticipated ecological implications.

- *Climate Change resistance*: As climate change exacerbates environmental stress, biostimulants may play an important role in increasing oilseed crop resistance to heavy metal toxicity as well as other abiotic stressors like drought and salt.

By addressing these future goals, biostimulants may be efficiently used to increase the productivity and sustainability of oilseed crops in heavy metal-contaminated soils, therefore contributing to food security and environmental health.

6.10 Conclusion

The use of biostimulants has emerged as a viable technique for reducing heavy metal stress in oilseed crops, improving growth, physiological resilience, and yield potential. Biostimulants such as humic compounds, protein hydrolysates, seaweed extracts, and microbial inoculants have been shown to lower metal toxicity by enhancing antioxidant defense, metal chelation, and nutrient absorption. These processes lead to improved plant tolerance and total production, making biostimulants an environmentally acceptable and sustainable method of cultivation in polluted soils. However, further study is required to optimize administration techniques, dose, and synergistic interactions with other agronomic practices to maximize their performance under a variety of environmental circumstances.

References

Abdel Latef AA (2013) Growth and some physiological activities of pepper (Capsicum annuum L.) in response to cadmium stress and mycorrhizal symbiosis. Journal of Agricultural Science and Technology. 15:1437–1448

Abeed AHA, Mahdy RE, Alshehri D, Hammami I, Eissa MA, Abdel Latef AA, Mahmoud GA (2022) Induction of resilience strategies against biochemical deteriorations prompted by severe cadmium stress in sunflower plant when Trichoderma and bacterial inoculation were used as biofertilizers. Frontiers in Plant Science 13:1004173

Abirami K (2023) Response of heavy metals stress during seed germination and early seedling growth in oil crop black sesame (Sesamum indicum L.). J Stress Physiol Biochem 19:28–42

Agliassa C, Mannino G, Molino D et al (2021) A new protein hydrolysate-based biostimulant applied by fertigation promotes relief from drought stress in Capsicum annuum L. Plant Physiol Biochem 166:1076–1086

Ahmad MSA, Ashraf M (2012) Essential roles and hazardous effects of nickel in plants. In: Whitacre DM (ed) Reviews of environmental contamination and toxicology. Springer, New York, pp 125–167

Akensous F-Z, Anli M, Meddich A (2022) Biostimulants as innovative tools to boost date palm (Phoenix dactylifera L.) performance under drought, salinity, and heavy metal (Oid) s' stresses: a concise review. Sustain For 14:15984

Ali B, Gill RA (2022) Heavy metal toxicity in plants: recent insights on physiological and molecular aspects, volume II. Front Plant Sci 13:1016257

Ali H, Khan E, Sajad MA (2013) Phytoremediation of heavy metals—concepts and applications. Chemosphere 91:869–881

Ali Q, Shehzad F, Waseem M et al (2020) Plant-based biostimulants and plant stress responses. In: Hasanuzzaman M (ed) Plant ecophysiology and adaptation under climate change: mechanisms and perspectives I. Springer, Singapore, pp 625–661

Ali S, Moon Y-S, Hamayun M et al (2022) Pragmatic role of microbial plant biostimulants in abiotic stress relief in crop plants. J Plant Interact 17:705–718. https://doi.org/10.1080/17429145.2022.2091801

Alloway BJ (2012) Heavy metals in soils: trace metals and metalloids in soils and their bioavailability. Springer Science & Business Media

Arioli T, Mattner SW, Islam MT et al (2024) Applications of seaweed extracts in agriculture: an Australian perspective. J Appl Phycol 36:713–726. https://doi.org/10.1007/s10811-023-03120-x

Asadi Aghbolaghi M, Sedghi M, Sharifi RS, Dedicova B (2022) Germination and the biochemical response of pumpkin seeds to different concentrations of humic acid under cadmium stress. Agriculture 12:374

Azeem W, Ashraf M, Shahzad SM et al (2017) Phosphate-arsenate relations to affect arsenic concentration in plant tissues, growth, and antioxidant efficiency of sunflower (Helianthus annuus L.) under arsenic stress. Environ Sci Pollut Res 24:24376–24386

Azevedo R, Rodriguez E (2012) Phytotoxicity of mercury in plants: a review. J Bot 2012:1–6. https://doi.org/10.1155/2012/848614

Bartucca ML, Cerri M, Del Buono D, Forni C (2022) Use of biostimulants as a new approach for the improvement of phytoremediation performance—a review. Plan Theory 11:1946

Bashir S, Qayyum MA, Husain A et al (2021) Efficiency of different types of biochars to mitigate cd stress and growth of sunflower (Helianthus; L.) in wastewater irrigated agricultural soil. Saudi J Biol Sci 28:2453–2459

Bilal K, Elahi NN, Imtiaz M (2024) Selection of canola (BRASSICA NAPUS) based on agronomic and physiological responses for better growth under cadmium toxicity. Pak J Bot 56:2139–2150

Boysan Canal S, Bozkurt M, Yılmaz H (2022) The effect of humic acid on plant growth, phytoremediation and oxidative stress in rapeseed (Brassica napus L.) grown under heavy metal stress. Yüzüncü Yıl Üniversitesi Ziraat Fakültesi Tarım Bilimleri Dergisi 32

Calvo P, Nelson L, Kloepper JW (2014) Agricultural uses of plant biostimulants. Plant Soil 383:3–41. https://doi.org/10.1007/s11104-014-2131-8

Carillo P, Papa S, Rouphael Y (2023) Multiple levels of crosstalks in biostimulant modulation of plant metal homeostasis. In: Biostimulants in alleviation of metal toxicity in plants. Elsevier, pp 123–147

Chandrasekaran VRM, Hsu D, Liu M (2014) Beneficial effect of sesame oil on heavy metal toxicity. J Parenter Enter Nutr 38:179–185. https://doi.org/10.1177/0148607113490960

Chekhova IV (2021) Socio-economic significance of oil crops products

Chen B, Fu J, Zheng Y et al (2024) Biostimulants in agriculture: innovations in seed treatment and coating technologies. Seed Sci Technol 52:227–249

Chen Y-H, Liu J, Liu Y, Liu Z-Q (2025) Biochemical sources of fulvic acid and its application in agriculture. MROC 22:400–410. https://doi.org/10.2174/0118756298321828240826074104

Chernikova O, Mazhaysky Y, Buryak S, et al (2021) Comparative analysis of the use of biostimulants on the main types of soil

Colla G, Rouphael Y, Di Mattia E et al (2015) Co-inoculation of *Glomus intraradices* and *Trichoderma atroviride* acts as a biostimulant to promote growth, yield and nutrient uptake of vegetable crops. J Sci Food Agric 95:1706–1715. https://doi.org/10.1002/jsfa.6875

Csótó A, Tóth G, Riczu P et al (2024) Foliar spraying with Endophytic Trichoderma biostimulant increases drought resilience of maize and sunflower. Agriculture 14:2360

de Abreu CA, Coscione AR, Pires AM, Paz-Ferreiro J (2012) Phytoremediation of a soil contaminated by heavy metals and boron using castor oil plants and organic matter amendments. J Geochem Explor 123:3–7

Demirbas A (2009) Biofuels securing the planet's future energy needs. Energy Convers Manag 50: 2239–2249

Deolu-Ajayi AO, Van Der Meer IM, Van Der Werf A, Karlova R (2022) The power of seaweeds as plant biostimulants to boost crop production under abiotic stress. Plant Cell Environ 45:2537–2553. https://doi.org/10.1111/pce.14391

Dotaniya ML, Dotaniya CK, Solanki P et al (2020) Lead Contamination and Its Dynamics in Soil–Plant System. In: Gupta DK, Chatterjee S, Walther C (eds) Lead in plants and the environment. Springer International Publishing, Cham, pp 83–98

Du Jardin P (2015) Plant biostimulants: definition, concept, main categories and regulation. Sci Hortic 196:3–14

Durrett TP, Benning C, Ohlrogge J (2008) Plant triacylglycerols as feedstocks for the production of biofuels. Plant J 54:593–607. https://doi.org/10.1111/j.1365-313X.2008.03442.x

Dutta SK, Layek J, Yadav A et al (2023) Improvement of rooting and growth in kiwifruit (Actinidia deliciosa) cuttings with organic biostimulants. Heliyon 9:e17815

Elango D, Devi KD, Jeyabalakrishnan HK et al (2022) Agronomic, breeding, and biotechnological interventions to mitigate heavy metal toxicity problems in agriculture. J Agric Food Res 10: 100374

Elsadig EH, Aljuburi HJ, Elamin AHB, Gafar MO (2017) Impact of organic manure and combination of NPKS, on yield, fruit quality and fruit mineral content of Khenazi date palm (Phoenix dactylifera L.) cultivar. J Sci Agric 1:10–25081

El-Shahir AA, El-Tayeh NA, Ali OM, Abdel Latef AA, Loutfy N (2021) The effect of Endophytic *Talaromyces pinophilus* on growth, absorption and accumulation of heavy metals of *Triticum aestivum* grown on Sandy soil amended by sewage sludge. Plan Theory 10(12):2659

Embaye WT, Bergtold JS, Archer D et al (2018) Examining farmers' willingness to grow and allocate land for oilseed crops for biofuel production. Energy Econ 71:311–320

Eynck C, Shrestha D, Vollmann J et al (2013) Sustainable oil crops production. In: Singh BP (ed) Biofuel crop sustainability, 1st edn. Wiley, pp 165–204

Fakhar A, Gul B, Rafique M, Ortas I (2022) Use of biostimulants to increase heavy metal tolerance in cereals. In: Abdel Latef AAH (ed) Sustainable remedies for abiotic stress in cereals. Springer Nature, Singapore, pp 575–598

Franzoni G, Cocetta G, Prinsi B et al (2022) Biostimulants on crops: their impact under abiotic stress conditions. Horticulturae 8:189

Friedt W, Tu J, Fu T (2018) Academic and economic importance of Brassica napus rapeseed. In: Liu S, Snowdon R, Chalhoub B (eds) The Brassica napus genome. Springer International Publishing, Cham, pp 1–20

Ghadbane M, Medjekal S, Benderradji L, et al (2021) Assessment of arbuscular mycorrhizal fungi status and rhizobium on date palm (Phoenix dactylifera L.) cultivated in a Pb contaminated soil. In: recent advances in environmental science from the Euro-Mediterranean and surrounding regions (2nd edition) proceedings of 2nd Euro-Mediterranean conference for environmental integration (EMCEI-2), Tunisia 2019. Springer, pp 703–707

Goldsmith PD (2008) Economics of soybean production, marketing, and utilization. In: Soybeans. Elsevier, pp 117–150

Gopal R, Khurana N (2011) Effect of heavy metal pollutants on sunflower. African J Plant Sci 5: 531–536

Grankina A, Bocharnikova E, Matichenkov V (2022) 6 silicon-based biostimulators for sustainable agriculture. Biostimulants for crop production and sustainable agriculture/Mirza Hasanuzzaman 85

Gu J, Hu C, Jia X et al (2022) Physiological and biochemical bases of spermidine-induced alleviation of cadmium and lead combined stress in rice. Plant Physiol Biochem 189:104–114

Gunstone F (2009) The chemistry of oils and fats: sources, composition, properties and uses. Wiley

Gupta DK, Vandenhove H, Inouhe M (2013) Role of Phytochelatins in heavy metal stress and detoxification mechanisms in plants. In: Gupta DK, Corpas FJ, Palma JM (eds) Heavy metal stress in plants. Springer, Berlin/Heidelberg, pp 73–94

Halpern M, Bar-Tal A, Ofek M et al (2015) The use of biostimulants for enhancing nutrient uptake. Adv Agron 130:141–174

Hasan MM, Uddin MN, Ara-Sharmeen I et al (2019) Assisting phytoremediation of heavy metals using chemical amendments. Plan Theory 8:295

He Z, Shentu J, Yang X, et al (2015) Heavy metal contamination of soils: sources, indicators and assessment

Houben D, Evrard L, Sonnet P (2013) Beneficial effects of biochar application to contaminated soils on the bioavailability of Cd, Pb and Zn and the biomass production of rapeseed (Brassica napus L.). Biomass Bioenergy 57:196–204

Ibraheim S (2014) Effect of foliar spray with some biostimulants on growth, yield and seeds quality of pea plants grown in sandy soil. J Appl Sci Res 10:400–407

Islam A, Yaakob Z, Anuar N (2011) Jatropha: a multipurpose plant with considerable potential for the tropics. Sci Res Essays 6:2597–2605

Kamran M, Malik Z, Parveen A et al (2020) Ameliorative effects of biochar on rapeseed (Brassica napus L.) growth and heavy metal immobilization in soil irrigated with untreated wastewater. J Plant Growth Regul 39:266–281

Kaur R, Das S, Bansal S et al (2021) Heavy metal stress in rice: uptake, transport, signaling, and tolerance mechanisms. Physiol Plant ppl.13491:430. https://doi.org/10.1111/ppl.13491

Khan W, Rayirath UP, Subramanian S et al (2009) Seaweed extracts as biostimulants of plant growth and development. J Plant Growth Regul 28:386–399

Kumar S, Shah SH, Vimala Y et al (2022) Abscisic acid: metabolism, transport, crosstalk with other plant growth regulators, and its role in heavy metal stress mitigation. Front Plant Sci 13:972856

Kumar U, Jha AK, Kumar N (2024) Cadmium toxicity in the environment: sources, issues, remediation, and challenges. In: Kumar N (ed) Cadmium toxicity. Springer Nature, Cham, pp 1–28

Lafarga T (2021) Production and consumption of oils and Oilseeds. In: Lafarga T, Bobo G, Aguiló-Aguayo I (eds) Oil and oilseed processing, 1st edn. Wiley, pp 1–21

Li C, Zhou K, Qin W et al (2019) A review on heavy metals contamination in soil: effects, sources, and remediation techniques. Soil Sediment Contam Int J 28:380–394. https://doi.org/10.1080/15320383.2019.1592108

Ma Y, Freitas H, Dias MC (2022) Strategies and prospects for biostimulants to alleviate abiotic stress in plants. Front Plant Sci 13:1024243

Makhaye G, Mofokeng MM, Tesfay S et al (2021) Influence of plant biostimulant application on seed germination. In: Biostimulants for crops from seed germination to plant development. Academic, Amsterdam, pp 109–135

Mani D, Kumar C (2014) Biotechnological advances in bioremediation of heavy metals contaminated ecosystems: an overview with special reference to phytoremediation. Int J Environ Sci Technol 11:843–872

Mannan MA, Yasmin A, Sarker U et al (2023) Biostimulant red seaweed (Gracilaria tenuistipitata var. liui) extracts spray improves yield and drought tolerance in soybean. PeerJ 11:e15588

Mansoor S, Ali A, Kour N et al (2023) Heavy metal induced oxidative stress mitigation and ROS scavenging in plants. Plan Theory 12:3003

Mearaji HS, Ansari A, Igdelou NKM et al (2021) Phytohormones and abiotic stresses: roles of phytohormones in plants under abiotic stresses. In: Handbook of plant and crop physiology. CRC Press, pp 175–213

Miransari M (2016) Soybean production and heavy metal stress. In: Abiotic and biotic stresses in soybean production. Elsevier, pp 197–216

Nagajyoti PC, Lee KD, Sreekanth TVM (2010) Heavy metals, occurrence and toxicity for plants: a review. Environ Chem Lett 8:199–216

Nazari F, Alizadeh AM, Alikord M et al (2023) Impacts of heavy metals in seed crops and oil seed on human health: a treat to food safety-review. Carpat J Food Sci Technol:106

Nephali L, Piater LA, Dubery IA et al (2020) Biostimulants for plant growth and mitigation of abiotic stresses: a metabolomics perspective. Meta 10:505

Neves FP, Leite AR, Mantovani LP et al (2023) The economic importance of the peanuts production chain. Revista Brasileira de Engenharia de Biossistemas 17

Okon Y, Baker R (1987) Microbial inoculants as crop-yield enhancers. Crit Rev Biotechnol 6:61–85. https://doi.org/10.3109/07388558709086985

Oliveira H (2012) Chromium as an environmental pollutant: insights on induced plant toxicity. J Bot 2012:1–8. https://doi.org/10.1155/2012/375843

Orzali L, Corsi B, Forni C, Riccioni L (2017) Chitosan in agriculture: a new challenge for managing plant disease. In: Biological activities and application of marine polysaccharides, vol 10. IntechOpen, pp 17–36

Pari L, Suardi A, Stefanoni W et al (2020) Environmental and economic assessment of castor oil supply chain: a case study. Sustain For 12:6339

Picos-Corrales LA, Morales-Burgos AM, Ruelas-Leyva JP et al (2023) Chitosan as an outstanding polysaccharide improving health-commodities of humans and environmental protection. Polymers 15:526

Rahman A, Shaikh NG, Naseer F (2024) The impact of heavy metal tolerant bacterium and Sargassum ilicifolium as organic fertilizer on growth and antioxidant enzymes of sunflower under lead (Pb) and nickel (Ni) stress

Rakow G (2004) Species origin and economic importance of Brassica. In: Pua E-C, Douglas CJ (eds) Brassica. Springer, Berlin/Heidelberg, pp 3–11

Ramazani SHR, Kargar M (2023) Germination and seedling growth of two canola (Brassica napus L.) cultivars under heavy metals. Environ Resour Res 11:283–290

Rathor P, Gorim LY, Thilakarathna MS (2023) Plant physiological and molecular responses triggered by humic based biostimulants-a way forward to sustainable agriculture. Plant Soil 492:31–60

Rizwan M, Ali S, Rizvi H et al (2016) Phytomanagement of heavy metals in contaminated soils using sunflower: a review. Crit Rev Environ Sci Technol 46:1498–1528. https://doi.org/10.1080/10643389.2016.1248199

Rouphael Y, Colla G (2018) Synergistic biostimulatory action: designing the next generation of plant biostimulants for sustainable agriculture. Front Plant Sci 9:1655

Rouphael Y, Colla G (2020a) Biostimulants in agriculture. Front Plant Sci 11:40

Rouphael Y, Colla G (2020b) Toward a sustainable agriculture through plant biostimulants: from experimental data to practical applications. Agronomy 10:1461

Santos MS, Nogueira MA, Hungria M (2019) Microbial inoculants: reviewing the past, discussing the present and previewing an outstanding future for the use of beneficial bacteria in agriculture. AMB Expr 9:205. https://doi.org/10.1186/s13568-019-0932-0

Savarese C, Cozzolino V, Verrillo M et al (2022) Combination of humic biostimulants with a microbial inoculum improves lettuce productivity, nutrient uptake, and primary and secondary metabolism. Plant Soil 481:285–314. https://doi.org/10.1007/s11104-022-05634-8

Sears ME (2013) Chelation: harnessing and enhancing heavy metal detoxification—a review. Sci World J 2013:219840. https://doi.org/10.1155/2013/219840

Shaffique S, Hussain S, Kang S-M et al (2023) Recent progress on the microbial mitigation of heavy metal stress in soybean: overview and implications. Front Plant Sci 14:1188856

Shahid M, Dumat C, Khalid S et al (2016) Cadmium bioavailability, uptake, toxicity and detoxification in soil-plant system. In: De Voogt P (ed) Reviews of environmental contamination and toxicology, vol 241. Springer International Publishing, Cham, pp 73–137

Shahrajabian MH, Sun W (2024) The importance of salicylic acid, humic acid and fulvic acid on crop production. LDDD 21:1465–1480. https://doi.org/10.2174/1570180820666230411102209

Sherameti I, Varma A (eds) (2015) Heavy metal contamination of soils: monitoring and remediation. Springer International Publishing, Cham

Singh AL, Singh S, Kurella A et al (2022) Plant bio-stimulants, their functions and use in enhancing stress tolerance in oilseeds. In: New and future developments in microbial biotechnology and bioengineering. Elsevier, pp 239–259

Siwik-Ziomek A, Szczepanek M (2018) Soil enzyme activity and Sulphur uptake by oilseed rape depending on fertilization and biostimulant application. Acta Agriculturae Scandinavica Sect B Soil Plant Sci 68:50–56. https://doi.org/10.1080/09064710.2017.1358762

Soomro AH, Marri A, Shaikh N (2023) Date palm (Phoenix dactylifera): a review of economic potential, industrial valorization, nutritional and health significance. In: Ismail T, Akhtar S, Lazarte CE (eds) Neglected plant foods of South Asia. Springer International Publishing, Cham, pp 319–350

Sun W, Shahrajabian MH, Kuang Y, Wang N (2024) Amino acids biostimulants and protein hydrolysates in agricultural sciences. Plan Theory 13:210

Tóth G, Hermann T, Da Silva MR, Montanarella L (2016) Heavy metals in agricultural soils of the European Union with implications for food safety. Environ Int 88:299–309

Tursunov N, Fayzullayev AS, Jurayeva SI, Azimova GA (2023) Characteristics of the sunflower plant, its importance in agriculture and its role in food security. Ethiop Int J Multidiscip Res 10: 167–174

Ul Hassan Z, Ali S, Ahmad R et al (2017) Biochemical and molecular responses of oilseed crops to heavy metal stress. In: Ahmad P (ed) Oilseed crops, 1st edn. Wiley, pp 236–248

uz Zaman Q, Guo L, He X et al (2025) Synergistic effect of arbuscular mycorrhizal fungi, biochar and seaweed extract for improving the copper tolerance in hemp. Ind Crop Prod 224:120358

Wadduwage J, Liu H, Egidi E et al (2023) Effects of biostimulant application on soil biological and physicochemical properties: a field study. J Sust Agric Env 2:285–300. https://doi.org/10.1002/sae2.12057

Yadav R, Kumar A, Saharan BS et al (2024) Arbuscular mycorrhizal fungi as biostimulant for plant growth and abiotic stress amelioration. In: Microbial biostimulants for plant growth and abiotic stress amelioration. Elsevier, pp 81–95

Yakhin OI, Lubyanov AA, Yakhin IA, Brown PH (2017) Biostimulants in plant science: a global perspective. Front Plant Sci 7:2049

Zainab N, Amna, Khan AA, Azeem MA, Ali B, Wang T, Shi F, Alghanem SM, Hussain Munis MF, Hashem M, Alamri S, Abdel Latef AA, Ali OM, Soliman MH, Chaudhary HJ (2021) PGPR-Mediated Plant Growth Attributes and Metal Extraction Ability of Sesbania sesban L. in Industrially Contaminated Soils. Agronomy. 11(9):1820.

Zhao F-J, McGrath SP, Meharg AA (2010) Arsenic as a food chain contaminant: mechanisms of plant uptake and metabolism and mitigation strategies. Annu Rev Plant Biol 61:535–559. https://doi.org/10.1146/annurev-arplant-042809-112152

Zouari I, Mechri B, Tekaya M et al (2020) Olive oil quality influenced by biostimulant foliar fertilizers. Braz J Biol Sci 7:3–18

Chapter 7
Biostimulants for Sustainable UV Stress Tolerance in Oilseed Crops

Seyedeh-Somayyeh Shafiei-Masouleh

Abstract Climate change has intensified environmental stresses, including ultraviolet (UV) radiation, posing significant challenges to global agriculture. A definitive role of oilseed crops in human nutrition and health is known, and these crops are particularly vulnerable to UV stress. This stress, alone or in combination with other environmental factors, causes morphological changes, physiological responses, and biochemical reactions in plants. Such alterations can adversely affect crop performance, leading to quantitative and qualitative reductions in oil yield. Plants have evolved diverse molecular and biochemical defense strategies to counteract these effects, including producing antioxidants, protective pigments, and stress-responsive proteins. However, these natural responses may not always be sufficient to mitigate UV-induced damage, especially under extreme climatic conditions. Therefore, farmers must adopt effective strategies to ensure the sustainable production of oilseed crops. One promising approach is the application of biostimulants, which enhance plant resilience against environmental stresses by improving growth, metabolism, and stress tolerance. Integrating the biostimulants into sustainable agricultural practices can remarkably contribute to maintaining the oilseed crop yield and quality against climate change. Understanding the intricate interactions between UV stress and plants, innovative solutions to promote crop sustainability could be developed. Further studies on plant responses to UV stress based on the molecular and physiological mechanisms will be essential for optimizing agronomic practices and improving global food security.

Keywords Oilseed crops · Oxidative stress tolerance · UV radiation · Biostimulants · Plant genetic

S.-S. Shafiei-Masouleh (✉)
Department of Genetics and Breeding, Ornamental Plants Research Center (OPRC), Horticultural Sciences Research Institute (HSRI), Agricultural Research, Education and Extension Organization (AREEO), Mahallat, Iran
e-mail: shafiee.masouleh@areeo.ac.ir

© The Author(s), under exclusive license to Springer Nature Singapore Pte Ltd. 2025

A. A. H. Abdel Latef (ed.), *Oilseed Crops Under Abiotic Stress*, Sustainability Sciences in Asia and Africa, https://doi.org/10.1007/978-981-96-8346-8_7

7.1 Introduction

Oilseeds, such as peanuts, rapeseeds, sesame, and sunflower seeds, are rich in fat (at least 15%) and protein (6–45%) and they are valuable sources of plant-based nutrition. Beyond producing vegetable oils, oilseeds offer significant dietary protein, with more than 200 varieties available worldwide. Commonly consumed oilseeds include soybeans, sunflower seeds, and sesame, while less common ones like chia and evening primrose are gaining attention (Zhang et al. 2023). Vegetable oils are vital commodities for both food and nonfood uses, and new oil types are needed due to the driving demand for specific quality requirements (Velasco and Fernández-Martínez 2002). Furthermore, oilseed proteins have various health benefits, including antioxidants, anti-inflammatory, and antibacterial properties for humans. Despite oilseeds' nutritional potential, protein-rich by-products are often used as animal feed, resulting in resource underutilization. Advances in research on oilseed proteins and their functional properties aim to improve processing, expand their application in food, and enhance human dietary protein sources, offering sustainable solutions to protein demand (Zhang et al. 2023).

Climate change, through its effects on temperature, carbon dioxide, solar radiation, and precipitation, restricts resources vital for oilseed crop growth, causing phenotypic shifts. Understanding *phenotypic plasticity* is crucial for managing these impacts (Jaradat 2016). Since the depletion of the ozone layer in the late 1970s, the level of UV-B radiation at the Earth's surface has increased. Although the Montreal Protocol has stabilized UV-B exposure since the mid-1990s, full ozone recovery to pre-1980 levels may need decades (Yoon et al. 2016). UV-B radiation is a plant stressor and damages deoxyribonucleic acid, proteins, and membranes. Plants counter UV stress by producing antioxidants like phenolics and flavonoids and UV-B screening molecules (Mareri et al. 2022). Plants, being sessile, cannot avoid environmental stresses like UV-B radiation. Mild stress triggers acclimation responses, while severe stress disrupts metabolism. UV-B affects various metabolites, including phenolics and alkaloids, with complex patterns of increase, decrease, or fluctuation (Jansen et al. 2008).

Agricultural practices are increasingly shifting toward organic, sustainable, and environment-friendly systems. Modern agriculture seeks to minimize inputs while maintaining both yield and quality. While selective breeding can contribute to these objectives, it is often species-specific and time-intensive. An alternative approach lies in identifying organic molecules capable of stimulating plant metabolism, offering a faster and more cost-effective means of enhancing crop performance. Biostimulants contain an array of bioactive compounds, many of which remain largely unexplored. These natural enhancers improve nutrient use efficiency as well as strengthen plant resilience against biotic and abiotic stresses, making them a promising tool for advancing sustainable agricultural productivity (Bulgari et al. 2015).

Using plant biostimulants has become an integral part of modern intensive agriculture, offering a sustainable approach to enhancing crop productivity. Farmers are increasingly incorporating biostimulants at key stages of plant development to optimize growth, improve nutrient efficiency, and bolster resilience against environmental stressors. These natural formulations enhance the plant's ability to withstand adverse conditions and contribute to superior crop quality and yield. Stimulating the physiological processes at the cellular level, biostimulants promote more vigorous root development, improved nutrient uptake, and greater metabolic efficiency. As agriculture continues to evolve toward more sustainable and resource-efficient practices, biostimulants are emerging as valuable tools that bridge the gap between high productivity and environmental responsibility, ensuring healthier crops and more resilient farming systems (Petrova et al. 2023). Biostimulants encompass various products, including amino acid derivatives, humic acids, seaweed extracts, and microbial inoculants. In Asia and Africa, where the oilseed sector faces challenges in sustaining a growing population with limited natural resources, biostimulants play a crucial role in enhancing nutrient yield and quality. Humic substances improve nutrient uptake by increasing soil cation exchange capacity, while compounds like glycine betaine and L-proline act as antioxidants, regulating ion fluxes and mitigating oxidative stress. Melatonin (MT) enhances seedling stress tolerance, while plant hormones such as brassinosteroids, abscisic acid, salicylic acid, gibberellins, and jasmonic acid contribute to various physiological functions. Seaweed and plant extracts enriched with auxins and cytokinins promote growth, while those containing jasmonic acid (JA), salicylic acid (SA), abscisic acid (ABA), and ethylene aid in stress adaptation. Additionally, induced systemic resistance (ISR) through plant growth-promoting rhizobacteria (PGPR) offers an effective strategy for managing plant diseases, reinforcing natural defense mechanisms against pathogen-induced stress (Singh et al. 2022).

This chapter discusses the effects of UV stress on various oilseed crops studied to date. UV radiation, as an environmental stressor, can significantly impact plant physiology, growth, and yield. The discussion delves into the diverse responses of oilseed crops to UV exposure, highlighting both the detrimental effects and potential adaptive mechanisms. Additionally, a key focus is placed on the role of biostimulants in enhancing sustainable tolerance to UV stress. Different categories of biostimulants, including microbial inoculants, protein hydrolysates, seaweed extracts, and humic substances, are examined for their ability to mitigate UV-induced damage and promote resilience. Understanding the interplay between biostimulants and plant defense mechanisms, this chapter provides insights into innovative strategies for safeguarding crop productivity under increasing environmental challenges. Ultimately, these findings contribute to developing more resilient agricultural systems in the face of global climate change.

7.2 UV Radiations and Plant Stress

7.2.1 Climate Changes and UV Radiation

In natural environments, plants receive various photonic stimuli, including UV radiation, alongside the radiation required for photosynthesis (Cano et al. 2023). UV radiation is categorized into three ranges: UV-A (315–400 nm), UV-B (280–315 nm), and UV-C (100–280 nm) (Dawood and Abdel Latef 2022; Cano et al. 2023). Though minor, UV significantly influences plant growth, metabolism, and development (Dawood and Abdel Latef 2022). While UV-C is entirely absorbed by the ozone layer, UV-A is less harmful and acts as a photomorphogenic signal. About 5% of UV-B radiation reaches the Earth's surface, serving as both a signal and a stressor. Excessive UV-B can damage proteins, lipids, and nucleic acids, leading to the accumulation of reactive oxygen species (ROS). To counteract UV-B-induced oxidative stress, terrestrial plants have evolved strategies to mitigate damage and acclimate to UV-B exposure, which remains a significant environmental challenge (Cano et al. 2023).

Global changes, including thinning of the stratospheric ozone layer, are expected to increase UV-B radiation reaching the Earth's surface. Meanwhile, tropospheric ozone levels have risen due to human activities. Future models predict that both ozone and UV-B radiation will increase simultaneously, affecting elevated regions (Langebartels et al. 2002). The anthropogenic release of chlorofluorocarbons has depleted the stratospheric ozone layer, increasing ultraviolet-B radiation on Earth's surface. This rise in UV-B poses significant threats to crop plants by damaging proteins, nucleic acids, and lipids, leading to mutagenesis and disruptions in critical physiological processes. Alongside heavy metals, UV-B radiation is considered a "novel" artificial abiotic stress factor, further highlighting the need for understanding and mitigating their combined impacts on plant health and productivity (Teramura et al. 1990; Li et al. 2012; Müller-Xing et al. 2014).

7.2.2 UV Stress on Oilseed Crops

With the decline in ozone and increased UV-B radiation, research has shifted from viewing UV-B as harmful to recognizing it as a regulatory factor in plants. On the other hand, while UV-B photoreceptor signaling is better understood, the role of UV-B-induced reactive oxygen species (ROS) remains unclear. Experiments suggest UV-B can trigger oxidative stress; the effects of pro-oxidants in complex biological systems are not fully understood. A hypothesis proposes that UV-B-induced the activation of peroxidase (POD) and superoxide dismutase (SOD) plays a key role in plant acclimation to UV-B (Czégény et al. 2016).

Although UV-B makes up less than 5% of total solar radiation, it has significant photobiological effects due to its strong absorption by macromolecules like proteins and nucleic acids. UV-B radiation impacts plant growth and physiology, causing damage to the photosynthetic apparatus, altering protein content and enzyme activity, and affecting membranes and leaf chemistry. Morphological effects include stunted growth, leaf discoloration (chlorosis or bronzing), and reduced biomass and seed yield. Numerous studies have demonstrated these effects across various plant species, highlighting the detrimental impact of increased UV-B exposure under different growing conditions (Teramura et al. 1990). Increased UV-B intensity threatens plant viability, causing physiological damage such as reduced photosynthesis and pollen fertility, as well as morphological effects. Most vascular plants struggle to adapt to high UV-B levels, making it a critical challenge for survival (Yoon et al. 2016). UV-B light significantly influences plant interactions with herbivorous arthropods by altering constitutive and inducible defenses. While UV-B can sometimes enhance plant adaptation to abiotic stresses like heat and drought, such responses may also be counteracted by changing environmental conditions, potentially reducing plant resistance. These complex interactions between UV-B, environmental factors, and plant biochemistry directly affect how plants defend against herbivorous pests, for instance, highlighting the need for a deeper understanding of these dynamics to optimize agricultural strategies (Escobar-Bravo et al. 2017). To mitigate damage, plants produce "sunscreen" flavonoids, with the UV-B receptor UVR8 playing a key role in promoting their biosynthesis. Recent research has uncovered UVR8-mediated and independent pathways, involving factors like HY5, BES1, MYB13, MKP1, and ATR, in UV-B stress tolerance. Additionally, visible light receptors and endogenous melatonin accumulation contribute significantly to UV-B signaling and stress tolerance, forming an integrated response to UV-B exposure (Shi and Liu 2021).

Future climate change scenarios predict increased UV-B radiation, potentially reducing crop productivity and oil quality in oilseed crops; however, the plants' responses depend on their genetic background and are different based on the genus, species, and even cultivars of a species. For instance, in a study, defense responses to elevated UV-B (+7.2 kJ m^{-2} day^{-1}) in two soybean (*Glycine max*) cultivars (JS-335, PS-1042), and their impact on seed nutrients and oil composition were examined. UV-B exposure increased fatty acid unsaturation, altering oil properties and reducing nutritional value, and with JS-335, it was affected more. Insufficient ROS quenching led to oxidative stress. PS-1042 performed better due to enhanced defense mechanisms, including higher lignin and flavonoids (quercetin, kaempferol) accumulation, and more wax deposition (Choudhary and Agrawal 2016). In a six-season field study, it is examined how supplemental UV-B, simulating 16–25% ozone depletion, affects soybean yield and seed quality. Two cultivars, Essex and Williams, showed contrasting responses. Essex was UV-B sensitive, experiencing a 20% yield reduction, while Williams was more tolerant. However, seasonal microclimate variations significantly influenced yield outcomes. Essex had the more significant yield loss in seasons with high water availability, whereas Williams' yield declined only under severe water stress. These findings highlight the need for long-term studies to understand UV-B interactions with other environmental factors. A deeper understanding of these interactions is essential for assessing the broader implications of stratospheric ozone depletion on crop productivity (Teramura et al. 1990). Nguyen et al. (2020) examined the effects of UV-B on phenolic accumulation, resveratrol content, and antioxidant activity in three peanut cultivars (Lac sen, L14, L27) during germination. Seeds were exposed to UV-B for 1 h daily, and sprouts were harvested on days 1, 3, and 5. Results showed that germination time and UV-B exposure significantly increased total phenolics, resveratrol, and antioxidant activity, and 5-day sprouts of Lac sen under UV-B showed the highest resveratrol content. Findings suggest UV-B treatment as a promising method to enhance the nutritional value of peanut sprouts for functional food production. Resveratrol, a phenolic compound naturally found in peanuts, is present in low amounts but can be induced by UV-B exposure (Nguyen et al. 2020).

The types of plant reactions to the intensity of UV radiation are different in oilseed crops. Costa et al. (2002) showed the antioxidant defense system in sunflower cotyledons exposed to UV-B radiation. UV-B doses (15.0 and 30.0 kJ m^{-2}) caused oxidative damage, increasing lipid peroxidation and decreasing chlorophyll and SOD activity. However, antioxidant enzyme activities like catalase (CAT) and glutathione dehydrogenase were enhanced, suggesting a protective role against oxidative stress. In a report by Cechin et al. (2007), the effects of enhanced UV-B radiation (4.0 Wm^{-2} for 7 h/day) on *Helianthus annuus* L. cv. Catissol-01 grown in a greenhouse after 12 or 21 days did not show any shoot dry matter, specific leaf weight, and UV-B absorbing compounds. They observed that photosynthesis was inhibited early, linked to reduced stomatal conductance and transpiration, though intercellular CO_2 remained stable. Chl a and chl a/b declined initially, while minimal fluorescence increased. Despite no biomass loss, findings suggest UV-B radiation influences sunflower performance. Based on a report (Mark and Tevini 1996), UV-B radiation at +30% intensity reduced growth parameters (height, leaf area, fresh and dry weight, stem elongation, and relative growth rate) by up to 35% at 28 °C on sunflower and maize seedlings. However, increasing temperature by 4 °C, with or without doubled CO_2, mitigated the UV-B effects, restoring growth to control levels. Chlorophyll content increased under intense UV-B, while photosynthetic rates decreased by 29%. CO_2 doubling enhanced photosynthesis in sunflowers, but not in maize. Transpiration rates increased with higher temperatures, while CO_2 reduction lowered transpiration and improved water use efficiency. UV-B radiation accelerates leaf senescence in cotton by reducing photosynthesis and chlorophyll content. Elevated atmospheric CO_2 enhances leaf photosynthesis by 32%, but UV-B exposure significantly lowers photosynthetic rates, especially in older leaves. While chlorophyll content is affected by UV-B, especially at higher levels, photosynthesis proves to be a more sensitive indicator of leaf aging. A study on cotton highlighted how UV-B stress accelerates leaf senescence and suggested that hyperspectral reflectance can effectively monitor these changes, offering insights into the impact of UV-B on cotton growth and health (Kakani et al. 2004). An experiment assessed the effects of enhanced UV-B radiation on cotton (*Gossypium hirsutum* L.) morphology and leaf anatomy in controlled environments. Plants were exposed to 0, 8, or 16 kJ m^{-2} d^{-1} UV-B for 8 h daily from emergence to harvest. UV-B exposure caused chlorosis and necrosis, with high UV-B reducing vegetative and reproductive parameters, resulting in a smaller canopy. UV-B increased epicuticular wax and stomatal index but reduced leaf thickness, especially in palisade and mesophyll tissues. Reproductive traits were affected at 8 and 16 kJ m^{-2} d^{-1}, while vegetative traits were only impacted at 16 kJ m^{-2} d^{-1}. Cotton is sensitive to UV-B at both the plant and anatomical levels (Kakani et al. 2003). Reddy et al. (2003) investigated the impact of increased UV-B radiation on cotton growth and development, aiming to develop UV-B functional algorithms for simulation models. Two experiments were conducted under optimal growing conditions with varying UV-B radiation levels. Results showed that UV-B exposure caused chlorotic and necrotic leaf patches, reduced canopy photosynthesis, and led to smaller leaf area and shorter plant height. Stem length was more sensitive to UV-B than leaf width. Critical UV-B limits for stem elongation, leaf expansion, canopy photosynthesis, and dry weight were identified, providing key parameters for predicting cotton yields under changing climatic conditions.

Furthermore, the responses of oilseed crops depend on the periods of UV stress exposure. For instance, in a report, sunflower seeds (*Helianthus annuus* cv. Toledo-2) were exposed to UV-C radiation for varying durations (0, 5, 10, 15, 30, and 60 min) in dry and hydrated states. Seeds germinated immediately or were stored in darkness for 7 days before germination. UV-C exposure reduced standard seedling percent and increased abnormal seedlings, with maximum inhibition at 60 min. Storage for 7 days after irradiation helped partially reverse this inhibition, especially in dry seeds. Hydrated seeds were more sensitive to UV-C radiation, showing 2.74 times greater sensitivity than dry seeds in germination performance (Torres et al. 1991). UV-B radiation is a known stressor that reduces plant biomass, impairs photosynthesis, and induces oxidative stress. Trigonelline, a secondary metabolite, is synthesized in response to such abiotic stress. A study investigated biochemical stress responses in four peanut (*Arachis hypogaea* L.) genotypes—Spanish, Virginia, Valencia, and Runner—following UV-B exposure at different durations (60, 120, and 180 min). It examined the influence of plant age on trigonelline biosynthesis. A significant positive correlation was observed between trigonelline levels and UV-B exposure time. No genotype-based differences were detected (Willmon et al. 2017). In a study, Zhu et al. (2021) examined the impact of light radiation on stilbene induction and ROS metabolism in peanut sprouts. Various light sources (white LED, UV-A, UV-B, and UV-C) were applied during germination, significantly enhancing stilbene and phenolic content, with different stilbenes induced by specific light types. UV-C was further tested at varying intensities, showing a dose-dependent increase in stilbenes, phenolics, flavonoids, antioxidant enzymes, and phenylalanine ammonia-lyase (PAL) activity. A strong negative correlation was observed between H_2O_2 and these compounds. Overall, light radiation, particularly UV-C, effectively boosts stilbene and antioxidant accumulation by regulating oxidant-antioxidant balance.

Similarly, plant responses to UV light components vary, and different UV spectral ranges uniquely influence physiological reactions and morphology in oilseed crops. A field experiment was conducted to evaluate the effects of solar UV-B and UV-A radiation on cotton (*G. hirsutum* var. Vikram) growth and antioxidant enzyme activity (Dehariya et al. 2011). UV components were filtered using polyester films that absorbed UV-B (<315 nm) and UV-B/A (<400 nm), with plants grown under unfiltered sunlight serving as controls. UV exclusion resulted in increased plant height, leaf area, and biomass. Antioxidant enzyme activity (superoxide dismutase, ascorbic acid peroxidase, glutathione reductase, and guaiacol peroxidase) was lower in UV-excluded plants, along with a decrease in ascorbic acid levels and UV-absorbing substances. These findings suggest that solar UV components limit cotton growth and induce stress, while the exclusion of UV enhances growth by improving primary metabolism, offering agricultural benefits (Dehariya et al. 2011). Research examined the effects of elevated UV-B irradiance on cotton (*G. hirsutum* L., Sukang 103) grown under natural and four supplemental UV-B regimes in the field. With a 9.5% increase in UV-B radiation over the growing season, significant negative impacts were observed. Cotton height, leaf area, and total biomass decreased by 14%, 29%, and 34%, respectively. Additionally, fiber quality was reduced, and the economic yield dropped by 72%, with the economic coefficient falling by 58%. The study discusses how these impacts on cotton compare to findings from other studies on different plant species, including trees, highlighting the potential risks of increased UV-B radiation on agricultural productivity (Gao et al. 2003). Ambient solar UV radiation can significantly impact cotton growth and yield, but excluding specific UV components may improve plant performance. By blocking UV-B or UV-B/A radiation using specialized filters, studies have shown that cotton plants grow taller, with larger leaves and increased biomass. Photosynthesis improves in UV-excluded plants, leading to higher yield parameters such as boll number, weight, and seed count. Fluorescence measurements reveal enhanced photosynthetic efficiency, while changes in stomatal conductance and carbon fixation further support these improvements. This suggests that reducing UV exposure could be a promising strategy to boost cotton production and optimize growth under environmental stressors (Dehariya et al. 2011).

Improvement of the plant growth conditions or removing the stress components with suitable tools and methods could effectively enhance the performance of oilseed crops. Understanding how crops respond to climate change is crucial for improving agricultural practices. Research shows that elevated atmospheric CO_2 increases cotton growth, leaf photosynthesis, and biomass, leading to larger leaves and better carbohydrate accumulation. However, high UV-B radiation significantly reduces growth, causing smaller plants, shorter stems, and smaller leaf areas. UV-B stress also impacts fruit development, causing fruit abscission and reduced dry weight. Interestingly, while higher CO_2 improves growth, it does not fully counteract the adverse effects of increased UV-B radiation, particularly on boll retention and fruit yield (Zhao et al. 2003). Excluding ambient solar UV-A/B radiation significantly boosts cotton plant growth, increasing height, leaf area, and dry weight. Different cotton varieties show varied sensitivity to UV, with Khandwa-2 being the most sensitive. The reduction in UV-B-absorbing substances and antioxidant enzyme activity upon UV exclusion suggests that solar UV-induced stress triggers the plant's defense mechanisms. By eliminating this stress, the plant's metabolism shifts toward primary growth, enhancing overall development. This highlights the impact of UV stress on cotton and the potential benefits of UV exclusion for improving crop yield (Kataria et al. 2012).

7.2.3 UV Stress with Other Abiotic Stresses on Oilseed Crops

Undoubtedly, as with other stresses, the response of oilseed crops varies when UV stress is combined with other environmental stresses. In the following, some of the ways that plants encounter these challenges will be explained.

UV Radiation and Heavy Metals In a study, the interactive effects of heavy metal contamination (Cd^{+2}) and UV-B exposure on nutrient uptake, biomass, and chlorophyll content in mustard (*Brassica campestris* L.) seedlings were examined. Plants were exposed to Cd^{+2} concentrations of 0.5, 1.0, 2.5, and 5.0 mg L^{-1} in a medium and UV-B radiation (0.4 mW cm^{-2}) for 30 min daily. After 5 and 10 days, the combined stresses led to a significant decline in biomass. At the highest Cd^{+2} concentration (5 mg L^{-1}), exposure to UV-B caused a notable reduction in chlorophyll a (43%), chlorophyll b (23%), and carotenoid (53%) levels. Additionally, calcium (Ca^{+2}), magnesium (Mg^{+2}), and potassium (K^{+}) uptakes in the roots decreased by 51%, 48%, and 62%, respectively, indicating a significant disruption in nutrient absorption. Both root and shoot leaf nutrient uptake decreased during the early growth phase under these conditions. Interestingly, Cd^{+2} uptake increased by 33% in the presence of UV-B. These findings highlight the impact of environmental stress on the early growth period of crops, which is crucial for overall productivity and yield. This study emphasizes the need to address the effects of combined environmental stresses on crop development (Shukla et al. 2008). In another study, exposure to elevated UV-B and arsenic (As) stress in *Helianthus annuus* seedlings reduced growth, including root and shoot lengths, fresh mass, and leaf area. Photosynthetic pigments, such as chlorophyll, decreased, and oxidative stress markers like superoxide radicals ($O_2{\bullet}^-$), hydrogen peroxide (H_2O_2), malondialdehyde (MDA), and reactive carbonyl groups (RCG) increased with both As and UV-B treatments. The combined stress further diminished growth and photosynthesis. However, photoprotective pigments, including carotenoids, anthocyanins, and flavonoids, were enhanced. Antioxidant enzyme activities (SOD, POD, CAT) and proline levels were also elevated, indicating a protective response to the combined stress (Yadav et al. 2016). This study assessed the effects of enhanced UV-B (0.4 W m^{-2}) and nickel (0.01, 0.10, 1.00 mM) on soybean seedlings, focusing on growth, photosynthesis, ROS accumulation, and antioxidant responses. Higher Ni concentrations (0.10, 1.00) and UV-B reduced growth and chlorophyll content, with even more substantial declines under combined exposure. While low Ni (0.01) stimulated growth, its combination with UV-B became inhibitory. Photosynthetic electron transport, mainly PS2 and whole-chain activity, was significantly impaired under stress, while PS1 was stimulated by Ni alone. UV-B and Ni elevated ROS (O_2^-, H_2O_2), lipid peroxidation, and electrolyte leakage while reducing ascorbate levels.

Antioxidant enzyme activities (CAT, SOD, POD) increased, but Ni1.00 inhibited SOD, and combined exposure further suppressed CAT. Overall, simultaneous UV-B and Ni stress caused severe oxidative damage and potent inhibition of photosynthesis (Prasad et al. 2005). Li et al. (2012) demonstrated that Cd and UV-B stress have detrimental effects on soybeans. However, the sequential application of Cd and UV-B stress is less harmful than their combined effects. Their study showed that root length and dry weight significantly decreased under Cd and Cd + UV-B stress. Additionally, each stressor alone led to reductions in pigment content, stomatal conductivity, net photosynthetic rate, transpiration, and SOD activity. They concluded that cross-adaptation between these two stresses can help protect soybeans against the combined effects of Cd and UV-B.

UV Radiation and Water Stress Soybeans (*G. max*cv. Essex) were grown under ambient and enhanced UV-B (5.1 kJ m^{-2} daily) with either water stress or supplemental irrigation. Under well-watered conditions, UV-B reduced leaf area, dry weight, and photosynthesis. However, in water-stressed plants, UV-B had no significant effects. This insensitivity was likely due to water stress-induced adaptations, such as increased UV-absorbing compounds and leaf thickening, which protected against UV-B damage (Murali and Teramura 1986). Cechin et al. (2008) investigated the effects of drought and UV-B radiation on sunflower plants. Plants exposed to UV-B (8.6 W m^{-2}) for 7 h per day under drought stress showed reduced shoot dry matter, photosynthesis, stomatal conductance, and transpiration. However, UV-B alleviated the adverse effects of drought on photosynthesis and transpiration as stress intensified. In sunflower plants, photosynthetic pigments were primarily unaffected by UV-B radiation, although water stress reduced chlorophyll content, especially chlorophyll a, under high UV-B. Fv/Fm, the measure of photosystem II efficiency, was reduced by UV-B but not by drought. Proline levels increased under drought stress, but UV-B did not affect their accumulation (Cechin et al. 2008). In a report by Moghadam et al. (2011), the combined effects of UV radiation, CO_2, and water stress on canola plants were examined. The plants were grown under twelve conditions, including varying UV radiation levels (UV-A, UV-B, UV-C), CO_2 concentrations, and irrigation treatments. The results highlighted significant impacts on plant traits and yield components. Water stress notably reduced most traits, except for the oil percent. High levels of CO_2, however, increased the 1000-seed weight, final yield, oil yield, specific leaf area, plant height, and the number of branches. UV radiation, in contrast, negatively affected all traits measured during the experiment. High CO_2 partially mitigated the negative effects of UV radiation, improving traits like seed weight, final yield, and oil yield. The best yield was observed in the "Talaye" cultivar, under full irrigation, high level of CO_2, and optimal sunlight conditions. The increase in yield was mainly attributed to higher seed weight and more significant number of branches (Moghadam et al. 2011).

UV Radiation and Ozone Stress A field study on two linseed cultivars, Padmini and T-397, assessed the effects of supplemental UV-B (+7.2 kJ m^{-2} d^{-1}) and elevated O_3 (+10 ppb). Both stresses increased ROS, membrane damage, and secondary metabolites while reducing biomass and yield. Padmini was more sensitive, while T-397 showed higher protective compounds. O_3 impacted growth more, whereas UV-B reduced yield. Combined exposure was less harmful (Tripathi and Agrawal 2013). The effects of supplemental UV-B (7.2 kJ m^{-2} d^{-1}) and elevated ozone (+10 ppb) on two *H. annuus* L. cultivars, Sungold and DRSF 108, under field conditions were examined in a report by Tripathi et al. (2019). Applying individually and in combination, these stressors induced changes in ultrastructure, growth, biomass, yield, and oil composition. Reduced leaf area in Sungold minimized UV-B perception and ozone uptake, limiting carbon loss compared to DRSF 108. Both stressors decreased the yield and oil content and increased the rancidity, making the oil less suitable for human consumption (Tripathi et al. 2019). Drought and UV-B radiation are major stressors affecting plant growth and productivity. This study compared their effects on maize and soybean seedlings. Reduced leaf water content was the primary cause of drought-induced growth decline, while UV-B exposure increased anthocyanin and phenol levels. Both stresses caused severe membrane damage, indicated by lipid peroxidation and osmolyte leakage. Antioxidant enzymes (CAT, POD, SOD) and hydrogen peroxide accumulated under stress. Proline increased mainly in drought-stressed seedlings, suggesting a protective role. Chlorophyll content, photosynthesis, and total dry matter decreased under single and combined stress conditions. Overall, UV-B had a more detrimental impact on growth than drought based on physiological and biochemical responses (Shen et al. 2015). The UV-A radiation on UV-B damaged peanut leaves using the laser-induced fluorescence (LIF) method was examined by Fukuchi et al. (2004). They assessed the vitality index through chlorophyll fluorescence kinetics alongside LIF spectra measurements. They observed that while visible light exposure helped recover UV-B-damaged leaves, UV-A radiation significantly reduced their vitality index, though it had no adverse effect on undamaged leaves. The decline in vitality likely resulted from UV-A-induced degradation of UV-absorbing pigments, which generally protect leaves by limiting UV penetration and preventing photosystem II damage. This suggests a synergistic adverse effect of UV-A on UV-B stressed leaves.

UV Radiation and Temperature Stress Cotton plants (*G. hirsutum* L. cv. NuCOTN 33B) exposed to high temperatures and enhanced UV-B radiation experience significant fruit abscission, a process heavily influenced by nonstructural carbohydrate (TNC) limitations. When plants are subjected to high daytime temperatures (36 °C) and UV-B radiation of 7 and 14 kJ m^{-2} per day, boll retention decreases drastically. While high temperatures did not directly affect leaf photosynthesis or square abscission, UV-B radiation reduced the leaf's nonstructural carbohydrates, particularly in floral buds and young bolls, which correlated with increased square and boll drop. This suggests that the depletion of carbohydrates in reproductive organs, rather than in leaves, plays a key role in the fruit drop observed under these stress conditions (Zhao et al. 2005). Rising temperatures and UV-B radiation threaten to speed up development but reduce boll production, especially under increased UV-B. Moderate UV-B (7 kJ m^{-2} day^{-1}) cuts yield drastically, while extreme levels (14 kJ m^{-2} day^{-1}) ultimately prevent boll formation. Photosynthesis, stomatal conductance, and chlorophyll levels decline under stress. Leaf reflectance changes with temperature, UV-B, and leaf age, which is a signal for plant stress. Future climate shifts may severely impact cotton, demanding adaptation through breeding and protective strategies (Reddy et al. 2004). In a controlled study on sunflower and maize seedlings, the effects of reduced UV-B radiation and increased temperature on growth and physiological processes were evaluated. Results showed that a 4 °C temperature increase enhanced growth parameters such as fresh and dry weight, plant size, and leaf area. Reduced UV-B radiation improved growth at lower temperatures, but no significant changes were observed at higher temperatures. Additionally, the increased temperature led to higher respiration rates in maize and increased transpiration in both species (Mark and Tevini 1996). Climate change imposes multiple abiotic stresses on plants, impacting growth and development. In *Brassica oleracea* and *B. juncea*, stressors like UV-B, high temperature, and their combinations reduced growth, except under elevated CO_2 levels, which mitigated the harmful effects of UV-B and temperature (Del Buono et al. 2023).

7.3 Plant Genetics to Cope with UV Stress

Breeding for resilience beyond seed and oil yield can enhance adaptation to unpredictable environments. Gene banks play a critical role in conserving and utilizing these resources. As classical breeding faces challenges, genomic tools and metabolic engineering provide efficient strategies for developing climate-resilient oilseed crops (Jaradat 2016). Over the past four decades, plant breeders have focused on improving seed oil quality, initially relying on natural variation within oilseed crops and related species. Since the 1970s, mutagenesis has been a key method for altering fatty acid profiles and creating novel oil types. Today, oil quality is defined not only by fatty acid composition but also by parameters like triacylglycerol and tocopherol content. Advances in molecular techniques now offer precise, efficient tools for improving seed oil quality (Velasco and Fernández-Martínez 2002).

UV-B radiation acts as a crucial environmental signal, triggering a wide range of plant responses that influence metabolism, development, and survival. These responses are shaped by the nature of UV-B exposure, the plant's level of adaptation, and interactions with other environmental factors. UV-B effects involve nonspecific pathways such as those triggered by DNA damage, ROS, and defense signals and specific photomorphogenic pathways activated by low UV-B levels. Photomorphogenic signaling promotes plant survival by inducing genes that provide UV protection. The UV-B-specific receptor *UV RESISTANCE LOCUS8* (*UVR8*) is central to photomorphogenic responses. UVR8, in cooperation with *CONSTITUTIVE PHOTOMORPHOGENESIS1* (*COP1*), regulates the expression of *ELONGATED HYPOCOTYL5* (*HY5*), a transcription factor that governs key UV-protective genes, enhancing plant resilience under UV-B stress (Jenkins 2009). Based on the report, the identified components regulating UV-B-specific photomorphogenic responses include the UVR8 photoreceptor, COP1, HY5, HYH, and genes like *CHS* involved in secondary metabolite biosynthesis. These elements enable nondamaging developmental responses to low-dose UV-B. However, high UV-B levels induce genotoxic stress, triggering ROS production and activating distinct stress defense genes. Unlike photomorphogenesis, stress responses to high UV-B levels remain poorly understood, particularly the intracellular signaling mechanisms involved. These findings highlight the dual role of UV-B radiation in plant development and stress response, underscoring the complexity of plant adaptation to UV-B fluctuations (Yoon et al. 2016).

Mareri et al. (2022) cited the results of two studies in their report. In one study, UV-B radiation interception in kale using 3D ray-tracing simulations revealed that UV-B-induced flavonoids and phenolics were more concentrated in younger leaves, with flavonoid content more influenced by UV-B interception than leaf age. In another study, *Arabidopsis mutants* lacked chlorophyll b due to a defect in chlorophyllide a oxygenase (*CAO*). Under low-light stress, the mutants adjusted photosynthetic apparatus structures (PSI and PSII), similar to high-light responses in ecotypes. These studies highlight mechanisms of UV protection and adaptive strategies in plants under varying light conditions (Mareri et al. 2022). Flavonoids are vital secondary metabolites in plants, aiding adaptation to UV radiation and terrestrial environments. The moss *Pohlia nutans* produces a type I flavone synthase (*PnFNSI*) that converts naringenin to apigenin, with optimal activity at 15–20 °C. Subcellular analysis showed *PnFNSI* is localized in the cytoplasm. Overexpressing PnFNSI in *Arabidopsis* improved tolerance to drought and UV-B radiation by enhancing the flavone biosynthesis pathway and increasing ROS scavenger activity, reducing oxidative stress. These findings reveal the role of *PnFNSI* in flavone metabolism and its protective function, highlighting its potential to improve plant resilience to environmental stresses (Wang et al. 2020). Increased UV-B radiation due to ozone depletion affects plant growth, including *Brassica oleracea* var. *italica* (broccoli). When low-to-moderate UV-B doses (0.3–1 kJ m^{-2} d^{-1}) were applied to broccoli sprouts, it led to an accumulation of nonvolatile secondary metabolites like flavonoids (kaempferol, quercetin) and glucosinolates (GS), particularly 4-methylsulfinylbutyl GS and 4-methoxy-indol-3-ylmethyl GS. It is reported that carotenoids and chlorophyll levels remained unchanged. This metabolic shift was linked to the activation of salicylate and jasmonic acid defense pathways and gene expression related to pathogen responses. Pre-exposure to UV-B also negatively impacted insect performance, such as the caterpillar *Pieris brassicae* and aphid *Myzus persicae*, with altered GS induction in response to insect presence (Mewis et al. 2012).

Ozone layer depletion has intensified UV-B radiation, imposing complex abiotic stresses such as heat, excessive light, and dehydration on plants. However, UV-B stress signaling remains poorly understood in soybeans. In a study, the transcriptional responses in two soybean genotypes, UV-B-sensitive *Cheongja*, and UV-B-resistant *Buseok*, using RNA-seq analysis under UV-B exposure for 0, 0.5, and 6 h were examined. Differentially expressed genes (DEGs) linked to UV-B stress were identified, with overrepresented categories including immune response, stress signaling, and ROS metabolism. Notably, *Buseok* exhibited a distinct response pathway, utilizing phosphatidic acid-dependent signaling over phospholipase D at high UV-B influence rates. Further analysis showed that *Buseok* upregulated genes involved in ABA signaling, MAPK cascades, and ROS management, contributing to its resistance. Defense-related DEGs for TIR-NBS-LRR proteins and heat shock proteins were also activated. These findings suggest that high-intensity UV-B defense mechanisms differ from low-level UV-B photomorphogenic adaptations. This study provides insights into UV-B stress tolerance and could aid in developing soybean genotypes resistant to extreme UV-B conditions, ensuring better growth and productivity under environmental stress (Yoon et al. 2016).

The BBX gene family encodes transcription factors that regulate plant growth, development, and stress responses. However, their role in soybean's UV-B stress defense remains poorly understood. A study by Azam et al. (2024) identified 184 *GmBBX* genes and analyzed for their conserved domains, including single and double B-box and CCT domains, indicating functional diversity. Physicochemical properties revealed variations in molecular size, hydrophobicity, and stability. Phylogenetic analysis classified *GmBBX* genes into five clades with nonuniform chromosomal distribution, with chromosome 13 containing the most genes (26). Structural analysis showed variations in exon-intron patterns, while motif components highlighted conserved sequences among *GmBBX* proteins. Gene duplication analysis indicated that *the GmBBX* gene expansion resulted from tandem and dispersed duplications. Gene ontology analysis linked *GmBBX* genes to light responses, abiotic stress adaptation, and molecular functions like zinc ion binding. Promoter analysis revealed cis-acting elements associated with hormones, light, and stress responses, suggesting regulatory roles in plant adaptation to environmental challenges. Expression analysis showed that *GmBBX* genes respond differentially to UV-B stress, indicating their potential role in soybean's defense mechanisms. This study provided essential insights into the molecular functions of *GmBBX* genes under UV-B stress and laid the groundwork for future research on stress-resilient soybean cultivars. Understanding these genes can aid in developing soybean varieties with enhanced tolerance to UV-B radiation and other environmental stresses (Azam et al. 2024).

7.4 Biostimulant-Induced UV Tolerance in Oilseeds

According to the European Biostimulants Industry Council (EBIC 2012), plant biostimulants are products containing substances and/or microorganisms that, when applied to plants or their rhizosphere, activate natural processes to enhance nutrient absorption, improve nutrient efficiency, increase tolerance to abiotic stress, and boost crop quality (Brown and Saa 2015). They can be categorized into nonmicrobial and microbial groups. Nonmicrobial biostimulants include humic and fulvic acids, protein hydrolysates, phosphites, seaweed extracts, chitosan derivatives, and silicon-based compounds. Microbial biostimulants include arbuscular mycorrhizal fungi (AMF), plant growth-promoting rhizobacteria (PGPR), and *Trichoderma* species. Biostimulants contribute to sustainable agriculture by optimizing plant performance without relying heavily on synthetic fertilizers (Rouphael and Colla 2020). Biostimulants offer numerous benefits to crops, enhancing their visual appeal, strengthening their immune systems, and stimulating the biosynthesis of defensive biomolecules. They aid in soil remediation by removing heavy metals, improving crop performance, reducing nutrient leaching, and promoting root development and seed germination. Additionally, they enhance resilience to abiotic and biotic stressors, facilitate crop establishment, and boost nutrient-use efficiency (Sun et al. 2024). This section reviews scientific studies on applying biostimulants for sustainable oilseed crop production under UV stress. Although research in this field has been limited, existing studies suggest that biostimulants can enhance stress tolerance and improve oilseed crop performance. The biostimulants examined on oilseed crops can be categorized into three main groups: microbial biostimulants, organic biostimulants, and inorganic and chemical biostimulants. These categories highlight the diverse mechanisms through which biostimulants aid plants in coping with UV stress, contributing to more sustainable agricultural practices.

7.4.1 Microbial Biostimulants

In a study, the impacts of UV exclusion and AMF inoculation on soybean photosynthetic parameters were investigated. UV exclusion was conducted in nursery bags under iron mesh with UV cut-off filters. The research revealed that plants subjected to UV exclusion combined with AMF inoculation (I) exhibited significantly enhanced photosynthetic pigments, carbonic anhydrase activity, and photosynthetic rate while displaying reduced internal CO_2 concentration, increased transpiration rate, and improved stomatal conductance. Notably, –UVB + I and –UVAB+I plants showed a higher performance index, enhanced activity of the water-splitting complex on the donor side of PSII, and an increased number of active PSII reaction centers per excited cross-section. These results underscore the synergistic benefits of UV exclusion and AMF in optimizing PSII quantum yield and overall photosynthetic efficiency. This study paves the way for future research on the role of these factors in improving soybean productivity, contributing to sustainable agricultural practices (Laad et al. 2023).

7.4.2 Organic Biostimulants

Plant-Derived Extracts Germinating soybeans under UV-B radiation is a promising approach to enhance isoflavone accumulation, yet the underlying mechanisms remain elusive. A study revealed that γ-Aminobutyric acid (GABA) plays a pivotal role in this process by acting as a signaling molecule that regulates secondary metabolite accumulation and plant development. Soybeans treated with GABA (5 mM) under UV-B exhibited significant improvements in sprout length and fresh weight, antioxidant enzyme activities (peroxidase and catalase), and Ca^{+2} influx. At the same time, oxidative stress markers such as malondialdehyde, H_2O_2, and $O_2^{\bullet-}$ fluorescence intensity were markedly reduced. Additionally, GABA treatment enhanced total flavonoid content and notably increased the activity of PAL and cinnamic acid-4-hydroxylase (C_4H)-key enzymes in isoflavone biosynthesisalongside the upregulation of related gene expression. Conversely, applying a GABA inhibitor (3-mercaptopropionic acid) suppressed these beneficial effects, confirming GABA's crucial role. These findings suggest that GABA acts as a vital mediator in soybean sprout growth and isoflavone enrichment under UV-B exposure, offering new insights into plant metabolic regulation (Gu et al. 2022). Stilbenes, a group of phytoalexins with diverse biological activities, are synthesized in plants under stress conditions. It is revealed that UV-C radiation enhances stilbene biosynthesis in peanut sprouts and induces oxidative damage (Zhu et al. 2023). Zhu et al. (2023) investigated the role of GABA in regulating stilbene production and oxidative stress under UV-C exposure during peanut germination. UV-C significantly increased GABA accumulation by upregulating genes and enzymes in its biosynthesis pathway. Exogenous GABA further affected stilbene synthesis, and enhanced phenolics, flavonoids, and antioxidant enzyme activities, while mitigating oxidative stress. Additionally, GABA influenced nitric oxide accumulation, suggesting its involvement in stress response mechanisms.

Plants have developed mechanisms to counter UV radiation damage, with free radicals triggering antioxidant defense systems. In a study, resveratrol accumulation and antioxidant enzyme activity in peanut seedlings were assessed after UV-C exposure. Resveratrol accumulation and stilbene synthase mRNA increased significantly following UV-C irradiation. Pretreatment with resveratrol mitigated leaf damage, while inhibition worsened the effects. Resveratrol treatment showed similar effects on malondialdehyde, H_2O_2, O_2^-, and antioxidant enzymes. Key enzymes like superoxide dismutases, catalase, and ascorbate peroxidase exhibited increased activity, suggesting resveratrol's role in managing oxidative stress induced by UV-C (Tang et al. 2010). Resveratrol is an antioxidant that is abundant in plants such as grapes and peanuts and even has potential therapeutic applications in human health (Costa et al. 2024).

Protein hydrolysates, derived from enzymatic or chemical hydrolysis of agro-industrial by-products, mitigate environmental stress, stimulate growth, and elevate productivity. Furthermore, amino acids, obtained through similar processes, enhance nutrient assimilation, increase yield and stress tolerance, and regulate key physiological activities by influencing enzymatic functions, gene expression, and redox homeostasis (Sun et al. 2024).

Hormone-Like Substances Abiotic stresses, including metals, ozone, salinity, extreme temperatures, UV-B radiation, and drought, threaten agricultural productivity. These stresses, whether alone or combined, negatively impact plants by disrupting biochemical, physiological, and molecular processes, leading to decreased growth and yield. Phytohormones, particularly salicylic acid (SA), have emerged as practical tools in enhancing plant tolerance to abiotic stresses (Abdelhameed et al. 2021; Dawood et al. 2022; Khan et al. 2015). Jasmonic acid (JA) and SA are crucial plant hormones that regulate growth, development, and defense mechanisms under various stresses. UV-B radiation, a significant abiotic stressor due to ozone depletion, adversely affects plant growth and yield, threatening food security. Understanding how JA and SA interact under UV-B stress can help maintain agricultural productivity. Studies show that UV-B exposure increases endogenous JA and SA levels, with JA enhancing plant resistance and SA contributing to oxidative stress. Their synergistic and antagonistic interplay plays a key role in plant defense, influencing ROS generation and overall stress responses (Choudhary et al. 2021). ROS are produced during normal oxidative processes in aerobic organisms, playing a role in cell signaling and defense. Environmental stresses, especially UV radiation, cause a significant increase in ROS levels, leading to oxidative stress. Anthropogenic pollutants, such as chlorofluorocarbons, contribute to ozone layer depletion, increasing UV-B exposure to Earth's surface. ROS cause lipid peroxidation of membranes, damages antioxidants, and alters plant gene expression. ROS includes nonradicals (H_2O_2, $^1O_2$) and free radicals ($OH^•$, $O_2^{•-}$, $RO^•$). In chloroplasts, PSI and PSII generate $^1O_2$ and $O_2^{•-}$. In mitochondria, complex I, ubiquinone, and complex III are key sources of $O_2^{•-}$. Plants have both enzymatic (e.g., superoxide dismutase, catalase) and nonenzymatic (e.g., ascorbic acid, glutathione) antioxidant systems to protect against ROS-induced damage. These defense systems scavenge ROS, preventing oxidative damage (Pathak et al. 2019).

Anthropogenic activities have led to an increase in harmful gaseous pollutants, depleting the stratospheric ozone layer and allowing more UV-B radiation to reach the Earth's surface. Plants rely on sunlight for energy. However, even a slight increase in UV-B radiation can be detrimental. In soybean plants, UV-B exposure significantly reduced biomass at juvenile and vegetative stages (Arsule et al. 2024). Based on a report by Arsule et al. (2024), among treatments, orange juice was the most effective, followed by ascorbic acid and salicylic acid, in mitigating UV-B damage, highlighting their potential in reducing UV-B stress on plants.

Abiotic stresses like drought, temperature extremes, and salinity create a challenge for plant growth and productivity. Salicylic acid plays a key role in enhancing stress tolerance. It activates plant defense mechanisms against heavy metal toxicity, UV light, and osmotic pressure. SA's effects depend on concentration, application method, and environmental factors (Salem et al. 2021). Li et al. (2014) investigated the effects of SA pretreatment on soybean seedlings under cadmium (Cd) and/or UV-B stress. They showed that both stresses reduced dry mass, pigment content, photosynthesis (PN), stomatal conductance (gs), and transpiration (E). SA alleviated growth, pigment, PN, and gs declines but did not improve E or Cd-induced chlorophyll fluorescence inhibition. Cd/UV-B increased oxidative stress and lipid peroxidation, which SA reduced. SA also stabilized membranes and acted as an antioxidant, enhancing soybean resistance to Cd and UV-B stress. Seed priming with SA has emerged as an effective strategy to enhance plant tolerance to abiotic stresses, including hexavalent chromium (Cr (VI)) and UV-B radiation. In maize seedlings, SA priming alleviates the toxic effects of these stresses by reducing the accumulation of Cr and ROS, which are responsible for cellular damage. This process boosts antioxidant activity, including enzymes like superoxide dismutase and ascorbate peroxidase, while improving chlorophyll fluorescence and growth. Ultimately, SA priming triggers a defensive response, helping plants mitigate oxidative damage, improve stress resilience, and enhance overall growth despite Cr (VI) and UV-B exposure (Singh et al. 2016). Salicylic acid pretreatment helped mitigate the harmful effects of Cd and UV-B stress on soybean seedlings. While Cd and UV-B exposure reduced growth, pigment content, and photosynthetic rates, SA alleviated these impacts but didn't fully restore transpiration or chlorophyll fluorescence. The stresses induced oxidative damage, increasing lipid peroxidation, but SA reduced this damage. The SA also modulated antioxidant enzyme activities, increasing SOD activity and decreasing POD activity, supporting plant resistance. These findings suggest that SA is an antioxidant and membrane stabilizer, enhancing plant tolerance to Cd and UV-B stresses (Li et al. 2014).

Salicylic Acid is a phenolic compound produced by many plant species, initially known for its medicinal use in humans. In plants, it participates in various physiological and metabolic reactions, particularly in responses to both biotic and abiotic stresses.SA plays a role in plant responses to abiotic stresses such as heavy metal toxicity, heat, chilling, drought, UV-light, and osmotic stress. Endogenous SA levels increase under abiotic stress. Applying SA induces stress tolerance in plants. Using genetically modified plants with altered SA synthesis helps to understand its molecular action.SA interacts with hormones like jasmonic acid, ethylene, abscisic acid, gibberellic acid, and cytokinin to regulate immune responses. The impact of SA depends on its application mode, concentration, environmental conditions, and the plant species and organs involved. The effects of SA may explain its role in enhancing stress tolerance (Shahrajabian and Sun 2024).

Jasmonates are lipid-derived compounds that play a crucial role in plant stress responses and development. They mediate defense against herbivores and necrotrophic pathogens, as well as responses to light, seasonal changes, circadian rhythms, cold, desiccation, salt, and UV stress. JAs inhibit the growth of roots, shoots, and leaves during development, while their precursor, 12-oxo-phytodienoic acid (OPDA), partially affects seed germination and flower development. Their involvement in various signal transduction pathways has led to an interest in horticultural and biotechnological applications, including intercropping, mycorrhization, induced resistance, insect resistance priming, and use in pharmaceutical production and cancer treatment (Wasternack 2014). The JAs, lipid derivatives, are key signaling compounds in plant stress responses and development. They mediate defense against herbivores, pathogens, nematodes, and various abiotic stresses such as UV, salt, and temperature. JA signaling interacts with other phytohormones like auxin, gibberellic acid, and salicylic acid in complex crosstalk networks. Additionally, JAs regulate beneficial plant–microbe interactions, including those with PGPR and AMF, enhancing plant resilience to stress (Dar et al. 2015).

UV-B radiation also plays a role in improving the color, nutritional quality, and postharvest disease resistance of fruits and herbs. Melatonin applications amplify these benefits, offering a promising tool for agricultural and postharvest practices. A study indicated that combining UV-B exposure with melatonin treatment could enhance phenolic compound production, antioxidant activity, and stress tolerance, ultimately improving crop quality. These findings underscore the importance of understanding and harnessing melatonin's protective effects to mitigate UV-B-induced stress, with potential implications for agricultural sustainability and crop productivity in changing environmental conditions (Cano et al. 2023). Exposure to UV-B radiation is a well-known strategy to enhance the isoflavone content in germinated soybeans, yet it significantly hampers their growth and biomass accumulation. Melatonin, as a novel plant biostimulant, not only mitigates abiotic stress but also modulates secondary metabolite production. In a study, the impact of exogenous melatonin (25 μM) on soybean germination under UV-B stress was reported. Compared to UV-B treatment alone, melatonin application significantly improved sprout length, fresh weight, Ca^{+2} influx, and peroxidase activity, while reducing malondialdehyde, H_2O_2 levels, and oxidative fluorescence signals (H_2O_2 and $O_2{}^{\bullet -}$). Notably, melatonin did not significantly alter SOD and CAT activity during germination. Moreover, melatonin substantially increased total flavonoid content and key isoflavone monomers (daidzein, genistein, daidzin, glycitin, and genistin) in 4-day-old soybean sprouts under UV-B stress. This enhancement was accompanied by a significant rise in PAL activity and the upregulation of PAL1, isoflavone reductase, chalcone synthase, and flavanone 3-hydroxylase—key enzymes in the pathway of isoflavone biosynthesis. Ultimately, exogenous melatonin alleviated the inhibitory impacts of UV-B stress on growth and development, while simultaneously boosting flavonoid and isoflavone accumulation by enhancing gene expression related to biosynthesis of isoflavone and enzyme activity (Yin et al. 2022).

7.4.3 Inorganic and Chemical Biostimulants

As mentioned above, UV-B radiation induces oxidative stress in plants by generating active oxygen species like singlet oxygen, hydrogen peroxide, superoxide anion, and hydroxyl radicals. Hydrogen peroxide acts as a signaling molecule in response to stress. In sunflower cotyledons, UV-B treatment altered the activities and isoforms of CAT, ascorbate peroxidase (APX), and POD. UV-B exposure increased the activities of CAT and guaiacol peroxidase (GPX) under dark recovery, but both returned to control levels under white light recovery. POD isoforms (POD1, POD2, and POD3) showed varying responses to UV-B and hydrogen peroxide treatments, suggesting a role for POD in oxidative stress management and polyphenol metabolism, enhancing antioxidant capacity (Yannarelli et al. 2006).

Silicon (Si) enhances plant defense against stress, but its role in mitigating UV-B radiation effects on field crops is less understood. This study examined how Si influences biomass, physiology, and antioxidant activity in soybean seedlings under UV-B stress. UV-B exposure significantly reduced leaf area (73.9–94.7%), total dry weight (11.8–36.6%), and root-to-shoot ratio (9.2–30.2%), while increasing antioxidant activity. Si supplementation alleviated these effects by improving growth parameters and reducing oxidative stress markers. In Si-treated plants, catalase, superoxide dismutase, and peroxidase activities decreased under UV-B, confirming their protective role in reducing stress-induced damage (Shen et al. 2010).

UV-B radiation can cause oxidative stress in plants, damaging tissues such as maize leaves. Nitric oxide (NO) protects against this stress by enhancing the plant's antioxidant defenses. When NO is applied through a donor like sodium nitroprusside (SNP), it helps to reduce the adverse effects of UV-B, such as growth suppression and chlorophyll degradation. NO boosts the accumulation of protective compounds like flavonoids and anthocyanins, which absorb UV-B rays. It also increases the activity of antioxidant enzymes, such as catalase and ascorbate peroxidase, helping to neutralize ROS. This highlights NO's function in activating a plant's defense mechanisms, improving its resistance to UV-B damage (Kim et al. 2010).

7.5 Conclusion and Future Studies

Biostimulants, including microbial, organic, inorganic, and chemical types, play a crucial role in enhancing the resilience of oilseed crops against environmental stresses. By influencing biochemical processes, boosting antioxidant activity, and improving physiological responses, these biostimulants help plants combat oxidative stress and mitigate the adverse effects of UV radiation. Their application enhances plant growth, stress tolerance, and productivity, contributing to sustainable agriculture. Integrating biostimulants into crop management strategies can optimize oilseed production, ensuring better yield and quality under challenging climatic conditions. Future research should focus on refining their use to maximize efficiency and long-term agricultural sustainability.

References

Abdelhameed RE, Abdel Latef AA, Shehata RS (2021) Physiological responses of salinized fenugreek (Trigonella foenumgraecum L.) plants to foliar application of salicylic acid. Plants 10:657

Arsule AA, Nigam B, Chaudhary IJ (2024) Comparative effectiveness of ascorbic acid, salicylic acid and orange juice on soybean cultivar (*Glycine Max* L.) under UV-B (ultraviolet radiation) stress. J Med Plants Stud 12(1):203–213

Azam M, Usman M, Manzoor MA, Yao L, Xiaohong M, Yan Z et al (2024) Comprehensive characterization and expression profiling of BBX gene family in soybean in response to UV-B stress. Plant Stress 13:100560

Brown P, Saa S (2015) Biostimulants in agriculture. Front Plant Sci 6:671

Bulgari R, Cocetta G, Trivellini A, Vernieri PAOLO, Ferrante A (2015) Biostimulants and crop responses: a review. Biol Agric Hortic 31(1):1–17

Cano A, Hernández-Ruiz J, Arnao MB (2023) Melatonin in plants under UV stress conditions. In: Melatonin: role in plant signaling, growth and stress tolerance: phytomelatonin in normal and challenging environments. Springer International Publishing, Cham, pp 263–277

Cechin I, Fumis TDF, Dokkedal AL (2007) Growth and physiological responses of sunflower plants exposed to ultraviolet-B radiation. Cienc Rural 37:85–90

Cechin I, Corniani N, de Fátima FT, Cataneo AC (2008) Ultraviolet-B and water stress effects on growth, gas exchange and oxidative stress in sunflower plants. Radiat Environ Biophys 47:405–413

Choudhary KK, Agrawal SB (2016) Assessment of fatty acid profile and seed mineral nutrients of two soybean (*Glycine max* L.) cultivars under elevated ultraviolet-B: role of ROS, pigments and antioxidants. Photochem Photobiol 92(1):134–143

Choudhary KK, Singh S, Agrawal M, Agrawal SB (2021) Role of jasmonic and salicylic acid signaling in plants under UV-B stress. In: Jasmonates and salicylates signaling in plants, pp 45–63

Costa H, Gallego SM, Tomaro ML (2002) Effect of UV-B radiation on antioxidant defense system in sunflower cotyledons. Plant Sci 162(6):939–945

Costa CE, Romaní A, Domingues L (2024) Overview of resveratrol properties, applications, and advances in microbial precision fermentation. Crit Rev Biotechnol:1–17

Czégény G, Mátai A, Hideg É (2016) UV-B effects on leaves—oxidative stress and acclimation in controlled environments. Plant Sci 248:57–63

Dar TA, Uddin M, Khan MMA, Hakeem KR, Jaleel H (2015) Jasmonates counter plant stress: a review. Environ Exp Bot 115:49–57

Dawood MF, Abdel Latef AAH (2022) Use of biostimulants to improve UV tolerance in cereals. In: Sustainable remedies for abiotic stress in cereals. Springer Nature Singapore, pp 599–623

Dawood MFA, Zaid A, Abdel Latef AA (2022) Salicylic Acid Spraying-Induced Resilience Strategies Against the Damaging Impacts of Drought and/or Salinity Stress in Two Varieties of Vicia faba L. Seedlings. Journal of Plant Growth Regulation 41:1919–1942

Dehariya P, Kataria S, Pandey GP et al (2011) Assessment of impact of solar UV components on growth and antioxidant enzyme activity in cotton plant. Physiol Mol Biol Plants 17:223–229. https://doi.org/10.1007/s12298-011-0071-9

Del Buono D, Regni L, Proietti P (2023) Abiotic stresses, biostimulants and plant activity. Agriculture 13(1):191

Escobar-Bravo R, Klinkhamer PG, Leiss KA (2017) Interactive effects of UV-B light with abiotic factors on plant growth and chemistry, and their consequences for defense against arthropod herbivores. Front Plant Sci 8:278

Fukuchi K, Takahashi K, Tatsumoto H (2004) Impacts of UV-A irradiation on laser-induced fluorescence spectra in UV-B damaged peanut (*Arachis hypogaea* L.) leaves. Environ Technol 25(11):1233–1240. https://doi.org/10.1080/09593332508618370

Gao W, Zheng Y, Slusser JR, Heisler GM (2003) Impact of enhanced ultraviolet-B irradiance on cotton growth, development, yield, and qualities under field conditions. Agric For Meteorol 120(1–4):241–248

Gu M, Yang J, Tian X, Fang W, Xu J, Yin Y (2022) Enhanced total flavonoid accumulation and alleviated growth inhibition of germinating soybeans by GABA under UV-B stress. RSC Adv 12(11):6619–6630

Jansen MA, Hectors K, O'Brien NM, Guisez Y, Potters G (2008) Plant stress and human health: do human consumers benefit from UV-B acclimated crops? Plant Sci 175(4):449–458

Jaradat AA (2016) Breeding oilseed crops for climate change. In: Breeding oilseed crops for sustainable production. Academic, pp 421–472

Jenkins GI (2009) Signal transduction in responses to UV-B radiation. Annu Rev Plant Biol 60(1): 407–443

Kakani VG, Reddy KR, Zhao D, Mohammed AR (2003) Effects of ultraviolet-B radiation on cotton (*Gossypium hirsutum* L.) morphology and anatomy. Ann Bot 91(7):817–826

Kakani VG, Reddy KR, Zhao D, Gao W (2004) Senescence and hyperspectral reflectance of cotton leaves exposed to ultraviolet-B radiation and carbon dioxide. Physiol Plant 121(2):250–257

Kataria S, Dehariya P, Guruprasad KN, Pandey GP (2012) Effect of exclusion of ambient solar UV-A/B components on growth and antioxidant response of cotton (*Gossypium hirsutum* L.). Acta Biol Crac Ser Bot 54(2):47

Khan MIR, Fatma M, Per TS, Anjum NA, Khan NA (2015) Salicylic acid-induced abiotic stress tolerance and underlying mechanisms in plants. Front Plant Sci 6:462

Kim TY, Jo MH, Hong JH (2010) Protective effect of nitric oxide against oxidative stress under UV-B radiation in maize leaves. J Environ Sci 19(12):1323. https://doi.org/10.5322/jes.2010.19.12.1323

Laad P, Patel P, Guruprasad KN, Sharma MP, Kataria S, Brestic M (2023) Effect of UV exclusion and AMF inoculation on photosynthetic parameters of Glycine max. Photosynthetica 61(2):236–243. https://doi.org/10.32615/ps.2023.014

Langebartels C, Schraudner M, Heller W, Ernst D, Sandermann H (2002) Oxidative stress and defense reactions in plants exposed to air pollutants and UV-B radiation. In: Oxidative stress in plants, pp 105–135

Li X, Zhang L, Li Y, Ma L, Bu N, Ma C (2012) Changes in photosynthesis, antioxidant enzymes and lipid peroxidation in soybean seedlings exposed to UV-B radiation and/or cd. Plant Soil 352:377–387

Li XM, Ma LJ, Bu N et al (2014) Effects of salicylic acid pre-treatment on cadmium and/or UV-B stress in soybean seedlings. Biol Plant 58:195–199. https://doi.org/10.1007/s10535-013-0375-4

Mareri L, Parrotta L, Cai G (2022) Environmental stress and plants. Int J Mol Sci 23(10):5416

Mark U, Tevini M (1996) Combination effects of UV-B radiation and temperature on sunflower (*Helianthus annuus* L., cv. Polstar) and maize (*Zea mays* L., cv. Zenit 2000) seedlings. J Plant Physiol 148(1–2):49–56

Mewis I, Schreiner M, Nguyen CN, Krumbein A, Ulrichs C, Lohse M, Zrenner R (2012) UV-B irradiation changes specifically the secondary metabolite profile in broccoli sprouts: induced signaling overlaps with defense response to biotic stressors. Plant Cell Physiol 53(9):1546–1560

Moghadam HRT, Ghooshchi F, Zahedi H (2011) Effect of UV radiation and evaluated CO2 on morphological traits, yield, and yield components of canola (*Brassica napus* L.) grown under water deficit stress. Not Bot Horti Agrobo 39(1):213–219

Müller-Xing R, Xing Q, Goodrich J (2014) Footprints of the sun: memory of UV and light stress in plants. Front Plant Sci 5:474

Murali NS, Teramura AH (1986) Effectiveness of UV-B radiation on the growth and physiology of field-grown soybean modified by water stress. Photochem Photobiol 44(2):215–219

Nguyen VL, Dinh TN, Nguyen HB, Nguyen TMA, Vu TY, Thieu TPL, Lai TNH, Nguyen VH (2020) UV-B exposure induced accumulation of phenolics and resveratrol and enhanced antioxidant activities in peanut sprout. Carpathian J Food Sci Technol 12(1):89–101

Pathak J, Rajneesh AH, Singh DK, Singh PR, Kumar D, Kannaujiya VK, Singh SP, Sinha RP (2019) Oxidative stress and antioxidant defense in plants exposed to ultraviolet radiation. In: Reactive oxygen, nitrogen and sulfur species in plants: production, metabolism, signaling and defense mechanisms, pp 371–342

Petrova I, Ivanova S, Stoyanova S, Mincheva R, Pavlova M (2023) Influence of biostimulants and humic extracts treatment on the fatty acid profile of the spring oilseed rape variety. Agric. Sci Technol 15:52

Prasad SM, Dwivedi R, Zeeshan M (2005) Growth, photosynthetic electron transport, and antioxidant responses of young soybean seedlings to simultaneous exposure of nickel and UV-B stress. Photosynthetica 43:177–118

Reddy KR, Kakani VG, Zhao D, Mohammed AR, Gao W (2003) Cotton responses to ultraviolet-B radiation: experimentation and algorithm development. Agric For Meteorol 120(1–4):249–265

Reddy KR, Kakani VG, Zhao D, Kotl S, Gao W (2004) Interactive effects of ultraviolet-B radiation and temperature on cotton physiology, growth, development, and hyperspectral reflectance. Photochem Photobiol 79(5):416–427

Rouphael Y, Colla G (2020) Biostimulants in agriculture. Front Plant Sci 11:40

Salem KFM, Saleh MM, Abu-Ellail FFB, Aldahak L, Alkuddsi YA (2021) The role of salicylic acid in crops to tolerate abiotic stresses. In: Hayat S, Siddiqui H, Damalas CA (eds) Salicylic acid – a versatile plant growth regulator. Springer, Cham. https://doi.org/10.1007/978-3-030-79229-9_7

Shahrajabian MH, Sun W (2024) The importance of salicylic acid, humic acid and fulvic acid on crop production. Lett Drug Des Discov 21(9):1465–1480

Shen X, Li X, Li Z, Li J, Duan L, Eneji AE (2010) Growth, physiological attributes and antioxidant enzyme activities in soybean seedlings treated with or without silicon under UV-B radiation stress. J Agron Crop Sci 196(6):431–439

Shen X, Dong Z, Chen Y (2015) Drought and UV-B radiation effect on photosynthesis and antioxidant parameters in soybean and maize. Acta Physiol Plant 37:1–8

Shi C, Liu H (2021) How plants protect themselves from ultraviolet-B radiation stress. Plant Physiol 187(3):1096–1103

Shukla UC, Murthy RC, Kakkar P (2008) Combined effect of ultraviolet-B radiation and cadmium contamination on nutrient uptake and photosynthetic pigments in Brassica campestris L. seedlings. Environ Toxicol 23(6):712–719

Singh VP, Kumar J, Singh M et al (2016) Role of salicylic acid-seed priming in the regulation of chromium (VI) and UV-B toxicity in maize seedlings. Plant Growth Regul 78:79–91. https://doi.org/10.1007/s10725-015-0076-4

Singh AL, Singh S, Kurella A, Verma A, Mahatama MK, Venkatesh I (2022) Plant bio-stimulants, their functions and use in enhancing stress tolerance in oilseeds. In: New and future developments in microbial biotechnology and bioengineering. Elsevier, pp 239–259

Sun W, Shahrajabian MH, Kuang Y, Wang N (2024) Amino acids biostimulants and protein hydrolysates in agricultural sciences. Plan Theory 13(2):210

Tang K, Zhan JC, Yang HR, Huang WD (2010) Changes of resveratrol and antioxidant enzymes during UV-induced plant defense response in peanut seedlings. J Plant Physiol 167(2):95–102

Teramura AH, Sullivan JH, Lydon J (1990) Effects of UV-B radiation on soybean yield and seed quality: a 6-year field study. Physiol Plant 80(1):5–11

Torres M, Frutos G, Duran JM (1991) Sunflower seed deterioration from exposure to UV-C radiation. Environ Exp Bot 31(2):201–207

Tripathi R, Agrawal SB (2013) Evaluation of changes in lipid peroxidation, ROS production, surface structures, secondary metabolites, and yield of linseed (*Linum usitatissimum* L.) under individual and combined stress of ultraviolet-B and ozone using open-top chambers. Indian J Biochem Biophys 50(4):318–325

Tripathi R, Rai K, Singh S et al (2019) Role of supplemental UV-B in changing the level of ozone toxicity in two cultivars of sunflower: growth, seed yield, and oil quality. Ecotoxicology 28: 277–293. https://doi.org/10.1007/s10646-019-02020-6

Velasco L, Fernández-Martínez JM (2002) Breeding oilseed crops for improved oil quality. J Crop Prod 5(1–2):309–344

Wang H, Liu S, Wang T, Liu H, Xu X, Chen K, Zhang P (2020) The moss flavone synthase I positively regulates the tolerance of plants to drought stress and UV-B radiation. Plant Sci 298: 110591

Wasternack C (2014) Action of jasmonates in plant stress responses and development—applied aspects. Biotechnol Adv 32(1):31–39

Willmon D, Devireddy AR, Inupakutika M, Puppala N, Cho Y (2017) Stress responses of peanut (*Arachis hypogaea* L.) genotypes as measured by trigonelline content after exposure to UV-B radiation. Am J Plant Sci 8(05):998

Yadav G, Srivastava PK, Parihar P, Tiwari S, Prasad SM (2016) Oxygen toxicity and antioxidative responses in arsenic-stressed *Helianthus annuus* L. seedlings against UV-B. J Photochem Photobiol B Biol 165:58–70

Yannarelli GG, Gallego SM, Tomaro ML (2006) Effect of UV-B radiation on the activity and isoforms of enzymes with peroxidase activity in sunflower cotyledons. Environ Exp Bot 56(2): 174–181

Yin Y, Tian X, Yang J, Yang Z, Tao J, Fang W (2022) Melatonin mediates isoflavone accumulation in germinated soybeans (*Glycine max* L.) under ultraviolet-B stress. Plant Physiol Biochem 175: 23–32

Yoon MY, Kim MY, Shim S, Kim KD, Ha J, Shin JH, Kang S, Lee SH (2016) Transcriptomic profiling of soybean in response to high-intensity UV-B irradiation reveals stress defense signaling. Front Plant Sci 7:1917

Zhang M, Wang O, Cai S, Zhao L, Zhao L (2023) Composition, functional properties, health benefits, and applications of oilseed proteins: a systematic review. Food Res Int 171:113061

Zhao D, Reddy KR, Kakani VG, Read JJ, Sullivan JH (2003) Growth and physiological responses of cotton (*Gossypium hirsutum* L.) to elevated carbon dioxide and ultraviolet-B radiation under controlled environmental conditions. Plant Cell Environ 26(5):771–782

Zhao D, Reddy KR, Kakani VG, Koti S, Gao W (2005) Physiological causes of cotton fruit abscission under conditions of high temperature and enhanced ultraviolet-B radiation. Physiol Plant 124(2):189–199

Zhu T, Yang J, Zhang D et al (2021) Light radiation promoted stilbene accumulation in peanut sprouts: a response of the reestablishment of oxidant-antioxidant homeostasis. Acta Physiol Plant 43:137. https://doi.org/10.1007/s11738-021-03305-2

Zhu T, Ruan S, Zhang J et al (2023) γ-Aminobutyric acid (GABA) mediated stilbene biosynthesis and alleviated oxidative stress induced by UV-C radiation during peanuts (*Arachis hypogaea* L.) germination. Acta Physiol Plant 45:102. https://doi.org/10.1007/s11738-023-03580-1

Chapter 8
Organic Fertilizers as a Solution to Abiotic Stress Challenges in Oilseed Crops

Sidra Ahmad, Nayab Ahmad, Muhammad Nauman Khan, Hülya Doğan, and Rashid Iqbal

Abstract In this chapter, the essential adaptive mechanisms and the critical challenges faced by oilseed crops under various abiotic stress conditions are discussed. These stresses have profound impacts on crop yield with long-range consequences on growth, yield and oil quality via complex physiological, biochemical and molecular disruptions. It describes how crops have adapted with the aid of osmotic adjustment, antioxidant defenses and changes in root architecture to survive and grow after suffering from environmental stress. In addition to supplying critical nutrients, organically based fertilizers improve the physical and chemical properties of soil, increase water retention and stimulate beneficial microbial activity, all of which together strengthen the defenses of the plant. These beneficial characteristics enable better stress tolerance mechanisms, which include antioxidant enhancement and root expansion and better nutrient acquisition. As a result, under difficult environmental conditions, oilseed crops maintain their development process and production levels. This chapter demonstrates that sustainable organic fertilizers hold great promise in stress mitigation, which leads to enhanced oilseed yields and resilience.

S. Ahmad · N. Ahmad
Institute of Biotechnology and Genetic Engineering (IBGE), The University of Agriculture, Peshawar, Peshawar, Pakistan

M. N. Khan (✉)
Department of Botany, Islamia College Peshawar, Peshawar, Pakistan

Biology Laboratory, University Public School, University of Peshawar, Peshawar, Pakistan
e-mail: nomiflora@uop.edu.pk

H. Doğan
Yozgat Bozok University, Hemp Research Institute, Department of Agriculture and Food, Yozgat, Türkiye

R. Iqbal
Department of Agronomy, Faculty of Agriculture and Environment, The Islamia University of Bahawalpur, Bahawalpur, Pakistan

Department of Life Sciences, Western Caspian University, Baku, Azerbaijan

© The Author(s), under exclusive license to Springer Nature Singapore Pte Ltd. 2025

A. A. H. Abdel Latef (ed.), *Oilseed Crops Under Abiotic Stress*, Sustainability Sciences in Asia and Africa, https://doi.org/10.1007/978-981-96-8346-8_8

Keywords Abiotic stress · Stress mitigation · Organic fertilizers · Oilseed crops

8.1 Introduction

Oilseeds represent an important part of global agricultural production, and they supply the human population with edible oils, potential biofuels and industrial products. Most relevant for food security and the economy, especially in agricultural-dependent regions, soybean, sunflower, rapeseed and groundnut are important oil crops. Oilseed crops are negatively impacted by several abiotic stressors, including drought, salt, elevated temperatures and nutrient deficiency (Ahmad et al. 2021a). These stresses are detrimental to the growth and productivity of the crops, which can directly affect the quality of the oil and the by-products derived from the plants. Organic fertilizers (e.g., compost and manure) supply nutrients to crops and improve soil health while sustaining soil microbial activity and enhancing crop resilience against environmental stresses (Vicente 2022).

Climate change and the increase in weather variability experienced over the last few decades have made abiotic stress an important issue in agricultural production and have enhanced the importance of the study of oilseed crops under abiotic stress conditions. These increasing hurdles and obstacles linked to water stress, soil salinization, as well as climate change increase the need to adapt agriculture into a more resistant one. To satisfy the nutrient requirements of crops, organic fertilizers have been widely used as a sustainable option that can also improve long-term soil fertility and environmental benefits (Khan et al. 2024). Organic inputs slowly release nutrients during the multiplication phase of the soil ecology process while also creating a more stable soil structure compared to synthetic fertilizers. Under abiotic stress conditions, plants need a constant supply of critical nutrients to sustain physiological activities and so this slow nutrient release could be advantageous (Liu 2024).

Oilseed crops have been grown for thousands of years since early agricultural societies realized the importance of such crops for dietary and industrial consumption. Oilseed crop domestication began with the early ancient civilizations of Mesopotamia, Egypt and the Indus Valley civilizations, which used oil-bearing plants for cooking and lighting (McIntosh 2017). Breeding approaches and the development of stress-tolerant varieties were the early steps in enhancing crop resilience. With the advancement of the field, molecular biology and biotechnology illuminated corresponding genes and pathways mediated by stress responses. Such a molecular understanding has, of course, been supplemented by agroecological work highlighting the importance of soil–plant interactions and organic matter as a buffer against environmental stresses on crops (Chugh et al. 2022).

Recent research trends show that there is more focus on integrated interventions that incorporate agronomic methods with biotechnological tools. Current studies have shown that there is a synergistic effect of using organic fertilizers along with drought-tolerant varieties. These combinations enhance water consumption efficiency and reduce the formation of reactive oxygen species (ROS), which are harmful by-products generated under stress conditions (Chakradhar et al. 2017). This research area has a wide range of methodologies, from greenhouse to field experiments. Soil and plant tissue analysis, gas exchange measurements and biochemical assays are among the most widely used techniques to assess the effect of abiotic stress and the efficacy of organic fertilizers. All methods have pros and cons. Greenhouse studies are an example of a controlled environment with the potential to isolate variables, but they can lose their field when it comes to complexity. On the other hand, field studies yield more naturalistic data but come with additional challenges in isolating variables in the environment for data interpretation (Patono et al. 2023).

By strengthening the resilience of important crops, we ensure more reliable vegetable oil supplies since oils are a major part of nutrition and a component of many products. Organic fertilizers promote sustainable agricultural practices in developing countries that comply with environmental policy objectives of lower carbon emissions, reinforcement of soil quality and high soil fertility. Implementation of such practices can support farmers in coping with the growing risks of climate variability while decreasing reliance on finite resources (Alcock et al. 2022). Nutrient availability and uptake can be influenced by temperature fluctuations, soil type, etc. These challenges need to be overcome and can be achieved in future studies, which must explore the heterogeneity of both organic and precision systems to tailor fertilizer usage. Finally, more long-term work is warranted to evaluate the total effects of organic additions on soil properties and plant response during subsequent years of production (Hedley 2015).

8.1.1 Overview of Oilseed Crops and Their Importance

Oilseed crops are one of the most important global crops, providing an array of food security, economic and industrial benefits. Soybean, sunflower, rapeseed (canola), groundnut (peanut) and sesame are the most grown oilseed crops across diverse climatic zones. Seeds from these plants are the principal sources of vegetable oils for cooking, food processing and the production of margarine and other food products. Apart from being directly involved in human nutrition, oilseed crops also offer high-quality protein meals that feed the animal sector. Oilseed crops can be used for nonfood purposes, such as eco-friendly biofuels, bioplastics and lubricants, which help diversify sources of energy and ultimately reduce dependency on fossil fuels (Lafarga 2021).

The significance of oilseed crops has increased over the last couple of decades, mainly due to the rising global need for oils and proteins of plant origin. Such trends are supported not only by the growing global population but also recommendations by health experts that favor unsaturated and plant-based fats over animal fats and by the expansion of bio-based materials (Czwartkowski et al. 2022). Since oilseeds are economically important cash crops, they have significant importance in the economies of some countries and regions where agriculture is one of the leading sectors. Oilseeds and their derivatives exports enhance trade while also promoting rural development by securing farmers a steady source of revenue (Mathur et al. 2023).

Furthermore, oilseed crops are integral to sustainable agriculture. Certain oilseed varieties, such as soybeans and groundnuts, can assimilate atmospheric nitrogen through symbiotic bacteria in the soil, enhancing soil fertility and reducing dependence on synthetic nitrogen fertilizers. Improving this feature could render cropping systems more sustainable by mitigating water pollution and greenhouse gas emissions associated with excessive chemical fertilizer application (Sindhu et al. 2019).

Oilseed crops are agronomically diverse, enabling the use of different climates to their advantage to provide a wide geographical coverage of the landscape suitable for growth. Abiotic stresses can have a detrimental impact on both yield and oil quality, which highlights the need for research into oilseed crop resilience (Ahmad et al. 2021b). The development of stress-resistant varieties via breeding programs and genetic transformation, along with the innovation of sustainable farming methods such as organic fertilizer application, is crucial for maintaining productivity to satisfy global demand (Panjabi et al. 2019).

8.1.2 Definition and Types of Abiotic Stress

Abiotic stress is an environmental condition that negatively impacts the growth, development and productivity of plants (Dawood et al. 2025; Umer et al. 2025). Drought is one of the most prevalent and damaging forms of abiotic stress and takes place when the soil does not hold sufficient water for the proper functioning of plant growth. Drought stress leads to a lack of water inside the plant tissue, which ultimately interrupts several physiological processes (e.g., photosynthesis and nutrient transport) (Ding et al. 2018). Salinity stress is a second major category where salt in the soil solution accumulates to levels that affect the ability of crops to extract water, due to leaching with high saline input water or as a result of natural soil conditions. Salt stress induces ion toxicity, osmotic stress and nutritional imbalance, which impede plant growth and productivity (Majeed and Muhammad 2019).

Another challenge arises from temperature stress, which includes heat and cold stress. It happens when temperatures reach a level above the range where plants can grow and develop optimally and disrupt the structures of the cell, enzymatic activities and photosynthesis, consequently impacting the yield and quality of the crops. Cold stress can reduce metabolism and induce damage to cell membranes (Tahjib-Ul-Arif et al. 2024; Waraich et al. 2012). Nutrient stress arises from either deficiency or excess of essential nutrients that affect plant growth. If a plant is short of nutrients, it will not be able to grow and respond well to other stressors, and if a plant experiences an overload of nutrients, this can lead to toxicity (Zaid et al. 2020).

Excessive or insufficient light causes light stress, which can also subsequently affect plant productivity. Photo inhibition due to high-intensity light leads to the destruction of chlorophyll and the destruction of photosystems involved in the

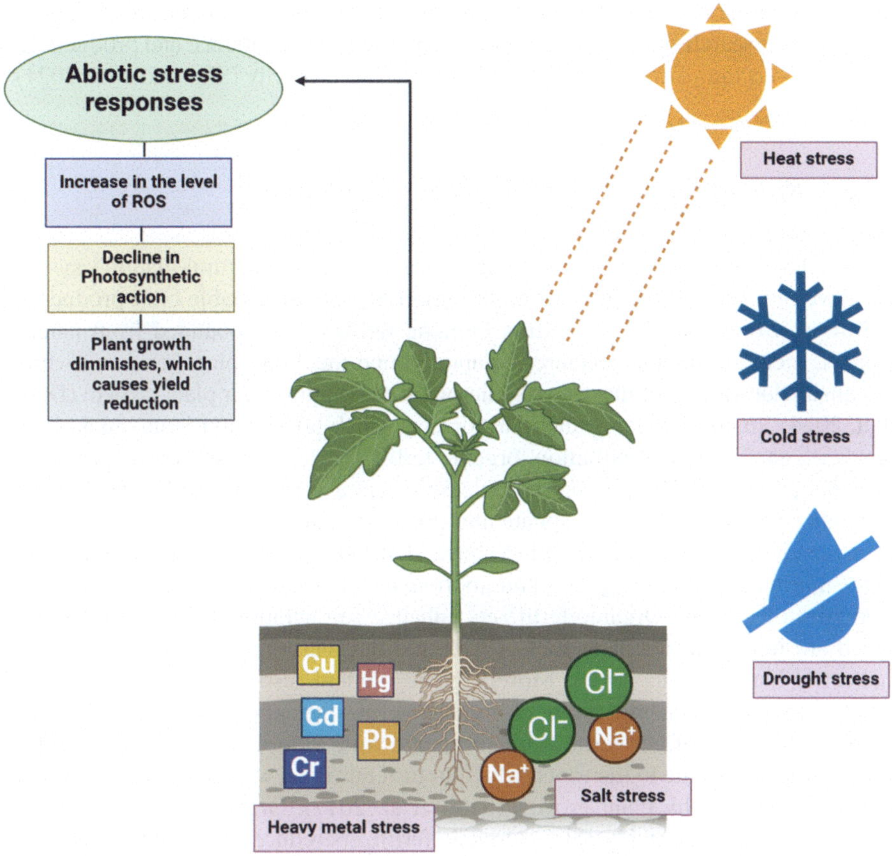

Fig. 8.1 Impact of abiotic on oilseed crops

process of photosynthesis. Conversely, low light decreases the rate of photosynthesis, which ultimately limits growth and biomass accumulation (Didaran et al. 2024). Oxidative stress is a typical secondary consequence of numerous abiotic stressors, characterized by the excessive production of ROS. Cellular dysfunction arises from the extremely reactive characteristics of these compounds, which can harm proteins, lipids and nucleic acids (Juan et al. 2021). High levels of metals such as cadmium, lead or mercury constitute heavy metal stress (Fig. 8.1), which also results in toxicity by interfering with metabolic functions and hindering healthy plant growth (Balali-Mood et al. 2021).

Flooding stress, also known as waterlogging, is a condition where soil is oversaturated and cannot supply enough oxygen to the roots (Nasrullah et al. 2022). This condition aggravates cellular respiration and nutrient uptake, which leads to further complications, inducing stress. Plants have developed different defense mechanisms to overcome these different types of abiotic stress, including osmotic adjustment, synthesis of stress-related proteins and activation of antioxidant defense systems. It is well known that abiotic stress is an important target area of agricultural research since it is a key factor that affects the resilience and productivity of crops and is thus directly implicated in global food security (Shahzad et al. 2021).

8.1.3 Role of Organic Fertilizers in Agriculture

Organic fertilizers are an important part of the current agricultural planting system and development and the greatest contribution toward sustainable crop production and the conservation of soil fertility. Organic fertilizers are obtained from natural sources, including animal manure, compost, bone meal and plant residues, which contain a wide variety of macro- and micronutrients required for plant growth (Dutta et al. 2024). In contrast to synthetic fertilizers, which provide nutrients quickly and can cause environmental pollution, organic fertilizers decompose slowly, providing nutrients gradually. The progressive release offers a consistent supply of essential nutrients such as nitrogen, phosphate and potassium, aligned with plant absorption needs, leading to enhanced growth (Yahaya et al. 2023). Organic fertilizers provide all the nutrients required by plants and improve the physical, chemical and biological properties of the soil. Organic fertilizers enhance soil aeration and water retention, which are important for plant root growth and nutrient uptake, by adding organic matter (Usharani et al. 2019). More organic content also boosts the activity of beneficial microorganisms, essential organisms for nutrient cycling and which promotes the process of converting organic matter into soluble forms. The microbes then further contribute to the roots, which in turn nurture a well-balanced ecosystem that promotes healthy plant growth (Sahu et al. 2017). Organic fertilizers have a number of environmental benefits. Organic fertilizers reduce the potential for nutrient loss through leaching and runoff, which are more prevalent with synthetic fertilizers. Thus, organic fertilizers play the role of helping carbon emission reduction by decreasing the use of chemical fertilizers while enhancing soil health, which also complies with sustainability practices based on sound environmental principles (Verma et al. 2020). Further, the plants trap carbon in soil, which is important for climate change mitigation as it removes CO_2 from the atmosphere (Lal et al. 2007). Apart from environmental benefits, organic fertilizers contribute to soil fertility, ensuring future generations can farm over the same land without net soil loss (Izaurralde et al. 2001). Organic fertilizers lessen the need for farmers to rely on nonrenewable resources needed to produce synthetic fertilizers. Organic fertilizers are useful for smallholders and resource-poor farmers who are looking for low-cost, sustainable ways to keep producing (Sarkar et al. 2020). An overview of abiotic stress impacts and the role of organic fertilizers in oilseed crops is given in Table 8.1.

Table 8.1 Overview of abiotic stress impacts and the role of organic fertilizers in oilseed crops

S. No	Parameter	Abiotic stress	Effect on specific oilseed crops	Physiological impact	Adaptation strategies	Role of organic fertilizers	Key benefits in mitigation	Reference
1	Growth performance	Drought	Soybean	Reduced biomass, stunted growth	Osmotic adjustment, deeper root development	Enhances soil water retention and root expansion	Maintains consistent growth and water use efficiency	Arya et al. (2021)
2	Seed yield and oil content	Salinity	Soybean	Decreased seed set and oil quality	Ion compartmentalization, synthesis of osmolytes	Improves nutrient balance and ion homeostasis	Supports better seed formation and stable oil composition	Anwar-ul-Haq et al. (2023)
3	Reproductive health	High temperature	Canola, soybean	Flower and pod abortion, decreased seed viability	Heat shock protein activation, membrane stabilization	Provides thermal buffering and supports root function	Reduces heat stress impacts, promotes reproductive success	Sita et al. (2017)
4	Nutrient uptake efficiency	Nutrient stress	Canola, sunflower	Limited nutrient absorption, chlorosis	Enhanced nutrient uptake through root adaptations	Supplies slow-releasing, balanced nutrients	Sustains plant nutrition and minimizes deficiency symptoms	Javeed et al. (2023)
5	Photosynthetic activity	Drought and salinity	Sunflower, canola	Reduced chlorophyll content, impaired photosynthesis	Stomatal regulation, antioxidant enzyme activity	Increases chlorophyll stability and enzymatic activity	Enhances photosynthesis and energy balance under stress	Saleem et al. (2024)
6	Root system adaptation	Abiotic stress	Soybean, canola	Shallow root systems under stress conditions	Root elongation, symbiotic microbial interactions	Promotes beneficial root–microbe relationships	Strengthens root systems for improved water/nutrient access	Vamerali et al. (2021)
7	Antioxidant response	Heat and drought	Sunflower, soybean	Accumulation of ROS, cellular damage	Upregulation of SOD, CAT and POD enzymes	Supplies micronutrients that	Protects cells from oxidative damage	Ma et al. (2023)

(continued)

Table 8.1 (continued)

S. No	Parameter	Abiotic stress	Effect on specific oilseed crops	Physiological impact	Adaptation strategies	Role of organic fertilizers	Key benefits in mitigation	Reference
						activate antioxidant pathways	and maintains integrity	
8	Soil health and microbial activity	Nutrient and salinity stress	Canola, sunflower	Nutrient cycling and microbial diversity	Interaction with rhizosphere microbes	Enhances microbial biodiversity and nutrient availability	Boosts overall soil health, supports plant vigor	Singh et al. (2022a)

8.1.4 *Importance of Studying Oilseed Crops Under Abiotic Stress Conditions*

Oilseed crops under various abiotic stress conditions are important to study which as the key to stabilizing and sustaining global food systems. From a human and animal nutritional view, oilseed crops, including soybean, sunflower, rapeseed and groundnut, are significant sources of oil and high-protein meals (Chugh et al. 2022). Nonetheless, abiotic factors, including drought, salinity, severe temperatures and nutrient deficits, can negatively impact the productivity and quality of these crops (Sabagh et al. 2019a). Exploring oilseed crop response to these environmental stressors is key to creating adaptive strategies to increase crop resilience and to maintain production in unfavorable situations (Sabagh et al. 2019b). Studying oilseed crops in abiotic stress conditions is also important for sustainable agriculture. Research in this field provides opportunities to identify critical physiological, biochemical and molecular components that allow plants to tolerate stress. Such information is critical for breeding programs targeting the development of stress-tolerant cultivars. These studies allow us to devise agronomic practices that will lessen the impact of stress on plants by improving irrigation systems, soil maintenance, or using organic fertilizers (Chugh et al. 2022). Different field studies on organic fertilizers for crops under abiotic stress are shown in Table 8.2. Oilseed crops comprise part of human dietary staples but also represent an essential source of livestock feed and industrial products, such as biofuels and biodegradable materials (Chugh et al. 2022). Because these crops generally act as the foundation for which supply chains are built, any decline in productivity caused by these stressful conditions can lead to higher production costs and contribute to the economic instability of agricultural production. Thus, understanding these stress conditions is beneficial for agricultural stakeholders such as farmers, policymakers and industry leaders to inform evidence-based decisions regarding the resilience and sustainability of oilseed production (Sabagh et al. 2019a). The other aspect of this research is that it not only provides knowledge for the further development of oilseed crops under abiotic stress, but it also brings additional value in terms of environmental protection through encouraging practices in accordance with eco-friendly and sustainable agriculture. The insights into the interaction mechanism between stress and plant growth provide valuable guidance for scientists and agricultural practitioners to adopt approaches that lower the use of chemical inputs and increase the regulation of water and nutrient use efficiency (Полушкина et al. 2020). Thereby reducing the overall impact of farming on land and promoting long-term soil health important for sustained agricultural yield in the future (Zewide 2021). Finally, biotechnological interventions and solutions to abiotic stressors in oilseed crops under abiotic stress conditions could hold much promise for better precision to increase crop production. Finding genes and metabolic pathways that respond positively to environmental stress can be used to produce transgenic crops with increased resistance to environmental stress (Chugh et al. 2022). Such scientific innovations may result in the

Table 8.2 Field studies on organic fertilizers for oilseed crops under abiotic stress

S. No	Crop	Abiotic stress	Fertilizer used	Observed benefits	Region	References
1	Castor bean	Drought stress	Humic acid	Grain yield, iron content and antioxidant activity	Iran	Rahbari et al. (2021)
2	Soybean	Water-deficit conditions	Vermicompost (humic acid, fulvic acid and humin)	Enhanced in plant growth	India	Maura Santos Reis de Andrade da et al. (2022)
3	Oilseed flax	Water-deficit conditions	Farm manure	Improved yield	China	Li et al. (2023)
4	Cotton	Heat stress	Salicylic acid (SA)-based Zn-Fe-B consortia (S-N consortia)	Improved yield attributes, better fiber quality, soil and plant health	Pakistan	Hannan et al. (2024)
5	Soybean	Heat stress & water deficit	Biosolids, Anaerobic digestate	Reduction in ethylene levels & nutrient support	Canada	McMillan et al. (2022)
6	*Brassica napus* (rapeseed)	Water deficit	Vermicompost	Increased grain yield, improved oil content and quality, enhanced biodiesel production, reduced chemical fertilizer usage by 67%, improved nutrient uptake	Iran	Nasrollahzadeh et al. (2023)
7	Canola (*Brassica napus* L.)	Salt stress	Calcium-fortified composted animal manure, ca-FCM	Improved growth and yield, higher nutrient levels (N, P, K) Reduced Na^+/K^+ ratio	China	Naveed et al. (2020)
8	Mustard	Drought stress	Ultragreen, home-grown natural vegetable foods	Improved growth, better water retention, increased membrane stability, higher chlorophyll content, improved nutrient content	USA	Geremew et al. (2021)
9	Sunflower	Drought stress	Cattle manure, zeocompost, vermicompost, zeolite-amended chicken manure	Higher grain yield under well-irrigated and drought stress conditions, reduced chemical fertilizer use	Iran	Khodael-Joghan et al. (2018)

10	Maize	Salinity stress	Compost, ca-fortified compost	Increased growth, yield, and physiological parameters, improved nutrient uptake (NPK), reduced Na^+/K^+ ratio in shoots	Pakistan	Niamat et al. (2019)
11	Linseed	Drought stress	*Eruca sativa* seed cake, chicken peat and *Moringa oleifera* leaves	Improved soil fertility (N, P, K, Zn, Fe), enhanced growth, better water retention (LRWC), higher seed yield, fiber length and fiber weight	Pakistan	Khan et al. (2020)

breakthroughs needed to change the paradigm of oilseed crop production, aiding in the worldwide transition to resilient climate-smart agriculture (Tian et al. 2021c).

8.2 Abiotic Stresses Affecting Oilseed Crops

8.2.1 Organic Fertilizers and Oilseed Crops Under Drought Stress

Oilseed crops are significantly impacted by many abiotic stressors during their growth, with drought stress being one of the most formidable, leading to diminished plant development, compromised physiology and lower output. Water is vital for plant growth, and its scarcity can adversely impact agricultural productivity and food security (Sehgal et al. 2018). As a result of climate change and adverse weather conditions, drought stress is becoming more common and more severe around the world, and therefore, it becomes essential to understand how oilseed crops react to this drought stress and cope with it (Ahmad et al. 2021b; Khayatnezhad et al. 2022). Oilseed crops such as soybean, sunflower, rapeseed and groundnut are especially threatened since they are often grown in regions where there is less water availability (Reddy and Suresh 2009; Altaf et al. 2023). Numerous physiological, biochemical and molecular responses in plants arise due to drought stress, ultimately resulting in reduced growth and production, as well as alterations in the quality of the produced oil (Fig. 8.2). Such effects can be direct (reduced uptake of water) and indirect (lower regulation in nutrient availability) (Kumar et al. 2018).

(a) *Effects on Crop Physiology and Yield*

Drought stress has a far-reaching impact on oilseed crops, from the cellular to the whole plant level. Initially, cells lose turgor pressure due to a lack of water. The consequence of this is a loss of turgor, cell expansion is affected, growth rates decline and plant parts become smaller (Raza et al. 2017). Drought-induced stress significantly impacts leaves, the primary organs responsible for photosynthesis, frequently resulting in reduced leaf area and accelerated leaf senescence in oilseed crops (Ahmad et al. 2021c). Collectively, the changes reduce the overall photosynthetic capacity of the plant, so limiting the energy accessible for net production and the carbon compounds necessary for growth and development (Raza et al. 2017). The closure of stomata, the microscopic structures located on leaf surfaces and other organs responsible for gas exchange, is one of the primary consequences of drought stress (Lv et al. 2023). Closure of stomata may limit transpiration and conserve water, but also restrict carbon dioxide (CO_2) influx necessary for photosynthesis (Kim et al. 2010). Such limitations lead to a reduction in carbon assimilation and consequently, a fall in the availability of photosynthates (Li et al. 2016). Photosynthetic activity depletion, which directly impacts the vegetative and reproductive growth stages, sequentially results in a series of implications that ultimately diminish

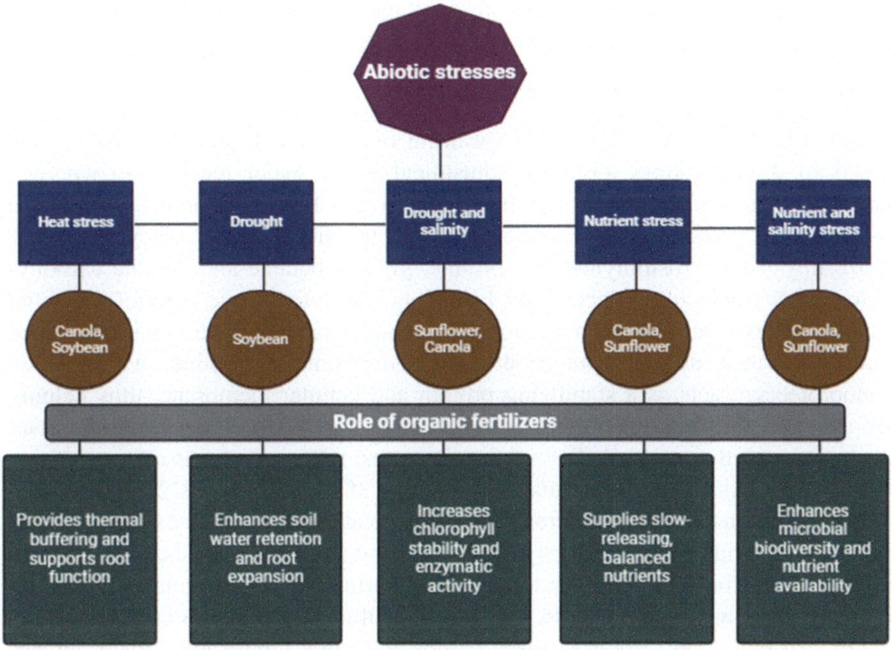

Fig. 8.2 Abiotic stress and the impact of organic fertilizers on some oilseed crops

yield (Sabagh et al. 2019b). Drought stress can also change chlorophyll content, decrease chlorophyll fluorescence and decrease the overall photochemical efficiency of the photosynthetic apparatus (Ahmad et al. 2021c). Such disruptions can lead to an altered state of energy metabolism, whereby the plant may prioritize mechanisms of survival and away from growth or reproduction (Raza et al. 2017). Drought stress in oilseed crops results in various impacts on yield reduction and is one of the most visible effects. The drought reduced the production and quality of the seeds (Si et al. 2022). Drought can inhibit normal processes during key developmental phases (e.g., flowering and seed filling). Severe drought stress at flowering can lead to poor pollination, flower abortion and pollen viability, which indirectly affects seed set and seed number (Ahmad et al. 2021c). When the seed-filling process is disrupted, oil content and seed quality may both be negatively affected under drought stress (Raza et al. 2017). For oilseed crops, this developmental stage is critical as it establishes the final oil content and quality that are important for nutritional and industrial applications (Ahmad et al. 2021c). Moreover, drought stress flexibly changes oil content in seeds (Amini et al. 2014). The changed metabolic environment during water deficit can modify the fatty acid synthesis balance (Hatzig et al. 2014). This change can affect the nutritional quality of the oil and render it undesirable for consumption due to its health impact (Amini et al. 2014). More favorable cardiovascular health effects are associated with oils that have higher levels of unsaturated fatty acids, oleic and linoleic acid. Thus, drought stress induces changes that not only affect yield but also both the economic and nutritional value of the oilseed crops (Hatzig et al. 2014).

(b) *Mechanisms of Drought Tolerance in Oilseed Crops*

Oilseed crops have developed a spectrum of adaptive strategies to reduce the effects of drought stress. Osmotic adjustment is the major mechanism whereby plants acquire osmotically compatible solutes (osmolytes) to compensate for the loss of turgor pressure and to continue with metabolic functions. They frequently utilize tiny solutes (osmolytes) like proline, glycine betaine and soluble carbohydrates. Their molecules function by balancing the cytoplasmic osmotic potential against the environmental osmotic potential, which enables cells to retain water and remain hydrated during osmotic drought. For example, proline, which is an osmoprotectant, acts as a stabilizing protein and cellular membrane, thus helping maintain structure and function under stress (You et al. 2019). The second important mechanism is the upregulation of many genes related to antioxidant defense responses to oxidative stress caused by drought (Chugh et al. 2022). Water deficiency conditions lead to an increase in the generation of reactive oxygen species, such as superoxide radicals, hydrogen peroxide and hydroxyl radicals. Such reactive oxygen species possess significant potential to harm cellular components, including lipids, proteins and nucleic acids, resulting in cellular malfunction and diminished viability of plants. In response, oilseed crops activate a robust antioxidant defense system comprising both enzymatic and nonenzymatic elements. Principal enzymatic antioxidants include superoxide dismutase (SOD), catalase (CAT) and peroxidase (POD), which collaboratively break down ROS to mitigate oxidative damage. The enzymes increase in expression to sustain redox equilibrium and facilitate the plant's fundamental physiological activities during drought stress (Amini et al. 2014). At the molecular level, the responses to drought stress are critical for the adaptation of oilseed crops to drought stress (Ahmad et al. 2021b). The expression of drought-responsive genes is a crucial factor for plants to implement defensive and adaptive responses against drought stress. Many of these genes are stimulated by transcription factors such as dehydration-responsive element-binding (DREB) proteins and ABA-responsive elements (ABREs). Abscisic acid (ABA) phytohormones are crucial in drought response and stress signal transduction. Accumulation of ABA due to drought leads to stomatal closure, osmotic correction and the production of defensive genes. Collectively, these genetic and hormonal characteristics facilitate a metabolic response that enhances the plant's capacity to endure water scarcity (Jaime et al. 2018). Over time, beneficial plant traits such as deep and expansive root systems to access water become increasingly important, especially when surface soil moisture is depleted due to excessive dryness. Root characteristics (e.g., length, density and branching pattern) enable plants to access more water stored in deeper levels of the soil profile. Genotypes of oilseed crops that have a higher capacity for root development also often cope better with drought through greater capacity for water uptake and physiological functioning at low soil moisture levels (Wu et al. 2018). Identification, availability and utilization of genetic variability in the root traits is the target for breeding programs to develop drought-resistant oilseed crop varieties (Hatzig et al. 2014).

Organic fertilizers help to improve soil conditions, which facilitates water retention capacity (Khan and Kwot 2024). Fertilization contributes to soil structure through its organic matter, which allows water to penetrate the soil and retain it until the plants need it in dry periods. Also, the organic inputs enhance the activity of soil microorganisms that are involved with nutrient cycling and the formation of beneficial plant root–microbe associations involved with plant stress tolerance (Pattnaik et al. 2021). Genetic and soil management strategies are the solutions to alleviate the effects of drought stress in oilseed crops (Hatzig et al. 2014).

8.2.2 Organic Fertilizers and Oilseed Crops Under Salinity Stress

Salinity is a critical abiotic element influencing plant growth and development, posing a significant agricultural constraint on sustainable crop production, especially in arid and semiarid environments (Ashraf et al. 2021; Ishrat et al. 2022). Soil salinity is caused by the build-up of soluble salts, mainly sodium chloride (NaCl), in the soil, which causes osmotic and ionic stress that limits water uptake and disturbs cell homeostasis (Arif et al. 2020).

(a) *Impact on Plant Growth and Oil Quality*

Salinity stress on plant growth is perceived in different ways, which starts from the root and spreads throughout the entire plant system (Eisa et al. 2017). The first major effect of high soil salinity is osmosis, reducing water availability to roots in soil despite the presence of water in the soil (Antonio Fabio da Silva et al. 2021). This condition is comparable to drought stress as it results in less water uptake, resulting in cellular-level dehydration (Sinthumule et al. 2022). Consequently, oilseed crops usually exhibit poor plant growth, limited leaf number and leaf senescence at an early stage (Hassanpouraghdam et al. 2022). Such morphological alterations reduce photosynthesis, which in turn impairs the rate of the essential process that allows biomass build-up and growth (Fetouh 2016).

At the cellular level, salinity stress results in toxic ion accumulation, especially sodium (Na^+) and chloride (Cl^-), which can displace critical nutrient ions like potassium (K^+) and calcium (Ca^{2+}) (Chaturvedi et al. 2017). Such ionic imbalance alters enzyme activity, protein synthesis and membrane stability, risking cell function (Agudelo et al. 2021). Due to salinity stress, oilseed crops may affect chlorophyll content and structural impairment of the chloroplast (Guimarães et al. 2019). This decreased photosynthetic activity results in reduced photosynthetic efficiency, which in turn contributes to the reduction in growth and yield (Ezzati and Karimi 2015). Oxidative stress generated by salt stress results in the excessive generation of ROS, which harm lipids, proteins and DNA, hence affecting cellular health and plant viability (Mbarki et al. 2016). Salinity stress not only exerts its adverse effects on the quantitative parameters of oilseed production but also on the qualitative characteristics of the oil (Arif et al. 2020). Salinity stress results in metabolic changes, and the production of fatty acids can be altered to modify the oil composition (Akhtar et al. 2015). According to studies, salt stress increases the ratio between saturated and unsaturated fatty acids (Boughalleb et al. 2017). The fatty acid profile changes result in reduced nutritional and industrial value of the oil due to the presence of unsaturated fat, especially oleic and linoleic acids, owing to their associated health benefits and importance in several industries (Coelho et al. 2018). Such a reduction in oil quality makes it difficult to meet the market standards and consumer preferences, highlighting the necessity of strategies to alleviate salinity stress in oilseed crops (Hütsch et al. 2016). Overall, the impact of salinity on oilseed crops is governed by the timing and severity of salt exposure. Salinity stress at sensitive growth stages like seed germination, flowering or seed filling would lead to yield reduction and oil quality deterioration. Seedlings are particularly sensitive to salt stress as salinity during the seedling stage can result in poor root development and lead to poor water and nutrient accessibility (Mehdi et al. 2018). In later stages of reproduction, high salinity may reduce pollen viability, disrupt fertilization and increase the number of empty seeds (Guo et al. 2017). It interferes with the seed-filling process in terms of oil deposition in seeds and this determines the quantity and also the quality of the oil produced (Faried et al. 2016).

(b) *Adaptation Strategies in Oilseed Crops*

Research has demonstrated how organic fertilizers protect oilseed crops from the adverse effects of salinity. Research demonstrates that safflower plants (*Carthamus tinctorius*) have more potential to handle salinity stress when compost is used as treatment to improve soil structure and enhance nutrient availability, thus promoting better plant development in saline environments (Keleş and Akbaş 2023). The seed germination and growth of sesame (*Sesamum indicum*) seedlings improved when researchers applied organic manure. A two-year research project examining organic fertilizer substitution on saline-sodic soil and *Helianthus annuus* plant growth occurred in northwestern China. A previous study demonstrated that organic fertilizer, when replaced by urea, resulted in better soil structural development while increasing water retention capacity and leading to higher nitrogen-use efficiency and also improved sunflower yield (Cheng et al. 2023). Under saline conditions, canola (*Brassica napus*) benefited substantially from the application of 3% calcium-fortified composted manure (Ca-FCM), which resulted in enhanced growth together with better yield and physiological characteristics. Research indicated that Ca-FCM treatment enhanced photosynthesis along with water retention and nutrient absorption, yet minimized Na^+/K^+ ratios up to 83.33%, thus making it a good method to reduce salt stress in canola cultivation (Naveed et al. 2020).

Salinity, together with drought stress, creates one of the primary obstacles that diminish agricultural production. Research showed how the combination of moringa leaf extract (MLE) at concentrations from 2% to 6% and seaweed extract (SWE) at levels between 1000 and 3000 ppm affected olive plants (*Olea europaea* cv. Kalamata) growing under salinity conditions. The application of MLE and SWE to olive plants resulted in enhanced leaf chlorophyll content as well as better results for fruit percentage and total yield and increased oil content. The combined application of MLE at 6% mixed with SWE at 3000 ppm under the combination demonstrated better results than the separate use of these treatments to combat salinity stress in perennial crops (Al-Saif et al. 2023).

Salinity stress damage in oilseed crops is mitigated through the implementation of diverse physiological, biochemical and molecular adaptation mechanisms. A crucial method for adaptation is osmotic adjustment, namely the buildup of suitable solutes or osmoprotectants. These chemicals (proline, glycine betaine and soluble sugars) establish osmotic equilibrium between the cell and its environment, reducing the cell's osmotic potential, facilitating water influx and stabilizing cellular structures (Punia et al. 2021; Luo et al. 2017). Proline, in particular, acts as an osmolyte and protein and cell membrane stabilizer to protect them from the denaturing effects of high salt concentrations (Hosseini et al. 2023). Ion homeostasis is the second most important adaptation strategy used by oilseed crops to tolerate salt stress. Ion uptake and compartmentalization are selective to avoid toxic sodium and chloride accumulation in the cytoplasm. Mechanisms such as plasma membrane sodium-proton antiporter (SOS1) or vacuolar localized Na^+/H^+ Exchangers such as NHX1 are responsible for sequestering excess Na + ions into vacuoles and preventing Na^+ ions from interfering in cellular metabolism (Luo et al. 2017). This compartmentalization allows for management of an optimal ratio of cytoplasmic ions, potassium, which is critical for enzyme activation, protein synthesis and many other metabolic activities (Lam et al. 2014). Oxidative stress generated by salinity is succeeded by the activation of the antioxidant defense mechanism. Reactive oxygen species are generated under salt stress, resulting in oxidative damage to oilseed crop cells, which may subsequently activate antioxidant defense mechanisms to eradicate these harmful metabolic by-products (Lv et al. 2013). Scavenging ROS to avoid cellular damage, enzymatic antioxidants such as SOD, CAT and POD function normally (Barman et al. 2012). Ascorbate (vitamin C) and glutathione are examples of nonenzymatic antioxidants that also assist in detoxifying ROS and maintaining cellular redox balance (Punia et al. 2021). Activation of these antioxidant pathways facilitates metabolic activities and protection of cellular structures under long-term salt exposure (Farhadi et al. 2022). The response of gene expressions in salinity stress at the molecular level in the adaptation process is an important step. MYB, NAC and DREB are examples of transcription factors that can regulate the expression of downstream genes in many stress response pathways (Salgado et al. 2021). They stimulate the transcription of genes for osmotic adjustment, ion transport and antioxidant proteins. An example is the Salt Overly Sensitive (SOS) pathway, comprising SOS1, SOS2 and SOS3 proteins, which plays a pivotal role in the maintenance of ion homeostasis during saline stress (Luo et al. 2017). The expression of stress-responsive genes that strengthen the adaptive capacity for the plant through the ABA-dependent as well as ABA-independent signaling pathways is also mediated. Among the salinity responses of plants, one is the phytohormone ABA, which induces stomatal closure for water conservation as well as expression of osmotic adjustment and antioxidant protection genes (Wang et al. 2022).

Another key area of adaptation of oilseed crops used to improve salinity stress tolerance is root system architecture. The root system of a halophyte plant is typically deep and extensive, so that it can reach water in the soil at greater depths and the salt tends to concentrate in the warmer upper portion of the soil (Farhadi et al. 2022). There exists good genetic variation among the oilseed crop species and their cultivars, which is one aspect of root architecture relevant to breeding for salt-tolerant varieties (Khan et al. 2016). Organic fertilizers and soil amendments can aid in an economical way in developing soil structure and physical qualities that can ultimately raise the soil saving ability of water and nutrients and help to reduce the challenge of salinity on plants (Farhadi et al. 2022). Organic matter helps to increase the cation exchange capacity of the soil, allowing excess salts to be leached while simultaneously improving the availability of essential nutrients (Lv et al. 2013). In addition, by applying organic fertilizers, beneficial soil microorganisms are promoted, which can help improve nutrient uptake and stress tolerance. Identification of salinity tolerance traits in oilseed crops and approaches through breeding and genetic engineering to develop varieties with improved salinity tolerance have been pursued. The development of modern tools such as marker-assisted selection (MAS) and genome editing tools like CRISPR/Cas9 has sped up the improvement of salt-tolerant crops by precise modification of stress-responsive genes (Khan et al. 2016). Overexpressing the genes responsible for osmotic adjustment, ion transport and antioxidant defense through a transgenic approach has the potential to improve the salinity tolerance of oilseed crops. This has led to the production of transgenic plants with enhanced growth and yield when grown in saline environments, for example, genes coding for major osmotic transporters and those for osmolyte biosynthetic enzymes were introduced (Hamwieh et al. 2011).

8.2.3 Organic Fertilizers and Oilseed Crops Under Temperature Stress

Temperature stress (both heat and cold stress) is one of the most important abiotic factors that negatively influence the growth, development and productivity of oilseed crops. Many crops, including soybean, sunflower, rapeseed and groundnut often exposed to fluctuations in temperature that affect their physiological and biochemical processes. Temperature stress influences every stage of plant development, including germination, seedling establishment, flowering and seed production, leading to considerable yield losses and quality reduction (Ahmad et al. 2021b).

(a) *Effects on Germination, Flowering and Seed Production*

Germination, the first stage of plant development, is where temperature stress begins affecting oilseed crops. The seed germination process would be unsuccessful if the seed germination temperature range is not correctly observed and maintained (Maulana et al. 2017). Heat stress is defined as temperatures that exceed the optimum range for a particular crop, which can negatively impact the germination process by lowering seed viability and modifying metabolic processes. However, high temperatures speed up metabolism, causing depletion of the energy of the seed and restricting seedling emergence (Rosental et al. 2014). In contrast, cold stress (temperatures below the optimal range) reduces enzymatic activity and metabolic processes so that germination is delayed or never completed (Bhattacharya 2022). Seeds that are exposed to low temperatures may not absorb enough water in order to kick-start the biochemical pathways for germination. Chronic exposure to less than-optimal temperatures can then decrease seedling vigor for reduce the capacity for growth of the plants (Ahmad et al. 2021b). Temperature stress can be especially damaging during flowering, a key time for reproductive success. Heat stress at flowering can adversely affect pollination and fertilization (Wu et al. 2020). Lower fertilization success directly impacts the number of pods or seeds and results in serious yield losses (Bennett et al. 2011). Sunflower and rapeseed are the most heat-sensitive crops and flowering is particularly susceptible to heat stress because excessive temperature leads to flower abortion and reduced seed set (Kourani et al. 2022). Further, heat stress often affects hormonal balance in plants, causing plants to flower early or change flowering time (Borghi et al. 2019). High or low temperatures can cause a delay in flowering and disrupt the flowering process. The damage due to cold to the reproductive organs can cause deformed flowers that do not yield viable seeds. Moreover, cold during the reproductive stage can decrease pollen viability and fertilization. In oilseed crops such as soybean and groundnut, flowering under cold stress can cause abnormal or uneven pod filling, leading to reduced yield and seed quality (Staniak et al. 2021). Seed production is also affected by temperature stress. If it encounters heat stress during the seed-filling stage, it may hasten the rate of development and produce tiny and poorly developed seeds. The genes that are responsible for filling seeds control the rates of bulk storage compound accumulation during seed filling, such as oils and proteins, which are important traits for both economic and nutritional value for oilseed crops (Maity et al. 2016). Consequently, the oil gets lower quality for nutritional and industrial processes. On the other hand, cold stress during seed development may inhibit metabolic processes and this may decrease oil and protein synthesis rates. The seed that is subject to low temperatures may never fully mature, favoring shrunken or immature seeds, which are less viable and have a lower oil content (Soualiou et al. 2022).

(b) *Heat and Cold Tolerance Mechanisms*

Oilseed crops have evolved different strategies to cope with and acclimate to heat and cold stress. Physiological, biochemical, or molecular adaptations are the most crucial mechanisms allowing plants to function and remain productive under temperature extremes (O'Neill et al. 2019).

(i) *Heat Tolerance Mechanisms*

There are several interlinked strategies to develop heat tolerance in oilseed crops. Heat shock proteins (HSPs) are one of the primary mechanisms through which environmental stress resistance is maintained (Richter et al. 2010; Xu et al. 2013). HSPs are intracellular chaperones responsible for maintaining the integrity of cellular proteins through denaturation and aggregation by heat (Yu et al. 2012). HSPs modulate cellular integrity under heat stress by stabilizing the proteins and helping them to fold for the active working of plant metabolism. Heat shock transcription factors (HSFs) regulate the expression of HSPs, activating upon elevated temperatures to induce transcription of the HSP gene (Bi et al. 2022). Elevated temperatures result in more inconsistency of cell membranes, which poses a risk to their structural integrity and function. This environmental challenge is met by oilseed crops through a change in the composition of membrane lipids that increases the level of saturated fatty acids, thereby stabilizing the membrane under heat stress. These changes are essential to prevent the loss of cytoplasm and protect membrane-associated proteins, including pigment-protein cofactors for photosynthetic activity and membrane transport proteins (Richter et al. 2010). Another important mechanism of heat tolerance is osmotic adjustment. Compatible solutes like proline, glycine and soluble sugars are accumulated by oilseed crops during drought to maintain osmotic balance and protect cellular structures from dehydration (Marcolino-Gomes et al. 2014). Besides stabilizing proteins and membranes, these osmolytes can also act as ROS scavengers, reducing the oxidative damage that may result from high temperatures. Heat stress elevates the production of antioxidants such as SOD, CAT and ascorbate peroxidase (APX) that are required to detoxify ROS and rescue cells from damage (Song et al. 2014; Marcolino-Gomes et al. 2014). At the molecular level, heat stress triggers the expression of genes responsible for initiating different protective pathways against stress. Heat shock factors (HSFs), MYB and bZIP transcription factors regulate the expression of genes involved with heat shock proteins, antioxidant defense and osmolyte synthesis, respectively (Waters et al. 2017). When activated, such genes improve the plant's response to heat stress through increased defense and repair systems. Besides, other phytohormones, such as ABA and ethylene, are also important in the regulation of heat stress responses (Tian et al. 2021b; Bi et al. 2022). ABA accumulation induces stomatal closure, which decreases water loss in the atmosphere through transpiration and regulates cellular water balance. In contrast, ethylene regulates multiple stress signaling pathways, i.e., growth and developmental pathways, to restore the plant affected by heat stress (Blair et al. 2019; Marcolino-Gomes et al. 2014).

(ii) *Cold Tolerance Mechanisms*

Cold tolerance mechanisms in oilseed crops are diverse and are at least at physiological and preferably biochemical and molecular levels (Cheong et al. 2022). Altering membrane lipid composition is one of the main physiological adaptations to cold stress. Whereas heat stress increases membrane fluidity, which can be detrimental to the function of membrane proteins and cellular transport-related processes, cold stress can increase membrane rigidity. In response to cold stress, the membranes of oilseed crops can alter the composition of unsaturated fatty acids to help them keep fluid and functional (Abeynayake et al. 2015). In response to cold, the accumulation of osmoprotectants such as proline and soluble sugars plays a dual role in osmotic adjustment as well as protection from oxidative damage (Dhaliwal et al. 2021; Wani et al. 2021). These solutes accumulate to decrease the freezing point of cellular fluids and stabilize proteins and cellular structures. Proline can preserve enzyme activity and promote the scavenging of ROS produced by cold stress (Li et al. 2018). One important response that helps to address the oxidative damage caused by cold stress is the activation of antioxidant defense systems. To eliminate ROS, the activity of detoxifying enzymes like SOD, CAT and POD was upregulated, preventing damage to cellular macromolecules (Dhaliwal et al. 2021).

In addition, cold stress triggers cold-responsive (COR) genes that are regulated differentially by transcription factors like C-repeat binding factor (CBF) or dehydration-responsive element-binding protein (DREB) (Maryan et al. 2023). They attach to the promoter regions of COR genes and trigger the expression of genes that produce proteins that offer freezing protection and keep cellular processes functioning. The CBF pathway is a typical cold tolerance regulatory mechanism, which enhances the expression of genes for osmotic adjustment, membrane stabilization and antioxidant defense through a chain of signal cascades (Liu et al. 2017; Maryan et al. 2023). The content of phytohormones accumulated at the time of cold stress, such as ABA, is also important for cold stress response. So, ABA triggers the expression of cold-responsive genes, thus aiding cold acclimatization in plants (Liu et al. 2017). Meanwhile, other phytohormones, such as gibberellins and cytokinins, also play important roles in the regulation of growth and development with respect to cold stress, with a focus on the partitioning of energy between a stress response and growth (Hassan et al. 2021). Some of these mechanisms also involve structural changes in the oilseed crops. For instance, alterations to leaf form, such as leaves being smaller and thicker to minimize heat loss and protect reproductive organs. For this, protective barriers like cuticular waxes are formed, which limit water loss and additionally can act as an insulator against cold temperatures. Higher root mass, length and density are root system adaptations that enhance the acquisition of water and nutrients and overall performance of plants growing in cold soils, alleviating this stress on metabolic needs (Li et al. 2022).

(iii) *Integrated Tolerance Strategies*

Natural mechanisms alone failed in adapting oilseed crops to temperature stress. These crops have inherent tolerance, but they are also complemented through agronomic practices and breeding strategies. Enhancing heat and cold-resistant cultivars via conventional breeding and gene engineering has become an urgent agricultural request. The integration of marker-assisted selection (MAS) and quantitative trait loci (QTL) mapping has facilitated the discovery of significant genes linked to temperature tolerance, thus enabling the breeding of stress and heat-tolerant cultivars for climate-resilient agriculture (Chugh et al. 2022; Raman et al. 2014). Technologies like CRISPR/Cas9 biotechnological tools have recently provided ease in the modification of stress-responsive genes to develop transgenic plants with higher acclimatization to drastic temperatures (Chugh et al. 2022).

The soil temperatures can be buffered to a certain extent by agronomic practices, especially through the application of mulching and the management of irrigation, which can alleviate the adverse effects of temperature stress (Kephe 2024). Organic fertilizers can promote better soil structure, facilitate moisture retention and stimulate beneficial microbial activity that adjusts the plant to high or low temperature (Stasnik et al. 2023). Antioxidants or other growth-enhancing materials can also be used as foliar sprays for immediate protection during the temperature stress period (Hwang et al. 2014). This set of natural plant mechanisms, breeding progress and strategic agronomic practices constitutes an integrated approach to temperature stress management in oilseeds (Chugh et al. 2022). Such an integrated approach can help to ensure productivity and quality seed production and maintain the economic viability of oilseed crops under a rapidly changing climatic scenario (Lamichhane et al. 2018).

8.2.4 Organic Fertilizers and Oilseed Crops Under Mineral Deficiency Stress

Nutrient stress is an important abiotic stress for the growth, development and productivity of oilseed crops. Nutritional stress occurs when a plant has a deficiency or imbalance of key nutrients that restrict physiological function within the plant. Nutrient absorption is crucial for the construction of different plant structures, initiation of enzyme action, energy transfer and metabolic regulation. Nutrient stress leads to impaired growth, reduced yield and lower oil quality when one or more nutrients are deficient or it is unevenly available to the plants (Ott et al. 2019). Various oilseed crops, including soybean, sunflower, rapeseed and groundnut, face stress due to nutrient deficit that results in impairing their growth but also drastically alters the nutritional and industrial quality of the oil extracted from their seeds (Ott et al. 2019; Pabuayon et al. 2019).

(a) *Nutrient Deficiency and Its Effects on Oilseed Crops*

Nutrient deficiency in oilseed crops depends on which essential nutrient is deficient. The most important nutrients for oilseed crops are nitrogen (N), phosphorus (P), potassium (K), sulfur (S) and micronutrients such as zinc (Zn), boron (B) and iron (Fe) (Liu et al. 2010; Si et al. 2022). Each of the nutrients performs a specific function within the plant during growth and development and when there is a deficiency, plants undergo distinct physiological and biochemical changes, which may lead to stunted growth. Nitrogen is undoubtedly one of the most essential macronutrients, a building block of amino acids, proteins, chlorophyll and nucleic acids. Nitrogen deficiency leads to a state of chlorosis (yellowing of the leaves) because of insufficient chlorophyll. It significantly diminishes the photosynthetic effectiveness of oilseed plants, resulting in limited growth, inferior biomass development and decreased seed production. Nitrogen deficiency significantly reduces the oil quality of oilseed crops by influencing the synthesis of storage proteins and lipids. These lead to a decrease in their protein quantity and a possible change in the fatty acid profile of the oil, negatively affecting the oil's nutritional value and also its industrial use (Si et al. 2022). Oilseed crops are also affected by deficient phosphorus. It is required to produce ATP, root growth and nucleic acid synthesis, which all require phosphorus for energy transfer processes. Marked stunted development and poorer root growth in oilseed crops affected by phosphorus deficiency restrict the crops' ability to access more water and other nutrients. This deficiency can also affect maturity and seed development, causing a smaller number of seeds and oil content. Phosphorus deficiency is especially at the reproductive stages, where the energy demand is high and this problem can severely limit yield (Ren et al. 2015a; Liu et al. 2010). Potassium, another macronutrient, is essential for oilseed crops, as it maintains osmotic balance, activates important enzymes and facilitates photosynthesis (Zheng et al. 2023). Potassium deficiency lowers the turgor pressure, impairs the uptake and transport of water in the plant (Tränkner et al. 2018). It may cause leaf curls as well as marginal chlorosis and necrosis. Potassium deficiency also limits the adaptability of plants to several abiotic stresses by regulating cellular osmotic balance, which is crucial during drought and salinity. Potassium is essential to produce oils and storage compounds in seeds and potassium-deficient oilseed crops can experience reduced seed and lower-quality oil (Zheng et al. 2023; Shiaa 2024). Sulfur is an important nutrient required for the production of amino acids, proteins and some vitamins. Sulfur is necessary for chlorophyll formation and its deficiency can cause chlorosis, especially in younger leaves. Therefore, in oilseed crops, sulfur deficiency generally reduces both the quantity and quality of oil due to its effect on the production of essential fatty acids. This deficiency also hinders glucosinolates biosynthesis in rapeseed and other members of the Brassicaceae family essential for plant defense and oil quality. Furthermore, it participates in the biosynthesis of coenzymes and secondary metabolites involved in stress-defense reactions (Nagaram 2020; Asif et al. 2023). Those nutrients that are required in smaller quantities as compared with macronutrients are often referred to as micronutrients and their deficiency can considerably affect oilseed crops. Zinc plays a vital role in the activation of enzymes and the synthesis of proteins, where its deficiency results in dwarfism and leaf deformation. Whereas, iron deficiency causes interveinal chlorosis, which limits photosynthesis and energy production (Dinh et al. 2021). Nutrient deficiency is also influenced by other stress factors, which aggravate the problem even more. For example, oilseed crops that are low in nutrients are more likely to suffer from drought, salinity and temperature fluctuations. Such a shock is particularly able to take place since these fundamental nutrients are critical to maintaining cellular homeostasis and stress response pathways. The plant becomes more susceptible to damage from abiotic stresses when there is a lack of an important nutrient, since this inhibits the plant from activating these stress response mechanisms (Dejoux et al. 2003).

(b) *Role of Balanced Nutrition in Stress Mitigation*

The provision of balanced nutrition is thus critical for offsetting the negative impact of nutrient stress while regulating the growth and development of oilseed crops (Elferjani and Soolanayakanahally 2018). Changes in nutrient supply strategically increase the abiotic resistance, growth and yield of crops while preserving the quality of the resulting oil (Gawęda and Haliniarz 2021). Balanced nutrition requires the best provision of both macro and micronutrients to ensure that all physiological and biochemical pathways work properly (Depotter et al. 2016). Balanced nutrition is mainly responsible for photosynthesis and energy metabolism. Sufficient nitrogen, phosphorus and potassium levels make possible chlorophyll synthesis, ATP synthesis and carbon fixation. This support is especially important for oilseed crops because, without photosynthesis, there is no energy or carbon skeletons for growth and seed (Zhu et al. 2023). Such a well-nourished plant is more capable of keeping up its photosynthetic rate during stresses like drought or high temperatures. However, correct nutrient management leads to an adequate supply of nitrogen and phosphorus, which may improve the capacity of plants to continue the photosynthetic process even under stress conditions (Zheng et al. 2023). Maintaining water balance and osmotic regulation are also critically dependent on balanced nutrition. Potassium, especially, is vital for the regulation of stomata function and osmotic adjustment in the cells. Sufficient potassium levels aid the maintenance of turgor pressure, essential for the uptake and mobilization of water in plant cells (Sardans and Peñuelas 2021). This regulation is very critical when limiting water availability or drought stresses. Sufficient potassium helps oilseed crops fight against water deficits during dry conditions and thus ensures a number of metabolic activities of the plants. Potassium also increases plant drought and salt stress tolerance by balancing ions within the cell (Patel et al. 2022).

Balanced nutrition also plays a role in the activation of stress response pathways and the synthesis of protective compounds (Bjørklund and Chirumbolo 2017). For instance, sulfur participates in the synthesis of an important antioxidant, glutathione, which plays a role in detoxifying reactive oxygen species arising from stress conditions. The role of zinc, copper and manganese are known as micronutrients acting as cofactors for enzymes and constituents of antioxidant defense systems in plants (Chandra and Pandey 2014). SOD, CAT and POD are some of these enzymes that eliminate reactive oxygen species to reduce oxidative damage to cellular components (Liu et al. 2014). Sufficient concentrations of these nutrients help to maintain proper antioxidant defense mechanisms for oilseed crops by relieving the oxidative stress induced by high temperatures, salinity and other abiotic stresses (Chandra and Pandey 2014). Another nutrient that plays a role in alleviating stress is phosphorus, which is involved in energy transfer and signaling pathways (Khan et al. 2023). Sufficient phosphorus allows the formation of ATP, necessary for a variety of cellular processes, including the activation of stress response signaling pathways (Damaris and Yang 2021). Phospholipid synthesis occurs in the cell membrane. Phosphorus is also involved in this process. The membranes are vital to maintaining cellular structure and function, particularly during changes in temperature or under desiccation conditions. Adequate phosphorus maintains the permeability of membranes in oilseed crops and plays a part in rapid recovery from stress due to harsh environments (Castrillo et al. 2017). Reduced nutrition also enables the synthesis of secondary metabolites and signaling molecules, which help in stress defense (Bartwal et al. 2013). For example, sulfur and nitrogen are essential nutrients that participate in the biosynthesis of osmoprotectants, such as amino acids and peptides. Osmoprotectants like proline and glycine betaine are known to accumulate in many plants when exposed to stress and exert a protective effect by stabilizing proteins and membranes. Their synthesis is coupled with nitrogen supply, which represents an intersection of nutrient supply and stress response mechanisms. These compounds contribute to maintaining the balanced cellular physiology to osmotic homeostasis of oilseed crops during environmental stress (Kujeke et al. 2017).

Balanced nutrition and nutrient management practices are also important for improving root development and function. An oilseed crop with a well-developed root system can better extract water and nutrients, allowing it to maintain growth even in less-than-ideal situations (Fageria and Moreira 2011). Phosphorus and potassium promote root growth and function. While phosphorus stimulates root elongation and branching, potassium generally aids the plant in its ability to conduct water and nutrients from the roots to the shoots. Maintenance of adequate levels of these nutrients in oilseed crops can enhance their stress tolerance and this is critical for improving yield potential under stress environments (Tang et al. 2019; Sefaoğlu 2021). Balanced nutrition can be ensured through agronomic practices such as the application of organic as well as inorganic fertilizers, soil testing for available nutrients and the use of nutrient-specific amendments. Natural fertilizers like compost and manure are slow-releasing sources of nutrients and improve soil structure, making them able to hold in more moisture and nutrients, which is especially useful for oilseeds cultivated on marginal soils or rain-fed conditions where the availability of water and nutrients vary. Foliar sprays containing micronutrients can also provide a rapid supply of these essential elements during critical stages of plant growth and in stress conditions (Shaheb and Nazrul 2017). Technologies associated with precision agriculture, like nutrient mapping and precision fertilization, can optimize nutrient management by ensuring balanced nutrition of oilseed crops based on their specific nutritional needs. With the help of data-driven approaches, these technologies can track soil nutrient levels, crop growth and environmental conditions and hence apply nutrients precisely. Precision agriculture reduces the chances of nutrient deficiency and improves plant adaptation to stress by aligning nutrient supply with crop demand (Ren et al. 2015b).

8.3 Effect of Organic Fertilizer on Oilseed Crops Under Heavy Metal Stress

Heavy metals contamination of the agricultural soil constitutes a major threat to crop productivity, soil health and food safety. Specifically, oilseed crops like soybean (*Glycine max*), rapeseed (*Brassica napus*), sunflower (*Helianthus annuus*) and flaxseed (*Linum usitatissimum*) are highly sensitive to heavy metal stress, which negatively affects physiological processes, biomass and seed yield and oil quality (Farooq et al. 2016). Application of organic fertilizers, compost, manure, biochar and vermicompost has been recently recommended as a sustainable means of reducing heavy metal toxicity by improving soil properties and increasing plant defense mechanisms (Iqbal et al. 2024; Younas et al. 2023).

Complexation, adsorption and precipitation of organic fertilizers are of great importance in the reduction of bioavailability of toxic metals like cadmium (Cd), lead (Pb) and arsenic (As). For example, immobilization of Cd and Pb by biochar and reduced Cd and Pb uptake into oilseed crops and reduced metal accumulation in seeds have been reported. Similarly, organic matter is added to the soil through compost and manure, which also enhances cation exchange capacity and metal sequestration. It is demonstrated that the application of compost to Cd-contaminated soils significantly reduces Cd concentration in rapeseed and this promotes growth and increases oil yield (Hamid et al. 2018).

Organic fertilizers not only immobilize metal but also improve plant tolerance by biochemical and physiological mechanisms. Improvement of plant vigor is due to the supply of essential nutrient elements, especially nitrogen, phosphorus and potassium, as well as organic amendments that trigger antioxidant enzyme activity, which minimizes heavy metals induced oxidative stress (Jha and Subramanian 2016). It has been reported that Vermicompost, enriched with humic substances and beneficial microbes, increased seed germination and root elongation in oilseed crops under Pb stress and such changes occurred through altering the expression of stress-related genes. In addition, organic amendments encourage microbial diversity of the rhizosphere, resulting in microbial chelation and enzymatic breakdown of heavy metals (Rigobelo 2024).

Although organic fertilizers can reduce heavy metal stress, further research should be carried out on the optimization of application rates of organic amendments, mixing different organic amendments and the synergistic effects of biofertilizers and organic fertilizers on oilseed crops under heavy metal stress. Along with these, long-term field studies are needed to evaluate the long-term influence of organic amendments on soil health and also metal bioaccumulation in edible plant parts to ensure food safety and sustainability (Alengebawy et al. 2021).

Heavy metals toxicity can be mitigated by means of organic fertilizers in oilseed crops like mustard (*Brassica juncea*) and rapeseed by immobilizing the heavy metals in the soil and by enhancing microbial activity. According to research, ammonia and sulfur-containing amendments increase mustard plant tolerance to cadmium stress by stimulation of the antioxidant enzyme activity, phosphorus fertilization on oilseed crops exposed to lead and chromium stress reduced metal toxicity and promoted plant growth. Thus, plant growth regulators such as salicylic acid and gibberellic acid have been applied to sunflower and mustard to enhance the resistance to heavy metal stress. In sunflower, chelating agents such as EDTA have been employed for the uptake of heavy metals, along with increasing the root and shoot growth. Through these studies, the possible role of organic fertilizers in reducing the toxicity of heavy metals and maintaining the productivity of oilseed crops under contaminated soil conditions is highlighted (Rizwan et al. 2017).

A study was conducted using organic fertilizers (chicken manure (CM), farmyard manure and compost) and cobalt (Co) on the growth, yield and nutritional composition of *Moringa oleifera*. It has been shown that the use of CM increases plant height, leaf number, leaf area and dry weight when combined with Co application. On the other hand, organic fertilization increased the accumulation of nutrients such as nitrogen (N), phosphorus (P), potassium (K), zinc (Zn) and copper (Cu) in moringa leaves, which are necessary for plant growth and productivity. Besides promoting mineral nutrition, organic fertilizers also positively influenced the biochemical composition of moringa by increasing the protein level, total carbohydrate content, total soluble solids, total phenolics, carotenoids and vitamin C. Thus, these findings showed the importance of organic fertilization strategies for increasing soil fertility and enhancing the economic and nutritional value of oilseed crops (Gad et al. 2019).

8.4 Effect of Organic Fertilizer on Oilseed Crops Under Waterlogging/Flooding

Oilseed crops are greatly affected by waterlogging, which is a major abiotic stress because of decreased availability of oxygen and alteration of plant processes such as nutrient uptake and changes in soil microbial activity. Severe physiological and biochemical damage is caused in plants due to prolonged flooding, which eventually leads to yield loss (Bridgemohan et al. 2020). It has been examined that organic fertilizers can act as potential agents for improving soil structure, nutrient availability and microbial function. While this effectiveness under more waterlogged conditions has not been well studied (Deng et al. 2022).

Oilseed crops are affected in various physiological and biochemical processes owing to waterlogging-induced hypoxia. Waterlogging for longer than three days usually caused a dramatic response in root starch concentration and subsequent inhibition of the plant growth and delayed postflooding recovery were observed in a study on oilseed rape (*Brassica napus*) (Liu 2021). The results of the study also showed that waterlogging resulted in changes to N and P accumulation in plant tissues and waterlogged soils contained lowered soil mineral N and available P concentration. Furthermore, flooding conditions suppressed activity of key soil enzymes like urease, phosphatase and invertase, which led to a decrease of available nutrients and microbial function. Such negative effects became more pronounced in longer flooding durations and there was a greater effect on phosphorus than on nitrogen availability. This suggests that nutrient imbalances and the suppression of critical enzymes have an essential role in the waterlogging stress responses of oil crop plants (Gu et al. 2019).

Composts, manure and biofertilizers are organic fertilizers that have proven to improve soil structure, aeration, as well as enhance microbial diversity. It is possible that through their application, they can counteract some of the adverse effects of waterlogging by increasing soil porosity, which improves drainage and oxygen diffusion (Ahmed et al. 2023). Organic fertilizers can also maintain microbiological activity, they can help enhance enzyme function, which is normally inhibited under waterlogged conditions. Flooding stress and resultant imbalance of nutrients could be mitigated by their ability to increase nutrient retention, particularly phosphorus (Kaur et al. 2020). Specific mechanisms by which organic fertilizers affect waterlogged soil environments, however, are poorly investigated in oilseed crops. Because waterlogging alters soil microbial dynamics and enzyme activities, inorganic amendments need to be investigated in association with these processes to maintain nutrient cycling and enhance plant resilience (Zhang et al. 2025).

Although the known benefits of organic fertilizers in sustaining agriculture are well accepted, however, their role in alleviating the waterlogging stress in oilseed crops has not been exploited yet. Most works on waterlogging effects on crop growth and soil nutrients dynamics that are available focus on the general effects and not organic fertilizer application under such conditions. Because organic amendments have the potential to improve oilseed crop performance in waterlogged soils in view of the growing frequency of extreme weather events such as heavy rainfall and flooding, it is important to explore this further with further research.

8.5 Effect of Organic Fertilizer on Oilseed Crops Under UV Stress

There has been great interest in the use of organic fertilizers in sustainable agriculture as it can lead to improved soil health and agronomic productivity (M. Tahat et al. 2020). Edible oils from oilseed crops such as *Brassica napus*, *Helianthus annuus* and *Linum usitatissimum* are of major importance for industrial uses. These crops, however, are increasingly being affected by environmental stressors such as elevated tropospheric ozone (O_3) and ultraviolet (UV) radiation that negatively affect their growth and yield (Tripathi and Agrawal 2016). It is also important because tropospheric ozone is a very potent phytotoxic air pollutant that is produced by photochemical reactions involving volatile organic compounds and nitrogen oxides. O_3 at elevated levels renders oxidative damage in plants, compromises photosynthesis, lowers chlorophyll content and results in reduced growth and yield (Tiwari et al. 2018). Similarly, plants can be exposed to elevated UV radiation, especially UVB

(280–315 nm) that could damage the cellular components, impair DNA and cause a decrease in photosynthetic efficiency. Consequently, elevated UV levels can lead to reduced biomass accumulation and seed quality of oilseed crops, affecting total yield (Gustavsson 2019).

Organic fertilizers contain organic matter and essential nutrients and are derived from plant or animal residues. The application of biochar to the soil improves the soil organic matter (SOM), increases soil structure, increases water retention and stimulates microbial activity, which ultimately increases soil fertility and crop resilience (Khan et al. 2024). Organic fertilization enhances soil microbial activity, which is very important in the mitigation of stress. The presence of a diverse and active microbial community is beneficial for nutrient cycling, plant growth-promoting activities and systemically induces plant resistance to an array of abiotic stresses such as ozone and UV radiation (Zhao et al. 2024).

In a study, the effect of two organic fertilizers (compost coupled with biofertilizer (CCB) and filter mud cake (FMC)) versus conventional inorganic fertilizers was examined on sunflower (*Helianthus annuus*). Inorganic fertilizers caused higher plant height, chlorophyll content and phenolic secondary metabolites in oil; on the other hand, organic fertilizers increased drastically both sunflower seed and oil yield. FMC showed the highest oil content of the organic treatments, followed by CCB and the inorganic treatment had the lowest. Organic fertilizers proved to be an excellent source of essential nutrients and a positive factor for sunflower oil production (Alzamel et al. 2022).

Organic fertilizers enhance the fertility of the soil, nutrient utilization efficiency, as well as general crop productivity in diverse oilseed crops, including *Helianthus annuus*. Application enhanced the soil's organic matter, microbial activity, nutrient uptake and supports plant growth. Besides, organic fertilizers can also enhance abiotic stress tolerance like droughts, salinity, heat and heavy metal stress by means of enhanced leaf water status, chlorophyll synthesis and antioxidant activity. However, due to these benefits, organic fertilizers are a sustainable solution to chemical fertilizers in enhancing the resilience of the oilseed crops (Geremew et al. 2021).

Chemical fertilizers tend to reduce soil quality, and their long-term use is dangerous, but the use of organic fertilizers maintains soil fertility and increases oilseed crop productivity. Organic fertilizers improve soil structure and nutrient availability, soil microbial diversity and better growth and yield in various oilseed crops such as *Helianthus annuus*, as observed through various studies. Moreover, organic amendments are helpful in deriving abiotic stress tolerance in plants through alterations in genetics related to hormones, osmolytes and antioxidants. This implies that organic fertilizers are very useful in conserving soil fertility and are very significant in helping sustainable oilseed crop production (Liu et al. 2024).

8.6 Oilseed Crops and Their Response to Abiotic Stress

*8.6.1 Canola (***Brassica napus***)*

Canola (*Brassica napus*) is one of the top oilseed crops in the world due to its high oil content and a good composition of unsaturated fatty acids, with the predominance of oleic acid (Silva 2024). Canola produces seeds that can be used to make biodiesel and cooking oil as well as livestock feed and its ability to thrive in diverse growing conditions has made it an important agricultural crop (Oh et al. 2018). Nevertheless, abiotic stress factors such as drought, salinity and extremely high or low temperatures affect both the productivity and quality of canola. Gaining insight into canola responses to these stresses is crucial for breeding canola varieties that are continuously high-yielding and high quality when exposed to such environmental extremes and for developing management practices that maintain high yield and quality (Lohani et al. 2020).

(i) *Responses to Drought Stress*

One of the biggest challenges to canola productivity is drought stress, especially in areas where rainfall is not stable, and water availability is an issue. Canola is an oilseed crop that requires a sufficient water supply for growth, flowering and seed fill, similar to many other oilseed crops (Jaliya et al. 2015). Canola is affected at different growth stages by drought stress, with flowering and seed-filling periods being the most important critical times for high-yielding environments (Zhang and Flottmann 2016). Drought stress induces physiological and morphological processes in canola plants that help conserve water and support basic, purified metabolic activity (Mota et al. 2021). Drought stress is immediately tackled through the closure of stomata. It also inhibits CO_2 uptake, which in turn may decrease photosynthesis. This leads to lesser energy production as photosynthesis is the main source of energy for plants and it hinders the production of assimilates that are critical for growth and seed development (Nowicka et al. 2018). Therefore, drought stress normally results in reduced plant height, decreased leaf area and an early onset of leaf senescence in canola plants (Tesfamariam et al. 2010). The drought stress response mechanism at the reproductive development stage of canola is especially important. When flowering occurs under reduced water availability, it can lead to incomplete fertilization or failed fertilization, leading to fewer pods and reduced seed set. Seed filling is a sensitive period and drought stress that occurs at this time may cause incomplete deposition of oil in the seed, which may lead to immature seeds that produce seeds with low oil content. Consequently, these effects lead to low yield and poor oil quality (Tian et al. 2021a). Biochemically, osmoprotectants such as proline, glycine betaine and soluble sugars are built up in canola under drought stress conditions (Vieira et al. 2018). The purposes of these compounds are twofold, as they assist in osmotic adjustment and provide stability of cellular structures under water deficit conditions. Proline also maintains cell turgor while functioning as an antioxidant to protect the cells from the damage done by ROS (Hu et al. 2011). During drought, the canola antioxidant defense system is triggered, leading to increased activity of important antioxidant enzymes to scavenge and detoxify ROS and alleviate oxidative stress (Akram et al. 2018). Molecular regulation of canola's response includes the expression of drought-responsive genes and transcription factors. As a major phytohormone involved in signaling for drought stress, ABA is one of the phytohormones that accumulate in canola under water deficit conditions (Muhammad Aslam et al. 2022). ABA closes stomata and induces the expression of stress-responsive genes, increasing stress tolerance under drought conditions. Genes involved in osmotic adjustment and antioxidant defense as well as aquaporins (water transporters), are unregulated to maintain cellular homeostasis and allow water flow (Zuber et al. 2013).

A remarkable reduction of the drought stress in avocado trees was alleviated using soil amendments such as *Ascophyllum nodosum* seaweed extract, humic acids and mycorrhizae because they increase soil water retention and root growth. On the other hand, zeolite, biopolymers and diazotrophic bacterial treatments proved to be less effective in improving water availability and alleviating stress (Beyá-Marshall et al. 2025). Another research conducted by Geremew et al. (2021) showed that organic fertilizers had a positive impact on the growth and physiological characteristics of mustard as affected by drought stress, thereby reducing the effect of water scarcity. The application of organic fertilizers had positive effects on nutrient uptake and physiological responses of castor bean plants under drought stress, which can be beneficial for enhanced drought tolerance (Rahbari et al. 2021). Another study showed (Khodaei-Joghan et al. 2018) that combining organic fertilizers with zeolite under drought conditions produced higher grain yields as a response to climatic drought resilience.

(ii) *Responses to Salinity Stress*

Salinity stress is also one of the most important abiotic stresses affecting the production of canola, especially under elevated soil salinity levels and irrigation-induced salinity. Maladaptive mechanisms of salinity stress include a drought-like osmotic force that restricts the water uptake from plant roots and the cationic toxicities induced through surrogate accumulation of Na^+ and Cl^- ions (Hidri et al. 2022). At the onset of salinity stress, canola reacts by first deploying osmotic adjustment mechanisms to sustain water absorption and cell turgor. Under high salinity stress, canola often accumulates osmolytes like proline and soluble sugars to overcome the osmotic imbalance. These compounds aid in protein and membrane stabilization, enabling the plant to maintain fundamental metabolic functions in stressful conditions. Canola also ameliorates its antioxidant defense system to cope with salinity-induced oxidative stress. Antioxidants are upregulated to remove excessive ROS and preserve the integrity of cellular structures from oxidative damage (Abdel Latef 2011; Abdel Latef et al. 2021; Dawood et al. 2024; Wang et al. 2019). Maintaining and managing ion balance is an essential feature of canola's response to salinity stress. The plant utilizes strategies to absorb necessary nutrients and exclude or sequester toxic ions. Na^+/H^+ antiporters in the plasma membrane and tonoplast serve to sequester Na^+ ions into vacuoles, thus maintaining low levels of these ions in the cytoplasm, where they could cause secondary toxic effects on many enzymes by disrupting enzymatic activity and other metabolic processes. Such compartmentalization prevents loss from a specific cytoplasmic environment, which is conducive to cell activity. Potassium (K^+) role in alleviating Na^+ toxicity is also equally important (Padan and Landau 2016; Chakraborty et al. 2018). Salinity response in canola is regulated by the expression of stress-responsive genes and regulatory proteins at the molecular level. Transcriptional regulators such as MYB, NAC and DREB control the expression of osmotic adjustment, ion transport and antioxidant defense genes. SOS (Salt Overly Sensitive) pathway encodes the SOS1, SOS2 and SOS3 proteins, which maintain ionic homeostasis. The regulatory salinity stress response genes are also mediated by phytohormones, ABA, which alters gene expression and enables cleaner adaptations to saline habitats (Mahajan et al. 2008).

(iii) *Responses to Temperature Stress*

The extremes of temperature, like heat and cold stress, have detrimental effects on canola growth, development and yield. Canola is grown in a variety of environments and is sensitive to extreme temperatures throughout its entire life cycle, from germination through seed development (Robertson and Lilley 2016).

(iv) *Heat Stress Response in Canola*

Heat stress occurs when temperatures are higher than the optimum for canola growth and affects different physiological and biochemical processes. Higher temperatures increase transpiration rates in plants and result in the quick loss of water, leading to heat-induced dehydration. Although canola plants close their stomata under heat stress to prevent water loss, which also restricts the amount of CO_2 absorbed, ultimately lowering the efficiency of photosynthesis. Reduced photosynthesis leads to lower energy production and reduced growth and development (Elferjani and Soolanayakanahally 2018). Heat stress during flowering and seed filling has devastating consequences for canola yield and oil quality. Extremely high temperatures can, however, often result in flower abortion, reduction in pollen viability or pollen germination on stigma, fertilization and consequently a reduction in pod formation and seed set. Heat stress imposed on the plant during the period of seed filling leads to increased metabolic rates that may affect the storage and accumulation of oil and other compounds, but therefore gives rise to seeds with lower fresh weight, lower oil content and alteration of fatty acid composition. It is essential for nutritional and industrial purposes that the oil produced under heat stress also contains a higher concentration of saturated fatty acids (Lohani et al. 2022). Canola plants synthesize heat shock proteins (HSPs), which are molecular chaperones that help mitigate the impacts of heat stress. Heat shock proteins (HSPs) stabilize denatured proteins to refold them and provide protection from heat-damaged proteins. Heat Shock Transcription Factors (HSFs) regulate the expression of HSPs during heat stress and are activated by increased temperatures to trigger the transcription of genes encoding heat shock proteins. This response helps maintain cellular function by avoiding protein aggregation and sustaining the action of enzymes (Gomez-Pastor et al. 2018). Under heat stress, canola also improves its antioxidant defense system. In high temperatures, the production of ROS escalates, and it can cause oxidative damage. Antioxidant enzymes such as SOD, CAT and APX are also upregulated to scavenge ROS and protect the structural components. Synthesis of osmoprotectants, like proline and glycine betaine, also plays a role in heat stress tolerance by stabilizing proteins and membranes and keeping the osmotic balance of cells (Dikilitas et al. 2020). Low temperatures experienced as cold stress can slow or even completely prevent metabolic processes, at the same time causing cellular damage. Cold stress makes enzymatic activity slow, perturbs photosynthesis and disrupts membrane fluidity. Canola plants under cold stress grow slowly, become chlorotic and delay flowering. Cold stress at flowering results in partial fertilization and a lower seed set. Membrane lipid composition is one of the key adaptive traits to cold stress in canola. Membranes get slightly stiff in cold temperatures, which can also impair other functions. In response, canola alters the total amount and types of unsaturated fatty acids in membrane lipids, thereby maintaining fluidity and membrane functionality under cold stress. This adaptation reinforces the functionality of cellular transport pathways and allows the movement of both water and nutrients (Soualiou et al. 2022; Ghassemi et al. 2021). Osmoprotectant accumulation and antioxidant defense system activation are also important components for canola cold stress tolerance. While solute production of proline and soluble sugars protects structures and lowers the freezing point of the fluids to prevent the propagation of ice crystals. Molecularly, cold-responsive (COR) genes are key to canola's ability to withstand cold temperatures. Transcription factors like CBF (C-repeat binding factor) bind to the promoter regions of the COR genes to regulate their expression. In response to cold stress, the CBF pathway activates the expression of genes that produce protective proteins to maintain cell function. So, the participation of ABA and phytohormones in cold stress signal transduction regulates the plant metabolic response to allow canola to acclimate to and recover from cold stress (Manasa S et al. 2022; Nakashima and Yamaguchi-Shinozaki 2006). Canola responds to drought, salinity and temperature stress through diverse physiological, biochemical and molecular mechanisms that underpin survival and yield production under stress. However, these innate responses can be improved upon with appropriate agronomic practices and breeding strategies (Gilliham et al. 2017). Developing canola varieties that are tolerant to stress by classical breeding and genetic transformation is a priority. Multiomics to identify and edit stress-responsive genes through marker-assisted selection (MAS) and genome editing techniques like CRISPR/Cas9 to develop more abiotic stress-resilient canola types (Raza et al. 2021b; Nawaz et al. 2019). By manipulating agronomic practices such as soil water replenishment, organic amendments through green manure crops and foliar spray of nutrients or antioxidants, we can enhance the stress tolerance of canola. Soil and crop health monitoring precision agriculture tools can provide interventions that are site-specific and crucial during growth stages (Rastogi et al. 2019).

8.6.2 Sunflower

Sunflower (*Helianthus annuus* L.) is one of the major oilseed crops with high oil and nutritional quality and wide environmental adaptability. Nevertheless, like many other crops, sunflower yield can be negatively impacted by abiotic stresses, particularly water deficit (drought) and salinity. They represent major stresses for sunflower growing in semiarid and saline-affected areas. The sunflower tolerance potential to grow and yield under these circumstances embraces a complex combination of physiological, biochemical and molecular mechanisms (Hussain et al. 2018).

(a) *Adaptation to Drought Conditions*

Sunflowers display numerous adaptive characteristics for surviving drought stress and is thus considered more drought-tolerant among a range of oilseed crops. A well-developed root system is an important adaptation of the sunflower to drought conditions. The deep and extensive root systems of sunflowers help them access water from deeper soil layers. This characteristic facilitates continuous water absorption during surface water deficit. Increased root development during drought enables the plant to continue to maintain its water balance and sustain the metabolic activities necessary for survival and productivity (Khaleghi et al. 2020). Sunflower plants experience physiological adaptations to conserve water and maintain cell homeostasis in response to water stress conditions. One of the most important types of response is stomatal regulation; stomata can close in sunflowers, reducing transpiration and hence water loss. This prevents water loss, but it also restricts CO_2 uptake and this, in turn, may limit the efficiency of photosynthesis (Lawson et al. 2011). Osmotic adjustment is also one of the important mechanisms of adaptation to drought in sunflower. Osmolytes, including proline, soluble sugars and glycine betaine accumulation, allow the plant to maintain cell turgor and stabilize cellular structures under drought stress. Notably, proline functions as an osmoprotectant and as an antioxidant, scavenging reactive oxygen species generated during drought situations to ameliorate oxidative damage (Dikilitas et al. 2020; Kosar et al. 2022). Sunflower also improves the antioxidant defense system during drought stress. Production of ROS under water-stress conditions can damage cellular components, including lipids, proteins and nucleic acids. In sunflower plants, antioxidant enzymes are upregulated to neutralize ROS and avoid oxidative damage to cellular structures. This antioxidant response assists in the conservation of cellular health and boosts general stress tolerance in the plant (Raza et al. 2021a). Drought stress induces the expression of several stress-responsive genes and regulatory proteins at the molecular level in sunflower. Transcription factors like DREB proteins and MYB are involved in the activation of genes with osmoregulation activity, antioxidant defense and stress signaling pathways. Accumulation of ABA, a principal phytohormone involved in drought stress response, induces the expression of ABA-responsive genes involved in stomatal closure and osmoprotectant synthesis.

Molecular regulation that aids the adaptation of sunflowers to prolonged water deficit through physiological and biochemical processes (Hussain et al. 2021).

(b) *Adaptation to Saline Conditions*

Several physiological and biochemical mechanisms of salinity tolerance have been elucidated that help alleviate the osmotic and ionic stresses experienced by sunflowers under high soil salinity. Under saline conditions, the initial response of sunflower plants is osmotic stress, given that the high concentration of soluble salts in the soil decreases the availability of water to the roots. In response, sunflowers use the same osmotic adjustment mechanisms that they develop during drought stress to counteract this. Osmolyte accumulation (proline and glycine betaine) was necessary to keep cell turgor and continue to absorb water despite the lower water potential of the soil (Singh et al. 2015). Na^+ and Cl^- pools, both of which result from salinity stress, can also accumulate in excess, interfering with nutrient balance and other metabolic processes. To cope with these toxic ions, sunflower plants use ion homeostasis mechanisms. Transporters and channels selectively transport ions; for example, Na^+ can be removed from the cytoplasm and sequestered into vacuoles or Na^+ uptake can be restricted in the root. This separation allows for the maintenance of an optimal cytoplasmic environment favorable to cell activity. One of the most relevant is the need to maintain a high K^+/Na^+ ratio since K^+ serves critical functions in the activation of numerous enzymes, osmotic regulation and cellular signaling. Sunflower plants regulate ion transport so that the K^+ level is enough to cope with Na^+ toxicity (Akram et al. 2009). This suggests that the antioxidant defense system is equally important in salinity stress adaptation in sunflower. Excess salts create oxidative stress by increasing the accumulation of ROS. Sunflower plants respond to oxidative stress with an increase in some protective enzymes such as SOD, CAT and glutathione peroxidase (GPX). NAC and WRKY transcription factors are the key players in regulating the expression of various stress-responsive genes that optimize plant performance under salinity stress (Punia et al. 2021). The role of ABA signaling in salinity response through the regulation of stomatal behavior and the induction of genes involved in osmoprotectant synthesis and ion transport. Collectively, the responses of these traits would enable sunflower plants to tolerate saline habitats and continue to grow and produce (Rajasheker et al. 2019).

*8.6.3 Soybean (*Glycine max*): Mechanisms of Stress Tolerance*

Glycine max is an oilseed crop with a high protein and oil yield, which is among the most cultivated oilseed crops. Soybean is an important oil crop, but it is susceptible to drought, salinity and high or low temperature abiotic stresses. These stresses can severely affect growth, yield and seed quality (Miransari 2016). More particularly, soybeans have evolved a wide variety of adaptive mechanisms to tolerate this abiotic stress, enabling these plants to maintain productivity (Joshi and Karan 2013).

(a) *Mechanisms of Drought Tolerance*

Soybean under drought stress initiates numerous physiological, biochemical and molecular responses that allow the plant to adapt to drought stress. This response is reflected physiologically by stomata closure, which is a mechanism to reduce the loss of water via transpiration. But this reaction also restricts CO_2 intake, thereby reducing the photosynthetic activity. In turn, soybean plants may increase their water-use efficiency through changes in their metabolic pathways and by reallocating resources to key growth processes. Drought stress may decrease the leaf area, which constrains transpiration while conserving water; however, it may also decrease photosynthetic surfaces (Liu et al. 2012). At the biochemical level, osmoprotectants such as proline, glycine, betaine and soluble sugars accumulate in soybean plants. They assist in osmotic adjustment, cellular stabilization and provide metabolic activity when exposed to drought conditions. Notably, proline functions both as an osmolyte and an antioxidant, scavenging ROS and limiting oxidative damage (Singh et al. 2015). Under drought stress, the antioxidant defense system in soybean is upregulated involving the enhanced activity of SOD, CAT and APX. This enzymatic activity contributes to the detoxification of ROS, an essential defense against oxidative stress, while facilitating metabolic processes (Hasanuzzaman et al. 2012). At the molecular level, the response to drought in soybean consists of the regulation of gene expression and activation of stress-responsive pathways. Recent advances promote the discovery and profiling of the major TFs, including DREB, MYB and bZIP that regulate genes associated with osmotic adjustment, scavenging of reactive oxygen species and signaling during abiotic stress. Under water-deficit conditions, the growth-responsive hormone, ABA, accumulates in soybean, which facilitates stomatal closure and the expression of many stress-responsive genes involved in drought response. These molecular mechanisms ensure their survival under drought conditions, enhancing cellular homeostasis and even adaptation to long-term drought (Hussain et al. 2021).

(b) *Mechanisms of Salinity Tolerance*

Soybean plants exert a range of physiological and biochemical strategies to cope with saline environments. These include the regulation of ions transport and homeostasis. Soybean plants utilize selective ion transporters that extrude Na^+ from the cytoplasm or compartmentalize it into vacuoles (Arif et al. 2020). It stops the cytoplasmic build-up of excess Na^+, enabling the plant to preserve a suitable K^+/Na^+ balance crucial for a variety of cellular processes and enzyme functions (Hertz et al. 2013). Under salinity stress, plants produce osmolytes like proline and glycine betaine and soybean plants also respond to salinity stress by osmotic accumulation. These molecules play osmoregulatory roles and serve to protect cellular structures from the harmful effects of high salinity. Osmolyte accumulation helps retain water and aids in the stabilization of proteins and membranes in stress conditions (Gomathi et al. 2013; Osman et al. 2021). Soybean antioxidant defense system is a key mechanism for coping with salinity-induced oxidative stress which includes SOD, CAT and GPX. The main enzymatic antioxidants are upregulated to eliminate ROS and protect the cellular machinery from oxidative impairment. This metabolic response aids in attenuating metabolic activity and maintenance of cellular structure integrity during salinity stress (Ighodaro and Akinloye 2018).

In soybean, molecular responses are characterized by the upregulation of stress-responsive genes and the reprogramming of regulatory pathways. TF, such as NAC and WRKY, are responsible for regulating the expression of genes related to osmotic adjustment, ion transport and antioxidant defense (Pandey and Gautam 2020). ABA-mediated signaling is also critical in salinity response, inducing the accumulation of osmoprotectants and modulating stomatal movement to avoid excessive water loss (Negin and Moshelion 2016). These responses include physiological, biochemical and molecular responses that are integrated to facilitate the growth and productivity of soybean plants under saline conditions (Gupta and Huang 2014).

(c) *Mechanisms of Temperature Stress Tolerance*

Temperature stress (heat and cold) is one of the most important abiotic stresses affecting soybean production. Metabolic rates may increase at higher temperatures and higher rates of dehydration occur due to heat stress, while lower temperatures may decrease enzymatic activity and damage cells (Djanaguiraman et al. 2018). In soybean, heat shock proteins (HSPs) act as molecular chaperones and are synthesized in response to heat stress. HSPs refold denatured proteins and also prevent protein aggregation to shield plants from heat-induced impairment (Kampinga and Bergink 2016). HSFs, which transactivate HSP expressions in response to increased temperatures, mediate the expression of HSPs (Fragkostefanakis et al. 2015). Moreover, in response to heat stress, soybean plants also strengthen their antioxidant defense system through increased activities of enzymes to scavenge ROS and avert oxidative damage (Azarabadi et al. 2017) (Fig. 8.3). Cold stress induces compositional changes in membrane lipids to help maintain membrane fluidity (Huang and Cheung 2021).

8.7 Role of Organic Fertilizers in Mitigating Abiotic Stress

Organic fertilizers are a key component of sustainable agriculture, benefiting crops facing abiotic stressors, such as drought, salinity and extreme temperatures. Organic fertilizers include sources such as compost, manure, green manure and bone meal and provide not just macro and micronutrients but also improve soil structure and health (Liu 2024). An important role is their contribution to the reduction of abiotic stress as they can improve crop vitality, root development and beneficial soil microorganisms while not only serving for nutrition. Organic fertilizers are the key to increasing productivity and quality for oilseed crops like canola, sunflower and soybean under stressful environmental conditions (Singh et al. 2022b). Organic fertilizers can reduce abiotic stress mainly by their effects on soil properties. They

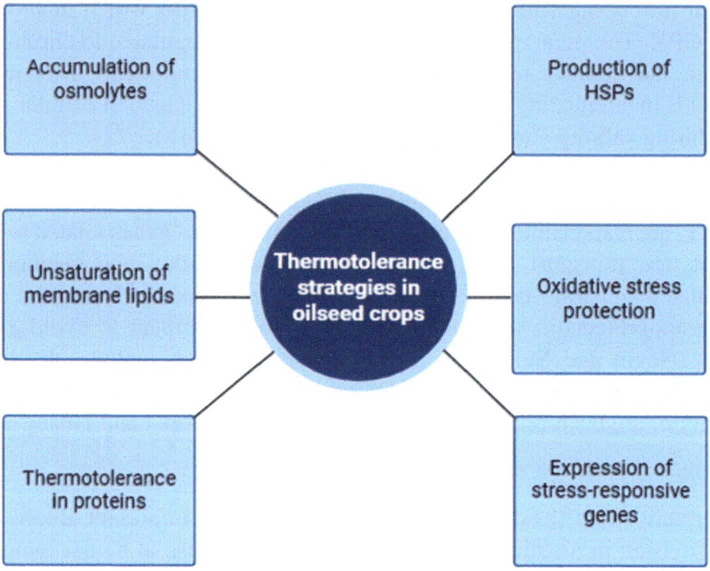

Fig. 8.3 Thermotolerance/temperature stress tolerance in oilseed crops

add organic matter into the soil, thus enhancing the structure, porosity and water retention capacity of the soil. This is especially useful when regions face drought stress when water scarcity can potentially hinder crop growth and production. The enhanced water holding capacity helps sustain plant hydration, supports physiological functions and alleviates water deprivation impact on crops (Ullah et al. 2021).

Fertilizers provide an additional source of organic matter, and this encourages a more extensive root system. Stress-resilient root systems help oilseed crops explore a greater amount of soil volume for water and nutrients to survive and remain productive under drought stress (Azemi-Ardakani et al. 2020). Balanced nutrients also support optimal root growth. Organic fertilizers provide essential elements to the plants that are required for the development of roots. During this period, they get a slow-release source of nutrients from the organic fertilizers that move inside the plant from the root zone (Shaji et al. 2021). In contrast to synthetic fertilizers, nutrients immediately cause the leaching of nutrients to the deeper subsoil and therefore destroy the environment and organic fertilizers decompose slowly. Controlled nutrient dispersion allows the plants to have a constant supply of all the necessary elements in their system, especially during stress conditions when the demand for nutrients increases significantly. The availability of nutrients is most critical to stress resilience. Balanced nutrition allows plants to maintain the function of photosynthesis, enzymes and other metabolic processes under suboptimal conditions (Sarkar et al. 2024; Araújo et al. 2015).

Several nutrients, including nitrogen, phosphorus, potassium and micronutrients (e.g., zinc and iron) contributed by organic fertilizers which play important roles in stress responses in plants. Sufficient nitrogen maintains chlorophyll content, upon which photosynthesis can persist during drought or temperature stress (Ahanger et al. 2016). Potassium helps transfer energy and establishes roots, which allows the plant to extend to a greater depth to access more water and nutrients. Potassium, due to its osmotic balance and water uptake regulation, is crucial during drought and salinity stress since it is required for maintaining tissue turgor and limiting excess water loss (Sardans and Peñuelas 2021). Since organic fertilizers contain micronutrients like zinc, copper and manganese that function as cofactors for antioxidant enzymes. Organic fertilizers make a significant contribution to the maintenance of the process of the antioxidant defense system, which counteracts stresses such as drought, salinity or extreme temperatures by protecting cellular structures and stabilizing metabolism in the crops (Chauhan 2023). Because organic fertilizers can enhance soil beneficial microorganisms, which have a significant contribution to increasing the plant tolerance toward abiotic stress events. These microbes, such as mycorrhizal fungi and nitrogen-fixing bacteria, form symbiotic associations with root systems that enhance nutrient absorption and promote overall plant growth (Soumare et al. 2015). A mycorrhizal fungus reaches beyond the root system with its hyphal networks, increasing the surface area for water and nutrient acquisition. These resources are beneficial, especially concerning drought and nutrient-deficient conditions (Chandrasekaran 2022).

Additionally, soil microbes help decompose organic matter and free up nutrients and other substances needed by plants in relatively available forms. This sequence provides important components when abiotic stress is detected, therefore fulfilling the metabolic demand of the plant (Zechmeister-Boltenstern et al. 2015). Moreover, certain microorganisms synthesize plant growth-promoting substances like phytohormones and diminish the effects of stress on plants by stimulating the action of natural stress response mechanisms in plants (Fahad et al. 2015).

Organic fertilizers help the plant to maintain osmotic equilibrium under the conditions of water deficit or salt stress. The gradual potassium release from the organic fertilizers supports stomatal regulation, which helps to retain water and minimizes transpiration loss (Liaqat et al. 2022). It helps maintain the osmotic potential of the plant cell, which allows for the retention of turgor even when external water is limited, keeping the cell hydrated. This function is essential for oilseed crops because it has been demonstrated that these species can carry out metabolic processes and growth under some stress conditions (Ali et al. 2014). However, in salinity stress, high concentrations of Na^+ and Cl^- ions can disturb tissue nutrient balance and density as well as disrupt proper function. Organic fertilizers promote nutrient balance and ion concentrations within the plant by providing necessary nutrients (potassium and calcium) in a balanced way (Mohamed et al. 2021). Potassium competes with sodium for uptake within the plant, and potassium competes with sodium for transport within the plant tissues, which reduces the toxic effect of sodium accumulation. An even supply of nutrients afforded by organic fertilizers aids the plant in keeping toxic ions out or sequestering them, which limits damage and allows growth to continue (Wakeel 2013). Excessive production of ROS under abiotic stress conditions can result in oxidative damage to proteins, lipids and nucleic acids. Organic fertilizers supply micronutrients that are part of the antioxidant defense system, which further increases the plants' ability to tolerate oxidative stress. Sodium, iron, zinc, copper and manganese are also needed for the activity of these ROS-scavenging enzymes, which protect the plant against oxidative damage commonly found in organic fertilizers (Tavanti et al. 2021). Under stress conditions, the upregulation of such enzymes maintains the homeostasis of the cell and preserves the activities of metabolic pathways. Organic fertilizers help in the synthesis of osmoprotectants (proline and glycine betaine) that aid in osmotic adjustment and cellular protection. These compounds also stabilize proteins and membranes, allowing the plant to maintain metabolic activity during severe drought or salinity-stressed conditions (Singh et al. 2015).

Soil with good health is also essential to provide thermal regulation against temperature stress, along with organic fertilizers, through promoting the root growth that facilitates water uptake. Enough moisture in the root zone during hot weather enables the transpiration process to cool the plant and also helps avoid heat injury (Sharma and Kumar 2023). By providing nutrients, organic fertilizers promote the synthesis of nutrients for the activation of HSPs and others that stabilize proteins and assist cellular function during heat stress (Bozaykut et al. 2014). Organic fertilizers help maintain soil fertility and environmental sustainability. Soil organic carbon increases due to the continuous application of organic fertilizers which subsequently enhances soil fertility, and physical properties and minimizes the erosion effects. Moreover, organic fertilizers are known as a key solution to chemical fertilizer use and its consequences in nutrient runoff and environmental degradation. By combining organic fertilizers with other methods of sustainable agriculture such as crop rotation and intercropping, oilseed crops become even more resilient to abiotic stress (Martínez-Mena et al. 2020).

8.8 Mechanisms of Stress Tolerance Induced by Organic Fertilizers

Organic fertilizers play a greater role in agriculture than just supplying nutrients, as they help plants overcome a range of abiotic stresses, including drought, salinity and temperature extremes. Organic fertilizers induce stress tolerance through complex and multifaceted mechanisms involving activities such as antioxidant activity, with stimulation of root system development and nutrient absorption (Liu 2024).

8.8.1 Enhancement of Antioxidant Activity

The improvement of stress tolerance in crops by the application of organic fertilizers can also be achieved by improving antioxidants. The overproduction of ROS in plant cells is one of the major responses of plants to abiotic stresses like drought, salinity and extreme temperature (Hasanuzzaman et al. 2021). ROS, like superoxide anions and hydrogen peroxide, are found near the site of the pathogen, in the cell cytoplasm in small quantities due to their important role in the cell signaling pathway. Their excessive amount causes oxidative stress and is harmful to the cell as they damage protein, lipids and nucleic acid and destroy normal cell functioning and leading to cell death (Silva and Coutinho 2010). Organic fertilizers play a significant role in reducing oxidative injury by providing micronutrients required as cofactors for antioxidant enzymes. Specific micronutrients, such as zinc, copper and manganese, are essential for the function of the SOD which catalyzes the dismutation of superoxide radicals to oxygen and hydrogen peroxide and is present in most organic fertilizers. Copper and manganese are also required in CAT and POD to decompose hydrogen peroxide to water and oxygen, to reduce the damage to cells (Gupta et al. 2024). Organic fertilizers can also stimulate nonenzymatic antioxidants such as ascorbate (vitamin C) and glutathione. These, together with enzymatic antioxidants, scavenge for ROS and create redox homeostasis. The organic part in fertilizers activates beneficial soil microbes that can function to synthesize metabolites and secrete enzymes able to modulate plant defense pathways. These actions cumulatively enhance the oxidative stress tolerance of the plant under abiotic stresses (Martínez and Araya 2010). Furthermore, organic fertilizers also modulate the expression of antioxidant-producing stress-responsive genes. It also increases the accessibility of essential phytohormones like ABA and salicylic acid (SA), which regulate stress response pathways mediated by these hormones (Kohli et al. 2018). It is well known that ABA and SA signaling pathways can induce the expression of genes that encode for antioxidant enzymes, which in turn increase the capacity of the cells to cope with oxidative stress (Vina et al. 2006).

8.8.2 Root System Development and Nutrient Uptake

Establishing a strong root system is a vital trait for plant survival under abiotic stress. Organic inputs are indispensable to ensure root structure and function, leading to better water and nutrient acquisition to support plant growth in conditions of drought, salinity or nutrient limitation (Koevoets et al. 2016). By adding organic matter content to the soil, organic fertilizers also improve the soil structure and porosity and thus better soil infiltration and retention quality. Taking these soil amendments develops a supportive habitat between the roots and plants, which can grow deeper with larger root development. However, deep root systems allow plants to retrieve moisture from deeper levels of the soil, which can be key to survival during long dry spells. Improved root proliferation enhances the plants' potential to appropriate vital nutrients to facilitate metabolic and physiological processes (Bodner et al. 2021; Pan et al. 2018). Organic fertilizers release good amounts of nitrogen, phosphorus and potassium, which are crucial for essential root development. Nitrogen helps in protein and nucleic acid synthesis required for cell division and root elongation. As part of ATP, a key compound for energy transfer, phosphorus must be present for root cell metabolism and growth. In water stress, potassium regulates osmotic adjustment and water uptake to sustain the turgor-needed cellular functions of the root (Karthika et al. 2018; Mostofa et al. 2022). The organic content helps in supplying microbial activity in the rhizosphere, so it also helps in making the required nutrients available to the plants. Organic fertilizers contain species of humic and fulvic acids and are well known for their effectiveness in stimulating root growth and nutrient uptake. These organic acids enhance the exchange capacity of the soil and therefore help in retaining and making the essential nutrients available. Humic substances can promote the elongation and branching of plant roots, which increases the root surface area for nutrition and water absorption. The enhanced root architecture promotes plant health and productivity, which increases the ability of plants to withstand abiotic stress conditions (Karimi et al. 2020).

8.8.3 Microbial Interactions

Soil microbial interactions enhance crop stress tolerance. Organic fertilizers trigger beneficial soil microbial interactions. The organic matter in fertilizers serves as food for beneficial soil microorganisms, which thrive on their activity. These microbes promote plant health by enhancing nutrient availability, production of growth-promoting substances and induction of systemic resistance to abiotic stress (Bhattacharjee et al. 2024). Among root-associated beneficial microorganisms, plant growth-promoting rhizobacteria (PGPR) and mycorrhizal fungi are key players. The PGPR help improve the nutrient uptake by fixing atmospheric nitrogen, solubilizing phosphorus and producing the siderophores that enhance the uptake of iron (Timofeeva et al. 2023). They additionally release phytohormones like auxins, cytokinins and gibberellins, which stimulate root growth or development. Such root architecture formed root systems to help the plant access water and nutrients, better allowing it to hedge against abiotic stressors such as drought and salinity (Kosakivska et al. 2022).

Organic fertilizers can enhance the population of microbes that synthesize metabolites (e.g., exopolysaccharides, antioxidant enzymes) that counteract stress. Soil aggregation and water-holding capacity are enhanced by exopolysaccharides, which facilitate soil moisture maintenance (Srividhya et al. 2020). Moreover, some PGPR also have the potential to induce systemic tolerance (IST) by regulating stress-responsive pathways in plants. These stress-related genes help plant cope with various stresses. Soil, like gut microbiome, is rich in microorganisms and organic fertilizers can support soil microorganisms and nutrient cycling, pathogen suppression and production of secondary metabolites to provide stress tolerance. Such metabolites as flavonoids, phenolics and other antioxidant properties in plants prevent oxidative damage from stress (Koza et al. 2022; Kapoor et al. 2024).

8.9 Conclusion

Considering the importance of oilseed crops for world agriculture and food security, the impact of abiotic stress on oilseed crops is very important to study. Oilseed crops like canola, sunflower and soybean are critical for edible oil, biofuel and animal feed production and their productivity is crucial to economic and nutritional requirements. Nevertheless, the cultivation of these crops is adversely affected by numerous abiotic stresses such as drought, salinity, extreme temperatures and nutrient imbalances, which significantly impair their growth, yield and oil quality. Characterizing these abiotic stressors and their effects on the physiology and gene expression profile of these plants is crucial for formulating specific adaptation and mitigation measures for these plants. Mechanisms of tolerance such as osmotic adjustment, antioxidant defenses and changes in root architecture, are essential to support plant crops to overcome such adverse environments. Organic fertilizers strengthen the resilience of crops by improving antioxidant activity, stimulating the root systems and helping host–plant beneficial microbes. These assist with the plant's natural stress responses to help oilseed crops continue to grow and produce despite the stresses. The future of organic fertilizers in alleviating abiotic stress seems optimistic as research, technology and policymaker support increase. Future research should focus on the combinatorial application of organic fertilizers with biostimulants and biofertilizers alongside precision agriculture strategies to fine-tune nutrient input and mediate oilseed crops' defense resilience.

References

Abeynayake S, Etzerodt T, Jonavičienė K, Byrne S, Asp T, Boelt B (2015) Fructan metabolism and changes in Fructan composition during cold acclimation in perennial ryegrass. Front Plant Sci 6. https://doi.org/10.3389/fpls.2015.00329

Abdel Latef AA (2011) Ameliorative effect of calcium chloride on growth, antioxidant enzymes, protein patterns and some metabolic activities of canola (Brassica napus L.) under seawater stress. J Plant Nutr 34:1303–1320

Abdel Latef AA, Omer AM, Badawy AA, Osman MS, Ragaey MM (2021) Strategy of Salt Tolerance and Interactive Impact of Azotobacter chroococcum and/or Alcaligenes faecalis Inoculation on Canola (Brassica napus L.) Plants Grown in Saline Soil. Plants 10(1):110

Agudelo A, Carvajal M, Martínez-Ballesta MC (2021) Halophytes of the Mediterranean Basin—underutilized species with the potential to be nutritious crops in the scenario of the climate change. Food Secur 10(1):119. https://doi.org/10.3390/foods10010119

Ahanger MA, Morad-Talab N, Abd-Allah EF, Ahmad P, Hajiboland R (2016) Plant growth under drought stress: significance of mineral nutrients. Water Stress Crop Plants 2:649–668

Ahmad M, Waraich EA, Skalicky M, Hussain S, Zulfiqar U, Anjum MZ, Habib ur Rahman M, Brestic M, Ratnasekera D, Lamilla-Tamayo L (2021a) Adaptation strategies to improve the resistance of oilseed crops to heat stress under a changing climate: an overview. Front Plant Sci 12:767150.

Ahmad MS, Skalický M, Hussain S, Anjum MZ, Brestič M, Ratnasekera D, Lamilla-Tamayo L, Al-Ashkar I, Sabagh AE (2021b) Adaptation strategies to improve the resistance of oilseed

crops to heat stress under a changing climate: an overview. Front Plant Sci 12. https://doi.org/10.3389/fpls.2021.767150
Ahmad Z, Anjum S, Skalický M, Waraich EA, Tariq RMS, Ayub MA, Hossain A, Hassan MM, Brestič M, Islam MS, Habib-ur-Rahman M, Wasaya A, Iqbal MA, Sabagh AE (2021c) Selenium alleviates the adverse effect of drought in oilseed crops Camelina (Camelina Sativa L.) and Canola (Brassica Napus L.). Molecules 26(6):1699. https://doi.org/10.3390/molecules26061699
Ahmed T, Noman M, Qi Y, Shahid M, Hussain S, Masood HA, Xu L, Ali HM, Negm S, El-Kott AF (2023) Fertilization of microbial composts: a technology for improving stress resilience in plants. Plan Theory 12(20):3550
Akhtar SS, Andersen MN, Liu F (2015) Biochar mitigates salinity stress in potato. J Agron Crop Sci 201(5):368–378. https://doi.org/10.1111/jac.12132
Akram MS, Ashraf M, Akram NA (2009) Effectiveness of potassium sulfate in mitigating salt-induced adverse effects on different physio-biochemical attributes in sunflower (Helianthus annuus L.). Flora 204(6):471–483
Akram NA, Iqbal M, Muhammad A, Ashraf M, Al-Qurainy F, Shafiq S (2018) Aminolevulinic acid and nitric oxide regulate oxidative defense and secondary metabolisms in canola (Brassica napus L.) under drought stress. Protoplasma 255:163–174
Alcock TD, Salt DE, Wilson P, Ramsden SJ (2022) More sustainable vegetable oil: balancing productivity with carbon storage opportunities. Sci Total Environ 829:154539
Alengebawy A, Abdelkhalek ST, Qureshi SR, Wang M-Q (2021) Heavy metals and pesticides toxicity in agricultural soil and plants: ecological risks and human health implications. Toxics 9(3):42
Ali M, Bakht J, Daraz Khan G (2014) Effect of water deficiency and potassium application on plant growth, osmolytes and grain yield of Brassica napus cultivars. Acta Bot Croat 73(2):299–314
Al-Saif AM, Ali MM, Ben Hifaa AB, Mosa WF (2023) Influence of spraying some biostimulants on yield, fruit quality, oil fruit content and nutritional status of olive (Olea europaea L.) under salinity. Horticulturae 9(7):825
Altaf M, Sarwar S, Iqbal J, ur Rahman S, Mustafa HSB, Nazir S, Habib I, Nawaz M (2023) Revealing the yield and quality responses of soybean advanced lines under semi-arid conditions. Adv Life Sci 10(1):54–60
Alzamel NM, Taha EM, Bakr AA, Loutfy N (2022) Effect of organic and inorganic fertilizers on soil properties, growth yield, and physiochemical properties of sunflower seeds and oils. Sustain For 14(19):12928
Amini H, Arzani A, Karami M (2014) Effect of water deficiency on seed quality and physiological traits of different safflower genotypes. Turk J Biol 38:271–282. https://doi.org/10.3906/biy-1308-22
Antonio Fabio da Silva L, Santos MF, Oliveira ML, Sousa GG, Filho PFM, Luz LND (2021) Physiological responses of inoculated and uninoculated peanuts under saline stress. Ambiente E Agua 16(1):1. https://doi.org/10.4136/ambi-agua.2643
Anwar-ul-Haq M, Iftikhar I, Akhtar J, Maqsood M (2023) Role of exogenous osmolyte supplementation in ameliorating osmotic and oxidative stress and promoting growth in salinity-stressed soybean genotypes. J Soil Sci Plant Nutr 23(3):3682–3694
Araújo R, Castro ACM, Fiúza A (2015) The use of nanoparticles in soil and water remediation processes. Mater Today Proc 2(1):315–320
Arif Y, Singh P, Siddiqui H, Bajguz A, Hayat S (2020) Salinity induced physiological and biochemical changes in plants: An omic approach towards salt stress tolerance. Plant Physiol Biochem 156:64–77
Arya H, Singh MB, Bhalla PL (2021) Towards developing drought-smart soybeans. Front Plant Sci 12:750664
Ashraf M, Naz U, Abid M, Shahzad SM, Aziz A, Akhtar N, Mühling KH (2021) Salinity resistance as a function of NH4+:NO3- ratio and its impact on yield and quality of tomato (Solanum

Lycopersicum L.). J Plant Nutr Soil Sci 184(2):246–254. https://doi.org/10.1002/jpln.202000257

Asif M, Safdar M, Akhtar N, Gul S, Javed MA, Raza N, Haq I, Saleem U, Aslam M (2023) Sulfur application improves the yield and quality of sunflower (Helianthus Annuus L.) hybrids. SABRAO J Breed Genet 55(3):907–916. https://doi.org/10.54910/sabrao2023.55.3.25

Azarabadi S, Abdollahi H, Torabi M, Salehi Z, Nasiri J (2017) ROS generation, oxidative burst and dynamic expression profiles of ROS-scavenging enzymes of superoxide dismutase (SOD), catalase (CAT) and ascorbate peroxidase (APX) in response to Erwinia amylovora in pear (Pyrus communis L). Eur J Plant Pathol 147:279–294

Azemi-Ardakani M, Dehestani-Ardakani M, Zarei A, Soltani-Gerdfaramarzi S (2020) Influence of different soil amendments on drought stress tolerance of Maclura pomifera. Plant Physiol Rep 25(3):405–417

Balali-Mood M, Naseri K, Tahergorabi Z, Khazdair MR, Sadeghi M (2021) Toxic mechanisms of five heavy metals: mercury, lead, chromium, cadmium, and arsenic. Front Pharmacol 12:643972

Barman HK, Patra SK, Das V, Mohapatra SD, Jayasankar P, Mohapatra C, Mohanta R, Panda RP, Rath SN (2012) Identification and characterization of differentially expressed transcripts in the Gills of freshwater prawn *Macrobrachium rosenbergii* under salt stress. Sci World J 2012:1–11. https://doi.org/10.1100/2012/149361

Bartwal A, Mall R, Lohani P, Guru S, Arora S (2013) Role of secondary metabolites and brassinosteroids in plant defense against environmental stresses. J Plant Growth Regul 32: 216–232

Bennett EJ, Roberts JA, Wagstaff C (2011) The role of the pod in seed development: strategies for manipulating yield. New Phytol 190(4):838–853

Beyá-Marshall V, Moraga F, Galaz B, Kremer C, Fichet T, Seguel Ó (2025) Avocado drought stress alleviation through different soil amendments. J Appl Phycol:1–20

Bhattacharjee P, Nath S, Bhattacharjee S, Yadav MR (2024) Exploring the synergy of microbes and plants: a promising strategy for managing moisture stress in medicinal and aromatic crops. In: Microbial biostimulants for plant growth and abiotic stress Amelioration. Elsevier, pp 367–395

Bhattacharya A (2022) Effect of low-temperature stress on germination, growth, and phenology of plants: a review. In: Physiological processes in plants under low temperature stress, pp 1–106

Bi H, Miao J, He J, Chen Q, Qian J, Li H, Xu Y, Ma D, Zhao Y, Tian X, Liu W (2022) Characterization of the wheat heat shock factor TaHsfA2e-5D conferring heat and drought tolerance in Arabidopsis. Int J Mol Sci 23(5):2784. https://doi.org/10.3390/ijms23052784

Bjørklund G, Chirumbolo S (2017) Role of oxidative stress and antioxidants in daily nutrition and human health. Nutrition 33:311–321

Blair EJ, Bonnot T, Hummel M, Hay E, Marzolino JM, Quijada IA, Nagel DH (2019) Contribution of time of day and the circadian clock to the heat stress responsive transcriptome in Arabidopsis. Sci Rep 9(1). https://doi.org/10.1038/s41598-019-41234-w

Bodner G, Mentler A, Keiblinger K (2021) Plant roots for sustainable soil structure management in cropping systems. In: The root systems in sustainable agricultural intensification, pp 45–90

Borghi M, Perez de Souza L, Yoshida T, Fernie AR (2019) Flowers and climate change: a metabolic perspective. New Phytol 224(4):1425–1441

Boughalleb F, Abdellaoui R, Nbiba N, Mahmoudi M, Neffati M (2017) Effect of NaCl stress on physiological, antioxidant enzymes and anatomical responses of Astragalus Gombiformis. Biologia 72(12):1454–1466. https://doi.org/10.1515/biolog-2017-0169

Bozaykut P, Ozer NK, Karademir B (2014) Regulation of protein turnover by heat shock proteins. Free Radic Biol Med 77:195–209

Bridgemohan P, Mohammed M, Bridgemohan RS, Mohammed Z (2020) Ecophysiology of anaerobiosis stress due to flood and waterlogging in rice. J Hortic Postharvest Res:113–128

Castrillo G, Paulo José Pereira Lima T, Paredes SH, Law TF, Lorenzo LD, Feltcher ME, Finkel OM, Breakfield NW, Mieczkowski PA, Jones CD, Paz-Ares J, Dangl JL (2017) Root microbiota drive direct integration of phosphate stress and immunity. Nature 543(7646):513–518. https://doi.org/10.1038/nature21417

Chakraborty K, Basak N, Bhaduri D, Ray S, Vijayan J, Chattopadhyay K, Sarkar RK (2018) Ionic basis of salt tolerance in plants: nutrient homeostasis and oxidative stress tolerance. Plant nutrients and abiotic stress tolerance, pp 325–362
Chakradhar T, Mahanty S, Reddy RA, Divya K, Reddy PS, Reddy MK (2017) Biotechnological perspective of reactive oxygen species (ROS)-mediated stress tolerance in plants. In: Reactive oxygen species and antioxidant systems in plants: role and regulation, pp 53–87
Chandra N, Pandey N (2014) Influence of sulfur induced stress on oxidative status and antioxidative machinery in leaves of Allium cepa L. Int Sch Res Not 2014:1–9. https://doi.org/10.1155/2014/568081
Chandrasekaran M (2022) Arbuscular mycorrhizal fungi mediated enhanced biomass, root morphological traits and nutrient uptake under drought stress: a meta-analysis. J Fungi 8(7):660
Chaturvedi S, Gudipati T, Shrivastav AN (2017) Influence of rhizobacterium inoculation on NaCl salinity tolerance in Pusa Sukomal and RC101 varieties of cowpea (Vigna Unguiculata L.). Rec Adv Biol Med 03:35. https://doi.org/10.18639/rabm.2017.03.454927
Chauhan DK (2023) Beneficial chemical elements of plants. Food Process 54:1215–1247
Cheng Y, Luo M, Zhang T, Yan S, Wang C, Feng H, Zhang T, Kisekka I (2023) Organic substitution improves soil structure and water and nitrogen status to promote sunflower (Helianthus annuus L.) growth in an arid saline area. Agric Water Manag 283:108320
Cheong BE, Yu D, Martinez-Seidel F, Ho WWH, Rupasinghe T, Roessner U (2022) The effect of cold stress on the root-specific Lipidome of two wheat varieties with contrasting cold tolerance. Plan Theory 11(10):1364. https://doi.org/10.3390/plants11101364
Chugh V, Kaur D, Purwar S, Kaushik P, Sharma V, Kumar H, Rai A, Singh CM, Kamaluddin, Dubey RB (2022) Applications of molecular markers for developing abiotic-stress-resilient oilseed crops. Life 13(1):88. https://doi.org/10.3390/life13010088
Coelho DS, Simões WL, Salviano AM, Mésquita AC, Alberto KC (2018) Gas exchange and organic solutes in forage sorghum genotypes grown under different salinity levels. Rev Bras Eng Agríc Ambient 22(4):231–236. https://doi.org/10.1590/1807-1929/agriambi.v22n4p231-236
Czwartkowski K, Wierzbic A, Golimowski W (2022) Quality, key production factors, and consumption volume of niche edible oils marketed in the European Union. Sustain For 14(3):1846
Damaris RN, Yang P (2021) Protein phosphorylation response to abiotic stress in plants. In: Plant phosphoproteomics: methods and protocols, pp 17–43
Dawood MFA, Tahjib-Ul-Arif M, Sohag AAM, Abdel Latef AA (2024) Role of Acetic Acid and Nitric Oxide against Salinity and Lithium Stress in Canola (Brassica napus L.). Plants 13(1):51
Dawood MFA, Mohamed HI, Abideen Z, Sheteiwy MS, Eissa MA, Abdel Latef AA (2025) Biochar amendments and drought tolerance of plants. In: Abdel Latef AAH. (Ed.). Biochar in Mitigating Abiotic Stress in Plants. Academic Press. Elsevier Inc. Pp 229–246.
Dejoux J-F, Meynard J-M, Reau R, Roche R, Saulas P (2003) Evaluation of environmentally-friendly crop management systems based on very early sowing dates for winter oilseed rape in France. Agronomie 23(8):725–736. https://doi.org/10.1051/agro:2003050
Deng L, Fu L, Zhang K, Shen Y, Feng G, Zhang L, Li H, Liu C (2022) Effects of fertilizer and waterlogging on the diversity and functioning of the microbial community in greenhouse cultivation soil. Chem Biol Technol Agric 9(1):31
Depotter JRL, Deketelaere S, Inderbitzin P, Tiedemann AV, Höfte M, Subbarao KV, Wood TA, Thomma BPHJ (2016) Verticillium longisporum, the invisible threat to oilseed rape and other brassicaceous plant hosts. Mol Plant Pathol 17(7):1004–1016. https://doi.org/10.1111/mpp.12350
Dhaliwal LK, Gannaban RB, Shrestha AK, Shim J, Mangat PK, Singleton JJ, Angeles-Shim RB (2021) Integrated morpho-biochemical and transcriptome analyses reveal multidimensional response of upland cotton (<i>Gossypium Hirsutum</I>L.) to low temperature stress during seedling establishment. Plant Environ Interact 2(6):290–302. https://doi.org/10.1002/pei3.10067

Didaran F, Kordrostami M, Ghasemi-Soloklui AA, Pashkovskiy P, Kreslavski V, Kuznetsov V, Allakhverdiev SI (2024) The mechanisms of photoinhibition and repair in plants under high light conditions and interplay with abiotic stressors. J Photochem Photobiol B Biol 259:113004

Dikilitas M, Simsek E, Roychoudhury A (2020) Role of proline and glycine betaine in overcoming abiotic stresses. In: Biochemical and molecular perspectives Prot Chem Agents Amelior Plant Abiotic Stress, pp 1–23

Ding L, Lu Z, Gao L, Guo S, Shen Q (2018) Is nitrogen a key determinant of water transport and photosynthesis in higher plants upon drought stress? Front Plant Sci 9:1143

Dinh AQ, Akhtar N, Sagervanshi A, Mühling KH (2021) One-time foliar application and continuous resupply via roots equally improved the growth and physiological response of B-deficient oilseed rape. Plan Theory 10(5):866. https://doi.org/10.3390/plants10050866

Djanaguiraman M, Boyle D, Welti R, Jagadish S, Prasad P (2018) Decreased photosynthetic rate under high temperature in wheat is due to lipid desaturation, oxidation, acylation, and damage of organelles. BMC Plant Biol 18:1–17

Dutta PK, Deka P, Kalita S, Nath RK, Dutta P (2024) Soil fertility management in organic farming. In: Advances in organic farming pp 263–276

Eisa SS, Eid MA, El-Samad EHA, Hussin S, Abdel-Ati AA, El-Bordeny NE, Ali SH, Al-Sayed HM, Lotfy M, Ahmadi M, El-Naggar AM, Ebrahim M (2017) Chenopodium Quinoa Willd. A new cash crop halophyte for saline regions of Egypt. Aust J Crop Sci 11(03):343–351. https://doi.org/10.21475/ajcs.17.11.03.pne316

Elferjani R, Soolanayakanahally R (2018) Canola responses to drought, heat, and combined stress: shared and specific effects on carbon assimilation, seed yield, and oil composition. Front Plant Sci 9. https://doi.org/10.3389/fpls.2018.01224

Ezzati R, Karimi M (2015) Effect of various soil salinity level on the antioxidant and physiological properties of corn plant (Zea Mays). J Exp Biol Agric Sci 3(5):448–452. https://doi.org/10.18006/2015.3(5).448.452

Fageria NK, Moreira A (2011) The role of mineral nutrition on root growth of crop plants. Adv Agron 110:251–331

Fahad S, Hussain S, Bano A, Saud S, Hassan S, Shan D, Khan FA, Khan F, Chen Y, Wu C (2015) Potential role of phytohormones and plant growth-promoting rhizobacteria in abiotic stresses: consequences for changing environment. Environ Sci Pollut Res 22:4907–4921

Farhadi A, Liu Y, Xu C, Wang X, Li E (2022) The role of the renin-angiotensin system (RAS) in salinity adaptation in Pacific white shrimp (Litopenaeus Vannamei). Front Endocrinol 13. https://doi.org/10.3389/fendo.2022.1089419

Faried HN, Ayyub CM, Muhammad A, Ahmed R (2016) Salinity impairs ionic, physiological and biochemical attributes in potato. Pak J Agric Sci 53(01):17–25. https://doi.org/10.21162/pakjas/16.4766

Farooq MA, Ali B, Gill RA, Islam F, Cui P, Zhou W (2016) Breeding oil crops for sustainable production: heavy metal tolerance. Breeding oilseed crops for sustainable production, Elsevier, pp 19–31

Fetouh MI (2016) Protective role of salicylic acid on Rosmarinus Officinalis L. plant in response to salinity. J Plant Prod 7(12):1287–1293. https://doi.org/10.21608/jpp.2016.47022

Fragkostefanakis S, Roeth S, Schleiff E, Scharf KD (2015) Prospects of engineering thermotolerance in crops through modulation of heat stress transcription factor and heat shock protein networks. Plant Cell Environ 38(9):1881–1895

Gad N, Sekara A, Abdelhamid MT (2019) The potential role of cobalt and/or organic fertilizers in improving the growth, yield, and nutritional composition of Moringa oleifera. Agronomy 9(12):862

Gawęda D, Haliniarz M (2021) Grain yield and quality of winter wheat depending on previous crop and tillage system. Agriculture 11(2):133. https://doi.org/10.3390/agriculture11020133

Geremew A, Carson L, Woldesenbet S, Carpenter C, Peace E, Weerasooriya A (2021) Interactive effects of organic fertilizers and drought stress on growth and nutrient content of Brassica juncea at vegetative stage. Sustain For 13(24):13948

Ghassemi S, Delangiz N, Lajayer BA, Saghafi D, Maggi F (2021) Review and future prospects on the mechanisms related to cold stress resistance and tolerance in medicinal plants. Acta Ecol Sinica 41(2):120–129

Gilliham M, Able JA, Roy SJ (2017) Translating knowledge about abiotic stress tolerance to breeding programmes. Plant J 90(5):898–917

Gomathi R, Shiyamala S, Vasantha S, Johnson DE, Janani P (2013) Kinetics of metabolism in sugarcane (Saccharum officinarum L.) under heat stress. Indian J Plant Physiol 18:41–47

Gomez-Pastor R, Burchfiel ET, Thiele DJ (2018) Regulation of heat shock transcription factors and their roles in physiology and disease. Nat Rev Mol Cell Biol 19(1):4–19

Gu C, Zhang S, Han P, Hu X, Xie L, Li Y, Brooks M, Liao X, Qin L (2019) Soil enzyme activity in soils subjected to flooding and the effect on nitrogen and phosphorus uptake by oilseed rape. Front Plant Sci 10. https://doi.org/10.3389/fpls.2019.00368

Guimarães MJM, Simões WL, Oliveira ARd, Gherman Garcia Leal de A, Ênio Farias de França e S, Willadino L (2019) Biometrics and grain yield of sorghum varieties irrigated with Salt water. Rev Bras Eng Agríc Ambient 23 (4):285–290. https://doi.org/10.1590/1807-1929/agriambi.v23n4p285-290

Guo J, Zhou Q, Li XJ, Bing-jun YU, Luo Q (2017) Enhancing NO3- supply confers NaCl tolerance by adjusting Cl- uptake and transport in G. Max & G Soja. J Soil Sci Plant Nutr (ahead). https://doi.org/10.4067/s0718-95162017005000015

Gupta B, Huang B (2014) Mechanism of salinity tolerance in plants: physiological, biochemical, and molecular characterization. Int J Genomics 2014(1):701596

Gupta R, Verma N, Tewari RK (2024) Micronutrient deficiency-induced oxidative stress in plants. Plant Cell Rep 43(9):213

Gustavsson M (2019) The effects of ultraviolet B radiation and high temperature on Stomal aperture and development in Arabidopsis Thaliana. University of Bristol

Hamid Y, Tang L, Wang X, Hussain B, Yaseen M, Aziz MZ, Yang X (2018) Immobilization of cadmium and lead in contaminated paddy field using inorganic and organic additives. Sci Rep 8(1):17839

Hamwieh A, Tuyen D, Huang C, Benitez ER, Takahashi R, Xu D (2011) Identification and validation of a major QTL for Salt tolerance in soybean. Euphytica 179(3):451–459. https://doi.org/10.1007/s10681-011-0347-8

Hannan M, Ahmed W, Akhtar M, Hussain M, Mubeen K (2024) Integration of compost and salicylic acid–micronutrients consortia with conventional fertilizers to alleviate heat stress on nutrient use efficiency, yield and quality of cotton. Sarhad J Agric 40(2):625–636

Hasanuzzaman M, Hossain MA, da Silva JAT, Fujita M (2012) Plant response and tolerance to abiotic oxidative stress: antioxidant defense is a key factor. In: Crop stress and its management: perspectives and strategies, pp 261–315

Hasanuzzaman M, Parvin K, Bardhan K, Nahar K, Anee TI, Masud AAC, Fotopoulos V (2021) Biostimulants for the regulation of reactive oxygen species metabolism in plants under abiotic stress. Cells 10(10):2537

Hassan M, Chen X, Farooq M, Muhammad N, Yan Z, Xu H, Ke Y-Y, Bruno AK, Zhang L, Li J (2021) Cold stress in wheat: plant acclimation responses and management strategies. Front Plant Sci 12. https://doi.org/10.3389/fpls.2021.676884

Hassanpouraghdam MB, Mehrabani LV, Kheiri MSA, Chrysargyris A, Tzortzakis N (2022) Physiological and biochemical responses of Tanacetum Balsamita L. to the foliar application of Dobogen biostimulant, glucose and KNO3 under salinity stress. Sci Rep 12(1). https://doi.org/10.1038/s41598-022-13150-z

Hatzig S, Zaharia LI, Abrams SR, Hohmann M, Legoahec L, Bouchereau A, Nési N, Snowdon RJ (2014) Early osmotic adjustment responses in drought-resistant and drought-sensitive oilseed rape. J Integr Plant Biol 56(8):797–809. https://doi.org/10.1111/jipb.12199

Hedley C (2015) The role of precision agriculture for improved nutrient management on farms. J Sci Food Agric 95(1):12–19

Hertz L, Xu J, Song D, Yan E, Gu L, Peng L (2013) Astrocytic and neuronal accumulation of elevated extracellular K+ with a 2/3 K+/Na+ flux ratio—consequences for energy metabolism, osmolarity and higher brain function. Front Comput Neurosci 7:114

Hidri R, Mahmoud OM-B, Zorrig W, Mahmoudi H, Smaoui A, Abdelly C, Azcon R, Debez A (2022) Plant growth-promoting rhizobacteria alleviate high salinity impact on the halophyte Suaeda fruticosa by modulating antioxidant defense and soil biological activity. Front Plant Sci 13:821475

Hosseini S, Ramezanpour SS, Soltanloo H, Seifati SE (2023) RNA-seq analysis and reconstruction of gene networks involved in response to salinity stress in Quinoa (Cv. Titicaca). Sci Rep 13(1). https://doi.org/10.1038/s41598-023-34534-9

Hu K, Zhu Z, Zang Y, Hussain SA (2011) Accumulation of glucosinolates and nutrients in Pakchoi (Brassica Campestris L. Ssp. Chinensis Var. Communis) two cultivar plants exposed to sulfur deficiency. Hortic Environ Biotechnol 52(2):121–127. https://doi.org/10.1007/s13580-011-0097-5

Huang JJ, Cheung PCK (2021) Cold stress treatment enhances production of metabolites and biodiesel feedstock in Porphyridium cruentum via adjustment of cell membrane fluidity. Sci Total Environ 780:146612

Hussain M, Farooq S, Hasan W, Ul-Allah S, Tanveer M, Farooq M, Nawaz A (2018) Drought stress in sunflower: physiological effects and its management through breeding and agronomic alternatives. Agric Water Manag 201:152–166

Hussain Q, Asim M, Zhang R, Khan R, Farooq S, Wu J (2021) Transcription factors interact with ABA through gene expression and signaling pathways to mitigate drought and salinity stress. Biomol Ther 11(8):1159

Hütsch BW, He W, Schubert S (2016) Nitrogen nutritional status of young maize plants (Zea Mays) is not limited by NaCl stress. J Plant Nutr Soil Sci 179(6):775–783. https://doi.org/10.1002/jpln.201500565

Hwang EY, Song Q, Jia G, Specht JE, Hyten DL, José Maria Correia da C, Cregan PB (2014) A genome-wide association study of seed protein and oil content in soybean. BMC Genomics 15(1). https://doi.org/10.1186/1471-2164-15-1

Ighodaro O, Akinloye O (2018) First line defence antioxidants-superoxide dismutase (SOD), catalase (CAT) and glutathione peroxidase (GPX): their fundamental role in the entire antioxidant defence grid. Alexandria J Med 54(4):287–293

Iqbal A, Khan R, Hussain Q, Imran M, Mo Z, Hua T, Adnan M, Abid I, Rizwana H, Soliman Elshikh M (2024) Vermicompost application enhances soil health and plant physiological and antioxidant defense to conferring heavy metals tolerance in fragrant rice. Front Sustain Food Syst 8:1418554

Ishrat F, Afrasiab H, Chaudhury FA (2022) Effects of salinity stress on growth and physio-biochemical parameters of three pea (Pisum sativum l.) cultivars of different maturity duration. Adv Life Sci 9(3):380–391

Izaurralde RC, Rosenberg NJ, Lal R (2001) Mitigation of climatic change by soil carbon sequestration: issues of science, monitoring, and degraded lands

Jaime R, Alcántara JM, Manzaneda AJ, Rey PJ (2018) Climate change decreases suitable areas for rapeseed cultivation in Europe but provides new opportunities for white mustard as an alternative oilseed for biofuel production. PLoS One 13(11):e0207124. https://doi.org/10.1371/journal.pone.0207124

Jaliya MM, Othman MK, Sani BM, Babaji BA, Sani YA, Zongoma AM (2015) Nitrogen and Sulfur absorption by quality protein maize (QPM) maize as affected by interaction between nitrogen and Sulfur fertilizers at Samaru, Zaria, Nigeria. Afr J Agric Res 10(28):2687–2694. https://doi.org/10.5897/ajar2015.9499

Javeed HMR, Ali M, Zamir MSI, Qamar R, Andleeb H, Qammar N, Kanwal S, Farooq ABU, Tariq M, Tahir M (2023) Biochar application to soil for mitigation of nutrients stress in plants. In: Sustain Agric Rev 61: biochar to improve crop production and decrease plant stress under a changing climate. Springer, pp 189–216

Jha Y, Subramanian R (2016) Regulation of plant physiology and antioxidant enzymes for alleviating salinity stress by potassium-mobilizing bacteria. In: Potassium solubilizing microorganisms for sustainable agriculture, pp 149–162

Joshi R, Karan R (2013) Physiological, biochemical and molecular mechanisms of drought tolerance in plants. In: Molecular approaches in plant abiotic stress, pp 209–231

Juan CA, Pérez de la Lastra JM, Plou FJ, Pérez-Lebeña E (2021) The chemistry of reactive oxygen species (ROS) revisited: outlining their role in biological macromolecules (DNA, lipids and proteins) and induced pathologies. Int J Mol Sci 22(9):4642

Kampinga HH, Bergink S (2016) Heat shock proteins as potential targets for protective strategies in neurodegeneration. Lancet Neurol 15(7):748–759

Kapoor D, Sharma P, Sharma MMM, Yadav S, Husen A (2024) Exploring soil microbiota and their role in plant growth, stress tolerance, disease control and nutrient immobilizer. Biocatal Agric Biotechnol:103358

Karimi E, Shirmardi M, Dehestani Ardakani M, Gholamnezhad J, Zarebanadkouki M (2020) The effect of humic acid and biochar on growth and nutrients uptake of calendula (Calendula officinalis L.). Commun Soil Sci Plant Anal 51(12):1658–1669

Karthika K, Rashmi I, Parvathi M (2018) Biological functions, uptake and transport of essential nutrients in relation to plant growth. Plant Nutr Abiotic Stress Toler:1–49

Kaur G, Singh G, Motavalli PP, Nelson KA, Orlowski JM, Golden BR (2020) Impacts and management strategies for crop production in waterlogged or flooded soils: a review. Agron J 112(3):1475–1501

Keleş B, Akbaş F (2023) Morphological, physiological and biochemical responses of safflower (Carthamus tinctorius L.) exposed to salinity stress. Kahramanmaraş Sütçü İmam Üniv Tarım Doğa Derg 26(6):1246–1252

Kephe PN (2024) Use of modelling tools to assess climate change impacts on smallholder oil seed yields in South Africa. PLoS One 19(5):e0301254. https://doi.org/10.1371/journal.pone.0301254

Khaleghi M, Hassanpour F, Karandish F, Shahnazari A (2020) Integrating partial root-zone drying and saline water irrigation to sustain sunflower production in freshwater-scarce regions. Agric Water Manag 234:106094

Khan JA, Kwot O (2024) The role of natural organic fertilizers on soil fertility management and crop production in Ethiopia. Agric Pract Ethiop:1

Khan S, Javed DMA, Jahan N, Manan DFA (2016) A short review on the development of salt tolerant cultivars in Rice. Int J Public Health Sci 5(2):201. https://doi.org/10.11591/ijphs.v5i2.4786

Khan AU, Ullah F, Khan N, Mehmood S, Fahad S, Datta R, Irshad I, Danish S, Saud S, Alaraidh IA (2020) Production of organic fertilizers from rocket seed (Eruca sativa L.), chicken peat and Moringa oleifera leaves for growing linseed under water deficit stress. Sustain For 13(1):59

Khan F, Siddique AB, Shabala S, Zhou M, Zhao C (2023) Phosphorus plays key roles in regulating plants' physiological responses to abiotic stresses. Plan Theory 12(15):2861

Khan MT, Aleinikovienė J, Butkevičienė L-M (2024) Innovative organic fertilizers and cover crops: perspectives for sustainable agriculture in the era of climate change and organic agriculture. Agronomy 14(12):2871

Khayatnezhad M, Gholamin R, Keynoos H (2022) Investigation of the effect of end season drought stress on morphological on Durum wheat genotypes. Adv Life Sci 9(3):284–288

Khodaei-Joghan A, Gholamhoseini M, Majid A-A, Habibzadeh F, Sorooshzadeh A, Ghalavand A (2018) Response of sunflower to organic and chemical fertilizers in different drought stress conditions. Acta Agric Slov 111(2):271–284

Kim TH, Böhmer M, Hu H, Nishimura N, Schroeder JI (2010) Guard cell signal transduction network: advances in understanding Abscisic acid, CO and Ca signaling. Annu Rev Plant Biol 61(1):561–591. https://doi.org/10.1146/annurev-arplant-042809-112226

Koevoets IT, Venema JH, Elzenga JTM, Testerink C (2016) Roots withstanding their environment: exploiting root system architecture responses to abiotic stress to improve crop tolerance. Front Plant Sci 7:1335

Kohli SK, Handa N, Sharma A, Gautam V, Arora S, Bhardwaj R, Wijaya L, Alyemeni MN, Ahmad P (2018) Interaction of 24-epibrassinolide and salicylic acid regulates pigment contents, antioxidative defense responses, and gene expression in Brassica juncea L. seedlings under Pb stress. Environ Sci Pollut Res 25:15159–15173

Kosakivska IV, Vedenicheva NP, Babenko LM, Voytenko LV, Romanenko KO, Vasyuk VA (2022) Exogenous phytohormones in the regulation of growth and development of cereals under abiotic stresses. Mol Biol Rep 49(1):617–628

Kosar F, Alshallash KS, Akram NA, Sadiq M, Ashraf M, Alkhalifah DHM, Abdel Latef AA, Elkelish A (2022) Trehalose-Induced Regulations in Nutrient Status and Secondary Metabolites of Drought-Stressed Sunflower (Helianthus annuus L.) Plants 11:2780.

Kourani M, Mohareb F, Rezwan FI, Anastasiadi M, Hammond JP (2022) Genetic and physiological responses to heat stress in Brassica napus. Front Plant Sci 13:832147

Koza NA, Adedayo AA, Babalola OO, Kappo AP (2022) Microorganisms in plant growth and development: roles in abiotic stress tolerance and secondary metabolites secretion. Microorganisms 10(8):1528

Kujeke G, Gonye E, Edziwa X, Ncube A, Masekesa RT, Icishahayo D, Matikiti A, Chabata I (2017) Field performance of spider plant (Cleome Gynandra L) under different agronomic practices. Afr J Food Agric Nutr Dev 17(03):12179–12197. https://doi.org/10.18697/ajfand.79.15985

Kumar S, Sachdeva S, Bhat K, Vats S (2018) Plant responses to drought stress: physiological, biochemical and molecular basis. Biotic Abiotic Stress Toler Plants:1–25

Lafarga T (2021) Production and consumption of oils and oilseeds. Oil Oilseed Process Oppor Chall:1–21

Lal R, Follett RF, Stewart BA, Kimble JM (2007) Soil carbon sequestration to mitigate climate change and advance food security. Soil Sci 172(12):943–956

Lam SH, Lui EY, Li Z, Cai S, Sung W-K, Mathavan S, Lam TJ, Ip YK (2014) Differential transcriptomic analyses revealed genes and Signaling pathways involved in Iono-osmoregulation and cellular remodeling in the gills of Euryhaline Mozambique Tilapia. Oreochromis Mossambicus BMC Genomics 15(1):921. https://doi.org/10.1186/1471-2164-15-921

Lamichhane JR, Arseniuk E, Boonekamp PM, Czembor JH, Decroocq V, Enjalbert J, Finckh MR, Korbin M, Koppel M, Kudsk P, Mesterházy Á, Sosnowska D, Zimnoch-Guzowska E, Messéan A (2018) Advocating a need for suitable breeding approaches to boost integrated pest management: a European perspective. Pest Manag Sci 74(6):1219–1227. https://doi.org/10.1002/ps.4818

Lawson T, von Caemmerer S, Baroli I (2011) Photosynthesis and stomatal behaviour. Prog Bot 72: 265–304

Li J, Yang L, Zhen Y, Jiang J, Zhang M, Guo X, Ye Z, Zhao Y, Xiong H, Zhang Z, Shao YH, Jiang C, Zhang H, An G, Paek NC, Ali J (2016) OsASR5 enhances drought tolerance through a stomatal closure pathway associated with ABA and signalling in rice. Plant Biotechnol J 15(2): 183–196. https://doi.org/10.1111/pbi.12601

Li X, Yang W, Liu S, Li X, Jia J, Zhao P, Cheng L, Qi D, Chen S, Liu G (2018) LcFIN2, a novel chloroplast protein gene from sheepgrass, enhances tolerance to low temperature in Arabidopsis and Rice. Physiol Plant 166(2):628–645. https://doi.org/10.1111/ppl.12811

Li Z, Li X, He F (2022) Drip irrigation depth alters root morphology and architecture and cold resistance of alfalfa. Agronomy 12(9):2192. https://doi.org/10.3390/agronomy12092192

Li Y, Wu B, Gao Y, Wu L, Zhao X, Wu L, Zhou H, Tang J (2023) Combination of organic and inorganic fertilizers to counteract climate change effects on cultivation of oilseed flax (Linum usitatissimum L.) using the APSIM model in arid and semiarid environments. Agronomy 13(12):2995

Liaqat S, Chhabra S, Saffeullah P, Iqbal N, Siddiqi TO (2022) Role of potassium in drought adaptation: insights into physiological and biochemical characteristics of plants. Role potassium abiotic stress:143–162

Liu M (2021) Role of Aquaporins in Brassica napus responses to root hypoxia and re-aeration

Liu Y (2024) Multifaceted ability of organic fertilizers to improve crop productivity and abiotic stress tolerance: review and perspectives. Agronomy 14(6):1141. https://doi.org/10.3390/agronomy14061141

Liu L, Gan Y, Bueckert R, Rees KV, Warkentin TD (2010) Fine root distributions in oilseed and pulse crops. Crop Sci 50(1):222–226. https://doi.org/10.2135/cropsci2009.03.0156

Liu B, Cheng L, Ma F, Zou Y, Liang D (2012) Growth, biomass allocation, and water use efficiency of 31 apple cultivars grown under two water regimes. Agrofor Syst 84:117–129

Liu N, Lin Z, Guan L, Gaughan G, Lin G (2014) Antioxidant enzymes regulate reactive oxygen species during pod elongation in Pisum sativum and Brassica chinensis. PLoS One 9(2):e87588

Liu Z, Jia Y, Ding Y, Shi Y, Li Z, Guo Y, Gong Z (2017) Plasma membrane CRPK1-mediated phosphorylation of 14-3-3 proteins induces their nuclear import to fine-tune CBF signaling during cold response. Mol Cell 66(1):117–128.e115. https://doi.org/10.1016/j.molcel.2017.02.016

Liu Y, Lan X, Hou H, Ji J, Liu X, Lv Z (2024) Multifaceted ability of organic fertilizers to improve crop productivity and abiotic stress tolerance: review and perspectives. Agronomy 14(6):1141

Lohani N, Jain D, Singh MB, Bhalla PL (2020) Engineering multiple abiotic stress tolerance in canola, Brassica napus. Front Plant Sci 11:3

Lohani N, Singh MB, Bhalla PL (2022) Short-term heat stress during flowering results in a decline in Canola seed productivity. J Agron Crop Sci 208(4):486–496

Luo Z, Dong H, Li W, Chen Y (2017) Non-uniform salinity in the root zone alleviates Salt damage by increasing sodium, water and nutrient transport genes expression in cotton. Sci Rep 7(1). https://doi.org/10.1038/s41598-017-03302-x

Lv J, Liu P, Wang Y, Gao B, Chen P, Li J (2013) Transcriptome analysis of Portunus Trituberculatus in response to salinity stress provides insights into the molecular basis of osmoregulation. PLoS One 8(12):e82155. https://doi.org/10.1371/journal.pone.0082155

Lv X, Li Y, Chen R, Rui M, Wang Y (2023) Stomatal responses of two drought-tolerant barley varieties with different ROS regulation strategies under drought conditions. Antioxidants 12(4):790. https://doi.org/10.3390/antiox12040790

M. Tahat M, M. lananbeh K, A. Othman Y, I. Leskovar D (2020) Soil health and sustainable agriculture. Sustainability 12(12):4859

Ma L, Luo Y, Chen C, Luo H, Wang S, Yuan Y, Yang W, Liu C, Cao X, Li N (2023) Bacterial strategies for improving the yield, quality, and adaptability of oil crops. Agriculture 14(1):80

Mahajan S, Pandey GK, Tuteja N (2008) Calcium-and salt-stress signaling in plants: shedding light on SOS pathway. Arch Biochem Biophys 471(2):146–158

Maity A, Vijay D, Mukherjee A, Lamichaney A (2016) Potential impacts of climate change on quality seed production: a perspective of hill agriculture. Conserv Agric:459–485

Majeed A, Muhammad Z (2019) Salinity: a major agricultural problem—causes, impacts on crop productivity and management strategies. Plant Abiotic Stress Toler:83–99

Manasa SL, Panigrahy M, Panigrahi KC, Rout GR (2022) Overview of cold stress regulation in plants. Bot Rev 88(3):359–387

Marcolino-Gomes J, Rodrigues FA, Fuganti-Pagliarini R, Bendix C, Nakayama TJ, Celaya B, Molinari HBC, Oliveira MCN, Harmon FG, Nepomuceno AL (2014) Diurnal oscillations of soybean circadian clock and drought responsive genes. PLoS One 9(1):e86402. https://doi.org/10.1371/journal.pone.0086402

Martínez JP, Araya H (2010) Ascorbate–glutathione cycle: enzymatic and non-enzymatic integrated mechanisms and its biomolecular regulation. In: Ascorbate Glutathione Pathw Stress Toler Plants, pp 303–322

Martínez-Mena M, Carrillo-López E, Boix-Fayos C, Almagro M, Franco NG, Díaz-Pereira E, Montoya I, De Vente J (2020) Long-term effectiveness of sustainable land management practices to control runoff, soil erosion, and nutrient loss and the role of rainfall intensity in Mediterranean rainfed agroecosystems. Catena 187:104352

Maryan KE, Farrokhi N, Lahiji HS (2023) Cold-responsive transcription factors in Arabidopsis and Rice: a regulatory network analysis using Array data and gene co-expression network. PLoS One 18(6):e0286324. https://doi.org/10.1371/journal.pone.0286324

Mathur R, Sujatha M, Bera S, Rai P, Babu BK, Suresh K, Babu SS, Sharma M, Rani K, Ajay B (2023) Oilseeds and oil palm. In: Trajectory 75 years Indian Agric Indep. Springer, pp 231–264

Maulana F, Weerasooriya DK, Tesso T (2017) Sorghum landrace collections from cooler regions of the world exhibit magnificent genetic differentiation and early season cold tolerance. Front Plant Sci 8. https://doi.org/10.3389/fpls.2017.00756

Maura Santos Reis de Andrade S, Lucas Amoroso de Lopes C, Braos LB, Luiz Fernando de Sousa A, Camilla Santos Reis de Andrade da S, Cleudison Gabriel Nascimento da S, Pinheiro DG, Correia MEF, Araújo EdS, Colnago LA, Desoignies N, Zonta E, Rigobelo EC (2022) Effect of the application of Vermicompost and Millicompost humic acids about the soybean microbiome under water restriction conditions. Front Microbiol 13. https://doi.org/10.3389/fmicb.2022.1000222

Mbarki S, Cerdà A, Brestič M, Rai M, Abdelly C, Pascual JA (2016) Vineyard compost supplemented with Trichoderma Harzianum T78 improve saline soil quality. Land Degrad Dev 28(3):1028–1037. https://doi.org/10.1002/ldr.2554

McIntosh JR (2017) Mesopotamia and the rise of civilization

McMillan M, Kallenbach CM, Whalen JK (2022) Soybean abiotic stress tolerance is improved by beneficial rhizobacteria in biosolids-amended soil. Appl Soil Ecol 174:104425

Mehdi Y, Sharafzadeh S, Bazrafshan F, Zare M, Amiri B (2018) Application of Jasmonic acid can mitigate water deficit stress in cotton through yield-related physiological properties. Acta Agrobot 71(2). https://doi.org/10.5586/aa.1741

Miransari M (2016) Soybean, protein, and oil production under stress. In: Environ Stress Soybean Prod. Elsevier, pp 157–176

Mohamed HI, El-Sayed AA, Rady MM, Caruso G, Sekara A, Abdelhamid MT (2021) Coupling effects of phosphorus fertilization source and rate on growth and ion accumulation of common bean under salinity stress. PeerJ 9:e11463

Mostofa MG, Rahman MM, Ghosh TK, Kabir AH, Abdelrahman M, Khan MAR, Mochida K, Tran L-SP (2022) Potassium in plant physiological adaptation to abiotic stresses. Plant Physiol Biochem 186:279–289

Mota L, Lourenço PEC, Motta LJM, Motta AM, Mattos JS, Cabral CHA, Cabral CEA (2021) Morphological characteristics and production of Xaraes and Zuri grass fertilized with combinations of sulfur and potassium. Bol Ind Anim 78:1–14. https://doi.org/10.17523/bia.2021.v78.e1502

Muhammad Aslam M, Waseem M, Jakada BH, Okal EJ, Lei Z, Saqib HSA, Yuan W, Xu W, Zhang Q (2022) Mechanisms of abscisic acid-mediated drought stress responses in plants. Int J Mol Sci 23(3):1084

Nagaram IP (2020) A review on role of sulphur nutrition in oilseed crops. J Pharmacogn Phytochem 9(6):1152–1156. https://doi.org/10.22271/phyto.2020.v9.i6q.13104

Nakashima K, Yamaguchi-Shinozaki K (2006) Regulons involved in osmotic stress-responsive and cold stress-responsive gene expression in plants. Physiol Plant 126(1):62–71

Nasrollahzadeh S, Mamnabi S, Ghassemi-Golezani K, Raei Y, Weisany W (2023) PGPR and vermicompost with reduced chemical fertilizer enhances biodiesel production, nutrient uptake and improve oil composition of rapeseed grown under water deficit stress. S Afr J Bot 159:17–25. https://doi.org/10.1016/j.sajb.2023.06.001

Nasrullah AS, Umar M, Sun L, Naeem M, Yasmin H, Khan N (2022) Flooding tolerance in plants: from physiological and molecular perspectives. Braz J Bot 45(4):1161–1176

Naveed M, Sajid H, Mustafa A, Niamat B, Ahmad Z, Yaseen M, Kamran M, Rafique M, Ahmar S, Chen J-T (2020) Alleviation of salinity-induced oxidative stress, improvement in growth, physiology and mineral nutrition of canola (Brassica napus L.) through calcium-fortified composted animal manure. Sustain For 12(3):846

Nawaz M, Iqbal MQ, Shoaib I, Ahmad S, Ullah F, Hussain A (2019) Advances in the CRISPR/Cas9 mediated genome editing: a review: CRISPR/Cas9 mediated genome editing Proc Pak Acad Sci B Life Environ Sci 56(1):1-17–11-17

Negin B, Moshelion M (2016) The evolution of the role of ABA in the regulation of water-use efficiency: from biochemical mechanisms to stomatal conductance. Plant Sci 251:82–89

Niamat B, Naveed M, Ahmad Z, Yaseen M, Ditta A, Mustafa A, Rafique M, Bibi R, Sun N, Xu M (2019) Calcium-enriched animal manure alleviates the adverse effects of salt stress on growth, physiology and nutrients homeostasis of Zea mays L. Plan Theory 8(11):480

Nowicka B, Ciura J, Szymańska R, Kruk J (2018) Improving photosynthesis, plant productivity and abiotic stress tolerance–current trends and future perspectives. J Plant Physiol 231:415–433

O'Neill CM, Lü X, Calderwood A, Tudor E, Robinson P, Wells R, Morris RJ, Penfield S (2019) Vernalization and floral transition in autumn drive winter annual life history in oilseed rape. Curr Biol 29(24):4300–4306.e4302. https://doi.org/10.1016/j.cub.2019.10.051

Oh S-H, Suganuma T, Gogol M, Workman JL (2018) Histone H3 threonine 11 phosphorylation by Sch9 and CK2 regulates chronological lifespan by controlling the nutritional stress response. elife 7. https://doi.org/10.7554/elife.36157

Ott M, Eberle CA, Thom MD, Archer DW, Forcella F, Gesch RW, Wyse DL (2019) Economics and agronomics of relay-cropping pennycress and camelina with soybean in Minnesota. Agron J 111(3):1281–1292. https://doi.org/10.2134/agronj2018.04.0277

Osman MS, Badawy AA, Osman AI, Abdel Latef AA (2021) Ameliorative impact of an extract of the halophyte Arthrocnemum macrostachyum on growth and biochemical parameters of soybean under salinity stress. J Plant Growth Regul 40:1245–1256

Pabuayon ILB, Singh S, Ritchie GL (2019) Effects of deficit irrigation on yield and oil content of sesame, safflower, and sunflower. Agron J 111(6):3091–3098. https://doi.org/10.2134/agronj2019.04.0316

Padan E, Landau M (2016) Sodium-proton (Na+/H+) antiporters: properties and roles in health and disease. Alkali Met Ions Their Role Life:391–458

Pan R, Martinez A, Brito T, Seidel E (2018) Processes of soil infiltration and water retention and strategies to increase their capacity. J Exp Agric Int 20(2):1–14

Pandey AK, Gautam A (2020) Stress responsive gene regulation in relation to hydrogen sulfide in plants under abiotic stress. Physiol Plant 168(2):511–525

Panjabi P, Yadava SK, Kumar N, Bangkim R, Ramchiary N (2019) Breeding Brassica juncea and B. rapa for sustainable oilseed production in the changing climate: progress and prospects. In: Genomic designing climate smart oilseed crops, pp 275–369

Patel M, Fatnani D, Parida AK (2022) Potassium deficiency stress tolerance in peanut (Arachis hypogaea) through ion homeostasis, activation of antioxidant defense, and metabolic dynamics: alleviatory role of silicon supplementation. Plant Physiol Biochem 182:55–75

Patono DL, Alcatrão LE, Dicembrini E, Ivaldi G, Aimonino DR, Lovisolo C (2023) Technical advances for measurement of gas exchange at the whole plant level: design solutions and prototype tests to carry out shoot and rootzone analyses in plants of different sizes. Plant Sci 326:111505

Pattnaik S, Mohapatra B, Gupta A (2021) Plant growth-promoting microbe mediated uptake of essential nutrients (Fe, P, K) for crop stress management: microbe–soil–plant continuum. Front Agron 3:689972

Punia H, Tokas J, Malik A, Sangwan S, Rani A, Yashveer S, Alansi S, Hashim MJ, El-Sheikh MA (2021) Genome-wide transcriptome profiling, characterization, and functional identification of NAC transcription factors in sorghum under Salt stress. Antioxidants 10(10):1605. https://doi.org/10.3390/antiox10101605

Rahbari A, Sinaki JM, Damavandi A, Rezvan S (2021) Castor bean (Ricinus communis L.) responses to drought stress and foliar application of Zn-nano fertilizer and humic acid: grain yield, oil content, antioxidant activity, and photosynthetic pigments. Not Bot Horti Agrobo Cluj Napoca 49(4):12003

Rajasheker G, Jawahar G, Jalaja N, Kumar SA, Kumari PH, Punita DL, Karumanchi AR, Reddy PS, Rathnagiri P, Sreenivasulu N (2019) Role and regulation of osmolytes and ABA interaction in salt and drought stress tolerance. In: Plant signaling molecules. Elsevier, pp 417–436

Raman H, Raman R, Kilian A, Detering F, Carling J, Coombes N, Diffey S, Kadkol G, Edwards D, McCully ME, Ruperao P, Parkin IAP, Batley J, Luckett DJ, Wratten N (2014) Genome-wide delineation of natural variation for pod shatter resistance in Brassica Napus. PLoS One 9(7): e101673. https://doi.org/10.1371/journal.pone.0101673

Rastogi A, Zivcak M, Tripathi D, Yadav S, Kalaji H (2019) Phytotoxic effect of silver nanoparticles in Triticum aestivum: improper regulation of photosystem I activity as the reason for oxidative damage in the chloroplast. Photosynthetica 57(1)

Raza MAS, Shahid AM, Saleem MF, Khan I, Ahmad S, Ali MU, Iqbal R (2017) Effects and management strategies to mitigate drought stress in oilseed rape (Brassica Napus L.): a review. Zemdirbyste-Agric 104(1):85–94. https://doi.org/10.13080/z-a.2017.104.012

Raza A, Hussain S, Javed R, Hafeez MB, Hasanuzzaman M (2021a) Antioxidant defense systems and remediation of metal toxicity in plants. In: Approaches to the remediation of inorganic pollutants, pp 91–124

Raza A, Razzaq A, Mehmood SS, Hussain MA, Wei S, He H, Zaman QU, Xuekun Z, Yong C, Hasanuzzaman M (2021b) Omics: the way forward to enhance abiotic stress tolerance in Brassica napus L. GM Crops Food 12(1):251–281

Reddy B, Suresh G (2009) Crop diversification with oilseed crops for-maximizing productivity, profitability and resource conservation. Indian J Agron 54(2):206–214

Ren T, Zou J, Lu J, Chen F, Wu J, Li X (2015a) On-farm trials of optimal fertilizer recommendations for the maintenance of high seed yields in winter oilseed rape (Brassica Napus L.) Production Soil Sci Plant Nutr 61(3):528–540. https://doi.org/10.1080/00380768.2014.1003964

Ren T, Zou J, Wang Y, Li XK, Cong R, Lu JW (2015b) Estimating nutrient requirements for winter oilseed rape based on QUEFTS analysis. J Agric Sci 154(3):425–437. https://doi.org/10.1017/s0021859615000301

Richter K, Haslbeck M, Büchner J (2010) The heat shock response: life on the verge of death. Mol Cell 40(2):253–266. https://doi.org/10.1016/j.molcel.2010.10.006

Rigobelo EC (2024) Humic substances in combination with PGPR. In: Microbial services for cereal crops: reducing costs and environmental impact. Springer, pp 153–169

Rizwan M, Ali S, Abbas F, Adrees M, Zia-ur-Rehman M, Farid M, Gill RA, Ali B (2017) Role of organic and inorganic amendments in alleviating heavy metal stress in oilseed crops. In: Oilseed crops: Yield and adaptations under environmental stress, pp 224–235

Robertson M, Lilley J (2016) Simulation of growth, development and yield of canola (Brassica napus) in APSIM. Crop Pasture Sci 67(4):332–344

Rosental L, Nonogaki H, Fait A (2014) Activation and regulation of primary metabolism during seed germination. Seed Sci Res 24(1):1–15

Sabagh AE, Hossain A, Barutçular C, Görmüş Ö, Ahmad Z, Hussain S, Islam MS, Alharby HF, Bamagoos AA, Kumar NN, Akdeniz H, Fahad S, Meena RS, Abdelhamid MT, Wasaya A, Hasanuzzaman M, Sorour S, Saneoka H (2019a) Effects of drought stress on the quality of major oilseed crops: implications and possible mitigation strategies—a review. Appl Ecol Environ Res 17(2):4019–4043. https://doi.org/10.15666/aeer/1702_40194043

Sabagh AE, Hossain A, Barutçular C, Islam MS, Ratnasekera D, Kumar NN, Meena RS, Gharib HS, Saneoka H, Jaime ATdS (2019b) Drought and salinity stress management for higher and sustainable Canola (Brassica Napus L.) production: a critical review Aust J Crop Sci 13(01): 88–97. https://doi.org/10.21475/ajcs.19.13.01.p1284

Sahu N, Vasu D, Sahu A, Lal N, Singh S (2017) Strength of microbes in nutrient cycling: a key to soil health. In: Agriculturally important microbes for sustainable Agriculture: Volume I: Plant-soil-microbe nexus, pp 69–86

Saleem N, Noreen S, Akhter MS, Alshaharni MO, H-u-R A, Alzuaibr FM, Al-zoubi OM, Mahmood S (2024) Ascorbic acid-mediated enhancement of antioxidants and photosynthetic efficiency: a strategy for enhancing canola yield under salt stress. S Afr J Bot 173:196–207. https://doi.org/10.1016/j.sajb.2024.08.018

Salgado FF, Vieira LR, Silva VNB, Leão AP, Grynberg P, Marcos Mota do Carmo C, Togawa RC, Carlos Antônio Ferreira de S, Júnior MTS (2021) Expression analysis of miRNAs and their putative target genes confirm a preponderant role of transcription factors in the early response of oil palm plants to salinity stress. BMC Plant Biol 21 (1). https://doi.org/10.1186/s12870-021-03296-9

Sardans J, Peñuelas J (2021) Potassium control of plant functions: ecological and agricultural implications. Plan Theory 10(2):419

Sarkar D, Kar SK, Chattopadhyay A, Rakshit A, Tripathi VK, Dubey PK, Abhilash PC (2020) Low input sustainable agriculture: a viable climate-smart option for boosting food production in a warming world. Ecol Indic 115:106412

Sarkar S, Jaswal A, Singh A (2024) Sources of inorganic nonmetallic contaminants (synthetic fertilizers, pesticides) in agricultural soil and their impacts on the adjacent ecosystems. In: Bioremediation of Emerging Contaminants from Soils. Elsevier, pp 135–161

Sefaoğlu F (2021) Effect of organic and inorganic fertilizers or their combinations on yield and quality components of oil seed sunflower in a semi-arid environment. Turk J Field Crops 26(1): 88–95. https://doi.org/10.17557/tjfc.869335

Sehgal A, Sita K, Siddique KH, Kumar R, Bhogireddy S, Varshney RK, HanumanthaRao B, Nair RM, Prasad PV, Nayyar H (2018) Drought or/and heat-stress effects on seed filling in food crops: impacts on functional biochemistry, seed yields, and nutritional quality. Front Plant Sci 9: 1705

Shaheb MR, Nazrul MI (2017) Effect of bioslurry on transplanted aus rice under rainfed condition in Sylhet. Bangladesh J Agric Res 42(3):467–474. https://doi.org/10.3329/bjar.v42i3.34505

Shahzad R, Jamil S, Ahmad S, Nisar A, Amina Z, Saleem S, Iqbal MZ, Atif RM, Wang X (2021) Harnessing the potential of plant transcription factors in developing climate resilient crops to improve global food security: current and future perspectives. Saudi J Biol Sci 28(4):2323–2341

Shaji H, Chandran V, Mathew L (2021) Organic fertilizers as a route to controlled release of nutrients. In: Controlled release fertilizers for sustainable agriculture. Elsevier, pp 231–245

Sharma PK, Kumar S (2023) Soil temperature and plant growth. In: Soil physical environment and plant growth: evaluation and management. Springer, pp 175–204

Shiaa HH (2024) The response of several sesame cultivars to foliar spraying with potassium on yield and quality. Iop Conf Ser Earth Environ Sci 1371(5):052045. https://doi.org/10.1088/1755-1315/1371/5/052045

Si T, Wang X, Zhou Y, Zhang K, Xie W, Yuan H, Wang Y, Sun Y (2022) Seed yield and quality responses of oilseed crops to simulated nitrogen deposition: a meta-analysis of field studies. GCB Bioenergy 14(8):959–971. https://doi.org/10.1111/gcbb.12977

Silva MA (2024) Effects of nitrogen sources on primary and secondary production from annual temperate and tropical pastures in southern Brazil. Nitrogen 5(2):483–497. https://doi.org/10.3390/nitrogen5020031

Silva JP, Coutinho OP (2010) Free radicals in the regulation of damage and cell death--basic mechanisms and prevention. Drug Discov Ther 4(3)

Sindhu SS, Sharma R, Sindhu S, Sehrawat A (2019) Soil fertility improvement by symbiotic rhizobia for sustainable agriculture. Soil Fertil Manag Sustain Dev:101–166

Singh M, Kumar J, Singh S, Singh VP, Prasad SM (2015) Roles of osmoprotectants in improving salinity and drought tolerance in plants: a review. Rev Environ Sci Biotechnol 14:407–426

Singh AL, Rani K, Jena R, Kona P, Reddy KK, Gangadhara K (2022a) Chapter 8 - Microbes-based bio-stimulants towards sustainable oilseeds production: nutrient recycling and genetics involved. In: Singh H, Vaishnav A (eds) New and future developments in microbial biotechnology and bioengineering. Elsevier, pp 111–130. https://doi.org/10.1016/B978-0-323-85163-3.00013-2

Singh P, Singh RK, Zhou Y, Wang J, Jiang Y, Shen N, Wang Y, Yang L, Jiang M (2022b) Unlocking the strength of plant growth promoting Pseudomonas in improving crop productivity in normal and challenging environments: a review. J Plant Interact 17(1):220–238

Sinthumule RR, Ruzvidzo O, Dikobe TB (2022) Elucidation of the morpho-physiological traits of maize (Zea Mays L.) under Salt stress. J Exp Biol Agric Sci 10(6):1441–1452. https://doi.org/10.18006/2022.10(6).1441.1452

Sita K, Sehgal A, HanumanthaRao B, Nair RM, Vara Prasad P, Kumar S, Gaur PM, Farooq M, Siddique KH, Varshney RK (2017) Food legumes and rising temperatures: effects, adaptive functional mechanisms specific to reproductive growth stage and strategies to improve heat tolerance. Front Plant Sci 8:1658

Song Y, Chen Q, Ci D, Shao X, Zhang D (2014) Effects of high temperature on photosynthesis and related gene expression in poplar. BMC Plant Biol 14(1). https://doi.org/10.1186/1471-2229-14-111

Soualiou S, Duan F, Li X, Zhou W (2022) Crop production under cold stress: An understanding of plant responses, acclimation processes, and management strategies. Plant Physiol Biochem 190: 47–61

Soumare A, Diop T, Manga A, Ndoye I (2015) Role of arbuscular mycorrhizal fungi and nitrogen fixing bacteria on legume growth under various environmental stresses. Int J Biosci 7(4):31–46

Srividhya S, Kumari N, Surendranath R, Jeyakumar P (2020) Role of rhizobacteria in alleviating salt stress. In: Molecular aspects of plant beneficial microbes in agriculture, pp 279–294

Staniak M, Czopek K, Stępień-Warda A, Kocira A, Przybyś M (2021) Cold stress during flowering alters plant structure, yield and seed quality of different soybean genotypes. Agronomy 11(10): 2059. https://doi.org/10.3390/agronomy11102059

Stasnik P, Vollmann J, Großkinsky DK, Jonak C (2023) Carbohydrate metabolism enzymes and phenotypic characterization of diverse lines of the climate-resilient food, feed, and bioenergy crop Camelina Sativa. Food Energy Secur 12:3. https://doi.org/10.1002/fes3.459

Tahjib-Ul-Arif M, Asaduzzaman M, Shirazy BJ, Khan MSU, Rahman AMS, Murata Y, et al. (2024) Seed Priming Improves Chilling Stress Tolerance in Rice (Oryza sativa L.) Seedlings. Phyton-Int J Exp Bot 93(11):3013–3027

Tang W, He X, Qian L, Wang F, Zhang Z, Sun C, Lin L, Guan C (2019) Comparative transcriptome analysis in oilseed rape (brassica Napus) reveals distinct gene expression details between nitrate and ammonium nutrition. Genes 10(5):391. https://doi.org/10.3390/genes10050391

Tavanti TR, de Melo AAR, Moreira LDK, Sanchez DEJ, dos Santos Silva R, da Silva RM, Dos Reis AR (2021) Micronutrient fertilization enhances ROS scavenging system for alleviation of abiotic stresses in plants. Plant Physiol Biochem 160:386–396

Tesfamariam EH, Annandale JG, Steyn JM (2010) Water stress effects on winter canola growth and yield. Agron J 102(2):658–666

Tian P, Huang T, Huang J (2021a) Noncoding-Rna-mediated regulation in response to macronutrient stress in plants. Int J Mol Sci 22(20):11205. https://doi.org/10.3390/ijms222011205

Tian X, Qin Z, Zhao Y, Wen J, Lan T, Zhang L, Wang F, Qin D, Yu K, Zhao A, Hu Z, Yao Y, Ni Z, Sun Q, Smet ID, Peng H, Xin M (2021b) Stress granule-associated TaMBF1c confers Thermotolerance through regulating specific mRNA translation in wheat (Triticum Aestivum). New Phytol 233(4):1719–1731. https://doi.org/10.1111/nph.17865

Tian Z, Wang J-W, Li J, Han B (2021c) Designing future crops: challenges and strategies for sustainable agriculture. Plant J 105(5):1165–1178. https://doi.org/10.1111/tpj.15107

Timofeeva AM, Galyamova MR, Sedykh SE (2023) Plant growth-promoting soil bacteria: nitrogen fixation, phosphate solubilization, siderophore production, and other biological activities. Plan Theory 12(24):4074

Tiwari S, Agrawal M, Tiwari S, Agrawal M (2018) Effect of ozone on physiological and biochemical processes of plants. In: Tropospheric ozone its impacts crop plants: a threat to future global food security, pp 65–113

Tränkner M, Tavakol E, Jákli B (2018) Functioning of potassium and magnesium in photosynthesis, photosynthate translocation and photoprotection. Physiol Plant 163(3):414–431. https://doi.org/10.1111/ppl.12747

Tripathi R, Agrawal S (2016) Effect of supplemental UV-B on yield, seed quality, oil content and fatty acid composition of Brassica campestris L. under natural field conditions. Qual Assur Saf Crops Foods 8(1):11–20

Ullah N, Ditta A, Imtiaz M, Li X, Jan AU, Mehmood S, Rizwan MS, Rizwan M (2021) Appraisal for organic amendments and plant growth-promoting rhizobacteria to enhance crop productivity under drought stress: a review. J Agron Crop Sci 207(5):783–802

Usharani K, Roopashree K, Naik D (2019) Role of soil physical, chemical and biological properties for soil health improvement and sustainable agriculture. J Pharmacogn Phytochem 8(5): 1256–1267

Umer S, Akram NA, Bukhari I, Abdel Latef AA (2025) Biochar for alleviation of salinity stress in plants. In: Abdel Latef AAH. (Ed.). Biochar in Mitigating Abiotic Stress in Plants. Academic Press. Elsevier Inc. pp 173–192.

Vamerali T, Panozzo A, Visioli G, Dal Cortivo C (2021) Root–microbe interactions influencing water and nutrient acquisition efficiency. Root Syst Sustain Agric Intensif:159–192

Verma B, Pramanik P, Bhaduri D (2020) Organic fertilizers for sustainable soil and environmental management. Nutr Dyn Sustain Crop Prod:289–313

Vicente O (2022) Improving agricultural production and food security under climate change conditions. AgroLife Sci J 11(1):241–252

Vieira ERD, Silva EDB, Soares MA, Júnior SLA, Barroso GM, Lemes PG (2018) Lack of macronutrients in Eucalyptus urophylla ST Blake (Myrtaceae) seedlings affects feed and development of Podisus nigrispinus (Hemiptera: Pentatomidae). Biosci J:42–48. https://doi.org/10.14393/bj-v34n1a2018-34484

Vina J, Borras C, Gomez-Cabrera M-C, Orr WC (2006) Part of the Series: From Dietary Antioxidants to Regulators in Cellular Signalling and Gene ExpressionRole of reactive oxygen species and (phyto) oestrogens in the modulation of adaptive response to stress. Free Radic Res 40(2): 111–119

Wakeel A (2013) Potassium–sodium interactions in soil and plant under saline-sodic conditions. J Plant Nutr Soil Sci 176(3):344–354

Wang X, Liu H, Yu F, Hu B, Jia Y, Sha H, Zhao H (2019) Differential activity of the antioxidant defence system and alterations in the accumulation of osmolyte and reactive oxygen species under drought stress and recovery in rice (Oryza sativa L.) tillering. Sci Rep 9(1):8543

Wang N, Lin Y, Qi F, Xiaoyang C, Peng Z, Yu Y, Liu Y, Zhang J, Qi X, Deyholos MK, Zhang J (2022) Comprehensive analysis of differentially expressed genes and epigenetic modification-related expression variation induced by saline stress at seedling stage in Fiber and oil flax, Linum Usitatissimum L. Plants 11(15):2053. https://doi.org/10.3390/plants11152053

Wani UM, Majeed ST, Raja V, Wani ZA, Jan N, Andrabi KI, John R (2021) Ectopic expression of a novel cold-resistance protein 1 from Brassica oleracea promotes tolerance to chilling stress in transgenic tomato. Sci Rep 11(1). https://doi.org/10.1038/s41598-021-96102-3

Waraich E, Ahmad R, Halim A, Aziz T (2012) Alleviation of temperature stress by nutrient management in crop plants: a review. J Soil Sci Plant Nutr 12(2):221–244

Waters AJ, Makarevitch I, Noshay JM, Burghardt LT, Hirsch CN, Hirsch CD, Springer NM (2017) Natural variation for gene expression responses to abiotic stress in maize. Plant J 89(4): 706–717. https://doi.org/10.1111/tpj.13414

Wu W, Ma BL, Whalen JK (2018) Enhancing rapeseed tolerance to heat and drought stresses in a changing climate: perspectives for stress adaptation from root system architecture. Adv Agron 151:87–157

Wu C, Cui K, Tang S, Li G, Wang S, Fahad S, Nie L, Huang J, Peng S, Ding Y (2020) Intensified pollination and fertilization ameliorate heat injury in rice (Oryza sativa L.) during the flowering stage. Field Crop Res 252:107795

Xu J, Xue C, Dong X, Zhao J, Gai J, Guo N (2013) Overexpression of GmHsp90s, a heat shock protein 90 (Hsp90) gene family cloning from soybean, decrease damage of abiotic stresses in Arabidopsis Thaliana. PLoS One 8(7):e69810. https://doi.org/10.1371/journal.pone.0069810

Yahaya SM, Mahmud AA, Abdullahi M, Haruna A (2023) Recent advances in the chemistry of nitrogen, phosphorus and potassium as fertilizers in soil: a review. Pedosphere 33(3):385–406

You J, Zhang Y, Liu A, Li D, Wang X, Dossa K, Zhou R, Yu J, Zhang Y, Wang L (2019) Transcriptomic and metabolomic profiling of drought-tolerant and susceptible sesame genotypes in response to drought stress. BMC Plant Biol 19(1). https://doi.org/10.1186/s12870-019-1880-1

Younas S, Tahira SA, Farooq U (2023) Assessment of heavy metal contamination in vegetables collected from selected localities of Okara, Pakistan. Adv Life Sci 10(2):167–173

Yu J, Yuan SS, Tang Z, Li DW, Zu YG (2012) Proteomics response and acclimation of Camptotheca Acuminata seedlings to water deficit. Adv Mater Res 518-523:460–464. https://doi.org/10.4028/www.scientific.net/amr.518-523.460

Zaid A, Ahmad B, Jaleel H, Wani SH, Hasanuzzaman M (2020) A critical review on iron toxicity and tolerance in plants: role of exogenous phytoprotectants. Plant Micronutrients Deficiency Tox Manag:83–99

Zechmeister-Boltenstern S, Keiblinger KM, Mooshammer M, Peñuelas J, Richter A, Sardans J, Wanek W (2015) The application of ecological stoichiometry to plant–microbial–soil organic matter transformations. Ecol Monogr 85(2):133–155

Zewide I (2021) Review on status of organic farming. Nutr Food Process 4(6):01–04. https://doi.org/10.31579/2637-8914/061

Zhang H, Flottmann S (2016) Seed yield of canola (Brassica napus L.) is determined primarily by biomass in a high-yielding environment. Crop Pasture Sci 67(4):369–380

Zhang Y, Chen X, Geng S, Zhang X (2025) A review of soil waterlogging impacts, mechanisms, and adaptive strategies. Front Plant Sci 16:1545912

Zhao Q, Dong J, Li S, Lei W, Liu A (2024) Effects of micro/nano-ozone bubble nutrient solutions on growth promotion and rhizosphere microbial community diversity in soilless cultivated lettuces. Front Plant Sci 15:1393905

Zheng L, Xie H, Liu Z, Peng H, Chen JY, Huang W, Yu F (2023) Foliar spraying of potassium gluconate promotes the synthesis of the seed oil of Styrax Tonkinensis. J Am Oil Chem Soc 100(8):623–634. https://doi.org/10.1002/aocs.12708

Zhu J, Dai W, Chen B, Cai G, Wu X, Yan G (2023) Research progress on the effect of nitrogen on rapeseed between seed yield and oil content and its regulation mechanism. Int J Mol Sci 24(19):14504

Zuber H, Poignavent G, Signor CL, Aimé D, Vieren E, Tadla C, Lugan R, Belghazi M, Labas V, Santoni A-L, Wipf D, Buitink J, Avice JC, Salon C, Gallardo K (2013) Legume adaptation to Sulfur deficiency revealed by comparing nutrient allocation and seed traits in Edicago Truncatula. Plant J 76(6):982–996. https://doi.org/10.1111/tpj.12350

Полушкина TM, Akimova YA, Коваленко ЕГ, Yakimova O (2020) Organic agriculture in the system of the sustainable use of natural resources. Bio Web Conf 17:00219. https://doi.org/10.1051/bioconf/20201700219

Chapter 9
Phytohormonal Mechanisms for Resilience Against Salinity Stress in Oilseed Crops

Nazima Wahid, Sana Wahab, Muhammad Nauman Khan, Barkat Ullah, Abdul Razzaq, Alevcan Kaplan, Khizar Hayat, Fethi Ahmet Ozdemir, and Quaid Ahmad

Abstract Salinity stress is a prominent abiotic limitation that reduces the yield and quality of oilseed crops globally. Phytohormones modulate plant salinity responses by coordinating complex physiological and molecular networks that improve stress tolerance. This chapter delves into the integrative function of important phytohormones-abscisic acid (ABA), salicylic acid (SA), jasmonic acid (JA), ethylene (ET), cytokinins (CK), gibberellins (GA), and auxins—in regulating ion homeostasis, osmotic balance, antioxidant defense, and gene expression of oilseed crops under saline circumstances. Recent research demonstrates how hormonal cross-talk fine-tunes adaptive responses, allowing oilseed crops to withstand salt-induced damage while maintaining development. Understanding these phytohormonal systems provides intriguing opportunities for generating salt-tolerant oilseed crops via breeding, genetic engineering, and exogenous hormone applications. This analysis

---

N. Wahid · S. Wahab
Department of Plant Sciences, Quaid-i-Azam University, Islamabad, Pakistan

M. N. Khan
Department of Botany, Islamia College Peshawar, Peshawar, Pakistan

Biology Laboratory, University Public School, University of Peshawar, Peshawar, Pakistan

B. Ullah · K. Hayat
Department of Botany, Islamia College Peshawar, Peshawar, Pakistan

A. Razzaq
Qarshi Herb Research Center at Qarshi Industries (Pvt.) Ltd, Lahore, Pakistan

Qarshi University, Lahore, Pakistan

A. Kaplan (✉)
Department of Crop and Animal Production, Sason Vocational School, Batman University, Batman, Turkey

F. A. Ozdemir
Department of Molecular Biology and Genetics, Faculty of Science and Art, Bingol University, Bingol, Turkey

Q. Ahmad
Department of Botany, University of Peshawar, Peshawar, Pakistan

© The Author(s), under exclusive license to Springer Nature Singapore Pte Ltd. 2025

A. A. H. Abdel Latef (ed.), *Oilseed Crops Under Abiotic Stress*, Sustainability Sciences in Asia and Africa, https://doi.org/10.1007/978-981-96-8346-8_9

emphasizes the need for coordinated solutions that use phytohormonal signaling to assure long-term oilseed production in the face of rising saline levels.

Keywords Agriculture yield · Agronomic technique · Oilseed crops · Salt tolerance

9.1 Introduction

Salt stress is one of the primary abiotic factors influencing agricultural output worldwide, particularly in arid and semiarid regions where soil salinity and limited water supplies are common (Abdel Latef et al. 2019a, b; Shrivastava and Kumar 2015). Among various crops, oilseed crops such as sunflower (*Helianthus annuus*), canola (*Brassica napus*), and soybean (*Glycine max*) are vital for global food security and industrial applications due to their rich oil content and economic value. However, the increasing salinization of agricultural land, estimated to affect over 20% of irrigated areas globally, constitutes a hazardous effect on its quality and production (Munns and Tester 2008). The main way that salt stress impacts plants is by causing osmotic and ionic imbalances, which result in decreased water intake, nutritional shortages, and oxidative stress (Hasegawa et al. 2000). For oilseed crops, these challenges manifest as stunted growth, lower seed germination rates, reduced photosynthesis, and compromised oil quantity and quality (Ahmad 2017). The sensitivity of oilseed crops to salinity varies; for instance, sunflower exhibits moderate tolerance, while soybean and canola are relatively sensitive. Phytohormones such as salicylic acid, gibberellin, cytokinin, and abscisic acid are essential for regulating how plants respond to salt stress. Complex physiological and biochemical mechanisms mediated by these signaling molecules enhance plants' ability to withstand and adjust to stressful situations (Jogawat 2019). For example, phytohormones play a crucial role in regulating how plants react to salt stress by regulating physiological and biochemical mechanisms that enhance stress tolerance. Abscisic acid (ABA) accumulation is particularly significant during salt stress because it regulates stomatal closure, which reduces transpiration and water loss. Additionally, because ABA controls the production of stress-responsive genes, plants can adapt to osmotic stress conditions (Cutler et al. 2010). Similarly, salicylic acid (SA) is known to lessen oxidative damage caused by reactive oxygen species (ROS) generated during salt stress. When SA boosts the antioxidant defense system, two enzymes that protect cells from oxidative damage, superoxide dismutase (SOD) and catalase (CAT), become more active (Khan et al. 2015). Furthermore, during salt stress, jasmonic acid (JA) is crucial for controlling ion transport and preserving ionic homeostasis. Better ion balance within plant tissues is ensured by JA signaling, which modulates the expression of genes involved in potassium and salt uptake (Wasternack and Hause 2013). Exogenous injection of these phytohormones is one potential strategy to lessen salt stress in oilseed crops. Studies have demonstrated that applying ABA or SA to salt-affected crops enhances root architecture, boosts antioxidant enzyme activity, and promotes seed germination, all of which enhance growth and productivity in saline settings (Fahad et al. 2015b). Therefore, phytohormonal management can be used to sustainably increase oilseed crops' resilience to salt stress and their capacity to adapt to adverse environmental conditions.

Phytohormones play a major role in plants' enhanced resistance to salt in several interconnected ways (Azooz et al. 2004a, b). One important strategy is osmotic adjustment, where phytohormones regulate the synthesis of osmolytes such as proline and glycine betaine. By preserving the water equilibrium within cells, these osmolytes lessen the negative consequences of salt-induced dehydration. Moreover, phytohormones maintain ion homeostasis via controlling ion transporters such as the high-affinity potassium transporter (HKT) and the salt overly sensitive (SOS) pathway that actively reduce the accumulation of toxic sodium ions in plant cells. The ability of this control to avoid ion toxicity and preserve proper ionic balance is essential for plant survival in saline settings. Moreover, phytohormones fortify the antioxidant defense system by increasing the activity of critical antioxidant enzymes such as CAT and SOD. These enzymes protect cellular components from oxidative damage by scavenging ROS. Another important mechanism is the phytohormones' integration of many stress signals through intricate hormonal connections, which activate genes and proteins that respond to stress. The ability of plants to tolerate and adjust to salt stress is strengthened by this coordinated response, which enhances growth and overall productivity (Peleg and Blumwald 2011). All these processes together demonstrate the basic function of phytohormones in regulating physiological and molecular reactions, which makes them essential targets for enhancing oilseed crops' and other economically significant plants' ability to withstand salt. Given the growing challenges posed by salinity, the integration of phytohormones into crop management practices holds immense potential. In combination with conventional breeding, molecular tools and agronomic techniques, phytohormones can significantly increase the resistance of oilseeds to salt stress.

9.2 General Overview of Phytohormones

Low-molecular-weight phytohormones are defense mechanisms that are more frequently adopted by plants to fully absorb external stimuli against biotic stresses (Iqbal et al. 2022). Complicated plant responses are coordinated by the complicated signaling pathways of phytohormones, which guarantee plant life and adaptability to shifting environmental conditions. Based on their biological roles and functions in the growth and development of plants, phytohormones are categorized into broad classes. Auxins, one of the main classes, control cell elongation, root initiation, and apical dominance. Gibberellins are responsible for stimulating seed germination, stem elongation, and flowering, whereas cytokinins improve nutrient mobilization, delay senescence, and encourage cell proliferation. Abscisic acid, which primarily functions as a stress hormone, is necessary for stomatal closure and seed dormancy under unfavorable climatic conditions. Because it influences stress signaling, fruit ripening, and leaf abscission, ethylene is a crucial hormone for plant adaptation. Other plant hormones, such as strigolactones, brassinosteroids, salicylic acid, and jasmonic acid, are crucial for controlling growth, development, and stress responses in addition to these conventional phytohormones, underscoring the complexity of hormonal.

In order to coordinate plant responses across several biological processes, phytohormones are essential. Auxins control the growth and development of shoots and roots, whereas gibberellins and cytokinins affect cell division, flowering, and seed germination. Because ABA controls water loss and osmotic balance, which in turn controls how plants respond to drought and salinity, phytohormones are also essential in stress responses. Similarly, by strengthening resistance to diseases and herbivory, JA and SA support plant defense systems (Peleg and Blumwald 2011). Senescence and abscission are also influenced by phytohormones; ethylene promotes fruit ripening and leaf shedding, whereas cytokinins postpone senescence by preserving cellular activity and nutrient mobilization. Phytohormonal interference, or the cooperation of several phytohormones, is crucial for combining plant responses to many environmental stressors. Plants can adjust their physiological and biochemical reactions for maximum survival thanks to this interaction. To ensure efficient adaptation to water-deficient situations, for example, ABA and JA work in concert during drought stress to activate genes that respond to stress. In the meantime, root architecture is determined by the ratio of auxin to cytokinin, which also affects meristem activity and lateral root development (Verma et al. 2016). The significance of phytohormones in sustaining plant growth, development, and stress adaptation is highlighted by the complex web of hormonal connections, which also demonstrates the intricacy of plant regulatory mechanisms (Table 9.1).

- *Significance of Phytohormones*

Phytohormones are required for maximizing plant productivity under both optimum and stressful conditions. It is important to determine how plants can survive in harsh conditions. According to the results of the above studies, plants develop various adaptive strategies to survive under these stressful conditions. These include defense mechanisms, osmotic adjustment, antioxidant ion homeostasis, and stress signaling pathways. In general, developing effective strategies to reduce the negative effects of soil salinity and drought stress in agriculture requires a deeper understanding of these factors. Plants use a variety of adaptation mechanisms to thrive in harsh environments. These include drought tolerance mechanisms such as reducing water loss through transpiration and extensive root systems. They also have mechanisms to withstand and adapt to severe temperatures, including thermotolerance and cold climate adaptation. Due to nutrient deficiencies, plants form symbiotic relationships with mycorrhizal fungi and develop specialized root systems (Muhammad et al. 2024). Technological and molecular biology developments have made it possible to manipulate phytohormonal pathways precisely, providing potential approaches to sustainable agriculture.

Table 9.1 Major phytohormones and their roles

S#	Phytohormones	Primary functions	Role in stress response
1	*Auxins*	Cell elongation, apical dominance, and root initiation	Modulate root architecture for improved water and nutrient uptake under stress.
2	*Gibberellins*	Stem elongation, seed germination, and flowering	Promote growth recovery poststress and regulate seed germination under unfavorable conditions.
3	*Cytokinins*	Cell division, delay of senescence, nutrient mobilization	Counteract oxidative stress and delay leaf senescence during drought.
4	*Abscisic acid*	Stomatal closure, seed dormancy, stress signaling	Mediates drought and salinity stress by controlling water loss and activating stress genes.
5	*Ethylene*	Fruit ripening, leaf abscission, and stress signaling	Triggers defense responses under pathogen attack and oxidative stress.
6	*Salicylic acid*	Defense against pathogens, thermotolerance	Enhances systemic acquired resistance and antioxidant activity under stress.
7	*Jasmonic acid*	Defense against herbivory, growth regulation	Regulates ion transport and ROS scavenging under abiotic stresses like salinity.
8	*Brassinosteroids*	Cell division, elongation, and reproductive development	Improve photosynthetic efficiency and enhance stress tolerance.
9	*Strigolactones*	Root-shoot signaling, seed germination	Promote root architecture changes for better nutrient acquisition under stress.

9.3 Impact of Oilseed Crops in Agriculture

Oilseed crops are vital to world agriculture because they are the primary source of vegetable oils and protein-rich byproducts for human consumption, animal feed, and industrial use. Their cultivation is integral to food security, economic development, and sustainable agricultural practices. Prominent oilseed crops include soybean (*Glycine max*), sunflower (*Helianthus annuus*), canola (*Brassica napus*), and groundnut (*Arachis hypogaea*), each contributing uniquely to the agricultural economy and ecosystem services (Table 9.2).

(i) *Economic Importance*

Oilseed crops provide a sizable amount of the edible oils eaten globally, making them an essential part of the world's vegetable oil output. These two types of oil, which satisfy the growing demand in both rich and developing countries, dominate the world's oil markets (Exports 2024). Oilseed production is essential to both food security and economic stability because of the growing demand for vegetable oils brought on by the world's population growth. In addition to serving as raw materials for the food industry, these crops are important commodities in international trade, with the USA, Brazil, and Indonesia leading the way in both production and export.

Table 9.2 Key impacts of oilseed crops in agriculture

S#	Category	Impact	Examples
1	*Economic*	Income generation, export revenue	Soybean exports from the USA; canola in Canada
2	*Nutritional*	Source of essential oils and proteins for the human diet	Soybean oil, sunflower oil
3	*Animal feed*	Protein-rich oilseed meals for livestock	Soybean meal, groundnut meal
4	*Industrial*	Biodiesel, cosmetics, and pharmaceuticals	Canola oil biodiesel, linseed oil varnishes
5	*Environmental*	Soil health improvement, crop rotation benefits, and carbon sequestration	Deep-rooting crops like sunflower

In addition to their role in oil production, oilseed crops provide millions of farmers with a significant income, especially in countries where agriculture is the main economic sector. Oilseed cultivation gives both smallholders and commercial growers in nations like Brazil, India, and the USA a means of subsistence and financial stability. Furthermore, exports bring in a sizable sum of money for the countries that produce oilseeds, with sunflower and soybean oil becoming the most important commodities in global trade. The agriculture sector's contribution to international markets is strengthened by this export potential, which also boosts national economies.

(ii) *Nutritional and Industrial Significance*

Because they provide necessary fatty acids and fat-soluble vitamins, which are critical for sustaining good health, oilseed crops are fundamental to human nutrition. These oils support heart health and cognitive function and are a major source of energy in diets all around the world. Furthermore, oilseed meals, like soybean meal, are high in protein and provide premium feed for the aquaculture and livestock sectors, guaranteeing sustainable animal production (Yaheliuk et al. 2024). The importance of oilseeds in the global food chain is further supported by the fact that their protein-rich byproducts are essential for the production of fish, cattle, and poultry. Oilseed crops have a number of industrial uses, including their nutritional benefits. Their oils are extensively utilized in the manufacturing of biodiesel, which provides a sustainable energy source and a renewable substitute for fossil fuels. Furthermore, oilseed derivatives are significant in the pharmaceutical and cosmetics sectors as essential components of lubricants, skincare products, and medication formulations. Oilseeds' adaptability in the food and nonfood industries emphasizes how important they are to the expanding bioeconomy.

(iii) *Environmental and Agronomic Benefits*

Additionally, oilseed crops provide several agronomic and environmental benefits. By creating deep root systems that strengthen soil structure and make nutrient recycling easier, crops like sunflower and canola promote soil health. The overall productivity of agricultural land is supported by the ability of these deep roots to improve aeration and water infiltration by breaking up compacted soil layers. Additionally, by lowering the accumulation of pests and illnesses carried by the soil, oilseeds improve soil fertility when incorporated into crop rotation systems. Oilseed rotational planting promotes sustainable farming methods by reducing the need for chemical pesticides and fertilizers. The ability of oilseed crops to sequester carbon is another important environmental advantage. By taking carbon dioxide out of the atmosphere, these crops aid in lowering greenhouse gas emissions and slowing down climate change. Oilseed production especially that produced in sustainable agricultural systems contributes to the reduction of carbon emissions and aids in the fight against climate change worldwide. Oilseed crops continue to be essential for maintaining food security, advancing sustainable agriculture, and propelling global economic growth due to their numerous economic, nutritional, industrial, and environmental advantages.

9.4 Effects of Salt Stress on Oilseed Crops

Abiotic variables that restrict agricultural productivity, especially in arid and semi-arid regions, include salt stress. Salinity severely affects oilseed crops, which are vital for industrial uses and global food security, jeopardizing their physiological, biochemical, and molecular functions. Ion toxicity, oxidative damage, osmotic stress, and nutritional imbalance are indicators of stress brought on by salinity. These conditions hinder growth, reduce yield, and deteriorate seed quality (Table 9.3).

(i) *Physiological Effects of Salt Stress on Oilseed Crops*

(a) *Osmotic Stress and Water Uptake:* High salt concentrations in the soil create osmotic stress, which impairs plant roots' capacity to absorb water by lowering the soil's water potential. This osmotic imbalance causes wilting and stunted growth by dehydrating cells and lowering turgor pressure. Under saline circumstances, oilseed crops like *Brassica napus* and *Glycine max* show delayed developmental rates and decreased biomass accumulation, making them especially vulnerable to osmotic stress (Huang

Table 9.3 Salt stress effects on oilseed crops

S#	Category	Effect	Example
1	*Physiological*	Reduced water uptake, ionic imbalance, and stunted growth	Sunflower, canola
2	*Biochemical*	Increased ROS production, oxidative damage	Soybean
3	*Growth and yield*	Fewer flowers, pods, and reduced seed oil content	Canola, groundnut
4	*Nutritional quality*	Altered nutrient composition, fatty acid imbalance	Sunflower

et al. 2012). When vital physiological processes, such as photosynthesis and nutrient intake, are hindered by ineffective water absorption, the negative effects of salinity are intensified.

(b) *Ion Toxicity:* Salinity stress causes toxic ions, particularly sodium (Na^+) and chloride (Cl^-), to accumulate in plant tissues. The overabundance of these ions destabilizes membranes, interferes with enzyme activity, and upsets cellular ionic homeostasis. For instance, elevated Na^+ levels in *Helianthus annuus* impair photosynthetic efficiency by destroying chloroplast structures and lowering photosystem II activity (Khan et al. 2019). Furthermore, an overabundance of Na^+ depletes vital nutrients like potassium (K^+), which further hinders metabolic processes and deteriorates the root system.

(c) *Nutrient Imbalance:* Salinity causes deficits that impact plant metabolism by interfering with the uptake and translocation of vital minerals like potassium, calcium, and magnesium. Under salt stress, competitive Na^+ absorption frequently inhibits potassium uptake, although potassium is essential for stomatal control and enzyme activity. Deficits in calcium and magnesium also impair protein synthesis and cellular signaling, which lowers plant growth and stress tolerance. Reduced seed growth, decreased enzyme activity, and compromised structural integrity are the outcomes of this nutritional imbalance in oilseed crops, and these factors all affect crop output and quality.

(ii) *Biochemical Effects of Salt Stress*

(a) *Oxidative Stress:* The overproduction of ROS, including superoxide radicals (O_2^-) and hydrogen peroxide (H_2O_2), is a significant effect of salinity. Proteins, nucleic acids, and membrane lipids are all oxidatively damaged by these extremely reactive chemicals, which results in cellular malfunction and programmed cell death. Under salt stress, oilseed crops such as soybeans exhibit markedly increased levels of lipid peroxidation, a hallmark of oxidative stress, which compromises membrane stability and lowers photosynthetic effectiveness. SOD, CAT, and peroxidase (POD) are antioxidant defense systems that plants activate to combat oxidative damage. However, these defenses are overpowered by extended exposure to salinity.

(b) *Altered Metabolic Pathways:* Oilseed crops undergo metabolic changes to strengthen their tolerance to salt stress. One of the basic adaptive mechanisms is the accumulation of osmoprotectants, such as proline and glycine betaine. In times of stress, these compounds help maintain the cellular osmotic balance and stabilize proteins and membranes (Ashraf and Foolad 2007). Glycine betaine stabilizes macromolecules and enzymatic processes, whereas proline serves as an osmolyte, preventing water loss and shielding cellular structures from oxidative damage. A higher synthesis of soluble sugars, which act as osmoprotectants and energy stores during stress, is another change in the metabolism of carbohydrates.

(iii) *Growth and Yield Reductions*

(a) *Impaired Seed Development:* Reproductive development is negatively impacted by salinity stress, which results in decreased flower initiation, decreased pod set, and poor seed formation. For example, salt stress in canola affects both nutritional quality and commercial value by lowering seed oil content and changing the mix of fatty acids (Shahbaz and Ashraf 2013). Salinity is a significant problem in the production of oilseed crops since it reduces the amount and quality of seed, which immediately affects output and causes financial losses for farmers.

(b) *Reduced Biomass and Harvest Index:* Plant biomass is decreased as a result of salinity stress, which inhibits both shoot and root growth. A reduced harvest index, a crucial indicator of crop yield, results from a decrease in the biomass allocated to reproductive organs. The yield potential of oilseed crops is further jeopardized by salt-induced stress conditions, which frequently cause resource allocation to shift from reproductive growth to stress tolerance mechanisms. Additionally, the detrimental effects of salinity on overall plant performance are intensified by structural anomalies such as shorter internodes, fewer lateral roots, and decreased leaf area development.

9.5 Phytohormones and Oilseed Crops Under Salt Stress

A major abiotic stressor impacting agricultural output worldwide is salt stress. Canola (*Brassica napus*), peanut (*Arachis hypogaea*), sunflower (*Helianthus annuus*), soybean (*Glycine max*), and other oilseed crops are especially vulnerable to salinity. Plant growth, development, and yield are hindered by oxidative damage, osmotic stress, and ionic imbalance caused by salt stress. Since phytohormones control the physiological, biochemical, and molecular reactions of oilseed crops, they are essential for reducing the negative effects of salt (Table 9.4).

(i) *Role of Phytohormones in Salt Stress Response*

Plants naturally produce phytohormones, which regulate growth and stress reactions. To improve oilseed crops' resistance to salt, they modify a number of metabolic and signaling pathways. Plant response to salt stress is mostly regulated by phytohormones, which modulate a wide range of physiological, biochemical, and molecular activities. Through processes like stomatal control, antioxidant defense, osmoprotectants accumulation, and metabolic reprogramming, several hormones aid in stress adaptation.

1. *Abscisic Acid (ABA): A Central Regulator of Salt Stress Responses*

ABA is a vital phytohormone that orchestrates plant responses to salt stress by controlling osmotic adjustment, gene expression, and water balance. One of its primary functions is stomatal regulation; ABA stimulates stomatal closure to minimize water loss during osmotic stress. ABA causes the

Table 9.4 Mechanisms of phytohormonal action in oilseed crops

S#	Phytohormones	Mechanism of action	Examples of oilseed crops
1	*Abscisic acid*	Stomatal closure, proline accumulation, activation of antioxidant genes	Canola, soybean
2	*Auxins*	Root hair development, lateral root formation, interaction with ethylene	Sunflower, groundnut
3	*Cytokinins*	Antioxidant enzyme activation, chlorophyll retention	Soybean, canola
4	*Gibberellins*	Growth promotion, carbohydrate metabolism	Groundnut, sunflower
5	*Salicylic acid*	ROS scavenging, ion homeostasis, nonenzymatic antioxidant production	Sunflower, soybean
6	*Ethylene*	Root growth modulation, interaction with auxins and ABA	Canola, sunflower
7	*Jasmonates*	Defense gene activation, reduction of lipid peroxidation	Sunflower, groundnut

activation of stress-responsive genes at the molecular level, including those that encode late embryogenesis abundant (LEA) proteins and antioxidant enzymes, which protect cellular components from damage (Christmann et al. 2006). Additionally, by stimulating the accumulation of proline and sugars, ABA stimulates the synthesis of osmoprotectants, which aid oilseed crops like canola in maintaining osmotic equilibrium and enhancing their resistance to salt.

2. *Auxins: Modulating Root Architecture for Salt Tolerance*

   Auxins drastically change the root architecture to enhance water and nutrient intake in saline conditions. They encourage the development of lateral roots and root hairs, which enhance the plant's ability to absorb water and nutrients even in harsh environments. Auxins also exhibit hormonal crosstalk with other phytohormones such as ABA and ethylene to control how plants respond to salt (Javid et al. 2011). This link ensures that stress signaling and root development are coordinated to optimize survival and growth under salt stress.

3. *Cytokinins (CKs): Delaying Senescence and Maintaining Shoot Growth*

   Delaying salt-induced leaf senescence and sustaining shoot growth in saline environments depend on cytokinins (CK*s*). By upregulating enzymes like SOD, which neutralize ROS and stop oxidative damage, they support antioxidant defense. Chlorophyll retention is another essential role of CKs; in salt-stressed crops like soybeans, they help maintain photosynthetic efficiency by preventing chlorophyll breakdown (Thu et al. 2017). CKs assist oilseed crops in producing energy and vital metabolites required for stress tolerance by sustaining photosynthetic activity.

4. *Gibberellins (GAs): Promoting Growth Under Salt Stress*
   GAs stimulate cell elongation and division, which induces the development of shoots and roots and helps counteract the growth inhibition caused by salt. Additionally, they are essential for metabolic reprogramming, which improves the metabolism of carbohydrates to provide the energy needed for stress tolerance systems. Even in high saline environments, GAs allow oilseed crops to retain biomass accumulation by preserving growth processes and energy supplies.
5. *Salicylic Acid (SA): Regulating Oxidative Stress and Ion Homeostasis*
   In salt-stressed oilseed crops, SA is essential for reducing oxidative stress and preserving ion homeostasis. It stimulates the production of nonenzymatic antioxidants like glutathione and ascorbic acid, which scavenge excess ROS and protect cells from oxidative damage. Furthermore, SA controls ion transporters to maintain a low Na^+/K^+ ratio, which is essential for oilseed crops to withstand salt (Khan et al. 2015). SA lessens salt toxicity and increases crop resilience by limiting excessive sodium buildup and preserving cellular ionic balance.
6. *Ethylene: A Dual Regulator of Salt Stress Responses*
   Ethylene regulates salt stress in two different ways, depending on its concentration. Ethylene promotes root development at low concentrations, improving nutrient and water intake. Ethylene, however, can suppress growth at high quantities, resulting in early senescence and decreased biomass accumulation. To help plants maintain the ideal balance between growth and stress adaptation, ethylene also interacts with other phytohormones, such as auxins, ABA, and CK*s*.
7. *Jasmonates (JAs): Enhancing Stress-Induced Defense Mechanisms*
   In oilseed crops like sunflower, JAs help to enhance salt tolerance by activating genes linked to stress. Preventing lipid peroxidation and minimizing membrane damage from salt-induced oxidative stress is one of their primary roles (Wasternack and Strnad 2019). JAs increase the plant's resistance to high-saline environments by strengthening defense responses and cell membrane integrity.

(ii) *Applications of Phytohormones in Salt Stress Mitigation*

Strategic application of phytohormones can help oilseed crops reduce the detrimental effects of salt stress. Phytohormones assist plants in adjusting to high salinity environments and sustain growth, yield, and overall productivity by modifying physiological and biochemical pathways. Hormonal modulation has been used in several ways to improve oilseed crops' resistance to salt.

(a) *Exogenous Application of Phytohormones*

It has been demonstrated that applying phytohormones externally, such as JAs, SA, and ABA, increases oilseed crops' resistance to salt stress. Plants can tolerate high salinity environments thanks to these treatments, which activate stress-responsive pathways such as osmoprotection, antioxidant defense, and ion homeostasis. For instance, SA improves ROS scavenging and ion balance, limiting harmful ion accumulation in plant tissues, whereas ABA administration helps control stomatal closure, reducing water loss under salt stress.

(b) *Hormonal Priming for Enhanced Stress Tolerance*

A pretreatment method known as "hormonal priming" involves exposing seeds or seedlings to trace amounts of phytohormones prior to their exposure to salt stress. By doing this, plants are better equipped to handle future salt exposure by triggering defense mechanisms in anticipation of stress. Hormonal priming improves growth and survival in saline environments by increasing osmoprotectant accumulation, antioxidant enzyme activity, and gene expression. For example, it has been demonstrated that priming soybean seeds with SA improves photosynthetic efficiency, root growth, and germination rates in saline soils.

(c) *Genetic Engineering for Salt-Resilient Oilseed Crops*

By modifying hormone signaling pathways, scientists have been able to create salt-tolerant oilseed crops thanks to advancements in genetic engineering. Researchers have successfully increased the salt resistance of crops like soybean and canola by overexpressing genes related to stress signaling or phytohormone production. By controlling ABA-dependent stress responses, transgenic canola plants that overexpress DREB1A (Dehydration-Responsive Element Binding 1A) under the influence of stress-inducible promoters, for instance, demonstrate improved tolerance to salinity (Nayidu et al. 2013). This technique ensures that plants only activate defense mechanisms in reaction to stress, reducing the risk of adverse effects on normal growth and development.

9.5.1 Role of Auxins Under Salt Stress

Auxins, a class of phytohormones that includes indole-3-acetic acid (IAA) in particular, control the growth and development of plants. They have a major effect on how the plant reacts to salt stress by altering root architecture, ion homeostasis, osmotic adjustment, and oxidative stress mitigation (Abdel Latef et al. 2021c). Salt stress reduces growth and output via interfering with auxin distribution and signaling. However, auxins' exogenous application or endogenous modulation can enhance plant resilience against salinity. In oilseed crops under salt stress, auxins are essential for controlling the architecture of the root system, ion homeostasis, osmotic adjustment, and antioxidant defense. Through altering several physiological and biochemical mechanisms, these phytohormones enhance plants' capacity to adjust to elevated salinity levels (Table 9.5).

According to Pashang et al. (2021), to study the effects of foliar IAA application on morphophysiological and biochemical traits of safflower (*Carthamus tinctorius*)

Table 9.5 Role of auxins in salt stress mitigation

S#	Mechanism	Description	Effect on oilseed crops
1	*Root development*	Promotes lateral root and root hair formation	Improved nutrient and water uptake
2	*Ion homeostasis*	Regulates Na^+/K^+ balance and Na^+ compartmentalization	Reduced ionic toxicity in crops like canola
3	*Osmotic adjustment*	Enhances proline and sugar accumulation	Maintains cellular turgor under saline conditions
4	*Antioxidant defense*	Activates enzymatic and nonenzymatic antioxidants	Mitigates ROS-induced oxidative damage
5	*Stress signaling*	Modulates auxin-responsive genes and cross-talks with other hormones	Integrates multiple stress response pathways
6	*Exogenous application*	Application of IAA or NAA improves growth and yield under salt stress	Increased oilseed productivity in saline environments

cultivars under water-deficit stress (WDS) condition at the end of the growing season, an experiment was conducted using a split-plot factorial randomized complete block design with three replicates in 2018–2019 and 2019–2020 crop seasons. Compared to the Golmehr cultivar, the Goldasht cultivar had more grain (13.27%), grain yield (32.42%) and biological yield (34.29%). Higher bush height (15%), oil percentage (9.58%), linoleic acid (30.76%) and oleic acid content (19.59%) were observed in the Golmehr cultivar. Under WDS, the activities of antioxidant enzymes and proline in the two cultivars did not change significantly; however, under WDS, Goldasht culture showed higher catalase (12%) and peroxidase (13.5%) activities and proline (11.04%). Since auxin use alleviated the side effects of WDS by increasing the SPAD index (2.51%), proline content (17.93%) and activities of catalase (30.53%) and peroxidase (14.72%) enzymes, auxin application could be recommended to alleviate the effects of WDS. Duca et al. (2003) conducted a 5-year study on different sunflower genotypes and found that changes in phytohormones in different organs of sunflower (*Helianthus annuus*) under environmental influence were significant. Auxin hormone showed the most variation in sunflower leaves under environmental conditions and they stated that the variations were significant at a rate of 33%. However, in their in vitro study, Özel and Sağlam (2022) stated that the auxin hormone applied to the cotyledon leaves of sunflower accelerated aging but delayed aging under salt stress. In another study, Saravanakumar and Samiyappan (2007) found that plant growth-promoting rhizobacteria (PGPR) application increased the salt tolerance of *Arachis hypogaea* by reducing ethylene production, auxin production and exopolysaccharide production.

1. *Root System Architecture and Salt Stress Adaptation*

   Water and nutrient intake, especially under situations of salt stress, depends on a robust root system. Auxins affect the growth of lateral roots by increasing the root surface area and improving the efficiency of nutrient absorption (Salehin 2024). They also encourage the formation of root hairs, which increases water intake and lessens the impacts of salinity-induced dryness. However, because salt stress produces too many ROS, it frequently interferes with auxin transport and signaling. By controlling gene expression linked to root elongation and division, auxins mitigate this disruption and support root growth in the face of adversity (Zhao 2010).

2. *Ion Homeostasis and Na^+/K^+ Balance*

   Preventing ion toxicity by maintaining a favorable Na^+/K^+ ratio is one of the main problems under salt stress. The high-affinity potassium transporter (HKT1), which promotes K^+ absorption while limiting Na^+ buildup, is one example of a transporter protein that auxins control. To avoid harmful effects on the cytoplasm, auxins also encourage ion compartmentalization, which sequesters excess Na^+ into vacuoles (Liu et al. 2014). All of these processes work together to support oilseed crops' development and ionic stability in salinity.
3. *Osmotic Adjustment and Metabolic Regulation*

   Osmotic stress results from salt stress's reduction of water supply. By increasing proline synthesis, a crucial osmoprotectant that stabilizes cellular structures and preserves turgor pressure, auxins lessen this effect. In order to ensure adequate energy reserves for stress adaptation, they also have an impact on the metabolism of carbohydrates (Ahammed and Yu 2016). Auxins assist oilseed crops in maintaining physiological activity and osmotically adjusting to saline soils by controlling sugar metabolism.
4. *Antioxidant Defense Against Oxidative Stress*

   Oxidative stress can result from increased ROS production under salt stress, which can harm cellular components. By triggering enzymatic antioxidants that neutralize dangerous ROS, such as POD, CAT, and SOD, auxins support antioxidant defense. To preserve cellular redox balance and lessen oxidative damage, they also encourage the synthesis of nonenzymatic antioxidants such as glutathione and ascorbic acid (Fahad et al. 2015a).
5. *Auxin Signaling Pathways and Stress Adaptation*

   Auxin response factors (ARFs) and the TIR1/AFB receptor system are two examples of auxin signaling pathways that control different stress-adaptive processes. Auxins enable plants to withstand salinity stress by activating genes related to cell wall remodeling, ion transport, and osmolyte production. Additionally, auxins combine many signaling networks with other phytohormones, such as ethylene, cytokinins, and ABA, to enhance tolerance to salt stress.
6. *Exogenous Application of Auxins for Salt Tolerance*

   It has been shown that applying synthetic auxins like naphthalene acetic acid (NAA) or exogenous auxins IAA improves salt tolerance. Exogenous auxins ensure improved water and nutrient intake in saline environments by encouraging root elongation and biomass development. Additionally, oilseed crops treated with auxin, like sunflower and soybean, show better physiological performance and higher seed yields under salt stress (Peleg and Blumwald 2011).

9.5.2 Role of Gibberellins Under Salt Stress

GAs are hormones found in plants that control a variety of physiological functions, including flowering, stem elongation, and seed germination. Through their impacts on development, ionic balance, osmotic adjustment, and stress-responsive gene expression, they play a crucial role in assisting plants in overcoming the negative effects of salinity during salt stress. One of the most noticeable consequences of salt stress is the suppression of seed germination, which is brought on by impaired water uptake and a rise in the buildup of the hormone ABA, which is hostile to gibberellins. By breaking seed dormancy and encouraging the manufacture of hydrolytic enzymes such α-amylase, which are essential for seed germination, GA*s* reverse these effects (Colebrook et al. 2014). Even in saline environments, GA*s* have been demonstrated to improve seed germination in oilseed crops such as sunflower and canola. Salt stress also prevents roots and shoots from growing because of ionic toxicity and osmotic stress. In these situations, gibberellins are crucial for promoting growth. They encourage shoot elongation, which is necessary to maintain plant growth, by encouraging cell division and elongation in the stem tissues. Gibberellins also promote the growth of lateral roots and increase the growth of primary roots, which helps with root development. Enhancing root development makes it easier for plants to absorb water and nutrients, which helps them survive in saltwater conditions (Table 9.6).

Gibberellins also help plants under salt stress maintain their ionic balance. In plant cells, they help maintain a desirable Na^+/K^+ ratio by controlling the expression of ion transporters such as the SOS1 (Salt Overly Sensitive 1) transporter (Ryu and Cho 2015). This control is essential to avoiding the harmful buildup of ions such as Na^+ and Cl^+ in delicate tissues, which could otherwise disrupt plant function. Gibberellins help in osmotic correction in addition to ionic equilibrium. Gibberellins

Table 9.6 Role of gibberellins in salt stress mitigation

S#	Mechanism	Description	Effect on oilseed crops
1	*Seed germination*	Overcomes ABA-induced dormancy and promotes enzyme synthesis	Improved germination in sunflower and canola
2	*Shoot and root growth*	Stimulates cell elongation and root development	Increased biomass and water uptake
3	*Ion homeostasis*	Regulates Na^+/K^+ transport and compartmentalization	Reduced ionic toxicity in sensitive tissues
4	*Osmotic adjustment*	Enhances proline and sugar accumulation	Maintains turgor and osmotic balance
5	*Oxidative stress mitigation*	Activates enzymatic and nonenzymatic antioxidants	Reduces ROS damage to cellular structures
6	*Stress gene expression*	Upregulates GA-responsive genes and interacts with ABA signaling pathways	Fine-tuned stress adaptation and growth processes
7	*Exogenous application*	Application of GA_3 enhances growth and yield under saline conditions	Increased productivity in soybean and rapeseed

stabilize the cellular osmotic potential, which helps plants maintain cellular integrity in saline environments by promoting the synthesis of compatible solutes like proline and promoting the metabolism of carbohydrates to produce osmoprotectants like soluble sugars and amino acids (Shaddad et al. 2013). Gibberellins also have a crucial role in lowering oxidative stress, which is triggered by salt stress and results in the production of ROS. These ROS have the potential to cause serious oxidative damage to plant cells. Gibberellins help to lessen this stress by activating antioxidant enzymes (Colebrook et al. 2014). They also promote the synthesis of nonenzymatic antioxidants such as ascorbate and glutathione, which further scavenge ROS and mitigate their harmful effects on plants.

By altering the expression of genes that react to stress, gibberellins also increase a plant's resistance to salt stress. By upregulating genes linked to metabolism and development, they lessen the effects of salt stress. Together with ABA, gibberellins balance growth and stress tolerance, enabling the plant to continue growing while simultaneously defending against salt-induced stressors. The optimal response of the organism to stress is maintained by this hormonal interplay (Llanes et al. 2016). Furthermore, it has been demonstrated that oilseed crops under salt stress can benefit greatly from the exogenous injection of gibberellins, such as GA_3. Gibberellins increase crop yield and performance under saline circumstances by increasing biomass production, encouraging shoot elongation, and improving seed filling and oil content. This has proven especially advantageous for crops like rapeseed and soybean, where gibberellin administration has increased yield under salt stress, indicating the hormones' ability to improve crop resilience to difficult growing circumstances (Haq et al. 2023).

Rashed et al. (2017) conducted a study on the alleviation of salt stress by gibberellic acid and rhizobacteria in *Nigella sativa* L. and as a result, they found that salinity treatments adversely affected the growth and yield characters of *Nigella sativa* plants. Under salinity treatments, relative water content, chlorophyll content and fixed oil percentage decreased. However, proline, membrane permeability, enzyme activities, and total soil salts increased. Meanwhile, GA_3 or PGPR treatments alleviated the harmful effects of salinity on the previously declared parameters. GA_3 or PGPR treatments increased proline content and CAT and POD activities, which could be considered as promising mechanisms for alleviating salinity in *Nigella sativa* L. Growth rate improvement by gibberellin may lead to leaf area expansion, stimulation of cell division and/or cell elongation, stimulation of photosynthesis rate, altered partitioning of photosynthetic products, or a combination of these. GA_3-mediated invertase activity in elongating shoots may lead to the accumulation of hexoses, thought to be important for primary cell wall biosynthesis and consequently enhance seedling growth under stress conditions (Rashed et al. 2017). Enhancement of plant growth by GA_3 under salinity stress has been previously reported by Khan et al. (2010) on wheat and *Linum usitatissimum* L (Rashed et al. 2017). According to Kalidas and Rajasekhar (2010), oil palm *Elaeis guineensis* Jacq is a perennial plant with the highest oil yield and in their study, the effect of Gibberellic acid on the weight of Oil palm fruit clusters was evaluated. They reported that the results showed a variable trend in the weight of Oil palm clusters in all treatments 15 days before fruit harvest.

9.5.3 Role of Cytokinins Under Salt Stress

Cell division, the initiation of shoots, and the regulation of plant growth and development are all influenced by plant hormones called cytokinins (CK*s*). Cytokinins help plants adapt to salt stress by changing physiological, biochemical, and molecular processes. Through a variety of processes, including the regulation of ionic balance, antioxidant activity, and stress-related gene expression, they alleviate the unfavorable effects of salt stress. The primary means that salt stress upsets ionic equilibrium in plants is by generating an excessive buildup of Na^+ ions, which can be harmful. For plant cells to maintain a favorable Na^+/K^+ ratio, which is essential for healthy cellular function in saline environments, cytokinins aid in controlling the activity of ion transporters, including K^+ channels and Na^+/H^+ antiporters. Furthermore, by promoting K^+ absorption and increasing the sequestration of Na^+ into vacuoles, cytokinins assist in minimizing ionic toxicity and lessen the negative effects of excessive Na^+ buildup (Yu et al. 2021). By enhancing selective ion transport and compartmentalization, cytokinins also lessen the toxicity of ions in sensitive tissues, especially in the shoots, improving plant survival under salt stress (Table 9.7).

During salt stress, more ROS are created, leading to oxidative stress and damage to biological components. Cytokinins aid in the reduction of oxidative stress by upregulating antioxidant enzymes that are necessary for neutralizing ROS, such as POD, CAT, and SOD. Furthermore, cytokinins protect plant cells from oxidative damage by reducing lipid peroxidation, which stops ROS-induced membrane damage (Abdel Latef et al. 2021a, b; Javid et al. 2011). In addition to their role in oxidative stress defense, cytokinins aid in osmotic adjustment during salt stress. In order to help plants better withstand osmotic stress, cytokinins contribute to the accumulation of osmoprotectants, including proline, soluble sugars, and glycine betaine. These substances stabilize cellular structures and maintain cell turgor. This is because salt reduces the quantity of water that plants can get. Furthermore, cytokinins have an impact on the expression of genes linked to salt stress response.

Table 9.7 Role of cytokinins in salt stress mitigation

S#	Mechanism	Description	Effect on oilseed crops
1	*Ionic homeostasis*	Maintains Na^+/K^+ balance by regulating ion transporters	Reduced ionic toxicity and better nutrient uptake
2	*Antioxidant defense*	Enhances enzymatic and nonenzymatic antioxidant activities	Mitigates oxidative damage caused by ROS
3	*Osmotic adjustment*	Promotes proline and soluble sugar accumulation	Maintains cell turgor and osmotic stability
4	*Gene regulation*	Upregulates stress-responsive genes and interacts with ABA and auxins	Fine-tuned stress adaptation and growth balance
5	*Growth promotion*	Stimulates shoot and root growth under saline conditions	Improved biomass and water absorption
6	*Exogenous application*	Application of CKs like kinetin improves photosynthesis and yield	Increased productivity in oilseed crops

They contribute to the plant's overall stress tolerance by modifying the CK signaling pathway, which is implicated in gene networks linked to oxidative stress defense, ion transport, and osmotic balance. Additionally, cytokinins work in concert with other hormones, including auxins and ABA, to regulate the plant's stress responses and successfully balance growth and stress adaptation (Thu et al. 2017). By encouraging shoot growth, which is frequently inhibited by salt stress, cytokinins promote growth under stressful conditions. They help plants thrive in saltwater environments by encouraging cell proliferation and growth in shoot tissues.

Additionally, cytokinins regulate root architecture and improve water and nutrient uptake in salt-stressed plants by encouraging the growth of lateral roots and root hairs, providing them with the resources they need to survive and thrive (Ryu and Cho 2015). It has been shown that the exogenous treatment of cytokinins such as kinetin and benzyl adenine increases the tolerance of oilseed crops to salt stress. These applications demonstrate how cytokinins improve plant growth under stress by boosting seed yield and biomass output in saline soils. Additionally, cytokinins boost chlorophyll content and photosynthetic efficiency, which aid plants in overcoming stress and boosting yield under trying circumstances (El Sabagh et al. 2022).

Kharkeshi et al. (2023) stated that drought is one of the important limiting factors for crops that seriously reduces plant growth and productivity and the application of CK and silicon (Si) fertilizers can help increase tolerance to drought stress in sesame (*Sesamum indicum* L.) plants and planned and carried out a study in this way. In the present study, they aimed to evaluate the effects of CK and Si fertilizers on seed yield, malondialdehyde (MDA) content, proline content and antioxidant enzyme activities in sesame plants under drought stress conditions. While the highest MDA content and proline accumulation were observed when plants were subjected to SDS (severe drought stress), they stated that higher activity of antioxidant enzymes occurred under MDS (moderate drought stress). It was observed that higher antioxidant enzyme activity and decreased MDA content were observed in plants treated with the combined application of Si and CK under MDS. However, higher seed yield, more proline content, and higher antioxidant enzyme activities were reported from plants treated with nano-Si rather than calcium silicate. Overall, the results of the present study revealed that foliar application of nano-Si + Ck could be a promising option to reduce the adverse effects of drought stress and increase sesame seed yield.

9.5.4 Role of Abscisic Acid Under Salt Stress

A vital plant hormone called ABA controls how plants react to a variety of abiotic stressors, including salt stress. Through the modification of numerous physiological, biochemical, and molecular processes, it plays a crucial role in mediating plant responses to high salt levels. Plant growth and production can be significantly impacted by salt stress, which causes oxidative damage, osmotic stress, and ion imbalance. ABA helps plants cope with a variety of stresses by regulating stomatal closure, promoting osmotic adjustment, altering gene expression, and bolstering antioxidant defense systems. ABA is significant for controlling how plants react to salt stress, especially when it comes to reducing water loss through stomatal closure. Salt stress lowers turgor pressure and dehydrates plant cells in saline environments. By causing stomatal closure and instructing guard cells to seal stomata, ABA lessens these effects by lowering transpiration and preventing additional water loss from the leaves (Cen et al. 2020). During times of salt stress, this procedure improves water retention and aids in preserving the plant's water balance. ABA is important for water intake and osmotic adjustment in addition to controlling water loss. To preserve cellular functioning and to retain water within cells, ABA encourages the buildup of osmolytes such as proline, sugars, and betaines during salt stress (Dar et al. 2017). In saline settings, this osmotic control is essential for maintaining cellular turgor and permitting water intake. Additionally, ABA promotes root development, which improves water absorption and the ability of plants to absorb water from saline situations (Table 9.8).

An imbalance between Na^+ and K^+ ions is frequently caused by salt stress, and this might interfere with cellular processes. By controlling the movement of Na^+ and K^+, ABA aids in the restoration of ion homeostasis. According to Munns and Tester (2008), ABA reduces toxicity via modifying the expression of Na^+/H^+ antiporters and Na^+ channels, which aid in excluding Na^+ ions from the cytoplasm or sequestering them into vacuoles. Additionally, by boosting the expression of K^+ transporters, ABA encourages K^+ absorption, improving the Na^+/K^+ ratio and being crucial for preserving healthy cellular activities under salt stress (Gómez-Cadenas et al. 2002). Additionally, ABA is essential for controlling the expression of genes that respond to stress. It helps to lessen oxidative stress brought on by salt by activating genes involved in antioxidant defense (Cen et al. 2020). Furthermore,

Table 9.8 Role of abscisic acid in salt stress adaptation

S#	Mechanism	Description	Effect on plant under salt stress
1	*Stomatal regulation*	Induces stomatal closure to reduce water loss	Conserves water by limiting transpiration
2	*Osmotic adjustment*	Promotes the accumulation of osmoprotectants like proline and sugars	Maintains cell turgor and prevents dehydration
3	*Ion homeostasis*	Regulates Na^+/K^+ balance through ion transporters and antiporters	Prevents Na^+ toxicity and maintains essential ion balance
4	*Gene expression*	Activates genes for antioxidant enzymes, stress proteins, and protective molecules	Reduces oxidative damage and enhances stress tolerance
5	*Seed germination*	Inhibits premature germination under saline conditions	Prevents seedling death by controlling germination timing
6	*Root growth*	Stimulates root elongation and lateral root formation under salt stress	Enhances water and nutrient uptake in saline soil
7	*Hormonal cross-talk*	Interacts with other hormones like auxins, gibberellins, and cytokinins	Balances growth and stress adaptation

ABA encourages the production of defense proteins that assist plants in dealing with environmental stress, such as heat shock proteins and dehydrins (Swamy and Smith 1999). The control of seed dormancy and germination in saline environments is one of the best-known functions of ABA. By encouraging seed dormancy, ABA suppresses seed germination and avoids early germination under adverse circumstances (Finkelstein 2013). By ensuring that seeds only sprout when there is an adequate supply of water for growth, this regulator aids in coordinating seed germination with environmental factors. For salt tolerance, ABA also affects root development, which is important because roots absorb water and vital nutrients like Na^+ and K^+ ions. When plants are experiencing stress via salt, ABA encourages the growth of deeper and more branched roots, which are better able to withstand the depletion of soil moisture caused by salinity. ABA then interacts with other plant hormones like auxins, gibberellins, and cytokinins to fine-tune the plant's response to salt stress. For instance, during salt stress, auxins and ABA cooperate to control stomatal closure and root development (Tuteja 2007). Furthermore, gibberellin production is inhibited by ABA, which lowers plant growth and encourages stress adaptation (Zhu 2001). Additionally, during stress, ABA counteracts cytokinin action to prevent cell division and development, directing resources toward survival rather than growth (Yu et al. 2021). These interactions show how ABA plays a key role in coordinating many physiological processes in the plant's response to salt stress.

Nitric oxide (NO) mediates posttranslational modification of target proteins via S-nitrosylation and nitration. Under water-deficit conditions induced or not by salt stress, ABA induces NO and ROS synthesis (Niu and Liao 2016). Salt stress was found to promote significant increases in lipid peroxidation and H_2O_2 content at all analyzed time points and plant tissues; however, NO pretreatment effectively alleviated oxidative damage and ROS accumulation in *Jatropha curcas* seedlings (Pompelli et al. 2022). Gadelha et al. (2017) reported that NO-pretreated *J. curcas* seedlings significantly alleviated salt stress due to low toxic ion and ROS accumulation through strong antioxidant systems, resulting in high growth of NO-treated seedlings under saline conditions. Differential proteomic analysis data revealed that the tolerance-like genotype showed proteins of different pathways related to salinity response, including the production of antioxidant enzymes and signaling and stress regulation pathways, especially high sensitivity and reactivity to ABA (Pompelli et al. 2022). While it is well documented that plant species using high-efficiency antioxidant systems have improved tolerance to salt stress, this response can also be signaled by some small molecules such as H_2O_2, oligochitosan, NaCl, proline, ABA and NO, as documented by Gadelha and Miranda (2017).

9.5.5 Role of Ethylene Under Salt Stress

A key plant hormone, ethylene, controls a number of physiological and biochemical functions in plants. Ethylene is a key mediator in how plants react to environmental stress and influences how plants grow, develop, and cope with stress when exposed to salt. The generation of ethylene is caused by salt stress, which causes osmotic and ionic stress. Then, to minimize harm and increase survival, this ethylene controls plant reactions. Plants' adaptive response to salinity includes a considerable rise in ethylene synthesis when they are under salt stress. 1-Aminocyclopropane-1-carboxylate (ACC) synthase is a crucial enzyme in the production of ethylene and is activated in response to salt stress. The precursor of ethylene, 1-aminocyclopropane-1-carboxylate (ACC), is produced when S-adenosylmethionine (SAM) is converted by this enzyme. Higher ACC levels cause plants to produce more ethylene, which is crucial for controlling some stress-responsive pathways and assisting plants in reducing the adverse impacts of salinity (Ku et al. 2018). The control of stomatal closure is one of ethylene's primary responses to salt stress. Plant survival depends on regulating transpiration since salt stress causes water loss and dehydration. Under salt stress, ethylene encourages stomatal closure, which reduces water loss and helps retain moisture. Under conditions of extreme salinity, where dehydration can significantly impair plant viability, this function is especially important. Ethylene interacts with other hormones, including ABA, to further strengthen its role in stomatal control. To minimize water loss and stop more desiccation, ethylene and ABA work in concert to ensure that stomata seal effectively (Riyazuddin et al. 2020; Amjad et al. 2014).

Furthermore, ethylene has a crucial role in root formation, which is necessary for plants under salt stress. Their ability to absorb water and essential nutrients from saline soils is greatly influenced by the size and form of their roots. Ethylene influences root architecture by promoting the formation of lateral roots and inhibiting primary root elongation. This leads to a wider root system, which enhances the plant's ability to take in water and nutrients in salinized conditions (Tao et al. 2015). Likewise, ethylene promotes the development of root hair, which enhances the plant's ability to absorb water and essential nutrients, especially in salty environments where nutrient absorption is limited. Maintaining ion homeostasis is one of ethylene's other crucial roles in salt stress. Salt stress, which results in an imbalance between toxic Na^+ ions and essential K^+ ions, commonly disrupts cellular activities. Ethylene regulates ion homeostasis by changing the expression of ion transporters and channels that control the uptake and transport of Na^+ and K^+. By boosting the activity of the Na^+/H^+ antiporter, which helps sequester Na^+ ions into vacuoles, it lessens the negative effects of Na^+ in the cytoplasm. Additionally, ethylene encourages the uptake of K^+, which helps plant cells maintain the Na^+/K^+ balance necessary for healthy cell activity and osmotic regulation (Tao et al. 2015). ROS build up under salt stress, causing oxidative damage to biological components. By triggering antioxidant defense systems, ethylene reduces oxidative stress. It increases the synthesis of antioxidant enzymes that neutralize ROS and stop oxidative cell damage (Munns and Tester 2008). Furthermore, ethylene helps plants better handle oxidative stress when exposed to salt by participating in redox signaling pathways that control the expression of genes involved in stress responses (Zhang et al. 2024). Ethylene interacts with several other plant hormones to regulate the plant's overall response to salt stress. For example, ABA and ethylene work together to increase salt tolerance; ABA regulates stomatal closure, and ethylene ensures that

Table 9.9 Role of ethylene in salt stress tolerance

S#	Mechanism	Description	Effect on plant under salt stress
1	*Ethylene biosynthesis*	Increased production of ethylene in response to salt stress	Ethylene mediates various stress responses such as stomatal closure and ion regulation
2	*Stomatal regulation*	Ethylene promotes stomatal closure to reduce water loss	Helps conserve water and prevent dehydration
3	*Root development*	Ethylene inhibits primary root growth and promotes lateral root and root hair formation	Enhances root system architecture to better absorb water and nutrients
4	*Ion homeostasis*	Regulates Na^+/K^+ balance by modulating ion transporter activity	Helps prevent Na^+ toxicity and maintain cellular ion balance
5	*Oxidative stress mitigation*	Upregulates antioxidant enzymes such as SOD, CAT, and POD	Reduces oxidative damage caused by ROS accumulation under salt stress
6	*Hormonal interactions*	Interacts with ABA, cytokinins, and gibberellins to fine-tune plant responses to salt stress	Coordinates growth and stress tolerance mechanisms
7	*Gene expression*	Ethylene modulates gene expression related to stress tolerance and growth	Enhances stress resistance and survival under saline conditions

water loss is minimized during salt stress (Amjad et al. 2014). Additionally, ethylene counteracts the actions of cytokinins, which typically encourage cell growth and division. Ethylene restricts excessive growth under salt stress, focusing resources on stress tolerance instead of growth (Sun et al. 2021). Ethylene also inhibits the production of gibberellins, which are hormones that promote growth. By slowing down growth, this decrease in gibberellin levels prioritizes stress adaptation over growth, which helps conserve energy and resources (Tao et al. 2015) as given in Table 9.9.

Cebrián et al. (2021) studied the effect of three gain-of-function mutations affecting the ethylene receptors CpETR1B, CpETR1A, and CpETR2B of *Cucurbita pepo* on the salt stress response during germination, seedling establishment and subsequent vegetative growth of plants. They showed that all mutations reduced ethylene sensitivity but increased salt tolerance during both germination and vegetative growth, suggesting a positive role of the three ethylene receptors in salt tolerance. They reported that under salt stress, etr1b, etr1a, and etr2b germinated earlier than WT, and root and shoot growth rates of both seedlings and plants were less affected in the mutant than in WT. The enhanced salt tolerance response of etr2b plants was associated with reduced Na + accumulation in shoots and leaves as well as increased accumulation of compatible solutes, including proline and total carbohydrates and antioxidant compounds such as anthocyanins. They noted that their data suggest that ethylene is a negative regulator of salt tolerance during germination and vegetative growth. Nevertheless, they noted that higher up-regulation of genes involved in Ca2+ signaling (CpCRCK2A and CpCRCK2B) and ABA biosynthesis (CpNCED3A and CpNCED3B) in etr2b leaves under salt stress, possibly indicating that the function of ethylene receptors in the salt stress response in *C. pepo* may be mediated by Ca^{2+} and ABA signaling pathways.

9.5.6 Role of Jasmonic Acid Under Salt Stress

A plant hormone called jasmonic acid (JA) is crucial for controlling how plants react to biotic and abiotic stressors, such as salt stress. Plant growth, development, reproduction, and reactions to environmental stressors are among the physiological and biochemical processes that are regulated by JA and its derivatives, collectively referred to as jasmonates. Through its modulation of growth regulation, ion homeostasis, stress-responsive genes, and antioxidative responses, JA plays a crucial role in changing the plant's capacity to withstand excessive salinity during salt stress. Plants under salt stress create JA, which is a component of a complex signaling network that helps plants respond to environmental stressors. Through a series of metabolic events involving lipoxygenases (LOX) and allene oxide synthase (AOS), the octadecanoid pathway converts linoleic acid into JA. Because JA and its methyl ester, methyl jasmonate (MeJA), accumulate during salt stress and function as signaling molecules that activate stress-responsive pathways, plants can adapt to salinity (Salehin 2024). JA controls ion homeostasis, which is necessary for plants to withstand salt stress. The imbalance in ionic concentrations brought on by salt stress, especially the accumulation of sodium ions (Na^+), may interfere with cellular activity. JA helps to maintain ion balance by influencing the activity of ion transporters, including Na^+/H^+ antiporters, which help sequester Na + ions into vacuoles and reduce their cytoplasmic toxicity. Additionally, JA regulates potassium (K^+) channel activity, which helps to maintain the Na^+/K^+ balance and encourages K^+ absorption. Osmotic control in saline settings and proper cellular function depend on this equilibrium (Salehin 2024) as given in Table 9.10.

By improving ion homeostasis through these processes, JA helps plants resist ionic stress brought on by too much salt. Oxidative stress, which is caused by the accumulation of ROS under salt stress, is another challenge that plants face. JA contributes significantly to lowering oxidative damage by triggering the plant's antioxidant defense system. It raises the production of antioxidant enzymes, which scavenge ROS and protect cellular constituents from oxidative damage. Additionally, JA boosts the plant's defenses against oxidative stress and ROS by increasing the expression of genes linked to the glutathione-ascorbate cycle (Singh and Gautam 2013). One important adaptive response to salt stress is stomatal control, and JA plays a major role in this process. To preserve water balance in saline environments, JA helps minimize water loss through transpiration by encouraging stomatal closure. ABA, another hormone involved in stomatal control, frequently coordinates this response. In order to ensure efficient stomatal closure in response to elevated salinity, JA and ABA cooperate to trigger the expression of genes that control guard cell behavior (Kim et al. 2021; Li et al. 2019). Plant life under salt stress depends on root development since roots are responsible for absorbing water and

Table 9.10 Role of jasmonic acid in salt stress tolerance

S#	Mechanism	Description	Effect on plant under salt stress
1	*Jasmonic acid production*	Increased JA biosynthesis under salt stress through the octadecanoid pathway	Activates stress-responsive genes and stress tolerance mechanisms
2	*Ion homeostasis regulation*	Modulates Na^+/K^+ balance through ion transporters and channels	Maintains cellular ion balance, preventing Na^+ toxicity
3	*Oxidative stress mitigation*	Activates antioxidant enzymes (SOD, CAT, POD) and the glutathione-ascorbate cycle	Reduces ROS accumulation and oxidative damage
4	*Stomatal regulation*	Promotes stomatal closure under salt stress, reducing water loss	Helps conserve water and maintain hydration
5	*Root architecture regulation*	Inhibits primary root elongation but promotes lateral root formation	Improves nutrient and water uptake under saline conditions
6	*Gene expression*	Upregulates stress-responsive genes (e.g., HSPs, WRKY, MYC transcription factors)	Enhances salt tolerance by activating stress-related pathways
7	*Hormonal interactions*	Synergizes with ABA for stress tolerance, antagonizes cytokinin effects	Balances growth and stress responses

nutrients from the soil. JA influences root architecture by regulating both lateral root development and primary root elongation. When under salt stress, JA negatively regulates primary root elongation, which helps plants conserve energy and resources for stress tolerance. Meanwhile, JA promotes the development of lateral roots, allowing the plant to create a more extensive root system that enhances nutrient and water uptake in salinized environments (Bali et al. 2018; Vanderauwera et al. 2012). JA also controls the expression of several stress-responsive genes, including those that encode heat shock proteins (HSP*s*), enzymes involved in proline biosynthesis, and stress-related transcription factors like WRKY and members of the MYC family. By triggering these genes, JA helps the plant respond effectively to oxidative damage and osmotic stress caused by salt (Wang et al. 2021; Alisofi et al. 2020). Through its interactions with other plant hormones, JA regulates the plant's response to salt stress. Thus, JA and ABA often work together to enhance stress responses such as stomatal closure and antioxidant defenses, which in turn improve the plant's ability to store water and tolerate salinity (Hoque et al. 2021). However, JA inhibits the effects of cytokinins under stressful conditions, inhibiting excessive growth and redirecting resources away from growth and toward stress tolerance (Delgado et al. 2021).

Sheteiwy et al. (2021) reported that priming and foliar application of JA and/or their combination significantly improved the water potential, osmotic potential, water use efficiency and relative water content of both cultivars under salinity stress. Similarly, seed priming with JA, foliar application of JA and/or their combination significantly improved the following traits over untreated soybean seedlings under salinity stress, namely net photosynthesis rate was 68.03%, 59.85% and 76.67%, respectively; transpiration rate was 74.85%, 55.10% and 80.26%, respectively; stomatal conductance was 69.88%, 78.25% and 26.24%, respectively; intercellular carbon dioxide concentration was 61.64%, 40.06% and 65.79%, respectively; and total chlorophyll content increased by 47.41%, 41.02% and 55.73%, respectively. Soybean (*Glycine max* L.) plants primed, sprayed or treated with a combination of JA increased chlorophyll fluorescence, which was damaged by salinity stress. JA applications improved the levels of abscisic acid, gibberellic acid and JA under salt stress by 60.57%, 62.50% and 52.25%, respectively, compared to those in the control condition. The transcriptional levels of FeSOD, POD, CAT and ascorbate peroxidase (APX) genes were significantly increased in seedlings under NaCl stress regardless of JA applications. In addition, JA application resulted in a decrease in sodium ion concentration and an increase in potassium ion concentration in leaves and roots of both cultivars, regardless of salinity stress. They reported that monodehydroascorbate reductase, dehydroascorbate reductase and proline contents decreased in JA-treated seedlings under salinity stress, while ascorbate content increased by combining JA application with NaCl stress.

9.5.7 Role of Salicylic Acid Under Salt Stress

A plant hormone called salicylic acid (SA) is essential for controlling how plants defend against biotic and abiotic stressors, such as salt stress. SA alters metabolic pathways and gene expression to affect a number of physiological processes, such as growth, development, and the stress response. By controlling ion homeostasis, lowering oxidative damage, encouraging stress-related gene expression, and interacting with other plant hormones, including JA and ABA, SA improves salt tolerance under salt stress. When plants experience salt stress, they increase their production of SA as a coping mechanism. SA biosynthesis begins with phenylalanine, which is produced by the shikimate pathway. Phenylalanine is converted to SA by enzymes such as isochorismate synthase (ICS) and phenylalanine ammonia-lyase (PAL). The genes that code for these enzymes are expressed more when plants are under salt stress, which boosts the levels of SA. This increase in SA synthesis helps plants overcome the detrimental effects of too much salt by triggering downstream stress-responsive pathways. Salicylic acid is necessary to maintain ion homeostasis, which is necessary for plants to tolerate salt stress. Salt stress-induced imbalance, particularly the dangerous accumulation of Na^+ ions, can disrupt biological functions. SA controls this imbalance by facilitating the sequestration of Na^+ into vacuoles through the activation of Na^+/H^+ antiporters in vacuolar membranes. This action reduces the negative effects of Na^+ in the cytoplasm and helps maintain the essential Na^+/K^+ balance required for cellular function. Additionally, SA encourages the consumption of K^+ ions (which are necessary to maintain cellular osmotic equilibrium and turgor pressure in saline situations (Singh and Gautam 2013; Haq et al. 2023).

In this way, SA lowers the ion toxicity caused by salt stress, which helps the plant tolerate salt more generally. Salt stress also produces ROS, which can oxidatively damage lipids, proteins, and DNA. SA plays a key role in reducing oxidative stress by upregulating antioxidant enzymes such glutathione reductase (GR), POD, CAT, and SOD, which scavenge ROS and prevent cellular damage. Furthermore, SA triggers the glutathione-ascorbate cycle, an essential antioxidant process that aids in ROS detoxification. By bolstering the antioxidant defense system, SA protects plant cells from oxidative stress, which is particularly important when salt stress is present (Ahmad et al. 2020; Emamverdian et al. 2020). In reaction to salt stress, plants can reduce transpiration water loss by closing their stomata. SA regulates the motility of stomata by interacting with ABA, another hormone that promotes stomatal closure. SA and ABA work together to initiate signaling pathways in guard cells during salt stress, which leads to stomatal closure and water conservation (Khan et al. 2012). Additionally, via modifying ion channels in guard cells, SA induces the efflux of K^+ and promotes the closure of stomata in response to salt stress (Wang et al. 2021). SA regulates stomatal closure in saltwater environments, assisting plants in maintaining hydration and conserving water. Salicylic acid is also necessary for regulating the expression of genes that react to stress. It promotes the synthesis of a number of transcription factors, including WRKY, MYB, and NAC, which are implicated in ion transport, antioxidant defense, and stress tolerance.

Moreover, SA encourages the synthesis of proline, an osmoprotectant that stabilizes proteins and cell structures under salt stress, hence preserving cellular integrity (Tao et al. 2015; Khan et al. 2022). By altering the expression of genes that react to stress, SA increases the plant's resistance to oxidative damage, osmotic stress, and salt-induced ion toxicity. Additionally, salicylic acid coordinates a comprehensive response to salt stress by interacting with GA, JA, and ABA, among other plant hormones. In order to improve stomatal closure and conserve water, SA and ABA frequently work in concert, ensuring that the plant can compensate for dehydration when exposed to salt stress (Khan et al. 2012). SA also modifies the effects of JA to balance stress tolerance and growth inhibition. The plant's response to salt stress is optimized by SA's balanced expression of defense-related genes, while JA is known to activate these genes and limit development (Khan et al. 2012). Together with other hormones, SA helps the plant develop a unified defense against the different dangers that come with high salt (Table 9.11).

Farhangi-Abriz and Ghassemi-Golezani (2016) conducted a factorial randomized complete block design-based greenhouse experiment with four replications in 2015 to investigate the effects of salicylic acid (SA) (1 mM) and jasmonic acid (JA) (0.5 mM) on protein accumulation and amino acid composition of soybean (*G. max*) seeds at different salinity levels (0, 4, 7, and 10 dS/m NaCl, respectively). SA treatment improved the protein percentage and amount per seed at different stages of seed development at all salinity levels. The highest protein yield was recorded for SA-treated plants due to higher seed yield and protein content. In

Table 9.11 Role of salicylic acid in salt stress tolerance

S#	Mechanism	Description	Effect on plant under salt stress
1	*Biosynthesis*	Increased SA biosynthesis via the shikimate pathway in response to salt stress	Activates stress-responsive pathways and enhances salt tolerance
2	*Ion homeostasis*	Regulates Na^+/K^+ balance and ion transporters under salt stress	Reduces Na^+ toxicity and maintains cellular osmotic balance
3	*Antioxidative defense*	Upregulates antioxidant enzymes (SOD, CAT, POD) and the glutathione-ascorbate cycle	Detoxifies ROS and reduces oxidative damage
4	*Stomatal regulation*	Modulates stomatal closure via interaction with ABA	Conserves water and reduces dehydration
5	*Gene expression*	Activates stress-responsive genes (WRKY, MYB, NAC) and proline biosynthesis genes	Enhances salt tolerance through the activation of protective pathways
6	*Hormonal interactions*	Synergizes with ABA and interacts with JA for coordinated stress response	Balances growth inhibition and stress tolerance mechanisms

contrast, JA treatment reduced seed protein content and yield, resulting in decreased seed yield. With the increase of salinity, the contents of lysine, methionine, phenylalanine, threonine, aspartic acid, glutamic acid, proline and tyrosine increased, while the contents of leucine and valine decreased. Foliar spray of SA improved the contents of isoleucine, leucine, lysine, methionine, valine, alanine, aspartic acid, glutamic acid, glycine and serine, but JA treatment increased the sulfur-containing amino acids such as methionine and aromatic amino acids such as phenylalanine and tyrosine in soybean seeds. It was concluded that the treatment with SA had the greatest effect on improving the protein quantity and quality at different salinity levels.

9.5.8 Role of Melatonin Under Salt Stress

Initially, it was discovered that a little bioactive chemical called melatonin regulated the sleep-wake cycles of mammals. It now plays a number of physiological functions in plants, particularly in response to abiotic stresses like salt. Melatonin, a strong antioxidant, is crucial for controlling plants' stress tolerance mechanisms when they are subjected to extreme weather conditions, such salt stress. To mitigate the adverse effects of elevated salinity, it reduces oxidative stress, improves cellular homeostasis, and regulates plant growth processes. Melatonin is crucial for boosting a plant's tolerance to salt stress because it is a strong antioxidant. When salt stress occurs, ROS are produced, including hydroxyl radicals, H_2O_2, and superoxide radicals. Oxidative stress may result from this, endangering specific cell components. In these situations, melatonin strengthens the antioxidant defenses of the plant It increases the activity of key antioxidant enzymes that help scavenge ROS and prevent oxidative damage to cellular components like lipids, proteins, and DNA, including glutathione peroxidase (GPX), POD, CAT, and SOD (Ye et al. 2012). Melatonin also enhances the ascorbate-glutathione cycle, an essential antioxidant pathway that detoxifies ROS and maintains cellular redox balance during salt stress (Sun et al. 2021). Therefore, melatonin plays a crucial role in protecting plant cells from the harm caused by oxidative stress caused by too much salt. Another significant function of melatonin in salt stress is its regulation of osmotic balance and ion homeostasis. An imbalance in ion distribution, particularly a buildup of Na^+ ions, is usually caused by salt stress and can interfere with cellular functions. By controlling the Na^+/K^+ ratio, which is essential for cellular activity in salinity, melatonin aids plants in maintaining ion balance. It increases ion transporter function, including K^+ transporters and Na^+/H^+ antiporters, guaranteeing that K^+ is absorbed in adequate levels to preserve cellular homeostasis and that Na^+ is sequestered into vacuoles (Li et al. 2019).

Melatonin also aids in osmotic adjustment by encouraging the accumulation of osmoprotectants such as proline and carbohydrates, which stabilize proteins and protect cellular structures from salt-induced dehydration. Under salt stress, this also helps the plant retain water and increase its water-use efficiency (Cen et al. 2020). These actions improve the plant's ability to withstand the osmotic stress caused by high salinity. Salt stress hurts plant growth and stunts development by inhibiting cell division and elongation. Melatonin helps mitigate these effects by promoting root growth and maintaining leaf and shoot development. By encouraging root elongation and lateral root development, it enhances water and nutrient uptake in saline environments (Cen et al. 2020). Furthermore, GA and IAA signaling genes, which govern shoot and leaf growth, are modulated by melatonin, which controls growth pathways. Melatonin helps avoid salt's inhibitory effects on leaf expansion and shoot elongation by maximizing these pathways (Li et al. 2019). Even under situations of salt stress, plants can sustain improved growth and production due to this growth regulation. Also, melatonin influences the production of stress-responsive genes, strengthening the plant's defenses against challenges caused by salt. It causes the synthesis of transcription factors such WRKY, bZIP, and MYB, which regulate other genes related to salt tolerance, including those for antioxidant defense, ion transport, and osmotic regulation (Li et al. 2019).

Further, melatonin stimulates the expression of genes involved in proline biosynthesis, including P5CS (Δ1-pyrroline-5-carboxylate synthase), which facilitates proline buildup as an osmoprotectant during salt stress (Parveen et al. 2021). By altering the transcriptional response to salt stress, melatonin increases a plant's overall tolerance to salt and ability to tolerate high salinity. Furthermore, melatonin interacts with several other phytohormones, including GA, CK, and ABA, to coordinate the plant's response to salt stress. Melatonin often works with ABA, a stress hormone that promotes stomatal closure, to reduce water loss during salt stress. This compound improves the plant's ability to store water and prevent dehydration (Li et al. 2019). Meanwhile, melatonin affects the auxin signaling pathway during salt stress, which regulates root growth and development and helps to optimize root architecture for better nutrient and water uptake

Table 9.12 Role of melatonin in salt stress tolerance

S#	Mechanism	Effect	Outcome under salt stress
1	*Antioxidant defense*	Increases the activity of antioxidant enzymes (SOD, CAT, POD, GPX)	Reduces ROS accumulation and oxidative damage
2	*Ion homeostasis*	Enhances Na^+/K^+ ratio and osmoprotectants accumulation (proline, sugars)	Maintains osmotic balance and reduces ion toxicity
3	*Growth promotion*	Stimulates root elongation, lateral root formation, and shoot growth	Improves growth and development in saline environments
4	*Gene expression*	Activates stress-responsive transcription factors (MYB, WRKY, bZIP) and proline biosynthesis genes	Enhances stress tolerance mechanisms through gene regulation
5	*Hormonal interactions*	Synergizes with ABA, modulates auxins and cytokinins	Coordinated regulation of water conservation and growth

(Vishwakarma et al. 2017). Additionally, melatonin interacts with cytokinins, which are hormones that regulate cell division and differentiation, to help plants thrive even in hostile conditions like salt (Li et al. 2019). By coordinating numerous development and stress response pathways through these hormonal connections, melatonin improves plants' resistance to salt stress (Table 9.12).

Bakyani et al. (2022) reported that salinity affects the yield and quality of oilseed products. They conducted a study in which they evaluated the effects of a single foliar application of melatonin (MEL) solutions at different concentrations (0, 30, 60, or 90 μM) to camelina (*Camelina sativa* (L.) Crantz) plants grown in soil in a greenhouse and irrigated at four salinity levels (0.5, 4, 8, and 16 dS m^{-1}). Increased salinity decreased leaf chlorophyll and photosynthetic rates, decreased K concentrations in roots and shoots, increased Na concentrations, and increased oxidative marker levels and activity of protective antioxidant enzymes in leaves. Under severe salinity stress, MEL90 application caused increases in chlorophyll, gas exchange qualities, leaf antioxidant enzyme activities and decreases in leaf oxidative markers and Na. Salinity reduced seed yield and they noted that no seed was produced at salinities above 8 dS m^{-1}. MEL90 application caused increases in seed yield and poly- and mono-unsaturated fatty acid contents and decreases in saturated fatty acid contents. They stated that the MEL90 application was more effective in alleviating salinity effects than lower MEL concentrations. The highest K and K/Na ratio concentrations were observed with MEL90 application under nonstress conditions. They stated that the data suggest that MEL foliar applications may increase salinity stress tolerance in camelina.

9.6 Melatonin: A Plant Master Regulator/An Emerging New Hormone

Apart from its antioxidant properties, melatonin is now recognized as a significant phytohormone that controls various physiological processes and boosts a plant's ability to withstand abiotic stresses such as heavy metals, drought, salinity, and severe temperatures. Although melatonin's function in controlling animal circadian rhythms has long been understood, it has now come to light as a crucial regulator in plants, especially in response to a range of environmental stressors. The several roles that melatonin plays in plants are discussed in this section, along with its growing importance as a "master regulator" of plant stress responses. The importance of melatonin in regulating plant stress responses, especially under adverse environmental conditions, is increasingly recognized. Unlike mammals, plants are sessile creatures that are always subject to a range of environmental stressors, including drought, salt, extremes in temperature, and heavy metals. To cope with these challenges, plants have evolved complex signaling systems, and melatonin plays a crucial role in regulating these stress responses. Melatonin has a wide range of functions, including regulating ion homeostasis, boosting antioxidant defense systems, and changing the expression of genes related to stress tolerance.

Additionally, by controlling ion transporters, namely the Na^+/K^+ balance, melatonin enhances ion homeostasis under stress. This is important for preserving cellular integrity during drought and salinity stress (Sheikhalipour et al. 2022). Additionally, osmotic control, ion transport, and stress-responsive transcription factors, including MYB, WRKY, and bZIP are all essential for plant adaptation to abiotic stress, and melatonin affects their expression. In recent years, melatonin's function in improving plants' resistance to abiotic stressors has attracted a lot of interest. Melatonin helps lessen the consequences of salt stress in a number of ways, including osmotic stress, ionic toxicity, and oxidative damage. In order to sustain cellular functioning, it controls ion transport to prevent excessive Na^+ accumulation and promotes K^+ absorption (Sheikhalipour et al. 2022). Melatonin improves the plant's antioxidant capacity, lowering ROS levels and shielding cellular constituents from oxidative damage in addition to controlling ion balance (Li et al. 2022). Similarly, melatonin promotes root elongation during drought stress, which increases the plant's ability to absorb water (Li et al. 2022). Through its interaction with ABA, it also controls stomatal closure, which minimizes water loss and aids in the plant's moisture conservation at times when water supplies are scarce. Melatonin helps plants respond to temperature stress by shielding them from extremes of heat and cold. By controlling osmotic regulators and antioxidant systems, it improves cold tolerance and protects plant cells from damage caused by cold by modulating HSP*s* and other stress-related proteins (Amjad et al. 2014). Melatonin functions as a chelating agent under heavy metal stress, attaching itself to harmful metals like lead, cadmium, and arsenic to decrease absorption and lessen oxidative damage from these metals (Ahmad et al. 2021). Melatonin interacts with several distinct phytohormones to help plants respond to stress in a coordinated manner. In response to drought and salt stress, the hormones cooperate to control stomatal closure, assisting plants in conserving water. When it comes to abscisic acid (ABA), this is one of the most significant interactions (Ahmad et al. 2021). Additionally, melatonin promotes root development by modulating IAA signaling pathways, which improves the plant's capacity to absorb nutrients and water, particularly when it is under salt stress (Amjad et al. 2014). Meanwhile, melatonin promotes tissue growth and cell division by interacting with cytokinins (CK*s*), which helps plants grow better under stress (Sun et al. 2021).

Additionally, it controls GA biosynthesis, which affects plant growth by regulating root development and shoot elongation, therefore improving stress tolerance (Zhang et al. 2016). Melatonin is becoming a viable tool in sustainable agriculture because of its many advantages. Under unfavorable environmental circumstances, it may be used to increase agricultural productivity. Melatonin, for example, has been demonstrated to enhance crop yields and stress tolerance when applied as a foliar spray (Huang et al. 2022). Melatonin treatment of seeds before planting might increase germination rates and the vigor of seedlings, particularly under stressful circumstances. Melatonin is a useful tool for sustainable agricultural operations since it can be added to soil as a soil amendment to promote soil health and increase plant resilience to biotic and abiotic stresses.

9.7 Exogenous Application of Phytohormones

A tried-and-true method in contemporary agriculture for boosting plant resistance, growth, and productivity, especially in the context of environmental stressors, is exogenous phytohormone therapy. Organic substances called phytohormones, sometimes referred to as plant hormones, control a number of physiological functions in plants, including growth, development, and reaction to environmental stimuli. Exogenously, or externally, these hormones can affect a plant's metabolic pathways and improve its resistance to environmental stressors such as infections, salt, drought, and severe temperatures.

1. *Phytohormones and Their Role in Plant Growth and Development:* Plant functions such as seed germination, root and shoot growth, blooming, and fruiting are all regulated by phytohormones. They are produced naturally by plants, but their exogenous application can enhance their effects. Here is a brief overview of the most commonly used phytohormones in agriculture.

Table 9.13 Application of phytohormones in enhancing plant stress tolerance

S#	Method	Description	Applications
1	*Foliar spray*	Phytohormones are applied directly to the leaves, allowing quick absorption and response.	Effective for short-term stress management and growth promotion.
2	*Soil application*	Phytohormones are applied to the soil where they are absorbed by the plant roots.	Best for long-term growth enhancement and root development.
3	*Seed priming*	Seeds are treated with phytohormones before planting to enhance germination and early stage growth.	Improves seedling vigor, root development, and stress tolerance.
4	*Root dipping*	Phytohormones are applied by dipping plant roots into a solution of the hormone before transplanting.	Useful for transplanting stressed seedlings to improve root establishment.
5	*Hydroponic application*	Phytohormones are added to nutrient solutions in hydroponic systems.	Used in controlled environments, promotes growth, and stress resilience.

9.8 Methods of Exogenous Phytohormone Application

Depending on the hormone type and the intended stress tolerance effect, there are various ways to apply exogenous phytohormones to plants (Table 9.13). Root dipping, seed priming, and foliar spraying are common application techniques. Hormones can be directly applied to plant leaves through foliar spraying, where they can be rapidly absorbed and act. Seed priming is the process of applying phytohormones to seeds before planting to enhance germination and seedling vigor. Another technique for delivering hormones directly to the root system is root dipping, which can enhance nutrient and water intake and encourage root growth. The type of phytohormone being employed and the particular stress circumstances determine how effective each strategy is.

9.9 Conclusion

Salinity stress is a major threat to worldwide oilseed crop yield, demanding novel techniques to improve plant resilience. Phytohormones are key regulators of stress perception and adaptive responses, influencing a variety of physiological, biochemical, and molecular pathways to reduce salinity-induced damage. The delicate interplay of hormones such as ABA, SA, JA, ET, CKs, GAs, and auxins allows oilseed plants to fine-tune their growth-defense balance, maintain ionic, and osmotic equilibrium, and activate antioxidant systems in the face of salt stress. While substantial progress has been made in comprehending these hormonal networks, further study is needed to determine the exact signaling cascades and how they interact with other stress-responsive pathways. Using phytohormonal information with biotechnological techniques, hormone priming, and breeding tactics has enormous promise for generating salt-tolerant oilseed cultivars. Finally, incorporating phytohormonal insights into crop enhancement projects may help to promote sustainable agriculture and ensure food and oil security in salt-affected areas.

References

Abdel Latef AAH, Kordrostam M, Zakir A, Zaki H and Saleh OM (2019a) Eustress with H_2O_2 facilitates plant growth by improving tolerance to salt stress in two wheat cultivars. Plants 8(9):303. (IF. 2.632).

Abdel Latef AAH, Mostafa MG, Rahman MM, Abdel-Farid AB, Tran L-SP (2019b) Extracts from yeast and carrot roots enhance maize performance under seawater-induced salt stress by altering physiobiochemical characteristics of stressed plants. J Plant Growth Regul 38:966–979. (IF. 2.073).

Abdel Latef AA, Akter A, Tahjib-Ul-Arif M (2021a) Foliar application of auxin or cytokinin can confer salinity stress tolerance in Vicia faba L. Agronomy 11:790

Abdel Latef AA, Hasanuzzaman, Tahjib-Ul-Arif M (2021b) Mitigation of salinity stress by exogenous application of cytokinin in faba bean (Vicia faba L.). Notulae Botanicae Horti Agrobotanici Cluj-Napoca 49(1):12192

Abdel Latef AA, Tahjib-Ul-Arif M, Rhaman MS (2021c) Exogenous Auxin-mediated salt stress alleviation in faba bean (Vicia faba L.). Agronomy 11:547

Ahammed GJ, Yu JQ (2016) Plant hormones under challenging environmental factors. Springer

Ahmad P (2017) Oilseed crops: yield and adaptations under environmental stress. John Wiley & Sons

Ahmad F, Kamal A, Singh A, Ashfaque F, Alamri S, Siddiqui MH (2020) Salicylic acid modulates antioxidant system, defense metabolites, and expression of salt transporter genes in Pisum sativum under salinity stress. J Plant Growth Regul:1–14

Ahmad S, Cui W, Kamran M, Ahmad I, Meng X, Wu X, Su W, Javed T, El-Serehy HA, Jia Z (2021) Exogenous application of melatonin induces tolerance to salt stress by improving the photosynthetic efficiency and antioxidant defense system of maize seedling. J Plant Growth Regul 40:1270–1283

Alisofi S, Einali A, Sangtarash MH (2020) Jasmonic acid-induced metabolic responses in bitter melon (*Momordica charantia*) seedlings under salt stress. J Hortic Sci Biotechnol 95(2): 247–259

Amjad M, Akhtar J, Anwar-ul-Haq M, Yang A, Akhtar SS, Jacobsen SE (2014) Integrating role of ethylene and ABA in tomato plants adaptation to salt stress. Sci Hortic 172:109–116

Ashraf M, Foolad MR (2007) Roles of glycine betaine and proline in improving plant abiotic stress resistance. Environ Exp Bot 59(2):206–216

Azooz MM, Shaddad MA, Abdel-Latef AA (2004a) Leaf growth and K^+/Na^+ ratio as an indication of the salt tolerance of three sorghum cultivars grown under salinity stress and IAA treatment. Acta Agron Hung 52:287–296

Azooz MM, Shaddad MA, Abdel-Latef AA (2004b) The accumulation and compartmentation of proline in relation to salt tolerance of three sorghum cultivars. Indian J Plant Physiol 9:1–8

Bakyani MRF, Alinia M, Kazemeini SA, Abadía J, Dadkhodaie A (2022) Foliar application of melatonin improves the salt tolerance, ion and redox homeostasis and seed oil fatty acid profile in *Camelina sativa*. Plan Theory 11:3113

Bali S, Kaur P, Sharma A, Ohri P, Bhardwaj R, Alyemeni M, Wijaya L, Ahmad P (2018) Jasmonic acid-induced tolerance to root-knot nematodes in tomato plants through altered photosynthetic and antioxidative defense mechanisms. Protoplasma 255:471–484

Cebrián G, Iglesias-Moya J, García A, Martínez J, Romero J, Regalado JJ, Martínez C, Valenzuela JL, Jamilena M (2021) Involvement of ethylene receptors in the salt tolerance response of *Cucurbita pepo*. Hortic Res 8(73):1–14
Cen H, Wang T, Liu H, Tian D, Zhang Y (2020) Melatonin application improves salt tolerance of alfalfa (*Medicago sativa* L.) by enhancing antioxidant capacity. Plan Theory 9(2):220
Christmann A, Moes D, Himmelbach A, Yang Y, Tang Y, Grill E (2006) Integration of abscisic acid signalling into plant responses. Plant Biol 8(03):314–325
Colebrook EH, Thomas SG, Phillips AL, Hedden P (2014) The role of gibberellin signalling in plant responses to abiotic stress. J Exp Biol 217(1):67–75
Cutler SR, Rodriguez PL, Finkelstein RR, Abrams SR (2010) Abscisic acid: emergence of a core signaling network. Annu Rev Plant Biol 61(1):651–679
Dar NA, Amin I, Wani W, Wani SA, Shikari AB, Wani SH, Masoodi KZ (2017) Abscisic acid: a key regulator of abiotic stress tolerance in plants. Plant Gene 11:106–111
Delgado C, Mora-Poblete F, Ahmar S, Chen JT, Figueroa CR (2021) Jasmonates and plant salt stress: molecular players, physiological effects, and improving tolerance by using genome-associated tools. Int J Mol Sci 22(6):3082
Duca M, Port A, Rotaru T (2003) Influence on the variability in auxin and gibberelin contents in *Helianthus annuus* L. Helia 26(38):121–126
EL Sabagh A, Islam MS, Hossain A, Iqbal MA, Mubeen M, Waleed M, Reginato M, Battaglia M, Ahmed S, Rehman A (2022) Phytohormones as growth regulators during abiotic stress tolerance in plants. Front Agron 4:765068
Emamverdian A, Ding Y, Mokhberdoran F (2020) The role of salicylic acid and gibberellin signaling in plant responses to abiotic stress with an emphasis on heavy metals. Plant Signal Behav 15(7):1777372
Exports CCE (2024) Oilseeds: world markets and trade
Fahad S, Hussain S, Matloob A, Khan FA, Khaliq A, Saud S, Hassan S, Shan D, Khan F, Ullah N (2015a) Phytohormones and plant responses to salinity stress: a review. Plant Growth Regul 75: 391–404
Fahad S, Nie L, Chen Y, Wu C, Xiong D, Saud S, Hongyan L, Cui K, Huang J (2015b) Crop plant hormones and environmental stress. Sustain Agric Rev 15:371–400
Farhangi-Abriz S, Ghassemi-Golezani K (2016) Improving amino acid composition of soybean under salt stress by salicylic acid and jasmonic acid. J Appl Bot Food Qual 89:243–248
Finkelstein R (2013) Abscisic acid synthesis and response. In: The *Arabidopsis* book/American society of plant biologists, p 11
Gadelha CG, Miranda RDS, Alencar NLM, Costa JH, Prisco JT, Gomes-Filho E (2017) Exogenous nitric oxide improves salt tolerance during establishment of *Jatropha curcas* seedlings by ameliorating oxidative damage and toxic ion accumulation. J Plant Physiol 212:69–79
Gómez-Cadenas A, Arbona V, Jacas J, Primo-Millo E, Talon M (2002) Abscisic acid reduces leaf abscission and increases salt tolerance in citrus plants. J Plant Growth Regul 21:234–240
Haq I-U, Gill NP, Yasin G, Memon S, Soomro Z, Ahmed F (2023) Gibberellic acid supplements mitigate the sodium chloride effects on onion seed germination and its physio-chemical attributes. Sains Malaysiana 52(4):1161–1171
Hasegawa PM, Bressan RA, Zhu J-K, Bohnert HJ (2000) Plant cellular and molecular responses to high salinity. Annu Rev Plant Biol 51(1):463–499
Hoque MN, Tahjib-Ul-Arif M, Hannan A, Sultana N, Akhter S, Hasanuzzaman M, Akter F, Hossain MS, Sayed MA, Hasan MT (2021) Melatonin modulates plant tolerance to heavy metal stress: morphological responses to molecular mechanisms. Int J Mol Sci 22(21):11445
Huang GT, Ma SL, Bai LP, Zhang L, Ma H, Jia P, Liu J, Zhong M, Guo ZF (2012) Signal transduction during cold, salt, and drought stresses in plants. Mol Biol Rep 39:969–987
Huang X, Tanveer M, Min Y, Shabala S (2022) Melatonin as a regulator of plant ionic homeostasis: implications for abiotic stress tolerance. J Exp Bot 73(17):5886–5902

Iqbal S, Wang X, Mubeen I, Kamran M, Kanwal I, Díaz GA, Abbas A, Parveen A, Atiq MN, Alshaya H, Zin El-Abedin TK, Fahad S (2022) Phytohormones trigger drought tolerance in crop plants: outlook and future perspectives. Front Plant Sci 12:799318

Javid MG, Sorooshzadeh A, Moradi F, Modarres Sanavy SAM, Allahdadi I (2011) The role of phytohormones in alleviating salt stress in crop plants. Aust J Crop Sci 5(6):726–734

Jogawat A (2019) Crosstalk among phytohormone signaling pathways during abiotic stress. In: Molecular plant abiotic stress: biology and biotechnology, pp 209–220

Kalidas P, Rajasekhar P (2010) Effect of gibbrellic acid on the yield of oil palm. Biosci Biotechnol Res Asia 7(1):433–436

Khan MN, Siddiqui MH, Mohammad F, Naeem M, Khan MMA (2010) Calcium chloride and gibberellic acid protect linseed (*Linum usitatissimum* L.) from NaCl stress by inducing antioxidative defence system and osmoprotectant accumulation. Acta Physiol Plant 32:121 132

Khan MIR, Syeed S, Nazar R, Anjum NA (2012) An insight into the role of salicylic acid and jasmonic acid in salt stress tolerance. In: Phytohormones and abiotic stress tolerance in plants, pp 277–300

Khan MIR, Fatma M, Per TS, Anjum NA, Khan NA (2015) Salicylic acid-induced abiotic stress tolerance and underlying mechanisms in plants. Front Plant Sci 6:462

Khan MIR, Poor P, Janda T (2022) Salicylic acid: a versatile signaling molecule in plants. J Plant Growth Regul 41(5):1887–1890

Kharkeshi AM, Petroudi ER, Shahmiri FS, Mobasser H, Daneshmand A (2023) Variations of yield, biochemical and antioxidative responses in sesame with silicon and cytokinin treatments under drought stress. Acta Sci Pol Hortorum Cultus 22(6):43–54

Kim H, Seomun S, Yoon Y, Jang G (2021) Jasmonic acid in plant abiotic stress tolerance and interaction with abscisic acid. Agronomy 11(9):1886

Ku Y-S, Sintaha M, Cheung M-Y, Lam H-M (2018) Plant hormone signaling crosstalks between biotic and abiotic stress responses. Int J Mol Sci 19(10):3206

Li J, Liu J, Zhu T, Zhao C, Li L, Chen M (2019) The role of melatonin in salt stress responses. Int J Mol Sci 20(7):1735

Li J, Liu Y, Zhang M, Xu H, Ning K, Wang B, Chen M (2022) Melatonin increases growth and salt tolerance of *Limonium bicolor* by improving photosynthetic and antioxidant capacity. BMC Plant Biol 22(1):16

Liu L, Guo G, Wang Z, Ji H, Mu F, Li X (2014) Auxin in plant growth and stress responses. In: Phytohormones: a window to metabolism, signaling and biotechnological applications. Springer, New York, pp 1–35

Llanes A, Varela C, Luna V (2016) Crosstalk between gibberellins and abscisic acid under drought and salinity. In: Mechanisms behind phytohormonal signalling and crop abiotic stress tolerance. Nova Science Pub, pp 1–10

Muhammad M, Waheed A, Wahab A, Majeed M, Nazim M, Yong-Hong L, Li L, Wen-Jun L (2024) Soil salinity and drought tolerance: an evaluation of plant growth, productivity, microbial diversity, and amelioration strategies. Plant Stress 11(1–16):100319

Munns R, Tester M (2008) Mechanisms of salinity tolerance. Annu Rev Plant Biol 59(1):651–681

Nayidu N, Bollina V, Kagale S (2013) Oilseed crop productivity under salt stress. In: Ecophysiology and responses of plants under salt stress, pp 249–265

Niu L, Liao W (2016) Hydrogen peroxide signaling in plant development and abiotic responses: crosstalk with nitric oxide and calcium. Front Plant Sci 7(230):1–14

Özel H, Sağlam S (2022) Strigolactone and auxin applications on cotyledon senescence in sunflower seedlings under salt stress. Eur J Biol 81(2):190–196

Parveen A, Ahmar S, Kamran M, Malik Z, Ali A, Riaz M, Abbasi GH, Khan M, Sohail AB, Rizwan M (2021) Abscisic acid signaling reduced transpiration flow, regulated Na+ ion homeostasis and antioxidant enzyme activities to induce salinity tolerance in wheat (*Triticum aestivum* L.) seedlings. Environ Technol Innov 24:101808

Pashang D, Weisany W, Ghajar FGK (2021) Changes in the fatty acid and morphophysiological traits of safflower (*Carthamus tinctorius* L.) cultivars as response to auxin under water-deficit stress. J Soil Sci Plant Nutr 21:2164–2177

Peleg Z, Blumwald E (2011) Hormone balance and abiotic stress tolerance in crop plants. Curr Opin Plant Biol 14(3):290–295

Pompelli MF, Jarma-Orozco A, Rodríguez-Páez LA (2022) Salinity in Jatropha curcas: a review of physiological, biochemical, and molecular factors involved. Agriculture 12(594):1–22

Rashed NM, Shala AY, Mahmoud MA (2017) Alleviation of salt stress in *Nigella sativa* L. by gibberellic acid and rhizobacteria. Alex Sci Exch J 38(4):1–16

Riyazuddin R, Verma R, Singh K, Nisha N, Keisham M, Bhati KK, Kim ST, Gupta R (2020) Ethylene: a master regulator of salinity stress tolerance in plants. Biomol Ther 10(6):959

Ryu H, Cho YG (2015) Plant hormones in salt stress tolerance. J Plant Biol 58:147–155

Salehin M (2024) Emerging roles of auxin in plant abiotic stress tolerance. Physiol Plant 176(6): e14601

Saravanakumar D, Samiyappan R (2007) ACC deaminase from Pseudomonas fluorescens mediated saline resistance in groundnut (*Arachis hypogea*) plants. J Appl Microbiol 102:1283–1292

Shaddad M, Abd El-Samad H, Mostafa D (2013) Role of gibberellic acid (GA3) in improving salt stress tolerance of two wheat cultivars. Int J Plant Physiol Biochem 5(4):50–57

Shahbaz M, Ashraf M (2013) Improving salinity tolerance in cereals. Crit Rev Plant Sci 32(4): 237–249

Sheikhalipour M, Mohammadi SA, Esmaielpour B, Zareei E, Kulak M, Ali S, Nouraein M, Bahrami MK, Gohari G, Fotopoulos V (2022) Exogenous melatonin increases salt tolerance in bitter melon by regulating ionic balance, antioxidant system and secondary metabolism-related genes. BMC Plant Biol 22(1):380

Sheteiwy MS, Shao H, Qi W, Daly P, Sharma A, Shaghaleh H, Hamoud YA, El-Esawi MA, Pan R, Wan Q, Lu H (2021) Seed priming and foliar application with jasmonic acid enhance salinity stress tolerance of soybean (*Glycine max* L.) seedlings. J Sci Food Agric 101(5):2027

Shrivastava P, Kumar R (2015) Soil salinity: a serious environmental issue and plant growth promoting bacteria as one of the tools for its alleviation. Saudi J Biol Sci 22(2):123–131

Singh PK, Gautam S (2013) Role of salicylic acid on physiological and biochemical mechanism of salinity stress tolerance in plants. Acta Physiol Plant 35:2345–2353

Sun C, Liu L, Wang L, Li B, Jin C, Lin X (2021) Melatonin: a master regulator of plant development and stress responses. J Integr Plant Biol 63(1):126–145

Swamy P, Smith BN (1999) Role of abscisic acid in plant stress tolerance. Curr Sci:1220–1227

Tao J-J, Chen H-W, Ma B, Zhang W-K, Chen S-Y, Zhang J-S (2015) The role of ethylene in plants under salinity stress. Front Plant Sci 6:1059

Thu NBA, Hoang XLT, Truc MT, Sulieman S, Thao NP, Tran LSP (2017) Cytokinin signaling in plant response to abiotic stresses. In: Mechanism of plant hormone signaling under stress, vol 1, pp 71–100

Tuteja N (2007) Abscisic acid and abiotic stress signaling. Plant Signal Behav 2(3):135–138

Vanderauwera S, Vandenbroucke K, Inzé A, Van De Cotte B, Mühlenbock P, De Rycke R, Naouar N, Van Gaever T, Van Montagu MC, Van Breusegem F (2012) AtWRKY15 perturbation abolishes the mitochondrial stress response that steers osmotic stress tolerance in Arabidopsis. Proc Natl Acad Sci 109(49):20113–20118

Verma V, Ravindran P, Kumar PP (2016) Plant hormone-mediated regulation of stress responses. BMC Plant Biol 16:1–10

Vishwakarma K, Upadhyay N, Kumar N, Yadav G, Singh J, Mishra RK, Kumar V, Verma R, Upadhyay R, Pandey M (2017) Abscisic acid signaling and abiotic stress tolerance in plants: a review on current knowledge and future prospects. Front Plant Sci 8:161

Wang Y, Mostafa S, Zeng W, Jin B (2021) Function and mechanism of jasmonic acid in plant responses to abiotic and biotic stresses. Int J Mol Sci 22(16):8568

Wasternack C, Hause B (2013) Jasmonates: biosynthesis, perception, signal transduction and action in plant stress response, growth and development. An update to the 2007 review in Annals of Botany. Ann Bot 111(6):1021–1058

Wasternack C, Strnad M (2019) Jasmonates are signals in the biosynthesis of secondary metabolites-pathways, transcription factors and applied aspects-a brief review. New Biotechnol 48:1–11

Yaheliuk S, Fomych M, Rechun O (2024) Global market trends of grain and industrial crops. For Comm Bull 1(17):134–145

Ye N, Jia L, Zhang J (2012) ABA signal in rice under stress conditions. Rice 5:1–9

Yu Y, Li Y, Yan Z, Duan X (2021) The role of cytokinins in plant under salt stress. J Plant Growth Regul:1–13

Zhang M, Smith JAC, Harberd NP, Jiang C (2016) The regulatory roles of ethylene and reactive oxygen species (ROS) in plant salt stress responses. Plant Mol Biol 91:651–659

Zhang X, Sun J, Dong C-H (2024) Molecular regulations of ethylene signaling in plant salt stress responses. Plant Stress 14:100583

Zhao Y (2010) Auxin biosynthesis and its role in plant development. Annu Rev Plant Biol 61(1): 49–64

Zhu J-K (2001) Cell signaling under salt, water and cold stresses. Curr Opin Plant Biol 4(5): 401–406

Chapter 10
Drought Resilience in Oilseed Crops: The Role of Phytohormones in Stress Mitigation

Khizar Hayat, Saeed Ahmad Asad, Nazima Wahid, Kanwal, Baber Ali, Alevcan Kaplan, Maryam Bibi, Fethi Ahmet Ozdemir, Tabarak Malik, Muhammad Nauman Khan, and Karishma

Abstract Oilseed crops, including sunflower, soybean, and groundnut, play a vital role in agricultural production, serving as a key source of essential edible oils, biofuels, and proteins. However, these plants face various abiotic challenges, including drought stress. Due to climate change, water scarcity has been identified as a significant environmental constraint that reduces the yield and quality of oilseed crops. Phytohormones play a pivotal role in adapting plants to drought stress. These phytohormones regulate drought stress through cellular function at the molecular

---

K. Hayat (✉)
Department of Botany, Islamia College Peshawar, Peshawar, Pakistan

S. A. Asad
Department of Biosciences, COMSATS University Islamabad, Islamabad, Pakistan

N. Wahid · Kanwal · B. Ali
Department of Plant Sciences, Quaid-i-Azam University, Islamabad, Pakistan

A. Kaplan
Department of Crop and Animal Production, Sason Vocational School, Batman University, Batman, Turkey

M. Bibi
Department of Botany, University of Malakand, Chakdara, Lower Dir, Khyber Pakhtunkhwa, Pakistan

F. A. Ozdemir
Department of Molecular Biology and Genetics, Faculty of Science and Art, Bingol University, Bingol, Turkey

T. Malik
Department of Biomedical Sciences, Institute of Health, Jimma University, Jimma, Ethiopia

M. N. Khan
Department of Botany, Islamia College Peshawar, Peshawar, Pakistan

Biology Laboratory, University Public School, University of Peshawar, Peshawar, Pakistan

Karishma
Department of Agricultural Sciences and Technology, Atta-ur-Rahman School of Applied Biosciences (ASAB), National University of Sciences and Technology (NUST), Islamabad, Pakistan

© The Author(s), under exclusive license to Springer Nature Singapore Pte Ltd. 2025

A. A. H. Abdel Latef (ed.), *Oilseed Crops Under Abiotic Stress*, Sustainability Sciences in Asia and Africa, https://doi.org/10.1007/978-981-96-8346-8_10

level via various cell signalings to maintain water balance, regulate stomatal movement, enhance root growth, and activate the antioxidant enzyme system. Abscisic acid (ABA), for instance, a major plant hormone, has a significant function in regulating drought tolerance by closing stomata and inducing stress-responsive genes, while cytokinin modulates root-to-shoot signaling to control water uptake. Ethylene (ET), gibberellins (GA), salicylic acid (SA), and jasmonic acid (JA) are other hormones that are crucial for helping plants adapt to drought stress. This chapter focuses on the physiobiochemical and molecular mechanisms involved in phytohormonal regulation under drought conditions in oilseed crops. These mechanisms are important for sustainable agriculture against increasing water stress and climate variability.

Keywords Drought stress · Oilseed crops · Phytohormones · Sustainable agriculture

10.1 Introduction

Drought is one of the most prevalent abiotic stresses impacting agricultural production globally, and it immensely impairs the yield and quality of numerous economically significant oilseed species such as *Glycine max, Brassica napus*, and *Helianthus annuus* (Jamil et al. 2018; Iqbal et al. 2022; Kosar et al. 2022). These crops are important sources of food oils, but are also used for biofuels and various other industrial processes (Farooq et al. 2012). Plants experience a sequence of physiological and biochemical alterations in response to drought and exhibit multiple adjustments, such as alterations in stomatal conductance, osmotic homeostasis, and antioxidant activity, for water conservation, cell homeostasis, and suppression of oxidative stress (Chaves et al. 2009; Batool et al. 2022). Phytohormones, or regulatory chemicals that orchestrate numerous growth and stress response mechanisms, are essential for the plant's response to drought (Peleg and Blumwald 2011). Abscisic acid (ABA), auxins, cytokinins, gibberellins, ethylene, and jasmonates are especially important types of hormones in drought resistance (Shinozaki and Yamaguchi-Shinozaki 2007). ABA, for example, is often considered the major phytohormone during the response to drought, since it plays a direct role in inducing stomatal closure to reduce water loss and stimulates osmolyte synthesis to maintain cellular osmotic homeostasis under drought (Zhang et al. 2006; Zhu 2016). In addition, ABA regulates the expression of stress-response genes associated with compatible solute accumulation and antioxidant defenses critical for protecting cellular structures from oxidative damage due to drought (Cutler et al. 2010).

As the plant reallocates its resources to favor survival over growth under the stress of drought, auxins and cytokinins that generally promote cell division and elongation show activity modulation (O'Brien and Benková 2013). For example, less cytokinin activity during drought time helps molecules of the plant to growth of shoots, but shoot growth enhances efficiency in water uptake (Mughal et al. 2024). Ethylene and jasmonates, both stress signaling hormones, yield pathways that cross-talk with ABA and modulate drought response. For instance, ethylene controls the stomatal behavior and elongation of roots while the jasmonates are involved in signaling pathways that prime plants to cope with abiotic stresses, including drought (Wilkinson and Davies 2010; Ali and Baek 2020). Recent research highlights the role of these phytohormones in drought tolerance mechanisms and illustrates how their regulation and cross-talk allow to triggering a well-coordinated response in oilseed crops (Pérez-Llorca et al. 2023). The role of phytohormones in signaling networks that underlie stomatal behavior, water-use efficiency, and metabolic adaptation (of which there is considerable evidence) is an area that will significantly inform breeding and biotechnological strategies to enhance drought resistance in oilseed crops. Understanding these hormonal interactions can help in developing crop varieties that are still productive when the water supply is limited; this is important for fulfilling future food and biofuel needs under climate variance (Fahad et al. 2015; Jamshidi Zinab et al. 2023).

10.2 Overview of Oilseed Crops and Their Importance

Oilseed crops such as *Glycine max* (bean), *Brassica napus* (rapeseed), *Helianthus annuus* (sunflower), and *Elaeis guineensis* (oil palm) make important contributions to world agriculture as major means of supplying proteins, edible fats and oils, and feedstocks for biofuels. These crops are vital for food security, renewable energy production, and economic development, particularly in resource-poor countries where they help to sustain local economies and create job opportunities (Mustafa and Iqbal 2021). This diversity in crop types results in an optimum crop choice based on specific agronomic traits, unique nutritional and industrial advantages. Soybeans are the highest oil-seed crops worldwide with a high oil yield (approximately 20%) and protein content (approximately 40%). Besides its use as cooking oil, soybean oil is also being used in biodiesel production, because of its favorable fatty acid composition and availability (Karasulu et al. 2011). Canola (*Brassica napus* L. or *B. rapa* L.) is a variety of rapeseed that has low levels of erucic acid, making it a much-loved plant since it is edible to consumers (Khan et al. 2021). Canola oil, high in monounsaturated fats, promotes heart health and is commonly used across food industries globally.

Sunflower is also a valuable oilseed crop, producing ω6 linoleic acid-rich oil, which is cardioprotective. This trait is also favorably adaptable; thus, it has the potential to be cultivated under various environmental and agronomical conditions (Adeleke and Babalola 2020; Abeed et al. 2022). Moreover, oil palm is the highest-yielding oil crop per hectare and an important energy resource in tropical countries, mainly in Malaysia and Indonesia. Widely used in food, cosmetics, and biofuels, palm oil cultivation has led to increased deforestation and environmental concerns, prompting sustainability initiatives (Meijaard et al. 2020). Apart from their significance in human nutrition, oilseed crops also have important environmental benefits. As potential sources of biofuels, they represent a sustainable alternative to fossil fuels that may reduce greenhouse gas emissions (Singh and Singh 2010). It is especially important as countries try to meet climate targets by using renewable energy. Mucilage from Oil Seed Sources Mucilage coagulated in mucilaginous residues from oilseed extraction also contributes to the global economy while providing an abundant source of mucilage for lubrication, retention, and viscosity during biorefinement. Protein-rich residues after oil extraction are largely dry-milled and serve as animal feed, promoting the multiuse of agriculture while preventing other land and water resource usage for protein feed crops (Kumar et al. 2024).

10.3 Impact of Drought Stress on Crop Productivity

Drought stress is a major environmental factor that negatively affects crop yield worldwide and contributes significantly to the current threats to agriculture sustainability and food security. Water vulnerability impacts most physiological processes in plants, including photosynthetic carbon reduction, nutrient absorption, and cytoplasmic expansion—all of which reduce agricultural productivity (Farooq et al. 2009). Drought causes plants to frequently close their stomata to minimize water loss, restricting the entry of CO_2 and reducing photosynthetic efficiency, impairing the energy produced for growth and development (Chaves et al. 2009). Drought stress also induces an oxidative response that leads to the generation of reactive oxygen species (ROS), which damage cellular structures and disrupt plant metabolism (Shao et al. 2009). Plants activate antioxidant defense mechanisms to counteract ROS damage, but this process usually takes energy and resources away from growth, thus affecting crop productivity (Mittler 2002). Plants subjected to drought stress show significant declines in biomass accumulation, root growth, and leaf area; the effects are most readily apparent in drought-sensitive species and types (Jaleel et al. 2009). Another key factor affected by drought stress is nutrient acquisition efficiency. When water levels decrease, the plant cannot take up essential nutrients like potassium, phosphorus, and nitrogen needed during development and yield formation. Nutrient mobility in the soil is also affected by water availability (Kramer and Boyer 1995). This limitation of nutrients combined with decreased photosynthetic activity, in turn, causes poor seed development and decreased yield quality in drought-sensitive crops such as maize, wheat, and rice (Lobell et al. 2013).

Drought tolerance does not have uniform effects on crop species; the varieties considered capable of surviving drought limit their root structures to access deeper reserves of water, alter their leaf structures, and activate mechanisms of osmotic adjustment to retain cellular water (Blum 2011). However, these adaptive practices may not be effective, especially under conditions of extreme or prolonged drought. Drought stress, therefore, has a significant impact on global crop production, and ameliorating water limitations will be crucial in achieving stable crop yields under global climate change.

10.4 Role of Phytohormones in Plant Stress Response

An integral component of this phenomenon, phytohormones regulate the nature of a plant's response to a range of environmental stresses, e.g., drought, salinity, and pathogen infection, by altering growth and defense mechanisms (Peleg and Blumwald 2011; Abdel Latef et al. 2021a, b, c). As multi-impact mediators, they consist of abscisic acid, gibberellins, cytokinin, ethylene, auxins, jasmonic acid, and salicylic acid, which coordinate the cellular processes and exhibit stress-specific responses, which in turn improve the resilience of plants (Zhang et al. 2006). Phytohormones often show synergistic or antagonistic effects under stress conditions, generating complex signaling networks that further modulate the physiological and biochemical responses of the plant. Abscisic acid is one of the key hormones involved in the response to drought stress. It triggers antioxidant mechanisms to combat oxidative damage, regulates osmoprotective-associated gene expression, and induces stomatal closure to conserve water (Cutler et al. 2010). Gibberellins, which promote growth under normal conditions, are typically downregulated by stress to conserve energy. In contrast, auxins promote root growth, which can help access more water and nutrients from deeper soil layers during drought (Taiz and Zeiger 2002).

Ethylene and jasmonic acid are commonly linked to responses to pathogen attacks and mechanical damage; however, they also contribute to abiotic stress tolerance through the modulation of antioxidant defenses, as well as the promotion of senescence in injured tissues to allow resource prioritization to healthier parts (Xiong and Zhu 2003). Adversely, cytokinin and salicylic acid are growth-promoting hormones, but tend to increase stress tolerance through redox balance and nutrient profile (Verma et al. 2016). And given that these phytohormones frequently work in concert, this allows plants to balance growth with stress resilience, Balance Between Growth and Stress Resilience Some notable interacting pairs include ABA and ethylene, which are important in response to water stress/ oxidative stress; and jasmonic acid/ salicylic acid, which mediate pathogen resistance and stress signal amplification (Yoon et al. 2009). This complex network of hormone interactions is an integral part of the plant's ability to acclimatize to changing environments, providing targets for biotechnological interventions aimed at developing stress-tolerant crops (Fig. 10.1).

10.5 Oilseed Crops and Drought Stress

Oilseeds—a diverse group of crops that include soybeans, canola, sunflower and safflower—are a critical component of global food and energy security, serving as key sources of edible oils, protein-rich feed and biofuel (Sharma et al. 2012).

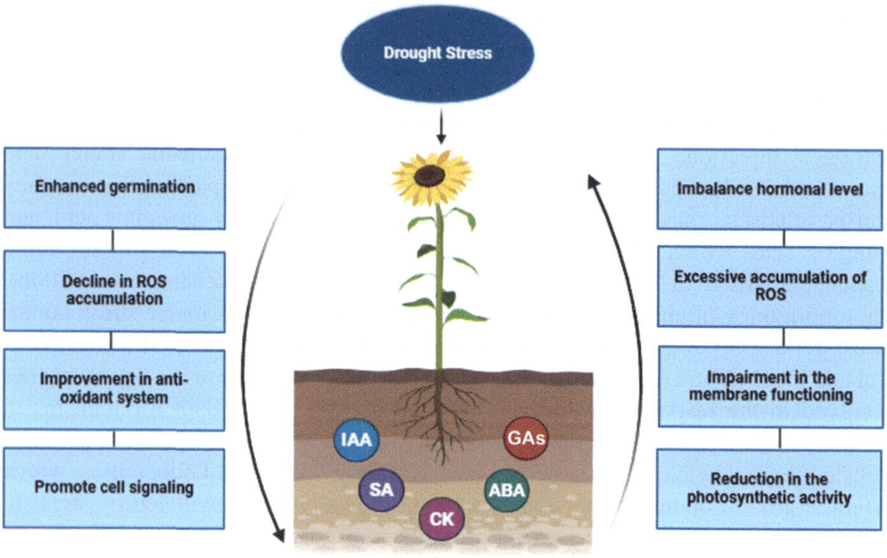

Fig. 10.1 Hormonal-mediated modulation of drought stress regulation in oilseed crops. SA salicylic acid, IAA indole acetic acid or auxin, ABA abscisic acid, CK cytokinin, GAs gibberellins

Drought stress, particularly in regions with low or erratic precipitation, is the most significant abiotic stress negatively affecting oilseed crop yield. Drought stress not only limits the water available for use but also induces oxidative stress, disrupts photosynthesis (Hussain et al. 2018). The impacts of drought on oilseed crops vary based on the crop type, growth stage, and duration of the water deficit. Drought at flowering and seed-filling stages is particularly damaging as it can trigger seed abortion and reduce oil content due to hormonal alterations and carbon allocation disturbances. In addition, the varying expression of traits associated with drought tolerance such as deep root systems, stomatal regulation, and antioxidant activity among various oilseed crops would contribute to their resilience to water stress (Farooq et al. 2009).

Adaptive responses of oilseed crops to drought stress are often modulated by phytohormones, including jasmonic acid, which initiates protective responses against cellular damage caused by drought, and abscisic acid, which plays a major role in controlling stomatal closing (Ahmad et al. 2018). Genetic and biotechnological approaches to improve drought tolerance have emerged as a promising part of agricultural research aimed at enabling oilseed crops to become more resilient with changing climatic conditions (Jogaiah et al. 2013). Sustainability of productivity, as well as oil quality, which is critical for the economic viability of these crops, is the focus of this research.

10.6 Physiological and Biochemical Effects of Drought on Oilseed Crops

The physiological and biochemical effects of drought stress on oilseed crops largely influence growth, production, and oil quality. On the physiological level, drought stress decreases leaf area, limits stomatal conductance and disturbs water and nutrient absorption, which then directly limits photosynthesis and inhibits plant growth in general (Farooq et al. 2009). Drought leads to less flowering, fewer seeds and less seed filling in oilseeds, including sunflower, soybean and canola which are particularly sensitive to water deficit during flowering. This decreases yield and oil content (Hussain et al. 2018). Based on biochemical findings, dryness induces oxidative stress in oilseed crops due to an excess of reactive oxygen species (ROS), which is detrimental to cellular components like proteins, lipids, and DNA (Hasanuzzaman et al. 2013). In response to the oxidative stress conditions caused by ROS, oilseed crops have been shown to upregulate numerous antioxidant enzymes such as superoxide dismutase (SOD), catalase (CAT), and peroxidase (POD), in addition to increased levels of nonenzymatic antioxidants (e.g., ascorbate and glutathione) (Jogaiah et al. 2013). Osmoprotectants, such as proline, glycine betaine, and sugars, are accumulated in many crops to maintain cellular turgor and to prevent membrane destabilization during drought environments. Drought stress also disrupts osmotic adjustment systems (Ahmad et al. 2018).

Moreover, drought stress was shown to influence lipid metabolism in oilseed crops that often leads to alterations in the composition and quality of stored oils. Drought, for instance, acts to modify the profiles of fatty acids, such as on sunflower and canola (Hussain et al. 2018). Yet, such changes may both compromise the nutritional quality and stability of the resulting oil, underscoring the necessity for deliberate water management and breeding programs to improve the drought tolerance of oilseed variants (Fig. 10.2).

10.6.1 Mechanisms of Drought Tolerance in Oilseed Crops

Drought stress perception and signaling pathways play a role in oilseed tolerance to drought stress, which ultimately can achieve survival and continued performance in conditions of water deficit through a complex network of physiological, biochemical, and molecular events. These mechanisms include changes in water uptake, osmotic balance, and the regulation of oxidative stress to adapt to drought stress in oilseed crops such as sunflower, soybean, and canola. Osmotic adjustment is a crucial mechanism by which crops accumulate compatible solutes such as proline, glycine betaine, and sugars to decrease osmotic potential for drought tolerance. As such, this enables cells to retain water and maintain turgor pressure in conditions of drought (Kumar et al. 2018). Stomatal regulation is also important for controlling water loss due to transpiration. Under drought stress, oilseed crops exhibit partial

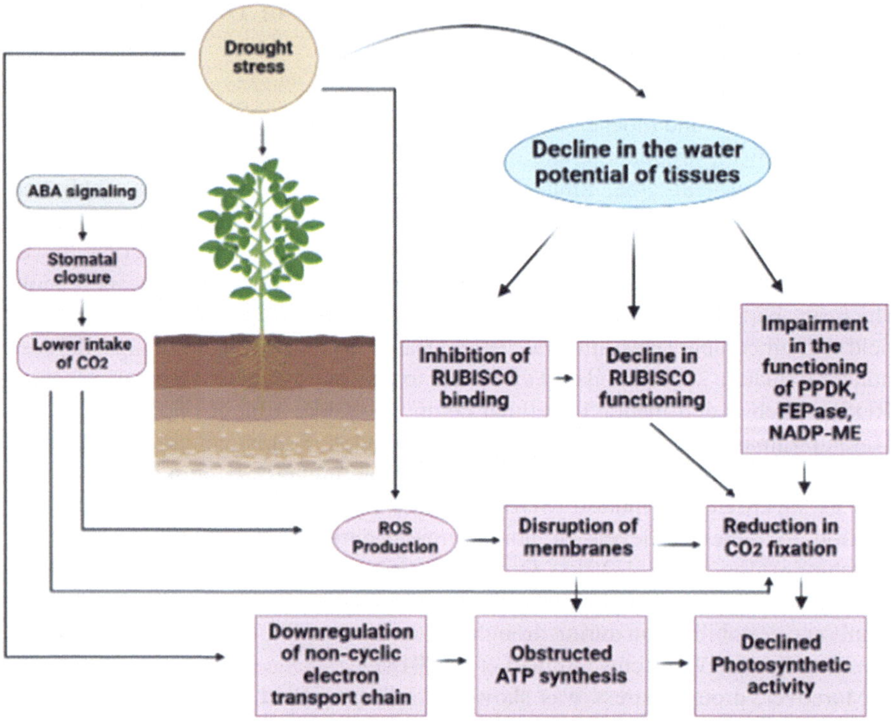

Fig. 10.2 Drought stress downregulates the biochemical pathways of various physiological mechanisms

closure of the stomata, which serves to reduce water loss but, consequently, can limit the entry of carbon dioxide into the leaves, ultimately reducing the photosynthetic efficacy of the plants. Another drought tolerance mechanism is the improved root system architecture: deeper and more extensive root systems are more efficient in extracting water and minerals from the soil (Luo et al. 2019).

Through the biochemical pathway, it triggers antioxidant defense systems in oilseed crops due to drought stress. Conversely, drought conditions cause the generation of ROS that can harm cellular structures. In response to this, plants induce antioxidant enzymes including SOD, CAT, and POD, which remove ROS and diminish oxidative damage (Hasanuzzaman et al. 2013). Some oilseed crops also accumulate nonenzymatic antioxidants (ascorbate and glutathione) that provide additional protection against oxidative stress (Farooq et al. 2009). The molecular responses to drought include upregulation of stress-responsive genes and signaling pathways, particularly those that involve ABA, a major phytohormone involved in regulating drought response. The activation of drought-responsiveness genes through ABA signaling contributes to osmotic adjustment, root development, and stomatal closing, hence improving drought tolerance (Xiong and Zhu 2003). Additionally, recent progress in genomics has revealed various transcription factors, including DREB (dehydration-responsive element binding), that regulate pathways associated with drought tolerance, and by extension, these are becoming potential targets for enhancement through genetically improved oilseed crop varieties (Yamaguchi-Shinozaki and Shinozaki 2006).

10.6.2 Examples of Oilseed Crops Affected by Drought (e.g., Sunflower, Canola, Soybean)

Drought stress is a major environmental factor that negatively affects the growth, physiology, and productivity of oilseed crops such as sunflower, canola, and soybean. Economically, they are important sources of oil and protein used in food, feed, and industrial applications. Drought effects are crop dependent and largely linked to the specific stress tolerance mechanisms existing in the individual plant species. The following section provides specific examples of how these oilseed crops are affected by drought.

10.6.2.1 Sunflower (*Helianthus annuus*)

Because sunflowers have a deep root system and can reach much deeper soil moisture, they are considered fairly drought tolerant. However, drought stress at crucial growth stages, especially during the reproductive phase, significantly reduces yield and oil content in sunflower. These negative effects of drought stress include the decrease in leaf area, reduction of chlorophyll content, and photosynthetic rate, affecting the overall health of the plant (Ghobadi et al. 2013). Moreover, drought can lead to early leaf senescence, which would shorten the grain filling period and reduce seed quality and oil percentage. Osmotic adjustment and increased antioxidant enzyme activity are part of the drought tolerance response of sunflowers; however, this is often not enough to completely counteract the effects of prolonged water deficit (Chimenti et al. 2002).

10.6.2.2 Canola (*Brassica napus*)

Canola responds poorly to drought stress, especially during flowering and pod-filling, both critical stages for yield and oil quality determination. Drought at these stages can lead to flower and pod abortion, reduction in seed number, and lower oil content owing to limited seed photosynthesis and nutrient transport (Qaderi et al. 2006). Moreover, drought stress induces a protein buildup in canola seed as well, but at the cost of oil, resulting in decreased oil yield. Stomatal regulation, osmotic adjustment and activation of antioxidant defenses are mechanisms adopted by canola to improve tolerance to drought stress conditions, and to reduce the oxidative damage (Khalili et al. 2012).

10.6.2.3 Soybean (*Glycine max*)

Soybean has moderate drought tolerance; however, considerable yield reduction occurs under water-deficit conditions. Drought stress is particularly detrimental at the pod formation and seed-filling stages because it can lead to a decrease in seed size, seed oil, and fatty acid composition, negatively affecting seed quality and market value (Desclaux et al. 2000). In response to drought, soybean plants may modify root architecture to explore deeper soil layers for moisture. Soybean plants also activate antioxidative enzymes and accumulate osmolytes to help maintain cellular functions under drought stress, though these adaptations may not fully prevent yield loss in severe drought conditions (Easwar Rao and Viswanatha Chaitanya 2020).

10.7 Role of Phytohormones in Drought Stress Tolerance

Plant hormones have a significant role in the plant developmental process as well as respond to environmental stresses, including drought. Endogenous plant hormones are considered key mitigators to tolerate climatic stress. However, water stress changes the biosynthetic pathways of a variety of hormones, i.e., cytokinins, auxin, gibberellins, ET, SA, JA, and ABA. Drought stress activates ABA expression after perception of the drought signal and is considered a major stress-responsive phytohormone. It is predominantly synthesized in roots and then transported to the aerial parts to regulate stomatal conductance, expression of stress-responsive genes, and channel activities. JA and SA also have a pivotal role in drought tolerance and to be synthesized in roots and then translocated to shoot. In shoot system, the JA and SA express ABA-responsive genes which serve to regulate stomatal aperture. Additions to the ABA, SA, and JA have been considered as drought stress-responsive phytohormones.

10.7.1 Abscisic Acid (ABA)

In drought conditions, ABA closes the stomata, allowing for maintenance of turgidity and osmoregulation of the cell (Montillet et al. 2013). ABA enhances potassium (K^+) and anion efflux from guard cells, which improves cytoplasmic calcium (second messenger), resulting in lowering turgidity of the cell and subsequently closing of stomata (Misra et al. 2015). The production of ABA analogs, such as xanthoxin and xanthoxic acid, produced by beneficial rhizobacteria, *Bacillus marisflavi* CRDT-EB-1, predicted that these compounds served as a precursor of ABA, thereby enhancing mustard (*Brassica juncea* L.) resistance against drought stress. Foliar spray of ABA at a concentration of 5 μM on sunflower improved the plant water relation, achene yields and accumulation of solute during bud and flowering initiation stage under drought stress (Hussain et al. 2020). Exogenous application of ABA at concentrations ranging from 10 μM to 100 μM improved the yield and growth of sesame plants. However, at concentration 100 μM, significantly increased seed yields, biomass production and water use efficiency as compared to 10 μM and 50 μM concentration (Darini et al. 2024). Tran et al. (2024) reported that the integrative effect of ABA and sodium nitroprusside priming (SNP) enhanced the drought tolerance in ground peanut (*Arachis hypogaea* L.) by improving the morphological, physiological and biochemical characteristics. The result showed that the SNP has a more significant effect than ABA alone. In addition, ROS facilitate stomata signaling to induce stomatal closure through stimulating ABA and generation of H_2O_2 in guard cells (Jha and Mohamed 2022). During the drought stress condition, the entry of CO_2 to the stomata is reduced adversely affecting photosynthetic mechanism and retarded growth of the plant (Mo et al. 2016). The Senescence-induced receptor-like kinase (OsSRLK) participates in hormone-mediated pathways that may degrade the chlorophyll molecule in rice, as reported by Shin et al. (2019). Plants evade drought stress by developing deep root systems. ABA regulates the root-shoot growth ratio by inhibiting the growth of the shoot and enhancing root growth through diverting mechanisms of photosynthesis in the root region (Rowe et al. 2016). Cotton transgenic tolerates prolonged drought stress by developing an elongated root system (Liu et al. 2014). Furthermore, ABA serves as a regulatory hormone to control the overexpression of stress-responsive genes and improve the biosynthesis of many important proteins, including LEA proteins, dehydrins, as well as protective proteins (Sreenivasulu et al. 2012; Wani et al. 2016). Biochemical pathways, such as ABA-dependent or ABA-independent, are essential to mediate signal transduction and ABA perception. ABA-dependent mechanisms have essential roles in expressing stress-responsive genes during osmotic stress conditions (Mehrotra et al. 2014). The ABA quantity is low during normal conditions, which inhibits the synthesis of protein like SnRK2 kinase through activation of PP2C phosphatases. The level of ABA is high under stress conditions which bind directly to the promoter of PYR/PYL/RCARs to stop the activity of PP2Cs. The inactivation of PP2Cs through the dissociation of SnRK2s can cause downstream phosphorylation targets and auto-activation, leading to ABA-induced physiological and molecular (Ullah et al. 2018a).The GmNAC019 is an important transcription factor found in soybean which was proposed to regulate drought tolerance in *Arabidopsis* through ABA ABA-mediated drought response mechanism (Hoang et al. 2019). Furthermore, genetically modified plants are typically designed to enhance and tolerate drought stress through the operation of additional regulatory mechanisms. Moreover, transgenic plants can be utilized to get more yields through some PHs-mediated drought tolerance.

10.7.2 Jasmonic Acid (JA)

Jasmonic acid (JA), a growth-regulating substance produced endogenously in many plant species and plays a positive role in plant response to stress conditions. Collectively, jasmonic acid includes fatty acid derivatives such as isoleucine conjugate and methyl ester (Ruan et al. 2019). Jasmonic acid is synthesized in cell organelles like peroxisomes, plastids and cytosol and produced in flower for its actions (Ullah et al. 2018a). Similar to ABA, jasmonic acid also induces the drought tolerance in plants by operating various mechanisms such as stomata closure, development of deep root system, scavenging of ROS, secondary metabolism as well as direct or indirect response mechanism (Riemann et al. 2015; Fang et al. 2016; Wani et al. 2016). In moisture-deficit conditions, JA enhanced soil water absorption through modulating root hydraulic conductivity (Sánchez-Romera et al. 2014). Exogenous application of JA at concentration of 40 ppm and 80 ppm on sunflower leaves increased the photosynthetic and transpiration rate, chlorophyll contents, root and shoot length as well as proline contents under drought stress conditions (Ashraf and Siddiqi 2024). The combined effect of brassinosteroid and jasmonic acid improved the drought tolerance in *Brassica rapa* L. by modulating the photosynthetic system, antioxidant defense, and osmolyte contents. JA enhanced drought tolerance in various *Brassica* species by improving the physiological performance, antioxidant enzyme activities and glyoxalase systems. Foliar spray of methyl jasmonate considerably increased the biochemical traits including saturated and unsaturated fatty acids, sugar fraction contents, flavonoids and phenolic acid in soybean in response to drought stress (Mohamed and Latif 2017). Foliar application of JA serves as an antioxidant in improving the content of dehydroascorbate reductase, ascorbate peroxidase, monodehydroascorbate reductase, and glutathione reductase in wheat seedlings during response to drought (Shan et al. 2015). Applying jasmonic acid exogenously helps to promote the growth of sugar beet and enhances antioxidant enzyme activities during water stress conditions (Ghaffari et al. 2019). In addition, the grapevine (*Vitis amurensis*) transcription factor (VaNAC17) plays an effective role in mediating drought tolerance through the biosynthesis of an endogenous JA in transgenic *Arabidopsis* (Su et al. 2020). Furthermore, the presence of jasmonic acid is a prerequisite for the stimulation of ABA activities in roots during drought stress conditions, as observed by de Ollas et al. (2013) and Sánchez-Romera et al. (2014). However, jasmonic acid also has a positive role in water conductivity while limited water is available in soils. Furthermore, the interaction of both ABA and JA signaling pathways determines the drought tolerance mechanism. It was observed that JA interacts with calcium and ABA, allowing for an increase in the hydraulic conductivity of the root system under drought conditions (de Ollas and Dodd 2016).

10.7.3 Salicylic Acid (SA)

Salicylic acid is another regulatory substance that is composed of seven-carbon compounds. It is derived from secondary metabolites–phenolic acid and is mostly synthesized in the chloroplast of the cell (Naeem et al. 2020; Abdelhameed et al. 2021; Rhaman et al. 2021; Dawood et al. 2022). SA has a critical role in the regulation of various plant physiological process including transpiration, photosynthesis, flowering and chlorophyll biosynthesis. It shows an antioxidant activity and is considered a potential enzymatic agent in enhancing the scavenging of ROS enzyme under water drought stress conditions (Arif et al. 2020; Wani et al. 2016). Generally, lower concentration of SA has a significant role in drought tolerance, while higher concentration causes oxidative stress. Kang et al. (2012) observed that the application of a high concentration of SA suppressed both plant growth and drought tolerance in wheat seedlings. Foliar application of SA improves drought tolerance in canola through various pathways such as enhancing antioxidant enzyme activities, improving water retention and modulating stress-related hormonal pathways. SA improves growth parameters including shoot and root length as well as fresh and dry weight of sunflower under drought stress (Nosherwan et al. 2024). The application of SA at high concentration enhanced chlorophyll contents, plant growth, seed and oil yield in sunflower underlying drought stress. Studies indicate that integrative exogenous application of silicon (Si) and SA effectively mitigates the drought stress and boosts cotton yield, increases biomass, as well as elevates photosynthetic efficiency and relative water content in cotton. During drought stress, SA improve nutrient uptake, physiological attributes and water relations, leading to better growth and yield of cotton (Mahdi et al. 2020) SA at concentration of 1.0 mM and 1.5 mM significantly enhances water use efficiency (WUE) and membrane stability index while reducing electrolyte leakage (Mahdi et al. 2020). Foliar application of SA stimulates physiological processes, reduces adverse effects, increases productivity, and promotes drought stress resistance in many important crops (Farooq et al. 2009; Hunter et al. 2013). Spraying of salicylic acid and methyl jasmonate on leaves and seeds of *Brassica rapa* has a significant effect on drought tolerance (Kareem et al. 2019; La et al. 2019).

10.7.4 Ethylene (ET)

A gaseous phytohormone performs various functions such as senescence, fruit ripening, as well as environmental stress responses. Methionine and S-adenosyl methionine are the main precursors to synthesize ethylene by the action of ACC synthase and ACC oxidase (ACO) (Tiwari et al. 2017). Ethylene signaling is an important mechanism that has been analyzed exclusively for its involvement in several important processes in plants, including root development, source-sink relationship, osmotic adjustment, and stress management through oxidative stress that plays a major role in improved crop productivity and drought stress tolerance in plants (Nazir et al. 2024). Drought stress: Overexpression of the HYR gene encoding ethylene responsive factor (AP2/ERF) improves yield quality and photosynthate assimilation in the leaves of rice (Ambavaram et al. 2014). In addition, studies evaluated on Jung et al. (2021) indicated that the ethylene signaling factor enhanced the yield quality traits, photosynthetic performance, and source sink dynamic in rice under the drought stress conditions. Jin et al. (2018) showed that the ERF gene expressed in the reproductive tissue was essential in pollen fertility and positively correlated with seed germination in response to drought stress. Moreover, walnut ethylene responsive factor (JeERF3) can activate the expression of genes related to C-metabolism and assimilation such as NtRub-SS, ChlGaPA, and NtRCA342, suggesting that JeERF3 plays a role in coordinating source-sink relationship which for example leads to enhanced plant performance under drought stress (Wu et al. 2008). To the best of our knowledge, AP2 and EREBP genes in cotton plants reported with increased drought and heat tolerance, respectively (Liu and Zhang 2017). The MAT gene expression produces Methionine Adenosyl Transferase enzyme (which is responsible for the biosynthesis of AdoMet) in soybean under water stress (Arraes et al. 2015). The ACO and ACS gene expression were improved in chickpea and soybean during drought (Tiwari et al. 2016; Arraes et al. 2015). A significant reduction in the level of CTR genes and ETR (ET receptors) were observed in soybean under drought conditions (Arraes et al. 2015).

10.7.5 Cytokinins (CKs)

Cytokinins (CKs) are recognized as a multifunctional hormone that performs various functions such as facilitating growth and development in plants. Cytokinins are vital in improving drought tolerance by contributing various mechanisms such as regulating water balance, promoting the antioxidant system, protecting the photosynthetic machinery, improving crosstalk with stress (i.e., PHs) and plant growth (Pavlů et al. 2018). Drought stress decreases the level of cytokinins in plants, which is compensated by ARR5 and AHP6 (negative regulators of cytokinin signaling) (Jang et al. 2017; Huang et al. 2018). The combined effect of silicon and cytokinin significantly improved oil yields, antioxidant enzyme activities and malondialdehyde contents, thereby strengthening the sesame plant against drought stress (Kharkeshi et al. 2023). The exogenous application of both BR and CK increased the drought tolerance in canola cultivars, resulting in an increase in photosynthetic rate, membrane stability and relative water content (Sima et al. 2024). Cytokinin and salicylic acid spray on sunflower plants increased the drought tolerance by improving the seed and oil contents; therefore, CK and SA spray are suggested as a solution to improve the plant yields and oil contents under drought stress conditions (Hossini et al. 2024). Golchin et al. (2022a) reported that cytokinin application minimized the drought stress in safflower and increased the physiological parameters including photosynthetic rate, chlorophyll contents, stomatal conductance, photosynthetic water use and seed oil yields under drought stress conditions. Cytokinin minimizes the negative effect of drought stress, thereby improving the photosynthetic partitioning of materials and seed oil yields in safflower (Golchin et al. 2022b). Cytokinins are important in priming the antioxidant system and protecting cells from the negative effects of ROS, as well as maintaining integrity of chloroplast (Pavlu et al. 2018). Cytokinins activate the expression of flavonoid biosynthetic genes, serving to improve drought resistance in barley (Vojta et al. 2016; Nakabayashi et al. 2014). Intercellular communication of the CKs signaling is crucial for crosstalk with ABA-regulating pathways and mediating stress tolerance during drought conditions (Hai et al. 2020).

10.7.6 Brassinosteroids (BRs)

Brassinosteroids are an essential plant regulatory hormone present in most parts of the plant. It is a polyhydroxy steroidal hormone that has a critical role in the formation of plant organs, including fruit and flower development, stem and root floral initiation (Wani et al. 2016). Drought tolerance of potato tubers was enhanced under the influence of brassinosteroids which were associated with the improvement of ABA, phenolics, terpenoid compounds and ethylene level (Akula and Ravishankar 2011). The exogenous application of BR and JA significantly enhanced the drought tolerance in *Brassica rapa* L. by improving several physiological parameters including photosynthetic rate, transpiration rate, stomatal conductance and antioxidant enzyme mechanisms. Brassinosteroids enhanced the quality of fatty acids in rapeseed which was evident that the level of oleic acid and linoleic acid were found to increase 0.7% and 1.1%, respectively, under BR application in response to drought stress. Naveen et al. (2021) reported that brassinolide enhanced the physiological and biochemical attributes in *Brassica juncea* by boosting the antioxidant compounds and reducing the oxidative stress indicators, thereby strengthening drought tolerance in the plant. The effect of 24-epibrassinolide (EBL) increased the composition of fatty acids in safflower under drought stress (Zafari et al. 2020). Results showed that the oleic acid was increased by 15.95%, linoleic (4.26%), linolenic (22.95%), and stearic acid (29.6%) as compared to the control. The use of brassinosteroids increases drought tolerance by 63% in sunflower as reported by Ghasemi et al. (2021), who studied that BR increased chlorophyll content, number of leaves which in turn improved yields, and oil contents as compared to the control. The drought stress reduced the oil yields and fatty acids in camelina (*Camelina sativa* L.), but foliar spray of both boran and 24-epibrassinolide significantly enhanced the oil contents by a range from 2% to 4% and polyunsaturated fatty acids. Brassinosteroids are involved in improving drought resistance mechanisms through interaction with transcription factors and Gibberellins to control the growth and production of the plant (Sánchez-Romera et al. 2014).

10.7.7 Gibberellins (GAs)

Gibberellin is a major plant hormone and it plays an important role in enhancing stem elongation, seed germination, leaf expansion, and fruit and flower development in a variety of plants (Yamaguchi 2008). Gibberellins are a group of tetracyclic diterpenoid carboxylic acids and have been extensively studied in water stress tolerance (Nir et al. 2014; Lo et al. 2017). Munteanu et al. (2014) investigated that the minimum concentration of GAs improved the drought stress in flowering plants, including transgenic potato and rice. Khan et al. (2020) demonstrated that treatment of canola seeds with GA_3 at 500 mg L^{-1} improved the drought stress tolerance by improving proline accumulation and antioxidant capacity as compared to water-sprayed control. GAs improved drought tolerance in rapeseed by modulating germination rate, maintaining osmotic adjustment and protecting the chloroplast from autophagy. Foliar spray of gibberellic acid increased chlorophyll contents, seed oil, seed content as well as relative water content in safflower during drought conditions (Barahouei et al. 2024). Foliar spray of GAs significantly increased total chlorophyll, seed oil and yields and decreased malondialdehyde in safflower under drought stress (Heidari et al. 2022). The study of Jan et al. (2019) showed that the combined effect of Zn, K, and GAs improved the biosynthesis of antioxidant compounds which in turn enhanced the drought tolerance in sunflower. However, GA interacts with ET significantly improved the stress tolerance mechanism as well as enhanced various developmental processes of the plants (Munteanu et al. 2014). This was confirmed that the AtGAMT1 gene encodes an enzyme that converts active GA to an inactive GA methyl ester by breaking methylation. Transgenic tomatoes were more resistant to drought stress and contained more water due to a low transpiration rate (Nir et al. 2014). DELLA proteins were reported as the main inducers of GAs to respond effectively to drought stress conditions. Gibberellin can be inactivated by the presence of such nuclear regulators. The 26S proteasome degrades DELLA proteins through the binding of gibberellins to the GID1 receptor which stimulates gibberellin responses (Li et al. 2012).

10.7.8 Auxins

Auxin is a multifunctional plant hormone and has a significant role in plant growth and development as well as responding to various environmental stresses (Singh et al. 2017). Auxin, especially IAA (indole-3-acetic acid), is an essential phytohormone, considered in enhancing drought resistance in a variety of crop species, including maize, popular, and white clover, as reported by many workers (Zhang et al. 2020; Quiroga et al. 2020; Salehin et al. 2019). Overexpression of IAA potentially enhanced the drought tolerance in potatoes by expressing the green properties (Ullah et al. 2018b). Pashang et al. (2021) reported that application of IAA at a concentration of 4 g L^{-1} improved proline content, antioxidant enzyme activities, and SPAD value in safflower under water-deficit stress. Mousavi et al. (2022) examined that auxin applied at a concentration of 30 ppm alleviated the adverse effect of drought by increasing the seed yields (18%) and palmitic acid of two safflower genotypes. In drought stress, the foliar application of α-naphthalene acetic acid (NAA) at a concentration of 100 μM induced ABA and IAA-dependent hydrogen peroxide accumulation in soybean which in turn improved the antioxidant enzyme activities. Auxin application mitigates the effect of water-deficit stress by improving proline content (17.93%), SPAD value (2.51%), activities of peroxidase (14.72%), and catalase (30.53%) in safflower (Pashang et al. 2021, 2023). Gadallah (2000) observed that the foliar spray of IAA at 10 mg L^{-1} enhanced the drought stress tolerance by increasing root length and maintaining the osmotic adjustment. Transcription factors of auxin-responsive genes such as ARFs (auxin response factors), directly bind to the responsive genes and can express or repress the genes, enabling them to increase or decrease the drought tolerance in tomato (Bouzroud et al. 2018). Gene such as TLD1/OsGH3.13 produces an enzyme indole-3-acetic acid-amido synthase, which has been extensively reported in minimizing drought. This gene is involved in expressing the late embryogenesis abundant (LEA) genes to boost drought resistance. The gene such as Aux/IAA was identified in rice and studied extensively in improving the drought tolerance (Iqbal et al. 2022). However, auxin also interacts with other phytohormones to further enhance drought tolerance. Auxins regulate various gene expression to produce enzymes such as 1-aminocyclopropane-1-carboxylate synthase (ACS) and ethylene (Table 10.1). This type of interaction further improves and strengthen the resistance mechanism of drought in plants (Colebrook et al. 2014).

10.8 Conclusions

In conclusion, drought stress intensively inhibits plant growth and development, resulting in reduced yields of crops. The increasing world population's scenario is compelling researchers to develop an efficient approach for securing crop yields to ensure food security under water stress conditions. To address this challenge, phytohormones offer a promising strategy and game-changing solution to enhance drought tolerance in crops, specifically in arid and semiarid regions. This current chapter provides remarkable information on utilizing plant hormones in alleviating the adverse effects of drought stress, with a special focus on oilseed crops. The major phytohormones such as ET, ABA, GA, CK, SA, JA, BRs, and auxin play a pivotal role in increasing plant response to drought stress. These remarkable molecules regulate drought stress through diverse biological processes such as modulating gene expression and biochemical pathways. However, up- or downregulation of phytohormone-associated genes controls the regulation of expression toward drought stress. Overexpression of phytohormone-associated genes takes part in synthesizing several biochemicals, evidently involved in controlling drought stress by modulating various biochemical pathways. Various techniques, such as

Table 10.1 Effect of phytohormones on the yields and growth of oilseed crops under drought stress

Phytohormones	Oilseed plants	Functions under drought stress	References
Salicylic acid	Sesame	Improved photosynthetic activity including rooting attributes and growth under drought stress	Yousefzadeh-Najafabadi and Ehsanzadeh (2021)
Kinetin	Sesame	Activated antioxidant systems	Hussein et al. (2016)
Salicylic acid and kinetin	Sesame	Induced drought stress tolerance by improving growth and physiology	Hussein et al. (2015)
Abscisic acid	Sunflower	Improved H_2O_2, anthocyanin, and cysteine contents. Reduced soluble sugars and antioxidant enzymes	Hussain et al. (2010), Hussain et al. (2013)
Abscisic acid	Soybean	Improved H_2O_2, anthocyanin, and cysteine contents. Reduced soluble sugars, antioxidant enzymes and photosynthetic pigments	Ruan YingHui et al. (2012)
Gibberellin	*Brassica napus* L.	Increased proline, proteins, sugars, and activities of antioxidant enzymes, whereas reduced MDA content and electrolyte leakage	Li et al. (2010)
Salicylic acid	*Brassica napus* L.	Improved growth and oil quality	Ullah et al. (2012)
Cytokinin	*Brassica napus* L.	Increased peroxidase and catalase activities and chlorophyll contents	Sima et al. (2024)
ABA	Sunflower	Stomatal closure, Increased growth rate and total biomass	Hussain et al. (2012a, b)
ABA	Sunflower	Increased oil yields, plant growth and development.	Hussain et al. (2010, 2012b, 2014)
Salicylic acid	Sunflower	Improved achene and oil yields	Hussain et al. (2008)
Gibberellic acid	*Brassica napus* L.	Improved biochemical and physiological activities	Elahi et al. (2022)
Gibberellic acid, salicylic acid, and ABA	Rapeseed-mustard	Increased plant growth and biochemical parameters	Nikita Nehal et al. (2017)
Salicylic acid	Rapeseed-mustard	Enhanced antioxidant defense and glyoxalase systems.	Alam et al. (2013)
Salicylic acid + nitric oxide	Safflower	Improved growth parameters and nonenzymatic defense mechanisms	Chavoushi et al. (2019)
Brassinolide	Safflower	Improved growth and oil yield	Roghiyeh et al. (2021)
GAs+ SA	Safflower	Enhanced agronomic traits including seed and oil yields	Heydari et al. (2021)
Salicylic acid	Safflower	Enhanced agronomic traits including seed and oil yields	Heydari et al. (2021)
Abscisic acid	Peanut	Improved leaf anatomy such as vascular bundle, palisade and spongy cells	Sankar et al. (2016b)
Gibberellic acid	Cotton seed	Enhanced ATP and 3-PGA content	Pandey et al. (2002)

(continued)

Table 10.1 (continued)

Phytohormones	Oilseed plants	Functions under drought stress	References
Auxin	Cotton seed	Improved 3-PGA and pyruvate content enhanced the photosynthetic efficiency	Pandey et al. (2002)
ABA	Cotton seed	Increased seed number and lint mass per plant	Pandey et al. (2003)
Salicylic acid	Peanut	Significantly increased quantitative traits	Karimian et al. (2015)
ABA	Peanut	Improved morphological parameters	Sankar et al. (2016a)
ABA	Peanut	Improved biochemical and physiological traits	Khan et al. (2011)
ABA	Sesame	Enhanced chlorophyll contents	Abraham et al. (2008)
ABA	Cotton seed	Reduced gaseous exchange and increased seed numbers	Pandey et al. (2003)
Gibberellic acid	Cotton seed	Improved ribulose-1,5-bisphosphate, gas exchange, carboxylase/oxygenase activity, pigment content and yields	Pandey et al. (2003)
Salicylic acid	Flaxseed	Increased relative water content, leaf area, phosphorus, and mycorrhizal colonization percentage	Ansari et al. (2016)
Auxin	Safflower	Improved biochemical traits such as SPAD index, proline content, activities of catalase and peroxidase enzymes	Pashang et al. (2021)
Brassinosteroids	Safflower	Increased oil and seed yields as well as SOD, CAT, PPO, and proline	Zafari et al. (2020)
ABA	*Ricinus communis* L	Influencing gene expression involved in drought stress responses	Jardim-Messeder et al. (2023)
Gibberellins	*Ricinus communis* L	Improved growth and oil yields	Metwally et al. (2014)
Gibberellins	Hemp	Improved seed germination, antioxidant enzyme, and osmotic regulating substance	Du et al. (2022)
Gibberellins	Flaxseed	Increased rate of seed germination	Heikal et al. (1982)
Gibberellins	Sesame	Increased rate of seed germination	Heikal et al. (1982)

engineered genes, are applicable in synthesizing endogenous as well as exogenous phytohormones are mainly responsible for efficient productivity under drought stress. However, there is much more attention required, and more work remains in alleviating drought stress in arid and semiarid regions of the globe. Advanced technological approaches, including genetic manipulation, should be applied in drought-sensitive oilseed crops that could contribute to overcoming drought stress through the development of drought-resistant varieties. Additionally, the integration of plant growth-promoting microbes alters endogenous phytohormones, osmoprotectants, and secondary metabolites, serving to enhance drought tolerance in many plant species. This mechanistic interaction is an environmentally friendly and low-cost technology and has a high chance of improving crop production. Lastly, gene manipulation and genetic engineering of responsive genes could assist drought tolerance in various crops.

References

Abdel Latef AA, Tahjib-Ul-Arif M, Rhaman MS (2021a) Exogenous Auxin-mediated salt stress alleviation in faba bean (*Vicia faba* L.). Agronomy 11:547

Abdel Latef AA, Hasanuzzaman, Tahjib-Ul-Arif M (2021b) Mitigation of salinity stress by exogenous application of cytokinin in faba bean (*Vicia faba* L.). Notulae Botanicae Horti Agrobotanici Cluj-Napoca 49(1):12192

Abdel Latef AA, Akter A, Tahjib-Ul-Arif M (2021c) Foliar application of auxin or cytokinin can confer salinity stress tolerance in *Vicia faba* L. Agronomy 11:790

Abdelhameed RE, Abdel Latef AA, Shehata RS (2021) Physiological responses of salinized fenugreek (*Trigonella foenum-graecum* L.) plants to foliar application of salicylic acid. Plan Theory 10:657

Abeed AHA, Mahdy RE, Alshehri D, Hammami I, Eissa MA, Abdel Latef AA, Mahmoud GA (2022) Induction of resilience strategies against biochemical deteriorations prompted by severe cadmium stress in sunflower plant when *Trichoderma* and bacterial inoculation were used as biofertilizers. Front Plant Sci 13:1004173

Abraham SS, Jaleel CA, Chang-Xing Z, Somasundaram R, Azooz M, Manivannan P, Panneerselvam R (2008) Regulation of growth and metabolism by paclobutrazol and ABA in Sesamum indicum L. under drought condition. Global Journal of Molecular Sciences 3(2): 57–66

Adeleke BS, Babalola OO (2020) Oilseed crop sunflower (Helianthus annuus) as a source of food: nutritional and health benefits. Food Sci Nutr 8(9):4666–4684

Ahmad Z, Waraich EA, Akhtar S, Anjum S, Ahmad T, Mahboob W, Rizwan M (2018) Physiological responses of wheat to drought stress and its mitigation approaches. Acta Physiol Plant 40:1–13

Akula R, Ravishankar GA (2011) Influence of abiotic stress signals on secondary metabolites in plants. Plant Signal Behav 6(11):1720–1731

Alam MM, Hasanuzzaman M, Nahar K, Fujita M (2013) Exogenous salicylic acid ameliorates short-term drought stress in mustard (Brassica juncea L.) seedlings by up-regulating the antioxidant defense and glyoxalase system. Aust J Crop Sci 7(7):1053–1063

Ali MS, Baek K-H (2020) Jasmonic acid signaling pathway in response to abiotic stresses in plants. Int J Mol Sci 21(2):621

Ambavaram MM, Basu S, Krishnan A et al (2014) Coordinated regulation of photosynthesis in rice increases yield and tolerance to environmental stress. Nat Commun 5:5302

Ansari A, Razmjoo J, Karimmojeni H (2016) Mycorrhizal colonization and seed treatment with salicylic acid to improve physiological traits and tolerance of flaxseed (Linum usitatissimum L.) plants grown under drought stress. Acta Physiol Plant 38:1–10

Arif Y, Sami F, Siddiqui H, Bajguz A, Hayat S (2020) Salicylic acid in relation to other phytohormones in plant: a study towards physiology and signal transduction under challenging environment. Environ Exp Bot 175:104040

Arraes FBM, Beneventi MA, Lisei de Sa ME, Paixao JFR, Albuquerque EVS, Marin SRR et al (2015) Implications of ethylene biosynthesis and signaling in soybean drought stress tolerance. BMC Plant Biol 15:1–20

Ashraf F, Siddiqi EH (2024) Mitigation of drought-induced stress in sunflower (Helianthus annuus L.) via foliar application of Jasmonic acid through the augmentation of growth, physiological, and biochemical attributes. BMC Plant Biol 24:592. https://doi.org/10.1186/s12870-024-05273-4

Barahouei M, Moosavi SG, Seghatoleslami MJ, Bradaran R, Javadzadeh SM (2024) Study on stress modulators and gibberellic acid on physiological, quantitative, and qualitative traits of safflower (Carthamus Tinctorius L.) under water deficit stress. In: Quantitative, and qualitative traits of safflower (Carthamus Tinctorius L) under water deficit stress

Batool M, El-Badri AM, Hassan MU, Haiyun Y, Chunyun W, Zhenkun Y et al (2022) Drought stress in Brassica napus: effects, tolerance mechanisms, and management strategies. J Plant Growth Regul:1–25

Blum A (2011) Drought resistance–is it really a complex trait? Funct Plant Biol 38(10):753–757

Bouzroud S, Gouiaa S, Hu N, Bernadac A, Mila I, Bendaou N et al (2018) Auxin response factors (ARFs) are potential mediators of auxin action in tomato response to biotic and abiotic stress (Solanum lycopersicum). PLoS One 13(2):e0193517

Chaves M, Flexas J, Pinheiro C (2009) Photosynthesis under drought and salt stress: regulation mechanisms from whole plant to cell. Ann Bot 103(4):551–560

Chavoushi M, Najafi F, Salimi A, Angaji S (2019) Improvement in drought stress tolerance of safflower during vegetative growth by exogenous application of salicylic acid and sodium nitroprusside. Ind Crop Prod 134:168–176

Chimenti C, Pearson J, Hall A (2002) Osmotic adjustment and yield maintenance under drought in sunflower. Field Crop Res 75(2–3):235–246

Colebrook EH, Thomas SG, Phillips AL, Hedden P (2014) The role of gibberellin signalling in plant responses to abiotic stress. J Exp Biol 217(1):67–75

Cutler SR, Rodriguez PL, Finkelstein RR, Abrams SR (2010) Abscisic acid: emergence of a core signaling network. Annu Rev Plant Biol 61(1):651–679

Darini S, Naser Alavi SM, Ghanbari J (2024) The effect of abscisic acid application on improving yield, yield components, and water use efficiency of sesame (Sesamum indicum L.) under water-limited conditions. Crop Sci Res Arid Regions 6:229–244

Dawood MFA, Zaid A, Abdel Latef AA (2022) Salicylic acid spraying-induced resilience strategies against the damaging impacts of drought and/or salinity stress in two varieties of *Vicia faba* L. seedlings. J Plant Growth Regul 41:1919–1942

de Ollas C, Dodd IC (2016) Physiological impacts of ABA–JA interactions under water-limitation. Plant Mol Biol 91:641–650

de Ollas C, Hernando B, Arbona V, Gómez-Cadenas A (2013) Jasmonic acid transient accumulation is needed for abscisic acid increase in citrus roots under drought stress conditions. Physiol Plant 147(3):296–306

Desclaux D, Huynh TT, Roumet P (2000) Identification of soybean plant characteristics that indicate the timing of drought stress. Crop Sci 40(3):716–722

Du G, Zhang H, Yang Y, Zhao Y, Tang K, Liu F (2022) Effects of gibberellin pre-treatment on seed germination and seedling physiology characteristics in industrial hemp under drought stress condition. Life 12(11):1907

Easwar Rao D, Viswanatha Chaitanya K (2020) Changes in the antioxidant intensities of seven different soybean (Glycine max (L.) Merr.) cultivars during drought. J Food Biochem 44(2): e13118

Elahi NN, Raza S, Rizwan MS, Albalawi BFA, Ishaq MZ, Ahmed HM, Mehmood S, Imtiaz M, Farooq U, Rashid M (2022) Foliar application of gibberellin alleviates adverse impacts of drought stress and improves growth, physiological and biochemical attributes of canola (Brassica napus L.). Sustainability 15(1):78

Fahad S, Hussain S, Bano A, Saud S, Hassan S, Shan D et al (2015) Potential role of phytohormones and plant growth-promoting rhizobacteria in abiotic stresses: consequences for changing environment. Environ Sci Pollut Res 22:4907–4921

Fang L, Su L, Sun X, Li X, Sun M, Karungo SK et al (2016) Expression of Vitis amurensis NAC26 in Arabidopsis enhances drought tolerance by modulating jasmonic acid synthesis. J Exp Bot 67(9):2829–2845

Farooq M, Wahid A, Kobayashi N, Fujita D, Basra SM (2009) Plant drought stress: effects, mechanisms and management. Sustain Agric:153–188

Farooq M, Hussain M, Wahid A, Siddique K (2012) Drought stress in plants: an overview. In: Plant responses to drought stress: from morphological to molecular features, pp 1–33

Gadallah M (2000) Effects of indole-3-acetic acid and zinc on the growth, osmotic potential and soluble carbon and nitrogen components of soybean plants growing under water deficit. J Arid Environ 44(4):451–467

Ghaffari H, Tadayon MR, Nadeem M, Razmjoo J, Cheema M (2019) Foliage applications of jasmonic acid modulate the antioxidant defense under water deficit growth in sugar beet. Span J Agric Res 17(4):e0805–e0805

Ghasemi M, Jahanbin S, Latifmanesh H, Farajee H, Mirshekari A (2021) Effect of brassinolide foliar application on some physiological and agronomic characteristics of sunflower (Helianthus annuus L.) under drought stress conditions. J Crop Prod 14 (1):fa31–fa48

Ghobadi M, Taherabadi S, Ghobadi M-E, Mohammadi G-R, Jalali-Honarmand S (2013) Antioxidant capacity, photosynthetic characteristics and water relations of sunflower (Helianthus annuus L.) cultivars in response to drought stress. Ind Crop Prod 50:29–38

Golchin L, Tavakoli A, Mohsenifard E (2022a) Effect of cytokinin application on photosynthesis, gas exchange and seed yield of safflower (Carthamus tinctorius) under drought stress condition. Iranian J Field Crop Sci 53:149–160

Golchin L, Tavakoli A, Mohsenifard E (2022b) Effects of cytokinin application on the photoassimilate partitioning and oil percentage in safflower cultivar under drought stress condition. Iranian J Field Crop Sci 53:249–261

Hai NN, Chuong NN, Tu NHC, Kisiala A, Hoang XLT, Thao NP (2020) Role and regulation of cytokinins in plant response to drought stress. Plan Theory 9(4):422

Hasanuzzaman M, Nahar K, Alam MM, Roychowdhury R, Fujita M (2013) Physiological, biochemical, and molecular mechanisms of heat stress tolerance in plants. Int J Mol Sci 14(5): 9643–9684

Heidari M, Tohidimoghadam HR, Ghooshchi F, Modares Sanavi SAM, Kasraie P (2022) Effect of seed priming and foliar application of growth regulators on morphophysiological changes and safflower grain and oil yield under drought stress. Environ Stresses Crop Sci 15(4):1073–1089

Heikal M, Shaddad M, Ahmed A (1982) Effect of water stress and gibberellic acid on germination of flax, sesame and onion seeds. Biol Plant 24:124–129

Heydari M, Moghadam HRT, Ghooshchi F, Modarres-Sanavy SAM, Kasraie P (2021) Foliar application of humic acid and some exo-and endophytic growth hormones on yield, yield components and fatty acid composition in safflower (Carthamus tinctorius L.) under drought stress. Journal of. Agric Sci 27(4):500–508

Hoang XLT, Nguyen NC, Nguyen Y-NH, Watanabe Y, Tran L-SP, Thao NP (2019) The soybean GmNAC019 transcription factor mediates drought tolerance in Arabidopsis in an abscisic acid-dependent manner. Int J Mol Sci 21(1):286

Hossini F, Meskarbashee M, Rahnama A (2024) The effect of foliar application of cytokinin and salicylic acid on yield and physiological characteristics of sunflower under drought stress conditions. J Plant Prod Res

Huang X, Hou L, Meng J, You H, Li Z, Gong Z et al (2018) The antagonistic action of abscisic acid and cytokinin signaling mediates drought stress response in Arabidopsis. Mol Plant 11(7): 970–982

Hunter LJ, Westwood JH, Heath G, Macaulay K, Smith AG, MacFarlane SA et al (2013) Regulation of RNA-dependent RNA polymerase 1 and isochorismate synthase gene expression in Arabidopsis. PLoS One 8(6):e66530

Hussain M, Malik M, Farooq M, Ashraf M, Cheema M (2008) Improving drought tolerance by exogenous application of glycinebetaine and salicylic acid in sunflower. J Agron Crop Sci 194(3):193–199
Hussain S, Saleem M, Ashraf M, Cheema M, Haq M (2010) Abscisic acid, a stress hormone helps in improving water relations and yield of sunflower (Helianthus annuus L.) hybrids under drought. Pak J Bot 42(3):2177–2189
Hussain S, Ali A, Ibrahim M, Saleem M, Bukhsh MAAHA (2012a) Exogenous application of abscisic acid for drought tolerance in sunflower (Helianthus annuus L.): a review. J Anim. Plant Sci 22(3):806–826
Hussain S, Ma B, Saleem M, Anjum SA, Saeed A, Iqbal J (2012b) Abscisic acid spray on sunflower acts differently under drought and irrigation conditions. Agron J 104(3):561–568
Hussain S, Ahmad M, Ahmad S, Iqbal J, Subhani MN, Nadeem SM et al (2013) Improvement of drought tolerance in sunflower (Helianthus annuus L.) by foliar application of abscisic acid and potassium chloride. Pak J Nutr 12(4):345
Hussain S, Saleem MF, Iqbal J, Ibrahim M, Atta S, Ahmed T, Rehmani M (2014) Exogenous application of abscisic acid may improve the growth and yield of sunflower hybrids under drought. Pak J Agric Sci 51(1)
Hussain M, Farooq S, Hasan W, Ul-Allah S, Tanveer M, Farooq M, Nawaz A (2018) Drought stress in sunflower: physiological effects and its management through breeding and agronomic alternatives. Agric Water Manag 201:152–166
Hussain S, Farooq S, Bashir MA et al (2020) 15. Abscisic Acid (ABA) mitigates drought stress in sunflower by enhancing water relations and osmotic adjustments. Pure Appl Biol (PAB) 10: 182–193
Hussein Y, Amin G, Azab A, Gahin H (2015) Induction of drought stress resistance in sesame (Sesamum indicum L.) plant by salicylic acid and kinetin. J Plant Sci 10(4):128
Hussein Y, Amin G, Gahin H (2016) Research article antioxidant activities during drought stress resistance of sesame (Sesamum indicum L.) plant by salicylic acid and kinetin. Res J Bot 11(1/3):1–8
Iqbal S, Wang X, Mubeen I, Kamran M, Kanwal I, Díaz GA et al (2022) Phytohormones trigger drought tolerance in crop plants: outlook and future perspectives. Front Plant Sci 12:799318
Jaleel CA, Manivannan P, Wahid A, Farooq M, Al-Juburi HJ, Somasundaram R, Panneerselvam R (2009) Drought stress in plants: a review on morphological characteristics and pigments composition. Int J Agric Biol 11(1):100–105
Jamil M, Ahamd M, Anwar F, Zahir ZA, Kharal MA, Nazli F (2018) Inducing drought tolerance in wheat through combined use of l-tryptophan and Pseudomonas fluorescens. Pak J Agric Sci 55(2):331
Jamshidi Zinab A, Hasanloo T, Naji AM, Delangiz N, Farhangi-Abriz S, Asgari Lajayer B, Farooq M (2023) Physiological and biochemical evaluation of commercial oilseed rape (Brassica napus L.) cultivars under drought stress. Gesunde Pflanzen 75(4):847–860
Jan AU, Hadi F, Akbar F, Shah A (2019) Role of potassium, zinc and gibberellic acid in increasing drought stress tolerance in sunflower (Helianthus annuus L.). Pak J Bot 51(3):809–815
Jang G, Chang SH, Um TY, Lee S, Kim J-K, Choi YD (2017) Antagonistic interaction between jasmonic acid and cytokinin in xylem development. Sci Rep 7(1):10212
Jardim-Messeder D, Cassol D, Souza-Vieira Y, Loureiro ME, Girke T, Boroni M, Sachetto-Martins G (2023) Genome-wide identification of core components of ABA signaling and transcriptome analysis reveals gene circuits involved in castor bean (Ricinus communis L.) response to drought. Gene 883:147668
Jha Y, Mohamed HI (2022) Plant secondary metabolites as a tool to investigate biotic stress tolerance in plants: a review. Gesunde Pflanzen 74(4):771–790
Jin Y, Pan W, Zheng X et al (2018) OsERF101, an ERF family transcription factor, regulates drought stress response in reproductive tissues. Plant Mol Biol 98:51–65
Jogaiah S, Govind SR, Tran L-SP (2013) Systems biology-based approaches toward understanding drought tolerance in food crops. Crit Rev Biotechnol 33(1):23–39

Jung SE, Bang SW, Kim SH et al (2021) Overexpression of OsERF83, a vascular tissue-specific transcription factor gene, confers drought tolerance in rice. Int J Mol Sci 22:7656

Kang G, Li G, Xu W, Peng X, Han Q, Zhu Y, Guo T (2012) Proteomics reveals the effects of salicylic acid on growth and tolerance to subsequent drought stress in wheat. J Proteome Res 11(12):6066–6079

Karasulu HY, Karasulu EM, Büyükhelvacıgil MT, Yıldız M, Ertugrul A, Büyükhelvacıgil K et al (2011) Soybean oil: production process, benefits and uses in pharmaceutical dosage form. In: Soybean and health. IntechOpen

Kareem F, Rihan H, Fuller MP (2019) The effect of exogenous applications of salicylic acid on drought tolerance and up-regulation of the drought response regulon of Iraqi wheat. J Crop Sci Biotechnol 22:37–45

Karimian MA, Dahmardeh M, Bidarnamani F, Forouzandeh M (2015) Assessment quantitative and qualitative factors of peanut (Arachis hypogaea L.) under drought stress and salicylic acid treatments. Paper presented at the Biological Forum

Khalili M, Aboughadareh AP, Naghavi MR, Talebzadeh S (2012) Response of spring canola (Brassica napus L.) genotypes to water deficit stress. Int J Agric Crop Sci 4:1579–1586

Khan A, Bakht J, Bano A, Malik NJ (2011) Effect of plant growth regulators and drought stress on groundnut (Arachis hypogaea L) genotypes. Pak J Bot 43(5):2397–2402

Khan MN, Khan Z, Luo T, Liu J, Rizwan M, Zhang J, Xu Z, Wu H, Hu L (2020) Seed priming with gibberellic acid and melatonin in rapeseed: consequences for improving yield and seed quality under drought and non-stress conditions. Ind Crop Prod 156:112850

Khan MAH, Bonifacio S, Clowes J, Foulds A, Holland R, Matthews JC et al (2021) Investigation of biofuel as a potential renewable energy source. Atmos 12(10):1289

Kharkeshi AM, Petroudi ER, Shahmiri FS, Mobasser H, Daneshmand A (2023) Variations of yield, biochemical and antioxidative responses in sesame with silicon and cytokinin treatments under drought stress. Acta Sci Pol Hortorum Cultus 22(6):43–54

Kosar F, Alshallash KS, Akram NA, Sadiq M, Ashraf M, Alkhalifah DHM, Abdel Latef AA, Elkelish A (2022) Trehalose-induced regulations in nutrient status and secondary metabolites of drought-stressed sunflower (*Helianthus annuus* L.). Plan Theory 11:2780

Kramer PJ, Boyer JS (1995) Water relations of plants and soils. Academic Press

Kumar S, Sachdeva S, Bhat K, Vats S (2018) Plant responses to drought stress: physiological, biochemical and molecular basis. In: Biotic and abiotic stress tolerance in plants, pp 1–25

Kumar M, Kumari N, Prakash S, Sharma N, Rajat R et al (2024) Oilseed meal as a source of protein: introductory remarks. In: Oilseed meal as a sustainable contributor to plant-based protein: paving the way towards circular economy and nutritional security. Springer, pp 1–29

La VH, Lee B-R, Zhang Q, Park S-H, Islam MT, Kim T-H (2019) Salicylic acid improves drought-stress tolerance by regulating the redox status and proline metabolism in Brassica rapa. Hortic Environ Biotechnol 60:31–40

Li Z, Lu G, Zhang X, Zou C, Cheng Y, Zheng P (2010) Improving drought tolerance of germinating seeds by exogenous application of gibberellic acid (GA3) in rapeseed (Brassica napus L.). Seed Sci Technol 38(2):432–440

Li Q-F, Wang C, Jiang L, Li S, Sun SS, He J-X (2012) An interaction between BZR1 and DELLAs mediates direct signaling crosstalk between brassinosteroids and gibberellins in Arabidopsis. Sci Signal 5(244):ra72-ra72

Liu C, Zhang T (2017) Expansion and stress responses of the AP2/EREBP superfamily in cotton. BMC Genomics 18:1–16

Liu G, Li X, Jin S, Liu X, Zhu L, Nie Y, Zhang X (2014) Overexpression of rice NAC gene SNAC1 improves drought and salt tolerance by enhancing root development and reducing transpiration rate in transgenic cotton. PLoS One 9(1):e86895

Lo SF, Ho THD, Liu YL, Jiang MJ, Hsieh KT, Chen KT et al (2017) Ectopic expression of specific GA 2 oxidase mutants promotes yield and stress tolerance in rice. Plant Biotechnol J 15(7): 850–864

Lobell DB, Hammer GL, McLean G, Messina C, Roberts MJ, Schlenker W (2013) The critical role of extreme heat for maize production in the United States. Nat Clim Chang 3(5):497–501

Luo L, Xia H, Lu B-R (2019) Crop breeding for drought resistance. In: In, vol 10. Frontiers Media SA, p 314

Mahdi A, Taha R, Emam SM (2020) Foliar applied salicylic acid improves water deficit-tolerance in Egyptian cotton. J Plant Prod 11(5):383–389

Mehrotra R, Bhalothia P, Bansal P, Basantani MK, Bharti V, Mehrotra S (2014) Abscisic acid and abiotic stress tolerance–different tiers of regulation. J Plant Physiol 171(7):486–496

Meijaard E, Brooks TM, Carlson KM, Slade EM, Garcia-Ulloa J, Gaveau DL et al (2020) The environmental impacts of palm oil in context. Nat Plants 6(12):1418–1426

Metwally SA, Mohamed SLM, Abou-Leila BH, Aly MS (2014) Effect of drought stress and helium neon (He-Ne) laser rays on growth, oil yield and fatty acids content in Caster bean (Ricinus communis L.). Agric For Fish 3(3):203–208

Misra BB, Acharya BR, Granot D, Assmann SM, Chen S (2015) The guard cell metabolome: functions in stomatal movement and global food security. Front Plant Sci 6:334

Mittler R (2002) Oxidative stress, antioxidants and stress tolerance. Trends Plant Sci 7(9):405–410

Mo Y, Wang Y, Yang R, Zheng J, Liu C, Li H et al (2016) Regulation of plant growth, photosynthesis, antioxidation and osmosis by an arbuscular mycorrhizal fungus in watermelon seedlings under well-watered and drought conditions. Front Plant Sci 7:644

Mohamed HI, Latif HH (2017) Improvement of drought tolerance of soybean plants by using methyl jasmonate. Physiol Mol Biol Plants 23:545–556

Montillet J-L, Leonhardt N, Mondy S, Tranchimand S, Rumeau D, Boudsocq M et al (2013) An abscisic acid-independent oxylipin pathway controls stomatal closure and immune defense in Arabidopsis. PLoS Biol 11(3):e1001513

Mousavi SB, Sayfzadeh S, Jabbari H, Valadabadi SA, Masouleh EH (2022) Effect of auxin foliar application on seed yield and fatty acids composition of two safflower genotypes under late-season drought. Plant Soil Environ 68(2):82

Mughal N, Shoaib N, Chen J, He Y, Fu M, Li X et al (2024) Adaptive roles of cytokinins in enhancing plant resilience and yield against environmental stressors. Chemosphere 143189: 143189

Munteanu V, Gordeev V, Martea R, Duca M (2014) Effect of gibberellin cross talk with other phytohormones on cellular growth and mitosis to endoreduplication transition. Int J Adv Res Biol Sci 1(6):136–153

Mustafa G, Iqbal A (2021) Economics of oil plants: demand, supply, and international trade. In: Oil Crop Genomics. Springer, Cham, pp 393–413

Naeem M, Basit A, Ahmad I, Mohamed HI, Wasila H (2020) Effect of salicylic acid and salinity stress on the performance of tomato plants. Gesunde Pflanzen 72(4):393

Nakabayashi R, Yonekura-Sakakibara K, Urano K, Suzuki M, Yamada Y, Nishizawa T et al (2014) Enhancement of oxidative and drought tolerance in Arabidopsis by overaccumulation of antioxidant flavonoids. Plant J 77(3):367–379

Naveen N, Kumari N, Avtar R, Jattan M, Ahlawat S, Rani B, Malik K, Sharma A, Singh M (2021) Evaluation of effect of brassinolide in Brassica juncea leaves under drought stress in field conditions. Horticulturae 7(11):514

Nazir F, Peter P, Gupta R et al (2024) Plant hormone ethylene: a leading edge in conferring drought stress tolerance. Physiol Plant 176:e14151. https://doi.org/10.1111/ppl.14151

Nikita Nehal NN, Nitish Sharma NS, Monika Singh MS, Prashansha Singh PS, Priyanka Rajpoot PR, Pandey AK et al (2017) Effect of plant growth regulators on growth, biochemical and yield of Indian mustard [Brassica juncea (L.) Czern. & Coss.] under drought stress condition. Plant Arch 17(1):580–584

Nir I, Moshelion M, Weiss D (2014) The Arabidopsis gibberellin methyl transferase 1 suppresses gibberellin activity, reduces whole-plant transpiration and promotes drought tolerance in transgenic tomato. Plant Cell Environ 37(1):113–123

Nosherwan M, Raza A, Ali MH, Noor HMA, Abbas J, Qureshi IS, Rehman A, Rana M, Moosa A, Ayub HM (2024) Mitigating the adverse effects of drought stress on sunflower plants via foliar application of salicylic acid. Plant Prot 8(4):759–768

O'Brien JA, Benková E (2013) Cytokinin cross-talking during biotic and abiotic stress responses. Front Plant Sci 4:451

Pandey D, Goswami C, Kumar B, Jain S (2002) Effect of growth regulators on photosynthetic metabolites in cotton under water stress. Biol Plant 45:445–448

Pandey D, Goswami C, Kumar B (2003) Physiological effects of plant hormones in cotton under drought. Biol Plant 47:535–540

Pashang D, Weisany W, Ghajar FG-K (2021) Changes in the fatty acid and morphophysiological traits of safflower (Carthamus tinctorius L.) cultivars as response to auxin under water-deficit stress. Journal of. Soil Sci Plant Nutr 21(3):2164–2177

Pashang D, Weisany W, Ghasem-Khan Ghajar F (2023) Effect of foliar application of auxin on morphophysiological and biochemical traits of safflower cultivars (spring and autumn) under drought stress. Env Stresses Crop Sci 16(1):19–33

Pavlů J, Novák J, Koukalová V, Luklová M, Brzobohatý B, Černý M (2018) Cytokinin at the crossroads of abiotic stress signalling pathways. Int J Mol Sci 19(8):2450

Peleg Z, Blumwald E (2011) Hormone balance and abiotic stress tolerance in crop plants. Curr Opin Plant Biol 14(3):290–295

Pérez-Llorca M, Pollmann S, Müller M (2023) Ethylene and jasmonates signaling network mediating secondary metabolites under abiotic stress. Int J Mol Sci 24(6):5990

Qaderi MM, Kurepin LV, Reid DM (2006) Growth and physiological responses of canola (Brassica napus) to three components of global climate change: temperature, carbon dioxide and drought. Physiol Plant 128(4):710–721

Quiroga G, Erice G, Aroca R, Zamarreño ÁM, García-Mina JM, Ruiz-Lozano JM (2020) Radial water transport in arbuscular mycorrhizal maize plants under drought stress conditions is affected by indole-acetic acid (IAA) application. J Plant Physiol 246:153115

Rhaman MS, Rauf F, Tania SS, Karim MM, Sagar A, Robin AHK, Abdel Latef AA, Murata Y (2021) Seed priming and exogenous application of salicylic acid enhance growth and productivity of okra (Abelmoschus esculentus L.) by regulating photosynthetic attributes. J Exp Biol Agric Sci 9(6):759–769

Riemann M, Dhakarey R, Hazman M, Miro B, Kohli A, Nick P (2015) Exploring jasmonates in the hormonal network of drought and salinity responses. Front Plant Sci 6:1077

Roghiyeh F-A, Kazem G-G, Safar N (2021) Grain and oil yields of safflower (Carthamus Tinctorius L.) affected by water deficit and growth regulators. Agriculture 67(2):87–94

Rowe JH, Topping JF, Liu J, Lindsey K (2016) Abscisic acid regulates root growth under osmotic stress conditions via an interacting hormonal network with cytokinin, ethylene and auxin. New Phytol 211(1):225–239

Hui RY, Kun DS, Jun LL, Shu SC, Bin WL, Qian GQ, Jia GZ (2012) Effect of exogenous abscisic acid (ABA) on the physiological characteristics during soybean flowering under drought stress. Soybean Science 31(3):385–388

Ruan J, Zhou Y, Zhou M, Yan J, Khurshid M, Weng W et al (2019) Jasmonic acid signaling pathway in plants. Int J Mol Sci 20(10):2479

Salehin M, Li B, Tang M, Katz E, Song L, Ecker JR et al (2019) Auxin-sensitive Aux/IAA proteins mediate drought tolerance in Arabidopsis by regulating glucosinolate levels. Nat Commun 10(1):4021

Sánchez-Romera B, Ruiz-Lozano JM, Li G, Luu DT, Martínez-Ballesta MDC, Carvajal M, Aroca R (2014) Enhancement of root hydraulic conductivity by methyl jasmonate and the role of calcium and abscisic acid in this process. Plant Cell Environ 37(4):995–1008

Sankar B, Gopinathan P, Karthishwaran K, Somasundaram R (2016a) Variation in growth of peanut plants under drought stress condition and in combination with paclobutrazol and ABA. Curr Bot 5:14–21

Sankar B, Karthishwaran K, Somasundaram R (2016b) Leaf anatomical changes in peanut plants in relation to drought stress with or without paclobutrazol and ABA. J Phytology 5:25–29
Shan C, Zhou Y, Liu M (2015) Nitric oxide participates in the regulation of the ascorbate-glutathione cycle by exogenous jasmonic acid in the leaves of wheat seedlings under drought stress. Protoplasma 252:1397–1405
Shao H-B, Chu L-Y, Jaleel CA, Manivannan P, Panneerselvam R, Shao M-A (2009) Understanding water deficit stress-induced changes in the basic metabolism of higher plants–biotechnologically and sustainably improving agriculture and the ecoenvironment in arid regions of the globe. Crit Rev Biotechnol 29(2):131–151
Sharma P, Jha AB, Dubey RS, Pessarakli M (2012) Reactive oxygen species, oxidative damage, and antioxidative defense mechanism in plants under stressful conditions. J Bot 2012(1):217037
Shin N-H, Trang DT, Hong W-J, Kang K, Chuluuntsetseg J, Moon J-K et al (2019) Rice senescence-induced receptor-like kinase (OsSRLK) is involved in phytohormone-mediated chlorophyll degradation. Int J Mol Sci 21(1):260
Shinozaki K, Yamaguchi-Shinozaki K (2007) Gene networks involved in drought stress response and tolerance. J Exp Bot 58(2):221–227
Sima NAK, Jabbari H, Ebadi A, Ghaffari MR, Koobaz P (2024) Comparative analysis of exogenous hormone application on contrasting canola (Brassica napus L.) genotypes under drought stress conditions. J Soil Sci Plant Nutr 24(1):308–317
Singh S, Singh D (2010) Biodiesel production through the use of different sources and characterization of oils and their esters as the substitute of diesel: a review. Renew Sust Energ Rev 14(1): 200–216
Singh VP, Prasad SM, Munné-Bosch S, Müller M (2017) Phytohormones and the regulation of stress tolerance in plants: current status and future directions. Front Plant Sci 8:1871
Sreenivasulu N, Harshavardhan VT, Govind G, Seiler C, Kohli A (2012) Contrapuntal role of ABA: does it mediate stress tolerance or plant growth retardation under long-term drought stress? Gene 506(2):265–273
Su L, Fang L, Zhu Z, Zhang L, Sun X, Wang Y et al (2020) The transcription factor VaNAC17 from grapevine (Vitis amurensis) enhances drought tolerance by modulating jasmonic acid biosynthesis in transgenic A rabidopsis. Plant Cell Rep 39:621–634
Taiz L, Zeiger E (2002) Plant physiology (Third Edition). Sinauer Associates, Inc., Sunderland, Publishers, 67–86
Tiwari S, Lata C, Chauhan PS, Nautiyal CS (2016) Pseudomonas putida attunes morphophysiological, biochemical and molecular responses in Cicer arietinum L. during drought stress and recovery. Plant Physiol Biochem 99:108–117
Tiwari S, Lata C, Singh Chauhan P, Prasad V, Prasad M (2017) A functional genomic perspective on drought signalling and its crosstalk with phytohormone-mediated signalling pathways in plants. Curr Genomics 18(6):469–482
Tran TT, Tran HT, Bui VT (2024) Effects of abscisic acid and sodium nitroprusside priming on yield and quality of peanut (Arachis hypogaea L.) under under drought stress. Hortic Sci 51:219
Ullah F, Bano A, Nosheen A (2012) Effects of plant growth regulators on growth and oil quality of canola (Brassica napus L.) under drought stress. Pak J Bot 44(6):1873–1880
Ullah A, Manghwar H, Shaban M, Khan AH, Akbar A, Ali U et al (2018a) Phytohormones enhanced drought tolerance in plants: a coping strategy. Environ Sci Pollut Res 25:33103–33118
Ullah R, ud Din S, Muhammad Z, Shah S, Jan SA (2018b) Biological efficacy of phyto-synthetic silver nanoparticles using ethanol extract of Euphorbia wallichii Hook Rhizome as bio-reductant and surfactant. Trop J Pharm Res 17(10):1903–1909
Verma V, Ravindran P, Kumar PP (2016) Plant hormone-mediated regulation of stress responses. BMC Plant Biol 16:1–10
Vojta P, Kokáš F, Husičková A, Grúz J, Bergougnoux V, Marchetti CF et al (2016) Whole transcriptome analysis of transgenic barley with altered cytokinin homeostasis and increased tolerance to drought stress. New Biotechnol 33(5):676–691

Wani SH, Kumar V, Shriram V, Sah SK (2016) Phytohormones and their metabolic engineering for abiotic stress tolerance in crop plants. Crop J 4(3):162–176

Wilkinson S, Davies WJ (2010) Drought, ozone, ABA and ethylene: new insights from cell to plant to community. Plant Cell Environ 33(4):510–525

Wu L, Zhang Z, Zhang H et al (2008) Transcriptional modulation of ethylene response factor protein JERF3 in the oxidative stress response enhances tolerance of tobacco seedlings to salt, drought, and freezing. Plant Physiol 148:1953–1963

Xiong L, Zhu J-K (2003) Regulation of abscisic acid biosynthesis. Plant Physiol 133(1):29–36

Yamaguchi S (2008) Gibberellin metabolism and its regulation. Annu Rev Plant Biol 59(1): 225–251

Yamaguchi-Shinozaki K, Shinozaki K (2006) Transcriptional regulatory networks in cellular responses and tolerance to dehydration and cold stresses. Annu Rev Plant Biol 57(1):781–803

Yoon JY, Hamayun M, Lee S-K, Lee I-J (2009) Methyl jasmonate alleviated salinity stress in soybean. J Crop Sci Biotechnol 12:63–68

Yousefzadeh-Najafabadi M, Ehsanzadeh P (2021) Correlative associations of photosynthetic and rooting attributes of sesame: drought-induced reversed associations are corrected upon salicylic acid exposure. S Afr J Bot 142:266–273

Zafari M, Ebadi A, Jahanbakhsh S, Sedghi M (2020) Safflower (Carthamus tinctorius) biochemical properties, yield, and oil content affected by 24-epibrassinosteroid and genotype under drought stress. J Agric Food Chem 68(22):6040–6047

Zhang J, Jia W, Yang J, Ismail AM (2006) Role of ABA in integrating plant responses to drought and salt stresses. Field Crop Res 97(1):111–119

Zhang Y, Li Y, Hassan MJ, Li Z, Peng Y (2020) Indole-3-acetic acid improves drought tolerance of white clover via activating auxin, abscisic acid and jasmonic acid related genes and inhibiting senescence genes. BMC Plant Biol 20:1–12

Zhu J-K (2016) Abiotic stress signaling and responses in plants. Cell 167(2):313–324

Chapter 11
Phytohormones in Waterlogging and Flooding Resilience of Oilseed Crops

Ali Bandehagh, Zahra Dehghanian, and Nicolas L. Taylor

Abstract Climate change is thoroughly linked to shifts in precipitation patterns and an increased frequency of flooding events. Flooding stress has emerged as a significant hazard to important oilseed crops, adversely impacting plant growth and development. Oxygen availability to plant roots is severely limited during waterlogged stress, resulting in hypoxic or anoxic environments. To manage such stresses, plants initiate a range of physiological, morphological, and biochemical adaptations, including the development of adventitious roots (ARs) and aerenchyma tissues, as well as the regulation of endogenous hormones and metabolites. Phytohormones are key modulators of plant adaptation to waterlogging stress, where their interactions strongly influence plant resilience. Critical genes and signaling pathways orchestrate these adaptive responses. Recent research has highlighted the intricacy of these phytohormone responses and the potential to exploit this knowledge to enhance crop tolerance. By understanding the precise roles of hormones such as ethylene, abscisic acid, and gibberellins, scientists are uncovering new strategies for mitigating the damaging impacts of waterlogging. Additionally, there is increasing interest in using phytohormones as exogenous treatments to strengthen plant defenses against flooding. Molecular studies have advanced our knowledge of the underlying metabolic pathways and target genes that enhance plant flooding tolerance. However, more research is required to fully elucidate the genetic regulation and functionally characterize all of the genes related to flooding tolerance. Additionally, innovative methods such as gene and genome editing and advanced breeding approaches hold promise for developing varieties with enhanced tolerance to waterlogging. This chapter provides a comprehensive review of the molecular mechanisms governing plant responses to waterlogging, focusing on the role of

A. Bandehagh (✉) · N. L. Taylor
School of Molecular Sciences, Australian Plant Phenomics Network, ARC Training Centre in Predictive Breeding for Agricultural Futures and The UWA Institute of Agriculture, The University of Western Australia, Boorloo, WA, Australia
e-mail: ali.bandehagh@uwa.edu.au; nicolas.taylor@uwa.edu.au

Z. Dehghanian
Department of Biotechnology, Faculty of Agriculture, Azarbaijan Shahid Madani University, Tabriz, Iran

© The Author(s), under exclusive license to Springer Nature Singapore Pte Ltd. 2025

A. A. H. Abdel Latef (ed.), *Oilseed Crops Under Abiotic Stress*, Sustainability Sciences in Asia and Africa, https://doi.org/10.1007/978-981-96-8346-8_11

phytohormones on plant growth, development, and productivity of oilseed crops. Future research challenges and opportunities are also discussed, particularly in developing climate-resilient agricultural systems.

Keywords Molecular mechanisms · Oilseed crops · Phytohormone regulation · Waterlogging stress · Plant adaptation · Flooding tolerance

11.1 Introduction

Oilseed crops play a crucial role in subtropical and tropical agriculture, primarily storing energy in the form of oil. As a highly nutritious food source for humans and animals, their economic significance has grown dramatically since domestication (Zandberg et al. 2022). Sesame (*Sesamum indicum* L.) is an annual oilseed crop that produces one of the most nutrient-rich edible oils compared with other major oilseeds cultivated globally, such as rapeseed (*Brassica napus*), soybean (*Glycine max* L.), sunflower (*Helianthus annuus*), and peanut (*Arachis hypogaea*) (Pathak et al. 2017).

Waterlogging results from excessive soil moisture and significantly negatively impacts crop development and yield (Tahjib-Ul-Arif et al. 2023). Recently, climate change has become a significant abiotic stress affecting vast regions across the globe (Aslam et al. 2023). Prolonged flooding reduces oxygen availability for plants, leading to oxygen scarcity (hypoxia or anoxia). This deficiency directly damages the root structure and indirectly causes chlorosis and leaf wilting (Srivastava et al. 2024). Prolonged waterlogging has significant adverse effects on root and shoot growth, potentially leading to plant wilting and even death, depending on the level of plant tolerance and the conditions of the surrounding environment (Topali et al. 2024). Hypoxic conditions suppress oxygen-dependent processes, inhibiting carbon assimilation, and reducing photosynthesis and oxygen transport between shoots and roots is adversely affected (de Souza et al. 2017). Oxygen limitation can also restrict plant growth due to changes in nutrient uptake and metabolism, resulting in reactive oxygen species (ROS) accumulation that initiates oxidative stress. Furthermore, flooding leads to significant losses in plant biomass and yield across different crops (Pais et al. 2022; Paul et al. 2021). Alternatively, plants have evolved several mechanisms to endure submergence conditions. One example is quiescence, where plants refrain from elongating their shoots during flooding to conserve energy and carbohydrates, allowing them to regrow once the water recedes. This approach is effectively employed by *Rumex acetosa* and soybean species, which can survive temporary deep floods (Parlanti et al. 2011; van Veen et al. 2024).

Escape is another strategy, where plants quickly elongate their stems or petioles to enable leaf aeration by reaching the water surface. This mechanism is employed by species such as *Rumex palustris* and deepwater rice (Jia et al. 2021; Nagai et al. 2020). These observations suggest that plants have evolved different strategies to manage the negative effects of waterlogging conditions. Studies show that under waterlogging stress, plants differentially express genes associated with signal transduction, phytohormone metabolism, transcriptional regulation, alcohol fermentation, sugar metabolism, suppression of the ROS-scavenging system and suffer from organelle dysfunction, cell wall loosening, and ubiquitin-mediated proteolysis (Komatsu et al. 2013b; Wang and Komatsu 2018). In certain instances, plants synthesize and translocate hormones to regulate their responses to waterlogging through a range of intricate signaling pathways (Yamauchi et al. 2014). These findings demonstrate that plants can adapt to flooding conditions by implementing a coordinated array of physiological adjustments governed by hormonal signaling. Phytohormones, including ethylene (ET), gibberellic acid (GA), and abscisic acid (ABA), are crucial in orchestrating the acclimation response to waterlogging stress (Komatsu et al. 2013a). Ethylene biosynthesis and physical entrapment have been shown to induce growth responses in plants during flooding stress. Ethylene stimulates the expression of Sub1A-1 at low concentrations, subsequently limiting further ethylene production during submergence (Kuroha and Ashikari 2020). A reduction in ABA abundance and a rise in GA synthesis have been observed under waterlogging stress, promoting shoot elongation to enable leaf emergence from floodwaters (Wang and Komatsu 2022). Additionally, ABA effectively delays cell death induced by ethylene and promoted by GA (Qi et al. 2023).

These findings suggest that phytohormones, including ABA, GA, and ET, work synergistically to facilitate plant adaptation to excess water. These hormones serve as key factors exhibiting significant crosstalk in their signaling responses under waterlogging conditions to control stress adaptation as well as the growth and development of plants. In recent decades, omics approaches and quantitative trait loci (QTL) mapping have been utilized to elucidate plant strategies to survive or adapt to flooding, emphasizing the pivotal role of plant hormones. This chapter summarizes the physiological regulations in plants under flooding stress and highlights how plant adaptations to flooding are mediated by hormonal signaling, particularly ABA, GA, ET, etc. Additionally, this review discusses the interactions among various phytohormone-mediated strategies that plants employ for adaptation and survival during flooding. It emphasizes the mechanisms of phytohormonal responses in plants under flooding conditions, paving the way for new agricultural applications to enhance plant growth in diverse flooding scenarios.

11.1.1 Overview of Waterlogging Stress in Oilseed Crops

Waterlogging stress has been progressively increasing globally, strongly impacting plant performance and yield (Hirabayashi et al. 2013). Almost all crop plants, including important oilseed crops, are susceptible to submersion of their roots and aerial tissues. When subjected to waterlogging, these tissues experience a substantial decrease in respiration and photosynthesis, ultimately reducing crop yield significantly (Yeung et al. 2018a, b), as has been observed in soybean (Linkemer et al. 1998), maize (Qiu et al. 2007), and canola (Wang et al. 2020a, b; Xu et al. 2015; Zou et al. 2014), among others.

11.1.2 Impact of Climate Change on Precipitation and Flooding

Adverse abiotic stresses, including excessive precipitation and flooding, have become highly unpredictable and intensive due to inadequate drainage systems, climate change, and enhanced land degradation. These can reduce plant growth and productivity, especially in areas susceptible to flooding (Bi et al. 2020; Combs-Giroir and Gschwend 2024; Tabari 2020; van Veen et al. 2013; Yeung et al. 2018a, b). Global warming is anticipated to intensify extreme rainfall through greater concentrations of atmospheric water vapor (Tabari 2020; Westra et al. 2014). Rainfall-induced floods are among the most harmful natural hazards globally (Hallegatte et al. 2013; Mastrantonas et al. 2020; Westra et al. 2014), affecting crop productivity (den Besten et al. 2021), human well-being (Ding et al. 2014), and urban infrastructure (Islam and Raja 2021).

11.1.3 Significance of Understanding Flooding Stress in Agriculture

Evidence shows that human-induced factors and their consequences intensify global waterlogging-related natural disasters. Under flooding stress, plants experience limitations in gas exchange, light intensity, and nutrient uptake, resulting in reduced photosynthetic activity and, ultimately, crop yield (Nishiuchi et al. 2012; Yan et al. 2018). Also, toxic compounds such as ROS, lactic acid, and aldehydes accumulate in plants under prolonged waterlogging, leading to cell senescence and death (Voesenek and Bailey-Serres 2013). Therefore, understanding the impacts of flooding stress on agriculture is paramount.

11.1.4 Historical Perspective on Waterlogging Stress

Waterlogging is an abiotic stress that plants experience even under natural conditions due to the overuse and/or poor management of irrigation. However, under excessive rainfall conditions in tropical and subtropical regions, crops are often exposed to waterlogging stress. In recent decades, climate change has drastically impacted the natural environment, causing unpredictable rainfall, flooding/waterlogging, and ultimately influencing plant growth and yield (Rupngam and Messiga 2024). It has been shown that the impacts of short-term waterlogging vary with plant developmental stage (Grassini et al. 2007; Yasumoto et al. 2011). However, all developmental stages, including seed emergence, root and shoot growth, and leaf expansion, are affected (Orchard and Jessop 1984; Singh 2015; Yasumoto et al. 2011) in sunflower.

11.2 Physiological Mechanisms Involved in Oilseed Crops Under Flooding Stress

Oxygen deprivation is the most important impact of waterlogging, resulting in hypoxic/anoxic conditions. Plants experience a variety of morphological, physiological, and biochemical alterations to deal with these conditions, including regulation of phytohormones and plant metabolism, the development of adventitious roots and aerenchyma, and adjustment of plant developmental and reproductive processes to escape stress conditions. In addition, flooding disrupts plant organelle homeostasis (Voesenek and Bailey-Serres 2013), reducing plant biomass and productivity (Kuroha et al. 2018; Ploschuk et al. 2021, 2023; Voesenek and Bailey-Serres 2013; Wollmer et al. 2018; Xu et al. 2015; Zhou and Lin 1995). Sunflower, maize, and soybean seeds all showed a significant decline in germination rate when exposed to excess water (Albuquerque and Carvalho 2003). Oilseed rape, which is commonly considered intolerant to waterlogging (Voesenek et al. 1999), has been studied intensively better to understand the adverse impacts on plant growth and productivity. Plants subjected to this stress experienced a significant reduction in seedling growth (Zou et al. 2014), root growth (Guo et al. 2020; Nabloussi et al. 2019; Pigliucci and Kolodynska 2002; Zou et al. 2014), plant height (Zhou and Lin 1995), adventitious root formation (Zou et al. 2014), primary branching (Ploschuk et al. 2020; Ploschuk et al. 2023; Zhou and Lin 1995), leaf area (Liu et al. 2020; Ploschuk et al. 2021), and seed yield (Ploschuk et al. 2020; Ploschuk et al. 2021, 2023; Wollmer et al. 2018; Xu et al. 2015; Zhou and Lin 1995; Zou et al. 2014). Studies have identified early senescence and maturation triggered by the regulation and signaling of ABA and ethylene as significant causes of these physiological changes (Zhu et al. 2023).

11.2.1 Oxygen Deprivation and Energy Crisis

Plants subjected to waterlogging experience significant decreases in gas diffusion, reducing photosynthesis and respiration, ultimately leading to reduced growth due to a carbon and energy crisis (Yeung et al. 2018a, b). This energy deficiency limits nutrient uptake and photosynthesis processes and can ultimately lead to cell death (Colmer and Greenway 2011; Malik et al. 2011). In sunflowers (*Helianthus annuus* L.), for instance, oxygen deprivation in the soil drives a metabolic flux reorganization, resulting in a rapid energy crisis and plant death (Paul et al. 2021). Additionally, flooding disrupts the oxidative decomposition of gases such as ethylene and carbon dioxide produced by roots and soil microorganisms. This accumulation of gases can negatively impact root growth and function. For instance, high levels of ethylene may restrict root extension, while increased carbon dioxide can severely damage the roots of certain species like soybean (*Glycine max*), but does not affect rice growth (Soares et al. 2021). Oilseed rape (*Brassica napus* L.) was subjected to 14 days of waterlogging at two key growth stages: stem elongation (BBCH 31) and floral bud formation (BBCH 55). Waterlogging during BBCH 31 hindered plant growth and nutrient absorption, resulting in decreased yields. Conversely, waterlogging at BBCH 55 harmed primary racemes but mitigated yield losses through enhanced production of secondary racemes (Wollmer et al. 2018).

11.2.1.1 Effects of Hypoxia on Root Metabolism

Hypoxia, involving both hypoxic events (waterlogging) and oxidative burst following re-aeration, has been shown to severely impact the root hydraulic system, the primary factor limiting gas exchange in leaves (León et al. 2021; Liu and Zwiazek 2022). Low oxygen conditions inhibit root hydraulic conductivity (Kamaluddin and Zwiazek 2002), and therefore, successful recovery from root hypoxia after waterlogging stress requires rapid water transport resumption to leaves (Maurel and Nacry 2020). In oilseed canola (*Brassica napus*), for instance, plants experienced a severe effect on their root hydraulic properties and gas exchange ability during re-aeration (Liu and Zwiazek 2022). Since such injuries involve both hypoxic events and oxidative burst, resistance to both stresses is required for successful waterlogging recovery (Liu and Zwiazek 2022). In another report on *B. napus*, hypoxia during flooding disrupted oxidative processes and reduced water transport to the leaves (Cutforth et al. 2013). *B. juncea* exhibits reduced tolerance to hypoxia primarily due to its smaller root system (effective depth: 120–130 cm) and longer exposure to waterlogging in certain soil profiles (Cutforth et al. 2013).

11.2.1.2 Adaptive Responses to Oxygen Limitation

Plants undergo several morphological, biochemical, and physiological modifications to adapt to oxygen limitation under hypoxic/anoxic conditions induced by flooding stress, including the formation of adventitious roots and aerenchyma (Eysholdt-Derzsó and Sauter 2019; Manghwar et al. 2024b). Additionally, it has been demonstrated that plants can alter their growth, reducing shoot growth and intensifying lateral root growth to enhance plant stability and nutrient uptake. They also exhibit physiological responses such as the stimulation of anaerobic mechanisms for energy production under oxygen-deficient conditions and a role in regulating phytohormones to coordinate adaptive processes to flooding stress (Pan et al. 2020). In maize (*Zea mays* L.), for instance, it has been shown that the formation of aerenchyma and adventitious roots is the most adaptive trait (Yao 2021; Zhai et al. 2013). Accordingly, it has been shown that elevation of oxygen availability to waterlogged organs is vital for maize to adapt to waterlogging conditions (Yu et al. 2019). Transcript profiling at the late stage of waterlogged maize plants under field conditions revealed the activation of several key genes involved in aerenchyma development, signal transduction, and carbon metabolism to initiate adaptive mechanisms during waterlogging, including aerenchyma development (Kaur et al. 2021).

11.2.1.3 Development of Adventitious Roots and Aerenchyma Tissues

One of the primary adaptive characteristics of plants to withstand flooding is the development of a gas space called aerenchyma in adventitious roots (Mui et al. 2021; Takahashi et al. 2014; Zhang et al. 2015). Aerenchyma creates an internal gas transport channel that helps roots thrive under waterlogged conditions by increasing internal oxygen diffusion to the root tips. This adaptation is vital for maintaining adequate oxygen levels in submerged roots in waterlogged soils, where oxygen availability is limited (Colmer 2003). Development of aerenchyma under waterlogged conditions has been demonstrated in several oilseed crops such as maize (*Zea mays* ssp. mays) (Abiko et al. 2012; Kaur et al. 2021), soybean (*Glycine max*) (Kondo et al. 2024), and canola (Ploschuk et al. 2018). It has been previously shown that in *Brassica napus*, high rates of radial oxygen loss from the root base due to the lack of aerenchyma formation are the main reason for the waterlogging sensitivity (Hussain et al. 2023).

11.2.2 ROS and Oxidative Stress

ROS, including hydroxyl radical, hydrogen peroxide, and superoxide, accumulate under waterlogging stress, leading to photooxidative damage and cell death (Pan et al. 2020; Yordanova et al. 2004). In soybean, it has been shown that stress priming can alleviate the damage triggered by waterlogging stress by reducing ROS accumulation (Agualongo et al. 2022).

11.2.2.1 Role of ROS in Stress Signaling

Excess ROS produced under waterlogging stress interferes with plant metabolism by oxidizing essential macromolecules, including DNA and protein. Plants use nonenzymatic and enzymatic antioxidant defense systems to scavenge excess ROS (Hasanuzzaman et al. 2017). The most important ROS-scavenging enzymes include superoxide dismutase (SOD), peroxidase (POD), and catalase (CAT), which are key components of biological defense against oxidative stress (Sharma et al. 2019; Wang et al. 2024).

11.2.2.2 Mechanisms of Oxidative Damage and Defense

Plants have evolved different strategies to adjust their metabolism to adapt to flooding conditions and to recover once it is removed, such as activation of the antioxidant defense system, development of distinct root structures, regulation of hormone levels, activation of anaerobic pathways to produce energy under oxygen-deficient conditions, and altering their growth and development patterns (Doupis et al. 2017; Eysholdt-Derzsó and Sauter 2019; Hasanuzzaman et al. 2020a, b; Manghwar et al. 2024b; Pan et al. 2020). In maize (*Zea mays* L.), for example, antioxidants are increased rapidly to alleviate oxidative damage under waterlogging stress (Yu et al. 2019). In soybean, the formation and function of secondary aerenchyma in waterlogged tissues is assumed to be an adaptive response to flooded conditions (Ploschuk et al. 2018).

11.2.2.3 Antioxidant Defense Systems in Flooded Conditions

Under waterlogging stress, antioxidant enzyme systems and other active antioxidants reduce oxidative stress damage by maintaining the dynamic balance of ROS (Doupis et al. 2017; Hasanuzzaman et al. 2020a, b). In maize, for instance, several genotypes showed higher POD, CAT, and SOD activities under waterlogging stress (Li et al. 2018). In soybean, the activity of antioxidant enzymes increased significantly under waterlogging conditions (Bester et al. 2024; Gangana Gowdra et al. 2025). Studies on rapeseed (*Brassica napus*) seedlings have shown that while CAT activity decreases, SOD and POD activities increase under waterlogging stress (Pan et al. 2021). In sunflower (*Helianthus annuus*), the activation of key antioxidant enzymes such as SOD and POD plays a crucial role in mitigating oxidative stress caused by ROS under waterlogging conditions (Hasanuzzaman et al. 2020a, b). The enhanced antioxidant activity is assumed to be an important mechanism for increasing plant tolerance under waterlogging stress. However, extended waterlogging may lead to mitochondrial dysfunction, decreasing the activity of antioxidant enzymes (Sharif et al. 2018).

11.2.3 Altered Root and Shoot Architecture

Flooding and waterlogging stress significantly alter oilseed crops' root and shoot architecture, impacting their growth and productivity. Roots often exhibit adaptive responses such as aerenchyma development, increased adventitious root formation, and reduced primary root elongation to enhance oxygen diffusion in hypoxic environments. Shoots may exhibit elongation, altered leaf growth, and reduced branching, driven by hormonal signals such as ethylene and gibberellins, to elevate foliage above water levels. While these structural adjustments aim to optimize plant survival, they often reduce biomass and yield, highlighting the importance of developing targeted strategies to mitigate these effects. In a study, waterlogging stress inhibited the growth of maize, resulting in reduced plant height, ear height, dry weight, leaf area index, and grain yield (Huang et al. 2022). In another study, soybean root growth was significantly inhibited by waterlogging stress (Tamang et al. 2014).

11.2.3.1 Changes in Root System Architecture

Since plants suffer from severe oxygen deprivation under waterlogging stress, one of their key adaptations following flooding is the alteration of root architecture and increasing lateral root and aerenchyma formation to improve the oxygen supply to roots for their survival (Visser and Voesenek 2005). As documented in maize, ethylene plays a key role in this process, reducing root growth and elongation (Geng et al. 2023). Rapeseed (*Brassica napus* L.) develops aerenchyma to enhance oxygen supply to roots under flooded conditions (Nabloussi et al. 2019).

11.2.3.2 Effects on Shoot Growth and Development

Shoot growth and development are reduced dramatically under waterlogging stress (Yeung et al. 2018a, b). As reported in sunflower, the reduction in shoot biomass is commonly associated with reduced plant height and leaf area (Islam et al. 2022; Orchard and Jessop 1984). However, plants can use different survival mechanisms to avoid a severe drop in shoot biomass during and after waterlogging stress (Yeung et al. 2018a, b). This includes forming radial oxygen loss (ROL) barriers, which help restrict oxygen diffusion from roots to the surrounding anaerobic soil, maintaining internal oxygen supply. Additionally, plants develop adventitious roots, which emerge from stem tissues above the anoxic soil and enhance oxygen uptake, facilitating nutrient and water absorption. Another critical adaptation is aerenchyma formation, where specialized air spaces develop within root and shoot tissues, improving internal oxygen transport to submerged organs. These mechanisms collectively support shoot growth and development by mitigating oxygen deficiency, maintaining metabolic activity, and reducing the negative effects of prolonged waterlogging stress. Flaxseed (*Linum usitatissimum*) develops secondary aerenchyma in stems and hypocotyls to maintain oxygen supply to submerged organs, supporting shoot growth (Yang et al. 2023). Canola cultivation under waterlogging conditions leads to increased lateral branching, which is a stress response but often negatively impacts yield (Schaefer et al. 2023).

11.2.3.3 Molecular Mechanisms of ROL Barriers, Adventitious Roots, and Aerenchyma Formation

Research on the molecular mechanisms that drive the production of flooding-induced ROL barriers remains limited. However, suberin is recognized as a significant contributor to the development of these barriers. Various genes involved in suberin biosynthesis, including KCS, CYP, GPAT, ABC, LTP, *ab*HD, and POD, have roles in this process (Shiono et al. 2019). Flooding or hypoxic stress can stimulate the expression of specific proteins involved in the formation of adventitious roots (ARs). Key genes include RBOHB, RBOHF3, various Ethylene

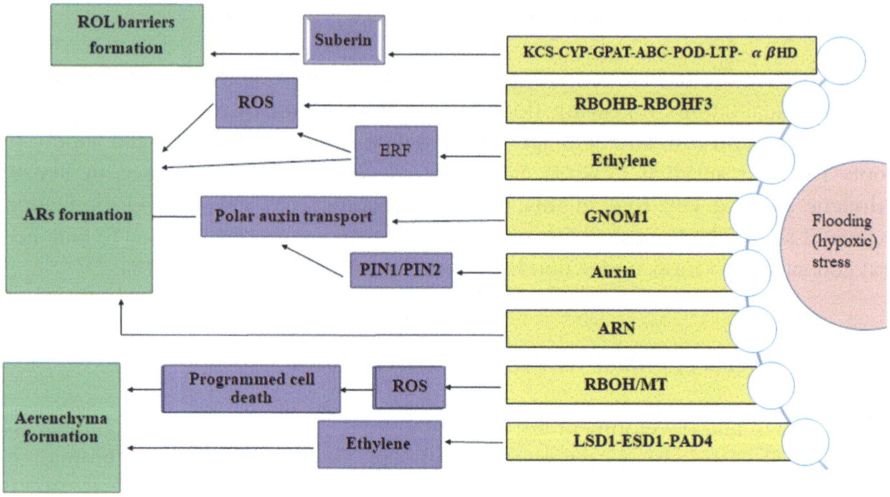

Fig. 11.1 Molecular mechanisms of flooding-induced ROL barrier formation and adventitious root development in plants. Abbreviations: ABC (ATP-binding cassette(, ARN (associated with adventitious root formation), *ab*HD (α/β-hydrolase domain-containing proteins), CYP (cytochrome P450), EDS1 (enhanced disease susceptibility 1), ERF (ethylene response factor), GPAT (glycerol phosphate acyltransferase), KCS (ketoacyl-CoA synthase), LTP (lipid transfer protein), LSD1 (lesion simulating disease 1), MT (metallothioneins), PAD4 (phytoalexin deficient 4), POD (peroxidases), *Respiratory burst oxidase homolog*)RBOH(, MT (metallothionein), ROS (reactive oxygen species), ROL (radial oxygen loss)

Response Factor (ERF) genes, and ARN. This gene induction affects AR development either directly or by regulating ROS (Qi et al. 2019; Xu et al. 2018) (Fig. 11.1). Plants can promote the formation ARs by enhancing auxin polar transport. Sunflower (*Helianthus annuus*) is sensitive to hypoxia and forms aerenchyma in response to waterlogging. Ethylene and ROS play crucial roles in this process, with programmed cell death (PCD) involved in aerenchyma formation in stems (Shao et al. 2025). This process is facilitated by activating several auxin-related genes, such as GNOM1, PIN1, and PIN2 (Dawood et al. 2016; Liu et al. 2009). Aerenchyma formation is triggered by the accumulation of ROS under waterlogging conditions. This adaptation enhances the transport of oxygen from the shoots to the roots, improving plant survival in saturated environments (Srivastava et al. 2024). *Respiratory burst oxidase homolog*)RBOH) is essential to produce ROS, whereas MT (metallothionein) plays a key role in scavenging ROS. Together, these are significant contributors to the regulation of ROS levels in plants (Rajhi et al. 2011). Maize also forms lysigenous aerenchyma under waterlogged conditions, with RBOH essential for generating ROS (Aslam et al. 2023).

The genes LSD1, EDS1, and PAD4 regulate the production of ethylene and ROS at an upstream level. These genes influence ROS signaling and ethylene biosynthesis, triggering downstream responses that promote adaptive structural modifications such as aerenchyma formation. These genes contribute to the development of aerenchyma, facilitating the plant's adaptation to waterlogged conditions (Bernacki et al. 2023). Typically, aerenchyma formation in most oilseed crops occurs after 5–7 days of waterlogging (Garcia et al. 2020). However, there are exceptions, particularly with oilseed rape (*Brassica napus* L.), which demonstrates a heightened vulnerability to flooding stress. The development of extensive lysigenous aerenchyma in the roots and the presence of aerenchymatous cells in the epidermis of the waterlogged ZZM25 sesame genotype provided it with tolerance to waterlogging stress. This contrasts with the disorganized aerenchyma observed in the susceptible genotype, Ezhi-2, which resulted from the collapse of parenchymatous cells (Wei et al. 2013). Kim et al. (2018) examined the impact of exogenous applications of ethephon, a synthetic plant growth regulator that releases ET upon decomposition and showed it significantly promoted the development of ARs and increased the root surface area in soybean plants, indicating its potential to enhance root growth and overall plant development.

11.3 Role of Plant Hormones in Modulating the Impacts of Flooding Stress

Plant hormones are pivotal in molecular, anatomical, biochemical, morphological, and signaling processes that allow plants to survive under oxygen-deficient stress conditions (Bashar et al. 2019). Several phytohormones, such as ethylene, gibberellin, auxin, brassinosteroids, and others, have been recognized as essential contributors to plant adaptation to flooding stress. These hormones serve as vital endogenous signals crucial for the induction of tolerance to waterlogging stress (Manik et al. 2022). Waterlogged conditions negatively impact plant growth in several ways. Waterlogging restricts the development of lateral roots and the elongation of primary roots. Waterlogging triggers changes in plant hormone levels (Fig. 11.2). Increased ET (via ACS7 and ACO2 gene expression) promotes primary root growth, adventitious root (AR) formation from stem nodes, and aerenchyma development. Concurrently, elevated gibberellin and indole-3-acetic acid (IAA) and reduced cytokinin and ABA correlate with AR emergence from waterlogged stems (Nguyen et al. 2018).

11.3.1 Ethylene (ET)

11.3.1.1 The Role of ET in Adventitious Root Formation and Aerenchyma Development

ET is a gaseous phytohormone essential for responding to waterlogging stress. However, its diffusion within plants occurs at a very slow rate (Hartman et al. 2019). When exposed to hypoxic environments, plants initially synthesize a

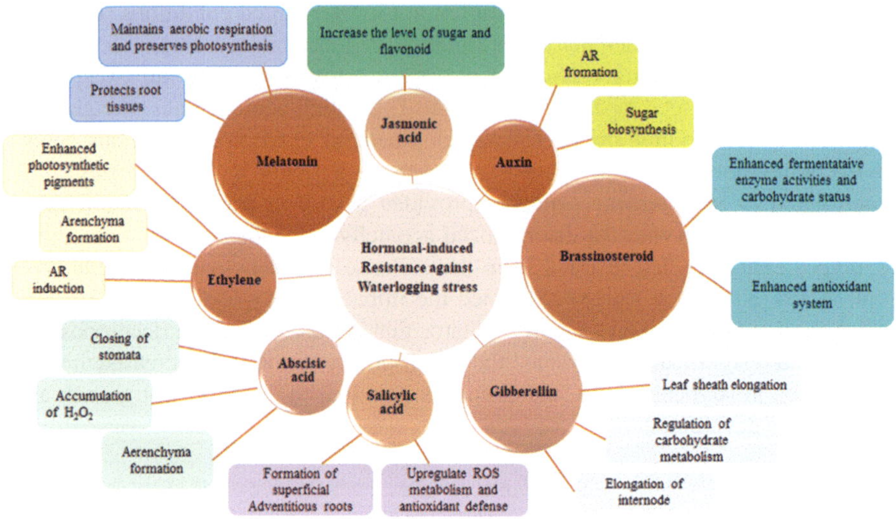

Fig. 11.2 Hormone-induced resistance to waterlogging stress in plants

significant quantity of 1-aminocyclopropane-1-carboxylic acid (ACC) to generate ethylene, which is facilitated by the enzyme ACC synthase (ACS). Subsequently, ACC is converted into ethylene by ACC oxidase (ACO) in a reaction that requires oxygen. Therefore, ACC must be transported from the roots to the plant's aerobic regions, where it can be oxidized to produce ethylene (Pan et al. 2021).

Previous research indicates that ethylene facilitates aerenchyma formation in plant roots under oxygen-deficient conditions by promoting cortical cell death (Nguyen et al. 2018; Yamauchi et al. 2014). To combat wilting caused by waterlogging, ethephon, a chemical that releases ethylene, has been shown to enhance aerenchyma development at root tips (Shiono et al. 2019). A study investigating the physiological mechanisms underlying flooding stress in soybean by applying different hormones, including ABA, GAs, IAA, jasmonic acid (JA), salicylic acid (SA), and kinetin (KN) revealed that ethephon treatment significantly alleviated waterlogging stress compared to untreated plants. Specifically, ethephon improved photosynthetic pigment levels, increased endogenous gibberellin content, enhanced root surface area, stimulated the initiation of ARs, and markedly boosted relative glutathione activity and the expression of glutathione transferases (Manghwar et al. 2024a). Ethylene has also been shown to mitigate water-logging stress by promoting the formation of aerenchyma cells. Its effects are influenced by interactions with other hormones, including ABA, GAs, IAA, and kinetin. These hormonal interactions contribute to the plant's ability to adapt to low oxygen conditions caused by waterlogging (Shimamura et al. 2016). Furthermore, ethylene interacts with JA and SA to enhance the formation of adventitious roots and aerenchyma cells in soybean (Yang et al. 2013). Ethylene plays a pivotal role in sunflowers' (*Helianthus annuus*) adaptation to flooding by inducing aerenchyma formation in roots. This adaptation enhances oxygen transport and alleviates hypoxia in submerged tissues. Exogenous ethylene treatments can accelerate this process, improving plant survival during prolonged waterlogging (Jia et al. 2021). Using a tolerant and a sensitive soybean line under waterlogging stress, the tolerant line exhibited superior formation of adventitious roots and aerenchyma and the tolerant line produced ET at a higher rate, which contributed to more vigorous growth under waterlogged conditions. These findings highlight the physiological advantages of the tolerant soybean line in coping with waterlogging stress (Kim et al. 2015). Molecular responses such as mitochondrial damage, cell wall loosening, protein degradation, and sugar accumulation have been identified as key factors contributing to the stunted growth observed in flooded soybean plants. These responses highlight the complex stress mechanisms negatively impacting plant development under flooding conditions (Komatsu et al. 2012). To withstand short-term flooding, soybean plants employ several collaborative mechanisms, including enhanced transcriptional regulation, protein balance maintenance, cell death suppression, and regulation of auxin-induced transcriptomic modifications. These adaptive strategies work together to improve the plants' resilience to flooding stress (Wang and Komatsu 2020). The findings indicate that auxin and ethylene are crucial for adapting soybean to flooding stress. These hormones regulate the formation of adventitious roots and promote energy preservation, facilitating the plant's ability to cope with waterlogged environments.

11.3.1.2 Regulation of ET Biosynthesis Under Flooding

ET is a multifunctional phytohormone that plays critical roles in plant growth and development, such as seedling emergence, root development, fruit ripening, and the abscission of flowers and leaves. In natural environments, plants facing prolonged stress employ strategies to avoid stress and develop mechanisms to cope with these challenges, thereby enhancing their overall stress tolerance. Research on higher plants shows that ethylene synthesis increases in response to biotic and abiotic stresses, highlighting its importance in plant adaptation to adverse environments (Marothia et al. 2021). Ethylene promotes plant growth and development at low levels, although elevated levels, typically occurring under stressful conditions, can cause detrimental effects, leading to abnormal plant growth and development (Fatma et al. 2022). In stressful conditions, an elevated abundance of ACC synthase (ACS) boosts the production of its substrate, ACC. This increase in ACC subsequently leads to enhanced ethylene synthesis within plant tissues, highlighting the role of ACS in the stress response of plants (Pattyn et al. 2021). Methionine is converted into S-adenosyl-L-methionine (SAM) by the SAM synthetase enzyme, which plays a crucial role in the ethylene production pathway in plants. Following this, ACC synthase (ACS) converts SAM into ACC, producing 5-methylthioadenosine (MTA) as a byproduct. MTA is subsequently recycled back to generate methionine via the Yang cycle. Finally, ACC is converted by ACC oxidase (ACO) into ethylene, completing the pathway (Adams and Yang 1979) (Fig. 11.3).

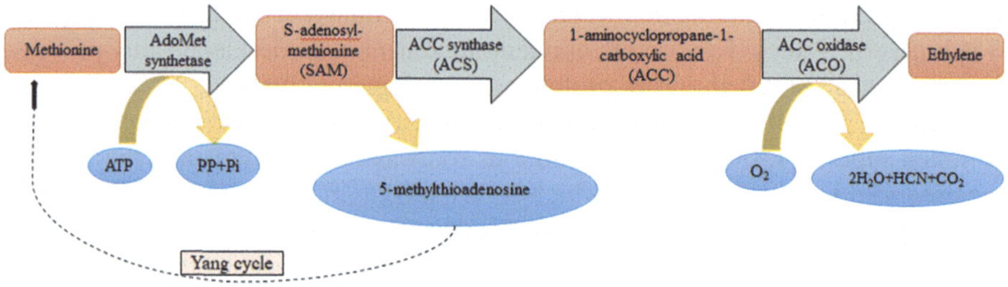

Fig. 11.3 Schematic diagram of the ethylene biosynthesis pathway

Early studies identified ACC synthase as a key enzyme in the ethylene biosynthesis pathway. This finding has prompted extensive research into the regulation of ACS enzyme activity and stability, highlighting its critical role in controlling ethylene synthesis in plants (Sato and Theologis 1989). Recent evidence indicates that ACC oxidase (ACO) may be the limiting enzyme in ethylene biosynthesis. This emerging understanding shifts the focus from ACS to ACO, suggesting that its activity can significantly influence ethylene production under specific conditions (Houben and Van de Poel 2019). Both ACS and ACO are essential for ethylene synthesis and the regulation of plant functions. ACS is known for its high turnover rate in the cytosol, and its enzymatic activity is highly dependent on its pyridoxal-5′-phosphate (PLP) cofactor. This highlights the importance of both enzymes in facilitating ethylene production and influencing various plant processes (Sato and Theologis 1989). Wang et al. (2023) demonstrated that the WRKY29 transcription factor controls ethylene synthesis by activating the expression of ACS and ACO. This regulation significantly impacts different aspects of plant development and growth, highlighting the importance of WRKY29 in ethylene production (Wang et al. 2023).

ACS is a polypeptide encoded by multiple genes that differ among species. Environmental factors regulate ACS expression, and its abundance varies throughout the plant life cycle. Research has shown that these genes contribute to plant tolerance response under different abiotic stresses. Specifically, ACS9, AtACS7, and ACS11 are crucial for maintaining the interaction between ROS, ethylene, and brassinosteroid hormones, highlighting their importance in plant stress adaptation (Lv et al. 2018). AtACS2 and AtACS5 participate in abscisic acid response pathways and play roles in regulating plant growth and development (Li et al. 2011). In hypoxic conditions, OsACS1 and OsACS3 are expressed in a tissue-specific manner in etiolated seedlings, whereas OsACS1 is mainly found in shoots, and OsACS3 accumulates in the roots. Conversely, OsACS2 is primarily expressed in roots but downregulated under hypoxia, while during submergence, OsACS5 mRNA accumulates in the vascular bundles of young shoots and leaves (Zhou et al. 2020). Earlier studies have shown that ACS upregulation enhances the production of defense proteins, leading to increased ACC production, which is then converted to ethylene (Eun et al. 2019).

Flooding stimulates increased ethylene production in canola (*Brassica napus*), which is crucial for root adaptation to hypoxic conditions. Ethylene facilitates the formation of aerenchyma, air-filled spaces within the roots that enhance oxygen diffusion and improve survival in waterlogged conditions. Research has shown that ethylene biosynthetic enzymes, specifically ACO and ACS, are upregulated in response to flooding. This rise in ethylene levels supports root development and triggers additional adaptive responses, such as leaf elongation, to help the plant escape submergence (Ruduś et al. 2013). When exposed to flooding, soybean plants also show elevated ethylene production, associated with improved adventitious root formation, enabling the plant to obtain oxygen from the atmosphere above the water. Higher levels of ACC synthase during flooding result in increased ethylene production, which supports physiological adaptations like improved seedling survival rates and enhanced antioxidant responses (Huang et al. 2019). Additionally, studies have shown that pretreating soybean seedlings with ethylene precursors can improve their tolerance to submergence stress by adjusting their physiological responses (Fatma et al. 2022). *Carthamus tinctorius* is another oilseed crop that positively responds to ethylene when subjected to stress conditions and flooding. During these stress conditions, ethylene signaling pathways in safflower are activated, resulting in enhanced antioxidant activity and better osmotic adjustment within the plant tissues. This adaptation allows safflower to sustain growth and productivity even in challenging environmental conditions (Huang et al. 2019). Research suggests that adjusting ethylene levels could improve stress tolerance in safflower.

11.3.2 Abscisic Acid (ABA)

11.3.2.1 The Function of ABA in Stress Response and Stomatal Regulation

Stomata are small pores on leaves that facilitate gas exchange. They allow for the uptake of CO_2 necessary for photosynthesis while also resulting in water loss. Additionally, stomata can serve as entry points for air pollutants and certain pathogens. Therefore, precisely regulating these pores is crucial for plant survival in changing environments. Environmental signals influence both the width and the number of stomata, affecting the plant's gas exchange efficiency (Stevens et al. 2021). Elevated CO_2 concentrations typically decrease the stomatal density across various plant species (Ayala-Ramos et al. 2024). Conversely, increased light intensity significantly increases stomatal density (Hernando et al. 2021). A key difference between the regulation of stomatal aperture and density is temporal, with stomata aperture adjusting rapidly, while stomatal density is modified throughout leaf development. The plant hormone ABA plays a crucial role in regulating stomatal aperture by controlling guard cell ion fluxes. Additionally, ABA influences stomatal density indirectly through transcriptional regulation of developmental factors such as SPEECHLESS (SPCH) (Kollist et al. 2014). ABA-induced stomatal closure begins when ABA binds to PYR/RCAR receptors, which in turn inactivate the type 2C protein phosphatases (PP2Cs) (Kollist et al. 2014). This leads to the activation of SNF-related kinases (SnRK2s) such as OPEN STOMATA1 (OST1), which phosphorylates and activates key ion channels, including SLOW ANION CHANNEL1 (SLAC1), triggering ion efflux and stomatal closure. The activation of these channels results in anion efflux, followed by the loss of K^+ and water, ultimately causing stomatal closure (Park and Kim 2021). ABA is a critical regulatory molecule for stomatal function, making precise adjustments to ABA levels and signaling essential for acclimation to abiotic stress. ABA levels in guard cells are adjusted through de novo synthesis, degradation, and transport from other tissues (Merilo et al. 2018). The protein OPEN STOMATA1 (OST1) plays a crucial role in ABA signaling; in ost1 mutants, the ABA/environmental-induced stomatal closure is significantly impaired (Merilo et al. 2013). In addition to its role in regulating stomata function, ABA also affects stomata development, which is also influenced by the environment, including CO_2 levels and light (Chater et al. 2015). Stomatal density in ABA-deficient mutants is higher than in wild-type plants (Brookbank et al. 2021). In contrast, the cyp707a1/a3 double mutant, which overaccumulates ABA, exhibits significantly lower stomatal densities (Tanaka et al. 2013), further supporting the role of ABA in stomata development. The ABA3 gene encodes a molybdenum cofactor sulfurase, which is essential for activating abscisic aldehyde oxidase. This enzyme catalyzes the final step in ABA biosynthesis by converting abscisic aldehyde to ABA. ABA3 expression increases in response to environmental stress and exogenous ABA (Jalakas et al. 2018). Under nonstressed conditions, the leaf ABA concentration in the aba3–1 mutant is around 45% of that found in wild-type plants (Merilo et al. 2018). The primary catabolic pathway for ABA degradation, ABA 8′-hydroxylation, is catalyzed by CYP707A family enzymes. Additionally, conjugation pathways, such as ABA-glucose ester formation, contribute to ABA homeostasis. The transcript levels of these genes are increased by abiotic stresses and ABA itself (Chen et al. 2020), particularly CYP707A1 and CYP707A3, which are two essential factors for postgermination plant growth. In mutants lacking these genes, seedling growth in response to exogenous ABA is significantly inhibited, with the effect being even more pronounced in double mutants, which accumulate higher ABA levels compared to single mutants (Boccaccini et al. 2016). Loss-of-function mutants for cyp707a1 and cyp707a3 exhibited decreased stomatal conductance, with the reduction being more pronounced in the cyp707a3 mutants (Merilo et al. 2013). Additionally, waterlogging for 5 to 10 days reduces ABA concentrations in soybean hypocotyls compared to the control, suggesting a negative correlation between ABA levels and the plant's ability to tolerate waterlogging (Manghwar et al. 2024a). Reduced ABA concentrations may impair a plant's ability to respond effectively to stress under flooded conditions. Conversely, ABA is known for promoting stomatal closure and increasing the accumulation of H_2O_2. This dual action can enhance a plant's tolerance to waterlogging and other adverse environmental conditions while simultaneously reducing water loss through transpiration (Hiron and Wright 1973). Flooding stress has been shown to reduce ABA accumulation in the leaves and roots of soybean seedlings, including flood-tolerant varieties. Notably, flood-tolerant varieties displayed lower ABA levels compared to sensitive lines (Tamang et al. 2021). Similar to rice and Arabidopsis, the role of ABA in water-logged soybeans seems complex; it negatively influences flooding tolerance. For example, ABA has been found to inhibit the growth of phellogen-derived cells during aerenchyma formation under flooding conditions (Shimamura et al. 2014). This finding supports previous observations that a reduction in ABA leads to the formation of well-structured aerenchyma cells (Kim et al. 2015). Conversely, applying ABA has enhanced soybean flood tolerance during the initial stress and recovery phases. While ABA levels remain unchanged during seedling establishment under flooding conditions (Komatsu et al. 2013a), it does influence the phosphorylation pattern of proteins (Yin and Komatsu 2015). A study showed that applying 10 μM ABA improved soybean tolerance in the early stages of flooding by activating enzymes associated with the regulation of transcription and translation (Yin et al. 2016). Additionally, the application of exogenous ABA at 10 μM enhanced plant survival during flooding by modulating energy conservation, promoting lignification, and inhibiting cytochrome P450 77A1 (Wang et al. 2018). ABA signaling and its role in waterlogging tolerance in soybean remains unresolved; however, applying ABA at a concentration of 10 μM may effectively enhance the plant's tolerance to short-term flooding.

11.3.2.2 Interactions Between ABA and Other Phytohormones

Plants under environmental stress must be able to recognize and react to stimuli in an intensely coordinated and interactive manner. As sessile organisms, plants must adjust to adverse and volatile conditions and exhibit growth plasticity through signaling networks to facilitate this adaptability. Abiotic stress sensing triggers signal transduction cascades that interact with core pathways regulated by phytohormones (Harrison 2012). Changes in stress-related plant hormones significantly impact cellular stability and are essential for the coordinated control of growth adaptations in stressful environments (Kohli et al. 2013). Crosstalk between phytohormones establishes a signaling network where hormones cooperate by triggering a phosphorylation cascade or a shared secondary messenger (Harrison 2012) Significant understanding of the biosynthesis routes and essential signaling elements of important phytohormones, including ABA, GAs, IAA, JA, Brassinosteroids (BRs), and ET, has been acquired throughout the last few decades (Vo et al. 2021). Despite scientific advancements, the molecular basis of abiotic stress tolerance remains mainly unresolved and complex. This complexity arises from diverse responses to varying stress thresholds, limited knowledge of tissue-specific stress responses, and incomplete understanding of genetic plasticity and its environmental adaptation (Kohli et al. 2013). As a result, these factors together lead to more misperception than clarity (Qiao et al. 2024). However, one of the key stress responses plants have evolved to cope with environmental challenges is alterations in phytohormone fluxes and their subsequent signal transduction cascades. Plant hormones collaboratively form a signaling network that enhances plant resistance to external stressors. The three most critical hormones in regulating these signaling pathways are ABA, JA, and SA. It has been shown that exogenous SA helps maize (*Zea mays*) seedlings and sunflowers (*Helianthus annuus*) to increase yield under waterlogging stress by protecting the photosynthetic system and antioxidant enzyme activities (Damalas and Koutroubas 2022; Xin et al. 2023). The exogenous application of SA has been demonstrated to effectively improve soybean (*Glycine max* L.) tolerance to waterlogging stress by enhancing antioxidant defense mechanisms and upregulating the glyoxalase system (Hasanuzzaman et al. 2022). A study systematically assessed the role of mixed salicylic acid (SA) and abscisic acid (ABA) spraying in reducing waterlogging stress and restoring yield and quality in winter oilseed rape (*Brassica napus* L.). The findings indicated that this approach effectively mitigates the adverse effects of waterlogging. Notably, the combined SA + ABA treatment increased peroxidase (POD), superoxide dismutase (SOD), and catalase (CAT) activities by 15–30%, thereby reducing oxidative damage from reactive oxygen species (ROS). SA amplified ABA's ability to upregulate antioxidant enzymes, mitigating ROS-induced damage. ABA reduced stomatal conductance to conserve water, while SA maintained photosynthetic efficiency (Shao et al. 2025).

Examining how phytohormones interact and how they coordinate to adapt to abiotic stressors has recently been the focus of research. These interactions are more frequent than previously thought, according to experimental data, and they are essential for integrating different signals, controlling development, and improving plants' ability to withstand stress (Kohli et al. 2013). Each phytohormone has several frequently redundant signaling intermediates, emphasizing their critical functions in crosstalk. Therefore, knowing how defense signaling networks and phytohormones interact is crucial as it is likely to identify novel targets for developing host resistance mechanisms and enhancing phytohormone efficacy (Grant and Jones 2009).

Stomatal closure is regulated by a complex network of signaling pathways, with ABA being the primary driver of this transient response by altering gene expression that promotes root development. ABA controls long-term growth responses and maximizes water intake (Muhammad Aslam et al. 2022). Through coordination with other plant hormones and signaling molecules like JA and nitric oxide (NO) to encourage stomatal closure, ABA plays a critical role in plant response to drought stress. Additionally, ABA controls gene expression by activating genes linked to auxin, cytokinin, and ethylene responses (Harrison 2012). It has been proposed that ABA-mediated stomatal closure and stress-induced JA production interact by increasing extracellular Ca^{2+} influx and/or activating signaling pathways involving NO and hydrogen peroxide (H_2O_2) (Harrison 2012). While ethylene regulates stomatal closure through its signal transduction pathway, it necessitates the creation of hydrogen peroxide (H_2O_2) in addition to stimulating ethylene biosynthesis (Wang et al. 2020a, b). This crosstalk includes JA, generated in reaction to adverse conditions, and regulates several genes linked to JA signaling (Li et al. 2021; Liu et al. 2018). JA influences ABA-regulated stomatal closure by enhancing Ca^{2+} influx, which stimulates the production of calcium-dependent protein kinases (CDPKs), triggering subsequent signaling cascades (Harrison 2012). Similarly, alterations in ABA-mediated gene regulation, primarily through transcription factors

(TFs) that bind to ABA-responsive elements (ABREs) in ABA-regulated genes, drive long-term physiological responses to abiotic stress. Notably, phosphorylation cascades modulate ABA-regulated transcription factors, including those involved in stomatal closure signaling. For instance, ABA-inducible kinases CPK4 or CPK11 phosphorylate ABA-responsive transcription factors ABF1 and ABF4, activating them (Atif et al. 2019; Harrison 2012). Isopentenyl transferase (IPT), a critical gene in the cytokinin biosynthesis pathway, was significantly inhibited by the exogenous application of abscisic acid (ABA). Concurrently, this treatment increased the expression of genes encoding cytokinin oxidases and dehydrogenases, suggesting a complex relationship between ABA and cytokinin signaling in plant adaptation (Hai et al. 2020). In addition to its well-known role in regulating plant growth and development alongside auxins and BRs (Qiu et al. 2020), GA plays a significant role in the hormonal crosstalk involved in responding to environmental stimuli (Nguyen et al. 2022).

11.3.3 Gibberellins (GAs)

11.3.3.1 Impact of GAs on Shoot Elongation and Stress Adaptation

During flooding conditions, ethylene levels increase due to limited diffusion, inhibiting ABA synthesis and promoting its degradation. This hormonal shift causes an increase in (GA) levels, which is essential for stem elongation. This elongation enables plants to grow above the water surface, enhancing their access to light and air. The interplay between ethylene and GA is crucial; higher ethylene levels boost GA content, supporting stem growth under submergence stress (Sasidharan and Voesenek 2015). GA is an essential plant hormone that regulates growth, development, and stress responses. During flash floods, plants tolerant to submergence often restrict their growth by activating the GA signaling pathway (Ritonga et al. 2023). Active gibberellin (specifically, GA_1) has been demonstrated to promote leaf sheath elongation in response to submergence stress (Dubois et al. 2011). Flooding activates the SD1 (SEMIDWARF1) gibberellin biosynthesis gene via an ethylene-responsive transcription factor, OsEIL1a. This increase in gibberellin abundance facilitates internode elongation (Kuroha et al. 2018). Two genes have been identified, ACE1 (Accelerator of Internode Elongation 1) and DEC1 (Decelerator of Internode Elongation 1), which contrastingly regulate internode elongation in relation to GA. The protein ACE1, whose function is currently unknown, enhances internode elongation in a GA-dependent manner. Both genes play a role in assisting plants in acclimating to flooding stress (Nagai et al. 2020). Plants utilize GA to mitigate flooding stress by regulating carbohydrate metabolism and stimulating internode elongation (Jia et al. 2021). Additionally, GA supports plant responses to both shade and flooding conditions (Li et al. 2017). A study investigating the impact of exogenous gibberellins ($GA_{4\ +\ 7}$) on soybean plants exposed to waterlogging stress revealed notable increases in chlorophyll content and elevated levels of endogenous bioactive gibberellins (GA_1 and GA_4) and JA (Khan et al. 2018).

11.3.3.2 GAs' Role in Plant Growth and Development During Flooding

GAs are pivotal phytohormones regulating diverse aspects of plant growth and development, particularly under stress conditions such as flooding. During flooding, the anaerobic environment disrupts normal physiological processes, but GAs play a critical role in mitigating these adverse impacts by promoting elongation, allowing plants to escape submergence. For instance, GA-mediated elongation of internodes and petioles enables plants to keep their leaves above water, ensuring photosynthesis and survival. Moreover, GAs interact with ethylene and abscisic acid to fine-tune growth responses, balancing escape strategies with energy conservation. The modulation of GA biosynthesis and signaling under flooding conditions highlights their adaptive significance, as plants prioritize energy-efficient growth to endure prolonged submergence. This dynamic interplay underscores the essential role of GAs in enhancing flood tolerance and maintaining developmental plasticity in plants (Colebrook et al. 2014).

GAs are crucial for regulating cell size and number, thereby playing a vital role in the secondary growth and development of plants (Gomez et al. 2018). During flooding stress, waterlogging-tolerant soybean cultivars exhibited a significant increase in GA concentration compared to stress-sensitive genotypes (Olorunwa et al. 2022). Additionally, applying GAs to the leaves of peanuts (*Arachis hypogaea*) under flooding stress enhanced physiological function, vegetative and reproductive growth, yield, and root biomass (Zeng et al. 2021). Also, Hong et al. (2023) observed that the exogenous application of GAs significantly enhanced flood resistance in rape plants by reducing malondialdehyde (MDA) content in their leaves and roots (Hong et al. 2023).

11.3.4 Brassinosteroids (BRs)

11.3.4.1 BRs Influence on Plant Growth and Stress Resilience

BRs are a class of polyhydroxy steroidal phytohormones that play a significant role in promoting plant growth (Bajguz and Piotrowska-Niczyporuk 2014). BRs were first isolated from the pollen of *Brassica napus* (Grove et al. 1979) and are classified based on their carbon structure into C27, C28, or C29 subtypes (Vardhini 2014). To date, over sixty related compounds have been identified (Basit et al. 2021), with three primary bioactive forms brassinolide (BL), 28-homobrassinolide (28-HomoBL), and 24-epibrassi-nolide (24-EpiBL) that are frequently employed in physiological and experimental research (Basit et al. 2021). BRs are essential for regulating various developmental and physiological functions in plants. BRs regulate numerous aspects of plant growth and development, including cell expansion and division, stem cell maintenance, vascular development, cell elongation, and the transition to flowering (Fàbregas et al. 2018). BRs contribute to hypocotyl elongation and other growth processes (Zou et al. 2018), including root and shoot growth (Mao and Li 2020; Ye et al. 2017), stomatal patterning (Wang et al. 2019), and the growth and development of pollen tubes (Ahammed et al. 2020). In addition to their other roles, BRs contribute to seed germination, the differentiation of tracheary elements (Amraee et al. 2019), xylem formation and differentiation (Sharma et al. 2019), photomorphogenesis, plant reproduction (Yin et al. 2019), and senescence (Nolan et al. 2020). BRs are essential for the growth and development of plants, and they play a significant role in regulating the cell cycle during seed germination (Martínez et al. 2018) by promoting endosperm rupture and cell elongation (Martins et al. 2017). BRs also help manage the proliferation of leaf cells (Sadura and Janeczko 2018) and are often used to enhance plant growth in plants grown in hydroponic systems (Ahammed and Yu 2016). BRs regulate plant responses to abiotic and biotic stresses, control stomatal development, regulate male and female fertility (Zhang et al. 2013), and stimulate stigma elongation (Tang et al. 2008). In addition to their other functions, BRs influence various aspects of plant metabolism (thermotolerance, proton transport) and architecture (tiller number, leaf angle, and leaf size) (Hussain et al. 2019). The external application of BRs or modifying their biosynthesis and signaling pathways can increase crop yields (Zhou et al. 2014). They are also crucial in regulating various plant genes. For example, Kagale et al. (2007) demonstrated that BRs activate the WRKY46, WRKY54, and WRKY70 genes, which positively promote plant growth regulated by BRs. Another study emphasized the role of the histone lysine methyltransferase SDG8 in regulating gene expression being affected by BRs. Additionally, brassinazole, a known inhibitor of BRs, was found to accelerate the transition to the heading stage in plants (Xi et al. 2013). Consequently, these substances are valuable biostimulants in agriculture, assisting crops in developing tolerance to abiotic stresses and improving overall plant efficiency (Fariduddin et al. 2009).

BRs are essential for mitigating various stresses by enhancing photosynthesis, and biomass accumulation, boosting antioxidant enzyme activity, promoting detoxification processes, and stimulating the expression of relevant genes (Ramakrishna and Rao 2013). BRs regulate multiple metabolic pathways and interact with various other plant growth regulators (Zhou et al. 2014). These regulatory functions underscore the critical role of BRs in assisting plants to adapt to environmental stresses (Yusuf et al. 2011). BR supplementation has been shown to enhance resistance to waterlogging stress in crops such as oilseed rape (*Brassica napus*) (Liang and Liang 2009) and promote seedling growth and resilience against similar stresses in soybean (Lu et al. 2006). A recent study demonstrated that the exogenous application of BRs significantly improves growth parameters and enhances antioxidant defense mechanisms in maize seedlings (*Zea maize* L.), effectively reducing the adverse effects of waterlogging stress. Furthermore, BRs were observed to reduce the area of root aerenchyma and decrease the production of reactive oxygen species (ROS) across various growth stages (Manghwar et al. 2024a). BRs are involved in regulating various physiological processes, including the management of ROS and metabolites, which are crucial for plant adaptation to environmental stresses (Zahra et al. 2020). These plant-specific ligands directly interact with cell surface receptors known as leucine-rich repeat receptor kinases, particularly BRASSINOSTEROID-INSENSITIVE 1 (BRI1) and BRI1-associated receptor kinase (BAK1). This interaction initiates a signaling cascade that activates various physiological responses, improving the plant's capacity to withstand environmental stresses. The signaling process involves phosphorylation in the cytoplasm, which includes the phosphorylation of the BSU1 protein and the proteasome-mediated degradation of BRASSINOSTEROID-INSENSITIVE 2 (BIN2) proteins. Inactivating BIN2, a GSK3-like kinase, is vital in the BR signaling pathway. The deactivation of specific regulatory pathways in plants facilitates the activation of critical transcription factors, particularly BRI1 EMS SUPPRESSOR 1 (BES1) and Brassinazole Resistant 1 (BZR1). These transcription factors are integral to the BRs signaling pathway, which is pivotal in regulating gene expression associated with plant growth and responses to environmental stresses. Exogenous BRs can effectively mitigate stress-induced physiological changes, attracting considerable interest in food production (Ahanger et al. 2020). While ongoing studies at physiological and molecular levels around the globe explore BRs' roles in alleviating stress effects, their precise mechanisms remain unclear. This uncertainty is thought to stem from both the presence of BRs and their complex interactions within signaling networks that ultimately help mitigate stress (Divi et al. 2016).

11.3.4.2 Interactions Between BRs and Other Hormonal Pathways

ABA is crucial for regulating plant growth, development, and adaptation to stress. The interaction between ABA and BRs is critical for enhancing plant resilience to environmental stress while promoting overall growth. Although ABA and BRs regulate different physiological processes, they engage in significant crosstalk that alters gene expression and protein activity. For example, BZR1, a key player in BR signaling, can suppress the expression of stress-related genes such as *ABI3* and *ABI5* by binding to their promoter regions. This mechanism helps to modulate the plant's stress responses (Xu et al. 2023). Conversely, ABA facilitates the degradation of BR-signaling components, including BRI1 and BES1/BZR1, through proteins such as PUB12/13 and SINA, inhibiting BR signaling (Ma et al. 2015). The synergistic effects of ABA and BRs on plant growth and stress resistance are more pronounced than the effects of either hormone alone. Additionally, BRs modulate ABA signaling by phosphorylating key regulators like SnRK2.2 and *ABI5*. Interactions between BES1 and *ABI5* also negatively impact *ABI5's* transcriptional activity (Fan et al. 2024). This interaction highlights the antagonistic regulation between BRs and ABA pathways, which is key in balancing plant growth and stress adaptation. Furthermore, BRs promote the synthesis and metabolism of endogenous ethylene, contributing to their inhibitory effects. By activating genes related to ethylene production, BRs enhance ethylene biosynthesis, increasing ethylene levels and carotenoid accumulation crucial for proper plant maturation. Recent research suggests that the growth regulation induced by BRs is facilitated through ethylene- and hydrogen peroxide-signaling pathways, which enhance stress tolerance. The interaction between BRs and auxin emphasizes their collaborative roles in regulating plant growth, development, and responses to stress. Notably, auxin signaling influences the expression of genes like *CPD* and *DWF4*, which are critical for brassinosteroid biosynthesis (Vert et al. 2008). Research shows that Brevis Radix (BRX) enhances the expression of *DWF4* and *CPD*, establishing a direct link between auxin signaling and BR synthesis. Additionally, BRs and auxin collaboratively regulate plant development through interactions involving BIN2 and ADP-ribosylation factor 2 (ARF2). A notable example of this interaction is seen in rapeseed, where the collaboration among BnaARF8, BnaIAA7s, and BnaBZR1 has been demonstrated to influence stem elongation (Zheng et al. 2019). The relationship between auxin and BRs is crucial for coordinating plant growth across physiological, biochemical, and genetic interactions. BRs also interact JAs through antimicrobial peptides such as thionine. Under optimal growth conditions, low concentrations of cytoplasmic JAs allow jasmonate-Zim domain (JAZ) proteins to inhibit JA signaling. They achieve this by binding to downstream transcription factors (TFs), preventing these factors from interacting with DNA and regulating gene expression. The relationship between BRs and JAs is characterized by mutual inhibition of their respective biosynthesis and signaling pathways. The external application of JAs can diminish the effectiveness of BR biosynthesis and modify the expression patterns of the *DWF4* gene. This demonstrates how JA can influence BR-related processes in plants (Nguyen et al. 2022). Similarly, methyl jasmonate (MeJA) application significantly downregulates genes related to BR biosynthesis while enhancing jasmonate-mediated immunity against root-knot nematodes. BRs collaborate with GAs to regulate numerous physiological processes through protein–DNA interactions involving BZR1/BES1. Importantly, a deficiency in GAs leads to a significant reduction in BZR1 activity, highlighting the interconnected roles of these hormones in plant growth and development (Retzer et al. 2019). BZR2 plays a crucial role in the BR response pathway by repressing the Ppy-dormancy-associated MADS-box gene 3 (*PpyDAM3*). When *PpyBZR2* is transiently overexpressed, it leads to elevated endogenous levels of GAs and JAs through its direct interaction with myelocytomatosis protein 2 (*PpyMYC2*). This interaction enhances the transcriptional activation of GA20 oxidase genes (*PpyGA20OX1L1* and *PpyGA20OX2L2*), resulting in increased levels of GA3 (Wang et al. 2024). In soybeans (*Glycine max*), shade-induced hypocotyl elongation involves a synergistic interaction between auxin and BRs. However, under shaded conditions, gibberellin biosynthesis is notably suppressed by inhibitors that target the IAA, GA, and BR pathways. This highlights the complex hormonal interplay that regulates plant responses to shading (Jiang et al. 2019). Exogenous application of high levels of BRs inhibits GA biosynthesis and increases the expression of the inactive *GA20ox-3* gene. This inhibition initiates a feedback mechanism that reduces endogenous GA levels, leading to growth retardation. These intricate interactions among ABA, BRs, auxin, JA, and GAs highlight a sophisticated regulatory network

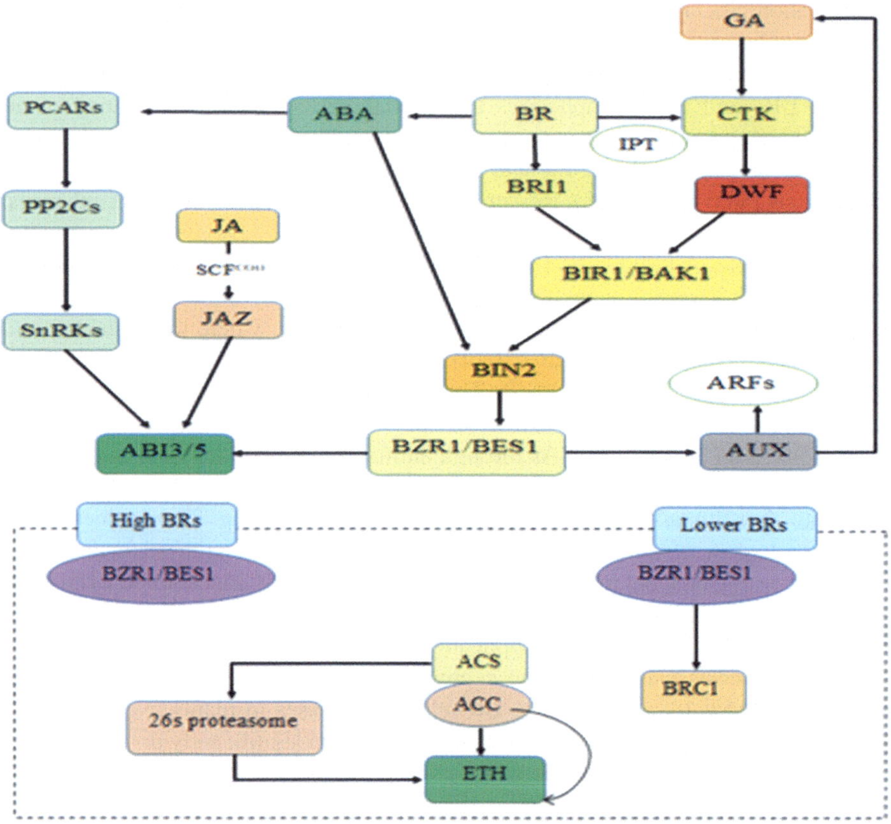

Fig. 11.4 Interactions among brassinosteroids (BRs), cytokinins (CTK), auxins (AUX), jasmonic acid (JA), gibberellins (GA), and ethylene (ETH) biosynthesis pathways are critical for plant growth regulation. Abscisic acid (ABA), JA, GA, BRI1 (brassinosteroid insensitive 1), BRs, BAK1 (BRI1-associated receptor kinase), BSU1 (BRI1-suppressor 1), BZR1 (brassinazole resistant 1), BES1 (BR-insensitive-EMS-suppressor 1), PP2C (protein phosphatases type 2C), BIN2 (brassinosteroid-insensitive 2), RCAR (regulation component of ABA receptors), SnRK (SNF-related protein kinase), and ABI3/5 (ABA-Insensitive 3/5) are key components in this network

essential for optimizing plant growth in response to changing environmental conditions (Fig. 11.4).

11.3.5 Other Phytohormones

Auxin (indole-3-acetic acid, IAA) is an essential hormone that significantly influences plant growth and development (Lv et al. 2019). In waterlogging conditions, the production of ethylene enhances the transport of auxin, and an elevation in auxin levels can stimulate the production of ethylene, which helps facilitate the transport of auxin to the flooded areas of the plant. This rise in auxin levels can promote the formation of ARs by initiating cell division, helping the plant adapt to waterlogged conditions (Lv et al. 2019). Numerous studies suggest that ethylene is a crucial signal for plant adaptation to flooding conditions. It orchestrates a hormonal cascade involving ABA and GA. This regulatory network is essential for helping plants cope with excess water and maintain their growth (Loreti et al. 2016). Moreover, the impacts of BR, JA, and SA on enhancing plant tolerance to flooding can occur either independently or dependently on ethylene signaling. These hormones support plant growth by encouraging the formation of adventitious roots, promoting shoot elongation, enhancing the production of photosynthetic pigments, and aiding in the scavenging of ROS (Koramutla et al. 2022). In waterlogged soybeans (*Glycine max* L.), the application of JA significantly mitigated water-logging-related damage and improved plant development compared to the control group. This treatment demonstrates JA's potential role in enhancing resilience and promoting growth under stressful conditions (Kamal and Komatsu 2016). Auxin has been identified as a key factor in promoting adventitious rooting in plants subjected to flooding (Nguyen et al. 2018). Interactions between auxin and ethylene are important for improving plant adaptation to flooding conditions. For instance, hypoxia decreases the abundance of the auxin efflux carrier PIN2. However, the overexpression of RAR2.12 exacerbates this reduction, indicating an antagonistic relationship between auxin and ethylene in response to flooding conditions (Eysholdt-Derzsó and Sauter 2017). ROS signaling is essential for facilitating adventitious rooting in waterlogged conditions. This is achieved by regulating the signaling pathways of auxin and ethylene, highlighting the importance of ROS in plant adaptation to excess water (Qi et al. 2019). These findings clarify how the interactions between auxin response, ethylene signaling, and ROS homeostasis affect root architecture and coleoptile growth in plants under hypoxic conditions. Additionally, the application of 24-epibrassinolide mitigated the adverse impacts of waterlogging of soybeans by enhancing root and leaf anatomy, increasing chlorophyll accumulation, and improving ROS scavenging (Pereira et al. 2020). BRs biosynthesis is regulated in a SUB1A-dependent manner. BRs play a dual role during submergence by limiting GA levels while promoting shoot elongation. This regulation helps plants adapt to submerged conditions effectively (Schmitz et al. 2013).

JA and SA are crucial factors in plant adaptation to flooding. They accomplish this by regulating antioxidant metabolism and shaping root architecture, improving the plant's capacity to withstand excess water conditions (Koramutla et al. 2022). The upregulation of RAP2–4 enhances seedling tolerance to waterlogging stress, with its expression activated by JA through its interaction with the JA response element (Phukan et al. 2018). The synthesis of JA in the root meristem is regulated by ZmPgb1.2, which positively enhances tolerance to anaerobic conditions. This regulation underscores the importance of JA in helping plants adapt to low-oxygen environments (Youssef et al. 2019). Additionally, SA promotes axial root formation independently of ethylene, whereas its influence on adventitious rooting relies on the presence of ethylene (Koramutla et al. 2022). Flooding induces oxidative stress in sunflowers (*Helianthus annuus*) due to reactive oxygen species (ROS) accumulation. Exogenous SA enhances the activity of antioxidant enzymes, such as superoxide dismutase (SOD) and catalase (CAT), reducing cellular damage and improving plant resilience under waterlogged conditions (Agüera and de la Haba 2021). These observations underscore the intricate interactions among plant hormones in boosting resilience to flooding stress. The activation of specific genes by JA and SA is crucial for promoting root growth and overall adaptation to waterlogged environments. These hormones and their metabolism provide an intriguing pathway for enhancing plant tolerance through targeted manipulation of hormonal signaling in response to flooding conditions.

11.4 Leveraging Natural Variation and Genomic Strategies to Enhance Stress Tolerance in Oilseed Crop Breeding

Stress-tolerant plants have evolved through natural selection, developing mechanisms for acclimation and adaptation to challenging environments. Traits linked to stress tolerance, which help sustain stable yields, are often conserved across various species. Moreover, exotic and wild relatives that thrive in diverse geographical regions and climates typically exhibit superior stress-tolerance traits compared to cultivated varieties. This underscores the potential for incorporating these resilient relatives into breeding programs to enhance crop resilience (Mickelbart et al. 2015). Integrating stress-tolerance traits into crop germplasm is vital for enhancing genetic diversity and ensuring sustainable crop production in the face of extreme climatic changes. This can be accomplished by identifying tolerant genetic resources from natural variation and integrating these traits into elite germplasm, thereby improving resilience and adaptability in agricultural systems (Foyer et al. 2016). Genomic strategies enable the identification of DNA elements, such as genes or markers associated with stress tolerance, which can be efficiently integrated into existing germplasm. Techniques such as marker-assisted selection (MAS) and genomic selection aid in the integration of desirable traits during the breeding process, thereby improving the resilience and adaptability of crops. Essential genetic elements linked to crop traits can be identified using various techniques, including quantitative trait locus (QTL) mapping, association mapping, and genome-wide association studies (GWAS). These approaches enable researchers to pinpoint specific genetic factors contributing to desirable traits, facilitating targeted breeding for improved crop performance (Varshney et al. 2016). Further characterization of genes associated with natural genetic variation in stress tolerance will enhance our understanding of the biological mechanisms governing this trait, thereby greatly influencing crop improvement programs. Advanced technologies such as whole-genome sequencing and high-throughput phenotyping enable thorough investigations of genetic diversity, aiding in identifying key loci and haplotypes linked to stress tolerance. This understanding is essential for the development of resilient crop varieties. Implementing next-generation breeding strategies, including breeding-by-design and genomic selection, will expedite the creation of stress-tolerant crop varieties with enhanced yields. Research into genetic variation for flooding tolerance in soybean germplasm has identified associated genomic regions that can be integrated into elite lines. This incorporation can help reduce significant yield losses caused by flooding stress, enhancing overall crop resilience (Adegoye et al. 2023). Shannon et al. (2005) evaluated a core collection of 350 soybean germplasm lines for their tolerance to flooding during early reproductive stages. Their study identified six lines that exhibited a strong correlation with flooding tolerance. The study found that susceptible lines experienced nearly double the yield loss compared to tolerant lines. Notably, exotic plant introductions (PIs) showed significantly better flooding tolerance than cultivated varieties. Six cultivated germplasm lines, including Archer, Misuzudaiz, PI 408105A, PI 561271, PI 567651, and PI 567343, were identified as potential donors for breeding programs focused on enhancing flooding tolerance in soybeans (Shannon et al. 2005). A screening of 84 Northern soybean cultivars demonstrated significant variation in their tolerance to waterlogging. This finding highlights the potential for selecting and breeding cultivars with improved resilience to flooding conditions (Abruzzi de Oliveira-Busatto et al. 2022). Two recombinant inbred lines (RIL) populations were created for QTL analysis, using the sensitive Noir 1 and Minsoy and the tolerant cultivar Archer as parents. This approach aims to identify genetic factors associated with flooding tolerance. The analysis conducted by Van Nguyen et al. (2017) identified a major locus, Sat_064, on chromosome 18 associated with flooding tolerance (Van Nguyen et al. 2017). However, no direct correlation with waterlogging tolerance was found when this locus was introduced into southern genotypes to create near-isogenic lines (NILs). Further investigation by Chandra et al. (2022) uncovered at least five additional markers on chromosomes 5 and 13 linked to stress tolerance in the cultivar Archer, indicating a more complex genetic basis for flooding resilience (Chandra et al. 2022). Although some RILs and NILs have demonstrated enhanced waterlogging tolerance, no specific gene has been definitively associated with the flooding tolerance observed in the cultivar Archer (Mustroph 2018). A QTL analysis involving two *Brassica oleracea* cultivars, a sensitive Chinese cultivar (A12DHd) and a more tolerant calabrese cultivar (GDDH33), identified three QTLs associated with germination ability under low-oxygen conditions (Prasad and Ranjan 2020). However, the specific genes linked to these QTLs remain unidentified. In addition, genetic analyses in rapeseed (*Brassica napus*) have pinpointed two genes related to waterlogging tolerance in the cultivar Zhongshuang 9, although no specific QTL has been documented for this trait (Cong Ye et al. 2009). Research on six different cultivars indicated the presence of both additive and nonadditive genetic effects, suggesting that multiple genes or alleles contribute to the observed traits. Among these cultivars, Zhongshuang 9 exhibited the highest level of tolerance (Guo et al. 2020). A study using a doubled haploid population derived from lines with differing waterlogging and drought tolerance levels identified at least 11 QTL associated with waterlogging tolerance, emphasizing the complex multigenic regulation of this trait. Research into the physiological mechanisms underlying waterlogging tolerance highlighted the importance of the antioxidant system (Mustroph 2018). Furthermore, differences in nitrate metabolism were observed between two Chinese *Brassica napus* cultivars under waterlogged conditions, establishing a connection between reduced ethanolic fermentation and improved waterlogging tolerance (Xu et al. 2016).

11.5 Conclusions and Future Perspectives

The complex interplay of phytohormones in oilseed crops during flooding stress highlights the critical need for a comprehensive understanding of plant physiological responses to environmental challenges. Waterlogging, primarily induced by hypoxic conditions, compels plants to activate various adaptive mechanisms essential for survival. These adaptations encompass the development of adventitious roots and aerenchyma tissues, crucial for enhancing oxygen uptake and sustaining vital metabolic functions. Phytohormones, including ethylene, gibberellins, and abscisic acid, are integral to these adaptive processes, as they modulate gene expression and signaling pathways that regulate plant growth and development under stress conditions. Recent studies have illuminated the intricate interactions among these phytohormones, emphasizing that their crosstalk is fundamental to effective stress responses. For example, ethylene not only initiates growth responses but also influences the synthesis of other hormones, such as gibberellins and abscisic acid, which are essential for mitigating the effects of waterlogging. This hormonal interplay enables plants to optimize energy utilization during flooding events, thereby enhancing their chances of survival until environmental conditions improve.

Furthermore, advancements in molecular biology techniques have significantly deepened our understanding of the genetic regulation underlying flooding tolerance. Identifying key genes involved in hormonal signaling pathways is crucial for developing crop varieties with improved resilience to waterlogging. The application of gene editing technologies presents promising opportunities for engineering oilseed crops capable of thriving in increasingly unpredictable climatic conditions. Future research endeavors should prioritize several key areas: first, elucidating the specific roles of lesser-studied phytohormones in enhancing flooding tolerance; second, investigating the synergistic effects of phytohormones when applied exogenously; and third, integrating omics approaches to gain a more comprehensive understanding of the metabolic pathways involved in stress adaptation. Addressing these research gaps will facilitate the development of innovative agricultural practices to bolster crop resilience against flooding.

In conclusion, as climate change continues to alter precipitation patterns and exacerbate flooding, a thorough understanding of the molecular mechanisms by which oilseed crops respond to waterlogging stress becomes increasingly imperative. Leveraging this knowledge through advanced breeding techniques and targeted phytohormonal applications could substantially enhance agricultural productivity and food security in regions susceptible to flooding. The ongoing investigation into phytohormonal interactions enriches our comprehension of plant biology and holds significant potential for devising sustainable solutions to address the agricultural challenges posed by climate change. In addition to these strategies, fostering collaborations between researchers, policymakers, and agricultural industries will be pivotal in translating scientific insights into practical applications. We can build more sustainable and adaptive agricultural systems by integrating innovative technologies with traditional farming practices. As research progresses, the focus must also shift toward educating farmers about the benefits of phytohormonal applications and stress-resistant crop varieties, ensuring that these advancements reach those who most need them. Together, these efforts will contribute to a resilient global food system capable of withstanding the challenges posed by climate change.

References

Abiko T, Kotula L, Shiono K, Malik AI, Colmer TD, Nakazono M (2012) Enhanced formation of aerenchyma and induction of a barrier to radial oxygen loss in adventitious roots of Zea nicaraguensis contribute to its waterlogging tolerance as compared with maize (Zea mays ssp mays). Plant Cell Environ 35(9):1618–1630

Abruzzi de Oliveira-Busatto L, da Silva P, Giordano C, Ferreira da Silva M, Uhry Junior D, Guzman F, Wiebke-Strohm B, Li Z, Bredemeier C, Bodanese-Zanettini MH (2022) Identification of functional genetic variations underlying flooding tolerance in Brazilian soybean genotypes. Int J Mol Sci 23(18):10611

Adams D, Yang S (1979) Ethylene biosynthesis: identification of 1-aminocyclopropane-1-carboxylic acid as an intermediate in the conversion of methionine to ethylene. Proc Natl Acad Sci 76(1):170–174

Adegoye GA, Olorunwa OJ, Alsajri FA, Walne CH, Wijewandana C, Kethireddy SR, Reddy KN, Reddy KR (2023) Waterlogging effects on soybean physiology and hyperspectral reflectance during the reproductive stage. Agriculture 13(4):844

Agualongo DAP, Da-Silva CJ, Garcia N, de Oliveira FK, Shimoia EP, Posso DA, de Oliveira ACB, de Oliveira D, do Amarante, L. (2022, Nov) Waterlogging priming alleviates the oxidative damage, carbohydrate consumption, and yield loss in soybean (Glycine max) plants exposed to waterlogging. Funct Plant Biol 49(12):1029–1042. https://doi.org/10.1071/fp22030

Agüera E, de la Haba P (2021) Climate change impacts on sunflower (Helianthus annus L.) plants. Plan Theory 10(12):2646

Ahammed GJ, Li X, Liu A, Chen S (2020) Brassinosteroids in plant tolerance to abiotic stress. J Plant Growth Regul 39(4):1451–1464

Ahammed GJ, Yu J-Q (2016) Plant hormones under challenging environmental factors. Springer

Ahanger MA, Mir RA, Alyemeni MN, Ahmad P (2020) Combined effects of brassinosteroid and kinetin mitigates salinity stress in tomato through the modulation of antioxidant and osmolyte metabolism. Plant Physiol Biochem 147:31–42

Albuquerque MCD, Carvalho NF (2003) Effect of the type of environmental stress on the emergence of sunflower (*Helianthus annus* L.), soybean (*Glycine max* (L.) Merril) and maize (*Zea mays* L.) seeds with different levels of vigor. Seed Sci Technol 31:465–479

Amraee L, Rahmani F, Mandoulakani BA (2019) 24-Epibrassinolide alters DNA cytosine methylation of Linum usitatissimum L. under salinity stress. Plant Physiol Biochem 139:478–484

Aslam A, Mahmood A, Ur-Rehman H, Li C, Liang X, Shao J, Negm S, Moustafa M, Aamer M, Hassan MU (2023) Plant adaptation to flooding stress under changing climate conditions: ongoing breakthroughs and future challenges. Plan Theory 12(22):3824

Atif RM, Shahid L, Waqas M, Ali B, Rashid MAR, Azeem F, Nawaz MA, Wani SH, Chung G (2019) Insights on calcium-dependent protein kinases (CPKs) signaling for abiotic stress tolerance in plants. Int J Mol Sci 20(21):5298

Ayala-Ramos MA, Montaño-Arias SA, Terrazas T, Grether R (2024) Ecological implications of stomatal density and stomatal index in the adult stage of Mimosa L.(Leguminosae, Caesalpinioideae). Protoplasma 261(3):477–486

Bajguz A, Piotrowska-Niczyporuk A (2014) Interactive effect of brassinosteroids and cytokinins on growth, chlorophyll, monosaccharide and protein content in the green alga Chlorella vulgaris (Trebouxiophyceae). Plant Physiol Biochem 80:176–183

Bashar KK, Tareq MZ, Amin MR, Honi U, Tahjib-Ul-Arif M, Sadat MA, Hossen QMM (2019) Phytohormone-mediated stomatal response, escape and quiescence strategies in plants under flooding stress. Agronomy 9(2):43

Basit F, Liu J, An J, Chen M, He C, Zhu X, Li Z, Hu J, Guan Y (2021) Brassinosteroids as a multidimensional regulator of plant physiological and molecular responses under various environmental stresses. Environ Sci Pollut Res 28:44768–44779

Bernacki MJ, Mielecki J, Antczak A, Drożdżek M, Witoń D, Dąbrowska-Bronk J, Gawroński P, Burdiak P, Marchwicka M, Rusaczonek A (2023) Biotechnological potential of the stress response and plant cell death regulators proteins in the biofuel industry. Cells 12(16):2018

Bester AU, Shimoia EP, Da-Silva CJ, Posso DA, Carvalho IR, Corrêa FM, de Oliveira AC, do Amarante L (2024) Enhancing stress resilience in soybeans (Glycine max): assessing the efficacy of priming and cross-priming for mitigating water deficit and waterlogging effects. Funct Plant Biol 51(9)

Bi W, Weng B, Yan D, Wang M, Wang H, Wang J, Yan H (2020) Effects of drought-flood abrupt alternation on phosphorus in summer maize farmland systems. Geoderma 363:114147

Boccaccini A, Lorrai R, Ruta V, Frey A, Mercey-Boutet S, Marion-Poll A, Tarkowská D, Strnad M, Costantino P, Vittorioso P (2016) The DAG1 transcription factor negatively regulates the seed-to-seedling transition in Arabidopsis acting on ABA and GA levels. BMC Plant Biol 16:1–11

Brookbank BP, Patel J, Gazzarrini S, Nambara E (2021) Role of basal ABA in plant growth and development. Genes 12(12):1936

Chandra S, Das RT, Nagar S, Satpute G, Kumawat G, Ratnaparkhe M, Gupta S, Rajesh V, Nataraj V, Shivakumar M (2022) Soybean improvement for waterlogging tolerance. In: Soybean improvement: physiological, molecular and genetic perspectives. Springer, pp 73–97

Chater C, Peng K, Movahedi M, Dunn JA, Walker HJ, Liang Y-K, McLachlan DH, Casson S, Isner JC, Wilson I (2015) Elevated CO2-induced responses in stomata require ABA and ABA signaling. Curr Biol 25(20):2709–2716

Chen T-T, Liu F-F, Xiao D-W, Jiang X-Y, Li P, Zhao S-M, Hou B-K, Li Y-J (2020) The Arabidopsis UDP-glycosyltransferase75B1, conjugates abscisic acid and affects plant response to abiotic stresses. Plant Mol Biol 102:389–401

Colebrook EH, Thomas SG, Phillips AL, Hedden P (2014) The role of gibberellin signalling in plant responses to abiotic stress. J Exp Biol 217(Pt 1):67–75. https://doi.org/10.1242/jeb.089938

Colmer TD (2003) Long-distance transport of gases in plants: a perspective on internal aeration and radial oxygen loss from roots. Plant Cell Environ 26(1):17–36. https://doi.org/10.1046/j.1365-3040.2003.00846.x

Colmer TD, Greenway H (2011, Jan) Ion transport in seminal and adventitious roots of cereals during O2 deficiency. J Exp Bot 62(1):39–57. https://doi.org/10.1093/jxb/erq271

Combs-Giroir R, Gschwend AR (2024) Physical and molecular responses to flooding in Brassicaceae. Environ Exp Bot 105664:105664

Cong Ye CY, Cheng Yong CY, Zou Chong Shun ZC, Zhang Xue Kun ZX, Wang Han Zhong WH (2009) Genetic analysis of waterlogging tolerance for germinated seeds of rapeseed (Brassica napus L.) with mixed model of major gene plus polygene. Agricultural and Food Sciences

Cutforth H, Angadi S, McConkey B, Miller P, Ulrich D, Gulden R, Volkmar K, Entz M, Brandt S (2013) Comparing rooting characteristics and soil water withdrawal patterns of wheat with alternative oilseed and pulse crops grown in the semiarid Canadian prairie. Can J Soil Sci 93(2): 147–160

Damalas CA, Koutroubas SD (2022) Exogenous application of salicylic acid for regulation of sunflower growth under abiotic stress: a systematic review. Biologia 77(7):1685–1697

Dawood T, Yang X, Visser EJ, Te Beek TA, Kensche PR, Cristescu SM, Lee S, Floková K, Nguyen D, Mariani C (2016) A co-opted hormonal cascade activates dormant adventitious root primordia upon flooding in Solanum dulcamara. Plant Physiol 170(4):2351–2364

de Souza KRD, de Oliveira Santos M, Andrade CA, da Silva DM, Campos NA, Alves JD (2017) Aerenchyma formation in the initial development of maize roots under waterlogging. Theor Exp Plant Physiol 29:165–175

den Besten N, Steele-Dunne S, de Jeu R, van der Zaag P (2021) Towards monitoring waterlogging with remote sensing for sustainable irrigated agriculture. Remote Sens 13(15):2929

Ding G, Gao L, Li X, Zhou M, Liu Q, Ren H, Jiang B (2014) A mixed method to evaluate burden of malaria due to flooding and waterlogging in Mengcheng County, China: a case study. PLoS One 9(5):e97520

Divi UK, Rahman T, Krishna P (2016) Gene expression and functional analyses in brassinosteroid-mediated stress tolerance. Plant Biotechnol J 14(1):419–432

Doupis G, Kavroulakis N, Psarras G, Papadakis IE (2017) Growth, photosynthetic performance and antioxidative response of 'Hass' and 'Fuerte' avocado (Persea americana Mill.) plants grown under high soil moisture. Photosynthetica 55(4):655–663. https://doi.org/10.1007/s11099-016-0679-7

Dubois V, Moritz T, García-Martínez JL (2011) Comparison of the role of gibberellins and ethylene in response to submergence of two lowland rice cultivars, Senia and Bomba. J Plant Physiol 168(3):233–241

Eun H-D, Ali S, Jung H, Kim K, Kim W-C (2019) Profiling of ACC synthase gene (ACS11) expression in Arabidopsis induced by abiotic stresses. Appl Biol Chem 62:1–11

Eysholdt-Derzsó E, Sauter M (2017) Root bending is antagonistically affected by hypoxia and ERF-mediated transcription via auxin signaling. Plant Physiol 175(1):412–423

Eysholdt-Derzsó E, Sauter M (2019, Jan) Hypoxia and the group VII ethylene response transcription factor HRE2 promote adventitious root elongation in Arabidopsis. Plant Biol (Stuttg) 21 Suppl 1:103–108. https://doi.org/10.1111/plb.12873

Fàbregas N, Lozano-Elena F, Blasco-Escámez D, Tohge T, Martínez-Andújar C, Albacete A, Osorio S, Bustamante M, Riechmann JL, Nomura T (2018) Overexpression of the vascular brassinosteroid receptor BRL3 confers drought resistance without penalizing plant growth. Nat Commun 9(1):4680

Fan Y, Chen Y, Li X, Chen H, Yu M, Miao L, Wei S, Li S, Qian M, Chang W (2024) The brassinosteroid-signaling gene BnaC04. BIL1 contributes to plant architecture in Brassica napus. Ind Crop Prod 210:118066

Fariduddin Q, Khanam S, Hasan S, Ali B, Hayat S, Ahmad A (2009) Effect of 28-homobrassinolide on the drought stress-induced changes in photosynthesis and antioxidant system of Brassica juncea L. Acta Physiol Plant 31:889–897

Fatma M, Asgher M, Iqbal N, Rasheed F, Sehar Z, Sofo A, Khan NA (2022) Ethylene signaling under stressful environments: analyzing collaborative knowledge. Plan Theory 11(17):2211

Foyer CH, Lam H-M, Nguyen HT, Siddique KH, Varshney RK, Colmer TD, Cowling W, Bramley H, Mori TA, Hodgson JM (2016) Neglecting legumes has compromised human health and sustainable food production. Nat Plant 2(8):1–10

Gangana Gowdra VM, Lalitha B, Halli HM, Senthamil E, Negi P, Jayadeva H, Basavaraj P, Harisha C, Boraiah K, Adavi SB (2025) Root growth, yield and stress tolerance of soybean to transient waterlogging under different climatic regimes. Sci Rep 15(1):6968

Garcia N, da Silva CJ, Cocco KLT, Pomagualli D, de Oliveira FK, da Silva JVL, de Oliveira ACB, do Amarante L (2020) Waterlogging tolerance of five soybean genotypes through different physiological and biochemical mechanisms. Environ Exp Bot 172:103975

Geng S, Lin Z, Xie S, Xiao J, Wang H, Zhao X, Zhou Y, Duan L (2023) Ethylene enhanced waterlogging tolerance by changing root architecture and inducing aerenchyma formation in maize seedlings. J Plant Physiol 287:154042. https://doi.org/10.1016/j.jplph.2023.154042

Gomez MD, Barro-Trastoy D, Escoms E, Saura-Sánchez M, Sánchez I, Briones-Moreno A, Vera-Sirera F, Carrera E, Ripoll J-J, Yanofsky MF (2018) Gibberellins negatively modulate ovule number in plants. Development 145(13):dev163865

Grant MR, Jones JD (2009) Hormone (dis) harmony moulds plant health and disease. Science 324(5928):750–752

Grassini P, Indaco GV, Pereira ML, Hall AJ, Trápani N (2007) Responses to short-term waterlogging during grain filling in sunflower. Field Crop Res 101:352–363

Grove MD, Spencer GF, Rohwedder WK, Mandava N, Worley JF, Warthen JD Jr, Steffens GL, Flippen-Anderson JL, Cook JC Jr (1979) Brassinolide, a plant growth-promoting steroid isolated from Brassica napus pollen. Nature 281(5728):216–217

Guo Y, Chen J, Kuang L, Wang N, Zhang G, Jiang L, Wu D (2020) Effects of waterlogging stress on early seedling development and transcriptomic responses in Brassica napus. Mol Breed 40: 1–14

Hai NN, Chuong NN, Tu NHC, Kisiala A, Hoang XLT, Thao NP (2020) Role and regulation of cytokinins in plant response to drought stress. Plan Theory 9(4):422

Hallegatte S, Green C, Nicholls RJ, Corfee-Morlot J (2013) Future flood losses in major coastal cities. Nat Clim Chang 3(9):802–806

Harrison MA (2012) Cross-talk between phytohormone signaling pathways under both optimal and stressful environmental conditions. In: Phytohormones and abiotic stress tolerance in plants. Springer, pp 49–76

Hartman S, Liu Z, Van Veen H, Vicente J, Reinen E, Martopawiro S, Zhang H, Van Dongen N, Bosman F, Bassel GW (2019) Ethylene-mediated nitric oxide depletion pre-adapts plants to hypoxia stress. Nat Commun 10(1):4020

Hasanuzzaman M, Ahmed N, Saha T, Rahman M, Rahman K, Alam MM, Rohman MM, Nahar K (2022) Exogenous salicylic acid and kinetin modulate reactive oxygen species metabolism and glyoxalase system to confer waterlogging stress tolerance in soybean (Glycine max L.). Plant Stress 3:100057

Hasanuzzaman M, Al Mahmud J, Nahar K, Anee TI, Inafuku M, Oku H, Fujita M (2017) Responses, adaptation, and ROS metabolism in plants exposed to waterlogging stress. In: Khan MIR, Khan NA (eds) Reactive oxygen species and antioxidant systems in plants: role and regulation under abiotic stress. Springer Singapore, pp 257–281. https://doi.org/10.1007/978-981-10-5254-5_10

Hasanuzzaman M, Bhuyan M, Zulfiqar F, Raza A, Mohsin SM, Mahmud JA, Fujita M, Fotopoulos V (2020a) Reactive oxygen species and antioxidant defense in plants under abiotic stress: revisiting the crucial role of a universal defense regulator. Antioxidants (Basel) 9(8). https://doi.org/10.3390/antiox9080681

Hasanuzzaman M, Bhuyan MB, Zulfiqar F, Raza A, Mohsin SM, Mahmud JA, Fujita M, Fotopoulos V (2020b) Reactive oxygen species and antioxidant defense in plants under abiotic stress: revisiting the crucial role of a universal defense regulator. Antioxidants 9(8):681

Hernando CE, Murcia MG, Pereyra ME, Sellaro R, Casal JJ (2021) Phytochrome B links the environment to transcription. J Exp Bot 72(11):4068–4084

Hirabayashi Y, Mahendran R, Koirala S, Konoshima L, Yamazaki D, Watanabe S, Kim H, Kanae S (2013) Global flood risk under climate change. Nat Clim Chang 3(9):816–821

Hiron R, Wright S (1973) The role of endogenous abscisic acid in the response of plants to stress. J Exp Bot 24(4):769–780

Hong B, Zhou B, Peng Z, Yao M, Wu J, Wu X, Guan C, Guan M (2023) Tissue-specific transcriptome and metabolome analysis reveals the response mechanism of Brassica napus to waterlogging stress. Int J Mol Sci 24(7):6015

Houben M, Van de Poel B (2019) 1-Aminocyclopropane-1-carboxylic acid oxidase (ACO): the enzyme that makes the plant hormone ethylene. Front Plant Sci 10:695

Huang C, Gao Y, Qin A, Liu Z, Zhao B, Ning D, Ma S, Duan A, Liu Z (2022) Effects of waterlogging at different stages and durations on maize growth and grain yields. Agric Water Manag 261:107334

Huang Y-C, Yeh T-H, Yang C-Y (2019) Ethylene signaling involves in seeds germination upon submergence and antioxidant response elicited confers submergence tolerance to rice seedlings. Rice 12:1–8

Hussain M, Khan TA, Yusuf M, Fariduddin Q (2019) Silicon-mediated role of 24-epibrassinolide in wheat under high-temperature stress. Environ Sci Pollut Res 26:17163–17172

Hussain MA, Naeem A, Pitann B, Mühling KH (2023) High sulfur (S) supplementation imparts waterlogging tolerance to oilseed rape (Brassica napus L.) through upregulating s metabolism and antioxidant pathways. J Plant Growth Regul 42(12):7591–7605

Islam MN, Bell RW, Barrett-Lennard EG, Maniruzzaman M (2022) Shallow surface and subsurface drains alleviate waterlogging and salinity in a clay-textured soil and improve the yield of sunflower in the Ganges Delta. Agron Sustain Dev 42(2):16. https://doi.org/10.1007/s13593-021-00746-4

Islam MR, Raja DR (2021) Waterlogging risk assessment: an undervalued disaster risk in coastal urban community of Chattogram, Bangladesh. Earth 2(1):151–173

Jalakas P, Merilo E, Kollist H, Brosché M (2018) Complexity of ABA signaling for stomatal development and aperture regulation. bioRxiv:335810

Jia W, Ma M, Chen J, Wu S (2021) Plant morphological, physiological and anatomical adaption to flooding stress and the underlying molecular mechanisms. Int J Mol Sci 22(3):1088

Jiang H, Tang B, Xie Z, Nolan T, Ye H, Song GY, Walley J, Yin Y (2019) GSK 3-like kinase BIN 2 phosphorylates RD 26 to potentiate drought signaling in Arabidopsis. Plant J 100(5):923–937

Kagale S, Divi UK, Krochko JE, Keller WA, Krishna P (2007) Brassinosteroid confers tolerance in Arabidopsis thaliana and Brassica napus to a range of abiotic stresses. Planta 225:353–364. http://dx.doi.org/10.1007/s00425-006-0361-6

Kamal AHM, Komatsu S (2016) Jasmonic acid induced protein response to biophoton emissions and flooding stress in soybean. J Proteome 133:33–47

Kamaluddin M, Zwiazek JJ (2002) Ethylene enhances water transport in hypoxic aspen. Plant Physiol 128(3):962–969. https://doi.org/10.1104/pp.010791

Kaur G, Vikal Y, Kaur L, Kalia A, Mittal A, Kaur D, Yadav I (2021) Elucidating the morpho-physiological adaptations and molecular responses under long-term waterlogging stress in maize through gene expression analysis. Plant Sci 304:110823

Khan MA, Hamayun M, Iqbal A, Khan SA, Hussain A, Asaf S, Khan AL, Yun B-W, Lee I-J (2018) Gibberellin application ameliorates the adverse impact of short-term flooding on Glycine max L. Biochem J 475(18):2893–2905

Kim Y-H, Hwang S-J, Waqas M, Khan AL, Lee J-H, Lee J-D, Nguyen HT, Lee I-J (2015) Comparative analysis of endogenous hormones level in two soybean (Glycine max L.) lines differing in waterlogging tolerance. Front Plant Sci 6:714

Kim Y, Seo C-W, Khan AL, Mun B-G, Shahzad R, Ko J-W, Yun B-W, Park S-K, Lee I-J (2018). Exo-ethylene application mitigates waterlogging stress in soybean (Glycine max L.). BMC Plant Biol 18:1–16. https://doi.org/10.1186/s12870-018-1457-4

Kohli A, Sreenivasulu N, Lakshmanan P, Kumar PP (2013) The phytohormone crosstalk paradigm takes center stage in understanding how plants respond to abiotic stresses. Plant Cell Rep 32: 945–957

Kollist H, Nuhkat M, Roelfsema MRG (2014) Closing gaps: linking elements that control stomatal movement. New Phytol 203(1):44–62

Komatsu S, Han C, Nanjo Y, Altaf-Un-Nahar M, Wang K, He D, Yang P (2013a) Label-free quantitative proteomic analysis of abscisic acid effect in early-stage soybean under flooding. J Proteome Res 12(11):4769–4784

Komatsu S, Hiraga S, Yanagawa Y (2012) Proteomics techniques for the development of flood tolerant crops. J Proteome Res 11(1):68–78

Komatsu S, Shirasaka N, Sakata K (2013b) 'Omics' techniques for identifying flooding–response mechanisms in soybean. J Proteome 93:169–178

Kondo R, Tanaka S, Fujisao K, Miyaji H (2024) Effects of depressions and drainage systems on soybean (Glycine max (L.) Merill) growth during the early growth stages under excess water stress conditions. Plant Prod Sci 27(4):283–293

Koramutla MK, Tuan PA, Ayele BT (2022) Salicylic acid enhances adventitious root and aerenchyma formation in wheat under waterlogged conditions. Int J Mol Sci 23(3):1243

Kuroha T, Ashikari M (2020) Molecular mechanisms and future improvement of submergence tolerance in rice. Mol Breed 40(4):41

Kuroha T, Nagai K, Gamuyao R, Wang DR, Furuta T, Nakamori M, Kitaoka T, Adachi K, Minami A, Mori Y (2018) Ethylene-gibberellin signaling underlies adaptation of rice to periodic flooding. Science 361(6398):181–186

León J, Castillo MC, Gayubas B (2021, Aug 11) The hypoxia-reoxygenation stress in plants. J Exp Bot 72(16):5841–5856. https://doi.org/10.1093/jxb/eraa591

Li M, Yu G, Cao C, Liu P (2021) Metabolism, signaling, and transport of jasmonates. Plant Commun 2(5):100231

Li W, Katin-Grazzini L, Gu X, Wang X, El-Tanbouly R, Yer H, Thammina C, Inguagiato J, Guillard K, McAvoy RJ (2017) Transcriptome analysis reveals differential gene expression and a possible role of gibberellins in a shade-tolerant mutant of perennial ryegrass. Front Plant Sci 8: 868

Li W, Mo W, Ashraf U, Li G, Wen T, Abrar M, Gao L, Liu J, Hu J (2018) Evaluation of physiological indices of waterlogging tolerance of different maize varieties in South China. Appl Ecol Environ Res 16:2059–2072. https://doi.org/10.15666/aeer/1602_20592072

Li Z, Zhang L, Yu Y, Quan R, Zhang Z, Zhang H, Huang R (2011) The ethylene response factor AtERF11 that is transcriptionally modulated by the bZIP transcription factor HY5 is a crucial repressor for ethylene biosynthesis in Arabidopsis. Plant J 68(1):88–99

Liang J, Liang Y (2009) Effects of plant growth substances on water-logging resistance of oilseed rape seedling. Xinan Shifan Daxue Xuebao, Ziran Kexueban 34:58–62

Linkemer G, Board JE, Musgrave ME (1998) Waterlogging effects on growth and yield components in late-planted soybean. Crop Sci 38(6):1576–1584. https://doi.org/10.2135/cropsci1998.0011183X003800060028x

Liu M, Tan X, Sun X, Zwiazek JJ (2020) Properties of root water transport in canola (Brassica napus) subjected to waterlogging at the seedling, flowering and podding growth stages. Plant Soil 454:431–445

Liu M, Zwiazek JJ (2022) Oxidative stress impedes recovery of canola (Brassica napus) plants from waterlogging by inhibiting aquaporin-mediated root water transport. Environ Exp Bot 200: 104931

Liu S, Lv Z, Liu Y, Li L, Zhang L (2018) Network analysis of ABA-dependent and ABA-independent drought responsive genes in Arabidopsis thaliana. Genet Mol Biol 41:624–637

Liu S, Wang J, Wang L, Wang X, Xue Y, Wu P, Shou H (2009) Adventitious root formation in rice requires OsGNOM1 and is mediated by the OsPINs family. Cell Res 19(9):1110–1119

Loreti E, van Veen H, Perata P (2016) Plant responses to flooding stress. Curr Opin Plant Biol 33: 64–71

Lu X, Chen Y, Gong W, Chen Y (2006) Effect of brassinolide on the seedling growth and waterlogging resistance of soybean. Chin Agric Sci Bull 23:37–38

Lv B, Tian H, Zhang F, Liu J, Lu S, Bai M, Li C, Ding Z (2018) Brassinosteroids regulate root growth by controlling reactive oxygen species homeostasis and dual effect on ethylene synthesis in Arabidopsis. PLoS Genet 14(1):e1007144

Lv B, Yan Z, Tian H, Zhang X, Ding Z (2019) Local auxin biosynthesis mediates plant growth and development. Trends Plant Sci 24(1):6–9

Ma X, Mo B, Cao X (2015) New players in ABA signaling: identification of PUB12/13 involved in degradation of ABA co-receptor ABI1. Sci China Life Sci 58(11):1173

Malik AI, Islam AKMR, Colmer TD (2011) Transfer of the barrier to radial oxygen loss in roots of Hordeum marinum to wheat (Triticum aestivum): evaluation of four H. Marinum—wheat amphiploids. New Phytol 190:499–508. https://doi.org/10.1111/j.1469-8137.2010.03519.x

Manghwar H, Hussain A, Alam I, Khoso MA, Ali Q, Liu F (2024a) Waterlogging stress in plants: unraveling the mechanisms and impacts on growth, development, and productivity. Environ Exp Bot 105824:105824

Manik SN, Quamruzzaman M, Livermore M, Zhao C, Johnson P, Hunt I, Shabala S, Zhou M (2022) Impacts of barley root cortical aerenchyma on growth, physiology, yield components, and grain quality under field waterlogging conditions. Field Crops Res 279:108461. https://doi.org/10.1016/j.fcr.2022.108461

Manghwar H, Hussain A, Alam I, Khoso MA, Ali Q, Liu F (2024b) Waterlogging stress in plants: unraveling the mechanisms and impacts on growth, development, and productivity. Environ Exp Bot 224:105824. https://doi.org/10.1016/j.envexpbot.2024.105824

Mao J, Li J (2020) Regulation of three key kinases of brassinosteroid signaling pathway. Int J Mol Sci 21(12):4340

Marothia D, Kaur N, Pati PK (2021) Abiotic stress responses in plants: current knowledge and future. Abiotic Stress in Plants 73. https://doi.org/10.5772/intechopen.93824

Martínez C, Espinosa-Ruíz A, de Lucas M, Bernardo-García S, Franco-Zorrilla JM, Prat S (2018) PIF 4-induced BR synthesis is critical to diurnal and thermomorphogenic growth. EMBO J 37(23):e99552

Martins S, Montiel-Jorda A, Cayrel A, Huguet S, Roux CP-L, Ljung K, Vert G (2017) Brassinosteroid signaling-dependent root responses to prolonged elevated ambient temperature. Nat Commun 8(1):309

Mastrantonas N, Herrera-Lormendez P, Magnusson L, Pappenberger F, Matschullat J (2020) Extreme precipitation events in the Mediterranean: spatiotemporal characteristics and connection to large-scale atmospheric flow patterns. arXiv preprint arXiv:2009.00912

Maurel C, Nacry P (2020, Jul) Root architecture and hydraulics converge for acclimation to changing water availability. Nat Plants 6(7):744–749. https://doi.org/10.1038/s41477-020-0684-5

Merilo E, Laanemets K, Hu H, Xue S, Jakobson L, Tulva I, Gonzalez-Guzman M, Rodriguez PL, Schroeder JI, Broschè M (2013) PYR/RCAR receptors contribute to ozone-, reduced air humidity-, darkness-, and CO2-induced stomatal regulation. Plant Physiol 162(3):1652–1668

Merilo E, Yarmolinsky D, Jalakas P, Parik H, Tulva I, Rasulov B, Kilk K, Kollist H (2018) Stomatal VPD response: there is more to the story than ABA. Plant Physiol 176(1):851–864

Mickelbart MV, Hasegawa PM, Bailey-Serres J (2015) Genetic mechanisms of abiotic stress tolerance that translate to crop yield stability. Nat Rev Genet 16(4):237–251

Muhammad Aslam M, Waseem M, Jakada BH, Okal EJ, Lei Z, Saqib HSA, Yuan W, Xu W, Zhang Q (2022) Mechanisms of abscisic acid-mediated drought stress responses in plants. Int J Mol Sci 23(3):1084

Mui NT, Zhou M, Parsons D, Smith RW (2021) Aerenchyma formation in adventitious roots of tall fescue and cocksfoot under waterlogged conditions. Agronomy 11(12):2487. https://www.mdpi.com/2073-4395/11/12/2487

Mustroph A (2018) Improving flooding tolerance of crop plants. Agronomy 8(9):160

Nabloussi A, Bahri H, Lakbir M, Moukane H, Kajji A, El Fechtali M (2019) Assessment of a set of rapeseed (Brassica napus L.) varieties under waterlogging stress at different plant growth stages. OCL 26:36

Nagai K, Mori Y, Ishikawa S, Furuta T, Gamuyao R, Niimi Y, Hobo T, Fukuda M, Kojima M, Takebayashi Y (2020) Antagonistic regulation of the gibberellic acid response during stem growth in rice. Nature 584(7819):109–114

Nguyen T-N, Tuan PA, Ayele BT (2022) Jasmonate regulates seed dormancy in wheat via modulating the balance between gibberellin and abscisic acid. J Exp Bot 73(8):2434–2453

Nguyen T-N, Tuan PA, Mukherjee S, Son S, Ayele BT (2018) Hormonal regulation in adventitious roots and during their emergence under waterlogged conditions in wheat. J Exp Bot 69(16): 4065–4082
Nishiuchi S, Yamauchi T, Takahashi H, Kotula L, Nakazono M (2012) Mechanisms for coping with submergence and waterlogging in rice. Rice 5:1–14
Nolan TM, Vukašinović N, Liu D, Russinova E, Yin Y (2020) Brassinosteroids: multidimensional regulators of plant growth, development, and stress responses. Plant Cell 32(2):295–318
Olorunwa OJ, Adhikari B, Brazel S, Popescu SC, Popescu GV, Shi A, Barickman TC (2022) Waterlogging during the reproductive growth stage causes physiological and biochemical modifications in the leaves of cowpea (Vigna unguiculata L.) genotypes with contrasting tolerance. Plant Physiol Biochem 190:133–144
Orchard PW, Jessop RS (1984) The response of sorghum and sunflower to short-term waterlogging. Effects of stage development and duraton of waterlogging on growth and yield. Plant Soil 81: 119–132
Pais IP, Moreira R, Semedo JN, Ramalho JC, Lidon FC, Coutinho J, Maçãs B, Scotti-Campos P (2022) Wheat crop under waterlogging: potential soil and plant effects. Plan Theory 12(1):149
Pan J, Sharif R, Xu X, Chen X (2020) Mechanisms of waterlogging tolerance in plants: research Progress and prospects. Front Plant Sci 11:627331. https://doi.org/10.3389/fpls.2020.627331
Pan J, Sharif R, Xu X, Chen X (2021) Mechanisms of waterlogging tolerance in plants: research progress and prospects. Front Plant Sci 11:627331
Park EJ, Kim T-H (2021) Thaumatin-like genes function in the control of both biotic stress signaling and ABA signaling pathways. Biochem Biophys Res Commun 567:17–21
Parlanti S, Kudahettige N, Lombardi L, Mensuali-Sodi A, Alpi A, Perata P, Pucciariello C (2011) Distinct mechanisms for aerenchyma formation in leaf sheaths of rice genotypes displaying a quiescence or escape strategy for flooding tolerance. Ann Bot 107(8):1335–1343
Pathak K, Rahman SW, Bhagawati S, Gogoi B (2017) Sesame (Sesamum indicum L.), an underexploited oil seed crop: current status, features and importance-A review. Agric Rev 38(3):223–227
Pattyn J, Vaughan-Hirsch J, Van de Poel B (2021) The regulation of ethylene biosynthesis: a complex multilevel control circuitry. New Phytol 229(2):770–782
Paul PLC, Bell RW, Barrett-Lennard EG, Kabir E, Mainuddin M, Sarker KK (2021) Short-term waterlogging depresses early growth of sunflower (Helianthus annuus L.) on saline soils with a shallow water table in the coastal zone of Bangladesh. Soil Systems 5(4):68
Pereira YC, Silva F Rd, Silva BRS, Cruz FJR, Marques DJ, Lobato AKS (2020) 24-epibrassinolide induces protection against waterlogging and alleviates impacts on the root structures, photosynthetic machinery and biomass in soybean. Plant Signal Behav 15(11):1805885
Phukan UJ, Jeena GS, Tripathi V, Shukla RK (2018) Ma RAP 2-4, a waterlogging-responsive ERF from Mentha, regulates bidirectional sugar transporter at SWEET 10 to modulate stress response in Arabidopsis. Plant Biotechnol J 16(1):221–233
Pigliucci M, Kolodynska A (2002) Phenotypic plasticity and integration in response to flooded conditions in natural accessions of Arabidopsis thaliana (L.) Heynh (Brassicaceae). Ann Bot 90(2):199–207
Ploschuk RA, Miralles DJ, Colmer TD, Ploschuk EL, Striker GG (2018) Waterlogging of winter crops at early and late stages: impacts on leaf physiology, growth and yield. Front Plant Sci 9: 1863
Ploschuk RA, Miralles DJ, Colmer TD, Striker GG (2020) Waterlogging differentially affects yield and its components in wheat, barley, rapeseed and field pea depending on the timing of occurrence. J Agron Crop Sci 206(3):363–375
Ploschuk RA, Miralles DJ, Striker GG (2021) Early-and late-waterlogging differentially affect the yield of wheat, barley, oilseed rape and field pea through changes in leaf area index, radiation interception and radiation use efficiency. J Agron Crop Sci 207(3):504–520
Ploschuk RA, Miralles DJ, Striker GG (2023) Waterlogging tolerance of winter crops: root mass density and canopy dynamics. Agron J 115(5):2506–2520

Prasad M, Ranjan R (2020) Brassicaceae plants response and tolerance to waterlogging and flood. The Plant Family Brassicaceae: Biology and Physiological Responses to Environmental Stresses, pp 311–335

Qi X, Hu Z, Chen X, Zhang M, Nakazono M (2023) Involvement of Phytohormones in flooding stress tolerance in plants. Plant Hormones and Climate Change:251–271

Qi X, Li Q, Ma X, Qian C, Wang H, Ren N, Shen C, Huang S, Xu X, Xu Q (2019) Waterlogging-induced adventitious root formation in cucumber is regulated by ethylene and auxin through reactive oxygen species signalling. Plant Cell Environ 42(5):1458–1470

Qiao M, Hong C, Jiao Y, Hou S, Gao H (2024) Impacts of drought on photosynthesis in major food crops and the related mechanisms of plant responses to drought. Plan Theory 13(13):1808

Qiu F, Zheng Y, Zhang Z, Xu S (2007) Mapping of QTL associated with waterlogging tolerance during the seedling stage in maize. Ann Bot 99(6):1067–1081. https://doi.org/10.1093/aob/mcm055

Qiu T, Qi M, Ding X, Zheng Y, Zhou T, Chen Y, Han N, Zhu M, Bian H, Wang J (2020) The SAUR41 subfamily of SMALL AUXIN UP RNA genes is abscisic acid inducible to modulate cell expansion and salt tolerance in Arabidopsis thaliana seedlings. Ann Bot 125(5):805–819

Rajhi I, Yamauchi T, Takahashi H, Nishiuchi S, Shiono K, Watanabe R, Mliki A, Nagamura Y, Tsutsumi N, Nishizawa NK (2011) Identification of genes expressed in maize root cortical cells during lysigenous aerenchyma formation using laser microdissection and microarray analyses. New Phytol 190(2):351–368

Ramakrishna B, Rao SSR (2013) Preliminary studies on the involvement of glutathione metabolism and redox status against zinc toxicity in radish seedlings by 28-homobrassinolide. Environ Exp Bot 96:52–58

Retzer K, Akhmanova M, Konstantinova N, Malínská K, Leitner J, Petrášek J, Luschnig C (2019) Brassinosteroid signaling delimits root gravitropism via sorting of the Arabidopsis PIN2 auxin transporter. Nat Commun 10(1):5516

Ritonga FN, Zhou D, Zhang Y, Song R, Li C, Li J, Gao J (2023) The roles of gibberellins in regulating leaf development. Plan Theory 12(6):1243

Ruduś I, Sasiak M, Kępczyński J (2013) Regulation of ethylene biosynthesis at the level of 1-aminocyclopropane-1-carboxylate oxidase (ACO) gene. Acta Physiol Plant 35:295–307

Rupngam T, Messiga AJ (2024) Unraveling the interactions between flooding dynamics and agricultural productivity in a changing climate. Sustain For 16(14):6141. https://www.mdpi.com/2071-1050/16/14/6141

Sadura I, Janeczko A (2018) Physiological and molecular mechanisms of brassinosteroid-induced tolerance to high and low temperature in plants. Biol Plant 62(4):601–616

Sasidharan R, Voesenek LA (2015) Ethylene-mediated acclimations to flooding stress. Plant Physiol 169(1):3–12

Sato T, Theologis A (1989) Cloning the mRNA encoding 1-aminocyclopropane-1-carboxylate synthase, the key enzyme for ethylene biosynthesis in plants. Proc Natl Acad Sci 86(17): 6621–6625

Schaefer PE, Nied AH, Heldwein AB, Dalmago GA, Pappis AC, Gouvêa JA, d. (2023) Modification of canola cultivation conditions in a waterlogging-susceptible subtropical environment. Acta Sci Agron 45:e61434

Schmitz AJ, Folsom JJ, Jikamaru Y, Ronald P, Walia H (2013) SUB 1 A-mediated submergence tolerance response in rice involves differential regulation of the brassinosteroid pathway. New Phytol 198(4):1060–1070

Shannon J, Stevens W, Wiebold W, McGraw R, Sleper D, Nguyen H (2005) Breeding soybeans for improved tolerance to flooding. In: Procedure of 30th soybean research conference. American Seed Trade Association, Chicago

Shao M, He Y, Han X, Qu H, Zhang X, Chen C, Zhang J, Song Q, Zhou J, Jie Y (2025) Effect of mixed spraying of SA and ABA on the growth and development of winter oilseed rape (Brassica napus L.) during the post-waterlogging Podding period. Agronomy 15(2):348

Sharif R, Xie C, Zhang H, Arnao MB, Ali M, Ali Q, Muhammad I, Shalmani A, Nawaz MA, Chen P, Li Y (2018) Melatonin and its effects on plant systems. Molecules 23(9). https://doi.org/10.3390/molecules23092352

Sharma A, Yuan H, Kumar V, Ramakrishnan M, Kohli SK, Kaur R, Thukral AK, Bhardwaj R, Zheng B (2019) Castasterone attenuates insecticide induced phytotoxicity in mustard. Ecotoxicol Environ Saf 179:50–61

Shimamura S, Nishimura T, Koshiba T, Yamamoto R, Hiraga S, Nakamura T, Komatsu S (2016) Effects of anti-auxins on secondary aerenchyma formation in flooded soybean hypocotyls. Plant Prod Sci 19(1):154–160

Shimamura S, Yoshioka T, Yamamoto R, Hiraga S, Nakamura T, Shimada S, Komatsu S (2014) Role of abscisic acid in flood-induced secondary aerenchyma formation in soybean (Glycine max) hypocotyls. Plant Prod Sci 17(2):131–137

Shiono K, Ejiri M, Shimizu K, Yamada S (2019) Improved waterlogging tolerance of barley (Hordeum vulgare) by pretreatment with ethephon. Plant Prod Sci 22(2):285–295

Singh A (2015) Soil salinization and waterlogging: a threat to environment and agricultural sustainability. Ecol Indic 57:128–130

Soares JC, Zimmermann L, Zendonadi dos Santos N, Muller O, Pintado M, Vasconcelos MW (2021) Genotypic variation in the response of soybean to elevated CO_2. Plant-Environ Interact 2(6):263–276. https://doi.org/10.1002/pei3.10065

Srivastava AK, Shukla SK, Singh A, Singh A, Jindal S, Malik P (2024) Biochemical, physiological and molecular aspects of waterlogging tolerance in economically important oilseed crops rapeseed, sesame and soybean. Genetika 56(1):221–241

Stevens J, Faralli M, Wall S, Stamford JD, Lawson T (2021) Chapter 2 stomatal responses to climate change. In: Photosynthesis, respiration, and climate change. Springer, pp 17–47

Tabari H (2020) Climate change impact on flood and extreme precipitation increases with water availability. Sci Rep 10(1):13768

Tahjib-Ul-Arif M, Hasan MT, Rahman MA, Nuruzzaman M, Rahman AS, Hasanuzzaman M, Haque MR, Hossain MA, Latef AAHA, Murata Y (2023) Plant response to combined salinity and waterlogging stress: Current research progress and future prospects. Plant Stress 7:100137. https://doi.org/10.1016/j.stress.2023.100137

Takahashi H, Yamauchi T, Colmer TD, Nakazono M (2014) Aerenchyma formation in plants. In: van Dongen JT, Licausi F (eds) Low-oxygen stress in plants: oxygen sensing and adaptive responses to hypoxia. Springer Vienna, pp 247–265. https://doi.org/10.1007/978-3-7091-1254-0_13

Tamang BG, Li S, Rajasundaram D, Lamichhane S, Fukao T (2021) Overlapping and stress-specific transcriptomic and hormonal responses to flooding and drought in soybean. Plant J 107(1):100–117

Tamang BG, Magliozzi JO, Maroof MS, Fukao T (2014) Physiological and transcriptomic characterization of submergence and reoxygenation responses in soybean seedlings. Plant Cell Environ 37(10):2350–2365

Tanaka Y, Nose T, Jikumaru Y, Kamiya Y (2013) ABA inhibits entry into stomatal-lineage development in A rabidopsis leaves. Plant J 74(3):448–457

Tang W, Kim T-W, Oses-Prieto JA, Sun Y, Deng Z, Zhu S, Wang R, Burlingame AL, Wang Z-Y (2008) BSKs mediate signal transduction from the receptor kinase BRI1 in Arabidopsis. Science 321(5888):557–560

Topali C, Antonopoulou C, Chatzissavvidis C (2024) Effect of waterlogging on growth and productivity of fruit crops. Horticulturae 10(6):623

Van Nguyen L, Takahashi R, Githiri SM, Rodriguez TO, Tsutsumi N, Kajihara S, Sayama T, Ishimoto M, Harada K, Suematsu K (2017) Mapping quantitative trait loci for root development under hypoxia conditions in soybean (Glycine max L. Merr.). Theor Appl Genet 130:743–755

van Veen H, Mustroph A, Barding GA, Vergeer-van Eijk M, Welschen-Evertman RA, Pedersen O, Visser EJ, Larive CK, Pierik R, Bailey-Serres J (2013) Two Rumex species from contrasting hydrological niches regulate flooding tolerance through distinct mechanisms. Plant Cell 25(11): 4691–4707

van Veen H, Triozzi PM, Loreti E (2024) Metabolic strategies in hypoxic plants. Plant Physiol kiae564

Vardhini BV (2014) Brassinosteroids' role for amino acids, peptides and amines modulation in stressed plants—a review. Plant adaptation to environmental change: Significance of amino acids and their derivatives, pp 300–316

Varshney RK, Singh VK, Hickey JM, Xun X, Marshall DF, Wang J, Edwards D, Ribaut J-M (2016) Analytical and decision support tools for genomics-assisted breeding. Trends Plant Sci 21(4): 354–363

Vert G, Walcher CL, Chory J, Nemhauser JL (2008) Integration of auxin and brassinosteroid pathways by auxin response factor 2. Proc Natl Acad Sci 105(28):9829–9834

Visser E, Voesenek L (2005) Acclimation to soil flooding sensing and signal-transduction. In: Lambers H, Colmer T (eds) Root physiology: from gene to function. Springer, pp 197–214

Vo KTX, Rahman MM, Rahman MM, Trinh KTT, Kim ST, Jeon J-S (2021) Proteomics and metabolomics studies on the biotic stress responses of rice: an update. Rice 14:1–16

Voesenek L, Armstrong W, Bögemann G, Colmer T, McDonald M (1999) A lack of aerenchyma and high rates of radial oxygen loss from the root base contribute to the waterlogging intolerance of Brassica napus. Funct Plant Biol 26(1):87–93. https://doi.org/10.1071/PP98086

Voesenek L, Bailey-Serres J (2013) Flooding tolerance: O2 sensing and survival strategies. Curr Opin Plant Biol 16(5):647–653

Wang H-Q, Sun L-P, Wang L-X, Fang X-W, Li Z-Q, Zhang F-F, Hu X, Qi C, He J-M (2020a) Ethylene mediates salicylic-acid-induced stomatal closure by controlling reactive oxygen species and nitric oxide production in Arabidopsis. Plant Sci 294:110464

Wang X, Komatsu S (2018) Proteomic approaches to uncover the flooding and drought stress response mechanisms in soybean. J Proteome 172:201–215

Wang X, Komatsu S (2020) Proteomic techniques for the development of flood-tolerant soybean. Int J Mol Sci 21(20):7497

Wang X, Komatsu S (2022) The role of phytohormones in plant response to flooding. Int J Mol Sci 23(12):6383

Wang X, Sakata K, Komatsu S (2018) An integrated approach of proteomics and computational genetic modification effectiveness analysis to uncover the mechanisms of flood tolerance in soybeans. Int J Mol Sci 19(5):1301

Wang X, Sun L, Li W, Peng M, Chen F, Zhang W, Sun C, Chen S, Hua W, Zhang J (2020b) Dissecting the genetic mechanisms of waterlogging tolerance in Brassica napus through linkage mapping and a genome-wide association study. Ind Crop Prod 147:112269. https://doi.org/10.1016/j.indcrop.2020.112269

Wang X, Wei J, Wu J, Shi B, Wang P, Alabd A, Wang D, Gao Y, Ni J, Bai S (2024) Transcription factors BZR2/MYC2 modulate brassinosteroid and jasmonic acid crosstalk during pear dormancy. Plant Physiol 194(3):1794–1814

Wang Y, Cao J-J, Wang K-X, Xia X-J, Shi K, Zhou Y-H, Yu J-Q, Zhou J (2019) BZR1 mediates brassinosteroid-induced autophagy and nitrogen starvation in tomato. Plant Physiol 179(2): 671–685

Wang Z, Wei X, Wang Y, Sun M, Zhao P, Wang Q, Yang B, Li J, Jiang Y-Q (2023) WRKY29 transcription factor regulates ethylene biosynthesis and response in Arabidopsis. Plant Physiol Biochem 194:134–145

Wei W, Li D, Wang L, Ding X, Zhang Y, Gao Y, Zhang X (2013) Morpho-anatomical and physiological responses to waterlogging of sesame (Sesamum indicum L.). Plant Sci 208: 102–111

Westra S, Fowler HJ, Evans JP, Alexander LV, Berg P, Johnson F, Kendon EJ, Lenderink G, Roberts N (2014) Future changes to the intensity and frequency of short-duration extreme rainfall. Rev Geophys 52(3):522–555

Wollmer AC, Pitann B, Mühling KH (2018) Waterlogging events during stem elongation or flowering affect yield of oilseed rape (Brassica napus L.) but not seed quality. J Agron Crop Sci 204(2):165–174

Xi Z, Wang Z, Fang Y, Hu Z, Hu Y, Deng M, Zhang Z (2013) Effects of 24-epibrassinolide on antioxidation defense and osmoregulation systems of young grapevines (V. Vinifera L.) under chilling stress. Plant Growth Regul 71:57–65

Xin L, Wang J, Yang Q (2023) Exogenous salicylic acid alleviates water deficit stress by protecting photosynthetic system in maize seedlings. Agronomy 13(9):2443

Xu B, Cheng Y, Zou X, Zhang X (2016) Ethanol content in plants of Brassica napus L. correlated with waterlogging tolerance index and regulated by lactate dehydrogenase and citrate synthase. Acta Physiol Plant 38:1–9

Xu G, Li L, Zhou J, He M, Lyu D, Zhao D, Qin S (2023) Integrated transcriptomics and metabolomics analyses reveal key genes and essential metabolic pathways for the acquisition of cold tolerance during dormancy in apple. Environ Exp Bot 213:105413

Xu M, Ma H, Zeng L, Cheng Y, Lu G, Xu J, Zhang X, Zou X (2015) The effect of waterlogging on yield and seed quality at the early flowering stage in Brassica napus L. Field Crop Res 180:238–245

Xu X, Ji J, Xu Q, Qi X, Weng Y, Chen X (2018) The major-effect quantitative trait locus Cs ARN 6.1 encodes an AAA ATP ase domain-containing protein that is associated with waterlogging stress tolerance by promoting adventitious root formation. Plant J 93(5):917–930

Yamauchi T, Watanabe K, Fukazawa A, Mori H, Abe F, Kawaguchi K, Oyanagi A, Nakazono M (2014) Ethylene and reactive oxygen species are involved in root aerenchyma formation and adaptation of wheat seedlings to oxygen-deficient conditions. J Exp Bot 65(1):261–273

Yan K, Zhao S, Cui M, Han G, Wen P (2018) Vulnerability of photosynthesis and photosystem I in Jerusalem artichoke (Helianthus tuberosus L.) exposed to waterlogging. Plant Physiol Biochem 125:239–246

Yang L, Li N, Liu Y, Miao P, Liu J, Wang Z (2023) Updates and prospects: morphological, physiological, and molecular regulation in crop response to waterlogging stress. Agronomy 13(10):2599

Yang W, Zhu C, Ma X, Li G, Gan L, Ng D, Xia K (2013) Hydrogen peroxide is a second messenger in the salicylic acid-triggered adventitious rooting process in mung bean seedlings. PLoS One 8(12):e84580

Yao Q (2021) Crucial waterlogging-responsive genes and pathways revealed by comparative physiology and transcriptome in tropical and temperate maize (Zea mays L.) inbred lines. J Plant Biol 64(4):313–325

Yasumoto S, Terakado Y, Matsuzaki M, Okada K (2011) Effects of high water table and short-term flooding on growth, yield, and seed quality of sunflower. Plant Produc Sci 14:233–248

Ye H, Liu S, Tang B, Chen J, Xie Z, Nolan TM, Jiang H, Guo H, Lin H-Y, Li L (2017) RD26 mediates crosstalk between drought and brassinosteroid signalling pathways. Nat Commun 8(1):14573

Yeung E, van Veen H, Vashisht D, Sobral Paiva AL, Hummel M, Rankenberg T, Steffens B, Steffen-Heins A, Sauter M, de Vries M (2018a) A stress recovery signaling network for enhanced flooding tolerance in Arabidopsis thaliana. Proc Natl Acad Sci 115(26):E6085–E6094

Yeung E, van Veen H, Vashisht D, Sobral Paiva AL, Hummel M, Rankenberg T, Steffens B, Steffen-Heins A, Sauter M, de Vries M, Schuurink RC, Bazin J, Bailey-Serres J, Voesenek L, Sasidharan R (2018b, Jun 26) A stress recovery signaling network for enhanced flooding tolerance in Arabidopsis thaliana. Proc Natl Acad Sci USA 115(26):E6085–e6094. https://doi.org/10.1073/pnas.1803841115

Yin W, Dong N, Niu M, Zhang X, Li L, Liu J, Liu B, Tong H (2019) Brassinosteroid-regulated plant growth and development and gene expression in soybean. Crop J 7(3):411–418

Yin X, Komatsu S (2015) Quantitative proteomics of nuclear phosphoproteins in the root tip of soybean during the initial stages of flooding stress. J Proteome 119:183–195

Yin X, Nishimura M, Hajika M, Komatsu S (2016) Quantitative proteomics reveals the flooding-tolerance mechanism in mutant and abscisic acid-treated soybean. J Proteome Res 15(6): 2008–2025

Yordanova RY, Christov KN, Popova LP (2004) Antioxidative enzymes in barley plants subjected to soil flooding. Environ Exp Bot 51(2):93–101. https://doi.org/10.1016/S0098-8472(03)00063-7

Youssef MS, Mira MM, Millar JL, Becker MG, Belmonte MF, Hill RD, Stasolla C (2019) Spatial identification of transcripts and biological processes in laser micro-dissected sub-regions of waterlogged corn roots with altered expression of phytoglobin. Plant Physiol Biochem 139: 350–365

Yu F, Liang K, Fang T, Zhao H, Han X, Cai M, Qiu F (2019) A group VII ethylene response factor gene, ZmEREB180, coordinates waterlogging tolerance in maize seedlings. Plant Biotechnol J 17(12):2286–2298

Yusuf M, Fariduddin Q, Ahmad A (2011) 28-Homobrassinolide mitigates boron induced toxicity through enhanced antioxidant system in Vigna radiata plants. Chemosphere 85(10):1574–1584

Zahra N, Wahid A, Shaukat K, Rasheed T (2020) Role of seed priming and foliar spray of calcium in improving flag leaf growth, grain filling and yield characteristics in wheat (Triticum aestivum)—a field appraisal. Int J Agric Biol 24:1591–1600

Zandberg JD, Fernandez CT, Danilevicz MF, Thomas WJ, Edwards D, Batley J (2022) The global assessment of oilseed Brassica crop species yield, yield stability and the underlying genetics. Plan Theory 11(20):2740

Zeng R, Chen T, Wang X, Cao J, Li X, Xu X, Chen L, Xia Q, Dong Y, Huang L (2021) Physiological and expressional regulation on photosynthesis, starch and sucrose metabolism response to waterlogging stress in peanut. Front Plant Sci 12:601771

Zhai L, Liu Z, Zou X, Jiang Y, Qiu F, Zheng Y, Zhang Z (2013) Genome-wide identification and analysis of microRNA responding to long-term waterlogging in crown roots of maize seedlings. Physiol Plant 147(2):181–193

Zhang, X., Shabala, S., Koutoulis, A., Shabala, L., Johnson, P., Hayes, D., Nichols, D. S., & Zhou, M. (2015, 2015/09/01). Waterlogging tolerance in barley is associated with faster aerenchyma formation in adventitious roots. Plant Soil, 394(1), 355–372. https://doi.org/10.1007/s11104-015-2536-z

Zhang Y, Zhu X, Ding H, Yang S, Chen Y (2013) Foliar application of 24-epibrassinolide alleviates high-temperature-induced inhibition of photosynthesis in seedlings of two melon cultivars. Photosynthetica 51:341–349

Zheng M, Hu M, Yang H, Tang M, Zhang L, Liu H, Li X, Liu J, Sun X, Fan S (2019) Three BnaIAA7 homologs are involved in auxin/brassinosteroid-mediated plant morphogenesis in rapeseed (Brassica napus L.). Plant Cell Rep 38:883–897

Zhou J, Wang J, Li X, Xia X-J, Zhou Y-H, Shi K, Chen Z, Yu J-Q (2014) H2O2 mediates the crosstalk of brassinosteroid and abscisic acid in tomato responses to heat and oxidative stresses. J Exp Bot 65(15):4371–4383

Zhou W, Lin X (1995) Effects of waterlogging at different growth stages on physiological characteristics and seed yield of winter rape (Brassica napus L.). Field Crop Res 44(2–3): 103–110

Zhou Y, Xiong Q, Yin CC, Ma B, Chen SY, Zhang JS (2020) Ethylene biosynthesis, signaling, and crosstalk with other hormones in rice. Small Methods 4(8):1900278

Zhu B, Yu J, Shi H, Yue K, Lu J, Zhang T (2023) Effects of waterlogging stress on rapeseed yield, oil content, fatty acid composition, and transcriptome differences. Plant Growth Regul 101(3): 769–779. https://doi.org/10.1007/s10725-023-01055-4

Zou LJ, Deng XG, Zhang L, Zhu T, Tan WR, Muhammad A, Zhu LJ, Zhang C, Zhang DW, Lin HH (2018) Nitric oxide as a signaling molecule in brassinosteroid-mediated virus resistance to cucumber mosaic virus in Arabidopsis thaliana. Physiol Plant 163(2):196–210

Zou X, Hu C, Zeng L, Cheng Y, Xu M, Zhang X (2014) A comparison of screening methods to identify waterlogging tolerance in the field in Brassica napus L. during plant ontogeny. PLoS One 9(3):e89731

Chapter 12
Phytohormone-Mediated Protection in Oilseed Crops: A Key Strategy Against Temperature Stress

Jalal Kassout, Kaoutar El Issaoui, Rachid Azenzem, Basma Chokrane, Soufiane Chakkour, Khalil Kadaoui, Abdelouahab Sahli, Sara Oulbi, Anas Chraka, and Mouaad Amine Mazri

Abstract Oilseed crops are vital to global agriculture, supplying edible oils, industrial materials, and biofuels. However, temperature stress increasingly threatens their productivity, which disrupts growth, physiological processes, and biochemical homeostasis. Phytohormones play a crucial role in mitigating these adverse effects by modulating morpho-physiological and biochemical responses, thereby enhancing stress tolerance and overall crop performance. This chapter explores the exogenous application of phytohormones as a strategy to improve temperature resilience in

J. Kassout (✉) · B. Chokrane · S. Oulbi
Regional Center of Agricultural Research of Marrakech, National Institute of Agricultural Research, Rabat, Morocco
e-mail: jalal.kassout@inra.ma; basma.choukrane@inra.ma; sara.oulbi@inra.ma

K. El Issaoui
Innovation City, Cadi Ayyad University, Marrakech, Morocco
e-mail: kaoutar.elissaoui@uca.ac.ma

R. Azenzem
Regional Center of Agricultural Research of Tanger, National Institute of Agricultural Research, Rabat, Morocco
e-mail: rachid.azenzem@inra.ma

S. Chakkour
Faculty of Sciences, Ibn Tofail University, Kenitra, Morocco
e-mail: soufian.chakkour@uit.ac.ma

K. Kadaoui · A. Sahli
Bio-Agrodiversity Team, Biology, Ecology, and Health Laboratory, Sciences Faculty, Abdelmalek Essaâdi University, BP, Tétouan, Morocco

A. Chraka
Laboratory of Engineering Materials and Sustainable Energy, Faculty of Science, Abdelmalek Essaadi University, Tetouan, Morocco

M. A. Mazri
Agro-Biotechnology Research Unit, Regional Center of Agricultural Research of Marrakech, National Institute of Agricultural Research, Rabat, Morocco
e-mail: mouaadamine.mazri@inra.ma

© The Author(s), under exclusive license to Springer Nature Singapore Pte Ltd. 2025
A. A. H. Abdel Latef (ed.), *Oilseed Crops Under Abiotic Stress*, Sustainability Sciences in Asia and Africa, https://doi.org/10.1007/978-981-96-8346-8_12

oilseed crops, highlighting their roles in promoting growth, photosynthetic efficiency, osmolyte accumulation, and antioxidant defense mechanisms. A bibliometric analysis reveals a growing research focus on hormonal crosstalk under stress conditions, underscoring the importance of integrating genomic, transcriptomic, proteomic, and metabolomic approaches to decipher the molecular networks governing phytohormone-mediated stress adaptation. Leveraging these insights in breeding and agronomic practices offers promising avenues for enhancing oilseed crop resilience and maintaining yield stability in the face of climate change.

Keywords Oilseed crops · Phytohormones · Heat stress tolerance · Temperature stress · Abiotic stress adaptation

12.1 Introduction

Oilseed crops, encompassing a diverse array of plant families, are globally recognized for their dual utility as a vital source of edible oils and as raw materials for various oleochemical industries. Among the myriad of oilseed species (Table 12.1), soybean, sesame, safflower, sunflower, groundnut, and castor stand out as the most economically and agriculturally significant (Sharma et al. 2012). Ranked by production, soybeans (398.21 million metric tons), rapeseed (87.44 mmt), sunflower seed (55.08 mmt), peanuts (50.46 mmt), cottonseed (41.46 mmt), palm kernel (20.71 mmt), and copra (6.05 mmt) were the major oilseed crop globally (USDA F 2018). These crops collectively occupy approximately 213 million hectares of the world's arable land, ranking fourth in importance among food commodities, trailing only cereals, vegetables and melons, and fruits and nuts (Salunkhe 1992; Rai et al. 2016). Notably, oilseed production experienced the highest growth rate among all crop types between 2000 and 2022, increasing by 121% (FAO 2023). This underscores their critical role in global agriculture and food security. The significance of oilseed crops extends beyond their nutritional value. They are a cornerstone of both human and animal diets, providing essential fats, proteins, and micronutrients (Damude and Kinney 2008; Wittkop et al. 2009; Sarwar et al. 2013). Moreover, their industrial applications are vast, ranging from the production of biodiesel to the manufacture of pharmaceuticals and other oleochemical products (Lu et al. 2011; Kotecka-Majchrzak et al. 2020). The food industry, in particular, heavily relies on oilseeds due to their enriched nutritional profiles, which have been further enhanced over time through advanced genetic engineering techniques (Maheshwari and Kovalchuk 2014). For instance, rapeseed, soybean, sunflower, and peanut have emerged as innovative plant-based protein sources, increasingly used as alternatives to animal-derived proteins (Usman et al. 2023; Zhang et al. 2023a). The primary product of oilseed plants—oil—serves as a fundamental ingredient in both food and feedstock industries (Harwood et al. 2013). In addition to their economic value,

Table 12.1 Impact of temperature stress on various oilseed crops at various growth stages

Oilseed crops	Temperature stress	Duration/period of study	Growth stage	Chemical composition changes	Yield impact	References
Rapeseed/Canola (*Brassica napus* L.)	36 °C	12 h	Reproductive stage	Reduced oil content, varying fatty acid composition, higher omega-6/omega-3 ratio	More than 50% decline in overall seed yield and seed weight.	Lohani et al. (2022)
Peanut (*Arachis hypogaea* L.)	33.8–43.5 °C	Full season	From sowing to maturity		Vegetative and pod yield decrease.	Kakani et al. (2014)
Cotton (*Gossypium* spp., 39 cotton genotypes)	40 °C	Full season	Seedlings	Increased electrolyte exchange	–	Guo et al. (2020)
Rapeseed/Canola (*Brassica napus* L.)	37 °C	2 days	Reproductive stage	Reduced seed oil accumulation. Increased seed sugar concentration. Downregulation of BnWRI1 and its downstream genes (involved in fatty acid production), as well as inhibition of sugar transporter genes	–	Huang et al. (2019)
Brassica species (various genotypes/ Canola)	35 °C	7 days	Reproductive stage	Changes in Vcmax (photosynthesis's maximal carboxylation rate) and TPU (triose phosphate utilization rate) were discovered.	–	Chen et al. (2019)
Soybean (cv. Pungsannamul)	45 °C	5 and 10 days	Reproductive phase	Increased abscisic acid (ABA), decreased salicylic acid (SA), and increased	–	Khan et al. (2020a, b)

(continued)

Table 12.1 (continued)

Oilseed crops	Temperature stress	Duration/period of study	Growth stage	Chemical composition changes	Yield impact	References
				Heat Shock Protein (HSP) expression initially (5 days) during heat stress.		
Soybean (*Glycine max* L.)	30/30 ° C day/night	5 days	At the initial seed filling stage	Increased fat content decreased protein content. Increased expression of lipid biosynthesis genes (GmBCCP2, GmKAS1, GmWRI1, GmDREBL), decreased expression of lipid degradation genes (GmACXs), downregulation of seed storage proteins (GmGy1, GmGy2, GmGy4, GmGy5, Gmβ-conglycinin), and upregulation of cysteine and aspartate proteinases.	Yield reduction associated with lower seed protein content	Nakagawa et al. (2020)
Sunflower (cv. HA89)	35, 37, and 40 °C	7 consecutive days during three different intervals of grain filling: 12–19, 19–26, and after 26 days after anthesis (daa).	Grain filling	Reduced oil grain percentage (up to −30%), decreased embryo oil percentage. Altered fatty acid content, especially when heat stress occurred between 19 and 26 daa. Activity of oleate desaturase (ODS) was specifically impacted.	Temperatures over 35 °C during early grain development (12–19 daa) result in lower final grain weight (up to −40%) due to shorter grain growth time.	Chimenti et al. (2001)

Sunflower (*Helianthus annuus* L.)	38 °C	21 days	Reproductive stage	Increased hydrogen peroxide, malondialdehyde, and electrolyte leakage (indicates oxidative stress).	–	Abdel Razik et al. (2021)
Camelina (*Camelina sativa* L.)	35 °C	14 days	Reproductive stage	Increased lipid peroxidation (malondialdehyde), hydrogen peroxide, and electrolyte leakage. Reduced chlorophyll content. Thiourea increased antioxidant activity (catalase, protease, ascorbate peroxidase). Thiourea increased osmolytes (total soluble sugars, total soluble proteins, proline, glycine betaine).	Reduced dry matter per plant (32.7%), and seed yield (60.9%).	Ahmad et al. (2021)
Camelina sativa	30–35 °C	3 h in a day	Reproductive stage	Rubisco activase activity was inhibited at temperatures above 35 °C and levels of cpn60 protein were elevated under higher growth temperature.	Growth under moderate heat stress caused a 63% reduction in seed yield.	Carmo-Silva and Salvucci (2012)
Camelina sativa L. (varieties 7126 and 8046)	32 °C	12 days	Reproductive stage	Heat stress reduced relative water content and photosynthetic rate	Heat stress negatively impacted crop growth and presumably yield	Waraich et al. (2021)
Camelina (*Camelina sativa* L.)	37 °C	14 days	Germination	Reduction in omega-3 α-linolenic acid	Low pods and seed size	Smith and Lu (2024)

oilseed crops play a vital role in agricultural systems by enhancing soil quality through nitrogen fixation—most oilseed crops form root nodules that enable nitrogen fixation through symbiotic interactions with Rhizobium bacteria—and acting as rotational crops that help control pest and disease cycles (Hegewald 2022; Zhang et al. 2023b). Beyond their culinary uses, oilseed crops are pivotal in the production of biodiesel, a renewable energy source that has gained prominence in the fight against climate change (Cruz et al. 2020), which positions oilseed crops as a sustainable alternative to fossil fuels, aligning with global efforts to mitigate atmospheric pollution and its detrimental effects. In addition to their nutritional and industrial roles, oilseed crops have carved a niche in the pharmaceutical sector. For example, olive oil, derived from olive fruit, is renowned for its medicinal properties, including anti-inflammatory and cardiovascular benefits (Kassout et al. 2023). Consequently, oilseed crops are a multifaceted resource with profound implications for nutrition, industry, and environmental sustainability.

In the context of climate change, agriculture is increasingly confronted with a multitude of environmental stresses that severely impact plant growth and development. These stresses, both biotic and abiotic, are largely driven by shifting climatic conditions and pose unprecedented challenges to global food security (Muluneh 2021). Among the most significant abiotic stresses are heat, drought, cold, waterlogging, and salinity, each of which disrupts physiological processes and reduces crop productivity (Ashraf et al. 2018; Kassout et al. 2023). Of these, elevated temperatures stand out as a critical concern, with climate models predicting a steady rise in atmospheric carbon dioxide (CO_2) levels, leading to global warming and its cascading effects on agricultural systems (Huang et al. 2017; Xu et al. 2019). Heat stress, in particular, has emerged as a formidable threat to crop production worldwide, often preventing the proper completion of developmental stages and resulting in immature or suboptimal yields (Fahad et al. 2017). This disruption in growth stages adversely affects seed yield and overall plant ecophysiological processes (Hasanuzzaman et al. 2013). For oilseed crops, which are highly sensitive to temperature fluctuations, heat stress poses a significant risk to both yield quantity and quality (Ahmad et al. 2021). Such losses highlight the vulnerability of oilseed crops to heat stress and their potential to undermine global oilseed production.

Plant responses to stressful conditions, such as temperature stress, involve complex ecophysiological processes and intricate interactions between hormones and other signaling molecules. A specific group of signaling molecules, known as phytohormones—including abscisic acid, gibberellic acid, ethylene, brassinosteroids, cytokinins, salicylic acid, jasmonates, and auxin—are present in small quantities within cells and play a crucial role in mediating responses to both external and internal stimuli. In this chapter, we provide an overview of the responses of oilseed crops to temperature stress, with a particular focus on the role of phytohormones in temperature stress mitigation.

12.2 Oilseed Crops and Phytohormones Under Temperature Stress—A Bibliometric Analysis

A bibliometric analysis was conducted on research related to oilseed crops and phytohormones under temperature stress between 2004 and 2024. The bibliometric analysis utilized the Scopus and Web of Science (WoS) databases, employing Boolean operators to refine search results and ensure comprehensive coverage (Pranckutė 2021). Publications were selected based on their relevance to the study objectives, applying inclusion criteria such as publication date (2004–2024) and document type (research articles). The analysis revealed a significant rise in the number of publications on this topic over the past decade, with an exponential increase observed from 2018 onward, peaking at 12 publications in 2021. A co-occurrence analysis of keywords used by authors facilitated the construction of a network visualization map comprising 75 keywords classified into four distinct clusters (Fig. 12.1). The most frequently occurring keywords included Phytohormone (42), Genetics (28), Metabolism (26), Plant (24), Gene expression regulation, and Physiological stress (22). Figure 12.1 illustrates the interactions between physiological, molecular, and genetic responses of oilseed crops to abiotic stress. Cluster 1 highlights the central role of phytohormones, particularly auxins, in regulating growth and responses to environmental stresses, including elevated temperatures.

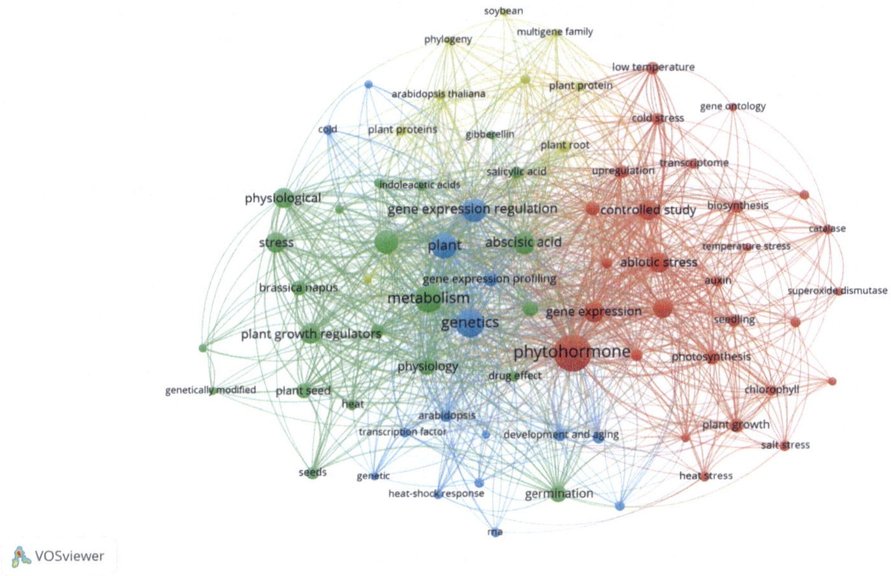

Fig. 12.1 Co-occurrence of keywords used for oilseed crops and phytohormones under temperature stress studies between 2004 and 2024

Cluster 2 focuses on plant physiology and growth regulators, emphasizing their role in metabolism and stress tolerance. Cluster 3 explores the regulation of gene expression and molecular mechanisms underlying stress responses, including transcription factors and heat shock responses. Finally, Cluster 4 links plant proteins and phylogeny to plant signaling pathways and adaptation mechanisms (Fig. 12.1).

12.3 Temperature Stress and Tolerance Mechanism in Oilseed Crops

Oilseed crops, like other plant species, regulate their growth and physiological performance in response to both external and internal stimuli, including elevated temperatures. One of the key regulatory mechanisms involves phytohormones, which play a crucial role in modulating plant responses to environmental stressors (Fig. 12.2) (Qu et al. 2013). Recent research has highlighted the significant role of phytohormones in mediating plant adaptation to heat stress, influencing a range of physiological and molecular processes that enhance stress tolerance (Grover et al. 2013; Hasanuzzaman et al. 2013). However, the regulatory network underlying phytohormone-mediated heat stress tolerance is highly complex. These signaling molecules can either act directly to mitigate heat-induced damage or indirectly by setting up broader defense mechanisms. This regulation often involves intricate coordination between multiple phytohormones and other signaling components, including reactive oxygen species (ROS), mitogen-activated protein (MAP) kinases, soluble sugars, and secondary messengers. Through these interconnected pathways,

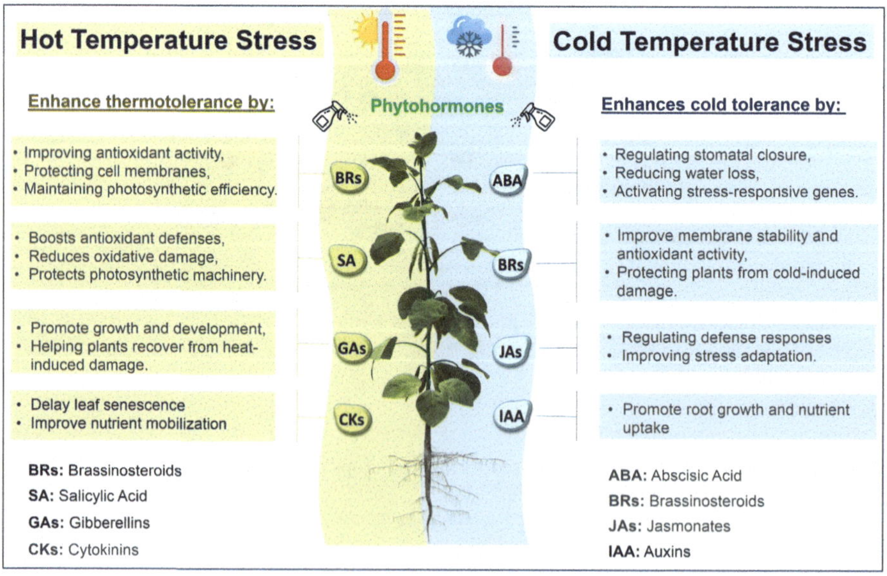

Fig. 12.2 Phytohormones are mostly integrated into oilseed crops to alleviate heat and cold stress

phytohormones integrate various stress signals, fine-tuning cellular responses to ensure optimal thermotolerance in oilseed crops (Smékalová et al. 2014; Ljung et al. 2015). Similarly, several key phytohormones, including auxins (IAA), abscisic acid (ABA), gibberellins (GA), ethylene, cytokinins (CKs), and salicylic acid (SA), coordinate a wide range of biochemical and physiological processes essential for plant adaptation (Li et al. 2021a). Among these, abscisic acid (ABA) is particularly well known for its role in mediating responses to biotic and abiotic stresses, significantly influencing plant growth and stress resilience (O'Brien and Benková 2013; Brunel-Muguet et al. 2015; Elferjani and Soolanayakanahally 2018). For instance, ABA has been found to enhance seed storage protein and amino acid accumulation in oilseed crops such as *Brassica napus* and *Medicago truncatula*, suggesting a pivotal role in seed quality and nutrient reserves under stress conditions (Xiong and Zhu 2003; Kishor et al. 2022). Tissue ABA levels have also been found to affect seed constituents during grain filling (Yang et al. 2014). Moreover, fluctuations in ABA concentration in seeds under high temperatures affected the expression of genes involved in the production of seed storage compounds in castor beans (*Ricinus communis* L.) (Chandrasekaran et al. 2014). In addition, ABA was found to improve linoleic acid accumulation with increased fatty acid biosynthesis in oil palm (Shi et al. 2021). Thus, it was found to enhance the linoleic acid and elaidic acid content in grape seed (*Vitis vinifera* L.) (Ju et al. 2016). In the olive tree, ABA was found to increase in the stomata guard cell under drought, which induces the expression of hexokinases, and consequently leads to the stomatal closure (Brito et al. 2018). Additionally, salicylic acid and abscisic acid were found to alleviate olive summer stress by controlling reactive oxygen species (ROS) and reducing the photooxidative effects (Brito et al. 2019).

Similarly, melatonin has been demonstrated to improve thermotolerance in soybean seedlings by promoting plant growth and development. This occurs through the accumulation of protective biochemicals, including phenolics, flavonoids, and proline, as well as an increase in endogenous melatonin and polyamines such as spermine, spermidine, and putrescine (Imran et al. 2021). Additionally, heat stress mitigation in soybean has been linked to the expression of heat shock protein 90 (*gmHsp90*) and heat shock transcription factor (*gmHsfA2*), which are involved in reactive oxygen species (ROS) detoxification via hydrogen peroxide-mediated signaling pathways (Imran et al. 2021). Auxins, particularly IAA, have also been shown to influence oilseed crop productivity. Studies indicate that IAA enhances oil yield in aromatic grasses such as *Cymbopogon martinii* and *Cymbopogon winterianus*, while also promoting seed production in linseed (*Linum usitatissimum*). GA, another crucial class of phytohormones, regulates key developmental processes, including floral induction, seed dormancy release, and vegetative growth. GA was found to regulate photosynthesis, stomatal conductance and transpiration under drought stress in rapeseed (Khan et al. 2020b). However, excessive GA concentrations could have negative effects on seed size and numbers (Silverstone et al. 2001). Similarly, in safflower (*Carthamus tinctorius*), GA treatment enhances oil biosynthesis, while in linseed, GA application stimulates stem elongation, cell wall thickening, and overall plant biomass accumulation (Arslan and Culpan 2017; Nejati et al. 2020).

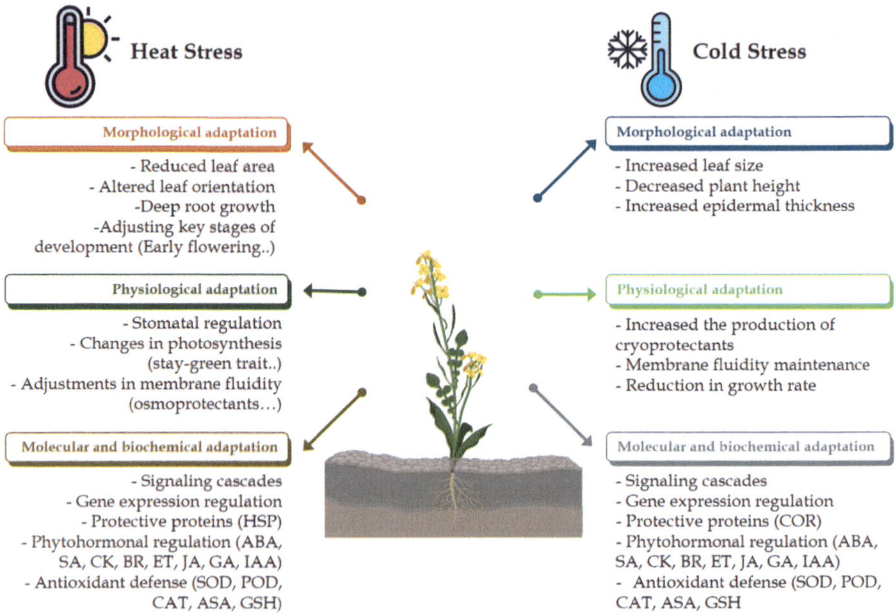

Fig. 12.3 Mean responses pathways of oilseed crops to temperature stress and adaptation strategies

As previously mentioned, plant responses to temperature stress are governed not by a single hormone but by a dynamic interplay among multiple phytohormones (Fig. 12.2). For instance, ABA and SA often act synergistically to enhance thermotolerance by upregulating stress-responsive genes and activating antioxidant defense mechanisms (Larkindale and Knight 2002; Dinler et al. 2014; Nazar et al. 2017). Additionally, ethylene and jasmonic acid (JA) coordinate growth and defense responses under both heat and cold stress conditions, further emphasizing the complexity of phytohormonal regulation in stress adaptation (Poór et al. 2022; Sharma et al. 2023). Moreover, auxins and cytokinins contribute to balancing root and shoot growth, ensuring optimal resource allocation under stressful conditions (Schwarz et al. 2020; Ahmad et al. 2021). Finally, it should be mentioned that along with internal mechanisms (e.g., phytohormone), oilseed crops adjust their morphological and phenological traits to improve tolerance to temperature stress (Fig. 12.3).

12.4 Biochemical Signaling and Temperature Stress in Oilseed Crops

Plants employ a range of biochemical mechanisms to mitigate the detrimental effects of heat stress, which can disrupt cellular homeostasis and impair growth (Fig. 12.2). Table 12.1 provides the mean occurring biochemical changes to enhance heat stress mitigation in major oilseed crops. Heat stress severely affects canola (*Brassica napus*), particularly during reproduction. As shown in Table 12.1, exposure to 36 °C for 12 h reduced oil content and altered fatty acid composition, increasing the omega-6/omega-3 ratio, indicating a stress-induced metabolic imbalance (Lohani et al. 2022). After 2 days at 37 °C, seed oil accumulation declined, while seed sugar content increased due to the downregulation of *BnWRI1* and its downstream genes involved in fatty acid biosynthesis. Heat stress also reduced sugar transporter gene activity, impairing energy storage and oil production (Huang et al. 2019). Short-term exposure during reproduction led to a 50% decline in seed yield and weight, demonstrating the vulnerability of reproductive development. Prolonged exposure (7 days at 35 °C) reduced photosynthetic carboxylation and triose phosphate utilization, further exacerbating stress effects (Chen et al. 2019). Soybean (*Glycine max*) undergoes substantial physiological and molecular changes under heat stress, particularly during the reproductive and seed-filling phases. Exposure to 45 °C for 5–10 days increased ABA levels while reducing SA, reflecting an adaptive response where ABA enhances osmotic regulation while SA suppression weakens defense mechanisms (Khan et al. 2020a).

Heat stress also induced heat shock proteins (HSPs), stabilizing cellular structures. During seed filling (30 °C/30 °C, day/night for 5 days), lipid content increased while protein content declined due to the upregulation of lipid biosynthesis genes (*GmBCCP2, GmKAS1, GmWRI1, GmDREBL*) and the downregulation of lipid degradation and seed storage protein genes, favoring lipid accumulation over protein synthesis (Nakagawa et al. 2020). Sunflower (*Helianthus annuus*) is highly sensitive to heat stress during reproduction and grain filling. Exposure to 35–40 °C for 7 days reduced oil content by up to 30%, particularly between 19 and 26 days after anthesis (daa), when decreased oleate desaturase (ODS) activity disrupted oleic acid conversion to linoleic acid (Chimenti et al. 2001). Exposure to 38 °C for 3 weeks increased oxidative stress markers such as hydrogen peroxide (H_2O_2), malondialdehyde (MDA), and electrolyte leakage, indicating cellular damage. These disruptions in lipid metabolism and oxidative balance resulted in significant yield and quality losses, with the 19–26 daa period being particularly critical for oil deposition (Abdel Razik et al. 2021). Similarly, *Camelina sativa* exhibits high sensitivity to heat stress, especially during reproduction (Ahmad et al. 2021; Smith and Lu 2024). Exposure to 35 °C for 14 days elevated lipid peroxidation, H_2O_2 levels, and electrolyte leakage, indicating oxidative damage (Carmo-Silva and Salvucci 2012). Thiourea treatment enhanced antioxidant enzyme activity (e.g., catalase, ascorbate peroxidase) and osmolyte accumulation (e.g., soluble sugars, proteins, proline, glycine betaine), partially mitigating stress effects. Short-term exposure (30–35 °C for 3 h per day) reduced Rubisco activase, impairing photosynthesis, while increasing *cpn60* protein levels, suggesting a protective response. Prolonged exposure (14 days at 35 °C) led to a 60.9% decline in seed yield and a 32.7% reduction in dry matter, emphasizing the vulnerability of carbon fixation and water regulation to heat stress (Waraich et al. 2021). In Peanut (*Arachis hypogaea* L.), heat stress during the full growing season was found to negatively impact the vegetative growth, thus a significant decrease in pod yield (Kakani et al. 2014). Exposure to high temperature (40 °C) in seedlings of 39 genotypes of cotton (Gossypium spp.) has greatly increased electrolyte leakage after heat stress, which impacted ecophysiological performances at the plant level (Guo et al. 2020). In peanut (*Arachis hypogaea* L.), heat stress during the full growing season was found to negatively impact the vegetative growth, thus a significant decrease in pod yield (Kakani et al. 2014). Exposure to high temperature (40 °C) in seedlings of 39 genotypes of cotton (Gossypium spp.) has greatly increased electrolyte leakage after heat stress, which impacted ecophysiological performances at the plant level (Guo et al. 2020).

12.5 In Vitro Culture Systems to Improve Temperature Stress Resilience in Oilseed Crops

Plant tissue culture, also known as plant micropropagation or in vitro culture, is a powerful technology widely used in modern agriculture, both commercially and in research (Tahiri et al. 2023). It serves various purposes, including rapid and large-scale propagation of healthy, uniform and disease-free plants, germplasm preservation, genetic improvement, and metabolite production (Shukla and Shukla 2024; Pérez et al. 2025). Micropropagation can be achieved through several methods, such as somatic embryogenesis, organogenesis, microcutting, and micrografting (Mazri et al. 2011; Mazri 2014; Koufan et al. 2020). Regardless of the method used, success is influenced by various factors, including species, genotypes, explant, culture medium composition, and environmental conditions. Among these, plant growth regulators (PGRs) are crucial. These compounds play a key role in the success of any micropropagation system, as they interact with endogenous phytohormones to promote cell division, differentiation, and growth. This ultimately leads to the formation of new organs and, eventually, fully developed plants (Gaspar et al. 1996, 2003).

In recent years, plant tissue culture has become a widely used strategy for enhancing plant resilience to abiotic and biotic stresses (Akhtar et al. 2024). Stress inducers can be incorporated during in vitro regeneration to create a controlled stressful environment, allowing the explants and resulting plants to develop improved resistance to specific stressors. For instance, sodium chloride (NaCl) has been added to the culture medium to enhance salt stress tolerance (Asthana et al. 2023; Radi et al. 2023), while mannitol, sorbitol, and polyethylene glycol (PEG) are commonly used to induce water stress and improve drought resilience (Sahu et al. 2023). Plant tissue culture also offers a valuable platform to assess plant responses to these stresses at multiple levels, including morphological, physiological, biochemical, and molecular. Morphological responses are most frequently evaluated under in vitro stress conditions. Additionally, changes in osmoprotectants (e.g., proline, glycine betaine), macromolecules (e.g., carbohydrates, proteins and chlorophylls), antioxidative enzymes (e.g., catalase, superoxide dismutase and peroxidase), and gene expression in explants have all been studied as indicators of stress responses under in vitro conditions (Hamooh et al. 2021; Radi et al. 2023; Vuksanović et al. 2023).

Temperature stress has also been studied under in vitro conditions, where plant cultures are exposed to elevated or low temperatures that simulate environmental extremes. Temperature stress affects plant growth and development by influencing morphological traits such as leaf and root structure, as well as biochemical markers such as proline content and superoxide dismutase activity (Zhang et al. 2022). Such stress conditions can also disrupt key cellular processes such as enzyme activity, membrane stability, and photosynthesis (Prasertthai et al. 2022; Ifeduba et al. 2024). Under in vitro conditions, both heat stress (high temperatures) and cold stress (low temperatures) have been used to assess plant tolerance mechanisms. Temperature stress can lead to an increase in reactive oxygen species (ROS), prompting antioxidant responses that mitigate oxidative damage (Mansinhos et al. 2022). Heat stress often triggers the accumulation of heat-shock proteins (HSPs), while cold stress promotes the production of cold-regulated proteins (CORs) and cryoprotectants, such as soluble sugars and polyols, which help protect cells from damage (Huang et al. 2016; Rudić et al. 2022). By evaluating in vitro responses of explants/plants to temperature stress, researchers can better understand the molecular mechanisms underlying temperature tolerance under a controlled environment, which is essential for developing crops resilient to temperature extremes in changing climates.

Although tissue culture systems have been widely used in various plant species to assess their tolerance to heat or cold stress, understand the mechanisms underlying their adaptation to these abiotic stresses under controlled conditions, and select or develop new genotypes with improved resilience to temperature stress, only a few studies have been conducted on oilseed crops. Burbulis et al. (2010) evaluated the role of the phytohormone ABA in cold acclimation of winter rapeseed (*Brassica napus* (L.) cv. Sunday. This phytohormone is known to play a key role in plant adaptation, growth and development under abiotic and biotic stress conditions (Klingler et al. 2010). It was found that exogenous ABA resulted in increased levels of proline and soluble sugars in winter rapeseed shoots, which were linked to enhanced cold tolerance. Besides, the combined effect of ABA and low acclimation temperature was more effective than ABA alone. This suggests that ABA might play a key regulatory role in the cold acclimation process of winter rapeseed, possibly by modulating the accumulation of protective solutes such as proline and sugars. Burbulis et al. (2011) further explored the complexities of winter hardiness by studying cold de-acclimation and re-acclimation in winter rapeseed (cv. Sunday) shoots. The findings of this research showed that when cold-acclimated shoots were exposed to normal growth temperatures (18/16 °C), their cold tolerance decreased, along with a reduction in proline and soluble sugar levels. However, reacclimating the shoots to 4 °C for 5 days allowed them to recover cold tolerance to levels comparable to or higher than before the de-acclimation treatment. This suggests that rapeseed plants possess a degree of plasticity in their cold tolerance mechanisms, which could be harnessed to improve winter hardiness through careful management of acclimation and re-acclimation cycles. In the case of spring canola (*Brassica napus*), McClinchey and Kott (2008) took a different approach by generating UV-induced mutations in microspores followed by a chemical in vitro selection to enhance cold tolerance. The study identified several mutant lines with elevated levels of key defense signaling molecules, including salicylic acid, jasmonic acid, and proline. These mutants showed improved cold tolerance both in controlled freezer tests and in field conditions. The study suggests that breeding canola cultivars with increased cold tolerance could significantly improve the crop's ability to survive in regions prone to late frosts, a critical issue in many areas of the world.

Burbulis et al. (2008) investigated the effects of low temperatures on winter rapeseed (cv. Valesca) shoots subjected to acclimation at 4 °C for 1 and 7 days, followed by freezing treatments at −9 °C for 24, 48, and 72 h. The study found that nonfreezing low temperatures induced an increase in proline content, which was linked to improved cold hardiness. However, prolonged freezing duration led to a decline in cold tolerance, suggesting that while proline accumulation is a response to cold stress, it does not directly correlate with shoot survival under extended freezing conditions. This result highlights the complexity of cold tolerance mechanisms, where biochemical markers such as proline do not always predict overall plant resilience. Jonytiene et al. (2012) evaluated the role of exogenous proline in enhancing cold tolerance in winter rapeseed cv. Sunday. The study showed that adding exogenous proline to culture media significantly increased the endogenous proline content of the shoots and improved their cold tolerance, as evidenced by lower electrolyte leakage in shoots subjected to freezing stress. This indicates that proline may play a protective role under cold stress. However, the relationship between proline accumulation and cold tolerance remains nuanced and may depend on complex and interlinked factors. Indeed, Janska et al. (2010) conducted an in vitro experiment to select twelve doubled haploid winter rapeseed plants. These plants had varying levels of proline and freezing tolerance and were selected for resistance to *trans*-4-hydroxy-l-proline (Hyp), which is a phytotoxic proline analogue for plants. According to these authors, the regenerants that survive in a Hyp-containing medium may have elevated proline levels, potentially resulting in enhanced freezing tolerance (Janska et al. 2010). Despite selecting for resistance, the study did not reveal significant changes in proline content or freezing tolerance when compared to control populations. This suggests a weak correlation between proline accumulation and cold tolerance in the selected plants, highlighting the potential limitations of relying solely on proline accumulation as a marker for freezing tolerance.

12.6 Omics-Based Strategies for Regulating Phytohormones to Enhance Heat Stress Tolerance in Oilseed Crop

Integrating genomic, transcriptomic, proteomic, and metabolomic approaches provides a powerful framework for regulating phytohormones to enhance heat stress tolerance in oilseed crops (Raza et al. 2021; Satrio et al. 2024). Genomic tools such as CRISPR/Cas9 and genome-wide association studies (GWAS) allow for the identification and targeted modification of key genes involved in phytohormone biosynthesis, signaling, and degradation (Subedi et al. 2020; Liao et al. 2024). For instance, the manipulation of ABA biosynthesis genes can enhance drought resistance by promoting stomatal closure and reducing transpirational water loss, whereas modifying auxin-related genes may improve root architecture and stress adaptation (Weselake et al. 2008; Kibret 2017). Moreover, oil biosynthesis and storage in rapeseed (*Brassica napus* L.) were found to be related to increasing ABA content as the upregulation of two members of *NCED6* (9-cis-epoxycarotenoid dioxygenase) gene family involved in ABA biosynthesis, and the upregulation of genes involved in ABA signal transduction (Li et al. 2021b). Similarly, transcriptomic analyses provide insights into heat-responsive gene networks, including transcription factors such as DREB, WRKY, and HSF, which regulate phytohormone-mediated stress responses (Gao et al. 2021; Su et al. 2022). These molecular insights facilitate the development of climate-resilient oilseed cultivars through precision breeding and genetic engineering, offering new avenues for mitigating yield losses under elevated temperatures (Zafar et al. 2019). Meanwhile, proteomic and metabolomic approaches further refine our understanding of heat stress adaptation by elucidating the posttranslational modifications and metabolic shifts that occur under high temperatures (Gu et al. 2019; Satrio et al. 2024).

Proteomic studies reveal alterations in key signaling proteins, such as phosphorylation and ubiquitination events in ABA and ethylene pathways, which play critical roles in cellular stress responses (Li et al. 2021b). Concurrently, metabolomic profiling allows for the quantification of phytohormone fluctuations under heat stress, identifying potential targets for metabolic engineering and exogenous hormone applications (Chauffour et al. 2019). For example, exogenous application of SA or BRs has been shown to enhance thermotolerance by modulating antioxidant enzyme activity and maintaining cellular homeostasis (Sahni et al. 2016; Ali et al. 2024). The integration of these omics-based strategies provides a comprehensive approach to improving oilseed crop resilience, enabling the development of tailored agronomic practices and biostimulant formulations to sustain productivity in the face of rising global temperatures.

12.7 Positive Role of Exogenous Phytohormones in Enhancing Morpho-Physiochemical Traits of Oilseed Crops Under Temperature Stress

As previously demonstrated, oilseed crops are highly susceptible to temperature stress, which disrupts growth, physiological functions, and biochemical homeostasis. To mitigate these adverse effects, the exogenous application of phytohormones has emerged as a promising strategy, modulating morpho-physiological and biochemical responses to enhance stress tolerance and overall crop performance.

Multiple studies have shown that exogenous phytohormones positively regulate growth-related traits in oilseed crops under temperature stress (Angon et al. 2024; Shehzad et al. 2024). For instance, in soybean, foliar application of auxins (2,4-D) and gibberellins (GA_3) significantly increased plant height, root length, nodulation, pod number, and seed yield. Notably, soybean seed treatment with 2,4-D (90 mg/L) resulted in the greatest growth improvements, highlighting its role in promoting cell elongation and division (Shehzad et al. 2024). Additionally, exogenous phytohormone treatments enhanced the leaf area index, stomatal conductance, and chlorophyll stability in soybean, ultimately improving photosynthetic performance (Shehzad et al. 2024). Furthermore, soybean plants treated with GA_3 and auxins exhibited elevated activities of ascorbate peroxidase (APX) and glutathione reductase (GR), contributing to improved redox balance and stress mitigation (Shehzad et al. 2024). Similarly, Lone et al. (2022) demonstrated that exogenous brassinosteroids and jasmonic acid improved drought tolerance in *Brassica rapa* by modulating osmolyte accumulation, antioxidant activity, and photosynthetic efficiency.

In canola, the application of exogenous cytokinins and epibrassinolide alleviated drought-induced stress, leading to increased plant biomass, shoot length, and root growth—traits essential for efficient water and nutrient uptake under stress conditions (Sima et al. 2024). Moreover, in heat-stressed canola, the application of gibberellins and brassinosteroids enhanced photosynthetic pigment retention and membrane stability, sustaining higher photosynthetic rates and carbon assimilation (Angon et al. 2024). These improvements translated into better energy production and resource allocation, crucial for maintaining yield under adverse environmental conditions. Additionally, exogenous abscisic acid (ABA) application in canola was found to increase proline accumulation, which functions as an osmoprotectant by stabilizing proteins and membranes (Iqbal et al. 2014; Yao et al. 2019). Similarly, cytokinins have been shown to enhance osmotic balance by regulating ion transport and improving root hydraulic conductivity (Sharma et al. 2022). Furthermore, cytokinins significantly stimulated antioxidant enzyme activities, including superoxide dismutase (SOD), catalase (CAT), and peroxidase (POD), thereby reducing oxidative damage and maintaining cellular homeostasis (Sima et al. 2024).

In maize, the application of exogenous phytohormones such as salicylic acid (SA) and moringa leaf extract induced chilling tolerance (Waqas et al. 2017). Moreover, Iqbal et al. (2020) reported that various exogenous phytohormones significantly improved heat stress tolerance in maize. In peanut (*Arachis hypogaea*), melatonin-treated plants exhibited enhanced photosynthetic efficiency, sucrose production, and nitrogen assimilation, particularly under reduced nitrogen (N) application (Li et al. 2024). Melatonin treatment also increased dry matter accumulation, leading to higher seed mass, pod yield, and improved fat and protein content (Li et al. 2024). Similarly, the application of SA and jasmonic acid in sunflower (*Helianthus annuus*) promoted catalase activity and sugar metabolism, stimulating seedling development and alleviating the decline in photosynthesis caused by chilling stress (Górnik and Lahuta 2017). Additionally, exogenous SA was found to regulate metabolic balance in sunflower seeds by influencing the expression of Zn^{2+} transport-related genes (Górnik et al. 2017), which in turn promoted the biosynthesis of endogenous SA and GA (Huang et al. 2021).

Overall, the exogenous application of phytohormones plays a crucial role in enhancing oilseed crop resilience to temperature stress by improving growth traits, photosynthetic efficiency, osmolyte accumulation, antioxidant enzyme activity, and membrane stability. These physiological and biochemical improvements contribute to sustaining higher yields and maintaining oil quality under challenging environmental conditions.

12.8 Conclusion and Perspectives

The increasing occurrence of temperature stress presents a major challenge to oilseed crop production, affecting both yield and quality. This chapter highlights the critical role of phytohormones in modulating oilseed crops' responses to heat stress, demonstrating their involvement in key physiological and molecular pathways. Advances in omics-based research have significantly improved our understanding of phytohormone-mediated stress adaptation, revealing intricate signaling networks that regulate plant tolerance to high temperatures.

The integration of genomic, transcriptomic, proteomic, and metabolomic studies has provided novel insights into stress-responsive genes, posttranslational modifications, and metabolic shifts that contribute to thermotolerance. Additionally, PGPR applications and in vitro tissue culture techniques present innovative strategies for enhancing oilseed crop resilience. Despite these advancements, future research should focus on translating laboratory findings into field applications, developing climate-resilient crop varieties through targeted breeding, and optimizing agronomic practices to enhance stress adaptation.

Exploring gene-editing technologies such as CRISPR/Cas9 to enhance specific stress-responsive traits, refining metabolomic profiling to identify key biomarkers of stress resilience, and expanding the use of bio-based stimulants to modulate phytohormone activity in oilseed crops could be relevant avenues for future research. Furthermore, interdisciplinary collaborations integrating molecular biology, agronomy, and environmental sciences will be crucial in developing sustainable strategies to mitigate heat stress impacts on oilseed production. By leveraging these scientific advancements, oilseed crops can be better adapted to future climatic conditions, ensuring food security and economic sustainability in agriculture.

References

Abdel Razik ES, Alharbi BM, Pirzadah TB et al (2021) γ-Aminobutyric acid (GABA) mitigates drought and heat stress in sunflower (Helianthus annuus L.) by regulating its physiological, biochemical and molecular pathways. Physiol Plant 172:505–527

Ahmad M, Waraich EA, Skalicky M et al (2021) Adaptation strategies to improve the resistance of oilseed crops to heat stress under a changing climate: an overview. Front Plant Sci 12:767150
Akhtar A, Javed SB, Ahmad A (2024) Stress biology in plant tissue culture: recent advances. In: Stress biology in photosynthetic organisms: molecular insights and cellular responses. Springer, Singapore, pp 185–211
Ali A, Kant K, Kaur N et al (2024) Salicylic acid: homeostasis, signalling and phytohormone crosstalk in plants under environmental challenges. S Afr J Bot 169:314–335
Angon PB, Das A, Roy AR, Khan JJ, Ahmad I, Biswas A, Pallob AT, Mondol M, Yeasmin ST (2024) Plant development and heat stress: role of exogenous nutrients and phytohormones in thermotolerance. Discover Plants 1(1). https://doi.org/10.1007/s44372-024-00020-3
Arslan B, Culpan E (2017) Effects of different gibberellic acid doses on seed yield, oil content and some quality traits of safflower (Carthamus tinctorius L.). J Glob Innov Agric Soc Sci 5:5–9
Ashraf MA, Akbar A, Askari SH et al (2018) Recent advances in abiotic stress tolerance of plants through chemical priming: an overview. In: Advances in seed priming. Springer, Singapore, pp 51–79
Asthana P, Rai MK, Jaiswal U (2023) In vitro selection, regeneration and characterization of NaCl-tolerant plants of Sapindus trifoliatus: an important multipurpose tree. Plant Cell Tissue Organ Cult 154:227–238
Brito C, Dinis LT, Meijón M, Ferreira H, Pinto G, Moutinho-Pereira J, Correia C (2018) Salicylic acid modulates olive tree physiological and growth responses to drought and rewatering events in a dose dependent manner. J Plant Physiol 230:21–32. https://doi.org/10.1016/j.jplph.2018.08.004
Brito C, Dinis LT, Ferreira H, Coutinho J, Moutinho-Pereira J, Correia CM (2019) Salicylic acid increases drought adaptability of young olive trees by changes on redox status and ionome. Plant Physiol Biochem 141:315–324. https://doi.org/10.1016/j.plaphy.2019.06.011
Brunel-Muguet S, d'Hooghe P, Bataillé M-P et al (2015) Heat stress during seed filling interferes with sulfur restriction on grain composition and seed germination in oilseed rape (Brassica napus L.). Front Plant Sci 6:213
Burbulis N, Kuprienė R, Blinstrubienė A (2008) Investigation of cold resistance of winter rapeseed in vitro
Burbulis N, Jonytiene V, Kupriene R et al (2010) Effect of abscisic acid on cold tolerance in Brassica napus shoots cultured in vitro. J Food Agric Environ 8:698–701
Burbulis N, Jonytienė V, Kuprienė R et al (2011) Biochemical and physiological factors related to cold de-acclimation and re-acclimation in rapeseed shoots in vitro. J Food Agri Environ 9:483–487
Carmo-Silva AE, Salvucci ME (2012) The temperature response of CO_2 assimilation, photochemical activities and Rubisco activation in Camelina sativa, a potential bioenergy crop with limited capacity for acclimation to heat stress. Planta 236:1433–1445
Chandrasekaran U, Xu W, Liu A (2014) Transcriptome profiling identifies ABA mediated regulatory changes towards storage filling in developing seeds of castor bean (Ricinus communis L.). Cell Biosci 4:1–12
Chauffour F, Bailly M, Perreau F et al (2019) Multi-omics analysis reveals sequential roles for ABA during seed maturation. Plant Physiol 180:1198–1218
Chen S, Guo Y, Sirault X et al (2019) Nondestructive phenomic tools for the prediction of heat and drought tolerance at anthesis in brassica species. Plant Phenomics, p 3264872
Chimenti CA, Hall AJ, Lopez MS (2001) Embryo-growth rate and duration in sunflower as affected by temperature. Field Crop Res 69:81–88
Cruz G, Silva AV, Da Silva JB et al (2020) Biofuels from oilseed fruits using different thermochemical processes: opportunities and challenges. Biofuels Bioprod Biorefin 14:696–719
Damude HG, Kinney AJ (2008) Enhancing plant seed oils for human nutrition. Plant Physiol 147:962–968

Dinler BS, Demir E, Kompe YO (2014) Regulation of auxin, abscisic acid and salicylic acid levels by ascorbate application under heat stress in sensitive and tolerant maize leaves. Acta Biol Hung 65:469–480

Elferjani R, Soolanayakanahally R (2018) Canola responses to drought, heat, and combined stress: shared and specific effects on carbon assimilation, seed yield, and oil composition. Front Plant Sci 9:1224

Fahad S, Bajwa AA, Nazir U et al (2017) Crop production under drought and heat stress: plant responses and management options. Front Plant Sci 8:1147

FAO (2023) Agricultural production statistics 2000–2022. FAOSTAT Analytical Briefs 79

Gao G, Hu J, Zhang X et al (2021) Transcriptome analysis reveals genes expression pattern of seed response to heat stress in Brassica napus L. Oil Crop Sci 6:87–96

Gaspar T, Kevers C, Penel C et al (1996) Plant hormones and plant growth regulators in plant tissue culture. In Vitro Cell Dev Biol Plant 32:272–289

Gaspar TH, Kevers C, Faivre-Rampant O et al (2003) Changing concepts in plant hormone action. In Vitro Cell Dev Biol Plant 39:85–106

Górnik K, Lahuta LB (2017) Application of phytohormones during seed hydropriming and heat shock treatment on sunflower (Helianthus annuus L.) chilling resistance and changes in soluble carbohydrates. Acta Physiologiae Plantarum 39(5). https://doi.org/10.1007/s11738-017-2413-x

Grover A, Mittal D, Negi M, Lavania D (2013) Generating high temperature tolerant transgenic plants: achievements and challenges. Plant Sci 205:38–47

Gu J, Hou D, Li Y et al (2019) Integration of proteomic and genomic approaches to dissect seed germination vigor in Brassica napus seeds differing in oil content. BMC Plant Biol 19:1–20

Guo Y, Fan R, Sun F, Qu Y, Zheng K, Chen Q, Chen Q (2020) Identification of exogenous ABA and heat stress tolerance in various cotton genotypes. Abstract Plant Genetic Resources: Characterization and Utilization 18(6):404–416. https://doi.org/10.1017/S147926212000043X

Hamooh BT, Sattar FA, Wellman G, Mousa MAA (2021) Metabolomic and biochemical analysis of two potato (Solanum tuberosum L.) cultivars exposed to in vitro osmotic and salt stresses. Plan Theory 10:98

Harwood JL, Ramli US, Tang M et al (2013) Regulation and enhancement of lipid accumulation in oil crops: the use of metabolic control analysis for informed genetic manipulation. Eur J Lipid Sci Technol 115:1239–1246

Hasanuzzaman M, Nahar K, Fujita M (2013) Extreme temperature responses, oxidative stress and antioxidant defense in plants. In: Abiotic stress-plant responses and applications in agriculture, vol 13. IntechOpen, Rijeka, pp 169–205

Hegewald H (2022) Oilseed rape in high-intensity crop rotations. Dissertation, Halle (Saale), Martin-Luther-Universität Halle-Wittenberg, 2022

Huang Y, Jin D, Lu C et al (2016) Proteomic responses associated with freezing tolerance in the callus of the Tibetan alpine plant Saussurea laniceps during cold acclimation. Plant Cell Tissue Organ Cult 124:81–95

Huang B, Thorne PW, Banzon VF, et al (2017) NOAA extended reconstructed sea surface temperature (ERSST), version 5. (No Title)

Huang R, Liu Z, Xing M et al (2019) Heat stress suppresses Brassica napus seed oil accumulation by inhibition of photosynthesis and BnWRI1 pathway. Plant Cell Physiol 60:1457–1470

Huang YT, Cai SY, Ruan XL, Chen SY, Mei GF, Ruan GH, Cao DD (2021) Salicylic acid enhances sunflower seed germination under Zn^{2+} stress via involvement in Zn^{2+} metabolic balance and phytohormone interactions. Sci. Hortic 275:109702. https://doi.org/10.1016/j.scienta.2020.109702

Ifeduba AM, Zhen S, Pandey J, Vales MI (2024) Leaf membrane stability under high temperatures as an indicator of heat tolerance in potatoes and genome-wide association studies to understand the underlying genetics. Plan Theory 13:2175

Imran M, Aaqil Khan M, Shahzad R et al (2021) Melatonin ameliorates thermotolerance in soybean seedling through balancing redox homeostasis and modulating antioxidant defense, phytohormones and polyamines biosynthesis. Molecules 26:5116

Iqbal N, Umar S, Khan NA, Khan MIR (2014) A new perspective of phytohormones in salinity tolerance: Regulation of proline metabolism. Environ Exp Bot 100:34-42. https://doi.org/10.1016/j.envexpbot.2013.12.006

Iqbal S, Hussain S, Qayyaum MA, Ashraf M, Hossain SA (2020) The response of maize physiology under salinity stress and its coping strategies. In: Plant Stress Physiology, IntechOpen

Janska A, Zelenkova S, Klima M et al (2010) Freezing tolerance and proline content of in vitro selected hydroxyproline resistant winter oilseed rape. Czech J Genet Plant Breed 46:35–40

Jonytiene V, Burbulis N, Kupriené R, Blinstrubiené A (2012) Effect of exogenous proline and de-acclimation treatment on cold tolerance in Brassica napus shoots cultured in vitro. J Food Agric Environ 10:327–330

Ju YL, Liu M, Zhao H, Meng JF, Fang YL (2016) Effect of exogenous Abscisic acid and methyl Jasmonate on anthocyanin composition fatty acids and volatile compounds of cabernet sauvignon (Vitis vinifera L.) grape berries. Molecules 21(10):1354. https://doi.org/10.3390/molecules21101354

Kakani VG, Wheeler TR, Craufurd PQ, Rachaputi RCN (2014) Effect of High Temperature and Water Stress on Groundnuts Under Field Conditions. In: Mahalingam R (ed) Combined stresses in plants physiological molecular and biochemical aspects. Springer International Publishing Cham, pp 159–180

Kassout J, Terral J-F, Azenzem R et al (2023) Olea europaea and stressful conditions. In: Medicinal plant responses to stressful conditions. CRC Press, pp 243–272

Khan MA, Asaf S, Khan AL et al (2020a) Thermotolerance effect of plant growth-promoting Bacillus cereus SA1 on soybean during heat stress. BMC Microbiol 20:1–14

Khan MN, Khan Z, Luo T et al (2020b) Seed priming with gibberellic acid and melatonin in rapeseed: consequences for improving yield and seed quality under drought and non-stress conditions. Ind Crop Prod 156:112850

Kibret KB (2017) Systems biological analysis of seedling vigour and osmotic stress tolerance in oilseed rape (Brassica napus L., Brassicaceae)

Kishor PBK, Tiozon RN, Fernie AR, Sreenivasulu N (2022) Abscisic acid and its role in the modulation of plant growth, development, and yield stability. Trends Plant Sci 27:1283–1295

Klingler JP, Batelli G, Zhu J-K (2010) ABA receptors: the START of a new paradigm in phytohormone signalling. J Exp Bot 61:3199–3210

Kotecka-Majchrzak K, Sumara A, Fornal E, Montowska M (2020) Oilseed proteins–properties and application as a food ingredient. Trends Food Sci Technol 106:160–170

Koufan M, Mazri MA, Essatte A et al (2020) A novel regeneration system through micrografting for Argania spinosa (L.) Skeels, and confirmation of successful rootstock-scion union by histological analysis. Plant Cell Tissue Organ Cult 142:369–378

Larkindale J, Knight MR (2002) Protection against heat stress-induced oxidative damage in Arabidopsis involves calcium, abscisic acid, ethylene, and salicylic acid. Plant Physiol 128:682–695

Li N, Euring D, Cha JY et al (2021a) Plant hormone-mediated regulation of heat tolerance in response to global climate change. Front Plant Sci 11:627969

Li Y, Zhang L, Hu S et al (2021b) Transcriptome and proteome analyses of the molecular mechanisms underlying changes in oil storage under drought stress in Brassica napus L. GCB Bioenergy 13:1071–1086

Li L, Cheng X, Zhang Y, Kohtz D, Wang X, Zhang X, Kong X, Xue H, Jia P, Bai N, Li Z, Xiao P, Pan X, Zhang Z, Zhang B (2024) Exogenous melatonin improves peanut field productivity and quality at reduced nitrogen application. Field Crops Res 319:109650. https://doi.org/10.1016/j.fcr.2024.109650

Liao W, Guo R, Li J et al (2024) CRISPR/Cas9-mediated mutagenesis of SEED FATTY ACID REDUCER genes significantly increased SEED oil content in soybean. Plant Cell Physiol pcae148

Ljung K, Nemhauser JL, Perata P (2015) New mechanistic links between sugar and hormone signalling networks. Curr Opin Plant Biol 25:130–137

Lohani N, Singh MB, Bhalla PL (2022) Short-term heat stress during flowering results in a decline in Canola seed productivity. J Agron Crop Sci 208:486–496

Lone WA, Majeed N, Yaqoob U, John R (2022) Exogenous brassinosteroid and jasmonic acid improve drought tolerance in Brassica rapa L. genotypes by modulating osmolytes antioxidants and photosynthetic system. Plant Cell Rep 41(3):603–617. https://doi.org/10.1007/s00299-021-02763-9

Lu C, Napier JA, Clemente TE, Cahoon EB (2011) New frontiers in oilseed biotechnology: meeting the global demand for vegetable oils for food, feed, biofuel, and industrial applications. Curr Opin Biotechnol 22:252–259

Maheshwari P, Kovalchuk I (2014) Genetic engineering of oilseed crops. Biocatal Agric Biotechnol 3:31–37

Mansinhos I, Gonçalves S, Rodríguez-Solana R et al (2022) Impact of temperature on phenolic and osmolyte contents in in vitro cultures and micropropagated plants of two mediterranean plant species, Lavandula viridis and Thymus lotocephalus. Plants 11:3516

Mazri MA (2014) Effects of plant growth regulators and carbon source on shoot proliferation and regeneration in date palm (Phoenix dactylifera L.)'16-bis'. J Hortic Sci Biotechnol 89:415–422

Mazri MA, Elbakkali A, Belkoura M, Belkoura I (2011) Embryogenic competence of calli and embryos regeneration from various explants of Dahbia cv, a Moroccan olive tree (Olea europaea L.). Afr J Biotechnol 10:19089–19095

McClinchey SL, Kott LS (2008) Production of mutants with high cold tolerance in spring canola (Brassica napus). Euphytica 162:51–67

Muluneh MG (2021) Impact of climate change on biodiversity and food security: a global perspective—a review article. Agric Food Secur 10:1–25

Nakagawa AC, Ario N, Tomita Y et al (2020) High temperature during soybean seed development differentially alters lipid and protein metabolism. Plant Prod Sci 23:504–512

Nazar R, Iqbal N, Umar S (2017) Heat stress tolerance in plants: action of salicylic acid. Salicylic acid: a multifaceted hormone 145–161

Nejati M, Alipour A, Yousefirad M et al (2020) Effects of seed germination with gibberellin on safflower germination under salinity stress. Iran J Plant Physiol 11:3443–3448

O'Brien JA, Benková E (2013) Cytokinin cross-talking during biotic and abiotic stress responses. Front Plant Sci 4:451

Pérez A, Gómez D, Acosta Y et al (2025) Mini-review on plant in vitro culture to obtain natural chemicals. In Vitro Cell Dev Biol Plant 61:1–9

Poór P, Nawaz K, Gupta R et al (2022) Ethylene involvement in the regulation of heat stress tolerance in plants. Plant Cell Rep 41:675–698

Pranckutė R (2021) Web of Science (WoS) and Scopus: the titans of bibliographic information in today's academic world. Publica 9:12

Prasertthai P, Paethaisong W, Theerakulpisut P, Dongsansuk A (2022) High temperature alters leaf lipid membrane composition associated with photochemistry of PSII and membrane thermo-stability in rice seedlings. Plan Theory 11:1454

Qu A-L, Ding Y-F, Jiang Q, Zhu C (2013) Molecular mechanisms of the plant heat stress response. Biochem Biophys Res Commun 432:203–207

Radi H, Bouchiha F, El Maataoui S et al (2023) Morphological and physio-biochemical responses of cactus pear (Opuntia ficus indica (L.) Mill.) organogenic cultures to salt and drought stresses induced in vitro. Plant Cell Tissue Organ Cult 154:337–350

Rai SK, Deeksha Charak DC, Rajeev Bharat RB (2016) Scenario of oilseed crops across the globe

Raza A, Razzaq A, Mehmood SS et al (2021) Omics: the way forward to enhance abiotic stress tolerance in Brassica napus L. GM Crops Food 12:251–281

Rudić J, Pantelić D, Oljača J, Momčilović I (2022) Effects of elevated temperature and salicylic acid on heat shock response and growth of potato microplants. Horticulturae 8:372

Sahni S, Prasad BD, Liu Q et al (2016) Overexpression of the brassinosteroid biosynthetic gene DWF4 in Brassica napus simultaneously increases seed yield and stress tolerance. Sci Rep 6: 28298

Sahu M, Maurya S, Jha Z (2023) In vitro selection for drought and salt stress tolerance in rice: an overview. Plant Physiol Rep 28:8–33

Salunkhe DK (1992) World oilseeds. Springer Science & Business Media

Sarwar MF, Sarwar MH, Sarwar M et al (2013) The role of oilseeds nutrition in human health: a critical. J Cereals Oilseeds 4:97–100

Satrio RD, Fendiyanto MH, Miftahudin M (2024) Tools and techniques used at global scale through genomics, transcriptomics, proteomics, and metabolomics to investigate plant stress responses at the molecular level. In: Molecular dynamics of plant stress and its management. Springer, pp 555–607

Schwarz I, Scheirlinck M-T, Otto E et al (2020) Cytokinin regulates the activity of the inflorescence meristem and components of seed yield in oilseed rape. J Exp Bot 71:7146–7159

Sharma M, Gupta SK, Mondal AK (2012) Production and trade of major world oil crops. In: Technological innovations in major world oil crops, volume 1: breeding. Springer, pp 1–15

Sharma S, Kaur P, Gaikwad K (2022) Role of cytokinins in seed development in pulses and oilseed crops: Current status and future perspective. Front Genet 13. https://doi.org/10.3389/fgene.2022.940660

Sharma M, Negi S, Kumar P et al (2023) Fruit ripening under heat stress: the intriguing role of ethylene-mediated signaling. Plant Sci:111820

Shehzad M, Sajid Munir M, Nazim M, Tahir MM, Maqbool M, Amjad Nawaz M, Muhammad Rashad Javeed H, Ahmed Al-Ghamdi A (2024) Evaluating the impact of phyto-hormones on the morpho-biochemical traits of soybean through seed treatment and foliar application. Journal of King Saud University - Science 36(10):103446. https://doi.org/10.1016/j.jksus.2024.103446

Shi P, Hua W, Htwe YM, Zhang D, Li J, Wang Y (2021) Abscisic acid improves linoleic acid accumulation possibly by promoting expression of EgFAD2 and other fatty acid biosynthesis genes in oil palm mesocarp. Front Plant Sci 12. https://doi.org/10.3389/fpls.2021.748130

Shukla S, Shukla SK (2024) Micropropagation for crop improvement and it's commercialization potential. In: The potential of microbes for a circular economy. Elsevier, pp 271–287

Silverstone AL, Jung H-S, Dill A et al (2001) Repressing a repressor: gibberellin-induced rapid reduction of the RGA protein in Arabidopsis. Plant Cell 13:1555–1566

Sima NAK, Jabbari H, Ebadi A, Ghaffari MR, Koobaz P (2024) Comparative analysis of exogenous hormone application on contrasting canola (Brassica napus L.) genotypes under drought stress conditions. J Soil Sci Plant Nutr 24(1):308–317. https://doi.org/10.1007/s42729-023-01527-6

Smékalová V, Doskočilová A, Komis G, Šamaj J (2014) Crosstalk between secondary messengers, hormones and MAPK modules during abiotic stress signalling in plants. Biotechnol Adv 32:2–11

Smith BE, Lu C (2024) Heat stress during reproductive stages reduces camelina seed productivity and changes seed composition. Heliyon 10:e26678

Su X, Gao T, Zhang P et al (2022) Comparative physiological and transcriptomic analysis of sesame cultivars with different tolerance responses to heat stress. Physiol Mol Biol Plants 28:1131–1146

Subedi U, Ozga JA, Chen G et al (2020) CRISPR/Cas-mediated genome editing for the improvement of oilseed crop productivity. Crit Rev Plant Sci 39:195–221

Tahiri A, Mazri MA, Karra Y et al (2023) Propagation of saffron (Crocus sativus L.) through tissue culture: a review. J Hortic Sci Biotechnol 98:10–30

USDA F (2018) Oilseeds: world markets and trade. Foreign Agricultural Service/USDA

Usman I, Saif H, Imran A et al (2023) Innovative applications and therapeutic potential of oilseeds and their by-products: an eco-friendly and sustainable approach. Food Sci Nutr 11:2599–2609

Vuksanović V, Kovačević B, Kebert M et al (2023) In vitro selection of drought-tolerant white poplar clones based on antioxidant activities and osmoprotectant content. Front Plant Sci 14: 1280794

Waraich EA, Ahmad M, Soufan W et al (2021) Seed priming with sulfhydral thiourea enhances the performance of Camelina sativa L. under heat stress conditions. Agronomy 11:1875

Waqas MA, Khan I, Akhter MJ, Noor MA, Ashraf U (2017) Exogenous application of plant growth regulators (PGRs) induces chilling tolerance in short-duration hybrid maize. Environ Sci Pollut Res 24(12):11459–11471. https://doi.org/10.1007/s11356-017-8768-0

Weselake RJ, Shah S, Tang M et al (2008) Metabolic control analysis is helpful for informed genetic manipulation of oilseed rape (Brassica napus) to increase seed oil content. J Exp Bot 59: 3543–3549

Wittkop B, Snowdon RJ, Friedt W (2009) Status and perspectives of breeding for enhanced yield and quality of oilseed crops for Europe. Euphytica 170:131–140

Xiong L, Zhu J-K (2003) Regulation of abscisic acid biosynthesis. Plant Physiol 133:29–36

Xu C, McDowell NG, Fisher RA et al (2019) Increasing impacts of extreme droughts on vegetation productivity under climate change. Nat Clim Chang 9:948–953

Yang DQ, Wang ZL, Ni YL et al (2014) Effect of high temperature stress and spraying exogenous ABA post-anthesis on grain filling and grain yield in different types of stay-green wheat. China Agric Sin 47:2109–2125

Yao C, Zhang F, Sun X, Shang D, He F, Li X, Zhang J, Jiang X (2019) Effects of S-Abscisic acid (S-ABA) on seed germination seedling growth and Asr1 gene expression under drought stress in maize. J Plant Growth Regul 38(4):1300–1313. https://doi.org/10.1007/s00344-019-09934-9

Zafar S, Li Y-L, Li N-N et al (2019) Recent advances in enhancement of oil content in oilseed crops. J Biotechnol 301:35–44

Zhang Y, Xu Z, Xie T et al (2022) In vitro selection and identification of a cold-tolerant variant in pineapple (Ananas comosus). Hortic Environ Biotechnol 63:275–286

Zhang M, Wang O, Cai S et al (2023a) Composition, functional properties, health benefits and applications of oilseed proteins: a systematic review. Food Res Int 171:113061

Zhang X, Yang Z, Wang L et al (2023b) The effects of plant growth-promoting rhizobacteria on plants under temperature stress: a meta-analysis. Rhizosphere:100788

Chapter 13
Phytohormones as Key Players: Boosting Oilseed Crop Tolerance to Nutrient Deficiency Stress

Nasir Assad, Marzia Batool Laila, Muhammad Naeem-Ul-Hassan, Baber Ali, Alevcan Kaplan, Muhammad Nauman Khan, Khizar Hayat, and Fethi Ahmet Ozdemir

Abstract Oilseed crops are vital for global food security, industrial applications and biofuel production. However, their productivity is stymied by nutrient deficiency stress, negatively affecting plant growth, metabolism and crop yield quality. Phytohormones such as auxins, cytokinins, gibberellins, abscisic acid, ethylene, jasmonic acid and salicylic acid are central in modulating plant responses to nutrient scarcity by regulating root growth, mechanisms of nutrient uptake and pathways of adaptation to stress. These hormones' coordination of complex signaling networks enhances nutrient transport, metabolic homeostasis and gene expression to strengthen plant resilience under nutrient-deficient conditions. There are several avenues to leverage, including genetic breeding, molecular biotechnology and precision agriculture, to enhance oilseed crops' nutrient-use efficiency. To that end, novel techniques such as CRISPR-Cas9 genome editing, transcriptomics and metabolomics allow more profound modifications within hormone biosynthesis and signaling pathways that support the development of nutrient-efficient and stress-tolerant cultivars. In addition, the combined use of phytohormonal treatments with microbial biofertilizers, *artificial intelligence (AI)-driven* nutrient management and site-specific fertilization strategies offers excellent potential for optimal crop

N. Assad (✉) · M. B. Laila · M. Naeem-Ul-Hassan
Institute of Chemistry, University of Sargodha, Sargodha, Pakistan

B. Ali
Department of Plant Sciences, Quaid-i-Azam University Islamabad, Islamabad, Pakistan

A. Kaplan
Department of Crop and Animal Production, Sason Vocational School, Batman University, Batman, Turkey

M. N. Khan · K. Hayat
Department of Botany, Islamia College Peshawar, Peshawar, Pakistan

F. A. Ozdemir
Department of Molecular Biology and Genetics, Faculty of Science and Art, Bingol University, Bingol, Turkey

© The Author(s), under exclusive license to Springer Nature Singapore Pte Ltd. 2025

A. A. H. Abdel Latef (ed.), *Oilseed Crops Under Abiotic Stress*, Sustainability Sciences in Asia and Africa, https://doi.org/10.1007/978-981-96-8346-8_13

performance with minimal environmental impacts. Future research should focus on deciphering the intricate hormonal crosstalk that controls nutrient acquisition and discovering novel transgenic and agronomic interventions for increasing resilience to nutrient stress. Formulating climate-resilient oilseed crops demands a multifaceted strategy in plant physiology, molecular biology and sustainable agricultural approaches. It is essential to navigate these advancements to enhance nutrient-use efficiency and increase crop productivity while ensuring sustainable agricultural systems can combat the global challenges of nutrient stress.

Keywords Nutrient deficiency stress · Nutrient uptake efficiency · Oilseed crops · Phytohormones · Stress tolerance mechanisms

13.1 Introduction

Worldwide, farmers gather oilseed crop seeds to press into vegetable oil, a versatile product with several uses in food and oleochemical industries. Oilseed crops rank as the third most significant category of crops, following cereals and legumes (Lafarga 2021). The food sector relies on oilseed crops such as soybean, safflower, sunflower, linseed, rapeseed, camelina, jatropha and castor bean, whose seeds are harvested mainly to produce oil (Bera et al. 2014). Soybeans, peanuts, sunflowers and rapeseeds are the four primary sources of edible oils. Although several factors affect oilseed crop development, phytohormones are among the most important. The diverse range of structures found in phytohormones allows plants to control their development and growth as well as respond to different types of stress. These allow plant cells to regulate cellular activities, including enzyme activity and also coordinate physiological processes like development, growth and metabolism (Iqbal et al. 2014). Phytohormones are biological signals that control plant development and growth; these signals may operate locally or be transported to other areas of the plant (Ashfaq and Khan 2017). A major negative impact of stress on agricultural output is the disruption of plant processes, which in turn impacts hormonal and nutritional imbalances. Plants are susceptible to several environmental threats, such as salt, dryness, heavy metals, malnutrition and illnesses (Nadeem et al. 2016). Here, we shall mostly focus on nutrient deficiency stress. Plants can't grow, among other physiological and metabolic processes, without the nutrients in the soil. The nutritional composition of a plant may tell a lot about its stage of growth (Vander Linden and Delvaux 2019). The many nutrient absorption, accumulation and translocation methods that plants have evolved support both soil nutrient mobilization and internal nutritional balance (Debona et al. 2017). An individual plant's growth and development are significantly affected by the nutrients it receives, including the kind, quantity, exposure time and origin of those nutrients (Bouain et al. 2019; Ali et al. 2020). On the other hand, if the plant lacks an adequate supply of a specific nutrient needed for its growth, metabolism and development, it can result in physiological and biochemical stress due to nutrient deficiency. The main micronutrients are manganese, iron and zinc and the main macronutrients are nitrogen, potassium and phosphorus. Nutritional stress may play a critical role in regulating various reproductive processes, spanning from root and shoot initiation to pollination, flowering and seed set (Khan et al. 2023). Another effect of nutritional stress for enhancing the susceptibility of plants to pests and diseases is the reduced production of secondary metabolites, which serve as a defense against biotic stress. The chronic effects of these causes are a reduction in biomass, a reduction in the number of fruits and a reduction in seed quality. They also cause significant losses in productivity and quality. Nutrient availability may be enhanced with time and nutrient deficiency stress may be treated via a combination of approaches.

13.2 Importance of Oilseed Crops in Global Agriculture

Oilseed crops are extremely valuable in the world's major crop production. Their relevance in human and industrial nutrition is obvious. Oilseed residues (i.e., sunflower hull, meal and oil) in a tertiary state are used for human nutrition and multiple industrial applications (Wrigley et al. 2015; Abiodun 2017). The wide-ranging and varied uses of oilseed crops highlight the necessity of multiple plant species to meet these different demands (Jaradat 2016). In a similar vein, the oil yields from oilseed crops are generally significant, though they vary by species and are influenced by the extraction methods employed; the high concentrations of polyunsaturated fatty acids also provide a safeguard against coronary heart disease. Beyond the nutritional advantages of oilseeds, a range of industrial uses has been recorded, such as biodiesel, fertilizers, pharmaceuticals, cosmetics, animal feed, fibers, paint and buttons, among others. Oilseed crops offer considerable potential to improve human diets, address malnutrition and food insecurity and generate employment through income opportunities within communities (Abiodun 2017). Soybean, sunflower, canola, cottonseed and peanut oilseed crops are among the most important ones being studied and grown right now. There are many different industrial uses for each, and they all have specific functions in human and animal nutrition. Also, major oil crops include olive, sesame, coconut, palm and safflower (Attia et al. 2021). Soybeans are the most widely produced legume in the world, and one-quarter of the world's edible oil supply is expected to come from this agricultural product. According to Patil et al. (2017) as cited by Santos Hansel et al. (2019), the oil's primary role is delivering protein, though they are also a product. The seed is widely harvested for human and animal consumption, due to its high protein, essential amino acid, oil and calorie content. Demand for soybean food products increased dramatically from $1 billion in 2000 to $4.5 billion in 2013, and so this crop is harvested on a massive scale, not just for humans, but now also for animals (Patil et al. 2017; Santos Hansel et al. 2019). Over the past few years, the growth of the market for soybean products has increased from 1 billion dollars in 2000 to 4.5 billion dollars in 2013 (Patil et al. 2017). Current evidence indicates that the commodity will continue to play a critical role on a global scale in the future. More than 34 million acres of land are reserved for the cultivation of sunflower

(*Helianthus annuus* L. as a source of high-quality oil, which is an oilseed for human consumption. This substance has numerous industrial applications, including its use in soaps and detergents, as well as in ruminant animal feed. Cottonseed oil consists of approximately 300 g kg^{-1} of saturated fatty acids, primarily derived from the seeds of *Gossypium hirsutum* or *Gossypium herbaceum* (Gunstone et al. 2007). Therefore, becoming a significant secondary ingredient in various food products like mayonnaise, cottonseed oil was regarded as a byproduct of the cotton fiber industry (Gunstone et al. 2007; Rashid et al. 2009). China stands as the foremost consumer and producer of cottonseed oil, with global consumption projected to approach around 5.2 million metric tons (List 2016). The main application of groundnuts, commonly referred to as peanuts or *Arachis hypogaea Fabr*., is in the production of oilseed crops. Approximately 500 g kg^{-1} of the seeds' weight consists of oil. This oil serves a dual purpose, being beneficial for soap production as well as for human consumption. Notably, it was among the pioneering vegetable oils demonstrated to be effective in biodiesel engines (Lewis 2011). Dominating the fatty acid profile are oleic acid (480 g kg^{-1}) and linoleic acid (320 g kg^{-1}) (USDA, 2020), which emphasizes its global importance (Attia et al. 2021).

13.3 Overview of Nutrient Deficiency Stress in Plants

Nutrient deficiency stress in plants arises when essential macronutrients (e.g., nitrogen, phosphorus and potassium) and micronutrients (e.g., iron, zinc and manganese) are insufficient, leading to impaired growth, metabolism and reproduction. Nitrogen deficiency, for instance, results in chlorosis (yellowing of leaves) and stunted growth due to its critical role in chlorophyll synthesis and protein formation. Phosphorus deficiency can cause dark green or purplish foliage and poor root development, as phosphorus is vital for energy transfer and nucleic acid synthesis (Mishra et al. 2024). Potassium deficiency often leads to necrosis (death of plant tissue) along leaf margins and increased susceptibility to drought, given potassium's role in osmotic regulation and enzyme activation.

Micronutrient deficiencies (e.g., iron or zinc) cause interveinal chlorosis and reduced activity of enzymes, respectively (Gupta et al. 2024). Plants have developed numerous adaptive mechanisms to nutrient limitation, such as modification of root systems to establish the highest possible uptake efficiency of nutrients, exudation from the rhizosphere to release nutrients in the soil water film and establishment of mycorrhizal symbiosis with mycorrhizal fungi (MF). Nutrient depletion at the molecular level can result in changes to gene expression, protein functions and metabolic response pathways as a compensatory response to remove the stress effect. The mechanism of these physiological and molecular responses is an indispensable step in designing the approach to improve nutrient-use efficiency and stress tolerance, which, in turn, can lead to an increase in crop production and sustainability (Pandey et al. 2021).

13.4 Role of Phytohormones in Plant Growth and Stress Response

One of the key controls of plant growth is the existence of a collection of small organic signaling molecules known as phytohormones. Phytohormones, at very low levels of concentration, have significant effects at multiple levels of plant development and stress resistance, mainly due to their marked physiological influence. These processes encompass cell division, elongation, differentiation, germination, root development, flowering, fruiting, sex determination, dormancy and organ shedding. One type of phytohormone may affect a broad spectrum of developmental phenomena, whereas a given set of phytohormones (possibly synergized) can affect a narrow set of selectable developmental events. Many phytohormones [auxins, cytokinins (CKs), gibberellins (GAs), abscisic acid (ABA), ethylene (ETH), brassinosteroids (BRs), jasmonate (JA) and salicylic acid (SA)] have been well known (Zhao et al. 2021). The phytohormones have played a critical role in the early stage of nutrient stress, which is to facilitate and drive the adaptive response capable of maximizing the ability to acquire and utilize such nutrients. ABA is a master controller of nutrient stress responses, particularly nitrogen and phosphorus starvation.

It controls stomatal closure for water saving and interacts with other hormones to control root growth and nutrient uptake. That is ethylene, which is synthesized in response to stress, works in concert with auxin to increase root hair formation and subsequent root surface area to improve nutrient uptake. Jasmonic acid and salicylic acid, which are best known for their involvement in defense, also play roles in nutrient stress responses by influencing secondary metabolism and scavenging reactive oxygen species (ROS), aiding in stress tolerance. Crosstalk among phytohormones is critical for controlling homeostasis in response to nutrient stress. For instance, the crosstalk between ABA, ethylene and cytokinin mediates the trade-off between growth and stress tolerance. Although predominantly described to have a stimulatory role in growth, gibberellins are downregulated under extreme nutrient deprivation, such as energy depletion for survival (Romera et al. 2021). Generally, phytohormones act as integrative signals, coordinating growth and stress responses, ensuring that plants adapt efficiently to nutrient-limiting conditions.

13.5 Phytohormones and Nutrient Deficiency in Oilseed Crops

13.5.1 Auxins

Indole-3-acetic acid (IAA) is a phytohormone that is structurally similar to tryptophan, and known as the main auxin responsible for regulating a variety of slow developmental processes in a wide range of plant processes, including cellular elongation, embryogenesis, transition to reproduction, vascular differentiation and postgermination development (Zažímalová et al. 2014; Lv et al. 2019). The studies in oilseed crops have demonstrated that either seed priming or foliar application of exogenous auxin improves morphophysiological traits. For instance, A. Ahmad et al. (2001) reported that foliar treatment with 10^{-8} M or 10^{-6} M IAA 29-day-old *Brassica juncea* plants resulted in 21.6% more dry biomass than aqueous controls. Similarly, Kulkarni et al. (2013) noted that the shoot and root dry mass of mustard plants treated with 5 μM indole-3-butyric acid (IBA) significantly increased, what stem height and primary branch length increase by 21.64% and 21.61%, respectively, compared to nontreated plants (Kulkarni et al. 2013).

In *Arachis hypogaea*, three foliar applications of 10 μM IAA under flowering have improved yield-related parameters (Peng et al. 2013). Furthermore, Chen et al. (2021) noted that 500 mg L^{-1} of IAA applied by spray significantly increased both shoot and root dry mass of *Helianthus annuus* seedlings aged 30 days (L. Chen et al. 2021). Recent work by Mir et al. (2020) found significantly stronger auxin-induced growth responses in *B. juncea*: Five doses of 10^{-8} M IAA, applied on 25-day-old seedlings, stimulated an 18.97% increase of root length, a 41% increase of shoot elongation, a 23.85% increase of leaf area expansion, a 47.80% increase of root fresh mass, a 26.80% increase of root dry mass and a 37.34% increase of shoot dry mass in relation to control cohorts. The results highlight the significance of utilizing auxins to enhance crop yields (Mir et al. 2020).

13.5.2 Cytokinins

As a precursor of cytokinins (CKs), a class of plant hormones involved in a wide array of physiological and developmental processes, adenine is also essential to plant function. These include modulating apical dominance, enabling apical meristem activity, promoting gametophyte growth, supporting vascular development and triggering bud dormancy release (Kieber and Schaller 2018; Wybouw and De Rybel 2019). Like other phytohormones, cytokinins play an important role in growth stimulation, photosynthesis and enzymatic processes, leading to a positive effect on yield and oilseed crops when applied exogenously. Cytokinins are also crucial for the enhancement of nutrient absorption, transport and utilization, thereby helping to avoid nutrient deficiency stress in oilseed crops. They improve nitrogen and phosphorus nutrient absorption by controlling root growth and gene expression of transporter genes. Cytokinins have a pivotal function in maintaining the proper function of source-sink relationships. This allows efficient delivery of nutrients to vital tissues, including developing seeds.

At 20 days old, safflower seedlings treated with 10 ppm kinetin (Kn) also showed higher means for branches and capitula per plant than those of the control as well (Dholekar et al. 2001). Moreover, Hamayun et al. (2015) reported that soybean plants of 17 days of age exhibited a shoot length improvement over control plants (shoot length) accompanied by an increase in fresh and dry cell mass/plant of the plant treated with Kn at 0.5, 1 and 5 μM. Tarkhorany et al. (2017) indicated that sesame seedlings 30 days of age, treated with 1 M Kn on the leaf surface, increased by 25.44% in height (Tarkhorany et al. 2017).

13.5.3 Gibberellins

The tetracyclic diterpenoid gibberellic acid (GA_3) plays an important role in plant development regulation. In particular, GA_3 is a valuable regulator of a number of physiological and developmental processes (stem elongation, flower initiation/ flowering commitment and pollen development as well as flowering) (Binenbaum et al. 2018; He et al. 2020; Hakla et al. 2021). Gibberellins, together with other phytohormones, exert their regulatory role to maintain growth and mortality homeostasis, as well as stress responses. For instance, GA_3 might be able to counteract the inhibitory action of ABA, a main stress messenger. In other words, growth mechanisms cannot be uniformly absent, even in the case of the complete absence of the nutrient. In stress, oil crops can balance reproductive, vegetative growth and maturation of seed in a way that maximizes net resource allocation via this interaction. Furthermore, GA_3 is also one of the key factors in the optimization of the biochemical and morpho-physiological response of crops in edible oilseeds. In a study involving 40-day-old soybean seedlings, it was found that a 100 ppm GA_3 treatment to the foliage of the seedlings two times at 20-day intervals resulted in a significant increase of leaf area, plant height, branch number and leaf number in comparison with the treatment that received water only (Sarkar et al. 2002). 50 and 100 mg L^{-1} GA_3 applied during the vegetative stage enhanced soybean growth by increasing the leaf area, plant height and stem diameter (Leite et al. 2003). Similarly, Khandagale et al. (2009) reported that the foliar application of sunflowers with 200 ppm GA_3 yielded effects similar to those obtained in mustard with the application of 75 mg L^{-1} GA_3 (Ahmad 2010). Additionally, Hamayun et al. (2010) found that the application of a 5.0 μM GA_3 solution to soybean foliage at the vegetative stage increased shoot elongation and the fresh and dry biomass of both shoots and root dry weight. Azizi et al. (2012) continue to confirm that a 125 ppm solution enhanced the parameters of the growth of the soybean (Azizi et al. 2012). Groundnut plants exhibited notable enhancements in plant height, total dry weight and branch number following treatment with 150 mg L^{-1} GA_3 applied twice (Hasan and Ismail 2018). Subsequently, Kumari et al. (2019) observed that mustard seedlings exposed to 50 ppm GA_3 at 30 days of age exhibited a 30.71% enhancement in plant height, a 30.05% improvement in leaf length and a 15.51% rise in leaf count compared to the control group.

13.5.4 Abscisic Acid (ABA)

Among the many morphological, physiological and biochemical adjustments that plants control, abscisic acid (ABA) is an important sesquiterpenoid hormone that is widely distributed. Yoshida et al. (2019) and Chen et al. (2020) reported its effect on a wide number of key species of agriculture, showing relevance on biomass yield, branching development, leaf area development, stomata closure, leaf senescence and osmotic regulation (Yoshida et al. 2019; Chen et al. 2020). Roots and shoots demonstrate enhanced coordination of systemic signals, resulting in improved resource efficiency. In reaction to nutritional stress, ABA induces the closure of stomata, thereby conserving water and ensuring the continued flow of nutrients to essential tissues. ABA enhances root elongation and lateral root development, thereby improving soil exploration and nutrient absorption. In their study, Abraham et al. (2008) described that seedlings of sesame (*Sesamum indicum* L. leaf) sprayed with 30 day–leafy sprayings of 10 μM ABA resulted in a significant increase root length (139.73% increase than nontreated) and an increase number of leaves (57.85% longer than nontreated) and number of branches per plant (74% increase than nontreated) (Abraham et al. 2008). When compared to the water-sprayed control, soybean seedlings treated with a foliar application of 300 mg L^{-1} ABA twice during the vegetative and blooming phases showed a 26.38% increase in shoot length and a 26.38% improvement in dry weight, as indicated in a different study (Travaglia et al. 2009). Furthermore, Hussain et al. (2014) showed that semidry soybean seedlings treated with 300 mg L^{-1} ABA twice (the first time at the vegetative and the second at the blooming stage), achieved 26.38% increased shoot length and dry weight compared to the control plants that only received water (Hussain et al. 2014). Moreover, Safari et al. (2018) revealed that sesame seeds subjected to 500 ppm ABA treatment before planting resulted in seedlings that exhibited greater weight compared to those in the control group (Safari et al. 2018).

13.5.5 Ethylene

In the pathways of ethylene (ET), e.g., seed germination, seedling/leaf growth, cell division/growth, flowering, senescence or other physiological processes, cell division/growth is regulated (Iqbal et al. 2017; Dubois et al. 2018). In stressful conditions, upregulation of ethylene production can be demonstrated in mature leaves and consequently in the remobilization of intracellular material at the cost of enhanced export of nutrients to the forming embryos. In other words, it is proposed that their oils extracted from crops (e.g., canola oil, soybean oil) should be further treated and oil extractions should be carried out. Results show that all the oilseed crop development, enzyme kinetics, yield and quality are significantly affected by the ET treatment. Under stress, mature leaves increase ethylene production, which breaks down cellular components and improves the transport of nutrients to developing seeds. So, it is suggested that canola and soybean crops should continue to be improved and their oil extracted. Results confirm that ET treatment significantly affects the growth of oilseed, the activity of enzymes, yield and quality (Mir et al. 2009). Further investigation conducted by Devi et al. (2011) revealed that 200 ppm of ethrel (ETR) promoted vegetative growth in soybeans (Devi et al. 2011). In addition, the study by Khan et al. (2019) found that treatment with 200 μl L^{-1} ETH resulted in a 27% increase in the leaf area and a 23% increase in the plant dry mass of 20-day-old mustard seedlings compared to the control treatment. A complementary investigation by Khan et al. (2020) found that ETH spray (200 μl L −1) application at seedling stage significantly increased relative growth rate (RGR) by 16.55% and plant dry mass (DM) by 6.33% in mustard seedlings over water treatment (control) during vegetative phase (Khan et al. 2020). Moreover, it was observed that when mustard seedlings at the vegetative stage were treated with 200 μL L^{-1} ETH, leaf area and plant weight increased by 27.91% and 2.89%, respectively (Fatma et al. 2021).

13.5.6 Jasmonic Acid and Salicylic Acid

Based on glycolipids, jasmonic acids (JAs) mainly regulate the various morphological and physio-biochemical responses. Edible oilseed crops show substantial improvements in seedling growth, increased enzymatic activity, higher photosynthetic efficiency, more chlorophyll and carotenoid and enhanced nutrient uptake and physiological performance along with improved stress tolerance (Huang et al. 2017; Li et al. 2018). In times of nutrient deficiency, jasmonic acid increases nutrient acquisition by radically altering root architecture. JA interaction with different hormones, including auxins and ethylene, plays an extremely key role in lateral roots and root hair development, thereby playing a major role in modulating root growth. These modifications to increase their root surface area also help these oilseed crops (like canola, sunflower and soybean) to absorb more soil minerals, including phosphorus and potassium. Yoon et al. (2009) found that in soybean seedlings, the root fresh weight was increased by 8.79% (compared to water-sprayed control) when treated with 20 μM methyl jasmonate (MeJA) and the root dry weight was increased by 0.46%, compared to water-sprayed control when treated with 30 μM MeJA (Yoon et al. 2009). Per et al. (2016) found that a 10 μM (1.5 mL per 1 L) MeJA foliar spray to 15-d-old mustard seedlings resulted in 2.46% more dry mass (Per et al. 2016).

Salicylic acid (SA) is a crucial regulator of a diverse range of biochemical and morpho-physiological processes in plants. Pluhařová et al. (2019) and Shaki et al. (2020) stated that key crops, particularly oilseed crops, are subject to numerous stages of changes, including seedling development, stomata regulation, leaf senescence, flowering and changes in the activity of various enzymes (Pluhařová et al. 2019; Shaki et al. 2020). In comparison, soybean seedlings treated with two 10^{-5} M SA solutions showed a significant increase of 26.98%, 22.73%, 0.49% and 0.22% in plant height, root length, shoot and root dry mass and total plant mass, respectively, compared with the control group receiving H_2O (Khan et al. 2003). This difference was reflected in the increase in dry mass of plants in treated seedlings compared to untreated ones, as the application of foliar solution of 10^{-5} M SA increased 2.39% of plant dry mass of 60-day-old mustard seedlings (Fariduddin et al. 2003). Moreover, Hayat et al. (2009) showed that 7-day-old mustard seedlings were given three applications of 10^{-5} M SA by the foliar method. Consequently, a significant improvement relative to the control group, namely a 13% improvement in root length, a 14.7% improvement in shoot length, a 14% improvement in plant fresh mass and a 35.5% improvement in plant dry mass. Further groundnut seeds presoaked with SA could increase plant growth traits at 5, 10, 25 and 50 ppm in a study by Jadhav and Bhamburdekar (2011). SA plays an important role in root development under nutrient deprivation stress. Acting upon SA contributes to the patterns of root development, allowing the plant to be more absorbent for nutrients. This increases the efficiency of nutrient usage and recycling. The role of phytohormones in enhancing oilseed crop tolerance to nutrient deficiency stress is summarized in Table 13.1, highlighting their impact on nutrient uptake, stress adaptation and overall plant resilience.

13.6 Molecular Mechanisms of Phytohormone Action in Nutrient Stress

The involvement of phytohormones is key in the molecular mechanisms that contribute to the ability of oilseed crops to cope with nutrient stress, which intricately orchestrates the crosstalk between signaling pathways, hormonal regulation and upregulation of transporters of the nutrients. Hormonal homeostasis is crucial for plants to respond to stresses, including auxins, cytokinins, abscisic acid (ABA), jasmonic acid (JA), salicylic acid (SA), ethylene and gibberellins, play important roles in regulating root architecture, nutrient uptake, root development, oxidative stresses and resource allocation. Collectively, these molecular mechanisms coordinate the optimized nutrient acquisition, resource reallocation and productivity, providing avenues to enhance oilseed crop resilience.

13.6.1 Hormonal Regulation of Root Architecture and Growth Under Nutrient Deficiency

When nutrients are limited, plants tend to reduce the amount of nutrients they require, and one of the earliest responses is often to suppress new shoot growth (Shin 2011). The decrease in inhibition of root growth leads to an increase in the root/shoot ratio, which allows the roots to better satisfy the demand for the depleted

Table 13.1 Phytohormones as key players in boosting oilseed crop tolerance to nutrient deficiency stress

Phytohormone	Associated oilseed crops	Role in nutrient stress tolerance	Key nutrients involved	Impact on oilseed crops	References
Auxins	Soybean, rapeseed, sunflower	Enhances root development, improving nutrient uptake efficiency	Nitrogen, phosphorus	Promotes root growth, leading to better nutrient acquisition	Ashfaq and Khan (2017)
Gibberellins	Peanut, cottonseed, corn	Stimulates seed germination and growth, aiding in nutrient mobilization	Potassium, calcium	Enhances seedling vigor, supporting early development under nutrient stress	Nowicka (2022)
Cytokinins	Sesame, flaxseed, mustard	Delays leaf senescence, maintaining nutrient remobilization	Nitrogen, zinc	Sustains photosynthetic activity, improving yield under nutrient deficiency	Raza et al. (2022)
Abscisic acid	Safflower, castor bean, palm kernel	Regulates stomatal closure, conserving water and nutrients	Sodium, chloride	Enhances drought tolerance, indirectly supporting nutrient uptake	Swain et al. (2023)
Ethylene	Rice bran, jatropha, camelina	Modulates stress-responsive genes, aiding in nutrient assimilation	Iron, sulfur	Improves stress resilience, supporting nutrient uptake mechanisms	Chen and Pang (2023)
Jasmonic acid	Nigella seed, pumpkin seed, hemp seed	Activates defense mechanisms, enhancing nutrient transport	Magnesium, boron	Improves stress tolerance, facilitating better nutrient utilization	Angon et al. (2024)
Salicylic acid	Evening primrose, Perilla seed, Tung oil tree	Enhances antioxidant defenses, mitigating oxidative stress	Copper, zinc	Boosts immunity, supporting overall plant health under nutrient stress	Sampedro-Guerrero et al. (2023)
Brassinosteroids	Moringa seed, Niger seed, neem	Promotes cell elongation, improving nutrient absorption	Phosphorus, magnesium	Enhances growth, leading to better nutrient uptake	Ashfaq and Khan (2017)

(continued)

Table 13.1 (continued)

Phytohormone	Associated oilseed crops	Role in nutrient stress tolerance	Key nutrients involved	Impact on oilseed crops	References
Strigolactones	Pongamia, babassu, mahua	Stimulates root hair elongation, enhancing phosphorus uptake	Phosphorus, nitrogen	Improves root architecture, facilitating nutrient acquisition	Nowicka (2022)
Melatonin	Sal seed, avocado, grape seed	Functions as an antioxidant, regulating stress-responsive genes	Zinc, copper	Mitigates oxidative stress, supporting plant metabolism under nutrient deficiency	Raza et al. (2022)

nutrient element (Marzec and Melzer 2018). When a sequel (of any nutrient) is detected, the plant activates physiological responses characterized by changes to its physiology, particularly at the root level, to mobilize and increase nutrient uptake. The most common physiological response is an increase in the ability of the plant to mobilize the deficient nutrient. Hormones play a critical function in modulating the activation and repression of these responses, particularly under conditions of nutrient limitation in plants. These two hormones are important in mediating many morphological and physiological responses to individual nutrient deficiencies (Lucena et al. 2015; Song and Liu 2015). Nutrient deficiencies often adapt root system architecture (RSA) to enhance the ability of roots to find nutrients (Wild et al. 2016; Lei et al. 2014). Nutrient deficiencies often change root system architecture (RSA) to enhance the ability of the roots to find nutrients (López-Bucio et al. 2019; Koprivova and Kopriva 2016). Czarnecki et al. (2013) and Marzec and Melzer (2018) showed that it is possible to control nutrient demand through shoot growth and branching manipulation (Czarnecki et al. 2013; Marzec and Melzer 2018). García et al. (2015) observed that nutrient inequality often influences the elongation of the tap root and the development of lateral roots (García et al. 2015). Developmental changes in root system architecture due to nutrient deprivation are mediated by several hormones, including auxin, ethylene, cytokinins, gibberellins, abscisic acid and strigolactones (López-Bucio et al. 2019; Marzec and Melzer 2018). Ethylene and abscisic acid are often associated with diminished root growth (Street et al. 2015; Vaseva et al. 2018), whereas auxins, cytokinins and strigolactones (SLs) contribute to lateral root formation (Jing and Strader 2019). Their interactions with other signaling molecules are complex (Kolbert 2019). In addition, under high iron conditions (Li et al. 2015) or elevated nitrate concentrations (Krouk 2016), the hormones respond via auxin, ethylene and cytokinins to changes in root system architecture and shoot adaptation (Krouk 2016).

13.6.2 Phytohormone-Induced Activation of Nutrient Transporters

Identifying nutritional deficits early is crucial for plants to respond appropriately to the deficiencies. By establishing a solid groundwork of knowledge, it becomes much easier to respond appropriately and reduce hazards. Plants have complex sensing systems that allow them to respond in a variety of ways before showing obvious symptoms of a deficiency (Khan et al. 2018). Sensors are also capable of detecting nutrients, peptides, amino acids, microRNAs, nutritional deficits and surplus (Zhang et al. 2018). Oilseed crops depend on the endogenous phytohormones for the expression of nutrient transporters toward nutrient availability under nutrient-deficient stages, for the effective utilization of nutrients. Proteins known as nutrient transporters facilitate the uptake of these vital elements—potassium, phosphorus and nitrogen by plant cells. The sensor can be placed on the surface of roots (Zheng et al. 2014; Li et al. 2017), or embedded in the shoots or roots (Khan et al. 2018). Transceptors, a specific class of membrane proteins, are also involved in signaling and nutrient transport (Dubeaux et al. 2018). There are two epidermal nitrate transporters, NRT2.1 and NRT1.1, potential for nitrate transport and sensing and acting as local sensors (Noguero and Lacombe 2016). Auxin transport by NRT1. 1 further confirms that this transporter behaves as a transceptor, engaged in nitrogen signaling (Gojon et al. 2011). IRTF1 is an epidermal iron transporter and cytoplasmic non-iron metal sensor that protects plants from reactive metals and facilitates better iron absorption (Dubeaux et al. 2018). IRT1 would therefore function as a transceptor (Dubeaux et al. 2018). Beyond the previously mentioned, SULTR1;2 that senses sulfur (Zheng et al. 2014) as well as AKT1 senses potassium (Li et al. 2017), another example of this class of nutrient sensor that modulates the function of the root surface. The internal sensors are represented by HRZs/BTS proteins, which play a role in the interactions with iron or iron-associated elements to modulate the reaction to iron deficiency and change the response to both iron and iron-associated compounds (Kobayashi and Nishizawa 2015). In addition, Puga et al. (2014) argue that SPX1 inhibits PHR1 DNA binding, which is a central regulator of the phosphate deprivation responses (Puga et al. 2014). Auxin is required for the activation of genes encoding some of these high-affinity nutrient transporters, e.g., IRT1 under iron starvation and NRT2 in response to a lack of nitrogen. Auxin is important for the activation of genes that encode certain high-affinity nutrient transporters, such as IRT1 in response to iron deficit and NRT2.

13.6.3 Signaling Pathways Involved in Phytohormonal Response to Nutrient Stress

In response to different nutritional deficits, both the plant and its hormones change their production and/or their reactivity (Song and Liu 2015; Marzec and Melzer

2018). Depending on which nutrients are deficient, the alterations that are seen are associated with changes in the expression of genes that are involved in hormone production and signaling. According to García et al. (2015), nutrient deficiencies lead to changes in gene expression involved in hormone production and signaling (Romera et al. 2017; Zeng et al. 2019). As noted by García et al. (2015), specific insufficiencies may lead to hormone levels that alter the response to other nutrient deficits via different signal transduction pathways. In nutrient-deficient environments, a hormone can mediate several responses (structural and functional changes) through a variety of signaling mechanisms (García et al. 2010). Some hormones trigger responses to nutrient deficits, while others inhibit them (Romera et al. 2017; Cui et al. 2018). Information on complex regulatory networks controlling responses to nutrient deficiency often involves multiple hormones, which makes it difficult to alter a single hormone in the system. Moreover, hormones can influence a wide range of target tissues via complex networks among themselves and with other signaling molecules, further complicating their interactions (Krouk 2016; Guan 2017). The manufacture, transport and action of other hormones may be influenced by certain hormones. In particular, auxin may enhance ethylene production, whereas ethylene can simultaneously influence auxin accumulation, polar auxin transport and the transcription of AUX/IAA and ARF genes (Romera et al. 2011).

13.7 Strategies for Enhancing Nutrient Deficiency Tolerance in Oilseed Crops

Improving nutrient deficiency tolerance in oilseed crops would require a combination of genetic, agronomic and biotechnological interventions. Genetic improvement by traditional breeding and molecular approaches can make crop varieties more efficient in nutrient uptake and tolerant to stress (Savadi et al. 2017). Deep root systems, higher root hair density and enhanced efficiency of nutrient transporter activity are traits that can be targeted to enhance nutrient acquisition under deficiency conditions. Key agronomic practices to sustain soil fertility and provide balanced nutrition to oilseed crops include soil testing, appropriate nutrient application and the use of organic amendments. Additionally, the introduction of biofertilizers that comprise nutrient-solubilizing microorganisms (NPS), phosphate-solubilizing bacteria (PSB) and mycorrhizal fungi (MF) should also be highlighted, as this approach can significantly contribute to increasing the availability of nutrients and uptake. Alternatively, "biotechnological" interventions including both the introduction of foreign genes into a native genome through "genetic engineering" and genome editing enable advanced solutions by incorporating or altering genes involved in nutrient acquisition, coordination and stress resistance. Some recent advances related to transgenic approaches and improved nutrient-use efficiency for oilseeds as an example include the overexpression of genes encoding nutrient transporters or for phytohormone biosynthesis pathways, shown to facilitate

Table 13.2 Strategies for enhancing nutrient deficiency tolerance

Strategy	Key features	Expected outcome
Genetic breeding	Developing nutrient-efficient varieties	Higher yield with improved nutrient-use efficiency
Biotechnological approaches	CRISPR-Cas9, gene editing for nutrient uptake improvement	Enhanced stress resistance and better nutrient uptake
Agronomic practices	Soil testing, balanced fertilizer application	Optimized soil fertility and reduced nutrient waste
Microbial biofertilizers	Use of phosphate-solubilizing bacteria, mycorrhizal fungi	Increased nutrient availability and uptake efficiency
Exogenous phytohormone application	Spraying auxins, gibberellins and cytokinins to enhance growth	Improved crop vigor and stress tolerance
Precision agriculture	Use of AI and IoT sensors for targeted nutrient application	Precision nutrient management and higher sustainability

the nutrient acquisition and stress resilience. Moreover, applying phytohormones such as auxins, cytokinins, gibberellins and salicylic acid exogenously is adopted to modulate plant growth and cope with the effects of nutrient stress (Ku and Ha 2020). For example, applications of omics technologies, genome editing (such as CRISPR/Cas9) and other advanced approaches now allow for precise identification and subsequent manipulation of genes associated with nutrient stress responses, leading to improved oilseed crop varieties with enhanced resiliency. These strategies need to be integrated into sustainable farming practices so that nutrient deficiencies in different agro-climatic conditions could be addressed without compromising productivity or resource use efficiency (Abawari et al. 2020). Strategies for enhancing nutrient deficiency tolerance are shown in Table 13.2.

13.7.1 Phytohormone-Based Approaches in Nutrient Management

In nutrient management with the phytohormone-based approaches, many plant hormones, including auxins, cytokinins, gibberellins, abscisic acid and ethylene, play essential roles in regulating nutrient uptake and distribution and the stress response. Auxins stimulate root elongation, lateral root formation and root hair development, which increase the root surface area and facilitate nutrient uptake, especially in the case of nitrogen and phosphorus deficiencies. It also controls the expression of genes of nutrient transporters to enhance nutrient efficiency (Choudhary et al. 2021). Cytokinins, produced in roots, are crucial for the redistribution of nutrients by inducing shoot expansion and accelerating chloroplast biogenesis to maintain photosynthetic activity. They play a crucial role in nitrogen signaling and maintaining root-to-shoot growth ratios under nutrient stress.

They influence root architecture and general plant growth, which indirectly enhances nutrient acquisition efficiency by including more surface area. Their interaction with auxins and cytokinins is critical for growth under nutrient-limited conditions (Shaukat et al. 2024). ABA plays a key role in the mediation of stress response and nutrient homeostasis, including changes in root architecture and modulation of stomatal closure which allows water conservation in response to drought stress with a subsequent parallel effect on nutrient uptake. ABA also regulates the expression of genes of specific nutrient transporters (phosphorus and nitrogen uptake transporter genes) under deficiency conditions. Ethylene is a gaseous hormone that mediates root hair development and alters root growth patterns, thereby increasing nutrient acquisition in nutrient-poor habitats. For example, having too much ethylene can prevent roots from elongating, demonstrating the need for a balanced hormone response. These phytohormones are dynamically cross-interactive, forming intricate signaling networks that allow plants to cope with nutrient availability. Root-to-shoot ratios highly depend on the concentration of auxin and cytokinin, which therefore greatly influence the exploration of the roots through soil to search for nutrients. Likewise, interactions between ABA and ethylene modulate the root growth responses to nutrient deficiencies (Chaffai et al. 2024). These hormonal pathways are explored through biotechnological and agronomic interventions for efficient nutrient use. For example, genetic engineering and genome editing technologies (e.g., CRISPR/Cas9) can be used to modify hormone biosynthesis or hormone signaling pathways to improve nutrient-use efficiency. Immediate benefits in nutrient acquisition and stress resistance can be achieved with exogenous treatment of phytohormones or their analogs. Among them, phytohormone-producing plant growth-promoting rhizobacteria (PGPR) can be applied directly to the soil as a biofertilizer to promote plant nutrient absorption. Consequently, combining this method with precision farming promotes sustainable nutrient management practices for addressing nutrient deficiencies, improving crop productivity and enhancing resilience to environmental stresses (Yang et al. 2024; Pillai et al. 2012).

13.7.2 Combining Fertilizers and Phytohormonal Treatments for Improved Efficiency

Phytohormone-signaling strategies fall under nutrient management approaches that reshape the roles of phytohormones in coordinating nutrient uptake, translocation and stress responses. Auxins stimulate the elongation, formation of lateral roots and development of root hair, thus increasing the surface area of roots and enhancing the absorption of materials, especially when nitrogen and phosphorus deficiencies are present. They also modulate the expression of nutrient transporter genes, driving a more efficient nutrient acquisition (Choudhary et al. 2021). Roots synthesize cytokinins that promote shoot growth and chloroplast growth to allocate nutrients to shoots for photosynthesis. They play a key role in nitrogen signaling and the adjustment of root-to-shoot growth ratios under nutrient stress.

Nutrient acquisition efficiency, particularly in the influence of gibberellins on root architecture and plant growth. They interact with auxins and cytokinins, which are key to the regulation of growth in nutrient-deprived conditions (Shaukat et al. 2024). The role of ABA in physiological stress responses and nutritional accessibility is paramount, as it generates root architecture changes and helps save water through stomatal closure during drought stress, where nutrient acquisition may be indirectly impacted. Under deficiency conditions, ABA also promotes the expression of several genes related to nutrient transporters, including phosphorus and nitrogen. Ethylene, an aerial hormone, regulates root hair formation and alters root growth behavior to improve nutrient uptake in nutrient-poor soils. On the contrary, excess ethylene retards the elongation of roots, emphasizing the need for a balanced hormonal response. These phytohormones act in concert with each other and their actions often create complex signaling networks that allow the plant to respond to nutrient availability. Plant resource allocation is still poorly comprehended, so the auxin-cytokinin balance, as an example, defines how the root-to-shoot ratio in a plant directly influences how good the plant can search for nutrients in the soil. In a similar way, interactions between ABA and ethylene fine-tunes responses of roots to nutrient deficiencies (Chaffai et al. 2024). These hormonal pathways are exploited both via biotechnological and agronomic interventions to manage nutrient application more efficiently. The nutrient-use efficiency can improve through genetic engineering or application of genome editing technologies like CRISPR/Cas9 to change hormone biosynthesis or signaling pathways. Supplementing with phytohormones or hormone analogs for exogenous use has immediate advantages for nutrient uptake and stress tolerance.

13.7.3 Breeding and Biotechnology Approaches for Enhanced Phytohormone Sensitivity

Breeding and biotechnological approaches to increase plants' sensitivities to phytohormones stand as an encouraging strategy that aims to enhance their growth, development and stress responses. Phytohormones play vital roles in regulating the physiological functioning of plants. This approach could allow for the improvement of agronomic traits more efficiently by boosting plant responsiveness to these hormones. To obtain different cultivars of a species responsive to a specific phytohormone, classic breeding approaches have been used. This includes screening germplasm collections for natural variants with favorable hormone sensitivity phenotypes. For example, increasing sensitivity to gibberellins as a target trait can create stem elongation and biomass. However, conventional breeding is usually a lengthy process and restricted to the genetic variation available within the species.

Biotechnology allows these phytohormone sensitivities to be manipulated with increased precision and seedlings to be selected that have improved traits. Higher levels of phytohormone receptors allow a better sensitivity to detect the presence in the plant. Overexpression of TIR1, a receptor part of the auxin signaling pathway, for example, increases sensitivity to auxin and enhances root architecture and nutrient uptake (such as GAI, a repressor in the gibberellin pathway). Modifying the gene related to hormone signaling transduction pathways, it has the potential to increase the gibberellin sensitivity of growth-related traits. Subcellular compartmentalization, passive diffusion or active transport via engineered transporters, such as PIN proteins for auxins or ABC transporters for abscisic acid, facilitates diffusion of the signaling molecules throughout the plant enhancing responsiveness. Specifically, CRISPR/Cas9 enables targeted and systematic alterations to important genes controlling hormone sensitivity. A loss of function in IAA7/AXR2 (negative regulators in the auxin pathway) will increase auxin responsiveness. To improve stress responses, it is important to increase the expression of positive regulators such as DREB (a transcription factor for ABA signaling) by altering promoter regions. Technologies involved in hormone signaling, enabling more focused interventions, study proteins responsive to hormones for identifying modification candidates, examines the role of phytohormones on metabolic pathways and suggests ways to increase sensitivity. RNAi can be used to silence negative regulators of hormone pathways to potentiate sensitivity to hormones. As an example, silenced transforming ethylene response inhibitor genes can improve stress tolerance. Synthetic phytohormones or analogs are designed to better bind to hormone receptors, effectively enhancing sensitivity. This allows for the amplification of plant responses by creating synthetic signaling networks that mimic the action of native hormone pathways, with the potential for more rapid and efficient downstream events (Pillai et al. 2012).

13.8 Future Perspectives

The integration of advanced genetic, agronomic and biotechnological strategies holds significant potential for enhancing the resilience of oilseed crops to nutrient deficiency stress. Future research directions and practical implementations include:

13.8.1 Precision Agriculture Integration

Leveraging precision agriculture tools, such as nutrient sensors, GPS-guided applications and data analytics, can optimize the application of fertilizers and phytohormonal treatments. This ensures targeted nutrient delivery, reducing wastage and environmental impact while improving nutrient-use efficiency.

13.8.2 CRISPR-Cas9 and Omics Technologies

Expanding the use of genome editing tools like CRISPR-Cas9 to precisely modify genes associated with nutrient transporter efficiency and hormone sensitivity is vital. Omics technologies, including transcriptomics, proteomics and metabolomics, can further identify key regulatory pathways and molecular targets for nutrient management in oilseed crops.

13.8.3 Development of Biofortified Varieties

Breeding and biotechnological advancements can focus on developing biofortified oilseed crop varieties with enhanced micronutrient content. This improves human nutrition while addressing nutrient stress challenges in agriculture.

13.8.4 Synergistic Use of Microbial Biofertilizers and Phytohormones

Future studies can explore the coapplication of biofertilizers containing nutrient-solubilizing microorganisms and phytohormones to synergistically enhance nutrient uptake, root development and stress tolerance.

13.8.5 Exogenous Phytohormonal Formulations

Designing cost-effective and environmentally friendly phytohormone formulations for exogenous applications, tailored to specific nutrient deficiencies and growth stages, can enhance crop productivity under stress conditions.

13.8.6 Climate-Resilient Oilseed Varieties

With climate change exacerbating nutrient stress, developing oilseed crop varieties with dual stress tolerance—nutrient deficiencies and abiotic factors like drought and salinity—is critical. This requires integrating advanced breeding techniques with high-throughput phenotyping.

13.8.7 Understanding Hormonal Crosstalk

Future research should focus on unraveling the complex hormonal crosstalk under nutrient-stress conditions. This includes interactions between auxins, cytokinins, ABA and ethylene, which regulate root architecture, nutrient acquisition and stress responses.

13.8.8 Policy Support and Sustainable Practices

Promoting policies that encourage sustainable farming practices and incentivize the adoption of nutrient-efficient and phytohormone-sensitive crop varieties will support long-term agricultural sustainability.

13.8.9 Artificial Intelligence (AI) and Machine Learning (ML)

Integrating AI and ML tools for predictive modeling of nutrient stress responses and optimization of phytohormone applications can revolutionize nutrient management strategies.

13.8.10 Educational Outreach and Farmer Training

Disseminating knowledge about advanced nutrient management practices, including the use of phytohormones and biofertilizers, can empower farmers to adopt these innovations effectively.

13.9 Conclusion

Effective management of nutrient deficiency stress in oilseed plants is critical for ensuring global food security and sustainable agriculture. Oilseed crops play an important role as a source of food oils, industrial materials and animal feed, with important implications for nutrition and economics. Phytohormones play the roles of highly influential regulators, conditioning plant growth and survival in conditions of limited nutrition by influencing root architecture, activity of a nutrient transporter and stress resistance. The use of advanced technologies (precision agriculture, genetic engineering and microbial biofertilizer) considerably increases the utilization efficiency of nutrients in oilseed crops. Technological advances such as CRISPR/Cas9 and omics technology tools offer precise tools to improve traits in crops, in addition to chemicals such as fertilizers and phytohormones, for which the uptake and accessibility of nutrients are optimized. In addition, the development of biofortified and climate-resistant genotypes ensures that oilseed crops will be able to withstand environmental and nutritional constraints. A multidisciplinary approach involving politicians, researchers and farmers is crucial to accomplish such goals. Policy support, in combination with information dissemination and farmer education, potentially contributes to sustainable practice adoption. Artificial intelligence and machine learning have the potential to increase both decision making and resource use in agriculture. The use of phytohormonal approaches, biotechnological developments and sustainable agricultural practices represent a pragmatic option for mitigating nutrient stress in oilseed crops. This integrated approach will contribute not only to increased crop yield and quality but also to a resilient and sustainable agricultural ecosystem.

References

Abawari RA, Tuji FA, Yadete DM (2020) Phosphate solubilizing bio-fertilizers and their role in bio-available P nutrient: an overview. Int J Appl Agric Sci 6(6):162

Abiodun OA (2017) The role of oilseed crops in human diet and industrial use. In: Ahmad P (ed) Oilseed crops, 1st edn. Wiley, pp 249–263. https://doi.org/10.1002/9781119048800.ch14

Abraham SS, Jaleel CA, Chang-Xing Z, Somasundaram R, Azooz MM, Manivannan P, Panneerselvam R (2008) Regulation of growth and metabolism by paclobutrazol and ABA in Sesamum indicum L. under drought condition. Global J Mol Sci 3(2):57–66

Ahmad P (2010) Growth and antioxidant responses in mustard (*Brassica juncea* L.) plants subjected to combined effect of gibberellic acid and salinity. Arch Agron Soil Sci 56(5): 575–588. https://doi.org/10.1080/03650340903164231

Ahmad A, Hayat S, Fariduddin Q, Ahmad I (2001) Photosynthetic efficiency of plants of Brassica juncea, treated with chlorosubstituted auxins. Photosynthetica 39(4):565–568

Ali N, Réthoré E, Yvin J-C, Hosseini SA (2020) The regulatory role of silicon in mitigating plant nutritional stresses. Plan Theory 9(12):1779

Angon PB, Das A, Roy AR, Khan JJ, Ahmad I, Biswas A, Pallob AT, Mondol M, Yeasmin ST (2024) Plant development and heat stress: role of exogenous nutrients and phytohormones in thermotolerance. Discover Plants 1(1):17. https://doi.org/10.1007/s44372-024-00020-3

Ashfaq M, Khan S (2017) Role of phytohormones in improving the yield of oilseed crops. Oilseed Crops: Yield and Adaptations under Environmental Stress, pp 165–183

Attia Z, Pogoda CS, Reinert S, Kane NC, Hulke BS (2021) Breeding for sustainable oilseed crop yield and quality in a changing climate. Theor Appl Genet 134(6):1817–1827

Azizi, K., Moradii, J., Heidari, S., Khalili, A., Feizian, M. (2012) Effect of different concentrations of gibberellic acid on seed yield and yield components of soybean genotypes in summer intercropping. https://www.cabidigitallibrary.org/doi/full/10.5555/20123239835

Bera AK, Pramanik K, Mandal B (2014) Response of biofertilizers and homo-brassinolide on growth, yield and oil content of sunflower (Helianthus annuus L.). Afr J Agric Res 9(48): 3494–3503

Binenbaum J, Weinstain R, Shani E (2018) Gibberellin localization and transport in plants. Trends Plant Sci 23(5):410–421

Bouain N, Krouk G, Lacombe B, Rouached H (2019) Getting to the root of plant mineral nutrition: combinatorial nutrient stresses reveal emergent properties. Trends Plant Sci 24(6):542–552
Chaffai R, Ganesan M, Cherif A (2024) Plant hormones and their regulatory roles in abiotic stress response. In: Plant adaptation to abiotic stress: from signaling pathways and microbiomes to molecular mechanisms. Springer, pp 249–274
Chen J, Pang X (2023) Phytohormones unlocking their potential role in tolerance of vegetable crops under drought and salinity stresses. Front Plant Sci 14:1121780
Chen K, Li G, Bressan RA, Song C, Zhu J, Zhao Y (2020) Abscisic acid dynamics, signaling, and functions in plants. J Integr Plant Biol 62(1):25–54. https://doi.org/10.1111/jipb.12899
Chen L, Hu W, Long C, Wang D (2021) Exogenous plant growth regulator alleviate the adverse effects of U and Cd stress in sunflower (Helianthus annuus L.) and improve the efficacy of U and Cd remediation. Chemosphere 262:127809
Choudhary K, Vivekanand V, Pareek N (2021) Phytohormones as fundamental regulators of plant–microbe associations under stress conditions. Plant, Soil and Microbes in Tropical Ecosystems 203–226
Cui Y, Chen C-L, Cui M, Zhou W-J, Wu H-L, Ling H-Q (2018) Four IVa bHLH transcription factors are novel interactors of FIT and mediate JA inhibition of iron uptake in Arabidopsis. Mol Plant 11(9):1166–1183
Czarnecki O, Yang J, Weston DJ, Tuskan GA, Chen J-G (2013) A dual role of strigolactones in phosphate acquisition and utilization in plants. Int J Mol Sci 14(4):7681–7701
Debona D, Rodrigues FA, Datnoff LE (2017) Silicon's role in abiotic and biotic plant stresses. Annu Rev Phytopathol 55(1):85–107
Devi KN, Vyas AK, Singh MS, Singh NG (2011) Effect of bioregulators on growth, yield and chemical constituents of soybean (Glycine max). J Agric Sci 3(4):151
Dholekar PO, Patil BN, Shivankar RS (2001) Effect of foliar spray of different growth regulators on yield and yield attributes of safflower. Agric Sci Dig 21(4):241–243
Dubeaux G, Neveu J, Zelazny E, Vert G (2018) Metal sensing by the IRT1 transporter-receptor orchestrates its own degradation and plant metal nutrition. Mol Cell 69(6):953–964
Dubois M, Van den Broeck L, Inzé D (2018) The pivotal role of ethylene in plant growth. Trends Plant Sci 23(4):311–323
Fariduddin Q, Hayat S, Ahmad A (2003) Salicylic acid influences net photosynthetic rate, carboxylation efficiency, nitrate reductase activity, and seed yield in Brassica juncea. Photosynthetica 41:281–284
Fatma M, Iqbal N, Gautam H, Sehar Z, Sofo A, D'Ippolito I, Khan NA (2021) Ethylene and sulfur coordinately modulate the antioxidant system and ABA accumulation in mustard plants under salt stress. Plan Theory 10(1):180
García MJ, Lucena C, Romera FJ, Alcantara E, Pérez-Vicente R (2010) Ethylene and nitric oxide involvement in the up-regulation of key genes related to iron acquisition and homeostasis in Arabidopsis. J Exp Bot 61(14):3885–3899
García MJ, Romera FJ, Lucena C, Alcántara E, Pérez-Vicente R (2015) Ethylene and the regulation of physiological and morphological responses to nutrient deficiencies. Plant Physiol 169(1): 51–60
Gojon A, Krouk G, Perrine-Walker F, Laugier E (2011) Nitrate transceptor (s) in plants. J Exp Bot 62(7):2299–2308
Guan P (2017) Dancing with hormones: a current perspective of nitrate signaling and regulation in Arabidopsis. Front Plant Sci 8:1697
Gunstone FD, Harwood JL, Dijkstra AJ (2007) The lipid handbook with CD-ROM. CRC Press. https://books.google.com/books?hl=en&lr=&id=INZa6WmqDA8C&oi=fnd&pg=PP1&dq=Gunstone+FD,+Harwood+JL,+Dijkstra+AJ+(2007)+The+lipid+handbook+with+CD-ROM.+CRC+Press,+Florida&ots=Sw4A1d9Cu3&sig=m2elFYtmxoHU8Qw9PDH_oWt5JRI
Gupta R, Verma N, Tewari RK (2024) Micronutrient deficiency-induced oxidative stress in plants. Plant Cell Rep 43(9):213

Hakla HR, Sharma S, Urfan M, Yadav NS, Rajput P, Kotwal D, Abdel Latef AA, Pal S (2021) Gibberellins target shoot-root growth, Morpho-physiological and molecular pathways to induce cadmium tolerance in *Vigna radiata* L. Agronomy 11:896

Hamayun M, Khan SA, Khan AL, Shin J-H, Ahmad B, Shin D-H, Lee I-J (2010) Exogenous gibberellic acid reprograms soybean to higher growth and salt stress tolerance. J Agric Food Chem 58(12):7226–7232. https://doi.org/10.1021/jf101221t

Hamayun M, Hussain A, Khan SA, Irshad M, Khan AL, Waqas M, Shahzad R, Iqbal A, Ullah N, Rehman G (2015) Kinetin modulates physio-hormonal attributes and isoflavone contents of soybean grown under salinity stress. Front Plant Sci 6:377

Hasan M, Ismail BS (2018) Effect of gibberellic acid on the growth and yield of groundnut (Arachis hypogaea L.). Sains Malaysiana 47(2):221–225

Hayat S, Masood A, Yusuf M, Fariduddin Q, Ahmad A (2009) Growth of Indian mustard (Brassica juncea L.) in response to salicylic acid under high-temperature stress. Braz J Plant Physiol 21: 187–195

He J, Xin P, Ma X, Chu J, Wang G (2020) Gibberellin metabolism in flowering plants: an update and perspectives. Front Plant Sci 11:532

Huang H, Liu B, Liu L, Song S (2017) Jasmonate action in plant growth and development. J Exp Bot 68(6):1349–1359

Hussain S, Saleem MF, Iqbal J, Ibrahim M, Atta S, Ahmed T, Rehmani MIA (2014) Exogenous application of abscisic acid may improve the growth and yield of sunflower hybrids under drought. Pak J Agric Sci 51(1) https://www.pakjas.com.pk/papers/2242.pdf

Iqbal N, Umar S, Khan NA, Khan MIR (2014) A new perspective of phytohormones in salinity tolerance: regulation of proline metabolism. Environ Exp Bot 100:34–42

Iqbal N, Khan NA, Ferrante A, Trivellini A, Francini A, Khan MIR (2017) Ethylene role in plant growth, development and senescence: interaction with other phytohormones. Front Plant Sci 8: 475

Jadhav SH, Bhamburdekar SB (2011) Effect of salicylic acid on germination performance in groundnut. Int J Appl Biol Pharm Technol 2(4):224–227

Jaradat AA (2016) Breeding oilseed crops for climate change. In: Breeding oilseed crops for sustainable production. Elsevier, pp 421–472. https://www.sciencedirect.com/science/article/pii/B9780128013090000185

Jing H, Strader LC (2019) Interplay of auxin and cytokinin in lateral root development. Int J Mol Sci 20(3):486

Khan W, Prithiviraj B, Smith DL (2003) Photosynthetic responses of corn and soybean to foliar application of salicylates. J Plant Physiol 160(5):485–492

Khan MA, Castro-Guerrero NA, McInturf SA, Nguyen NT, Dame AN, Wang J, Bindbeutel RK, Joshi T, Jurisson SS, Nusinow DA (2018) Changes in iron availability in Arabidopsis are rapidly sensed in the leaf vasculature and impaired sensing leads to opposite transcriptional programs in leaves and roots. Plant Cell Environ 41(10):2263–2276

Khan MIR, Jahan B, Alajmi MF, Rehman MT, Khan NA (2019) Exogenously-sourced ethylene modulates defense mechanisms and promotes tolerance to zinc stress in mustard (Brassica juncea L.). Plants 8(12):540

Khan MIR, Trivellini A, Chhillar H, Chopra P, Ferrante A, Khan NA, Ismail AM (2020) The significance and functions of ethylene in flooding stress tolerance in plants. Environ Exp Bot 179:104188

Khan I, Awan SA, Rizwan M, Brestic M, Xie W (2023) Silicon: an essential element for plant nutrition and phytohormones signaling mechanism under stressful conditions. Plant Growth Regul 100(2):301–319

Khandagale, G. B., Pawar, G. S., & Shinde, A. B. (2009). Influence of foliar spray of plant growth regulators on physiological parameter and yield in safflower. https://www.cabidigitallibrary.org/doi/full/10.5555/20103165244

Kieber JJ, Schaller GE (2018) Cytokinin signaling in plant development. Development 145(4): dev149344

Kobayashi T, Nishizawa NK (2015) Intracellular iron sensing by the direct binding of iron to regulators. In: Frontiers in plant science, vol 6. Frontiers Media SA, p 155

Kolbert Z (2019) Strigolactone-nitric oxide interplay in plants: the story has just begun. Physiol Plant 165(3):487–497

Koprivova A, Kopriva S (2016) Hormonal control of sulfate uptake and assimilation. Plant Mol Biol 91(6):617–627

Krouk G (2016) Hormones and nitrate: a two-way connection. Plant Mol Biol 91(6):599–606

Ku H-K, Ha S-H (2020) Improving nutritional and functional quality by genome editing of crops: status and perspectives. Front Plant Sci 11:577313

Kulkarni MG, Stirk WA, Southway C, Papenfus HB, Swart PA, Lux A, Vaculík M, Martinka M, Staden JV (2013) Plant growth regulators enhance gold uptake in *Brassica juncea*. Int J Phytoremediation 15(2):117–126. https://doi.org/10.1080/15226514.2012.683207

Kumari K, Singh SK, Mehmi V, Kumar U, Kaur K (2019) Influence of plant growth regulators and biofertilizers on growth and yield of Broccoli (Brassica oleracea L. var. Italica) under central region of Punjab. *Agriways* 7(1):44–49

Lafarga T (2021) Production and consumption of oils and oilseeds. In: Lafarga T, Bobo G, Aguiló-Aguayo I (eds) Oil and oilseed processing, 1st edn. Wiley, pp 1–21. https://doi.org/10.1002/9781119575313.ch1

Lei GJ, Zhu XF, Wang ZW, Dong F, Dong NY, Zheng SJ (2014) Abscisic acid alleviates iron deficiency by promoting root iron reutilization and transport from root to shoot in A rabidopsis. Plant Cell Environ 37(4):852–863

Leite VM, Rosolem CA, Rodrigues JD (2003) Gibberellin and cytokinin effects on soybean growth. Sci Agric 60:537–541

Lewis J (2011) Analysis of the performance of peanut fame in a single cylinder Idi engine and investigations of Neat Methyl Ester influence on fuel quality. https://digitalcommons.georgiasouthern.edu/etd/1041/

Li G, Song H, Li B, Kronzucker HJ, Shi W (2015) Auxin resistant1 and PIN-FORMED2 protect lateral root formation in Arabidopsis under iron stress. Plant Physiol 169(4):2608–2623

Li J, Wu W-H, Wang Y (2017) Potassium channel AKT1 is involved in the auxin-mediated root growth inhibition in Arabidopsis response to low K+ stress. J Integr Plant Biol 59(12):895–909

Li C, Wang P, Menzies NW, Lombi E, Kopittke PM (2018) Effects of methyl jasmonate on plant growth and leaf properties. J Plant Nutr Soil Sci 181(3):409–418. https://doi.org/10.1002/jpln.201700373

List GR (2016) Oilseed composition and modification for health and nutrition. In: Functional dietary lipids. Elsevier, pp 23–46. https://www.sciencedirect.com/science/article/pii/B9781782422471000028

López-Bucio JS, Salmerón-Barrera GJ, Ravelo-Ortega G, Raya-González J, León P, de la Cruz HR, Campos-García J, López-Bucio J, Guevara-García ÁA (2019) Mitogen-activated protein kinase 6 integrates phosphate and iron responses for indeterminate root growth in Arabidopsis thaliana. Planta 250:1177–1189

Lucena C, Romera FJ, García MJ, Alcántara E, Pérez-Vicente R (2015) Ethylene participates in the regulation of Fe deficiency responses in strategy I plants and in rice. Front Plant Sci 6:1056

Lv B, Yan Z, Tian H, Zhang X, Ding Z (2019) Local auxin biosynthesis mediates plant growth and development. Trends Plant Sci 24(1):6–9

Marzec M, Melzer M (2018) Regulation of root development and architecture by strigolactones under optimal and nutrient deficiency conditions. Int J Mol Sci 19(7):1887

Mir MR, Lone NA, Khan NA (2009) Impact of exogenously applied ethephon on physiological and yield attributes of two mustard cultivars under rainfed conditions. Appl Biol Res 11(1):44–46

Mir AR, Siddiqui H, Alam P, Hayat S (2020) Foliar spray of Auxin/IAA modulates photosynthesis, elemental composition, ROS localization and antioxidant machinery to promote growth of Brassica juncea. Physiol Mol Biol Plants 26(12):2503–2520

Mishra S, Levengood H, Fan J, Zhang C (2024) Plants under stress: exploring physiological and molecular responses to nitrogen and phosphorus deficiency. Plan Theory 13(22):3144

Nadeem SM, Ahmad M, Zahir ZA, Kharal MA (2016) Role of phytohormones in stress tolerance of plants. Plant, Soil and Microbes: Volume 2: Mechanisms and Molecular Interactions:385–421
Noguero M, Lacombe B (2016) Transporters involved in root nitrate uptake and sensing by Arabidopsis. Front Plant Sci 7:1391
Nowicka B (2022) Modifications of phytohormone metabolism aimed at stimulation of plant growth, improving their productivity and tolerance to abiotic and biotic stress factors. Plan Theory 11(24):3430
Pandey R, Vengavasi K, Hawkesford MJ (2021) Plant adaptation to nutrient stress. Plant Physiol Reports 26(4):583–586
Patil G, Mian R, Vuong T, Pantalone V, Song Q, Chen P, Shannon GJ, Carter TC, Nguyen HT (2017) Molecular mapping and genomics of soybean seed protein: a review and perspective for the future. Theor Appl Genet 130(10):1975–1991. https://doi.org/10.1007/s00122-017-2955-8
Peng Q, Wang H, Tong J, Kabir MH, Huang Z, Xiao L (2013) Effects of indole-3-acetic acid and auxin transport inhibitor on auxin distribution and development of peanut at pegging stage. Sci Hortic 162:76–81
Per TS, Khan NA, Masood A, Fatma M (2016) Methyl jasmonate alleviates cadmium-induced photosynthetic damages through increased S-assimilation and glutathione production in mustard. Front Plant Sci 7:1933
Pillai BVS, Kagale S, Chellamma S (2012) Enhancing productivity and performance of oil seed crops under environmental stresses. In: Venkateswarlu B, Shanker AK, Shanker C, Maheswari M (eds) Crop stress and its management: perspectives and strategies. Springer, pp 139–161. https://doi.org/10.1007/978-94-007-2220-0_4
Pluhařová K, Leontovyčová H, Stoudková V, Pospíchalová R, Maršík P, Klouček P, Starodubtseva A, Iakovenko O, Krčková Z, Valentová O (2019) "Salicylic acid mutant collection" as a tool to explore the role of salicylic acid in regulation of plant growth under a changing environment. Int J Mol Sci 20(24):6365
Puga MI, Mateos I, Charukesi R, Wang Z, Franco-Zorrilla JM, de Lorenzo L, Irigoyen ML, Masiero S, Bustos R, Rodríguez J (2014) SPX1 is a phosphate-dependent inhibitor of phosphate starvation response 1 in Arabidopsis. Proc Natl Acad Sci 111(41):14947–14952
Rashid U, Anwar F, Knothe G (2009) Evaluation of biodiesel obtained from cottonseed oil. Fuel Process Technol 90(9):1157–1163
Raza A, Salehi H, Rahman MA, Zahid Z, Madadkar Haghjou M, Najafi-Kakavand S, Charagh S, Osman HS, Albaqami M, Zhuang Y (2022) Plant hormones and neurotransmitter interactions mediate antioxidant defenses under induced oxidative stress in plants. Front Plant Sci 13: 961872
Romera FJ, García MJ, Alcántara E, Pérez-Vicente R (2011) Latest findings about the interplay of auxin, ethylene and nitric oxide in the regulation of Fe deficiency responses by strategy I plants. Plant Signal Behav 6(1):167–170
Romera FJ, Lucena C, García MJ, Alcántara E, Pérez-Vicente R (2017) The role of ethylene and other signals in the regulation of Fe deficiency responses by dicot plants. Stress Signaling in Plants: Genomics and Proteomics Perspective 2:277–300
Romera FJ, Lucena C, García MJ, Alcántara E, Angulo M, Aparicio MÁ, Pérez-Vicente R (2021) Plant hormones and nutrient deficiency responses. In: Gupta DK, Corpas FJ (eds) Hormones and plant response, vol 2. Springer International Publishing, pp 29–65. https://doi.org/10.1007/978-3-030-77477-6_2
Safari H, Hosseini SM, Azari A, Rafsanjani MH (2018) Effects of seed priming with ABA and SA on seed germination and seedling growth of sesame (Sesamum indicum L.) under saline condition. Aust J Crop Sci 12(9):1385–1392
Sampedro-Guerrero J, Vives-Peris V, Gomez-Cadenas A, Clausell-Terol C (2023) Efficient strategies for controlled release of nanoencapsulated phytohormones to improve plant stress tolerance. Plant Methods 19(1):47. https://doi.org/10.1186/s13007-023-01025-x
Santos Hansel DS, Schwalbert RA, Shoup DE, Holshouser DL, Parvej R, Prasad PVV, Ciampitti IA (2019) A review of soybean yield when double-cropped after wheat. Agron J 111(2):677–685. https://doi.org/10.2134/agronj2018.06.0371

Sarkar PK, Haque MS, Abdul Karim M (2002) Effects of GA and IAA and their frequency of application on morphology, yield 3. Pak J Agron 1(4):119–122

Savadi S, Lambani N, Kashyap PL, Bisht DS (2017) Genetic engineering approaches to enhance oil content in oilseed crops. Plant Growth Regul 83:207–222

Shaki F, Maboud HE, Niknam V (2020) Differential proteomics: effect of growth regulators on salt stress responses in safflower seedlings. Pestic Biochem Physiol 164:149–155

Shaukat S, Mustafa G, Cheng K (2024) Phytohormones for combating global challenges: an ecofriendly approach. In: Plant growth regulators: resilience for sustainable agriculture. Springer, pp 267–284

Shin R (2011) Transcriptional regulatory components responding to macronutrient limitation. J Plant Biol 54:286–293

Song L, Liu D (2015) Ethylene and plant responses to phosphate deficiency. Front Plant Sci 6:796

Street IH, Aman S, Zubo Y, Ramzan A, Wang X, Shakeel SN, Kieber JJ, Schaller GE (2015) Ethylene inhibits cell proliferation of the Arabidopsis root meristem. Plant Physiol 169(1): 338–350

Swain R, Sahoo S, Behera M, Rout GR (2023) Instigating prevalent abiotic stress resilience in crop by exogenous application of phytohormones and nutrient. Front Plant Sci 14:1104874

Tarkhorany T, Madani H, Sharif Abad HH (2017) Evaluating the effect of kinetin application on sesame cultivars. Scientific Papers. Series A Agronomy 60. https://agronomyjournal.usamv.ro/pdf/2017/Art65.pdf

Travaglia C, Reinoso H, Bottini R (2009) Application of abscisic acid promotes yield in field-cultured soybean by enhancing production of carbohydrates and their allocation in seed. Crop Pasture Sci 60(12):1131–1136

Vander Linden C, Delvaux B (2019) The weathering stage of tropical soils affects the soil-plant cycle of silicon, but depending on land use. Geoderma 351:209–220

Vaseva II, Qudeimat E, Potuschak T, Du Y, Genschik P, Vandenbussche F, Van Der Straeten D (2018) The plant hormone ethylene restricts Arabidopsis growth via the epidermis. Proc Natl Acad Sci 115(17):E4130–E4139

Wild M, Davière J-M, Regnault T, Sakvarelidze-Achard L, Carrera E, Diaz IL, Cayrel A, Dubeaux G, Vert G, Achard P (2016) Tissue-specific regulation of gibberellin signaling fine-tunes Arabidopsis iron-deficiency responses. Dev Cell 37(2):190–200

Wrigley CW, Corke H, Seetharaman K, Faubion J (2015) Encyclopedia of food grains. Academic Press

Wybouw B, De Rybel B (2019) Cytokinin—a developing story. Trends Plant Sci 24(2):177–185

Yang P, Condrich A, Scranton S, Hebner C, Lu L, Ali MA (2024) Utilizing plant growth-promoting Rhizobacteria (PGPR) to advance sustainable agriculture. Bacteria 3(4):434–451

Yoon JY, Hamayun M, Lee S-K, Lee I-J (2009) Methyl jasmonate alleviated salinity stress in soybean. J Crop Sci Biotechnol 12:63–68

Yoshida T, Obata T, Feil R, Lunn JE, Fujita Y, Yamaguchi-Shinozaki K, Fernie AR (2019) The role of abscisic acid signaling in maintaining the metabolic balance required for Arabidopsis growth under nonstress conditions. Plant Cell 31(1):84–105

Zažímalová E, Petrášek J, Benková E (eds) (2014) Auxin and its role in plant development. Springer Vienna. https://doi.org/10.1007/978-3-7091-1526-8

Zeng H, Zhang X, Ding M, Zhang X, Zhu Y (2019) Transcriptome profiles of soybean leaves and roots in response to zinc deficiency. Physiol Plant 167(3):330–351

Zhang L, Li G, Wang M, Di D, Sun L, Kronzucker HJ, Shi W (2018) Excess iron stress reduces root tip zone growth through nitric oxide-mediated repression of potassium homeostasis in Arabidopsis. New Phytol 219(1):259–274

Zhao B, Liu Q, Wang B, Yuan F (2021) Roles of phytohormones and their signaling pathways in leaf development and stress responses. J Agric Food Chem 69(12):3566–3584

Zheng Z-L, Zhang B, Leustek T (2014) Transceptors at the boundary of nutrient transporters and receptors: a new role for Arabidopsis SULTR1; 2 in sulfur sensing. Front Plant Sci 5:710

Chapter 14
Oilseed Crops and Phytohormones Under UV Stress

Sk Naim Aktar, Soumya Shephalika Dash, Sagun Hembram, Siddharth Shankar Patre, Soham Barik, Santanu Kundu, Maksud Hasan Shah, Mazharul Islam, Kalipada Pramanik, Akbar Hossain, and Arafat Abdel Hamed Abdel Latef

Abstract Ultraviolet (UV) radiation is a segment of the electromagnetic spectrum spanning from 10 to 400 nm and is categorized into three primary types: UV-A (320–400 nm), UV-B (280–320 nm), and UV-C (100–280 nm). The UV radiation that penetrates the Earth's atmosphere primarily consists of UV-A (95%) and a smaller proportion of UV-B (5%), as much of the latter is absorbed by the stratospheric ozone layer. This specific wavelength of radiation significantly influences plant growth and development by modifying key organic compounds essential for life. Research has shown that UV-B exposure leads to alterations in plant morphology, physiology, and both primary and secondary metabolism, ultimately impacting

Sk N. N. Aktar · S. Hembram · S. S. Patre · K. Pramanik
Department of Agronomy, Palli Siksha Bhavana (Institute of Agriculture), Birbhum, West Bengal, India

S. S. Dash
Department of Agricultural Entomology, Palli Siksha Bhavana (Institute of Agriculture), Birbhum, West Bengal, India

S. Barik
Department of Agronomy, Maharana Pratap University of Agriculture and Technology, Udaipur, Rajasthan, India

S. Kundu
Division of Agronomy, ICAR-Indian Agricultural Research Institute, New Delhi, Delhi, India

M. H. Shah (✉)
Department of Agronomy, Bidhan Chandra Krishi Viswavidyalaya, Nadia, West Bengal, India

M. Islam
Department of Agronomy, Kanas State University, Manhattan, KS, USA

A. Hossain
Division of Soil Science, Bangladesh Wheat and Maize Research Institute, Dinajpur, Bangladesh

A. A. H. A. Latef
Botany and Microbiology Department, Faculty of Science, South Valley University, Qena, Egypt

© The Author(s), under exclusive license to Springer Nature Singapore Pte Ltd. 2025
A. A. H. Abdel Latef (ed.), *Oilseed Crops Under Abiotic Stress*, Sustainability Sciences in Asia and Africa, https://doi.org/10.1007/978-981-96-8346-8_14

crop yield. However, the potential effects of fluctuating UV-B levels in response to ozone layer changes on agricultural ecosystems remain uncertain. This chapter aims to provide an overview of current findings on how UV radiation alters the major phytohormones and affects the morphological, biochemical, and genetic characteristics of plants, with a particular focus on major oilseed crops.

Keywords Ozone layer · Morphology · Secondary metabolism · Phytohormones · Oilseed crops

14.1 Introduction

Global population growth and the rising demand for food necessitate a significant increase in agricultural productivity. By 2050, it is projected that food production must increase by approximately 70% to support an additional 2.3 billion people (Tian et al. 2021). However, this goal is challenged by several environmental stressors, including ultraviolet (UV) radiation, which is an increasingly critical factor affecting crop growth and productivity. UV radiation accounts for approximately 9% of the solar spectrum and significantly influences the growth, development, and metabolic processes of plants (Dawood et al. 2021; Bais et al. 2019). While the ozone layer filters most harmful UV-C and a substantial portion of UV-B, human activities such as the use of hydrochlorofluorocarbons (HCFCs), halons, methyl bromide, and industrial solvents have led to ozone depletion, allowing increased penetration of UV-B radiation (Egamberdieva et al. 2017). Although UV-B radiation constitutes only a small fraction of sunlight, it significantly impacts plants by inducing oxidative stress, damaging cellular structures, and disrupting essential physiological functions (Carvalho and Castillo 2018). In contrast, UV-A plays a regulatory role in photosynthesis, pigment biosynthesis, and antioxidant metabolism, contributing to overall plant adaptation and response to light conditions (Correa et al. 2023; Barnes et al. 2019).

Oilseed crops, including soybean, sunflower, rapeseed, and linseed, are integral to global agriculture, providing essential edible oils, biofuels, and industrial products. Over the past decade, annual oilseed crops like safflower, sunflower, linseed, niger, and groundnut have shown declining growth in area and production (Mathur et al. 2023). India, the largest vegetable oil importer globally, heavily depends on imports, primarily palm, soybean, and sunflower oils, to meet its rising edible oil demand. Edible oil demand in India is expected to reach 17.84 million tonnes by 2050 (Mittal 2006). Despite their economic and nutritional importance, oilseed crops are highly susceptible to UV radiation, which negatively impacts their physiological and metabolic processes (Shan et al. 2022). It alters biomolecular structures, suppresses photosynthesis, reduces nitrogen assimilation, and diminishes overall biomass, leading to significant yield losses. These adverse effects emphasize the need for innovative strategies to enhance oilseed resilience to UV stress.

Phytohormones, as central regulators of plant growth and development, offer promising avenues for mitigating UV-induced stress. These endogenous signalling molecules coordinate cellular activities, modulate stress responses, and regulate metabolic pathways, enabling plants to adapt to environmental challenges (Ashfaq and Khan 2017). For oilseed crops, phytohormones such as auxins, cytokinins, gibberellins, abscisic acid (ABA), ethylene, and salicylic acid (SA) play critical roles in enhancing tolerance to abiotic stresses, including UV radiation. Auxins influence cell elongation, root development, and reproductive processes, while cytokinins promote cell division and nutrient assimilation. Gibberellins regulate stem elongation and flowering, and ABA mediates stomatal closure and osmotic adjustments under stress (Vishal and Kumar 2018). Ethylene and SA, meanwhile, are involved in antioxidant defence and photo protection (Vanhaelewyn et al. 2016). The manipulation of phytohormonal pathways represents a viable strategy to enhance the UV stress tolerance of oilseed crops. Exogenous applications of phytohormones, genetic engineering approaches, and hormonal priming have shown potential in mitigating oxidative stress, stabilizing cellular structures, and improving crop resilience (Chen et al. 2022). For instance, the application of salicylic acid has been demonstrated to enhance antioxidant enzyme activities and reduce UV-induced DNA damage (Vanhaelewyn et al. 2016). Similarly, jasmonic acid modulates flavonoid biosynthesis and promotes ROS scavenging, while ABA improves cellular water retention and stress signalling (Quan et al. 2018; Choudhary et al. 2021). These findings highlight the multifaceted role of phytohormones in adapting to and mitigating the deleterious effects of UV radiation.

In this chapter, we explore the intricate mechanisms by which phytohormones regulate UV stress responses in oilseed crops. Emphasis is placed on understanding the molecular and physiological interactions between UV radiation, plant hormones, and metabolic pathways. We also discuss potential biotechnological and agronomic interventions to enhance the UV tolerance of oilseeds, ensuring sustainable production amidst climate change-induced challenges. By delving into the roles of phytohormones and leveraging advanced research tools, this study aims to contribute to the development of climate-resilient oilseed varieties that can sustain productivity under adverse environmental conditions.

14.2 Overview of Oilseed Crops and Their Economic Importance

Oilseed crops are an integral part of global agriculture, playing a vital role in food security, industrial applications, and rural livelihoods. These crops, including soybean, sunflower, canola, flax, peanut, cottonseed, coconut, sesame, mustard, and niger seed, serve as primary sources of edible oils while also providing protein-rich byproducts for animal feed. Their economic significance is immense, as they support millions of farmers, drive agribusinesses, and contribute to diverse industries such as food processing, cosmetics, pharmaceuticals, and biofuels. The global oilseed market is witnessing substantial growth, valued at approximately USD 305.16 billion in

2024 and projected to reach USD 432.26 billion by 2031, with a compound annual growth rate (CAGR) of 5.1% (Coherent Market Insights 2024). This surge is primarily driven by increasing population and rising income levels, leading to greater demand for edible oils. In developing economies like India, China, and Indonesia, changing dietary habits and improving living standards have significantly boosted per capita oil consumption. Globally, India ranks as the fourth-largest producer of oilseeds after the United States, China, and Brazil, with cultivation spanning 26.77 million hectares and yielding 33.55 million tons, with an average productivity of 2549 kg per hectare (Kamaleshwaran and Elayaraja 2024).

Oilseed crops thrive across diverse agro-climatic regions, from tropical zones where coconut and peanut flourish to temperate regions that favour canola and sunflower (Asan 2024). This adaptability ensures a steady supply of oils essential for cooking, snacks, and processed foods while also supporting industrial applications. The biofuel sector, for instance, heavily relies on oilseed crops like canola, soybean, and sunflower to produce sustainable energy sources. Similarly, industries manufacturing paints, cosmetics, and lubricants benefit from the versatility of oilseeds. Additionally, the protein-rich residues from oil extraction serve as critical components of livestock and poultry feed, thereby enhancing global food production systems. Economically, oilseed cultivation sustains millions of smallholder farmers, especially in developing countries where crops like sesame, mustard, and niger seed provide crucial income sources. In India, mustard farming is deeply intertwined with cultural traditions, as seen in Rajasthan and Punjab, where its golden fields define the winter landscape. Mustard oil remains a staple in Indian households, flavouring regional delicacies like 'sarson ka saag'. Similarly, sesame seeds are deeply embedded in rituals and cuisines across Tamil Nadu and Odisha. Peanuts, predominantly grown in Gujarat and Andhra Pradesh, offer a drought-resilient income to farmers in rain-fed regions. Trade and geopolitical factors also influence the global oilseed market. Soybean and canola dominate international trade, fuelling agribusiness giants in North and South America, while sunflower oil from Ukraine and rapeseed from Canada underscore strategic economic and sustainability concerns (Kastakova et al. 2023). The rising demand for plant-based protein and sustainable oils has further propelled research into high-yield, climate-resilient oilseed varieties, making these crops even more valuable in the face of global challenges like climate change and food security. Each oilseed crop holds unique significance. Soybean, widely cultivated for its high protein and oil content, serves as a cornerstone of the global vegetable oil and livestock feed industries (Fang and Kong 2022). Sunflower oil, known for its light texture and high nutritional value, finds extensive use in cooking and skincare. Canola, derived from rapeseed, is prized for its heart-healthy oil and biofuel potential. Flaxseed, rich in omega-3 fatty acids, is in high demand for health foods and linseed oil production (Langyan et al. 2025). Peanuts contribute both as a nutritious snack and an essential oil source. Cottonseed oil, extracted as a byproduct of cotton farming, supports both culinary and industrial needs. Coconut oil remains a staple in tropical diets, with wide applications in food, beauty, and wellness industries. Sesame oil, known for its distinct aroma and medicinal benefits, is cherished in the culinary and pharmaceutical sectors. Mustard, with its pungent oil, is a fundamental ingredient in many traditional cuisines and condiments. Niger seed, though less recognized, produces an oil rich in essential fatty acids, supporting nutrition and income for small-scale farmers. Beyond economic significance, oilseeds play a crucial role in sustainable agriculture and nutrition security. They offer affordable dietary fats, promote soil health through crop rotation practices, and provide drought-resistant options for farmers facing erratic weather patterns. As the global demand for edible oils continues to rise, oilseed crops will remain at the heart of agricultural economies, supporting industries, farmers, and consumers alike.

14.3 Overview of UV Stress in Plants

Among various environmental stressors threatening the food security goal, ultraviolet (UV) radiation is emerging as a significant factor affecting plant growth and crop yields. It constitutes about 9% of the solar spectrum and has profound effects on plant physiology, influencing growth, metabolism, and overall productivity (Bais et al. 2019).

UV radiation is classified into categorized into UV-A (320–400 nm), UV-B (280–320 nm), and UV-C (100–280 nm) (Mutschlechner et al. 2024). The Earth's ozone layer effectively blocks most UV-C and a substantial portion of UV-B radiation. However, due to anthropogenic activities such as emissions of chlorofluorocarbons (CFCs) and carbon tetrachloride (CCl_4), ozone depletion has intensified, leading to increased penetration of harmful UV-B radiation (Egamberdieva et al. 2017). Despite constituting only a small fraction of sunlight, UV-B radiation induces oxidative stress, damages cellular structures, and disrupts key physiological processes in plants (Carvalho and Castillo 2018). UV-A, while less harmful, still has a crucial role in regulating photosynthesis, pigment synthesis, and antioxidant metabolism (Correa et al. 2023; Barnes et al. 2019).

UV stress exposure negatively affects plants, primarily by generating reactive oxygen species (ROS). These ROS lead to oxidative damage in cellular components, including lipids, proteins, and DNA. This oxidative stress leads to membrane peroxidation, enzyme inactivation, and disruptions in cellular homeostasis, ultimately impairing plant growth and yield (Shan et al. 2022). Additionally, UV-B radiation affects the photosynthetic apparatus by degrading chlorophyll, damaging the D1 protein of photosystem II, and reducing the efficiency of carbon fixation. As a result, plants experience decreased biomass accumulation and lower productivity (Carvalho and Castillo 2018). Beyond physiological damage, UV stress also alters plant morphology and developmental patterns. Prolonged exposure leads to stunted growth, reduced leaf expansion, and altered root architecture. These changes are often accompanied by a decrease in nitrogen assimilation and disruptions in secondary metabolism, affecting essential compounds such as flavonoids and phenolics. While some of these metabolites serve as protective UV filters, their overproduction often comes at the expense of growth and yield (Shan et al. 2022).

At the molecular level, UV radiation influences gene expression and signal transduction pathways. Key regulatory components such as the UV RESISTANCE LOCUS 8 (UVR8) photoreceptor and the CONSTITUTIVELY PHOTOMORPHOGENIC 1 (COP1) protein are central to UV-induced responses. Upon UV-B exposure, UVR8 undergoes a conformational change, triggering downstream signalling cascades that regulate photoprotection, antioxidant defence, and DNA repair mechanisms. Transcription factors such as ELONGATED HYPOCOTYL 5 (HY5) play a pivotal role in mediating these responses by controlling the expression of genes involved in flavonoid biosynthesis and stress adaptation (Vanhaelewyn et al. 2016). Furthermore, microRNAs (miRNAs) have been identified as key regulators of UV stress responses, modulating gene expression under high UV-B conditions to enhance stress tolerance (Quan et al. 2018; Choudhary et al. 2021). Despite these adaptive mechanisms, excessive UV radiation remains a major threat to crop productivity, particularly for oilseed crops such as soybean, sunflower, rapeseed, and linseed. These crops are highly susceptible to UV-induced stress, experiencing significant reductions in biomass, seed yield, and oil content due to impaired metabolic and physiological functions. Given the economic and nutritional importance of oilseeds, there is an urgent need to develop innovative strategies to enhance their resilience to UV stress.

Understanding the intricate mechanisms of UV stress and its impact on plant systems is essential for developing sustainable agricultural practices. As climate change continues to exacerbate environmental challenges, integrating advanced research tools and biotechnological interventions will be crucial in safeguarding global food security and ensuring resilient crop production in the future.

14.4 Overview of Major Phytohormones

Phytohormones are crucial regulators of plant growth, development, and stress adaptation. These signalling molecules control a wide range of physiological processes, from cell division and elongation to seed development and defence against environmental stressors. The primary phytohormones—auxins, gibberellins, cytokinins, abscisic acid (ABA), ethylene, brassinosteroids, jasmonic acid, and salicylic acid—interact in complex networks to maintain plant homeostasis and enhance resilience. This section provides an overview of their functions, particularly in oilseed crops, emphasizing their role in growth and productivity enhancement.

14.4.1 Auxin

Auxins, primarily represented by indole-3-acetic acid (IAA), are pivotal in plant growth regulation. They stimulate cell elongation, root initiation, and apical dominance, ensuring coordinated development. In oilseed crops, auxins contribute to seed and pod formation, enhance root architecture, and improve nutrient uptake efficiency. Studies have demonstrated that exogenous application of auxins can increase oil yield in *Cymbopogon martinii* and *Cymbopogon winterianus* while improving seed number and fibre strength in linseed (Saleem et al. 2024). Their role in modulating gene expression and enzymatic pathways further underscores their importance in optimizing crop yield.

14.4.2 Ethylene

Ethylene is a gaseous hormone that governs fruit ripening, seed germination, and stress responses. It acts in coordination with other phytohormones, viz., SA, ABA, and jasmonic acid to regulate abiotic and biotic stress adaptation. Ethylene signalling enhances resistance to pathogen attacks, facilitates leaf senescence, and modulates responses to mechanical stress (Mohammadizadeh et al. 2024). In oilseed crops, ethylene has a key role in pod formation and seed maturation, influencing yield and quality.

14.4.3 Abscisic Acid

ABA is often associated with stress responses, particularly drought tolerance, but it also has an important role in seed development and maturation. In oilseed crops, ABA regulates osmotic adjustments, facilitates seed storage protein accumulation, and influences grain filling rates. Elevated ABA levels enhance drought resilience in *Brassica napus* and *Medicago truncatula*, promoting seed viability under water-limiting conditions (Datta et al. 2024). However, ABA can also inhibit growth under certain stress conditions, highlighting its dual role in plant physiology.

14.4.4 Gibberellin

Gibberellins (GA) regulate stem elongation, seed germination, flowering, and fruit development. These hormones break seed dormancy and promote pod elongation in oilseed crops. However, excessive GA concentrations can reduce seed number per pod and negatively impact yield (Singh et al. 2024). Research indicates that GA treatment enhances safflower oil synthesis and improves stem elongation and flowering in mustard. By influencing metabolic pathways, GA plays a crucial role in determining the growth potential and productivity of oilseed crops.

14.4.5 Cytokinins

Cytokinins are primarily involved in cell division, nutrient mobilization, and delayed senescence. They are synthesized in roots and translocated to other plant parts, where they regulate shoot growth and seed development. In oilseed crops, cytokinins contribute to pod development by promoting cell differentiation and nutrient assimilation (Xing et al. 2024). Increased cytokinin levels have been linked to delayed leaf senescence and enhanced seed yield in various crops. Furthermore, their interaction with other hormones, particularly auxins and gibberellins, balances growth and developmental processes.

14.4.6 Jasmonic Acid (JA)

Jasmonic acid is a crucial regulator of plant defence mechanisms and secondary metabolite biosynthesis. It mediates responses to herbivory, pathogen attack, and mechanical injury by activating defence-related gene expression. Additionally, JA influences seed germination, pod formation, and oil accumulation in oilseed crops (Han et al. 2024; Zhou et al. 2024). Its interaction with ABA and ethylene plays a significant role in coordinating stress tolerance pathways, thereby enhancing crop resilience under adverse conditions.

14.4.7 Salicylic Acid (SA)

Salicylic acid functions as a key player in plant defence, regulating systemic acquired resistance (SAR) against pathogens. It also influences seed development, flowering, and metabolic processes. In oilseed crops, SA enhances antioxidant enzyme activities, mitigates oxidative stress, and improves pod formation (Sun et al. 2024). Studies have demonstrated that optimal SA concentrations enhance seed composition and branching in *Arabidopsis* (family: Brassicaceae) mutants, reinforcing its role in maximizing oilseed productivity.

14.4.8 Brassinosteroids

Brassinosteroids (BRs) are essential for cell elongation, photomorphogenesis, and reproductive development. They enhance plant growth by improving nutrient uptake, increasing biomass, and strengthening stress tolerance mechanisms (Al-Taey et al. 2024). BRs regulate auxin transport and contribute to root and shoot development (Wu et al. 2024), making them indispensable in oilseed crops where optimal growth is critical for productivity. Furthermore, they protect against environmental stresses, including temperature fluctuations and oxidative damage.

14.5 Significance of Studying UV Stress in Oilseed Crops

Oilseed crops play a crucial role in global agriculture, providing essential oils, proteins, and biofuels. However, climate change and environmental stressors, particularly UV radiation, which accounts for approximately 9% of the solar spectrum, pose a significant threat to their growth, yield, and quality (Bais et al. 2019). Understanding the impact of UV stress on oilseed crops is vital for developing resilient cultivars and sustainable agricultural practices. Research on UV stress in oilseeds is pivotal for several reasons:

Climate change and ozone layer depletion: The thinning of the stratospheric ozone layer, exacerbated by anthropogenic activities, has increased ground-level UV-B radiation. Oilseed crops, cultivated across diverse agro-climatic zones, are directly exposed to these shifts, which impair physiological processes such as photosynthesis, nutrient uptake, and flowering. For instance, studies on *Brassica* species (e.g. mustard, canola) reveal that UV-B exposure reduces chlorophyll content and stomatal conductance, directly compromising yield and oil quality (Ochoa et al. 2025).

Yield and quality implications: UV stress disrupts lipid biosynthesis (Liu et al. 2024), a key process in oilseed crops. In soybeans, prolonged UV-B exposure alters fatty acid composition, reducing polyunsaturated fats critical for both human nutrition and industrial applications (Wang et al. 2024).

Oxidative damage and defence mechanisms: UV radiation triggers ROS accumulation, leading to oxidative damage in plant tissues (Mahdavian 2024). Oilseeds like flax and sesame respond by upregulating antioxidant enzymes (e.g. superoxide dismutase (SOD), catalase (CAT)) and synthesizing UV-absorbing compounds such as flavonoids and phenolic acids. Understanding these adaptive mechanisms can aid in breeding resilient varieties. For example, research on sesame genotypes identified UV-tolerant lines with enhanced phenolic content, offering a genetic blueprint for stress resilience.

Ecological and agricultural sustainability: Smallholder farmers in tropical and subtropical regions, where UV radiation is intense, often lack the resources to mitigate such abiotic stresses. Studying UV stress helps develop low-cost agronomic practices, such as shade nets or bio-fertilizers that enhance UV tolerance.

Global food security: Oilseeds are dual-purpose crops, providing both oil and a protein-rich meal. UV-induced yield losses exacerbate gaps in edible oil supply, particularly in import-dependent nations like India. Research on UV stress aligns with Sustainable Development Goals (SDGs) by promoting climate-resilient agriculture and reducing reliance on costly imports.

Human health and nutrition: UV stress not only reduces crop productivity but also influences the composition of bioactive compounds. For instance, UV-B irradiation has been shown to decrease biomass and stem length in *Linum usitatissimum* L. (flax seed) sprouts, which are rich in dietary fibres, phenolic acids, lignins, carotenoids, and fat-soluble vitamins. However, while UV-induced stress can negatively impact growth, it also triggers metabolic changes that may enhance the nutraceutical properties of oilseeds. Understanding and balancing these biochemical alterations could provide opportunities for improving the functional food value of oilseed crops, making them more beneficial for health-focused markets.

14.6 Impacts of UV Stress on Plants

UV-B radiation from the sun significantly impacts terrestrial plants, leading to molecular damage and alterations in plant physiological processes, ultimately reducing yield (Fig. 14.1). Understanding plant responses to UV-B exposure is essential for assessing its broader effects on terrestrial habitat (Bornman et al. 2019). Although UV-B constitutes only a small fraction of the solar spectrum, it has a profound photo-biological influence on both plants and animals due to its interaction with essential biomolecules such as DNA/RNA, lipids, proteins, phytohormones, etc. (Singh et al. 2012).

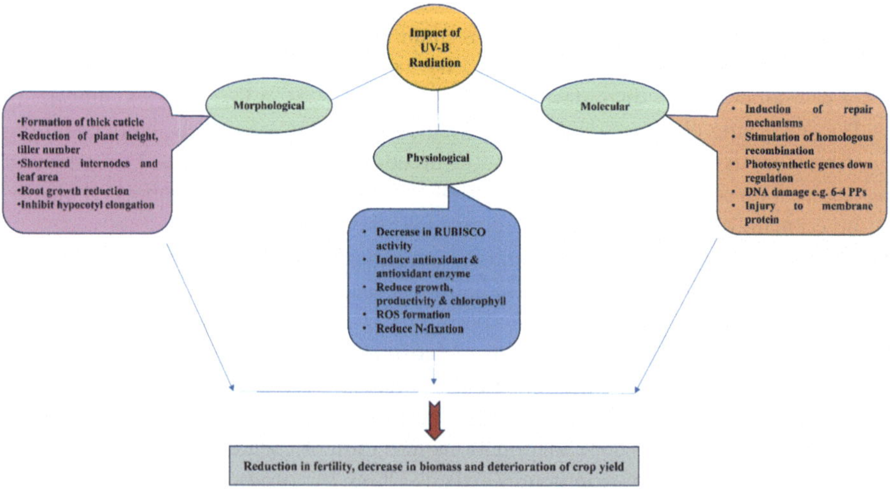

Fig. 14.1 Effects of UV-B on plant morphology, physiology, molecular characteristics, and yield

14.6.1 UV Stress and Plant Morphology

Photo-morphogenesis, a process where plants develop their organs in the presence of UV rays, has been thoroughly examined. Plants of various species modify their development, with stems and roots becoming longer and more abundant. Germination, a crucial stage, is greatly influenced by UV-B light. For example, germination of *Arabidopsis thaliana* seedlings with UV light showed slower growth of the hypocotyl. Krasylenko et al. (2013) found that UV radiation can also cause depolymerization/fragmentation of microtubules in plants, reorganizing the cytoskeleton and cell, reducing elongation and cell division, and causing shorter leaves. Some species of plants bounce back the UV radiation by forming a waxy cuticle on the epidermis, which increases in thickness and accumulates molecules like phenolic acids and flavonoids (Liu et al. 2022).

Krasylenko et al. (2013) found that UV radiation can also cause depolymerization/fragmentation of microtubules in plants, reorganizing the cytoskeleton and cell, reducing elongation and cell division, and causing shorter leaves. Some species of plants bounce back the UV radiation by forming a waxy cuticle on the epidermis, which increases in thickness and accumulates molecules like phenolic acids and flavonoids (Liu et al. 2022).

UV-B radiation not only affects plant growth and development, but also affects root-shoot ratio (Bussell et al. 2012) and flowering timing in various species (Yan et al. 2012; Rajendiran and Ramanujam 2004). Inflorescence elongation and axillary growth are also retarded due to its exposure, indicating a notable role in plant growth retardation (Hectors et al. 2007). Studies on UV-B also indicated its effects on flower development and secondary metabolites (Alonso et al. 2016). Higher UV-B exposure reduces growth rate and nitrogen productivity, affecting biomass and dry matter production (Eissa et al. 2011). Additionally, UV-B radiation can alter the taste and aromatic properties of plants by increasing sesquiterpene content.

14.6.2 Impacts of UV Stress on Plant Physiology

UV-B exposure can harm plants' physiological activities. Dicotyledons are more susceptible to UV-B as it directly impacts DNA, proteins, and lipids (Kataria et al. 2014; Sah et al. 2022). Mathur et al. (2024) revealed that reduced chlorophyll content, decreased chlorophyll a/b binding protein expression (Sah et al. 2022), and reduced photosynthetic pigment function. UV-B also interferes with radiation use efficiency in rice (Sah et al. 2022).

Photosynthesis, a light-dependent process, is significantly impacted by UV radiation (Liu et al. 2022). The effect of UV-B radiation on photosynthetic activities depends on various factors like crop species, UV-B amount, PAR to UV-B ratio, and cultivars. The photosynthetic apparatus is the most susceptible because of the demolition of photosynthetic pigments (Jenkins 2009). UV-B reduces chlorophyll a content (Ayash et al. 2017; Sebastian et al. 2018), leading to a decrease in net photosynthetic rate and stomatal conductance. Moreover, in some cases, UV-A radiation stimulates stomatal opening; however, UV-B radiation is known for its closure (Brestic et al. 2023).

UV light degrades pigments, particularly chlorophyll b and carotenoids, leading to imbalances in pigment proportions and altered photosynthetic apparatus. Recent studies show decreased fluorescence and chlorophyll concentration in maize lines after 19 days of exposure (Jovanić et al. 2022). Gray et al. (2024) demonstrated that plants can manipulate their floral pigmentation to protect the pollen grains from the adverse effects of UV-B doses.

Photosystem II, the site of photosynthesis, is reactive to UV radiation due to the degradation of pigments and proteins in the electron transport chain (Kataria et al. 2014). This leads to reduced levels of essential elements, such as oil, amino acids, and protein in sunflowers.

14.6.3 Impacts of UV Stress on Plant Oxidative Stress

UV radiation, a significant environmental stressor for plants, triggers protective mechanisms like accumulation of secondary metabolites (Fig. 14.2) (Gogo et al. 2018; Kim et al. 2018; Zhao et al. 2020) and increased leaf thickness (Liu et al.

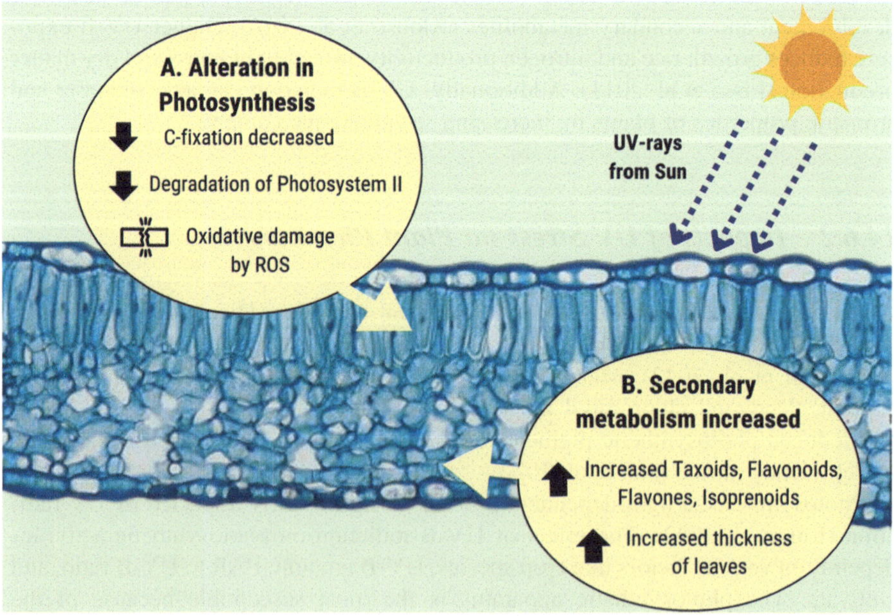

Fig. 14.2 UV radiation's effects on plants, including how it alters the photosynthetic pathway

2020). Interestingly, UV radiation acts as a hermetic stimulus, meaning that while high doses can be detrimental, moderate exposure can enhance protective responses (Duarte-Sierra et al. 2020). With repeated exposure, plants develop resilience to UV stress (Klem et al. 2022).

Numerous studies have documented changes in secondary metabolism across different plant species in response to UV radiation. For example, in the moss *Pohlia nutans*, UV-B exposure stimulated flavone biosynthesis by increasing type I flavone synthase activity (Wang et al. 2020). Similarly, in *Taxus cuspidata*, treatment with UV-B radiation (3 W/m^2) led to the accumulation of toxoids and flavonoids (Jiao et al. 2022). In *Scutellaria baicalensis*, flavonoid levels peaked at 41.86 mg g^{-1} after 7 days of UV-A exposure (Miao et al. 2022).

In *Pinus radiata*, UV exposure triggered an early response by reducing photosystem activity and promoting the collection of photo-protective compounds. Additionally, the primary metabolism adjusted to limit ROS production, while levels of isoprenoid-derived compounds such as carotenoids, tocopherols, phytol, and gibberellins decreased (Pascual et al. 2017). In *Clematis terniflora*, UV-B exposure followed by darkness led to a significant rise in flavonoid and coumarin levels, while proteins associated with photorespiration exhibited differential expression, suggesting a shift towards reduced energy consumption to maintain cellular balance (Tao et al. 2022). The marine plant *Nymphoides humboldtiana* illustrated increased antioxidant activity and flavonoid production, adding phloroglucinol, chlorogenic acid, epicatechin, quercetin, and ferulic acid, after 13 days of UV-B exposure. Likewise, *Colobanthus quitensis* responded to UV-B stress by enhancing flavonoid biosynthesis, particularly flavone C-glycosides, which were concentrated in the most metabolically active cells (Contreras et al. 2019). In *Melissa officinalis*, UV radiation affected glycolysis and the phenylpropanoid pathway, with variations in recovery time depending on the extent of stress (Kim et al. 2018).

UV-radiation stress affects medicinal plants. *Morus alba* reduces growth and secondary metabolism, *Gingko biloba* leaves increase flavonoid biosynthesis, and *Astragalus mongholicus* accumulates myricitrin and isoflavones (Liu et al. 2020). *Lonicera japonica* increases oxidative pentose phosphate and secondary metabolites (Zhu et al. 2017), *Adhatoda vasica* reduces superoxide radical production (Pandey et al. 2021), and *Centella asiatica* accumulates saponins and epidermal flavanols in younger leaves (Müller et al. 2013). As shown in these studies, UV light has forced the plants to modify their metabolism, especially the secondary metabolism, to intensify their ecological success rate.

14.6.4 The Role of ROS in Plant Responses to UV Radiation

Wild plants rely on sunlight for energy, which is crucial for photosynthesis within the 400–700 nm spectrum. UV light is categorized into UV-A (315–400 nm), UV-B (280–315 nm), and UV-C (100–280 nm). The ozone layer blocks UV-C entirely and most UV-B, but the small amount that reaches Earth can still impact plant cells, potentially disrupting photosynthesis. ROS are produced during oxidative stress conditions.

14.6.4.1 UV Radiation and ROS Production *vis-à-vis* Plant Antioxidant Systems

Excessive UV-B radiation can lead to the generation of ROS in plants, resulting in increased malondialdehyde (MDA) accumulation. To counteract ROS-induced stress, plants have developed various defence mechanisms, including enzymatic antioxidants such as catalase (CAT), peroxidase (POD), ascorbate peroxidase (APX), glutathione reductase (GR), and superoxide dismutase (SOD). Additionally, non-enzymatic antioxidants like ascorbic acid (AsA), glutathione (GSH), flavonoids, and alpha-tocopherol contribute to ROS detoxification (Kataria et al. 2014). The efficiency of these antioxidant responses depends on radiation intensity, exposure duration, and plant type.

MDA, a byproduct of ROS-induced lipid peroxidation, is commonly used as an indicator of oxidative stress. Hideg and Vass (1996) reported elevated MDA levels in broad bean leaves subjected to short-term UV-B exposure (0.46 kJ/m^2), indicating increased free radical activity. Similarly, Dai et al. found that UV-B radiation increased MDA and hydrogen peroxide (H_2O_2) concentrations in rice (*Oryza sativa*), along with enhanced superoxide (O_2^-) production, membrane leakage, which resulted in cell damage. Kalbin et al. (1997) observed a significant rise in ROS levels in pea (*Pisum sativum*) tissues following intense UV-B exposure by analysing oxidized and total glutathione (GSSG and GSH). Changes in the redox state of antioxidants can serve as an indicator of ROS fluctuations. Mishra et al. (2009) explore the method to estimate superoxide levels by tracing nitrite formation in UV-B-exposed seedling leaf homogenates (Biever et al. 2014; Jiao et al. 2022).

ROS accumulation in plants exposed to UV-B radiation is associated with both metabolic and photophysical responses. Disruptions in metabolism, such as damage to electron transfer in photosystem II (PSII) and disturbances in quinone components, can lead to ROS generation in photosystem I (PSI). Additionally, UV-B radiation can facilitate the photoconversion of H_2O_2 into highly reactive hydroxyl radicals ($OH^{\cdot}$), intensifying oxidative damage.

$$2O_2 + 2H + H_2O_2 + O$$

$$H_2O_2 + O_2 \;\; OH_- + O_2 + OH$$

ROS production due to UV exposure also resulted from higher mitochondrial respiration. Research indicates that UV-B exposure influences the activity of antioxidant enzymes, such as superoxide dismutase (SOD), in *Arabidopsis thaliana* and other plant species, with responses varying over time. For example, Rengin et al. found that after 3 h of UV-B treatment, SOD activity declined in both *Arabis alpina* and *Arabidopsis thaliana*, but after 6 h, it significantly increased, suggesting an adaptive response (Gogo et al. 2018).

It has been proposed that ROS may serve as signalling molecules, modulating antioxidant system responses to UV-B stress. However, further research is required for a proper understanding of the molecular mechanisms that regulate this process.

14.6.4.2 Effect of UV Rays on Metabolites Production via ROS

UV-B radiation excites the production of ROS and various secondary metabolites, including ascorbic acid, carotenoids, anthocyanins, phenols, flavonoids, terpenoids, and alkaloids (Nocchi et al. 2020; Britt 2004). The plant cuticle contains UV-filtering compounds, which make tissue-dependent ROS production. When wounding happens before UV exposure, magnifying radiation penetration amplifies ROS signalling, triggering ethylene (ET) and jasmonic acid (JA) biosynthesis, which further build up phenylalanine ammonia-lyase (PAL) activity and phenolic accumulation (Duarte-Sierra et al. 2020).

Hydrogen peroxide (H_2O_2), a key signalling molecule, has been linked to the boost of enzymes involved in isoflavone biosynthesis, such as PAL, chalcone synthase (CHS), and isoflavone synthase (IFS) (Zhao et al. 2020). Given its role in UV-B-induced signalling, further study is needed to explore additional genes and proteins involved in ROS-mediated metabolic regulation, particularly in alkaloid, terpene, carotenoid, and thio-glucoside biosynthesis.

14.6.4.3 Effect of UV Rays on Photosynthesis via ROS

The impact of UV-B rays on photosynthetic organisms has been extensively studied with scientific findings. It indicated that UV-exposure impaired photosynthetic activity, primarily due to increased ROS levels. These ROS contribute to lipid peroxidation, damaging the thylakoid membrane and disrupting the metabolic conditions essential for photosynthesis (Dukowic-Schulze et al. 2022; Quail 2002; Yi and Deng 2005). Moreover, UV-B exposure has been linked to increased ion permeability in plasma membranes (Lockhart 2014; Oravekz et al. 2006; Favory et al. 2009).

Photosystem II (PS II), a pigment-protein complex in the thylakoid membrane, captures light energy for photosynthesis. Research by Swarna et al. (2012) indicated that UV-B radiation could reduce efficiency of PS II up to 68% in maize leaves due to ROS accumulation (Wang et al. 2020). Moreover, UV light absorption by the manganese (Mn) in the oxygen-evolving complex (OEC) of PS II results in primary photodamage. Several researchers have stated that hydroxyl radical ($OH^{\cdot}$) accumulation following UV-B exposure is considered a pivotal factor in UV-induced damage.

ROS not only act as signalling molecules but also cause structural disintegration, which affects proteins within the light-harvesting complexes and the electron transport chain. Their accumulation can degrade Rubisco enzyme, leading to decreased atmospheric carbon fixation and reduced plant biomass (Nasibova 2022).

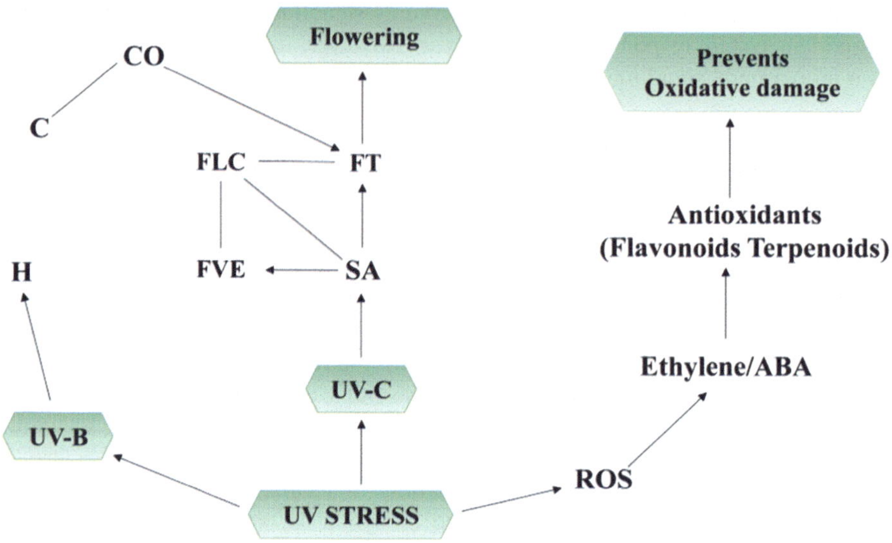

Fig. 14.3 Schematic representation of UV-stress pathway

Furthermore, oxidation of membrane fatty acids, combined with peroxidation and photo-oxidation, compromises thylakoid integrity, disrupting membrane-embedded photosynthetic complexes and lowering photosynthetic capacity (Swarna et al. 2012). Some studies also indicate that ROS can break the large subunit (LSU) of Rubisco under UV exposure.

In summary, UV-B rays accentuate oxidative damage in plant cells, which indicates a negative effect on membrane lipids, proteins, and photosynthetic pigment content. It also decreases PS II efficiency and impairs Rubisco activity. These disruptions ultimately lower crop productivity and yield (Fig. 14.3).

14.7 Role of Phytohormones in UV Stress Tolerance in Oilseed Crops

Phytohormones, a class of signalling chemicals in cells, play a vital part in controlling the growth and development of plants in response to various stimuli. They govern growth, evolution, source-sink relationship, and nutrient distribution to different parts, facilitating plant acclimatization to changing environments. Phytohormones are essential for plant survival as stable and static creatures, and their function can vary depending on the plant's environment. Several phytohormones, including ethylene, cytokinins, abscisic acid, gibberellins, auxin, brassinosteroids, salicylic acid and jasmonates, significantly influence plant development and plastic growth. The influence of UV stress on several oilseed crops and their adaptive mechanisms regulated by phytohormones is summarized in Table 14.1 and Fig. 14.4.

Table 14.1 Effects of UV stress on different oilseed crops

Oilseed crop	Physiological response	Adaptive mechanisms	Role of phytohormones
Soybean	Reduced chlorophyll, impaired nitrogen fixation	Increased flavonoid synthesis, ROS scavenging	Auxins enhance root growth and nutrient uptake (Mir et al. 2022)
Sunflower	Decreased biomass, altered lipid composition	Enhanced cuticle thickness, antioxidant enzyme activity	Cytokinins delay senescence and improve chlorophyll retention (Zwack and Rashotte 2015)
Canola	Disrupted membrane integrity, reduced seed oil quality	Upregulated stress-responsive genes, ABA-mediated stomatal closure	ABA regulates stomatal closure and enhances drought tolerance (Mukherjee et al. 2023)
Flaxseed	Minor reduction in photosynthesis and growth	High lignin and phenolic accumulation for UV protection	Jasmonic acid increases flavonoid and lignin biosynthesis (Choudhary et al. 2021)
Peanut	Altered root morphology, reduced nutrient uptake	Increased proline content, modulation of ethylene pathways	Ethylene modulates stress responses and improves tolerance (Yang et al. 2014)
Sesame	Enhanced oxidative stress, protein degradation	Overproduction of secondary metabolites like sesamin	Brassinosteroids enhance stress tolerance and antioxidant activity (Al-Taey et al. 2024)
Mustard	Reduced seed weight, delayed flowering	Activation of jasmonic acid pathways, enhanced leaf wax production	Gibberellins promote flowering and seed development (Kumar et al. 2024)

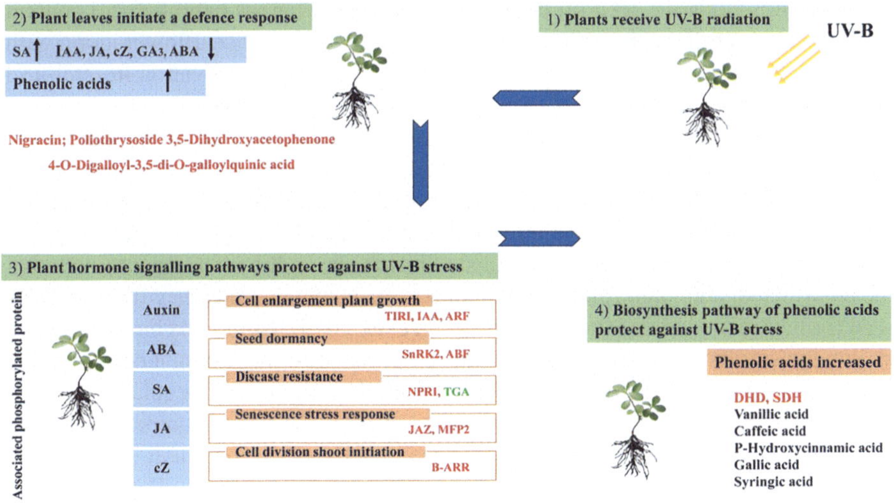

Fig. 14.4 Plant hormones response to UV-B stress

14.7.1 Role of Auxins in the Alleviation of UV Stress in Oilseed Crops

Auxin, the first identified phytohormone, plays a crucial part in plant maturation processes like root development, cell lengthening, vascular segregation and dormancy. They encourage plant growth under normal and stress conditions. In higher plants, IAAs exist as IAA conjugates and the primary free IAA involved in lateral root formation (Lymperopoulos et al. 2018). Auxin affects *Brassica juncea*'s capacity for phytoremediation, stress tolerance, and growth in Cu-contaminated soils. The results demonstrate that by increasing antioxidant activities and lowering ROS, MDA, and electrolyte leakage (EL) levels, Auxin treatment reduces harmful reactions under UV stress. Additionally, it keeps stomata and chloroplasts organized, lowers the mortality rate of root cells, and enhances sugar metabolism, nutrient enrichment, photosynthetic and chlorophyll fluorescence indices, as well as plant biomass. In Cu-contaminated soils, the research suggests adding 10–8 M of IAA to above-ground plant parts to boost growth, stress endurance, and plant detoxification capacity (Mir et al. 2022). Also, oilseed crops can be protected against UV-C radiation by the foreign application of auxin-like chemicals TA-12 and TA-14. This protection includes reduced membrane damage, restored protein, hydrogen peroxide, and proline content, and preserved antioxidants. The action of non-enzymatic antioxidants and ROS detoxifying enzymes combination may lead to positive and protective measures on oilseed plants (Sergiev et al. 2017). Foliar application of auxin promotes the undesirable effects of UV stress, which led to an 18% increase in seed yield and components of oilseeds. However, the application of auxin did not affect the seed oil content of safflower, whereas palmitic acid content was affected by foliar application of auxin. Generally, growing safflower along with auxin application can be recommended to develop safflower cultivation in semi-arid areas (Mousavi et al. 2022).

Seed formation and starch buildup in grain legumes are critical for human meals, and we must learn as much as we can about the variables controlling these processes. Our knowledge of how plants control their carbohydrate economy has improved as a result of the new information we offer in this study about the signalling relationships between auxin, seed size, and starch in starch-rich seeds. It's interesting to note that the chlorinated, abundant form of auxin that is exclusive to seeds may be crucial for seed formation in peas and their cousins (Lam et al. 2015). Numerous bioassays have shown the significant auxin activity of this chlorinated form. Auxin, a hormone that interacts with ABA and GA, is not considered a significant regulator of seed germination. However, it plays a critical role in response to environmental stimuli, as demonstrated by studies showing that exogenous auxin treatment significantly decreased Arabidopsis seed germination under stress conditions. The exact processes behind auxin's impact on seed germination remain unclear (Wang 2011).

14.7.2 Role of Gibberellins in the Alleviation of UV Stress in Oilseed Crops

UV-B radiation is known to be an aptitude stressor for organisms, causing not only morphological changes but also influencing the physiological and genetic characteristics of oilseed plants. GA stimulates cell elongation, which can help plants 'escape' intense UV radiation by increasing the distance between leaves and the sun. Under UV stress, GA can induce the production of antioxidants like flavonoids and phenolic compounds, which help neutralize harmful ROS created by UV radiation. The exogenous application of momilactone-rich bran fractions (MRBF), along with the GA_3 combination, helps in mitigating physio-biochemical stress responses induced by UV-B stress. GA_3 produced maximum levels of chlorophyll, carotenoid, and proline contents. Additionally, it improved the antioxidant property of oilseed and led to increased concentration of secondary metabolites, including salicylic acid. These metabolites might be helpful in UV-B stress mitigation. GA_3 plays important roles in building up physio-biochemical status and UV-B stress resilience in oilseeds. By maintaining healthy plant growth under UV stress, GA indirectly helps protect the photosynthetic machinery, ensuring optimal light capture and energy conversion. However, further research is needed to understand the proper role of GA_3 under UV-B stress for the oilseed crop (Kumar et al. 2024; Sindhu and Singh 2024).

14.7.3 Role of Cytokinins in the Alleviation of UV Stress in Oilseed Crops

Cytokinin is the primary plant hormone that controls a number of features of growth and development. Although cytokinin's role in several developmental stages has been well described, its impact on plants' ability to withstand stress is still unclear. The complex interaction between cytokinin and stress signalling, particularly the response to UV stress, may be the source of this (Zwack and Rashotte 2015). However, cytokinin application plays a remarkable role in flower production and has a positive effect on the yield of soybean and other oilseed crops. Cytokines play a defence function against high air temperature during flower formation, potentially lowering endogenous GA levels to improve ultraviolet tolerance, thus increasing the total seed production. Cytokinin has also been reported to release dormancy and enhance germination and chlorophyll concentration under UV stress. The external application of cytokinin has been shown to substitute for the physiological influence of roots on the growth of the soybean seedlings. Kinetin used as both seed treatment and foliar sprays, individually or in combination, increased the seed yield over the control condition. However, maximum favourable effects of kinetin were obtained with the combined application of seed treatment plus foliar spray, which increased seed yield over untreated plants. Studies noted that individual treatment of UV-B radiation in *Brassica campestris* PT-303 inhibited the chlorophyll a and chlorophyll b, respectively. Therefore, plant growth hormones would show the maximum mitigation against UV-B induced deleterious or negative impacts on the Chlorophyll-a and Chlorophyll-b of *Brassica campestris* PT-303 (Lal and Kaur 2016).

14.7.4 Role of Abscisic Acid in the Alleviation of UV Stress in Oilseed Crops

The plant hormone ABA is essential for starting defence mechanisms in oilseed crops under UV stress. It mainly does so by causing stomatal closure, which reduces water loss and shields the plant from the adverse effects of excessive UV radiation. In essence, ABA acts as a protective response to the stress. Elevated ABA levels under UV stress can also affect root architecture, encouraging deeper root growth to access more water and nutrients (Mukherjee et al. 2023).

ABA's capacity to signal the closing of stomata, which are the microscopic pores on leaves, is the most well-known function of ABA when it is subjected to UV stress. Stomata are responsible for reducing the amount of water that is lost through transpiration and protecting the plant from excessive exposure to UV light. The reaction mechanism of ABA includes the inhibition of ethylene synthesis in plants and intensifies plant growth (Zhang 2014). Ethylene production in plants is increased by UV-B as well as in water-lacking conditions. The plant species that were exposed to the UV-B radiations were found to be more tolerant to drought conditions, and therefore, ABA shows an effective way towards the plant's reaction under drought environment (Li et al. 2014; Shoaib et al. 2024).

In oilseed, UV-B increased flavonoid and carotenoid concentrations, albeit without significant ABA involvement. Conversely, in soybean, ABA amplified flavonoid production and altered gibberellin profiles, indicating species-specific regulation. Flavonoids, which act as natural sunscreens, are often induced by UV via ABA-dependent pathways, as demonstrated in sunflower, where ABA enhanced phenolic accumulation under high irradiance.

UV radiation often disrupts photosynthesis by degrading chlorophyll and impairing electron transport. In soybean, ABA counteracted UV-C-induced chlorophyll loss and PSII inhibition, aligning with studies in Arabidopsis where ABA preserved photosynthetic efficiency under UV stress by stabilizing chloroplast membranes. Similarly, in canola (*Brassica napus*), UV-B lessens the leaf area and dry matter, but ambient UV-B levels optimize growth. Although ABA's direct impact here was minimal, its interaction with watering regimes subtly influenced leaf dry matter and wax content, suggesting context-dependent roles in photoprotection.

Several studies found to relationship between the interaction incorporating UV-B and ABA, but some of them conclude that the presence of ABA increases the tolerance of oilseed to UV-B. The same effect was observed in leaves of Arabidopsis, concluding a positive effect of UV-B on ABA synthesis in tissues during high exposure to UV-B. Whereas additional studies reported that the interaction between water stress conditions to UV-B radiation shows less sensitivity to UV-B in various plant species during drought conditions (Berli and Bottini 2013).

14.7.5 Role of Ethylene in the Alleviation of UV Stress in Oilseed Crops

When oilseed crops are subjected to UV stress, they frequently experience an increase in the production of ethylene, which functions as a stress signal. This causes the plant to undergo physiological changes such as leaf senescence, reduced growth, and altered plant architecture. These changes are a means of mitigating the harmful effects of excessive UV radiation. In essence, ethylene assists the plant in adapting to the stress by activating mechanisms such as leaf shedding, which helps to minimize further UV damage. The simultaneous experiment demonstrated that increased UV-B radiation may stimulate the synthesis of ethylene, which was used as a signal molecule in the soybean seedlings. The amount of ethylene in the soybean seedlings was connected with ROS levels rather than Pn levels. There was a correlation between the change in ACC synthase activity and the change in ethylene content of the soybean seedlings exposed to the increased UV-B radiation (Yang et al. 2014). The concentration of both JA and ethylene increased suddenly on UV-B exposure, indicating a feasible role for these two compounds in UV-B-mediated responses in the oilseed crop. In pathogen defence, the entire JA and ethylene signal pathways are needed for maximum defence against UV-B-induced tissue damage. Increases in both JA and ethylene, consequently, are essential for the utmost induction of the *PDF1·2* gene by UV-B radiation. When exposed to top UV levels, plants produce more ethylene, a gaseous plant hormone that acts as a signal to initiate stress responses.

Therefore, although only ethylene cannot induce PR-1 transcripts, unlike in pathogen responses, it must participate in increases in PR-1 transcripts following UV-B treatment. Ethylene is mostly located upstream of SA in the signal pathway involved in the regulation of PR-1 transcripts and results in changes in gene expression in response to UV-B radiation. After the moderate-dose irradiation, there is a dramatic rise in ethylene evolution in oilseed. Hence, the mobilization of plant defence responses during plant adaptation was accompanied by an intense release of ethylene. One primary effect of ethylene under UV stress is the acceleration of leaf senescence, where older leaves naturally shed to conserve resources and minimize further damage. It appears that, unlike the short-term rise in ethylene content, the long-term ethylene production and drag out the action of comparatively high ethylene concentrations do not promote plant defence but exert an opposite effect, i.e., degradation of overstressed plants. Evidence for the involvement of ethylene, ABA, and polyamines in adaptation of oilseed crop to UV-B radiation and supports the existence of at least two pathways for the phytohormone-mediated transduction of UV-B signal. The primary pathway is inspired by low doses and secures plant adaptation to the UV-B radiation. The pathway reactions are accompanied by the temporary improvement of ethylene production, sustained collection of ABA, increased synthesis of putrescine, and by maintenance of repeatedly high levels of spermidine and spermine. The second pathway initiated by high-dose irradiation causes a significant increase in ethylene production, the reduced accumulation of ABA, the drastic negative changes of polyamine content and finally the plant death (Rakitin et al. 2008).

14.7.6 Role of Brassinosteroids in the Alleviation of UV Stress in Oilseed Crops

By improving antioxidant defence mechanisms, encouraging cell wall thickening, and controlling the expression of genes involved in UV protection, brassinosteroids (BRs) significantly contribute to oilseed crops in tolerating UV stress, which in turn improves plant growth and yield even in highly UV environments. In order to protect plant tissues from oxidative damage, BRs increase the formation of antioxidants such as phenolic compounds and flavonoids. These antioxidants help scavenge damaging ROS that are formed by exposed plant tissues to ultraviolet light (Hasanuzzaman et al. 2020). It is possible for the use of BRs to result in cell walls that are thicker and more robust, making them more resistant to the damage that is caused by ultraviolet light to cellular components. Brassinosteroids play a crucial part in controlling the expression of genes that are involved in various UV-protective mechanisms within organisms. Specifically, BRs can modulate the activity of genes responsible for producing enzymes that play a key role in the synthesis of UV-absorbing compounds, such as flavonoids in soybean and other oilseeds. Through this regulation, BRs help organisms to respond effectively to UV radiation exposure and protect themselves from its potentially harmful effects (Sen Mandi and Sen Mandi 2016). On the other hand, Brassinosteroids mitigating the UV stress in oilseed crops can help maintain optimal photosynthetic activity, leading to improved plant growth and yield.

14.7.7 Role of Jasmonic Acid in the Alleviation of UV Stress in Oilseed Crops

Jasmonic acid (JA), one of several plant growth regulators, plays a crucial and intricate role in oilseed plants under UV-B exposure. These signalling molecules can control cellular functions and are found in extremely small amounts. When plants' defence systems are activated, JA acts as an essential part of signalling pathways (Choudhary et al. 2021). JA, as well as its derivatives like methyl jasmonate (MeJA), facilitates well-coordinated responses to a variety of abiotic and biotic stressors. The synthesis of several defensive substances, like as phenolic acids, flavonoids, and anthocyanins, which function as antioxidants and UV-absorbing pigments, is triggered by the output of JA by soybean plants in response to UV exposure (Lingakumar 2023). The results, which are consistent with JA's well-established positive regulatory function in anthocyanin production, imply that JA could also be crucial in determining how rapeseed seedlings react to intense light. All of these findings suggest that JA and anthocyanin biosynthesis-related pathways mediate UV responses in rapeseed. Together, our results offer self-propelled insights into the mechanisms behind rapeseed's response to UV stress, and the main genes found may be utilized in breeding or genetic engineering to increase rapeseed cultivars' UV tolerance. Subsequent investigations revealed that the anthocyanin-biosynthesis and JA-biosynthesis pathways were particularly enriched in the UV-induced differently expressed genes (DEGs). Particularly, UV stress significantly increased the expression of genes linked to late anthocyanin biosynthesis, like those encoding BnANS and BnDFR, and genes implicated in the control of anthocyanin biosynthesis, such as those encoding BnPAP2 and BnGL3 TFs (Luo et al. 2021).

14.7.8 Role of Salicylic Acid in the Alleviation of UV Stress in Oilseed Crops

Acting as a signalling molecule, salicylic acid (SA) reduces UV stress in oilseed crops by activating several protective mechanisms, including boosting antioxidant production, controlling photosynthetic activity, and shielding cellular components from UV radiation, so improving plant tolerance to UV stress and perhaps increasing crop yield. SA promotes the synthesis of antioxidant enzymes like peroxidase, catalase, and superoxide dismutase, which scavenge damaging ROS produced by UV radiation, therefore reducing reactive damage to macromolecules and cell membranes. According to studies, SA may help plants repair UV-induced DNA damage and preserve genomic integrity by activating DNA repair mechanisms (Strzałka et al. 2020). In addition to providing extra protection to plant tissues, SA can increase the formation of phenolic compounds, which are naturally occurring chemicals that absorb ultraviolet light. Through the activation of certain genes, SA can modify the response of the plant to UV damage, which ultimately leads to the production of proteins and metabolites that are protective against stress (Jan et al. 2021). Exogenous administration of SA as a foliar spray has the potential to improve the tolerance of oilseed crops such as canola, sunflower, and soybean to the effects of UV stress, which might lead to an increase in the production of these crops when exposed to high levels of UV radiation. Efforts are currently being made to produce oilseed crops that have increased endogenous SA production in order to improve their resistance to the effects of ultraviolet stress (Damalas and Koutroubas 2022).

14.8 Conclusion and Future Perspective

Oilseeds play a pivotal role in global agriculture, providing essential nutrients and supporting food security while contributing significantly to the economy. However, the increasing prevalence of environmental stressors, particularly ultraviolet-B (UV-B) radiation, poses significant challenges to oilseed cultivation. With the depletion of the ozone layer, UV-B radiation reaching the Earth's surface has intensified, resulting in detrimental effects on plant physiology. UV-B stress leads to the generation of ROS, disruption of antioxidant enzyme activity, damage to cellular structures, and a reduction in photosynthetic efficiency. These physiological changes ultimately compromise the growth, yield, and seed quality of oilseed crops, raising concerns over their productivity and sustainability. Understanding the mechanisms of UV-B tolerance is crucial for mitigating these adverse effects and ensuring the resilience of oilseed crops. Plants employ various physiological responses to combat UV-B stress, many of which are regulated by phytohormones. Hormones such as abscisic acid, jasmonic acid, salicylic acid, and nitric oxide play essential roles in enhancing stress tolerance and photoprotection. These hormones help plants manage the harmful effects of UV-B by modulating antioxidant defence systems, improving cellular resilience, and promoting adaptive growth changes. For instance, ethylene may be upregulated as part of a stress response or downregulated under certain conditions, influencing morphogenic changes. Additionally, phytohormones like brassinosteroids, auxins, and strigolactones have shown functional associations with the UVR8 pathway, which is central to UV-B photomorphogenesis. Their roles in regulating UV-B stress tolerance remain a promising area for further research. Incorporating pheromones, which are naturally occurring chemical signals, offers another promising strategy for mitigating UV stress in oilseed crops. Pheromones have traditionally been associated with plant-insect interactions, but recent research highlights their potential role in plant stress management. Certain plant pheromones may enhance UV-B stress tolerance by modulating antioxidant systems, improving photosynthetic efficiency, and stimulating the production of protective secondary metabolites. Integrating pheromones into crop management strategies provides dual benefits: enhancing crop resilience to UV-B stress and reducing reliance on synthetic agrochemicals. This eco-friendly approach aligns with sustainable agricultural practices and helps preserve environmental health. Future research should focus on identifying specific pheromones that can enhance UV-B tolerance and understanding their underlying mechanisms. By combining pheromone-based interventions with advanced biotechnological tools like CRISPR-Cas9, breeding programs can explore the genetic incorporation of UV-B tolerance traits in oilseeds. Such strategies will contribute to the development of oilseed varieties that can thrive under increasing UV stress, ensuring more stable production and food security in the face of climate change. In conclusion, the integration of phytohormones and pheromones in managing UV stress represents a promising synergy for oilseed crops. Understanding their roles in UV-B tolerance and developing sustainable interventions will be crucial in protecting oilseed production from the adverse effects of environmental stressors. By advancing this research, we can enhance the resilience and sustainability of oilseed crops, ensuring their vital contribution to global food security and agricultural sustainability.

References

Alonso R, Berli FJ, Fontana A, Piccoli P, Bottini R (2016) Malbec grape (*Vitis vinifera* L.) responses to the environment: berry phenolics as influenced by solar UV-B, water deficit and sprayed abscisic acid. Plant Physiol Biochem 109:84–90

Al-Taey DK, Al-Musawi ZJ, Kadium SMA, Abbas AK, Alsaffar MF, Mahmood SS (2024) Brassinolides' function and involvement in salt stress response: a review. IOP Conf Ser Earth Environ Sci 1371(4):42032

Asan D (2024) Farming systems for sustainable agriculture. Educohack Press

Ashfaq M, Khan S (2017) Role of phytohormones in improving the yield of oilseed crops. In: Oilseed crops: yield and adaptations under environmental stress, pp 165–183

Ayash A, Mohammad I, Nassour R (2017) The effect of ultraviolet radiation on chlorophyll in Chlamydomonas reinhardtii. Int J Agric Environ Sci 4(6):23–27

Bais AF, Bernhard G, McKenzie RL, Aucamp PJ, Young PJ, Ilyas M, Patrick J, Makoto D (2019) Ozone–climate interactions and effects on solar ultraviolet radiation. Photochem Photobiol Sci 18(3):602–640

Barnes PW, Williamson CE, Lucas RM, Robinson SA, Madronich S, Paul ND, Bornman JF, Bais AF, Sulzberger B, Wilson SR, Andrady AL (2019) Ozone depletion, ultraviolet radiation, climate change and prospects for a sustainable future. Nat Sustain 2(7):569–579

Berli FJ, Bottini R (2013) UV-B and abscisic acid effects on grape berry maturation and quality. J Berry Res 3:1–14

Biever JJ, Brinkman D, Gardner G (2014) UV-B inhibition of hypocotyl growth in etiolated Arabidopsis thaliana seedlings is a consequence of cell cycle arrest initiated by photodimer accumulation. J Exp Bot 65(11):2949–2961

Bornman JF, Barnes PW, Robson TM, Robinson SA, Jansen MAK, Ballare CL (2019) Linkages between stratospheric ozone, UV radiation and climate change and their implications for terrestrial ecosystems. Photochem Photobiol Sci 18(3):681–716

Brestic M, Zivcak M, Vysoka DM, Barboricova M, Gasparovic K, Yang X, Kataria S (2023) Acclimation of photosynthetic apparatus to UV-B radiation. In: UV-B radiation and crop growth. Springer

Britt AB (2004) Repair of DNA damage induced by solar UV. Photosynth Res 81:105–112

Bussell JS, Gwynn-Jones D, Griffith GW, Scullion J (2012) Above- andbelow-ground responses of Calamagrostis purpurea UV-B radiation and elevated CO2 under phosphorus limitation. Physiol Plant 145:619–628

Carvalho SD, Castillo JA (2018) Influence of light on plant–phyllosphere interaction. Front Plant Sci 9:1482

Chen Z, Dong Y, Huang X (2022) Plant responses to UV-B radiation: signaling, acclimation and stress tolerance. Stress Biol 2(1):51

Choudhary KK, Singh S, Agrawal M, Agrawal SB (2021) Role of Jasmonic and salicylic acid signaling in plants under UV-B stress. In: Jasmonates and salicylates signaling in plants, pp 45–63

Coherent Market Insights (2024) Oil seed market size and share analysis-growth trends and forecasts 2014–2031

Contreras RA, Marisol P, Hans K, Zamora P, Zúñiga GE (2019) UV-B shock induces photoprotective flavonoids but not antioxidant activity in Antarctic Colobanthus quitensis (Kunth) Bart. Environ Exp Bot 159:179–190

Correa MDSS, Saavedra MERR, Parra EAE, Ontiveros EN, Flores JDCB, Montiel JGO, Contreras JEC, Urrutia EL, Acevedo JGÁ, Nopala GEJ, González AME (2023) Ultraviolet radiation and its effects on plants. In: Oliveira M, Fernandes-Silva A (eds) Abiotic stress in plants-adaptations to climate change. IntechOpen

Damalas CA, Koutroubas SD (2022) Exogenous application of salicylic acid for regulation of sunflower growth under abiotic stress: a systematic review. Biologia 77(7):1685–1697

Datta S, Mandal D, Mitra S, Chakraborty S, Nag CR (2024) ABI3 regulates ABI1 function to control cell length in primary root elongation zone. Plant J 120(6):2437–2455

Dawood MFA, Tahjib-Ul-Arif M, Sohag AAM, Abdel Latef AAH, Ragaey MM (2021) Mechanistic Insight of Allantoin in Protecting Tomato Plants Against Ultraviolet C Stress. Plants. 10(1):11.

Duarte-Sierra A, Munzoor Hasan SM, Angers P, Arul J (2020) UV-B radiation hormesis in broccoli florets: glucosinolates and hydroxy-cinnamates are enhanced by UV-B in florets during storage. Postharvest Biol Technol 168:111278

Dukowic-Schulze S, Harvey A, Garcia N, Chen C, Gardner G (2022) UV-B irradiation results in inhibition of hypocotyl elongation, cell cycle arrest, and decreased endoreduplication mediated by miR5642. Photochem Photobiol 98(5):1084–1099

Egamberdieva D, Wirth SJ, Alqarawi AA, Abd-Allah EF, Hashem A (2017) Phytohormones and beneficial microbes: essential components for plants to balance stress and fitness. Front Microbiol 8:2104

Eissa P, Mahdi B, Abolfazl T, Yasser E (2011) Effects of UV irradiation on plants. Afr J Microbiol Res 5(14):1710–1716

Fang C, Kong F (2022) Soybean. Curr Biol 32(17):R902–R904

Favory JJ, Stec A, Gruber H, Rizzini L, Oravekz A, Funk M (2009) Interaction of COP1 and UVR8 regulates UV-B induced photomorphogenesis and stress acclimation in Arabidopsis. EMBO J 28:591–601

Gogo EO, Förster N, Dannehl D, Frommherz L, Trierweiler B, Opiyo AM (2018) Postharvest UV-C application to improve health promoting secondary plant compound pattern in vegetable amaranth. Innovative Food Sci Emerg Technol 45:426–437

Gray LA, Sandra V, Carl DS (2024) Increased UV intensity reduces pollen viability in Brassica rapa. Flora 319(2024):152582

Han Q, Zhu G, Qiu H, Li M, Zhang J, Wu X, Xiao R, Zhang Y, Yang W, Tian B, Xu L (2024) Quality traits drive the enrichment of Massilia in the rhizosphere to improve soybean oil content. Microbiome 12(1):224

Hasanuzzaman M, Bhuyan MB, Zulfiqar F, Raza A, Mohsin SM, Mahmud JA, Fotopoulos V (2020) Reactive oxygen species and antioxidant defense in plants under abiotic stress: revisiting the crucial role of a universal defense regulator. Antioxidants 9(8):681

Hectors K, Prinsen E, De Coen W, Jansen MAK, Guisez Y (2007) Arabidopsis thaliana plants acclimated to low dose rates of ultraviolet B radiation show specific changes in morphology and gene expression in the absence of stress symptoms. New Phytol 175(2):255–270

Hideg E, Vass I (1996) UV-B induced free radical production in plant leaves and isolated thylakoids. Plant Sci 115:251–260

Jan R, Asaf S, Numan M, Lubna KKM (2021) Plant secondary metabolite biosynthesis and transcriptional regulation in response to biotic and abiotic stress conditions. Agronomy 11(5): 968

Jenkins GI (2009) Signal transduction in responses to UV-B RADIATION. Annu Rev Plant Biol 60(1):407–431

Jiao J, Xu XJ, Lu Y, Liu J, Fu YJ, Fu JX (2022) Identification of genes associated with biosynthesis of bioactive flavonoids and taxoids in Taxus cuspidata Sieb. Et Zucc. Plantlets exposed to UV-B radiation. Gene 823:146384

Jovanić B, Radenković B, Despotović-Zrakić M, Bogdanović Z, Barać D (2022) Effect of UV-B radiation on chlorophyll fluorescence, photosynthetic activity and relative chlorophyll content of five different corn hybrids. J Photochem Photobiol 10:100115

Kalbin G, Ohlsson AB, Berglund T, Rydström J, Strid A (1997) Ultraviolet-B-radiation induced changes in nicotinamide and glutathione metabolism and gene expression in plants. Eur J Biochem 249:465–472

Kamaleshwaran R, Elayaraja D (2024) Enhancement of oilseeds crop production through micronutrient fertilization: a review. Asian J Soil Sci Plant Nutr 10(3):126–130

Kastakova E, Bato V, Dubenova D (2023) Competitiveness on the global oilseeds market. J Econ Res Bus Admin 144(2):3–13

Kataria S, Jajoo A, Guruprasad KN (2014) Impact of increasing ultraviolet-B (UV-B) radiation on photosynthetic processes. J Photochem Photobiol B Biol 137:55–66

Kim S, Lee H, Kim KH (2018) Metabolomic elucidation of recovery of Melissa officinalis from UV-B irradiation stress. Ind Crop Prod 121:428–433

Klem K, Oravec M, Holub P, Šimor J, Findurová H, Surá K (2022) Interactive effects of nitrogen, UV and PAR on barley morphology and biochemistry are associated with the leaf C:N balance. Plant Physiol Biochem 172:111–124

Krasylenko YA, Yemets AI, Blume YB (2013) Plant microtubules reorganization under the indirect UV-B exposure and during UV-B-induced programmed cell death. Plant Signaling & Behavior 8(5):e24031

Kumar P, Ashique S, Kumar N, Jain A, Sharma H, Pandey SN, Singh A (2024) Regulation of plant hormones under abiotic stress conditions in plants. In: Plant secondary metabolites and abiotic stress, pp 243–276

Lal S, Kaur R (2016) Deleterious impacts of supplemental ultraviolet-B radiation on the chlorophyll contents of Brassica campestris Pt-303 and counteracted by some plant growth hormone. Imp J Interdiscip Res 2(8):1601

Lam HK, McAdam SAM, McAdam EL, Ross JJ (2015) Evidence that chlorinated auxin is restricted to the Fabaceae but not to the Fabeae. Plant Physiol 168:798–803

Langyan S, Yadava P, Khan FN, Sharma S, Singh R, Bana RS, Singh N, Kaur V, Kalia S, Kumar A (2025) Trends and advances in pre-and post-harvest processing of linseed oil for quality food and health products. Crit Rev Food Sci Nutr 65(4):746–769

Li C, Yue J, Wu X, Xu C, Yu J (2014) An ABA-responsive DRE-binding protein gene from Setaria italica, SiARDP, the target gene of SiAREB, plays a critical role under drought stress. J Exp Bot 65:5415–5427

Lingakumar K (2023) Interaction of salicylate and Jasmonate on the UV-B induced changes in physiological and biochemical activities of crop plants. In: UV-B radiation and crop growth, pp 129–181

Liu L, Wang X, Chang C (2022) Toward a smart skin: harnessing cuticle biosynthesis for crop adaptation to drought, salinity, temperature, and ultraviolet stress. Front Plant Sci 13:961829

Liu Q, Wang T, Ke M, Qian C, Li J, Huang X, Gao Z, Chen X, Tu T (2024) UV-B radiation disrupts membrane lipid organization and suppresses protein mobility of GmNARK in Arabidopsis. Plan Theory 13(11):1536

Liu Y, Liu J, Wang HZ, Wu KX, Guo XR, Mu LQ (2020) Comparison of the global metabolic responses to UV-B radiation between two medicinal Astragalus species: an integrated metabolomics strategy. Environ Exp Bot 176:104094

Lockhart J (2014) How ELONGATED HYPOCOTYL5 helps protect Plants from UV-B rays. Plant Cell 26(10):3826–3826

Luo Y, Teng S, Yin H, Zhang S, Tuo X, Tran LSP (2021) Transcriptome analysis reveals roles of anthocyanin-and jasmonic acid-biosynthetic pathways in rapeseed in response to high light stress. Int J Mol Sci 22(23):13027

Lymperopoulos P, Msanne J, Rabara R (2018) Phytochrome and phytohormones: working in tandem for plant growth and development. Front Plant Sci 9:1037–1042

Mahdavian K (2024) Effects of ultraviolet radiation on plants and their protective mechanisms. Russ J Plant Physiol 71(6):184

Mathur S, Bheemanahalli R, Jumaa SH, Kakar N, Reddy VR, Gao W, Reddy KR (2024) Impact of ultraviolet-B radiation on early-season morpho-physiological traits of indica and japonica rice genotypes. Front Plant Sci 15

Mathur RK, Sujatha M, Bera SK, Rai PK, Babu BK, Suresh K, Babu SS, Sharma M, Rani K, Ajay BC, Kumar M (2023) Oilseeds and oil palm. In: Trajectory of 75 years of Indian agriculture after independence. Springer, Singapore, pp 231–264

Miao N, Yun C, Shi Y, Gao Y, Wu S, Zhang Z (2022) Enhancement of flavonoid synthesis and antioxidant activity in Scutellaria baicalensis aerial parts by UV-A radiation. Ind Crop Prod 187 (Part B):115532

Mir AR, Alam P, Hayat S (2022) Auxin regulates growth, photosynthetic efficiency and mitigates copper induced toxicity via modulation of nutrient status, sugar metabolism and antioxidant potential in Brassica juncea. Plant Physiol Biochem 185:244–259

Mishra V, Srivastava G, Prasad SM (2009) Antioxidant response of bitter gourd (Momordica charantia L.) seedlings to interactive effect of dimethoate and UV-B irradiation. Sci Hortic 120:373–378

Mittal S (2006) Structural shift in demand for food: projections for 2020.; Working Paper No. 184. Indian Council for Research on International Economic Relations (ICRIER), New Delhi, India

Mohammadizadeh HN, Tohidfar M, Maleki ZB, Mohsenpour M, Ghanbari MSR, Esmaeilzadeh SK (2024) Co-overexpression of chitinase and β-1, 3-glucanase significantly enhanced the resistance of Iranian wheat cultivars to Fusarium. BMC Biotechnol 24(1):35

Mousavi SB, Sayfzadeh S, Jabbari H, Valadabadi SA, Masouleh EH (2022) Effect of auxin foliar application on seed yield and fatty acids composition of two safflower genotypes under late-season drought. Plant Soil Environ 68(2):82–88

Mukherjee A, Dwivedi S, Bhagavatula L, Datta S (2023) Integration of light and ABA signaling pathways to combat drought stress in plants. Plant Cell Rep 42(5):829–841

Müller V, Albert A, Winkler J, Lankes C, Noga G, Hunsche M (2013) Ecologically relevant UV-B dose combined with high PAR intensity distinctly affect plant growth and accumulation of secondary metabolites in leaves of Centella asiatica L. Urban. J Photochem Photobiol B Biol 127:161–169

Mutschlechner M, Chisté D, Hauptmann D, Schöbel H (2024, 2028) Ultraviolet (Spot) light on water treatment: targeting inactivation efficiency and stress responses of gram-positive and gram-negative bacteria using UV-B and UV-C LEDs. Water 16(14)

Nasibova AN (2022) Advances in biology and earth sciences. Jomard Publishing, Azerbaijan, p 84. ISSN 2520-2847

Nocchi N, Duarte HM, Pereira RC, Konno TUP, Soares AR (2020) Effects of UV-B radiation on secondary metabolite production, antioxidant activity, photosynthesis and herbivory interactions in Nymphoides humboldtiana (Menyanthaceae). J Photochem Photobiol B Biol 12:112021

Ochoa L, Shrivastava M, Srivastava S, Cota-Ruiz K, Zhao L, White JC, Hernandez-Viezcas JA, Gardea-Torresdey JL (2025) Nanomaterials for managing abiotic and biotic stress in the soil–plant system for sustainable agriculture. Environ Sci Nano 12:1037

Oravekz A, Baumann A, Máté Z, Brzezinska A, Molinier J, Oakeley EJ (2006) CONSTITUTIVELY PHOTOMORPHOGENIC is required for the UV-B response in Arabidopsis. Plant Cell 18:1975–1990

Pandey A, Jaiswal D, Agrawal SB (2021) Ultraviolet-B mediated biochemical and metabolic responses of a medicinal plant Adhatoda vasica Nees. at different growth stages. J Photochem Photobiol B Biol 216:112142

Pascual J, Cañal MJ, Escandón M, Meijón M, Weckwerth W, Valledor L (2017) Integrated physiological, proteomic, and metabolomic analysis of ultraviolet (UV) stress responses and adaptation mechanisms in Pinus radiata. Mol Cell Proteomics 16(3):485–501

Quail PH (2002) Photosensory perception and signalling in plant cells: new paradigms? Curr Opin Cell Biol 14(2):180–188

Quan J, Song S, Abdulrashid K, Chai Y, Yue M, Liu X (2018) Separate and combined response to UV-B radiation and jasmonic acid on photosynthesis and growth characteristics of Scutellaria baicalensis. Int J Mol Sci 19(4):1194

Rajendiran K, Ramanujam MP (2004) Improvement of biomass partitioning, flowering and yield by Triadimefon in UV-B stressed Vigna radiata (L.) Wilczek. Biol Plant 48(1):145–148

Rakitin VY, Prudnikova ON, Karyagin VV, Rakitina TY, Vlasov PV, Borisova TA, Novikova GV, Moshkov IE (2008) Ethylene evolution and ABA and polyamine contents in Arabidopsis thaliana during UV-B stress. Russ J Plant Physiol 55(3):321–327

Sah SK, Reddy KR, Li J (2022) Silicon enhances plant vegetative growth and soil water retention of soybean (*Glycine max*) plants under water-limiting conditions. Plan Theory 11(13):1687

Saleem MS, Khan SH, Ahmad A, Rana IA, Naveed ZA, Khan AI (2024) The 4Fs of cotton: genome editing of cotton for fiber, food, feed, and fuel to achieve zero hunger. Front Genome Ed 6: 1401088

Sebastian A, Kumari R, Kiran BR, Prasad MNV (2018) Ultraviolet B induced bioactive changes of enzymatic and non-enzymatic antioxidants and lipids in Trigonella foenum-graecum L. (Fenugreek). Euro Biotech J 2(1):64–71

Sen Mandi S, Sen Mandi S (2016) UV acclimation strategy in plants. In: Natural UV radiation in enhancing survival value and quality of plants, pp 73–133

Sergiev I, Todorova D, Katerova Z, Shopova E, Jankauskiene J, Jurkoniene S (2017) Beneficial effects of auxin-like compounds on pea plants irradiated with uv-c. Genet Plant Physiol 7(3–4): 135–146

Shan T, Gul B, Rafique M, Jabeen A, Gurmani AR, Ortas I (2022) Cereals and phytohormones under UV stress. In: Sustainable remedies for abiotic stress in cereals, pp 425–441

Shoaib N, Pan K, Mughal N, Raza A, Liu L, Zhang J, Wu X, Sun X, Lin Z, Pan Z (2024) Potential of UV-B radiation in drought stress resilience: a multidimensional approach to plant adaptation and future implications. Plant Cell Environ 47(2):387–407

Sindhu V, Singh SS (2024) Photosynthetic resilience under stress: unravelling the role of phytohormones and phytometabolites. In: Stress biology in photosynthetic organisms: molecular insights and cellular responses, pp 147–184

Singh V, Kumar P, Pandey A, Hallan V, Pati PK (2024) Functional characterization of WsPR-1 reveals its interplay with cytokinin and gibberellin signalling pathways. Int J Biol Macromol 278:134691

Singh VP, Srivastava PK, Prasad SM (2012) Differential effects of UV-B radiation fluence rates on growth, photosynthesis, and phosphate metabolism in two cyanobacteria under copper toxicity. Toxicol Environ Chem 94(8):1511–1535

Strzałka W, Zgłobicki P, Kowalska E, Dziga D, Banaś AK (2020) The dark side of UV-induced DNA lesion repair. Genes 11(12):1450

Sun Q, Li X, Sun L, Sun M, Xu H, Zhou X (2024) Plant hormones and phenolic acids response to UV-B stress in Rhododendron chrysanthum pall. Biol Direct 19(1):40

Swarna S, Lorente C, Thomas AH, Martin CB (2012) Rate constants of quenching of the fluorescence of pterins by the iodide anion in aqueous solution. Chem Phys Lett 542:62–65

Tao M, Zhu W, Han H, Liu S, Liu A, Li S (2022) Mitochondrial proteomic analysis reveals the regulation of energy metabolism and reactive oxygen species production in Clematis terniflora DC. Leaves under high-level UV-B radiation followed by dark treatment. J Proteome 254: 104410

Tian X, Engel BA, Qian H, Hua E, Sun S, Wang Y (2021) Will reaching the maximum achievable yield potential meet future global food demand? J Clean Prod 294:126285
Vanhaelewyn L, Prinsen E, Van Der Straeten D, Vandenbussche F (2016) Hormone-controlled UV-B responses in plants. J Exp Bot 67(15):4469–4482
Vishal B, Kumar PP (2018) Regulation of seed germination and abiotic stresses by gibberellins and abscisic acid. Front Plant Sci 9:838
Wang H, Liu S, Wang T, Liu H, Xu X, Chen K (2020) The moss flavone synthase I positively regulates the tolerance of plants to drought stress and UV-B radiation. Plant Sci 298:110591
Wang L (2011) Auxin Response Factor2 (ARF2) and its regulated homeodomain gene HB33 mediate abscisic acid response in Arabidopsis. PLoS Genet 7:123–129
Wang M, Wang Y, Bilal M, Xie C, Wang P, Rui X, Yang R (2024) UV-B radiation exhibited tissue-specific regulation of isoflavone biosynthesis in soybean cell suspension cultures. Food Secur 13(15):2385
Wu M, Zhou J, Li Q, Quan D, Wang Q, Gao Y (2024) Auxin–brassinosteroid crosstalk: regulating rice plant architecture and grain shape. Crop J 12:953
Xing Q, Wang K, Yin H, Yang C, Wang M (2024) Transcriptome and miRNA reveal the key factor regulating the somatic embryogenesis of camellia oleifera. Horticulturae 10(12):1291
Yan A, Pan J, An L, Gan Y, Feng H (2012) The responses of trichome mutants to enhanced Ultraviolet-B radiation in Arabidopsis thaliana. J Photochem Photobiol B Biol 113:29–35
Yang Q, Li Y, Wang L, Zhou Q (2014) Effect of lanthanum (III) on the production of ethylene and reactive oxygen species in soybean seedlings exposed to enhanced ultraviolet-B radiation. Ecotoxicol Environ Saf 104(1):152–159
Yi C, Deng XW (2005) COP1—From plant photomorphogenesis to mammalian tumorigenesis. Trends Cell Biol 15:618–625
Zhang D (2014) Abscisic acid: metabolism, transport and signaling. Springer, New York, NY
Zhao B, Wang L, Pang S, Jia Z, Wang L, Li W et al (2020) UV-B promotes flavonoid synthesis in Ginkgo biloba leaves. Ind Crop Prod 151:112483
Zhou C, Mughal N, Zhang X, Chen J, Shoaib N, Wang X, Yong T, Yang F, Liu W, Wu X, Wu Y (2024) Soybean plants enhance growth through metabolic regulation under heterogeneous drought stress. Agric Water Manag 303:109029
Zhu W, Zheng W, Hu X, Xu X, Zhang L, Tian J (2017) Variations of metabolites and proteome in Lonicera japonica Thunb. Buds and flowers under UV radiation. Biochim Biophys Acta Proteins Proteom 1865(4):404–413
Zwack PJ, Rashotte AM (2015) Interactions between cytokinin signaling and abiotic stress responses. J Exp Bot 6(16):4863–4871

Chapter 15
Integrative Omics Strategies for Enhancing Abiotic Stress Tolerance in Oilseed Crops

Alevcan Kaplan, Barkat Ullah, Muhammad Nauman Khan, Abdul Razzaq, Nazima Wahid, Baber Ali, Karishma, Nabeel Ahmad, and Rokayya Sami

Abstract Oilseed crops are critical to world agriculture, producing essential oils, proteins, and biofuels. However, abiotic factors, including drought, salt, severe temperatures, and nutritional deficits, considerably impact their production. To improve stress tolerance, omics approaches such as genomics, transcriptomics, proteomics, metabolomics, and phenomics provide valuable tools for studying plant responses at various cellular levels. High-throughput sequencing, gene expression profiling, and proteomic investigations have discovered critical stress-responsive genes, proteins, and metabolites that participate in adaptive processes. Integrating multiomics data with systems biology and modern breeding methods, such as genome editing and marker-assisted selection, allows for the production of

A. Kaplan (✉)
Department of Crop and Animal Production, Sason Vocational School, Batman University, Batman, Turkey

B. Ullah
Department of Botany, Islamia College Peshawar, Peshawar, Pakistan

M. N. Khan
Department of Botany, Islamia College Peshawar, Peshawar, Pakistan

Biology Laboratory, University Public School, University of Peshawar, Peshawar, Pakistan

A. Razzaq
Qarshi Herb Research Center, Qarshi Industries (Pvt.) Ltd, Hattar, Haripur, Pakistan

Qarshi University, Lahore, Punjab, Pakistan

N. Wahid · B. Ali
Department of Plant Sciences, Quaid-i-Azam University, Islamabad, Pakistan

Karishma
Department of Plant Biotechnology, Atta-Ur-Rehman School of Applied Biosciences, National University of Sciences and Technology, Islamabad, Pakistan

N. Ahmad
School of Allied Sciences Dev Bhoomi Uttarakhand University, Dehradun, Uttarakhand, India

R. Sami
Department of Food Science and Nutrition, College of Sciences, Taif University, Taif, Saudi Arabia

© The Author(s), under exclusive license to Springer Nature Singapore Pte Ltd. 2025
A. A. H. Abdel Latef (ed.), *Oilseed Crops Under Abiotic Stress*, Sustainability Sciences in Asia and Africa, https://doi.org/10.1007/978-981-96-8346-8_15

stress-resistant oilseed cultivars. This chapter focuses on current advances in omics-driven research and their potential applications to improve abiotic stress tolerance in important oilseed crops. A better knowledge of plant stress responses using omics technology will pave the path for sustainable agriculture and increased crop yield under changing environmental circumstances.

Keywords Omics · Abiotic stress · Genomics · Metabolomics · Oilseed crops

15.1 Introduction

One of the biggest obstacles to agricultural output is abiotic stressors, such as drought, salt, high temperatures, and nutrient shortages, particularly in oilseed crops. The availability of essential oil and protein sources for human consumption and industrial purposes may be impacted by these conditions, which can also lower growth, yield, and seed quality (Basu et al. 2018; Farooq et al. 2021). Therefore, creating crop types that can withstand stress has emerged as a major objective to guarantee consistent yields under harsh environmental circumstances. Plant stress research has been transformed by omics techniques, which include transcriptomics, proteomics, metabolomics, phenomics, and genomics. These techniques allow for high-throughput investigations of intricate biological systems (Langridge and Fleury 2011; Singh et al. 2021). These technologies facilitate the discovery of important genes, regulatory networks, proteins, metabolites, and phenotypic traits involved in stress tolerance by offering a thorough understanding of biological responses to environmental pressures. Researchers can identify biomarkers and molecular pathways that are essential for stress adaption through omics-based studies. These can then be used to improve crops through genetic engineering and breeding.

1. *Significance of Abiotic Stress Tolerance in Oilseed Crops*

   Normal physiological and biochemical processes are upset by abiotic stressors, which lower productivity and make people more vulnerable to secondary stressors. For example, salt stress results in ion toxicity and osmotic imbalance, whereas dryness causes cellular dehydration, oxidative damage, and reduced photosynthesis (Ashraf et al. 2021). Figure 15.1 portrays the impact of abiotic stress on plants. As oilseed crops are cultivated globally, the ability to withstand these environmental stresses is crucial for stabilizing yields. In canola, for example, drought and heat stresses significantly reduce seed yield and oil content, while salt stress can limit plant growth and productivity in soybean (Kumar et al. 2019). The development of stress-tolerant oilseed crops, therefore, represents a critical strategy for improving agricultural resilience and meeting the growing demand for edible oils and protein.

2. *Overview of Omics Approaches in Abiotic Stress Research*

   The last few decades have witnessed the emergence of multiple "omics" approaches as successful technologies for plant systems. Advances in

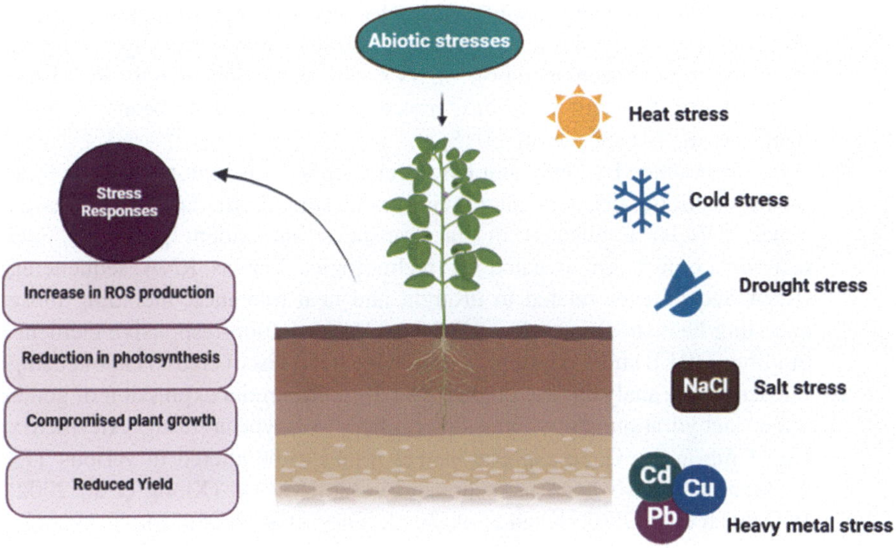

Fig. 15.1 Abiotic stress and its response in oilseed crops. ROS reactive oxygen species, NaCl sodium chloride, Cd cadmium, Cu copper, Pb lead

next-generation sequencing (NGS) have opened up new horizons by paving the way for a new generation of different omics such as genomics, transcriptomics and proteomics. However, metabolomics, ionomics and phenomics are also well documented in crop science. Multiomics approaches with high-throughput techniques have played a significant role in elucidating growth, senescence, yield and responses to biotic and abiotic stresses in a large number of crops. These omics approaches have been applied in several important crops, including soybean (*Glycine max* L.), maize (*Zea mays* L.), millet (*Setaria italica* L.), and cotton (*Gossypium hirsutum* L.). The integration of functional genomics with other omics highlights the relationships between crop genomes and phenotypes under specific physiological and environmental conditions (Yang et al. 2021a, b).

(a) *Genomics*: Finding genes linked to abiotic stress tolerance is part of the field of genomics, which studies an organism's whole genome. Stress-responsive genes and alleles in oilseed crops are found using genomic techniques like next-generation sequencing (NGS), quantitative trait locus (QTL) mapping, and genome-wide association studies (GWAS). For instance, GWAS has discovered soybean drought-tolerance QTLs, which offer useful markers for creating drought-resistant cultivars (Yue et al. 2020). GWAS is one of the effective methods used to evaluate how multiple abiotic stress factors affect oil content in sunflower (*Helianthus annuus* L.) (Mangin et al. 2017; Yang et al. 2021a, b). Yang et al. (2021a, b) reported that previously identified 48 QTLs related to maize (*Z. mays*) crop yield under heat and water stress through GWAS. Yang et al. (2021a, b) reported that structural variants (SVs) play a vital role in the genetic control of agronomically important traits in crops. The association of SVs with agronomical traits has been reported in GWAS of *B. napus*, maize (*Z. mays*) and soybean (*G. max*) (Gabur et al. 2018; Lu et al. 2015).

(b) *Transcriptomics:* By analyzing RNA transcripts, transcriptomics sheds light on how abiotic stressors alter gene expression. Differentially expressed genes (DEGs) implicated in ion transport, antioxidant pathways, and osmotic control are revealed by technologies like as RNA sequencing (RNA-Seq). Genes related to drought and heat tolerance, including those encoding heat shock proteins (HSPs) and dehydration-responsive element-binding (DREB) transcription factors, have been discovered in canola using transcriptome analysis (Liu et al. 2018a, b). Differential expression of genes under dehydration stress was determined in soybean using Affymetrix GeneChip array. Gene expression was found to be altered in various TFs in Arabidopsis, soybean in response to abiotic stress (Xiong et al. 2002; Wohlbach et al. 2008; Khan et al. 2017; Yang et al. 2021a, b).

(c) *Proteomics:* By examining the complete protein complement, proteomics makes it possible to identify stress-responsive proteins that are involved in metabolic, defensive, and signaling processes. Proteomics based on mass spectrometry (MS) has identified proteins essential for sunflowers' resistance to salt and drought, including those involved in ion transport and reactive oxygen species (ROS) scavenging. Proteomics complements transcriptomics by revealing posttranscriptional regulation and protein modifications. Functional proteomics approaches have identified ROS scavengers in sunflower crop, including quinone reductase, γ-glutamylcysteine synthetase, dehydrins, and dehydroascorbate reductase (Di-Baccio et al. 2004; Yang et al. 2021a, b). Recently, iTRAQ-based comparative proteomic analysis of two coconut cultivars identified numerous stress-responsive DEPs in two coconut cultivars (Yang et al. 2020). In addition, iTRAQ-based proteomic analysis provided new insights into somatic embryogenesis in cotton crop (Zhu et al. 2018). Therefore, iTRAQ-based quantitative proteomic studies also play important roles in the protection of crops against abiotic stresses (Yang et al. 2021a, b).

(d) *Metabolomics:* Metabolomics profiles small molecules and metabolites, offering insights into the metabolic adjustments plants make under stress conditions of salt and drought, metabolites such as proline, glycine betaine, and antioxidants build up in oilseed crops and are essential for oxidative defense and osmotic adjustment (Obata and Fernie 2019). Identifying such metabolites allows for the selection of traits that enhance stress tolerance. Various metabolomics techniques, including LC/GC-MS, GC/EI-TOF-MS, HPLC, and NMR, have been widely used in crop species such as corn and soybean in response to abiotic (drought, salt, oxidative and temperature) and biotic stress conditions (Ghatak et al. 2018; Yang et al. 2021a, b).

(e) *Phenomics*: The study of visible characteristics (phenotypes) of plants and how they react to their surroundings is known as phenomics. High-throughput phenotyping platforms now make it possible to assess traits like canopy temperature, leaf area, and root architecture under stress conditions. Stress-resistant phenotypes of oilseeds can be linked to biological responses by combining phenomics data with other omics data (Tardieu et al. 2017).

(f) *Ionomics*: The ionome concept was introduced about 10 years ago, and since then, significant progress has been made in the ionome field, mainly due to the integration of genetics and other high-throughput tools to identify genes that regulate the ionome (Lahner et al. 2003; Danku et al. 2013; Ali et al. 2021). The advances in ionomics to identify qualitative trait loci and element uptake and transport within plant tissues using genome-wide association analysis and genetic variation within species are intriguing (Ali et al. 2021). Ionomy has been used to investigate the role of minerals in different economically important crops such as *Glycine max*, *Zea mays*, *Brassica napus*, and *Olea europaea* (Ziegler et al. 2013; Baxter et al. 2014; Thomas et al. 2016; D'Attoma et al. 2019; Ali et al. 2021). Comprehensive elemental profiling has revealed QTLs responsible for grain mineral accumulation and yield in the maize crop (Yang et al. 2021a, b).

3. *Integrating Omics Approaches for Comprehensive Stress Tolerance Analysis*

   Integrating multiple omics layers provides a more holistic understanding of abiotic stress responses by linking genes to proteins, metabolites, and phenotypic traits. For example, combining genomics and transcriptomics can reveal how genetic variations influence gene expression patterns under stress. Similarly, integrating proteomics and metabolomics data might clarify how gene expression affects metabolic pathways and stress-related physiological reactions. These multiomics approaches allow researchers to identify key regulatory pathways and potential biomarkers for breeding stress-tolerant crops. Moreover, Fig. 15.2 represents the integrated multiomics strategies for improving plant resilience to abiotic stress.

15.2 Abiotic Stresses Affecting Oilseed Crops

Abiotic stresses are significant limiting variables that affect oilseed crop growth, yield, and quality. These pressures include drought, salinity, severe temperatures, and nutritional deficits. These stresses are particularly damaging to crops which are critical sources of oils and proteins for human consumption and industrial applications (Farooq et al. 2021; Kumar et al. 2020). It is essential to comprehend how oilseed crops react to these environmental difficulties in light of the growing climatic variability. Omics technologies, including transcriptomics, proteomics,

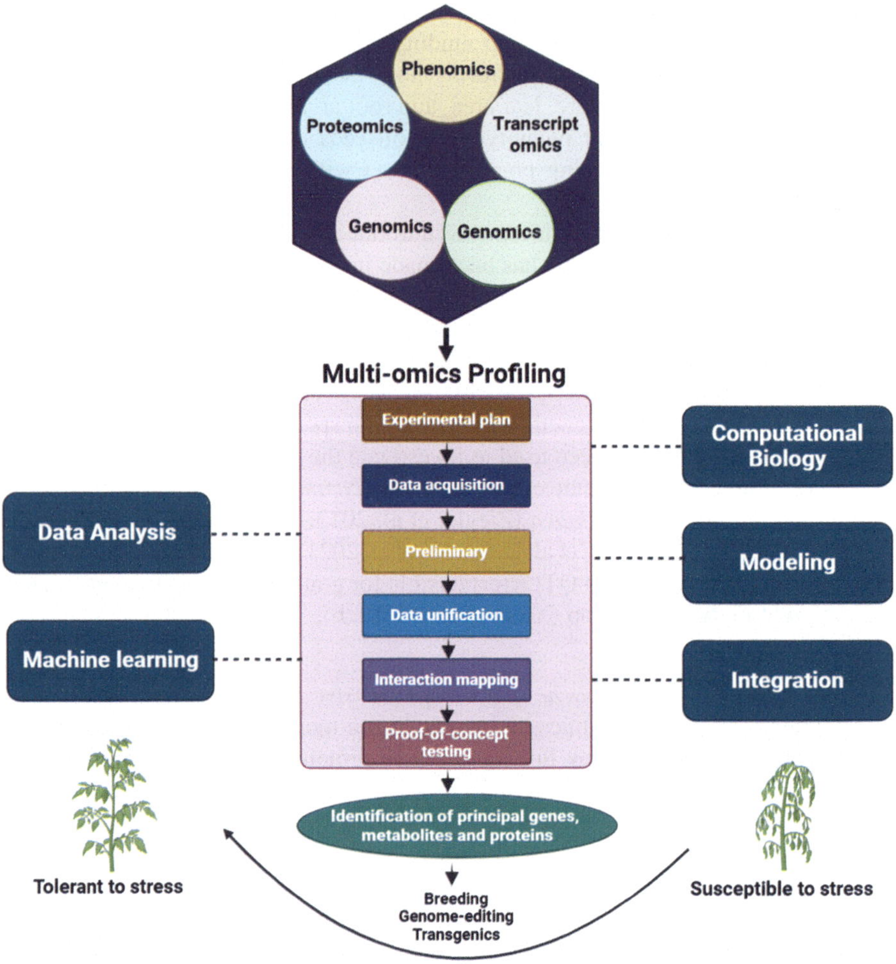

Fig. 15.2 Holistic multiomics frameworks for enhancing abiotic stress tolerance

metabolomics, phenomics, and genomics, provide effective instruments for analyzing the intricate processes that underlie oilseed crop stress responses (Langridge and Fleury 2011; Singh et al. 2021).

(a) *Genomics for Identifying Stress-Responsive Genes*

Stress-responsive genes in oilseed crops can be found using genomics techniques including next-generation sequencing (NGS), quantitative trait loci (QTL) mapping, and genome-wide association studies (GWAS). In soybean, for example, QTL mapping and GWAS have been utilized to pinpoint drought tolerance-associated loci, helping breeders identify genetic markers for improving drought resilience in breeding programs (Yue et al. 2020). In canola, candidate genes related to heat and drought tolerance, including those coding for transcription factors and heat shock proteins, have been identified using whole-genome sequencing (Liu et al. 2018a, b). The genetic foundation of abiotic stress tolerance must be discovered through genomics-based research to be used for crop improvement via genetic engineering and marker-assisted selection.

(b) *Transcriptomics for Gene Expression Profiling*

Transcriptomics provides insights into gene expression changes that occur under abiotic stresses. RNA sequencing (RNA-Seq) is commonly used to profile stress-induced transcriptional changes and identify differentially expressed genes (DEGs) associated with tolerance mechanisms. For example, in drought-stressed sunflower, transcriptomic analysis has identified genes involved in osmotic adjustment, antioxidant defense, and ion transport, which contribute to drought tolerance (Sharma et al. 2020). The increase of genes linked to heat shock proteins (HSPs) and dehydration-responsive element-binding (DREB) transcription factors, which are critical for drought and heat stress tolerance, was also found by RNA-Seq in canola (Khan et al. 2019). By identifying stress-responsive DEGs, transcriptomics facilitates the discovery of candidate genes that could be targeted in breeding programs to enhance stress tolerance.

(c) *Proteomics for Posttranslational Modifications and Functional Proteins*

Proteomics studies focus on the entire protein profile, identifying proteins that play direct roles in stress responses and providing insights into posttranslational modifications that regulate protein activity under stress. In oilseed crops, proteomics approaches have identified stress-responsive proteins involved in key pathways such as ROS scavenging, osmotic adjustment, and membrane stabilization. For example, proteomic analysis in peanut under drought stress revealed the upregulation of antioxidative enzymes, including superoxide dismutase and catalase, which mitigate oxidative damage (Ahmad et al. 2018). Proteomics complements transcriptomics by revealing functional proteins and regulatory pathways that are not always apparent from transcriptomic data, as protein abundance does not always correlate directly with mRNA levels due to posttranslational control mechanisms.

(d) *Metabolomics for Identifying Key Metabolites*

A distinct perspective on the biochemical alterations brought about by abiotic stressors is provided by metabolomics, which focuses on profiling small molecules and metabolites. Proline, glycine betaine, and sugars are examples of compatible solutes that oilseed crops accumulate during drought stress. These solutes help preserve cellular osmotic equilibrium and guard against dehydration (Obata and Fernie 2019). According to metabolomic research, drought stress causes soybeans to accumulate antioxidants and osmoprotectants, which increase drought tolerance. Furthermore, canola responds to salinity stress by raising levels of metabolites such as ascorbate and glutathione, which help detoxify ROS produced under salt stress (Shelden et al. 2020). By identifying stress-associated metabolites, metabolomics helps elucidate the metabolic adjustments that are critical for abiotic stress tolerance.

(e) *Phenomics for Stress-Associated Trait Analysis*

Phenomics integrates high-throughput phenotyping methods to quantify plant traits associated with abiotic stress responses. It includes measuring traits such as root architecture, canopy temperature, chlorophyll fluorescence, and water-use efficiency under stress conditions (Tardieu et al. 2017). In oilseed crops, phenomics has been applied to assess drought-tolerant traits in soybean, such as deeper root systems and greater water-use efficiency, which enable the plant to sustain productivity under water-limited conditions (Furbank and Tester 2011). Phenomics data can be integrated with other omics data to establish correlations between molecular responses and phenotypic traits, facilitating the identification of stress-tolerant phenotypes.

(f) *Integration of Multiomics Approaches*

Combining multiple omics layers provides a more comprehensive understanding of stress responses in oilseed crops by linking genetic and molecular data with physiological traits. For instance, integrating genomics, transcriptomics, and proteomics has revealed how genetic variations influence gene expression and protein abundance under drought and salinity stress in canola and sunflower. Similarly, integrating metabolomics with transcriptomics enables the identification of metabolic pathways affected by stress-responsive gene expression changes, providing insights into metabolic adjustments under stress conditions (Singh et al. 2021). The multiomics approach is thus instrumental in identifying candidate genes, proteins, and metabolites for developing stress-tolerant oilseed crops.

15.3 Genomics Approaches

Understanding abiotic stress tolerance in oilseed crops has benefited greatly from genomic research, which has made it possible to pinpoint important genes, metabolic pathways, and genetic variants that support resilience in the face of harsh environmental factors like drought, salinity, temperature extremes, and nutrient shortages. Through genomics, researchers aim to not only increase crop productivity but also ensure stability under stress conditions, which is essential given the rising climate challenges and global food demands (Singh et al. 2021). The discussion that follows focuses on key genomics techniques, such as next-generation sequencing (NGS), quantitative trait loci (QTL) mapping, and genome-wide association studies (GWAS), that are supporting abiotic stress tolerance research in oilseed crops.

(a) *Genome-Wide Association Studies (GWAS)*

In oilseed crops, GWAS has shown itself to be an effective method for locating genetic loci linked to abiotic stress tolerance. This method looks for single-nucleotide polymorphisms (SNPs) in the genome that are associated with particular phenotypic features, such as tolerance to drought or salinity (Varshney et al. 2014). For instance, GWAS has discovered SNP markers in soybean that are connected to drought resistance characteristics, including water-use efficiency and deeper root architecture (Yue et al. 2020). For complicated qualities like drought tolerance, where several genes may interact to impart resilience, this method is particularly helpful. Genes like DREB (Dehydration-Responsive Element-Binding protein) and LEA (Late Embryogenesis Abundant protein), which are essential for oilseed crops' stress response, have been found thanks to GWAS findings (Manavalan et al. 2009). Hassani et al. (2024) performed a genome-wide association study (GWAS) using DArTseq markers to identify marker-trait associations (MTAs) associated with drought tolerance across 90 globally diverse saffron (*Carthamus tinctorius* L.) genotypes. A total of 66 MTAs were identified for key agronomic traits, including seed yield, 1000 seed weight, number of spikes per plant, and days to maturity, and 45 MTAs were found to have significant associations with stress tolerance index (STI). Twelve MTAs were consistently identified under both stress and nonstress conditions, indicating stable loci for breeding practices. BLAST searches revealed that several identified markers corresponded to genes involved in abiotic stress response, including biotin carboxylase, serine/threonine protein kinase, and zinc finger domain proteins. Genotype clustering based on drought tolerance coefficient (DC) and STI values grouped genotypes into drought-susceptible and drought-tolerant categories without a strong correlation with geographical origin. These findings provide valuable insights into the genetic architecture of drought tolerance in safflower and support the development of drought-tolerant cultivars by highlighting candidate loci for marker-assisted selection.

(b) *Quantitative Trait Loci (QTL) Mapping*

To find genomic areas connected to abiotic stress tolerance features, QTL mapping has been widely used. By mapping the loci linked to quantifiable phenotypic qualities, this method sheds light on how complex features like heat and drought tolerance are genetically controlled. For example, QTL mapping in *Brassica napus* has shown drought-resistant areas that contain genes for antioxidant activity and osmotic control, which lessen the cellular impacts of stress (Shi et al. 2017). Similarly, QTL mapping in sunflower has revealed loci related to salinity tolerance, enabling researchers to link these regions with specific traits like ion homeostasis and cellular osmotic adjustment (Hussain et al. 2018). QTL mapping, thus, aids in pinpointing specific genomic regions that can be targeted for improving stress resilience in breeding programs.

(c) *Next-Generation Sequencing (NGS)*

NGS technologies, such as RNA sequencing (RNA-Seq) and whole-genome sequencing, have revolutionized genomics research in oilseed crops by providing large-scale, high-resolution data on gene expression, SNPs, and genetic variations. These technologies allow for detailed analysis of gene networks and pathways involved in stress responses. In canola, RNA-Seq under heat and drought stress has highlighted the upregulation of heat shock proteins (HSPs) and transcription factors (TFs), such as *HSF* (Heat Shock Factor) and *WRKY*, which are known to confer tolerance by stabilizing proteins and regulating stress-responsive genes (Wang et al. 2021). In soybeans, whole-genome sequencing has also made it possible to identify important genes linked to drought resilience, bolstering genome editing strategies meant to improve these characteristics (Zhang et al. 2016).

(d) *CRISPR-Cas9 and Genome Editing*

Genome editing tools like CRISPR-Cas9 are becoming invaluable in abiotic stress tolerance research. By enabling precise gene editing, CRISPR can be used to modify specific genes involved in stress responses. For example, CRISPR has been successfully used to edit the *DREB* gene in soybean, enhancing its expression and subsequently improving drought tolerance by boosting the plant's ability to retain water (Bao et al. 2019). Targeted gene editing is a significant advantage in oilseed crop breeding, as it reduces reliance on traditional breeding methods and enables the creation of stress-tolerant varieties more rapidly and precisely.

(e) *Gene Expression and Functional Genomics Studies*

The molecular mechanisms underlying stress tolerance are clarified by functional genomics research, which includes transcriptomics and gene expression analysis. Gene expression profiling during abiotic stress conditions allows researchers to identify candidate genes for enhancing resilience. In sunflower, gene expression studies under drought stress have shown increased expression of genes related to osmotic adjustment, such as *P5CS* (Pyrroline-5-carboxylate synthase), a key enzyme in proline synthesis (Jha et al. 2020). Functional studies also allow researchers to confirm the role of specific genes by overexpressing them in transgenic plants, helping validate their potential for improving abiotic stress tolerance in crops.

(f) *Genomic Selection (GS)*

Genomic selection is an emerging approach that uses genome-wide marker data to predict the performance of breeding lines under stress conditions. This approach allows breeders to select plants with the desired traits without waiting for phenotypic evaluations, making breeding for stress tolerance more efficient. By combining genetic information from GWAS and QTL research with phenotypic performance under controlled conditions, genomic selection has been used to predict drought tolerance in oilseed crops like canola (Golicz et al. 2016). This method accelerates the development of stress-tolerant cultivars by focusing on individuals with favorable genetic profiles.

15.3.1 Role of Genomics in Understanding Plant Stress Responses

Because genomics has made it possible to identify and characterize the genes, processes, and molecular networks that underpin stress resilience, it has completely changed our understanding of how plants respond to environmental pressures. The intricate genetic and molecular processes that plants use to withstand or lessen the negative impacts of biotic and abiotic stresses, such as drought, salinity, high temperatures, infections, and pests, can be uncovered by researchers using genomics. This section examines how genomics, including its applications in gene identification, functional genomics, transcriptional profiling, and genomic selection, is contributing to our growing understanding of plant stress responses.

(a) *Gene Identification and Functional Genomics*

The identification of stress-responsive genes in many plant species has been made easier by genomics, which has also shed light on the functions of these genes and their possible uses in enhancing plant resilience. For example, several plant species, especially oilseed crops, have been found to have drought-responsive genes like DREB (Dehydration-Responsive Element-Binding proteins) and HSP (Heat Shock Proteins), which are essential for controlling gene expression in situations of water scarcity (Zhang et al. 2020). The involvement of these genes in stress tolerance is supported by functional genomics research, which uses overexpression lines, mutants, and gene knockouts. For instance, it has been demonstrated that overexpressing the DREB1A gene in tomatoes and Arabidopsis improves drought tolerance by upregulating the expression of genes linked to downstream stress, which in turn improves cellular water retention and ion balance (Mizoi et al. 2012).

Stress-responsive pathways, such as the abscisic acid (ABA) signaling system, which is implicated in drought and salinity tolerance, are also revealed by gene identification and functional analysis. To lessen the consequences of stress, ABA controls a series of reactions that include osmotic adjustment, stomatal closure, and the activation of antioxidant enzymes (Cutler et al. 2010). By using genomics, scientists have discovered important elements of the ABA signaling cascade, like the SnRK2 kinase family, which interact with downstream targets like stress-responsive genes and transcription factors.

(b) *Transcriptional Profiling and Gene Expression Studies*

RNA sequencing (RNA-Seq) and other high-throughput sequencing technologies have made it possible for researchers to identify stress-induced genes through thorough transcriptional profiling. Understanding the temporal and geographical expression of genes in response to stressors is possible through transcriptome analysis. For instance, RNA-Seq research on rice under salinity stress has shown that the plant prioritizes ion homeostasis above development, as seen by the overexpression of ion transporters and the downregulation of genes linked to photosynthesis (Zhu 2016).

Likewise, transcriptomics has revealed the intricacy of drought response in oilseed crops such as soybeans, where exposure to drought triggers the expression of genes linked to antioxidant activity, cell membrane stability, and osmolyte production (Hu et al. 2019). The identification of regulatory networks involving transcription factors like MYB, WRKY, and NAC, which are essential in regulating downstream gene expression under stress-is another benefit of transcriptional profiling. These results are crucial for breeding efforts that select or alter particular stress-response genes to create crops that are tolerant to stress.

(c) *Genome-Wide Association Studies (GWAS)*

A potent genomics technique for locating genetic loci linked to stress tolerance characteristics in various plant populations is genome-wide association studies (GWAS). In a variety of populations, GWAS analyses identify correlations between particular genetic markers, including single-nucleotide polymorphisms (SNPs), and stress response characteristics. SNPs linked to heat tolerance in rice have been effectively discovered by GWAS, exposing potential genes connected to membrane stability and cellular heat-shock reactions (Yamaguchi et al. 2018). GWAS has been used to find SNPs linked to drought tolerance features in crops like canola and soybean, including osmotic adjustment capacity, water-use efficiency, and root architecture. For instance, GWAS found genes in soybeans associated with deep root growth, a characteristic that facilitates plants' ability to draw water from deeper soil layers during dry spells (Prince et al. 2017). By identifying the alleles linked to stress tolerance, this method offers useful markers for use in breeding and marker-assisted selection initiatives.

(d) *Genomic Selection (GS)*

In plant genomics, genomic selection is a new technique that predicts a plant's breeding value using genomic data without the need for significant phenotyping under stress. Plants with advantageous alleles for stress tolerance can be chosen by genomic selection, which estimates breeding values using genome-wide markers. When breeding complex traits that include several genes and environmental interactions, like as drought and heat tolerance, this strategy is very advantageous (Crossa et al. 2017). In wheat, genomic selection has been successfully used to increase yield forecast accuracy in hot and dry circumstances (Juliana et al. 2019). Breeding efforts are accelerated and crop resilience to climate variability is improved by genomic selection, which integrates genomic data from GWAS and QTL research to enable the quick selection of stress-tolerant cultivars.

(e) *CRISPR-Cas9 and Genome Editing*

Because genome editing techniques like CRISPR-Cas9 enable precise manipulation of stress-responsive genes, they have added a new dimension to genomics-based research on stress tolerance. Certain genes have been knocked out or enhanced using CRISPR-Cas9, which has shed light on their functional involvement in stress tolerance. For example, the OsRR22 gene in rice, which negatively controls drought tolerance, has been deleted using CRISPR-Cas9; the modified plants showed increased drought-tolerant survival and water-use efficiency (Zafar et al. 2020). CRISPR-based editing has also been applied in oilseed crops like soybean to modify genes associated with drought tolerance pathways. This technology allows researchers to fine-tune the expression of key genes, such as transcription factors and ion transporters, that are crucial for stress response, thus providing new avenues for developing resilient crops.

(f) *Transcriptomics and Functional Networks in Stress Tolerance*

Transcriptomics, combined with genomics, allows for the study of functional networks involved in stress tolerance. Transcriptomics reveals how different genes and pathways interact under stress conditions, providing insights into complex gene regulation networks. In maize, for example, transcriptomics has shown that drought induces genes related to ABA biosynthesis, secondary metabolism, and ROS scavenging. ABA regulates numerous developmental processes and adaptive stress responses in plants. Many ABA signaling components have been identified, but their interconnections and a consensus on the structure of the ABA signaling network have eluded researchers. Recently, several advances have led to the identification of ABA receptors and their three-dimensional structures, and an understanding of how key regulatory phosphatase and kinase activities are controlled by ABA. A new model for ABA action has been proposed and validated, in which the soluble PYR/PYL/RCAR receptors function at the apex of a negative regulatory pathway to directly regulate PP2C phosphatases, which in turn directly regulate SnRK2 kinases. This model unifies many previously defined signaling components and highlights the importance of future work focused on defining the direct targets of SnRK2s and PP2Cs, dissecting the mechanisms of hormone interactions (i.e., cross talk) and defining connections between this new negative regulatory pathway and other factors implicated in ABA signaling (Fábián et al. 2019). Functional networks derived from transcriptomics data also reveal key interactions between hormonal pathways (e.g., ABA, ethylene) and transcription factors that regulate stress-responsive genes.

15.3.2 Genome Sequencing and Identification of Stress-Responsive Genes

In plant biology, genome sequencing has proved revolutionary, particularly for comprehending how plants react to environmental stressors including salt, drought, and extremely high temperatures. Researchers can improve crops and increase their tolerance to abiotic stress by cataloging genes, regulatory elements, and structural changes that are crucial to stress responses through the sequencing of plant genomes. By identifying stress-responsive genes, genome sequencing contributes to the understanding of the molecular mechanisms behind stress tolerance and opens up new avenues for the development of crops that can withstand stress.

(a) *Role of Genome Sequencing in Plant Stress Research*

Genomic-based research on stress tolerance has been made possible by the whole genome sequences of model plants like *Arabidopsis thaliana* and staple crops like *Oryza sativa*. Large plant genomes can now be sequenced quickly and affordably thanks to high-throughput sequencing technologies like next-generation sequencing (NGS), which allows researchers to investigate the genetic underpinnings of stress responses in a variety of species (Michael and Jackson 2013). With reference genomes available, researchers can conduct comparative analyses to identify conserved and unique stress-response genes among different plant species. For instance, comparisons between stress-tolerant and stress-sensitive genotypes within species, like *Glycine max*, have led to the discovery of genetic variations linked to drought and heat tolerance (Prince et al. 2015).

Additionally, single-nucleotide polymorphisms (SNPs) and structural differences linked to stress tolerance can be found by genome sequencing. Different types of plants may have different levels of stress tolerance due to the effects of these genetic differences on gene expression, protein function and biochemical metabolic pathways. Genome sequencing has connected certain genes with stress-tolerant features through quantitative trait loci (QTL) mapping and genome-wide association studies (GWAS), allowing for focused breeding initiatives (Huang et al. 2010).

(b) *Identification of Stress-Responsive Genes through Genomics*

Numerous stress-responsive genes implicated in ion homeostasis, hormone signaling, ROS scavenging, and osmotic adjustment have been found by genome sequencing. For instance, it is well known that genes belonging to the families DREB (dehydration-responsive element-binding) and LEA (late embryogenesis abundant) are essential for responding to cold stress and drought. When plants experience drought, DREB transcription factors control downstream genes involved in osmolyte production, which aids in maintaining cellular water balance (Sakuma et al. 2006). Similarly, *LEA* proteins, first identified in seeds, protect cellular structures from dehydration damage by stabilizing proteins and membranes under stress conditions.

Heat shock proteins (HSPs) are another important class of stress-responsive genes. HSPs act as molecular chaperones, protecting proteins from aggregation and denaturation under high-temperature conditions. Genome sequencing in heat-tolerant plants has shown the upregulation of HSPs, which aid in protecting cellular structures from heat-induced damage (Gupta et al. 2010). For example, HSP70 and HSP90 are highly expressed in wheat and rice during heat stress, helping these plants maintain cellular functions at elevated temperatures.

(c) *Omics Integration for Identification of Stress-Responsive Genes*

Genome sequencing, combined with transcriptomics, proteomics, and metabolomics, has furthered the identification of stress-responsive genes and regulatory networks. Integrative omics approaches help link genome information to functional responses under stress conditions. Transcriptomics studies, for example, have shown how stress-responsive genes identified through genome sequencing are activated or suppressed in response to various abiotic stresses. For instance, RNA sequencing (RNA-Seq) of drought-stressed maize (*Zea mays*) revealed upregulation of genes related to ABA biosynthesis and secondary metabolism, both of which are critical for drought resilience (Fábián et al. 2019).

Additionally, metabolomics data have provided insights into stress-induced metabolic pathways regulated by identified stress-responsive genes. During salinity stress, for instance, the accumulation of osmolytes such as proline, which is encoded by *P5CS* genes, helps plants mitigate osmotic stress. In rice, genome sequencing has identified *P5CS* gene variants associated with higher proline accumulation, providing a genetic marker for salinity tolerance (Ashraf and Foolad 2007).

(d) *Functional Validation of Stress-Responsive Genes*

Once identified, stress-responsive genes undergo functional validation through techniques such as gene knockout, overexpression, and CRISPR-based genome editing. For example, in Arabidopsis, overexpression of *DREB1A*, a well-known drought-responsive gene, leads to improved drought tolerance due to the activation of downstream stress-protective genes (Sakuma et al. 2006). Similarly, CRISPR-Cas9 has been employed to knock out negative regulators of stress responses, enhancing tolerance to stresses like drought and salinity (Zafar et al. 2020).

Functional validation is essential for translating genomic information into practical applications. By verifying the roles of stress-responsive genes, researchers can focus on the most promising candidates for crop improvement through breeding or biotechnological approaches. For instance, wheat varieties with overexpressed *HSP70* genes show improved growth and yield under heat stress, highlighting the potential for genetic enhancement of stress tolerance (Qin et al. 2020).

(e) *Applications in Marker-Assisted Breeding and Genetic Engineering*

To speed up breeding for stress tolerance, marker-assisted selection (MAS) uses genetic markers connected to stress-responsive genes that have been found by genome sequencing. The time needed to generate stress-resilient crops is shortened by MAS, which allows breeders to choose plants with advantageous alleles of stress-responsive genes even when stress is not present. For example, markers linked to drought tolerance in maize and soybean have facilitated the development of varieties that can withstand water stress conditions, improving yields in drought-prone regions (Prince et al. 2015).

Genetic engineering also benefits from insights provided through genome sequencing. Improved resistance to a variety of pressures has been demonstrated by transgenic crops that produce genes that respond to stress, such as the DREB, LEA, and HSP families. For example, overexpression of the DREB1A gene in cotton increases survival in water-limited environments and improves drought tolerance (Lata and Prasad 2011).

15.3.3 Marker-Assisted Selection (MAS) and Its Role in Breeding Stress-Tolerant Oilseed Varieties

In plant breeding, marker-assisted selection (MAS) is a potent genetic method that makes it possible to choose particular features based on molecular markers linked to desired qualities. MAS are being utilized more and more in oilseed crops to increase resistance to a range of abiotic stressors, such as salinity, drought, and severe temperatures. By identifying stress-tolerant genotypes early in the growth process, even before plants display phenotypic features under field conditions, this method improves the efficiency and accuracy of breeding operations. In the wake of climate change, MAS has become crucial for creating resilient oilseed varieties that guarantee food security and sustainable farming methods.

(a) *Overview of Marker-Assisted Selection (MAS)*

In MAS, particular DNA markers associated with genes or quantitative trait loci (QTLs) that bestow desired qualities in plants are found and chosen. In contrast to traditional breeding, which depends on observable characteristics, MAS makes use of genetic markers that are strongly linked to stress-tolerance genes, such as microsatellites and single-nucleotide polymorphisms (SNPs). These markers are used as proxies for selecting plants with favorable genotypes, significantly reducing the time and effort needed to achieve desired outcomes (Collard and Mackill 2007). The MAS-guided approach is vital for crops such as soybeans, canola and sunflowers, which have been increasingly grown at the mercy of climate change and the stress factors that come with it in recent years (Azad and Zakaria 2024).

MAS is especially useful for complex, polygenic traits like stress tolerance, which involve multiple genes and intricate gene–environment interactions. Through MAS, breeders can target specific QTLs that contribute to stress responses, making it feasible to develop oilseed varieties with enhanced tolerance to drought, heat, and other environmental stresses (Varshney et al. 2014). This precision allows MAS to outperform traditional breeding methods, which are limited by long growth cycles, high resource demands, and the low heritability of certain stress-related traits. Kadirvel et al. (2020) reported that a series of high oleic cultivars were found to carry the same mutation in fatty acid desaturase 2-1 gene CtFAD2-1, the "ol" allele, which is assumed to be associated with high oleic acid content in safflower. The MAS-derived lines showed consistent expression of high oleic acid content across seasons and comparable seed/oil yield performance to the local control cultivars. The genotypic analyses reported in this study accurately predicted the robust, nondestructive, codominant, and high oleic acid trait in segregating populations; therefore, they are recommended for rapid breeding of high oleic acid cultivars in safflower. Spasibionek et al. (2020) obtained mutant winter rapeseed breeding lines capable of producing oil with high oleic acid (C18:1, more than 75%) and low linolenic acid (C18:3, less than 3%) content in Brassica napus by chemical mutagenesis using ethyl methanesulfonate. However, they reported that the mutant lines exhibited low agronomic value as they were characterized by low seed yield, low wintering, and high glucosinolate content in seed meal. They improved the mutant lines and developed high oleic and low linolenic recombinants showing both good oil quality and high agronomic value. They investigated the genotype × environment interaction for C18:1 and C18:3 acid content in seed oil. Genotyping was performed using allele-specific CAPS markers and SNaPshot assay for the selection of homozygous high oleic and low linolenic lines, respectively. They have obtained new high oleic and low linolenic winter rapeseed recombinant lines to be used as starting material for the development of new varieties that could have high value in the oil crop market. Feng et al. (2020) conducted a three-generation (G0, G1, and G2) diallelic mating and selection experiment in Camellia oleifera and used Sequence Associated Amplified Polymorphism (SRAP) to investigate their utility in a MAS framework. SRAP markers were effective in determining hybrid specificity and the presence of matrilineal or patrilineal inheritance, thus guiding mating pair choice and orientation (role as male or female). Over the 3 generations studied, estimates of genetic diversity parameters ((G0 to G1 and G1 to G2) 9.25 and 9.05: number of observed alleles; 3.12 and 7.80: mean effective alleles; 12.35 and 22.34: Nei's gene diversity; and 14.21 and 21.77: Shannon's index) showed a steady increase with percentile increase, indicating the absence of selection-associated diversity reduction. Genetic distance estimates and their correlation with heterosis were reported to be useful in guiding the selection of mating pairs to obtain desired yield and fruit attributes (fruit diameter, height, weight and index, rind thickness, number of seeds per fruit, seed weight per fruit and seed ratio). Most yield and fruit attributes exhibited high sense heritability with an increasing trend across generation intervals, indicating an increased potential for hybrid breeding for this species. Moreover, many studies on industrially important oilseed plants continue.

(b) *MAS in Breeding Drought-Tolerant Oilseed Crops*

For oilseed crops, drought tolerance is an essential characteristic, especially in areas where agricultural yield is constrained by water constraints. A complicated characteristic, drought resistance is impacted by a number of morphological, physiological, and biochemical changes. The selection of important genes linked to drought resistance, including those governing root architecture, osmotic adjustment, and antioxidant activity, is made possible by MAS.

In sunflower (*Helianthus annuus*), for instance, MAS has been used to select for QTLs associated with deep root systems, which improve water uptake during drought conditions. Studies have identified specific markers linked to root depth and density traits that correlate with enhanced drought tolerance, allowing breeders to develop varieties better adapted to water-limited environments (Haddadi et al. 2018). Similarly, in canola (*Brassica napus*), QTLs associated with drought tolerance traits like leaf water content and proline accumulation have been mapped, enabling the use of MAS for improving drought resilience (Basunanda et al. 2010).

MAS also helps identify markers for traits that contribute to drought avoidance. For example, QTLs related to stomatal conductance and leaf surface area are critical for reducing water loss under drought conditions in crops like soybean (*Glycine max*). Breeders have used MAS to choose soybean cultivars that have smaller leaves and a slower rate of transpiration, enabling the plants to retain water during drought stress (Prince et al. 2015). This strategy has sped up the creation of oilseed cultivars that can withstand drought, increasing the stability of their output in situations where water is scarce.

Studies on breeding drought stress resistant varieties of various oilseeds using the MAS technique and the techniques used are continuing on important species such as safflower (AFLPs), soybean (SSRs), sesame (SNPs), and groundnut (SSRs), respectively (Poodineh et al. 2021; Chen et al. 2021; Li et al. 2018; Gautami et al. 2012; Azad and Zakaria 2024).

(c) *MAS in Breeding Salinity-Tolerant Oilseed Crops*

Salinity tolerance is another vital trait for oilseed crops grown in saline soils, such as those found in coastal areas and irrigated regions. High soil salinity disrupts cellular homeostasis and causes ionic imbalances, leading to reduced crop productivity. MAS have facilitated the breeding of salinity-tolerant varieties by targeting genes and QTLs involved in ion homeostasis, osmotic regulation, and antioxidant defense.

In canola, QTL mapping has identified markers linked to traits associated with salinity tolerance, such as sodium exclusion, potassium retention, and high proline accumulation, which help maintain osmotic balance under salt stress. MAS enables breeders to select canola genotypes with these traits, leading to enhanced salinity tolerance without the need for extensive field testing. Similarly, sunflower breeding has benefited from the identification of QTLs associated with salt tolerance, such as those linked to higher root Na^+ exclusion and enhanced K^+ retention, which mitigate the adverse effects of salt on cellular processes.

In soybean, the development of salinity-tolerant varieties has been achieved by selecting markers linked to specific QTLs controlling salt-induced ion accumulation. For example, the *GmSALT3* locus, associated with salt tolerance, has been identified and incorporated into breeding programs through MAS. Soybean varieties carrying the *GmSALT3* allele exhibit reduced Na^+ accumulation in leaves, improving growth and yield under saline conditions (Do et al. 2016).

Studies on the breeding of species resistant to salinity stress in various oilseeds using the MAS technique and the techniques used are ongoing on important species such as sesame (SNPs) and soybean (SSR) (Li et al. 2018; Chen et al. 2008; Azad and Zakaria 2024).

(d) *MAS in Breeding Temperature-Stress-Tolerant Oilseed Crops*

The growth and yield of oilseed crops are greatly impacted by temperature extremes, including both heat and cold stress. Heat shock proteins (HSPs), antioxidant enzymes, and membrane stability under temperature changes are examples of features that MAS has shown to be useful in selecting. For example, QTLs linked to heat tolerance have been found in sunflower, which aids the plant in preserving enzyme activity and cellular stability in the face of high temperatures. MAS targeting these QTLs has facilitated the breeding of sunflower varieties capable of flowering and setting seeds even under heat stress, thus ensuring higher yield stability (Jung and Müller 2009). In canola, QTLs linked to cold tolerance at the seedling stage have been mapped, allowing breeders to select for genotypes that can germinate and grow in cooler temperatures, thus extending the growing season in temperate regions (Ebrahimi et al. 2018).

Soybean breeding for heat tolerance has also benefited from MAS by identifying markers linked to traits such as membrane stability, chlorophyll content, and ROS-scavenging enzyme activity. Heat-tolerant soybean varieties selected through MAS exhibit less cellular damage and retain higher photosynthetic efficiency under high temperatures, leading to improved productivity in warm climates (Board et al. 2011).

Studies and techniques used with the MAS technique related to the breeding of heat (cold and hot) stress resistant species of various oilseeds have been continuing from the past to the present on important species such as mustard and canola (RFLPs, AFLPs), soybean (SSRs), sunflower (ESTs), and rapeseed (AFLPs, SCARs) (Kole et al. 2002; Ikeda et al. 2009; Fernandez et al. 2008; Zeng et al. 2014; Azad and Zakaria 2024).

(e) *Advantages of MAS over Conventional Breeding*

MAS offers several advantages over traditional breeding methods, particularly for stress-tolerant oilseed crops:

(i) *Precision:* MAS allows for precise selection of favorable alleles based on molecular markers, ensuring that only desired traits are passed on to subsequent generations.
(ii) *Efficiency:* MAS accelerates the breeding process by enabling early selection, reducing the need for time-consuming and resource-intensive field trials under stress conditions.
(iii) *Targeting Polygenic Traits:* MAS is well-suited for polygenic traits like stress tolerance, where multiple genes contribute to a single trait, allowing for the cumulative selection of multiple favorable alleles (Collard and Mackill 2007).
(iv) *Marker Pyramidization:* MAS enables the combination (pyramiding) of multiple stress-resistance genes, leading to varieties with broader tolerance to multiple stresses, such as drought and salinity.

15.4 Transcriptomics Approaches

A crucial method for examining how genes are expressed in oilseed crops in response to abiotic stress is transcriptomics, which is the study of all RNA transcripts expressed in a cell or tissue under particular conditions. Through the identification of regulatory networks and differentially expressed genes (DEGs) under stress, transcriptomics contributes to the understanding of molecular mechanisms that facilitate plant adaptation to environmental stressors. This knowledge can guide breeding and genetic engineering efforts aimed at enhancing abiotic stress tolerance in important oilseed crops like *Glycine max*, *Brassica napus*, and *Helianthus annuus*.

1. *Role of Transcriptomics in Oilseed Crop Research Under Abiotic Stress*
   Extreme temperatures, salt, drought, and nutrient shortages are examples of abiotic stressors that interfere with regular physiological processes and necessitate reprogramming gene expression to preserve cellular homeostasis. RNA sequencing (RNA-Seq), microarrays, and quantitative real-time PCR (qRT-PCR) are examples of transcriptomics tools that have enabled genome-wide analysis of these variations in gene expression. RNA-Seq is a recommended technique for comprehending plant stress responses because it provides great resolution, sensitivity, and accuracy in detecting transcriptome changes .
2. *Key Findings from Transcriptomics Studies in Oilseed Crops Under Abiotic Stress*

   (a) *Drought Stress Response*
   Oilseed crops undergo significant transcriptional reprogramming in response to drought stress, affecting genes linked to oxidative stress tolerance, osmotic adjustment, and water conservation. Transcriptomics studies in soybeans have revealed overexpression of genes encoding aquaporins, dehydration-responsive element-binding (DREB) transcription factors, and late embryogenesis abundant (LEA) proteins, all of which contribute to improved drought tolerance and cell turgor. It has also been discovered that DREB and MYB transcription factors control stress-responsive pathways in canola, allowing adaptation to water-limited environments by modifying the expression of genes related to antioxidant defenses and osmotic balance (Liu et al. 2018a, b).

   (b) *Salt Stress Response*
   Osmotic and ionic imbalances brought on by salinity stress cause a variety of transcriptome alterations in oilseed crops. DEGs related to ion transport, such as those encoding potassium transporters and sodium/hydrogen exchangers (NHXs), have been found in canola by RNA-Seq research. These DEGs are essential for preserving ion homeostasis. Transcriptomics has shown that soybeans have higher expression of genes involved in the salt excessively sensitive (SOS) pathway, which reduces harm from salt by controlling ion fluxes across cell membranes. Additionally, the activation of abscisic acid (ABA) signaling pathways under salt stress has been well-documented, as ABA signaling modulates gene expression related to water retention and stress adaptation.

   (c) *Heat Stress Response*
   High temperatures disrupt protein stability and cellular function, requiring an effective transcriptional response for protection. Heat shock proteins (HSPs) and heat shock transcription factors (HSFs) are upregulated in sunflower during heat stress, according to transcriptomics research. These proteins help fold proteins and stop denatured proteins from aggregating. Heat stress has been observed to increase the expression of genes related to

antioxidant defense in canola, including those that encode catalase (CAT) and superoxide dismutase (SOD), which guard against oxidative damage brought on by heat (Jiang et al. 2019).

(d) *Cold Stress Response*

Cold stress impacts membrane fluidity, enzymatic reactions, and photosynthetic efficiency in plants. The potential of cold-responsive (COR) genes, which are involved in lipid modification and cell wall remodeling, to preserve membrane stability and cellular structure at low temperatures has been brought to light by transcriptomics investigations conducted in canola (Shi et al. 2020). By modifying the expression of cold-responsive genes that guard against cellular damage at low temperatures, C-repeat binding factors (CBFs) and cold-regulated (COR) genes have been demonstrated to improve freezing tolerance in soybeans.

(e) *Nutrient Deficiency Stress Response*

Transcriptomics has also been employed to study responses to nutrient deficiencies, such as phosphorus (P) and nitrogen (N) deficiencies, which are critical for oilseed productivity. In canola, low phosphorus availability has been found to upregulate genes related to phosphorus transporters and phosphate starvation response (PHR) proteins, which enhance phosphorus uptake and recycling under deficiency conditions (Tian et al. 2021). Similarly, in soybean, transcriptomics analyses have revealed that nitrogen deficiency triggers the expression of nitrate transporters and nitrogen assimilation genes, improving nitrogen use efficiency in low-fertility soils (Liu et al. 2020).

3. *Regulatory Mechanisms Uncovered by Transcriptomics*

Transcriptomics studies in oilseed crops (Table 15.1) have also uncovered important regulatory mechanisms, including:

Table 15.1 Key transcriptional responses in oilseed crops under abiotic stress

Crop	Stress type	Key genes/transcription factors	Role in stress tolerance	Reference(s)
Soybean	Drought	LEA, DREB, MYB	Osmotic adjustment, water conservation	Zhou et al. (2020)
Canola	Salt	NHX, SOS pathway genes	Ion homeostasis	Wang et al. (2022)
Sunflower	Heat	HSPs, HSFs	Protein stability, heat protection	Ceylan et al. (2023)
Canola	Cold	COR, CBF	Membrane stability	Shi et al. (2020)
Canola	Phosphorus deficiency	PHR, phosphate transporters	Nutrient acquisition	Tian et al. (2021)
Soybean	Nitrogen deficiency	Nitrate transporters, nitrogen assimilation genes	Nitrogen use efficiency	Liu et al. (2020)

(a) *Transcription Factors*: It has been determined that transcription factors such DREB, MYB, WRKY, and NAC are important modulators of abiotic stress reactions. Under conditions of drought and salinity stress, WRKY transcription factors are activated in canola, where they regulate genes related to water retention and antioxidant defense.
(b) *Noncoding RNAs*: Stress response genes are posttranscriptionally regulated by both long noncoding RNAs (lncRNAs) and small RNAs. It has been discovered that miRNAs such as miR398 that target superoxide dismutase genes in soybeans control oxidative stress responses, underscoring the function of RNA silencing in adjusting stress tolerance.
(c) *Alternative Splicing*: Stress conditions often induce alternative splicing, leading to different protein isoforms that can function in stress adaptation. In sunflower, alternative splicing has been observed in genes related to antioxidant enzymes, providing functional diversity in response to oxidative stress.

4. *Integrative Omics for Comprehensive Stress Response Analysis:* Combining transcriptomics with other omics approaches, such as genomics, proteomics, and metabolomics, enhances the understanding of stress responses in oilseed crops.

   (a) *Genomics and Transcriptomics*: Genome-wide association studies (GWAS) combined with transcriptomics can link genetic variations with specific gene expression patterns under stress, identifying genetic loci associated with abiotic stress tolerance.
   (b) *Proteomics and Transcriptomics*: By comparing transcriptomic data with protein expression data, researchers can study posttranscriptional regulations and identify proteins that are functionally relevant in stress tolerance.
   (c) *Metabolomics and Transcriptomics*: Studying the relationship between gene expression and metabolite levels reveals metabolic adjustments under stress, connecting biochemical changes to transcriptional regulation.

15.5 Proteomics Approaches

One effective method for comprehending the molecular pathways behind abiotic stress tolerance is proteomics, which is the study of the complete protein set expressed by an organism, tissue, or cell under specific conditions. Proteomics sheds light on the intricate biological mechanisms used by oilseed crops, including *Helianthus annuus*, *Brassica napus*, and *Glycine max*, to withstand harsh environmental factors like salinity, drought, temperature fluctuations, and nutrient shortages. By identifying key proteins and pathways responsive to these stresses, proteomics can facilitate the development of more resilient oilseed varieties.

1. *Overview of Proteomics in Oilseed Crops Under Abiotic Stress*

   Using techniques such as tandem mass spectrometry (MS/MS), liquid chromatography-mass spectrometry (LC-MS), and two-dimensional gel electrophoresis (2D-GE), proteomics entails profiling proteins that are expressed under particular environmental conditions. A thorough examination of protein abundance, posttranslational modifications (PTMs), and protein–protein interactions is made possible by these methods. Finding important enzymes, transporters, stress-responsive proteins, and signaling molecules involved in tolerance processes in oilseed crops has been made possible thanks in large part to proteomic techniques (Jorrín-Novo et al. 2019; Xu et al. 2019).
2. *Key Findings from Proteomics Studies in Oilseed Crops*

   (a) *Antioxidant Defense Mechanisms*

   ROS buildup brought on by abiotic stress frequently results in cellular damage. Proteins involved in antioxidant defense, such as ascorbate peroxidase (APX), catalase (CAT) and superoxide dismutase (SOD), have been found through proteomic investigations in oilseed crops. These proteins aid in the neutralization of ROS under stress. For instance, proteomics analyses conducted in canola under drought stress showed that antioxidant proteins, which prevent oxidative damage and preserve cellular homeostasis, were upregulated (Liu et al. 2018a, b). This antioxidant response is a key target for improving stress tolerance and is necessary for shielding cellular components from harm.

   (b) *Osmotic Adjustment and Compatible Solutes*

   Proline, glycine betaine, and trehalose are examples of suitable solutes that plants collect under drought or salinity stress. These solutes help preserve cell turgor and safeguard cellular proteins. According to Mahmood et al. (2019), proteomic investigations in soybeans have revealed an increase in enzymes involved in the manufacture of suitable solutes during drought, suggesting their function in osmotic adjustment. Finding potential genes to target in the breeding of drought-tolerant crops is made easier by these results.

   (c) *Heat Shock Proteins (HSPs) and Chaperones*

   Chaperones and heat shock proteins are essential for preserving the stability and functionality of proteins under stress. According to proteomics research on oilseed crops, HSPs are markedly elevated in response to drought and heat stress, where they serve to stop protein aggregation and misfolding. Proteomic research of sunflower, for example, has revealed that HSPs help stabilize cellular proteins and membranes, increasing the plant's resistance to high temperatures (Ghaffari et al. 2020).

   (d) *Signal Transduction Proteins*

   By sending stress signals to activate downstream pathways, proteins involved in signal transduction such as phosphatases, calcium-dependent protein kinases, and mitogen-activated protein kinases (MAPKs) play crucial roles in the abiotic stress response. Proteomics techniques have revealed that MAPKs and CDPKs are crucial regulators of salt and drought tolerance in

Table 15.2 Key proteins involved in abiotic stress tolerance in oilseed crops

Crop	Stress type	Key proteins/enzymes	Role in stress tolerance	Reference (s)
Canola	Drought	Superoxide dismutase (SOD), ascorbate peroxidase (APX)	Reactive oxygen species (ROS) scavenging	Liu et al. (2018a, b)
Soybean	Drought, salinity	Proline biosynthesis enzymes	Osmotic adjustment	Mahmood et al. (2019)
Sunflower	Heat	Heat shock proteins (HSPs)	Protein stability	Ghaffari et al. (2020)
Canola	Drought, salt	MAPKs, CDPKs	Signal transduction	Zhang et al. (2014)
Sunflower	Drought	Photosynthetic proteins, ATPases	Energy production	Basal et al. (2021)
Canola	Drought, heat	Phosphorylated metabolic enzymes	PTM-mediated stress regulation	Ali et al. (2020)

canola, allowing plants to activate genes and pathways that respond to stress (Yang et al. 2021a, b). In order to engineer stress tolerance in oilseed crops and to react to changing environmental conditions, this signaling network is essential.

(e) *Proteins Involved in Energy Metabolism*

Abiotic stress frequently interferes with plants' ability to produce and use energy. Stress-induced changes in the enzymes involved in glycolysis, the tricarboxylic acid cycle (TCA cycle), and photosynthesis have been found in oilseed crops by proteomic research. For instance, sunflower demonstrates alterations in the quantity of photosynthetic proteins and energy metabolism-related enzymes when subjected to drought stress, which reflects the plant's effort to sustain ATP synthesis in the face of water scarcity (Basal et al. 2021). The creation of cultivars that preserve energy balance under stress can be guided by this knowledge of metabolic adaptations (Table 15.2).

(f) *Posttranslational Modifications (PTMs)*

PTMs that control protein activity under stress include phosphorylation, ubiquitination, and glycosylation. Protein phosphorylation events are essential for triggering stress-response pathways in oilseed crops, according to phosphoproteomics, a subset of proteomics. PTMs are crucial for modifying metabolic pathways in response to stress, as evidenced by the observed phosphorylation of enzymes involved in carbon and nitrogen metabolism in canola during drought and heat stress (Ali et al. 2020). Understanding PTMs provides an additional layer of insight into how stress responses are regulated at the protein level.

3. *Integrative Omics for Comprehensive Understanding*

   Although proteomics provides useful information, a more complete picture of the mechanisms behind abiotic stress tolerance can be obtained by combining it with other omics technologies, such as transcriptomics, metabolomics, genomics, and epigenomics.

   (a) *Genomics and Proteomics*: Proteomics in conjunction with genome-wide association studies (GWAS) can help choose advantageous alleles for breeding programs by identifying genes and proteins associated with stress tolerance.
   (b) *Transcriptomics and Proteomics*: Transcriptomics provides insights into gene expression, while proteomics reveals posttranscriptional regulation and protein modifications, offering a more detailed view of stress response.
   (c) *Metabolomics and Proteomics*: Metabolomics identifies stress-related metabolites, while proteomics reveals the enzymes responsible for their biosynthesis, connecting metabolic changes with protein function.

15.6 Metabolomics Approaches

A key strategy for comprehending the biochemical reactions of plants to environmental stressors is metabolomics, which is the extensive study of metabolites within a biological system. Because metabolites are the byproducts of gene expression and represent the physiological state of the plant under stress, metabolomics is especially helpful in oilseed crops. These discoveries have the potential to inspire the creation of more robust cultivars and pinpoint biochemical indicators for abiotic stress tolerance breeding in oilseed crops.

1. *Importance of Metabolomics in Oilseed Crops Under Abiotic Stress*

   Significant alterations in primary and secondary metabolites result from the disruption of plant metabolism caused by abiotic conditions such as drought, salt, high temperatures, and nutrient shortages. Metabolomics methods like as nuclear magnetic resonance (NMR) spectroscopy, liquid chromatography-mass spectrometry (LC-MS), and gas chromatography-mass spectrometry (GC-MS) aid in profiling these compounds and identifying metabolic pathways linked to stress. Researchers can find important chemicals that improve stress tolerance and possible metabolic markers by tracking metabolite changes under stress (Farooq et al. 2021; Obata and Fernie 2019).
2. *Key Metabolomic Responses to Abiotic Stress in Oilseed Crops*

   (a) *Osmoprotectants and Compatible Solutes*

   Proline, glycine betaine, and polyols are examples of osmoprotectants that oilseed crops accumulate under drought and salinity stress. These substances aid in preserving the cellular osmotic balance and guard against the denaturation of proteins and membranes. Proline and other suitable solutes, which are essential for osmotic adjustment and stress tolerance, are elevated in soybeans during drought stress, according to metabolomics research (Luo et al. 2019). Glycine betaine and polyols like mannitol have also been identified as essential for mitigating osmotic stress in canola, showing that these metabolites serve as indicators and agents of stress adaptation (Jia et al. 2020).

(b) *Antioxidant Metabolites*

ROS build up because of oxidative stress brought on by abiotic stress, which can harm DNA, proteins, and lipids. Under drought and heat stress, metabolomic investigations in oilseed crops have found an increase in antioxidants like ascorbate, glutathione, flavonoids, and phenolic compounds. For instance, in sunflower, metabolomic profiling has revealed an increase in flavonoid and phenolic content under water stress, helping to neutralize ROS and protect cellular integrity. Such antioxidant responses provide valuable insights into the biochemical pathways that enhance oxidative stress tolerance.

(c) *Amino Acids and Nitrogen Metabolism*

As osmoprotectants and building blocks for secondary metabolites involved in stress defense, amino acids contribute to stress tolerance in two ways. Increases in amino acids like glutamate, aspartate, and valine have been linked to stress-induced metabolic changes in canola under salt stress, according to metabolomics research (Ahmad et al. 2018). Further emphasizing the significance of nitrogen-containing compounds in stress resilience, arginine and ornithine have also been connected to polyamine production, which stabilizes membranes and DNA under stress.

(d) *Sugars and Carbohydrate Metabolism*

Carbohydrates like sucrose, fructose, and trehalose build up during stressful situations in order to stabilize cellular structures and supply energy. According to metabolomic research, soybeans under drought stress contain higher levels of sucrose and raffinose, which help with osmotic adjustment and cell membrane stabilization (Awada et al. 2020). The idea that carbohydrates are essential for stress adaptation is supported by the discovery of trehalose, a sugar associated with desiccation tolerance, in oilseed crops under heat and drought stress.

(e) *Lipid Metabolism and Membrane Stability*

Abiotic stresses, particularly drought and salinity, affect lipid composition and membrane integrity in oilseed crops. Studies on canola metabolomics have revealed that drought stress modifies the amounts of lipid-related metabolites, including sterols, phospholipids, and fatty acids, which are important for preserving the fluidity and integrity of membranes. Furthermore, the regulatory significance of lipid metabolism in stress tolerance is highlighted by the fact that lipid signaling molecules, such as jasmonates made from linolenic acid, function as signaling agents in stress responses.

(f) *Secondary Metabolites and Stress-Induced Pathways*

Table 15.3 Key metabolites and their roles in abiotic stress tolerance in oilseed crops

Crop	Stress type	Key metabolites	Role in stress tolerance	Reference(s)
Soybean	Drought	Proline, glycine betaine	Osmotic adjustment and protection	Luo et al. (2019)
Canola	Salinity	Glutamate, aspartate	Nitrogen metabolism and polyamine synthesis	Ahmad et al. (2018)
Sunflower	Drought, heat	Flavonoids, phenolics	Antioxidant defense	Gök et al. (2022)
Soybean	Drought	Sucrose, trehalose	Osmotic adjustment, energy source	Awada et al. (2020)
Canola	Drought, salinity	Fatty acids, phospholipids	Membrane stability	Rana et al. (2021)
Sunflower	Drought	Alkaloids, terpenoids	ROS scavenging and osmotic regulation	Kefale et al. (2024)

Secondary metabolites (Table 15.3), including flavonoids, alkaloids, and terpenoids, play essential roles in stress defense by scavenging ROS, regulating osmotic pressure, and modulating signaling pathways. Drought stress causes sunflowers to accumulate flavonoids and phenolic chemicals, which help to prevent oxidative damage and provide antioxidative protection. Similarly, glucosinolates in canola act as defense compounds under salinity stress, showcasing the diversity of secondary metabolites involved in abiotic stress adaptation (Jia et al. 2020).

3. *Integrating Metabolomics with Other Omics Approaches*

   A more comprehensive understanding of stress responses is obtained by integrating metabolomics with other omics approaches, like as genomes, transcriptomics, and proteomics, even though metabolomics offers important insights into the biochemical pathways and metabolites linked to stress tolerance

   (a) *Genomics and Metabolomics*: Genome-wide association studies (GWAS) can help link metabolite levels with genetic loci, identifying candidate genes that regulate metabolite synthesis under stress.
   (b) *Transcriptomics and Metabolomics*: By linking gene expression with metabolite levels, researchers can uncover how transcriptional changes translate into metabolic responses, revealing the regulatory networks driving stress adaptation.
   (c) *Proteomics and Metabolomics*: Studying proteins involved in metabolite biosynthesis alongside the metabolites themselves provides a complete picture of biochemical pathways under stress.

15.7 Advances in Biotechnology for Abiotic Stress Tolerance

Extreme temperatures, salinity, drought, and heavy metal toxicity are examples of abiotic stressors that significantly reduce oilseed crop productivity. The creation of oilseed cultivars resistant to abiotic stress has gained importance because of the changing global climate and the growing demand for sustainable farming methods. Because stress-tolerance features are complicated and polygenic, traditional breeding methods frequently fail to produce stress-tolerant cultivars quickly. On the other hand, abiotic stress tolerance in oilseed crops has been better understood and enhanced thanks to contemporary biotechnology methods. These technologies, which include genome editing, genomic selection (GS), marker-assisted selection (MAS), and genetic engineering, are changing how breeders and researchers approach creating robust oilseed crops.

(a) *Genetic Engineering for Abiotic Stress Tolerance*

To increase oilseed crops' resistance to stress, genetic engineering entails introducing or changing particular genes. Genetically modified (GM) crops with improved drought, salinity, and heat tolerance have been created as a result of the use of genetic engineering to increase abiotic stress tolerance.

(i) *Drought Tolerance*: Osmotic control and water-use efficiency genes have been successfully inserted into *Helianthus annuus*. For instance, it has been demonstrated that the expression of the Arabidopsis dehydration-responsive element-binding protein AtDREB1A confers drought tolerance in sunflower by enhancing osmotic balance and stress-responsive gene expression. Similarly, it has been demonstrated that crop species like *Brassica napus* that overexpress P5CS (pyrroline-5-carboxylate synthetase), an enzyme involved in proline production, are more drought-tolerant (Sivamani et al. 2000).

(ii) *Salinity Tolerance*: Additionally, genetic engineering has been used to increase oilseeds' resistance to salinity. It has been demonstrated that introducing ion transport genes canola and soybean, such as SOS1 (Salt Overly Sensitive 1) and NHX1 (Na^+/H^+ exchanger), enhances their capacity to preserve cellular ion homeostasis in saline environments (Zhu 2001). By controlling the absorption and compartmentalization of sodium ions, these genes help plants thrive in high-salinity conditions by avoiding toxicity.

(iii) *Heat Tolerance*: Heat shock proteins (HSPs) play an essential role in protecting plants from heat-induced damage by stabilizing cellular structures and proteins. The overexpression of HSPs such as *HSP70* in sunflower and canola has been shown to improve heat tolerance by enhancing cellular stability during high-temperature stress (Scharf et al. 2012). While genetic engineering has shown great promise, it is also accompanied by challenges, including regulatory concerns, environmental impacts, and the acceptance of genetically modified crops by consumers and policymakers.

(b) *Marker-Assisted Selection (MAS)*

A molecular breeding method called marker-assisted selection (MAS) makes use of molecular markers connected to the genes or quantitative trait loci (QTLs) that provide desired characteristics like resistance to abiotic stress. By allowing the early selection of stress-tolerant plants without the requirement for time-consuming and resource-intensive field-based stress testing, MAS speeds up the breeding process.

(i) *Drought Tolerance:* QTLs pertaining to drought tolerance in canola, such as those governing root architecture and osmotic adjustment, have been found using MAS. For instance, the mapping of QTLs associated with root length and leaf water content made it easier to create canola varieties that can withstand drought (Khan et al. 2015). Similar to this, QTLs linked to characteristics such as osmotic adjustment, stomatal conductance, and root biomass have been found in soybeans and utilized in marker-assisted breeding to create drought-tolerant cultivars (Tuberosa et al. 2002).

(ii) *Salinity Tolerance*: MAS is also widely used in the selection of salt-tolerant oilseed crops. In sunflower, QTLs associated with sodium exclusion and potassium retention have been identified through MAS, allowing breeders to select varieties that exhibit better tolerance to saline environments. In canola, QTLs controlling salt-induced ion accumulation and osmotic regulation have been used to improve the plant's tolerance to salinity.

(iii) *Heat Tolerance*: Characteristics like membrane stability, chlorophyll content, and antioxidative enzyme activity have been the focus of marker-assisted selection for heat tolerance in oilseeds. To find QTLs that enhance membrane integrity and preserve photosynthetic efficiency in sunflowers under heat stress, MAS has been applied (Luo et al. 2013). Heat-tolerant soybean varieties were selected using MAS identification of genes related to the regulation of heat shock proteins (Wahid et al. 2007). MAS has proven to be an effective technique for improving the resistance of oilseeds to abiotic stress and enables the precise and efficient selection of the desired traits. However, the genetic diversity among the crop species, high-density markers, and reliable genetic maps are necessary for MAS to be successful.

(c) *Genome Editing Technologies*

The targeting and modification of genes in plants has been completely transformed by genome editing tools, especially CRISPR/Cas9. Researchers can change particular genes that are in charge of abiotic stress tolerance by using CRISPR/Cas9 to make precise, targeted changes in the plant genome.

(i) *Drought Tolerance*: CRISPR/Cas9 genome editing has been used to change genes linked to drought in oilseed crops. For instance, it has been possible to successfully modify the AtDREB1A gene, which is essential for drought tolerance, in canola to increase drought resistance. Without using transgenic techniques, this specific alteration can increase the plant's capacity to endure in environments with inadequate water.

(ii) *Salinity Tolerance*: In oilseeds like soybean and canola, CRISPR/Cas9 has been used to alter genes like NHX1 and SOS1 that are involved in salt tolerance. It has been demonstrated that altering these genes improves ion homeostasis and inhibits sodium ion toxicity in plant cells, hence increasing salt tolerance (Jiang et al. 2020).

(iii) *Heat Tolerance*: By altering genes that respond to heat, CRISPR/Cas9 has also been utilized to improve heat tolerance. For example, canola plants that have had their HSP17.8 gene altered have shown increased yield and survival in the face of high temperatures. A major benefit over conventional genetic engineering is the ability to accurately alter the plant genome using genome editing technologies like CRISPR/Cas9, which can enhance abiotic stress tolerance features without introducing foreign DNA. Public acceptance and regulatory issues, however, continue to be barriers to broad implementation.

(d) *Synthetic Biology and Abiotic Stress Tolerance*

Synthetic biology is an emerging field that integrates biological engineering, genomics, and biotechnology to design novel biological systems that can help plants cope with abiotic stresses. Synthetic biology can produce novel osmoprotectants, heat shock proteins, or stress-responsive enzymes, among other mechanisms for stress tolerance, by building synthetic pathways.

(i) *Osmoprotectants*: Through synthetic biology, researchers have engineered oilseed crops to produce high levels of osmoprotectants like glycine betaine, trehalose, or mannitol, which help plants maintain cellular integrity under drought and salinity stress (Gómez-Galera et al. 2010). These osmoprotectants prevent cellular dehydration and osmotic imbalance, improving stress tolerance.

(ii) *Enhanced Antioxidant Systems*: Enhancing plant antioxidant systems to reduce oxidative stress brought on by abiotic variables like heat and drought is another use of synthetic biology. Researchers can increase a plant's resistance to oxidative damage during abiotic stress by creating synthetic biosynthetic pathways for important antioxidants including glutathione, catalase, and superoxide dismutase (Kumar et al. 2014). By developing new biological systems that increase plant resilience, synthetic biology provides a novel way to increase abiotic stress tolerance. Nevertheless, the use of synthetic biology in crops is still in its infancy, and more study is required to convert lab results into practical uses.

15.8 Future Perspectives

- The integration of omics techniques offers great prospects for improving abiotic stress tolerance in oilseed crops. Advances in multiomics technologies, such as genome-wide association studies (GWAS), transcriptomics, proteomics, and metabolomics, will allow for a more detailed knowledge of stress-responsive systems at the molecular level. The use of machine learning and artificial intelligence (AI) in data analysis will improve stress tolerance prediction models, allowing more accurate trait selection in breeding programs.
- Genome editing techniques like CRISPR/Cas9 has enormous promise for targeted change of critical stress-responsive genes, permitting the production of climate-resilient oilseed varieties with higher yield stability. Furthermore, combining omics with systems biology and synthetic biology may lead to the development of crops with improved stress tolerance.
- Future research should concentrate on field validation of omics-driven discoveries in order to close the gap between laboratory studies and actual agricultural applications. Furthermore, cooperation among academics, breeders, and policymakers will be critical for converting omics advances into sustainable agriculture practices. As climate change continues to present problems, harnessing omics technology will be critical for guaranteeing food security and increasing the resilience of oilseed crops in a variety of environmental situations.

15.9 Conclusion

Omics methods have transformed our knowledge of abiotic stress tolerance in oilseed crops by giving detailed information on genetic, transcriptomic, proteomic, and metabolic responses. These technologies have allowed the discovery of critical stress-responsive genes, proteins, and metabolites, opening the path for the production of more robust crop types. The combination of multiomics data with modern breeding approaches, like as marker-assisted selection and genome editing, presents exciting opportunities for enhancing stress tolerance. Despite substantial advances, there are still hurdles in transferring laboratory results into field applications. Future research should combine omics with systems biology, artificial intelligence, and precision agriculture to improve stress resistance in oilseed crops. Collaboration among academics, breeders, and policymakers will be critical in using omics technology for sustainable agriculture. By capitalizing on these advances, we may create climate-resilient oilseed crops that assure food security and agricultural sustainability in the face of environmental problems.

References

Ali S, Zafar Z, Ahmad R, Zafar Z, Ahmad R (2020) Phosphoproteomics: a tool to decipher regulatory phosphorylation events during abiotic stress in plants. 176:104094

Ali S, Tyagi A, Bae H (2021) Ionomic approaches for discovery of novel stress-resilient genes in plants. Int J Mol Sci 22(13):7182

Ashraf M, Foolad MR (2007) Roles of glycine betaine and proline in improving plant abiotic stress resistance. Environ Exp Bot 59(2):206–216

Ashraf M, Akram NA, Al-Qurainy F (2021) Genetic resources of oilseed crops: a global perspective on their use for abiotic stress tolerance. 68(7):1871–1885

Awada SA, Younes JA, Othman RT (2020) Drought tolerance mechanisms in soybean. 167:103852

Azad HN, Zakaria AR (2024) Molecular marker utilization in oilseed crop breeding: a review. J Plant Mol Breed 12(1):107–120

Bao A, Chen H, Chen L, Chen S, Hao Q (2019) CRISPR/Cas9-mediated targeted mutagenesis of DREB enhances drought tolerance in soybean. 145:120–128

Basal H, Topcuoglu A, Uygun D (2021) Proteomic and metabolomic adjustments in sunflower (Helianthus annuus) under drought stress. 168:353–362

Basu S, Roychoudhury A, Majumder AL (2018) Abiotic stress adaptation and "omics" approaches for crop improvement. 155:143–165

Basunanda P, Spiller TH, Hasan M, Geiger HH, Freytag S, Snowdon RJ, Wittkop B (2010) Marker-assisted increase of genetic diversity improves drought tolerance in canola (Brassica napus L). 176(1):285–297

Baxter IR, Ziegler G, Lahner B, Mickelbart MV, Foley R, Danku J, Armstrong P, Salt DE, Hoekenga OA (2014) Single-Kernel Ionomic profiles are highly heritable indicators of genetic and environmental influences on elemental accumulation in maize grain (*Zea mays*). PLoS One 9:e87628

Board JE, Kahlon CS, Board JE, Kahlon CS (2011) Soybean yield formation: what controls it and how it can be improved. In: Soybean physiology and biochemistry. IntechOpen

Ceylan Y, Altunoglu YC, Horuz E (2023) HSF and Hsp Gene Families in sunflower: a comprehensive genome-wide determination survey and expression patterns under abiotic stress conditions. Protoplasma 260(6):1473–1491

Chen H, Cui S, Fu S, Gai J, Yu D (2008) Identification of quantitative trait loci associated with salt tolerance during seedling growth in soybean (*Glycine max* L.). Aust J Agric Res 59(12): 1086–1091

Chen H, Kumawat G, Yan Y, Fan B, Xu D (2021) Mapping and validation of a major QTL for primary root length of soybean seedlings grown in hydroponic conditions. BMC Genomics 22: 1–9

Collard BCY, Mackill DJ (2007) Marker-assisted selection: an approach for precision plant breeding in the twenty-first century. Philos Trans R Soc Lond B Biol Sci 363(1491):557–572

Crossa J, Pérez-Rodríguez P, Cuevas J, Montesinos-López O, Jarquín D, de los Campos G, Burgueño J, González-Camacho JM, Pérez-Elizalde S, Beyene Y, Dreisigacker S, Singh R, Zhang X, Gowda M, Roorkiwal M, Rutkoski J, Varshney RK (2017) Genomic selection in plant breeding: methods, models, and perspectives. Trends Plant Sci 22(11):961–975

Cutler SR, Rodriguez PL, Finkelstein RR, Abrams SR (2010) Abscisic acid: emergence of a core signaling network. Annu Rev Plant Biol 61(1):651–679. https://doi.org/10.1146/annurev-arplant-042809-112122

D'Attoma G, Morelli M, Saldarelli P, Saponari M, Giampetruzzi A, Boscia D, Savino VN, la Fuente LD, Cobine PA (2019) Ionomic differences between susceptible and resistant olive cultivars infected by Xylella Fastidiosa in the outbreak area of Salento, Italy. Pathogens 8:272

Danku JMC, Lahner B, Yakubova E, Salt DE (2013) Large-scale plant ionomics. Methods Mol Biol 953:255–276

Di-Baccio D, Navari-Izzo F, Izzo R (2004) Seawater irrigation: antioxidant defence responses in leaves and roots of a sunflower (*Helianthus annuus* L.) ecotype. J Plant Physiol 161:1359–1366

Do TD, Chen H, Hien VTT, Hamwieh A, Yamada T, Sato T et al (2016) Ncl synchronously regulates Na^+, K^+, and H^+ accumulations in soybean and greatly increases the grain yield in saline field conditions. Sci Rep 6:19147

Ebrahimi A, Karami E, Kahrizi D, Mohammadi R (2018) Genetic analysis of seed yield and related traits in canola (Brassica napus L) under drought and non-drought conditions. 21(3):257–266

Fábián A, Sáfrán E, Szabó-Eitel G, Barnabás B, Jäger K (2019) Physiological responses of wheat to drought stress: current status and future breeding prospects. 288:110–115

Farooq M, Hussain M, Wakeel A, Siddique KH (2021) Salt stress in oilseed crops: a review. 191: 104347

Feng JL, Jiang Y, Yang ZJ, Chen SP, El-Kassaby YA, Chen H (2020) Marker-assisted selection in *C. oleifera* hybrid population. Silvae Genetica 69(1):63–72

Fernandez P, Di Rienzo J, Fernandez L, Hopp HE, Paniego N, Heinz RA (2008) Transcriptomic identification of candidate genes involved in sunflower responses to chilling and salt stresses based on cDNA microarray analysis. BMC Plant Biol 8:1–18

Furbank RT, Tester M (2011) Phenomics – technologies to relieve the phenotyping bottleneck. Trends Plant Sci 16(12):635–644

Gabur I, Chawla HS, Liu X, Kumar V, Faure S, von Tiedemann S et al (2018) Finding invisible quantitative trait loci with missing data. Plant Biotechnol J 16:2102–2112

Gautami B, Pandey MK, Vadez V, Nigam SN, Ratnakumar P, Krishnamurthy L et al (2012) Quantitative trait locus analysis and construction of consensus genetic map for drought tolerance traits based on three recombinant inbred line populations in cultivated groundnut (*Arachis hypogaea* L.). Mol Breed 30:757–772

Ghaffari H, Nayyar H, Tuteja N (2020) Heat shock proteins: plant responses and functions under abiotic stresses. 11:215

Ghatak A, Chaturvedi P, Weckwerth W (2018) Metabolomics in plant stress physiology. Adv Biochem Eng Biotechnol 164:187–236

Gök SB, Erdoğdu Y, Göçmen E, Salbaş B, Erdem T (2022). Evaluation of the phenolic content and antioxidant activity of sunflower seeds under deficit irrigation conditions. Rivista Italiana delle Sostanze Grasse, 99(4).

Golicz AA, Bayer PE, Edwards D (2016) Recent advances in the understanding of the genetic basis of drought tolerance in canola. 30:112–118

Gómez-Galera S, Miralpeix B, Prat S (2010) Transgenic plants as bioreactors for osmoprotectants. 8(5):524–538

Gupta SC, Sharma A, Mishra M, Mishra RK, Chowdhuri DK (2010) Heat shock proteins in toxicology: how close and how far? Life Sci 86(11):377–384. https://doi.org/10.1016/j.lfs.2009.12.015

Haddadi P, Majidi MM, Saeidi G (2018) Identification of drought tolerance mechanisms in canola (Brassica napus L) cultivars under water stress. 20(2):327–338

Hassani SMR, Pourdad SS, Naji AM et al (2024) Genome-wide association mapping in safflower (*Carthamus tinctorius* L.) for genetic dissection of drought tolerance using DArTseq markers. Sci Rep 14:31490

Hu H, Xie W, Li J (2019) Stress responses in plants: mechanisms of drought and cold stress responses in Arabidopsis and rice. 10:112

Huang X, Wei X, Sang T, Zhao Q, Feng Q, Zhao Y, Li C, Zhu C, Lu T, Zhang Z, Li M, Fan D, Guo Y, Wang A, Wang L, Deng L, Li W, Lu Y, Weng Q et al (2010) Genome-wide association studies of 14 agronomic traits in rice landraces. Nat Genet 42(11):961–967. https://doi.org/10.1038/ng.695

Hussain SS, Kayani MA, Amjad M (2018) QTL mapping for salinity tolerance in sunflower. 17(4): gmr18069

Ikeda T, Ohnishi S, Senda M, Miyoshi T, Ishimoto M, Kitamura K, Funatsuki H (2009) A novel major quantitative trait locus controlling seed development at low temperature in soybean (*Glycine max*). Theor Appl Genet 118:1477–1488

Jha UC, Bohra A, Jha R (2020) Genetic engineering for abiotic stress tolerance in oilseed crops: present status and future perspectives. 60(1):1–20
Jia H, Wang X, Han Y, Liang J (2020) Salinity-induced osmotic and ionic stress tolerance in oilseed crops. 252:53258
Jiang Y, Qiu Y, Xu X (2019) Transcriptome analysis of heat stress-responsive genes in canola. 19: 459
Jiang WZ, Zhou H, Fromm M (2020) A CRISPR/Cas9 system for targeted mutagenesis in plant genomes (nature communications). Nat Commun 2735
Jorrín-Novo JV, Komatsu S, Weckwerth W, Wienkoop S, Plant Proteomics and Environmental Stress Group (2019) Plant proteomics and environmental stress: An overview. Proteomics 19(6):e1800042
Juliana P, Singh RP, Poland J, Mondal S, Crossa J, Montesinos-López OA (2019) Genomic selection in wheat breeding: current status and future prospects. Crop Sci 59(2):627–645
Jung C, Müller AE (2009) Flowering time control and applications in plant breeding. Trends Plant Sci 14(10):563–573
Kadirvel P, Veerraju C, Senthilvel S, Yadav P, Kiran BU, Shaik M et al (2020) Marker-assisted selection for fast-track breeding of high oleic lines in safflower (*Carthamus tinctorious* L.). Ind Crop Prod 158:112983
Kefale H, You J, Zhang Y, Getahun S, Berhe M, Abbas, ... Wang L (2024) Metabolomic insights into the multiple stress responses of metabolites in major oilseed crops. Physiol Plant 176(6): e14596
Khan AA, Singh A, Shahnaz S (2015) Marker-assisted selection for drought tolerance in canola. 199(3):411–424
Khan F, Chai HH, Ajmera I, Hodgman C, Mayes S, Lu C (2017) A transcriptomic comparisonof two bambara groundnut landraces under dehydration stress. Genes 8:1–19
Khan AS, Ali S, Farooq M, Ullah S (2019) Improving drought tolerance in canola by foliar application of mineral nutrients. 111(1):248–255
Kole C, Thormann CE, Karlsson BH, Palta JP, Gaffney P, Yandell B, Osborn TC (2002) Comparative mapping of loci controlling winter survival and related traits in oilseed *Brassica rapa* and *B. napus*. Mol Breed 9:201–210
Kumar S, Sharma P, Bansal N (2014) Engineering abiotic stress tolerance in plants using synthetic biology. 150(3):341–352
Kumar S, Thakur K, Kaushal N (2019) Impact of drought and heat stress on performance of oilseed crops and adaptation strategies. 11(2):180–185
Kumar S, Thakur K, Kaushal N (2020) Impact of drought and heat stress on performance of oilseed crops and adaptation strategies. 11(2):180–185
Lahner B, Gong J, Mahmoudian M, Smith EL, Abid KB, Rogers EE, Guerinot ML, Harper JF, Ward JM, McIntyre L et al (2003) Genomic scale profiling of nutrient and trace elements in *Arabidopsis thaliana*. Nat Biotechnol 21:1215–1221
Langridge P, Fleury D (2011) Genomics-assisted breeding for drought tolerance. 14(3):318–324
Lata C, Prasad M (2011) Role of DREBs in regulation of abiotic stress responses in plants. 62(14): 4731–4748
Li D, Dossa K, Zhang Y, Wei X, Wang L, Zhang Y et al (2018) GWAS uncovers differential genetic bases for drought and salt tolerances in sesame at the germination stage. Genes 9(2):87
Liu T, Zhu S, Tang Q, Chen P, Zhang L (2018a) Proteomic analysis of drought-stressed canola leaves using iTRAQ. 9:826
Liu Y, Zhang W, Xu H (2018b) Transcriptome profiling reveals nitrogen deficiency response in soybean. 62(6):888–900
Liu Y, Zhang W, Xu H (2020) Transcriptome profiling reveals nitrogen deficiency response in soybean. 62(6):888–900
Lu F, Romay MC, Glaubitz JC, Bradbury PJ, Elshire RJ, Wang T et al (2015) High-resolution genetic mapping of maize pan-genome sequence anchors. Nat Commun 6:6914

Luo X, Zhu X, Shen L (2013) Mapping and identification of QTLs for heat tolerance in sunflower (Helianthus annuus L) under field conditions. 191(3):339–348

Luo M, Liu X, Sun X, He Y (2019) Role of compatible solutes in osmotic adjustment under drought stress in soybean. 140:215–226

Mahmood T, Islam F, Hussain S (2019) Exploring drought-induced responses in soybean through proteomics. 518(4):728–735

Manavalan LP, Guttikonda SK, Tran LS, Nguyen HT (2009) Physiological and molecular approaches to improve drought resistance in soybean. 50(7):1260–1276

Mangin B, Casadebaig P, Cadic E, Blanchet N, Boniface MC, Carrère S et al (2017) Genetic control of plasticity of oil yield for combined abiotic stresses using a joint approach of crop modelling and genome-wide association. Plant Cell Environ 40:2276–2291

Michael TP, Jackson S (2013) The first 50 plant genomes. 14(6):113

Mizoi J, Shinozaki K, Yamaguchi-Shinozaki K (2012) AP2/ERF family transcription factors in plant abiotic stress responses. Biochim Biophys Acta (BBA) 1819(2):86–96

Obata T, Fernie AR (2019) Metabolomics and photosynthesis in plants under environmental stress. 70:123–150

Poodineh M, Nezhad NM, Mohammadi-Nejad G, Fakheri BA, Ebrahimi F (2021) Identification of safflower (*Carthamus tinctorius* L.) QTL under drought stress and normal conditions. Ind Crop Prod 171:113889

Prince SJ, Murphy M, Mutava RN, Nguyen HT (2015) Genomics and breeding for drought tolerance in soybean. 38(9):1702–1721

Prince SJ, Mutava RN, Peiris KHS, Chen P (2017) Genomic analysis of abiotic stress tolerance in soybean. 10(1)

Qin F, Shinozaki K, Yamaguchi-Shinozaki K (2020) Molecular mechanisms of drought tolerance in plants. 33(4):484–500

Rana S, Ali S, Zhang H (2021) Drought-induced alterations in lipid metabolism in oilseed crops

Sakuma Y, Maruyama K, Qin F, Osakabe Y, Shinozaki K, Yamaguchi-Shinozaki K (2006) Functional analysis of an Arabidopsis transcription factor, DREB2A, involved in drought-responsive gene expression. 18(5):1292–1309

Scharf KD, Hecker A, Jäger K (2012) Heat shock transcription factors: structure and function. 17(1):1–14

Sharma D, Aggarwal P, Kaur J (2020) Genetic and molecular insights into drought tolerance in sunflower. 29(3):429–441

Shelden MC, Roessner U, Sharp RE, Tester M (2020) Metabolomics for improving plant abiotic stress tolerance. 154:283–296

Shi J, Li R, Qiu D (2017) Genome-wide association analysis for drought tolerance and related traits in canola. 8:1050

Shi Y, Ding Y, Yang S (2020) Cold stress responses in plants: transcriptomic insights. 252:153258

Singh D, Singh CK, Shukla S, Sharma S (2021) Role of omics technologies in improving abiotic stress tolerance in oilseed crops. 140(5):587–602

Sivamani E, Murata N, Wada S (2000) Molecular breeding of drought-tolerant crops in the era of functional genomics. 3(2):149–155

Spasibionek S, Mikołajczyk K, Ćwiek–Kupczyńska H, Piętka T, Krótka K, Matuszczak M et al (2020) Marker assisted selection of new high oleic and low linolenic winter oilseed rape (*Brassica napus* L.) inbred lines revealing good agricultural value. PLoS One 15(6):e0233959

Tardieu F, Cabrera-Bosquet L, Pridmore TP, Bennett MJ (2017) Plant phenomics, from sensors to knowledge. Curr Biol 27(15):R770. https://nottingham-repository.worktribe.com/output/877029

Thomas CL, Alcock TD, Graham NS, Hayden R, Matterson S, Wilson L, Young SD, Dupuy LX, White PJ, Hammond JP et al (2016) Root morphology and seed and leaf ionomic traits in a Brassica napus L: diversity panel show wide phenotypic variation and are characteristic of crop habit. BMC Plant Biol 16:214

Tian J, Venkatachalam P, Zhu X (2021) Transcriptomic analysis reveals phosphorus deficiency responses in Brassica napus. 40(6):1073–1087

Tuberosa R, Salvi S, Sanguineti MC (2002) Mapping QTLs for drought tolerance in maize: an overview. 53(389):697–709

Varshney RK et al (2014) Integrated genomic approaches for improving drought tolerance in crop plants. 30:24–30

Wahid A, Gelani S, Ashraf M (2007) Heat tolerance in plants: an overview. Environ Exp Bot 61(3): 199–223

Wang J, Lu C, Zhang Y (2021) Genome-wide transcriptome analysis of drought stress response in canola. 302:110685

Wang W, Pang J, Zhang F, Sun L, Yang L, Fu T, ... Siddique KH (2022) Salt responsive transcriptome analysis of canola roots reveals candidate genes involved in the key metabolic pathway in response to salt stress. Sci. Rep 12(1):1666

Wohlbach DJ, Quirino BF, Sussman MR (2008) Analysis of the Arabidopsis histidine kinase ATHK1 reveals a connection between vegetative osmotic stress sensing and seed maturation. Plant Cell 20:1101–1117

Xiong L, Schumaker KS, Zhu JK (2002) Cell signaling during cold, drought, and salt stress. Plant Cell 14:65–S183

Xu G, Yuan M, Ai C, Liu L, Zhuang E (2019) Proteomic analysis reveals differentially expressed proteins under salt stress in canola. 202:103373

Yamaguchi M, Fujii N, Yamakawa H (2018) Association of the OsSIZ1 gene with heat tolerance in rice. 178(1):469–482

Yang Y, Saand MA, Abdelaal WB, Zhang J, Wu YI, Li J et al (2020) TRAQ-based comparative proteomic analysis of two coconut varieties reveals aromatic coconut cold-sensitive in response to low temperature. J Proteome 220:103766

Yang X, Lu M, Wang Y, Wang Y, Liu Z, Chen S (2021a) Response mechanism of plants to drought stress. Horticulturae 7(3):50

Yang Y, Saand MA, Huang L, Abdelaal WB, Zhang J, Wu Y, Li J, Sirohi MH, Wang F (2021b) Applications of multi-omics technologies for crop improvement. Front Plant Sci 12:563953

Yue X, Li X, Chen X, Zhang Z (2020) Mapping drought tolerance QTLs in soybean using GWAS. 133(6):1715–1727

Zafar SA, Zaidi SS, Gaba Y, Singla-Pareek SL, Pareek A (2020) CRISPR/Cas9 mediated targeted mutagenesis in rice for abiotic stress tolerance. 61:138–145

Zeng X, Li W, Wu Y, Liu F, Luo J, Cao Y et al (2014) Fine mapping of a dominant thermo-sensitive genic male sterility gene (BntsMs) in rapeseed (*Brassica napus*) with AFLP-and *Brassica rapa*-derived PCR markers. Theor Appl Genet 127:1733–1740

Zhang H, Liu W, Zhang Z, Deng Y, Niu M, Yang F, … Jiang YQ (2014) Identification, expression and interaction analyses of calcium-dependent protein kinase (CPK) genes in canola (Brassica napus L.). BMC genomics 15:1–17

Zhang H, Yang Y, Wang C, Liu M (2016) Whole-genome resequencing and SNPs identification in soybean. 17(1):1–12

Zhang H, Liu Y, Zhang Y (2020) Molecular mechanisms for improving drought tolerance in plants. 71(10):2829–2834

Zhou Y, Zhou W, Liu W, Liu H, Li Z (2020) Genome-wide analysis of the soybean DREB gene family: Identification, genomic organization and expression profiles in response to drought stress. Plant Breed 139(6):1158–1167

Zhu JK (2001) Plant salt tolerance. 6(2):66–71

Zhu JK (2016) Abiotic stress signaling and responses in plants. 167(2):313–324

Zhu HG, Cheng WH, Tian WG, Li YJ, Liu F, Xue F et al (2018) iTRAQ-based comparative proteomic analysis provides insights into somatic embryogenesis in *Gossypium hirsutum* L. Plant Mol Biol 96:89–102

Ziegler G, Terauchi A, Becker A, Armstrong P, Hudson K, Baxter I (2013) Ionomic screening of field-grown soybean identifies mutants with altered seed elemental composition. Plant Genome 6

Chapter 16
Nanotechnology in Oilseed Crops: A New Frontier for Abiotic Stress Adaptation

Nasir Assad, Ansa, Marzia Batool Laila, Sher Ullah, Muhammad Adnan, Sana Wahab, Alevcan Kaplan, Muhammad Nauman Khan, and Barkat Ullah

Abstract Oilseed crops like soybean (*Glycine max*), rapeseed (*Brassica napus*), and sunflower (*Helianthus annuus*) are essential in global agriculture, offering edible oils, protein sources, and raw materials for various industrial applications. Nonetheless, abiotic stresses frequently undermine their productivity, encompassing factors such as salinity, drought, heavy metal toxicity, radiation, and ozone stress. Conventional techniques for improving crop resilience face constraints, highlighting the need for novel strategies like nanotechnology. Nanoparticles (NPs) demonstrate considerable promise in alleviating abiotic stresses through the enhancement of nutrient uptake, the improvement of photosynthetic efficiency, and the activation of plant defense mechanisms. Metal-based nanoparticles such as ZnONPs, TiO_2NPs, and Fe_3O_4NPs play an important role in alleviating stress through the reduction of oxidative damage and the enhancement of metabolic stability. Moreover, nanoparticles derived from biopolymers, like chitosan NPs, have shown efficacy in scavenging reactive oxygen species (ROS), thereby mitigating cellular damage during stress conditions. Recent developments suggest that nanotechnology has the potential to significantly transform sustainable oilseed crop production, particularly in challenging environmental circumstances. While the results are encouraging, there are still significant challenges in comprehending the long-term environmental effects and the safe use of nanoparticles in agriculture. Further investigation is essential to enhance NP formulations, implement extensive field trials, and develop

N. Assad (✉) · Ansa · M. B. Laila · S. Ullah
Institute of Chemistry, University of Sargodha, Sargodha, Pakistan

M. Adnan
Iqbal Ahmad Qarshi Laboratory, Qarshi Industries (Pvt.), Ltd, Hattar Haripur, Pakistan

S. Wahab
Department of Plant Sciences, Quaid-i-Azam University, Islamabad, Pakistan

A. Kaplan
Department of Crop and Animal Production, Sason Vocational School, Batman University, Batman, Turkey

M. N. Khan · B. Ullah
Department of Botany, Islamia College Peshawar, Peshawar, Pakistan

© The Author(s), under exclusive license to Springer Nature Singapore Pte Ltd. 2025
A. A. H. Abdel Latef (ed.), *Oilseed Crops Under Abiotic Stress*, Sustainability Sciences in Asia and Africa, https://doi.org/10.1007/978-981-96-8346-8_16

regulatory frameworks for their responsible application. The potential for integrating nanotechnology with precision agriculture and smart delivery systems holds the promise of transforming crop management. This advancement could enhance yield and stress tolerance, all while reducing environmental risks. This investigation emphasizes the possible uses of nanotechnology in oilseed crops and its significance in promoting sustainable agricultural practices.

Keywords Nanotechnology · Oilseed crops · Abiotic stress · Heavy metals stress · Drought stress · Salinity stress · Radiation stress · Ozone stress

16.1 Introduction

Nanotechnology is an engineering field at the molecular scale, typically 1 to 100 nanometers (Assad et al. 2023; Hussain et al. 2025). In the recent decade, nanotechnology has emerged as a game changer in many sectors, including environmental remediation (Jabbar et al. 2023; Afzal et al. 2024), medicine (Hashmi et al. 2024; Khan et al. 2023), photocatalysis (Ghaffar et al. 2023; Assad et al. 2024), and agriculture (Paliani and Ratiani 2025). The ability to engineer materials at the molecular or atomic level makes it a new tool for tackling longstanding agricultural problems, from improving crop yields to enhancing nutrient use and reducing the impact on the environment (Thakur and Yadav 2025). Nanotechnology plays a significant role in agriculture by helping the development of products such as nanofertilizers, nanopesticides, and nanosensors that support precision farming through the targeted use of nutrients and agrochemicals for reducing wastes and contamination (Salem et al. 2022; Abdel Latef et al. 2020; Abdelhameed et al. 2021; El-Sayed et al. 2025). For example, nanofertilizers can provide nutrients in a controlled manner, based on the requirement of the crop's growth, so that losses due to percolation or volatilization can be minimized. Nanopesticides are more target-specific in their application and therefore lead to a lower application rate of the pesticides and thus reduce the nontarget effects. Furthermore, the nanosensors give real-time information on the soil, the plant itself and the environment, which in turn leads to early intervention and optimal use of resources. All these changes are combined to support sustainable farming and food safety with the increasing population of the world at present (Dev et al. 2025).

Oilseed crops such as rapeseed (*Brassica napus*), soybean (*Glycine max*), sunflower (*Helianthus annuus*), sesame (*Sesamum indicum*), and mustard (*Brassica juncea*), etc. are dominating global agriculture because they contribute to vegetable oils, proteins, and biofuels. These crops add essential fatty acids and are used as feedstock for multiple industrial applications, including biofuels (biodiesel and jet fuels), lubricants, and cosmetics. At the same time, the importance of oilseed crops goes beyond human nutrition, livestock feed and economic development. Soybean and other oilseed crops provide high protein food items and edible oils and are used in food production and livestock feeding. Rapeseed oil has low saturated fat, is a popular oil for cooking and as a feedstock for biodiesel. Sunflower oil has a neutral

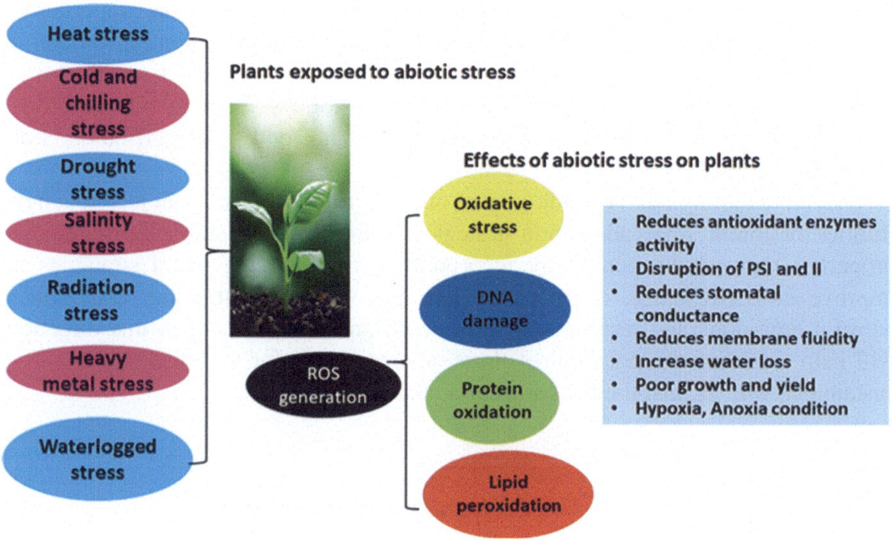

Fig. 16.1 Potential effects of abiotic stress on oilseed crops along with the major types of abiotic stress to which the plant is exposed

taste coupled with high levels of vitamin E, commonly used in cooking. Sesame oil has a unique flavor that is used in different cuisines, and it has antioxidant properties. These crops' cultivation and processing are essential components of the agricultural economies in many countries, impacting both trade balances and rural livelihoods.

However, the productivity and quality of oilseed crops are often negatively affected by abiotic stresses (drought, salinity, extreme temperatures and nutrient deficiency). By disturbing physiological and biochemical processes these environmental factors negatively affect plant growth, development, and yield as shown in Fig. 16.1. For example, drought stress decreases the availability of water, reduces photosynthesis, nutrient transport and subsequently decreases seed yield and oil content. Ionic and osmotic imbalances, such as ion toxicity and osmotic stress due to salinity stress, greatly arrest the growth and productivity of oilseeds crops. Extreme temperatures can affect flowering, pollination and seed development due to inhibiting membrane stability and enzymes activity. Nutrient deficiencies restraining their fundamental metabolic activities causing retarded plant growth and yield. Mitigating these issues is important for maintaining profitable and sustainable oilseed crop production (Ahmad et al. 2021a, b; Hladni et al. 2022).

The recent developments in nanotechnology provide novel potential strategies to improve oilseed crops' resilience against the abiotic stresses. Nanomaterials have a high surface area to volume ratio and unique physicochemical properties and can interact with plant systems at the cellular and molecular levels promoting improved stress tolerance. For instance, nanofertilizers effectively increased nutrient uptake efficiency and the resistance of plants to adverse stress conditions. Controlled release of plant growth regulators is achieved using nanoencapsulated forms of hormones, which help to modulate stress-responsive pathways. Also, antioxidant properties of nanoparticles can scavenge reactive oxygen species (ROS), mitigating oxidative damage of cellular components. The use of nanotechnology-based solutions has the potential to revolutionize the mitigation of abiotic stresses in oilseed crops, which has implications for sustainable agriculture and food security (Shukla et al. 2024; Hofmann et al. 2020).

Overall, coupling nanotechnology and oilseed crop production can lead to a new frontier for abiotic stress adaptation strategies. Nanomaterials exhibit several unique properties that can be potentially incorporated into enabling novel strategies to improve crop resistance, minimize energy and water consumption, and reduce harmful emissions and residues. Such research and development in this direction will help to utilize the potential of these advancements for the oilseed crops in a sustainable way against the backdrop of a changing climate.

16.2 Nanotechnology for Mitigating Abiotic Stress in Oilseed Crops

In recent years, nanotechnology has emerged as a promising approach to enhance the resilience of oilseed crops to abiotic stresses by improving their morphological, physiological, biochemical, and molecular responses to abiotic stresses. Many abiotic stresses negatively impact oilseed crops, including salinity, drought, extreme temperatures, heavy metal contamination, mineral deficiencies, waterlogging, and radiation. A key mechanism by which nanoparticles (NPs) mitigate salinity stress and inhibit oxidative stress is modulating ion homeostasis, inhibiting Na^+ absorption, and increasing K^+ absorption (Abdel Latef et al. 2017, 2018; Rajput et al. 2021). The use of nanoparticles enhances soil water retention, increases photosynthesis, and extends root systems that enable plants to get their water from deeper reservoirs during dry periods. TiO_2NPs (nanopack) protect cell membranes and increase heat shock proteins, which also aid in the upregulation of antioxidant defense machinery against both heat and cold stress.

NPs reduce heavy metal toxicity by altering metal bioavailability and increasing antioxidant activity (Sharifi et al. 2021; Al-Khayri et al. 2023). In addition to improving nutrient absorption, NPs control nutrient delivery throughout the plant's growth cycles (Khalid et al. 2022). Under waterlogged conditions, doping NPs allows plants to thrive in low-oxygen environments by enhancing root porosity and enabling anaerobic respiration. Meanwhile, UV light-protective and ozone-detoxifying nanoparticles diminish the oxidative stress corollary to radiation on plants (Aguirre-Becerra et al. 2022). Farming with nanoparticles has been shown to improve yields, stress resistance, and sustainability. It is important, however, to establish the viability of these technologies through field-based research in order to assess their potential for long-term agricultural resilience. In this chapter, we discuss abiotic stresses that affect oilseed crops and highlight the importance of NPs in alleviating these stressors, leading to a better yield for the crops.

16.2.1 Salinity Stress

Salinity is an important abiotic stress factor that drastically limits the growth and productivity of oilseed crops globally. It poses a significant challenge to agricultural sustainability, especially in arid and semiarid areas with widespread soil salinization. Salinity stress is affecting the oilseed crops, and thus, knowledge is required by utilizing the salinity stress-affected plant system to address this problem and to ensure food security. Salinity stress of plants occurs because of the aggregation of soluble salts, mainly sodium chloride (NaCl) in the soil, which has a harmful effect on plants. This accumulation causes osmotic stress, ion toxicity and oxidative stress, which somehow reduce plant growth and development. Soil salinity, the accumulation of soluble salts, affects more than 800 million hectares of land worldwide—over 6% of the world's total agricultural land. Salinity has affected 6.28 million hectares in Pakistan alone, thus making the problem severe (Maryum et al. 2022).

Plants affected by salinity stress act in different interconnected ways. High concentrations of salt in the soil reduce the soil water potential and make it difficult for plants to absorb the available water; thus, physiological drought occurs even when soil moisture is abundant. The excessive absorption of sodium (Na^+) and chloride (Cl^-) ions causes ion homeostasis disturbances, enzymatic inhibition, membrane instability, and metabolic disruption. Furthermore, salinity-induced ionic imbalances lead to the excessive production of ROS, which damage proteins, lipids, and nucleic acids via oxidative processes.

In oilseed crops, salinity stress causes negative effects during vegetative and reproductive stages, resulting in yield reduction and poor oil quality. In soybean (*Glycine max*), Na^+ accumulation disrupts nutrient absorption and hinders metabolic functions, resulting in lower seed yields, reduced germination rates, and stunted growth. A higher salt content in rapeseed (*Brassica napus, Brassica rapa*) causes a reduction in chlorophyll content and photosynthetic efficiency, as well as low growth (Ilyas et al. 2024). Under salinity stressful conditions, seed germination and oil content in sunflower (*Helianthus annuus*) are decreased due to a disruption in K^+ uptake resulting in ionic imbalances and reduced metabolic activities. Osmotic stress leads to reduced growth and yield of peanut (*Arachis hypogaea*), as it restricts water absorption, leading to wilting, chlorosis, and reduced pod formation. Salinity stress also has negative impacts on the Cottonseed (*Gossypium* spp.) plant stature, fresh and dry weight of root and shoot, node number, leaf area index, stomatal conductance, transpiration rate, photosynthesis, yield, canopy and root development and fiber quality (Maryum et al. 2022). Oil production from palm kernel (*Elaeis guineensis*) is heavily affected by salinity stress, especially in partially fertile lands, which can cause declines in standard yields. Studies have shown that salinity has the potential to reduce fresh fruit bunch (FFB) production and oil extraction rates, with an affected area producing on average 16.5 t ha^{-1} as opposed to the 23.5 t ha^{-1} seen in nonsaline areas. Moreover, higher salinity (>5 mS cm^{-1}) has been reported to negatively affect oil-to-bunch ratio, normally varying from 16.5% to 18.7% in saline areas and 22.2% to 22.4% in nonsaline areas (Henry and Wan HoongHak 2012).

Salinity stress also results in reduced growth and nut production in coconut (*Cocos nucifera*), accompanied by leaf chlorosis, necrosis, and premature nut abortion; these contribute significantly to loss in yield (Marinho et al. 2021).

Physiological and biochemical responses to salinity stress include osmotic adjustment, mediated by the accumulation of compatible solutes such as proline, glycine betaine, and soluble sugars that prevent loss of turgor and thereby allow osmotic balance. Selective uptake, compartmentalization of ions, and activation of ion transporters is important to maintain ionic balance and minimize Na^+ toxicity. Moreover, increased activity of antioxidant enzymes including superoxide dismutase (SOD), peroxidase (POD) and catalase (CAT) plays an important role in the scavenging of ROS and reducing oxidative damage (Liu et al. 2024; Shelar et al. 2024). Plants have several strategies at the molecular level to enable salinity tolerance.

Moreover, the genes responsible for boosting osmoprotective and ion transport as well as antioxidant functions are turned on by salinity stress. The expression of stress responsive genes is controlled by calcium ions (Ca_2^+), Mitogen Activated Protein Kinases (MAPKs), and signal transduction pathways involving a transcription factor. While responding to salinity stress, phytohormones like abscisic acid (ABA), salicylic acid (SA), and ethylene assume a vital role (Acharya et al. 2024).

The application of nanotechnology represents a method to lower the negative impacts of salt stress on oilseed crops. Nanoparticles provide different salvaging techniques to improve plant tolerance through both enhanced nutrient absorption and antioxidant defense mechanisms, with reduced salt damage effects. Multidimensional effects of ZnONPs on salt stress tolerance stem from their ability to modify stress-responsive genes, resulting in higher proline levels with balanced antioxidant enzyme activity regulation (Singh et al. 2022). The use of cerium oxide nanoparticles (CeO_2NPs) has been shown to reduce excessive accumulation of ROS and to enhance plant tolerance to salinity stress (Liu et al. 2021). The polymer-coated manganese oxide nanoparticles (Mn_3O_4) have also been shown to improve soybean seedling germination and growth under salinity stress (Mao et al. 2024). In addition, zinc oxide nanoparticles (ZnO-NPs) have been shown to increase chlorophyll content and photosynthesis efficiency in groundnuts under salinity stress (Kumari et al. 2024). Chitosan nanoparticles (CS-NPs) play a critical role in facilitating potassium uptake in *Brassica juncea* and limiting sodium influx by maintaining an ionic balance. In *Brassica juncea* grown in saline conditions, chitosan nanoparticles (CS-NPs) significantly enhanced antioxidant defenses, reducing oxidative damage (Junedi et al. 2023). In a similar manner, nanosilicon dioxide (SiO_2NPs) reduced the effects of oxidative stress while improving seed germination, growth, and photosynthesis rates in *Cucurbita pepo* (Siddiqui et al. 2014). Under salinity stress conditions, silver nanoparticles (AgNPs) enhanced photosynthetic pigments, increased soluble sugars, proline, and protein levels, while reducing hydrogen peroxide (H_2O_2) and malondialdehyde (MDA) (Khalofah et al. 2021).

In summary, salinity stress is a significant limitation to oilseed crop growth, yield, and oil quality. It is necessary to understand the physiological, biochemical, and molecular mechanisms in oilseed crops responding to salinity stress for their effective mitigation. Nanotechnology provides a new and efficient strategy to boost salt tolerance in oilseed crops through facilitation of the nutrient acquisition, osmotic balance homeostasis, and activation of the defense response cascades. However, NPs formulations and application methods would require further research and innovation to maximize their prospects to enhance salinity tolerance in oilseed crops.

16.2.2 Drought Stress

Drought stress, as an important abiotic factor, severely limits the growth, development, and yield of oilseed crops worldwide. Drought stress characterized by a lack of availability of water, can disrupt many physiological and biochemical processes in plants, significantly reducing crop productivity. The osmotic stress, caused by lower soil moisture, lowers the soil's water potential and limits water absorption, leading to cellular dehydration and diminished turgor pressure. Plants counter by closing stomata to limit water loss, but this also restricts atmospheric CO_2 entry, hampering photosynthesis and the energy production process. Upon this disruption of the photosynthetic equilibrium, an overexcretion of ROS occurs in such a way that oxidative damage, enzyme inhibition, and diminished membrane stability arise. Additionally, it affects the uptake and transport of nutrients that can lead to nutrient imbalances that can further impair plant growth and development. Oilseed crops face severe challenges in drought stress, especially during vegetative and reproductive phases, resulting in reduced seed yield and oil quality.

Soybean (*Glycine max* (L.) *Merrill*) is among the top 10 most widely grown crop species; it is a globally important food and feed crop (Pagano and Miransari 2016). On the other hand, it is also one of the most drought-susceptible crop plants (Valliyodan et al. 2017), with water deficit accounting for as much as 40% of annual yield losses. It's an even more serious problem in tropical countries like Vietnam, where drought has become more common and lasted longer over the years. The need for alternative approaches to conventional plant breeding and biotechnological approaches for micronutrient modification has been previously studied (Waraich et al. 2011; Tripathi et al. 2018). Mineral microelements, including iron, copper, cobalt, magnesium, manganese, nickel and zinc, also improve the crop yield and can significantly alleviate the stress of adverse environmental conditions and all have a positive impact on the growth and development of crops (Naveed 2012; Jaleel et al. 2009).

In soybeans (*Glycine max*), pollination is markedly reduced due to drought stress during the reproductive stage, resulting in greatly decreased stomatal conductance and elevated canopy temperature that contribute to reduced seed yield by decreasing both seed number and average seed weight (Poudel et al. 2023). Related to this, drought stress reduces fresh and dry weight, seedling length, and seed yield in rapeseed (*Brassica napus, Brassica rapa*). However, applications of proline externally have been found to promote seedling dry-matter accumulation under drought conditions (Jurkonienė et al. 2023). Drought conditions lower the overall productivity of sunflower (*Helianthus annuus*) by decreasing grain yield, grain weight, and number of grains per head (Ghaffari et al. 2023). Peanut (*Arachis hypogaea*), likewise, suffers from reduced seed yields and impaired commercial traits, including seed weight and quality, under drought stress (Sehgal et al. 2018). Cottonseed (*Gossypium* spp.) suffers from stunted growth of its plants, reduced number of bolls, and fiber quality, leading to a reduced economic value during prolonged water-deficit periods (Fahad et al. 2017). Drought stress significantly decreases oil yield in palm kernel (*Elaeis guineensis*) and selenium has been used as a means of increasing tolerance to drought in such crop plants (Ahmad et al. 2021b). Drought stress affects coconut (*Cocos nucifera*) by reducing nut production and affecting oil yield, and in olive (*Olea europaea*) long periods of drought adversely affect oil composition and fruit quality.

Sesame (*Sesamum indicum*) and flaxseed (*Linum usitatissimum*) demonstrate decreased seed yield, photosynthetic activity, and biomass accumulation due to drought. Another oilseed, mustard (*Brassica juncea*), as well as safflower (*Carthamus tinctorius*) and castor bean (*Ricinus communis*) show a decrease in seed yield and oil content when individuals are water-stressed, especially during flowering and seed-filling periods. Drought causes substantial reductions in kernel weight and oil yield in maize (*Zea mays*) due to impaired nutrient translocation and photosynthesis. Limited availability of water reduces yield and oil composition in Jatropha (*Jatropha curcas*), camelina (*Camelina sativa*) and other oilseed crops, such as nigella (*Nigella sativa*), pumpkin (*Cucurbita pepo*), hemp (*Cannabis sativa*), and moringa (*Moringa oleifera*). In summary, similar to all agricultural crops, drought stress represents the main constraint to oilseed crop production.

Among the sustainable agronomic strategies that incorporate drought-resistant varieties, biostimulants, promising plant protection and stimulation tools are needed for enhancing plant adaptation to mitigate drought impacts and ensure stable yields in a water-limited environment. Therefore, in response, recent advances in nanotechnology present integrative approaches for alleviating the deleterious impacts of desiccation stress of these crops. Drought can be mitigated by a multifactorial effect of NPs, including better water and nutrient uptake, alterations in stomatal conductance, activation of antioxidant defense mechanisms, and changes in the expression of genes involved in stress tolerance (Chandrashekar et al. 2023).

The responses of plants exposed to iron, copper, cobalt, and zinc oxide NPs were studied in soybean plants under drought-stress conditions in previous research. The results suggested that these metal-based NPs played a role in enhancing the drought tolerance of the NP-treated plants. The targeted physiological parameters: relative water content, drought tolerance index, and biomass reduction rate were significantly improved, in particular in the iron NP-treated plants. At the molecular level, quantitative PCR analysis of a few drought-responsive genes showed a tissue and NP-dependent upregulation of gene expression. Iron NP treatment enhanced the expression levels of all tested genes in roots; the expression of three drought-responsive genes was also higher in leaves of all NP-treated plants, and *GmERD1* (Early Responsive to Dehydration 1) expression was induced in roots and shoots as well in all NP treatments tested. These results indicate that NPs application enhances drought stress tolerance in soybean plants by inducing the expression of drought-related genes (Linh et al. 2020).

In another study, FH-770 sunflower variety was grown in pots under three stress levels (100% FC, 75% FC, and 50% FC) and four Se application rates (0 ppm, 30 ppm, 60 ppm, and 90 ppm). Foliar Se decreases the accumulation of H_2O_2 (hydrogen peroxide; ROS) (20.89%) that significantly increased glycine betaine (GB) (74.46%) and total soluble protein (Pro) (68.63%), improved the accumulation of ascorbic acid (AsA) (25.51%), total phenolics (TP) (39.34%), flavonoids (Flv) (73.16%), and anthocyanin (Ant) (83.73%), and improved the activity of antioxidant system superoxide dismutase (SOD) (157.63%), peroxidase (POD) (100.20%), and catalase (CAT) (49.87%), which ultimately improved sunflower growth (36.65%) in drought stress conditions. With Supplemental Se, shoot Se content was increased significantly (93.86%), as well as the improvement in calcium (Ca^{2+}), potassium (K^+), and sodium (Na^+) ions in roots by 36.16%, 42.68%, and 63.40%, respectively. In controlled and water-deficient conditions, Se supplements at low concentrations (60 and 90 ppm) enhanced the growth, development, and biochemical traits of sunflowers (Ameen et al. 2024).

In another study, the beneficial effects of exogenously foliar application of Fe_2O_3 modified carbon nitride nanostructures (control, $FeSO_4$, C_3N_4, and Fe_2O_3/g-C_3N_4) on pomological attributes, as well as physiological and biochemical changes occurring in the leaves and fruits of treated "Shengeh" olive cultivars were reported, when grown at various watering levels (100, 75, and 50% ET), in two experimental years (2022 and 2023). The results showed that pomological characteristics significantly decreased under drought stress, and the treatments could significantly compensate for this damage. In contrast, the fruits of olive suffered from severe drought showed more oil content percent in the fresh matter and water use efficiency (WUE) in oil with 30% and 52.5%, respectively, based on the average results of 2 years, and the two traits in olive deprived of severe drought can improve about 35% on average in 2 years after Fe_2O_3/g-C_3N_4 (Gholami et al. 2024).

In summary, drought stress continues to be a major constraint in oilseed crop productivity, adversely affecting yield, oil quality, and plant health. Water deprivation affects important physiological and biochemical mechanisms, such as photosynthesis, the uptake of nutrients and the control of ROS. Different oilseed crops have different degrees of drought tolerance, which negatively affects seed yield and oil content, as well as other commercial traits in soybean, sunflower, rapeseed, and olive. To overcome these challenges, multiple strategies including micronutrients supplementation, nanotechnology-based approaches, and biostimulant application have been researched. The synthesized NPs, including iron, zinc oxide, selenium, and Fe_2O_3-modified carbon nitride, have a vital role in enhancing drought tolerance by preventing water loss, regulating stomatal conductance, inducing antioxidant defense mechanisms, and upregulating stress-responsive genes. Nanotechnology-driven solutions provide a new avenue for enhancing drought tolerance in oilseed crops. Further studies should investigate nutrient solubilization mechanisms in response to NP formulations, their environmental consequences over the long term, and their complementation with sustainable agronomic practices for consistent.

16.2.3 Temperature (Cold and Heat) Stress

Both cold and heat stress have a strong impact on oilseed crops and their growth, development and productivity. Cold stress includes chilling (<20 °C) and freezing (<0 °C) conditions, both of which disturb cellular functions and result in decreased germination rates, stunted growth, and decreased seed yields. Rapeseed (Brassica napus), for instance, demonstrates inhibited seedling growth under cold stress, and reciprocal varieties of soybean (Glycine) positively correlate with their sensitivity to cold stress, as cold-sensitive cultivars show greater lipid peroxidation and greater damage to membrane integrity than tolerant ones (Jian et al. 2020).

Oilseed crops respond to cold stress through adaptive mechanisms, including modulation of gene expression, metabolic adjustments and exogenous treatments (e.g., agricultural treatments). Differential expression of homeologous genes under cold stress in *Camelina sativa* suggests a complex regulatory mechanism (Fang et al. 2023). Likewise, dramatic shifts in the metabolism of amino acids, organic acids, sugars, and other processes drive the increased tolerance to cold in rapeseed ecotypes (Jian et al. 2020). Exogenic agents, including trehalose and melatonin have been shown to strengthen rapeseed cold tolerance through increasing antioxidant enzyme activities and repairing damage caused by oxidative stress (Lei et al. 2022; Raza et al. 2022).

In addition, heat stress is also a serious challenge for oilseed crops, resulting in the inhibition of photosynthesis, membrane damage, and increased accumulation of ROS. All this plays a role in lower pollen viability, poor seed set, and lower oil quality. The extreme heat during seed fill causes extreme yield reductions and affects the ratio of fatty acids in seeds of *Camelina sativa* (Nadakuduti et al. 2023). Likewise, in oilseed rape (*Brassica napus*), at reproductive-stage heat stress results in diminished seed yield and quality (Magno et al. 2022).

Oilseed crops resist heat stress through antioxidant defense systems, physiological and molecular breeding. Two physiological traits, canopy temperature depression (CTD) and membrane thermostability have been associated with heat tolerance, and CTD is strongly correlated with yield (Y) in warm environments. The activities of antioxidant enzymes significantly contribute to alleviating oxidative damage induced by heat stress, and in rapeseed, trehalose treatment is reported to enhance antioxidant defenses (Raza et al. 2022). This is possible due to significant advances in molecular breeding and genomics that can couse heat-tolerance quantitative trait loci (QTLs) with recent genome-wide association analyses revealing loci controlling plasticity of oil yield under combined abiotic stresses in sunflower (Mangin et al. 2017). The extreme climate changes demand adaptive agricultural practices for sustainable oilseed production since they intensify temperature fluctuations. Three methods that increase plant tolerance against environmental stressors comprise breeding programs for thermotolerant characteristics alongside exogenous applications of both melatonin and trehalose and the deployment of cold plasma treatments. Scientific research indicates that exposing oilseed rape seeds to cold plasma leads to better plant growth and improved seed development even under dry conditions. The combined implementation of physiological expertise and genetic discoveries and field-based farming practices creates a complete solution for sustainable oilseed crop cultivation under severe environmental stress.

Scientists have been exploring nanotechnology as a developing technology to improve plant tolerance against high and low temperatures. Research has shown metal oxide nanoparticles (ZnO, TiO_2, Fe_3O_4) and carbon-based nanoparticles and nanocomposites enhance oilseed crop heat stress tolerance through their regulatory mechanism on antioxidants and their water retention capabilities and metabolic stability maintaining functions (Yashveer et al. 2022). By enhancing the activity of SOD, CAT, and POD, nanoparticles can decrease the accumulation of ROS in plants growing at high temperatures (Salaheldin et al. 2024). It has been demonstrated that nanoparticles can enhance chlorophyll retention and carbon assimilation efficiency under heat stress conditions and facilitate sustained photosynthesis (Guo et al. 2024). NPs increase the stability of lipid peroxidation, ensuring the preservation of cellular membranes at higher temperatures (Prasad et al. 2023). NPs containing silica and titanium dioxide promote water retention and regulate plant stomatal conductance (Yashveer et al. 2022). Oilseed crops can be improved in thermotolerance through nanoparticles that regulate heat-responsive genes and heat-sensitive proteins (Guo et al. 2024). Several studies have demonstrated that nano silica and nano selenium have been used as ameliorant agents for soybeans under stress conditions to improve physiological conditions, stabilize the membrane, and modulate the transcription of genes that respond to oxidative stress (Omar et al. 2023). The primary experimental investigation evaluated salinity stress, but researchers should find parallel mechanisms useful for understanding heat stress conditions also. Research shows metal oxide nanoparticles produce substantial effects that influence both peanut physiological state and nutritional characteristics. The application of these nanoparticles creates no growth restriction for plants but results in weight reduction alongside amino acid variations because of extreme environmental conditions (Rui et al. 2018). CuNPs synthesized by plant extracts have enhanced seed germination, root length, shoot length, and protease activity, demonstrating a tolerance to stress (Santos-Espinoza et al. 2020). A foliar spray of SiO_2NPs applied to peanuts has also been reported to improve water use efficiency, photosynthetic rate, and pod yield, proving that these nanoparticles are useful for mitigating heat and drought stress (Prasad et al. 2023).

Heat stress causes negative effects on both soybean seed harvest volume and oil qualitative characteristics. The application of TiO_2 nanoparticles to heat-stressed soybean plants enhances the operation of photosynthetic and antioxidant enzymes which leads to improved heat defense capabilities (Salaheldin et al. 2024). Reports suggest that ZnO and SiO_2 nanoparticles can stabilize cellular membranes and facilitate water homeostasis, which could enhance pod filling and oil synthesis in heat-stressed conditions. Heat stress in flowering inhibits the pollen viability and seed set of canola. Silicon (Si) and selenium (Se) nanoparticles modulate antioxidant defense mechanisms, leading to increased reproductive success (Yashveer et al. 2022). Incorporating iron oxide (Fe_3O_4) nanoparticles helps to stabilize chlorophyll in the exposed system, thus improving drought tolerance, which is significant in order to mitigate the combined heat and drought stress in canola. The combined treatment of silver (Ag) and cerium oxide (CeO_2) NPs enhances stress tolerance in lettuce and among the nanostructures, Ag and CeO_2 NPs could increase chlorophyll fluorescence and root water uptake (Guo et al. 2024). Nanosilica applications have increased heat tolerance by regulating osmotic balance. Flower abortion and impaired kernel filling occur due to heat stress on peanuts. Copper (Cu) and zinc (Zn) nanoparticles have been shown to improve antioxidant boosts and reproductive efficiency, increasing yield stability under heat stress (Salaheldin et al. 2024).

Overall, nanoparticles offer a revolutionary approach to enhancing heat tolerance in oilseed crops. Various nanoparticles can alleviate the negative effects of heat stress on oil yield and quality by promoting photosynthesis, balancing antioxidant routes, and protecting cell surfaces. More investigation is needed to tune nanoparticle formulations and to evaluate overarching impacts environment over time. The ongoing advancements in nanobioengineering may facilitate the design of nanomaterials for crop species-specific approaches in various oilseed species to the sustainable productivity in a changing climate.

16.2.4 Heavy Metals Stress

The oilseed crops are considerably affected by heavy metal stress in terms of their growth, development, and yield. Arsenic (As), lead (Pb), cadmium (Cd), mercury (Hg), nickel (Ni), and chromium (Cr) are the most common heavy metals that can accumulate in these crops, causing numerous physiological and biochemical disturbances as shown in Fig. 16.2 (Nazari et al. 2023). Natural processes and anthropogenic activities introduce heavy metals into agricultural systems. Natural sources include volcanic eruptions and forest fires, whereas man-made sources include industrial emissions, mining activities, the use of fertilizers, pesticides, and irrigation with contaminated water. These practices lead to higher levels of heavy metals in the soil, which are then absorbed by oilseed crops. Heavy metals also can interrupt morphological and anatomical processes, causing smaller root and shoot lengths and decreased fresh and dry weights. This effect in growth is generally associated with the symptoms of chlorosis, wilting, and necrosis (Majhi and Sikdar 2023). Heavy metals disrupt chlorophyll synthesis and activity, leading to reduced efficiency of photosynthesis. This inability disrupts the plant from generating particular key vitamins that are important for the total wellbeing of the plant, which could lead to a decrease in yield (Nasircilar et al. 2024). Heavy metals trigger the production of ROS in plants. The ROS cause oxidative damage to cellular components, including lipids, proteins, and nucleic acid, further compromising cell integrity and function (Ul Hassan et al. 2017). Heavy metals interfere with the absorption and transport of essential nutrients, causing deficiencies or toxicities that aggravate further plant growth and development (Nasircilar et al. 2024).

For instance, heavy metal accumulation in oilseed crops can affect plant health and pose health hazards to humans consuming the crops. It is well-known that Cd,

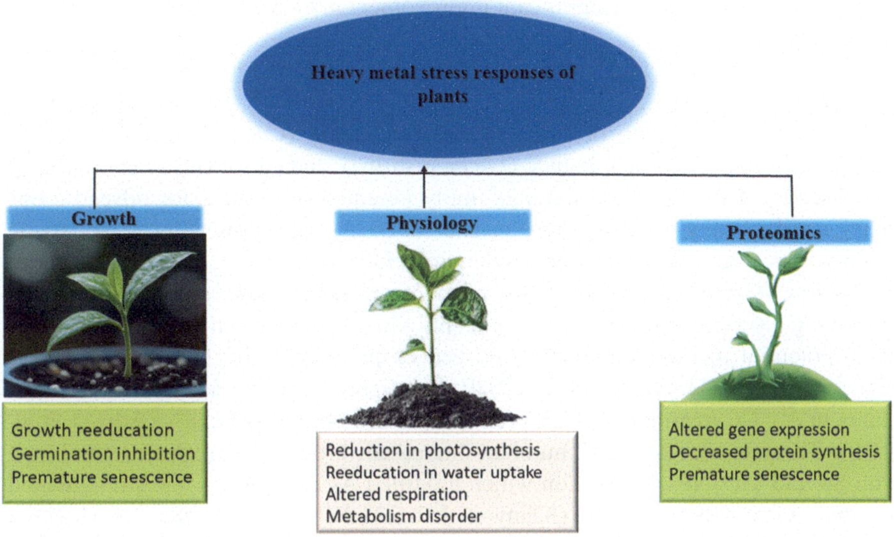

Fig. 16.2 Graphical representation of heavy metal effects on plant growth and development

Pb, and Hg are toxic to humans, seriously harming almost all organs and systems (Khan et al. 2025). Long-term exposure can cause serious health problems, among them neurological disorders and kidney damage (Ehsan et al. 2024). Heavy metals detected in edible oils made from such contaminated oilseed crops amplify food safety issues. Consumers need to be protected by regularly observing maximum allowable levels of heavy metals (Nazari et al. 2023).

Various methods can be used to reduce heavy metal stress in oilseed crops. Phytoremediation is a green technology that employs plants possessing great heavy metal absorption ability to restore polluted soil. For example, some Brassica species have been efficient in accumulating heavy metals and reducing soil pollution (Nasircilar et al. 2024). Plant growth-promoting rhizobacteria (PGPR) and endophytes are beneficial microbes that can enhance plant tolerance to toxic heavy metals. They act through multiple mechanisms such as phytohormones, antioxidant enzymes and metal toxins (Shaffique et al. 2023). Heavy metals in the soil may be immobilized when organic and inorganic amendments are applied, making them unavailable for plant uptake. That strategy assists de facto in reducing heavy metals in the food chain, but their efficiency is generally to a limited extent by slow metallic uptake and constrained bioavailability. Nanotechnology offers an alternative approach through nanoparticle-based therapies that improve these conventional treatments. Some NPs, including Fe_3O_4, TiO_2, and ZnO, have been shown to enhance phytoremediation through enhancing metal bioavailability, increasing root elongation, and reducing metal toxicity. With high surface area and reactivity, NPs tend to bind with heavy metals, the binding of which leads to transformation of heavy metals into less toxic forms, which could favor their uptake and removal from contaminated soils. Thus, a significant improvement in heavy metal detoxification of oilseed crops.

Sulfur nanoparticles had a positive effect on Cu stress and toxicity in this oilseed rape (*Brassica napus* L.) (Ghorbani et al. 2024). Iron oxide (Fe_3O_4) nanoparticles enhanced the root growth of crops such as cucumber (*Cucumis sativus*), tomato (*Solanum lycopersicum* L.), lettuce (*Lactuca sativa*), and carrot (*Daucus carota*) in response to Cd stress. These findings imply potential usefulness for oilseed crops facing similar heavy metal issues (Faizan et al. 2023). The application of zinc oxide nanoparticles through leaves helps salicylic acid reduce arsenic toxicity effects in rice plants (*Oryza sativa* L.). The treatment enhanced plant growth while also improving gas exchange parameters and antioxidant enzyme activities, thus providing a potential method for treating oilseed crops affected by heavy metal stress (Faizan et al. 2021). Adrees et al. (2020) highlighted that the adaptation of wheat plants under Cd treatment was improved by the addition of FeNPs (Adrees et al. 2020). The total rice seedling numbers increased through As treatment combined with Si- and TiO_2NPs application which modified As transporters and strengthened the antioxidant defense system (Kiany et al. 2022). NPs are now widely used with a lot of advantages, enhancing growth and adaptation under different abiotic stress conditions in plants like wheat (Adrees et al. 2021), corn (Oliveira et al. 2016), barley (Ghorbanpour et al. 2020), rice (Ahmed et al. 2021), as well as tomato (Hernández-Hernández et al. 2018). Owing to their excellent chemical and physical stability, low toxicity, corrosion resistance and low-cost properties, TiO_2NPs are now widely considered to have a great capacity for application in environmental conservation (Samanta and Roychoudhury 2021). TiO_2NPs have proven their potential for ecological restitution due to their strong oxidizing properties, remarkable resistance to photo-degradation, and selectivity toward redox reactions (Wu et al. 2021). As adsorbents have a higher affinity for AsV than AsIII in most cases of As removal, preoxidation is required to maximize the utilization of adsorbent potential by converting AsIII to AsV completely. Additionally, TiO_2NPs possess intrinsic photocatalytic activity disrupting the balance of As^{III} and As^{V} (Samanta and Roychoudhury 2021). Recent studies on TiO_2NPs were recently conducted to understand how they buffered As^{III} accumulation. Compared with control groups, TiO_2NPs significantly reduced As accumulation by 90% (rutile) and 40% (anatase). Anatase and rutile reduced the accumulation in seedlings by 14% and 90%, respectively, compared to control groups (Wu et al. 2021). In addition, the TiO_2NPs attenuated As-induced increase of MDA and ROS through upregulating expression levels of antioxidant genes. These data imply that TiO_2NPs, especially those synthesized through environmentally benign methods, promote the antioxidant power of plants, which helps minimize excessive levels of ROS and alleviate As toxicity. Thus, the application of greener prepared NPs appears as a safe, cost-effective, and environmentally friendly approach that can be justified as an advantageous alternative to chemically synthesized NPs. In contrast, various other research conducted 2000 mg/kg TiO_2NPs along with soil amendments and found minimal influence on the properties of soil (You et al. 2018). TiO_2 (2.5 mg/L) was tested on *Abelmoschus esculentus* and *Helianthus annuus* in a recent study. Higher yield, fruit length, plant height, chlorophyll content and antioxidant enzyme activity were observed. Treatment with TiO_2 also reduced the accumulation of Cr in the fruit, root, and stem (Kumar et al. 2022).

Overall, nanotechnology emerges as a viable strategy to alleviate heavy metal stress in oilseed crops, improving plants' resilience, growth, and yield under toxic metal exposure. Several studies have shown that these nanoparticles (sulfur, iron oxide, zinc oxide, and silicon nanoparticles) relieve the stress effect of metals (cadmium, lead, arsenic, and copper). The action of these nanoparticles includes increasing nutrient uptake, enhancing antioxidant defense mechanisms, and decreasing metal bioavailability in plants. While most of the studies highlighted were conducted on nonoilseed crops, the insights gleaned from these studies form a basis for further exploration of the application of nanotechnology for the protection of oilseed crops. Further studies are needed to optimize nanoparticle formulations, evaluate long-term environmental consequences, and establish the safety of these materials for use in sustainable agriculture. It is also worth mentioning that incorporation of nanoparticles with phytoremediation/microbial-based strategies is also a good target for improving the tolerance of oilseed crops toward heavy metal stress without compromising food security and agricultural sustainability.

16.2.5 Minerals Deficiency Stress

Mineral stress in oilseed crops, as well as many other plants, arises through the inadequate supply or excess presence of important nutrients that can disturb normal physiological processes. Mineral stress refers to either deficiency stress, as in the case of macronutrient minerals such as nitrogen, phosphorus, potassium, or micronutrient minerals such as zinc, boron, and iron, or toxicity stress, which is due to the excessive accumulation of minerals such as sodium or heavy metals within plants, resulting in toxicity. The deficiencies can cause impaired enzyme function as well as impacts on protein synthesis, chlorophyll production, and energy transfer which ultimately, reduce yield and oil quality. As an example, sulfur deficiency in oilseed rape (*Brassica napus*) interrupts amino acid and protein synthesis, resulting in stunted growth and chlorotic leaves, and boron deficiency is linked to reduced pollen viability and seed filling. Excessive minerals are known to cause oxidative stress and to interfere with absorption of normal nutrients—for example, sodium has toxic effects by disrupting ion balances within cells and damaging membranes. Factors such as bad soil quality, pH imbalance, and climate (e.g., drought) lower the accessibility and uptake, resulting in mineral stress. Reducing mineral stress can be achieved via balanced fertilization, soil amendments, biofertilizers, and using nanotechnology-based nutrient-delivery systems that can enhance nutrient absorption efficiency (Safdar et al. 2023). Likewise, biofortification strategies have been proposed to increase the availability of minerals in deficient soils to increase crop resilience and yield. Addressing mineral stress is crucial to sustaining oilseed productivity and ensuring high-quality oil production for commercial and nutritional purposes.

To mitigate mineral stress in oilseed crops, a dual approach comprising traditional and high-tech agronomic practices should play a key role in improving the physicochemical availability of nutrients, their uptake and utilization. The paramount method of fertilization is balanced fertilization as it helps to provide adequate macronutrients and in sufficient amounts of micronutrients (such as zinc, boron, and iron). Adding organic matter, lime or gypsum to the soil can enhance nutrient retention and pH balance, reducing nutrient deficiencies that can occur in soils that are not suitable for the plants. Soil microorganisms play a pivotal role in soil fertility management through biofertilizers, especially PGPR and mycorrhizal fungi, that promote root growth and solubilization of nutrients ensuring minimal reliance on chemical fertilizers. Precision agricultural practices such as site-specific fertilization and condition testing maximize nutrient application and minimize excess or deficiency stress. Otherwise, foliar sprays work well, particularly for all those micronutrients, where you have limited root uptake because the soils aren't in good condition.

An emerging solution to mitigate mineral stress in oilseed crops is the application of NPs as advanced nutrient carriers and stress alleviators. Nanoparticles improve nutrient efficiency due to their high surface area, controlled release properties, and ability to enhance plant metabolism. Studies show that zinc oxide (ZnO) NPs improve zinc bioavailability, leading to enhanced chlorophyll production and seed quality in oilseed crops under deficiency conditions (Safdar et al. 2023). Similarly, silicon (Si) NPs have been found to enhance root architecture and increase the uptake efficiency of essential minerals, reducing stress effects from nutrient deficiencies and imbalanced soil conditions. NPs made from Fe_3O_4 improve the absorption of iron and counteract chlorosis resulting from iron deficiency. The unavailability of iron for root absorption occurs frequently in alkaline soil environments. Chitosan-based NPs work as biostimulants to provide nutritional elements while strengthening plant resistance against stress through better antioxidant performance and osmotic control. The combination of oilseed crop nanotechnology applications with regular nutrient management systems helps decrease mineral stress problems. Nanotechnology represents an environmentally friendly, precise method for addressing nutrient deficiencies because it increases nutrient dissolution rates while cutting down losses through leaching and optimizing plant nutrient uptake. Nanotechnology functions as an ultimate tool because it serves as a practical method together with standard nutrient management to reduce oilseed crop mineral stress. Nutrient deficiencies receive precise environmentally friendly treatment using nanoparticles. Plants acquire nourishment optimally when nutrient solubility increases while leaching losses decrease and nutrient solubility reaches its highest level. Research indicates that NPs successfully enhance the resistance of oilseed crops against mineral stress which limits agricultural yield production. The enhancement of nutrient absorption by NPs enables better plant uptake of nutrients and greater tolerance against environmental stress factors which supports plant development in such conditions. The introduction of defending factors through NPs boosts oilseed crops to withstand both nutritional deficiencies and toxic effects in order to produce better agricultural yields (Cao et al. 2025). Nanoparticles improve nutrient uptake efficiency and activate stress tolerance mechanisms, enhancing oilseed crops' tolerance to mineral stress. This model responds to nutrient deficiency and results in mastery of abiotic stresses through yield and quality (Zulfiqar et al. 2024). Zinc oxide together with iron oxide nanoparticles show two beneficial effects for oilseed crops by enhancing mineral uptake and lowering the stress that results from mineral deficiencies. Plants gain better nutrient access because of these enhancements that lead to successful growth in low-energy soil environments. Plant resistance against stress factors becomes stronger through nanoparticle treatment because this leads to both enhanced antioxidant capabilities and modified gene networks resulting in better growth rates, increased stress tolerance and increased crop production (Kumari et al. 2024). In cases of mineral stress, nanoparticles, like ZnO and graphene quantum dots, induce positive effects for oilseed crops through improved nutrient uptake and translocation while also stimulating soil microbial communities, reducing plant growth and yield problems caused by mineral toxicity (Nawaz et al. 2023).

Overall, oilseed crop productivity is severely challenged by mineral stress because it disrupts nutrient balance, impairs physiological processes, and reduces yield. The use of nanotechnology is a powerful tool that helps to increase the biological availability of nutrients and also helps the crops to recover more effectively from stress, which increases crop resilience. Several nanoparticles, such as ZnO, Fe_3O_4, and silicon-based formulations, have shown the potential to alleviate nutritional deficiencies and reverse the effects of toxicity. Further studies should be directed toward nanoparticle formulations, a better understanding of long-term environmental impacts, and applications in safe agriculture. Conventional and advanced practices must be integrated to develop sustainable nutrient management plans that increase oilseed crops productivity and quality under different agroecological environments.

16.2.6 Waterlogging Stress/Flooding Stress

Brassica napus (rapeseed/canola), *Helianthus annuus* (sunflower), and *Glycine max* (soybean) experience susceptibility to waterlogging during their lifecycle. Waterlogging in the root zone creates hypoxia or anoxia conditions that disrupt aerobic respiration and disrupt energy production processes. The damage inflicted upon root systems by this condition produces chlorosis symptoms together with leaf wilting. Plants under waterlogging might wilt until death happens although this outcome depends on the plant type and ecological setting. The deficiency of oxygen blocks essential biological functions that include carbon adsorption along with photosynthesis. Recognition deficits in oxygen transmission from roots to shoots impair both metabolic processes and nutritional uptake functions. The creation of ROS as well as oxidative stress develops from these interruptions. A wide selection of crops experiences reduced dry matter accumulation and seed yield because of the impact (Nabloussi et al. 2019). Several anatomical as well as morphological adaptations have developed across plants because of waterlogging in their environments. Plants develop three different adaptations to waterlogging conditions including the formation of aerenchyma in roots and the development of adventitious roots at the soil surface and leaf modifications. Plants have evolved modifications that facilitate oxygen transfer to roots under flood conditions. Plants manipulate ethylene hormones together with other phytohormones so they can alter their responses to stress factors (Aslam et al. 2023).

Waterlogging stress can be eased with various agronomic practices. Water accumulation can be prevented by using proper drainage systems and soil management techniques. Waterlogging-tolerant crop varieties can be developed through breeding programs that select traits like efficient aerenchyma formation and robust adventitious rooting. In addition, compatible solutes and growth regulators may enhance tolerance to waterlogging (Raza et al. 2017; Ploschuk et al. 2018). The numerous physiological and biochemical mechanisms operated by NPs enable waterlogging stress mitigation in oilseed crop plants. Plant flood resistance improves through the combined efforts of silver nanoparticles (AgNPs) and silicon nanoparticles (SiNPs) and zinc oxide nanoparticles (ZnONPs) which regulate antioxidant mechanisms together with phytohormone functions and nutrient intake efficiency. Experts have recently evaluated how NPs can reduce detrimental effects in research studies. Soybean (*Glycine max*) plants receive relief from flooding stress through silver nanoparticles (AgNPs) which control amino acid output and protein manufacturing and glycolytic process and wax formation (Al-Khayri et al. 2023). Similarly, O-carboxymethyl chitosan nanoparticles (O-CMC-NPs) improve maize tolerance to water stress. By improving cellular stability, oxygen diffusion, and anaerobic respiration mechanisms, nanoparticles could improve oilseed crops' tolerance to waterlogging stress (Wu et al. 2024).

Under waterlogged environments reactive oxygen species (ROS) generate extreme amounts of reactive oxygen that degrade cellular structures alongside impairing metabolic activities. Plants benefit from engineered nanoparticles as the chemicals activate three antioxidant enzymes: SOD, CAT and POD to protect against oxidative damage. Reforms in antioxidant-related genes occur because of SiNPs treatment which results in lower lipid peroxidation levels and increased root and shoot biomass. The flooding conditions lead to better nutrient access because NPs enhance their availability and uptake processes. The incorporation of iron oxide nanoparticles together with zinc oxide nanoparticles create stronger root membranes which decrease the effects of waterlogged soil deficiencies. NPs make changes to three phytohormones: ethylene together with abscisic acid (ABA) and gibberellins because these hormones manage hypoxia adaptive responses. Through their influence NPs stimulate the natural formation of aerenchyma tissues and induce the development of adventitious roots. When applied to plants, silicon-based nanoparticles enhance root porosity so oxygen moves from shoots to roots, and silver-based nanoparticles accelerate the growth of adventitious roots which in turn increases waterlogged survival rates (Al-Khayri et al. 2023).

The practical application of NPs consists of using them either through foliar delivery or by using them in seed priming solutions to boost oilseed crops against waterlogging. The effectiveness of NPs depends on achieving optimal formulation parameters, yet this requires preventing any phytotoxic reactions that may occur. The combination of NPs based strategies together with conventional drainage practices and crop breeding programs aimed at selecting stress-resistant varieties represents a method to reduce waterlogging damage in oilseed plants (El-Saadony et al. 2022).

16.2.7 Radiation and Ozone Stress

The main source of radiation stress in oilseed crops stems from ultraviolet rays together with two types of ionizing radiation that include gamma rays and X-rays. DNA receives direct harm from UV-B and UV-C light which leads to mutations and collapses proteins and organ failure in cells. The photosynthetic process suffers from UV radiation because the ultraviolet rays decompose chlorophyll molecules leading to decreased light absorption ability and disrupted electron transport. The reduction of biomass production and seed yield alongside growth stunting occurs as a result of these events. UV radiation creates reactive oxygen species that function as oxidative stress agents to damage cell membranes, besides proteins and nucleic acids. Although plants have protective responses to UV exposure that produce more flavonoids and antioxidant enzymes plants still suffer negative effects on their overall productivity from extended light exposure. The DNA modifications induced by ionizing radiation like gamma rays together with gene expression alterations and chromosomal mutation cause additional damage to fundamental growth and development processes.

Oilseed crops experience major detrimental effects from ozone (O_3) stress because they grow in areas with heavy air pollution. Leaf tissues gain entry to tropospheric ozone via stomata which produces ROS that results in oxidative stress. The damaging effects of oxidizing membrane lipids generate chlorophyll breakdown and alter stomatal functions which lead to decreased photosynthetic performance and impaired transpiration rates. The stress from ozone exposure causes Brassica napus and Glycine max to experience early leaf death followed by reduced carbon absorption leading to lower seed mass production. Plants exposed to O_3 develop modifications in their defensive systems along with seed oil traits because O_3 modifies secondary metabolite synthesis of glucosinolates and flavonoids. Research validates that high ozone concentrations create three observable effects on seeds: they reduce weight while diminishing oil volume and protein elements while heightening markers of oxidative stress which lessens plant resistance to environmental conditions and biological threats.

The growth and yield of oilseed crops including soybeans, rapeseed and sunflower are reduced when they experience strokes due to combinations of radiation and ozone exposure. Different resolution approaches have been examined for these stresses. Heat-resistant plant cultivars require both contemporary genetic approaches and conventional seed breeding as essential development tools. Heat resistance exists as QTLs that affect sesame and groundnut crops, thus demonstrating genetic tolerance to stress. Agronomic practices exist that make crops better adapted to environmental stress exposures. Scientists confirmed that applying sulfur through leaves can benefit canola plants during heat exposure by improving photosynthetic functions and yield attributes. Application of thiourea under conditions of plant growth regulation demonstrates its ability to reduce the heat stress impact on sunflower and canola crops. Further, advanced biotechnological tools, including CRISPR/Cas9 genome editing, offer exciting opportunities to develop stress-tolerant oilseed crops. This would enhance the radiation-and-ozone stress tolerance and subsequently improve crop productivity under adverse environmental conditions by regulating potentially stress-responsive genes. To minimize the effects of radiation and ozone stresses on oilseed crops like soybean, rapeseed, and sunflower, new NPs are likely to offer tremendous support. With their unparalleled physicochemical properties, NPs may impart interaction at the molecular level with the plant systems to increase tolerance toward stress, preserving cellular structures, and supporting the health of plants in broader terms. Other examples of engineered nanomaterials demonstration in potential for improved physiological function against abiotic stresses relate to such types of stress as radiation and ozone: zinc oxide (ZnO), titanium dioxide (TiO_2), iron oxide (Fe_3O_4), and chitosan-based NPs.

Excessive UV (type B and C) radiation destroys the plant cells by imposing oxidative stress, degrading chlorophyll, and erratically inducing photosynthetic processes. Thus far, extensive studies on the use of nanoparticles like ZnO and TiO_2 describe their effectiveness in reducing the UV-induced damage to crops. ZnO NPs work by increasing the production of antioxidant enzymes-SOD and CAT, which consequently scavenge the harmful ROS that build up during radiation stress. Studies show that ZnO application in soybean stabilizes chlorophyll and promotes stress tolerance through the modifying activity of important ameliorating enzymes which confer better growth and yield under UV exposure (Cao et al. 2025). Similarly, TiO_2 NPs have been reported to absorb harmful UV radiation becoming a shield against UV-induced damage in sunflowers. These NPs improve photosynthesis efficiency by maintaining integrity along the chloroplast membranes and by decreasing lipid peroxidation and thus preventing the degradation of the fundamental photosynthetic pigments. Fe_3O_4 NPs contribute to improved stress tolerance by encouraging rooting and nutrient acquisition in rapeseed, particularly under radiation stress. Research evidence suggests that the Fe_3O_4 nanoparticles increased the seed germination percentage and seedling vigor, allowing them to withstand environmental fluctuation. In addition, applying these nanoparticles improves iron availability in the soil, augmented by the plant capacity to synthesize chlorophyll, an important pigment for maintaining the photosynthetic efficiency (Al-Khayri et al. 2023).

Air pollutants such as tropospheric O_3 are significant sources of oxidative stress for plants due to the generation of ROS. ROS attack cellular membranes, degrade chlorophylls, and impair photosynthetic efficiency, resulting in early leaf senescence and lower seed yields. A chitosan-based nanoparticle can effectively alleviate ozone stress in crops, particularly when loaded with antioxidants such as N-acetyl cysteine. Plant cells are protected from oxidative damage by these nanoparticles, which strengthen the antioxidant system. Chitosan nanoparticles have been shown to reduce lipid peroxidation levels in wheat and enhance ROS-scavenging enzyme activity, protecting the plants from ozone stress (Nowroz et al. 2024). ZnO nanoparticles have been reported to increase soybean tolerance to ozone stress by upregulating antioxidant enzyme activities and increasing proline accumulation as an osmoprotectant. Under polluted atmospheric conditions, ZnO nanoparticles applied foliarly protect soybean leaves from oxidative damage and increase seed yield. Furthermore, Fe_3O_4 nanoparticles showed enhanced cell defense and efficient nutrient uptake in rapeseed seedlings under ozone stress. Under an ozone-prompted environment of oxidative stress, these nanoparticles assist in maintaining metabolic homeostasis. Ozone-stressed sunflower plants generally show low photosynthetic rates and decreased chlorophyll content, which contributes to their inferior growth. Photocatalyst properties of TiO_2NPs minimize oxidative damage and enhance photosynthetic activity by acting as a photocatalyst. Furthermore, TiO_2 nanoparticles increase stomatal regulation and decrease ROS accumulation in sunflower, thus maintaining normal physiological functions even at high ozone levels.

The ability of nanoparticles to enhance antioxidant defenses, improve photosynthetic efficiency, and protect cellular structures from oxidative damage makes them an important source of protection from radiation and ozone stress in oilseed crops. Crops such as soybean, rapeseed, and sunflower have shown positive impacts on stress tolerance and yield sustainability by using nanoparticles. Prior to large-scale application in agricultural systems, more research is needed to optimize formulations, delivery methods, and potential environmental impacts. In various oilseed crops, nanoparticles play a crucial role in mitigating abiotic stress by enhancing nutrient uptake, improving physiological attributes, and reducing oxidative stress, as shown in Table 16.1.

16.3 Future Perspectives

The integration of nanotechnology into oilseed crop production presents a promising frontier for enhancing abiotic stress tolerance, improving nutrient efficiency, and ensuring sustainable agricultural practices. However, several key areas require further research and development to fully harness the potential of nanomaterials in crop improvement.

16.3.1 Optimization of Nanoparticle Formulations

Future studies in this field should focus on developing NP formulations specific to particular oilseed crops and given conditions in the environment. The development

Table 16.1 Role of nanoparticles in mitigating abiotic stresses in oilseeds crops

Abiotic stress	Oilseed crop	Nanoparticle type	Mechanism of action	Reference
Drought	*Glycine max*	Selenium nanoparticles (Se NPs)	Enhance drought tolerance by improving antioxidant enzyme activities, reducing oxidative stress, and maintaining higher relative water content.	Zeeshan et al. (2024)
Drought	*Brassica napus*	Calcium nanoparticles (Ca-NPs)	Improve drought stress tolerance by enhancing water uptake and maintaining cellular functions.	Ayyaz et al. (2022)
Heavy metal contamination	*Brassica napus*	Sulfur nanoparticles (SNPs)	Alleviate copper (Cu) stress by reducing oxidative stress and enhancing plant growth.	Yuan et al. (2023)
Salinity	*Helianthus annuus* L.	Urea-functionalized hydroxyapatite nanoparticles (Urea-HANPs)	Enhance agronomic and physiobiochemical attributes, elevate chlorophyll content, boost antioxidant systems, and reduce osmolyte accumulation under salt stress.	Ullah et al. (2022)
Salinity	*Olea europaea*	Zinc oxide (ZnO) nanoparticles	Increase chlorophyll a and b content, enhancing photosynthetic efficiency and overall plant growth under saline conditions	Kumari et al. (2024)
Salinity	*Arachis hypogaea*	Silicon dioxide (SiO_2) NPs	Enhance plant growth under salt stress by improving nutrient uptake and reducing sodium accumulation	El-Saadony et al. (2022)
Nutrient deficiency	*Glycine max*	Iron oxide nanoparticles (Fe_3O_4 NPs)	Improve nutrient uptake efficiency, enhance chlorophyll content, and promote overall plant growth under nutrient-deficient conditions	Singh et al. (2024)
Waterlogging	*Helianthus annuus* L.	Silver (Ag) and aluminum oxide (Al_2O_3) NPs	Improve oxygen diffusion in roots, induce root porosity, and facilitate anaerobic respiration under hypoxic conditions.	Thabet and Alqudah (2024)
Drought	*Arachis hypogaea*	Zinc oxide (ZnO) NPs	Improve drought tolerance by enhancing antioxidant enzyme activities and maintaining higher relative water content	Tripathi et al. (2022)
Drought	*Olea europaea*	Iron oxide/graphitic carbon nitride (Fe_2O_3/g-C_3N_4) nanocomposites	Enhance water use efficiency, improve relative water content, stabilize cell membranes, and boost antioxidant capacity, leading to better growth and fruit quality under severe drought conditions	Gholami et al. (2024)
Salinity	*Brassica napus*	Calcium nanoparticles (Ca-NPs)	Enhance nutrient uptake and improve plant growth under saline conditions.	Ayyaz et al. (2022)

of biocompatible, eco-friendly, and biodegradable nanoparticles will curtail environmental concerns while enhancing efficiency. Green synthesis methods using plant extracts, fungi, and bacteria offer sustainable solutions to conventional NP production and minimize the risks of toxicity.

16.3.2 Mechanistic Understanding of NP–Plant Interactions

Most studies offer a clear assessment of the advantages of nanomaterials for oilseed crops, while an in-depth knowledge of their interaction at the level of molecular, physiological, and biochemical statuses is greatly solicited. Sophisticated imaging and molecular techniques on NPs like transcriptomics, proteomics, and metabolomics will clarify how nanomaterials interact with gene expression, enzyme activity, and metabolic pathways under conditions of stress.

16.3.3 Large-Scale Field Trials and Commercial Applications

Most recent research on the application of NPs on oilseed crops is mostly carried under controlled greenhouse conditions. Large-scale field trials must be conducted to confirm the efficacy of NPs in real agricultural conditions. Researchers, together with agricultural industries and policymakers, will have to collaborate to bridge the gap between lab results and their practical applications on farms.

16.3.4 Nanoenabled Smart Delivery Systems

The future direction should aim at developing nano-based intelligent delivery systems for fertilizers, pesticides, and stress-relieving agents. Mathematical models for controlled release through nanocarriers, polymeric nanoparticles, and nanoencapsulated agrochemicals can help improve the efficiency of nutrient uptake, reduce the frequency of application, and minimize environmental pollution.

16.3.5 Environmental and Food Safety Assessments

With the rapidly expanding use of nanotechnology in agriculture, a general consideration must be given to evaluate the environmental fate and bioaccumulation and potential toxicity of nanoparticles. Regulatory frameworks are to be made to ensure the safe and responsible NPs use in crop production without compromising food safety or soil health.

16.3.6 *Integration with Precision Agriculture and Digital Technologies*

In oilseed crops, the future of nanotechnology lies in the full development of integration into precision agriculture technologies like remote sensing, artificial intelligence (AI), and the Internet of Things (IoT). Nanosensors providing real-time monitoring of soil moisture levels, nutrient deficiencies, and abiotic stress levels would allow data-driven decisions, minimal resource use, and superior resilience to be made by farmers.

16.3.7 *Climate-Resilient Nanotechnology Strategies*

Although climate change poses an increasing challenge to global agriculture, future climate-resilient oilseed crop varieties should certainly be developed using nanotechnology-derived tools. Chief among the strategies will be the nano-based stress modulators, NP-assisted gene editing (CRISPR-Cas9), and engineered nanomaterials, which will improve photosynthesis efficiency under extreme conditions and are critical to sustaining productivity.

16.4 Conclusion

Nanotechnology has emerged as a powerful tool for improving oilseed crops' abiotic tolerance. As a result of the application of NPs in agriculture, plant resilience has been enhanced against salinity, drought, heavy metal toxicity, radiation, and ozone stress. Nanoparticles enhance nutrient uptake, regulate antioxidant defense mechanisms and stabilize cellular metabolism to sustain crop productivity in adverse environmental conditions. As a key advantage of nanotechnology, it can provide targeted and controlled nutrient delivery through nanofertilizers and nanoencapsulated growth regulators. As a result, nutrient absorption is maximized, losses are minimized and contamination of the environment is reduced. Furthermore, metal-based nanoparticles have been found to improve plants' health and yields of oil. The use of chitosan-based nanoparticles has also been shown to reduce oxidative stress caused by pollutants such as tropospheric ozone by scavenging ROS. Despite these promising benefits, there are still challenges in implementing nanotechnology on a large scale. For nanoparticles to be used safely, their long-term environmental and ecological effects need to be thoroughly investigated. A regulatory framework should be established to evaluate the toxic effects, bioaccumulation and soil health impacts of nanoparticles. Additionally, extensive field trials are needed to validate laboratory findings and optimize nanoparticle formulations. Precision agriculture, artificial intelligence and climate-smart farming strategies can revolutionize oilseed production in the future with nanotechnology. To achieve sustainable implementation, eco-friendly, biodegradable nanoparticles should be developed, and smart delivery systems should be enhanced. Nanotechnology can play an important role in ensuring food security and improving oilseed crops productivity in a rapidly changing climate by overcoming current challenges.

References

Abdel Latef AA, Alhmad MF, Abdelfattah KE (2017) The possible roles of priming with ZnO nanoparticles in mitigation of salinity stress in Lupine (*Lupinus termis*) plants. J Plant Growth Regul 36:60–70

Abdel Latef AA, Srivastava AK, Abdel-sadek MS, Kordrostam M, Tran L-SP (2018) Titanium dioxide nanoparticles improve growth and enhance tolerance of broad bean plants under saline conditions. Land Degrad Dev 29:1065–1073

Abdel Latef AA, Zaid A, Abu Alhmad MF, Abdelfattah KE (2020) The impact of priming with Al_2O_3 nanoparticles on growth, pigments, osmolytes, and antioxidant enzymes of Egyptian roselle (*Hibiscus sabdariffa* L.) cultivar. Agronomy 10:681

Abdelhameed RE, Abu-Elsaad NI, Abdel Latef AA, Metwally RA (2021) Tracking of zinc ferrite nanoparticle effects on pea (*Pisum sativum* L.) plant growth, pigments, mineral content and arbuscular Mmcorrhizal colonization. Plan Theory 10:583

Acharya BR, Gill SP, Kaundal A, Sandhu D (2024) Strategies for combating plant salinity stress: the potential of plant growth-promoting microorganisms. Front Plant Sci 15:1406913

Adrees M, Khan ZS, Ali S et al (2020) Simultaneous mitigation of cadmium and drought stress in wheat by soil application of iron nanoparticles. Chemosphere 238:124681

Adrees M, Khan ZS, Hafeez M et al (2021) Foliar exposure of zinc oxide nanoparticles improved the growth of wheat (Triticum aestivum L.) and decreased cadmium concentration in grains under simultaneous Cd and water deficient stress. Ecotoxicol Environ Saf 208:111627

Afzal M, Ullah S, Assad N et al (2024) Photocatalytic degradation of organic dyes and biological potentials of biogenic zinc oxide nanoparticles synthesized using the polar extract of *Cyperus scariosus* R.Br. (Cyperaceae). Green Process Synth 13:20240038. https://doi.org/10.1515/gps-2024-0038

Aguirre-Becerra H, Feregrino-Perez AA, Esquivel K et al (2022) Nanomaterials as an alternative to increase plant resistance to abiotic stresses. Front Plant Sci 13:1023636

Ahmad M, Waraich EA, Skalicky M et al (2021a) Adaptation strategies to improve the resistance of oilseed crops to heat stress under a changing climate: an overview. Front Plant Sci 12:767150

Ahmad Z, Anjum S, Skalicky M et al (2021b) Selenium alleviates the adverse effect of drought in oilseed crops camelina (Camelina sativa L.) and canola (Brassica napus L.). Molecules 26:1699

Ahmed T, Noman M, Manzoor N et al (2021) Nanoparticle-based amelioration of drought stress and cadmium toxicity in rice via triggering the stress responsive genetic mechanisms and nutrient acquisition. Ecotoxicol Environ Saf 209:111829

Al-Khayri JM, Rashmi R, Surya Ulhas R et al (2023) The role of nanoparticles in response of plants to abiotic stress at physiological, biochemical, and molecular levels. Plan Theory 12:292

Ameen M, Zia MA, Najeeb Alawadi HF et al (2024) Exogenous application of selenium on sunflower (Helianthus annuus L.) to enhance drought stress tolerance by morpho-physiological and biochemical adaptations. Frontiers. Plant Sci 15:1427420

Aslam A, Mahmood A, Ur-Rehman H et al (2023) Plant adaptation to flooding stress under changing climate conditions: ongoing breakthroughs and future challenges. Plan Theory 12: 3824

Assad N, Naeem-ul-Hassan M, Ajaz Hussain M et al (2023) Diffused sunlight assisted green synthesis of silver nanoparticles using *Cotoneaster nummularia* polar extract for antimicrobial and wound healing applications. Nat Prod Res:1–15. https://doi.org/10.1080/14786419.2023.2295936

Assad N, Abbas A, ur Rehman MF, Naeem-ul-Hassan M (2024) Photo-catalytic and biological applications of phyto-functionalized zinc oxide nanoparticles synthesized using a polar extract of Equisetum diffusum D. RSC Adv 14:22344–22358

Ayyaz A, Fang R, Ma J et al (2022) Calcium nanoparticles (Ca-NPs) improve drought stress tolerance in Brassica napus by modulating the photosystem II, nutrient acquisition and antioxidant performance. NanoImpact 28:100423

Cao Y, Turk K, Bibi N et al (2025) Nanoparticles as catalysts of agricultural revolution: enhancing crop tolerance to abiotic stress: a review. Front Plant Sci 15:1510482

Chandrashekar HK, Singh G, Kaniyassery A et al (2023) Nanoparticle-mediated amelioration of drought stress in plants: a systematic review. 3 Biotech 13:336. https://doi.org/10.1007/s13205-023-03751-4

Dev P, Jaiswal S, Goswami S, Singh SK (2025) Comparative study of various nanofertilizers and their impact on crop yield. In: Kumar P, Dubey RC (eds) Nanofertilizers for sustainable agriculture. Springer Nature Switzerland, Cham, pp 451–471

Ehsan A, Sultana N, Munir B et al (2024) Assessment of some heavy metals for their potential health implications in the skin whitening creams available in Pakistani cosmetic market. Int J Environ Anal Chem:1–13. https://doi.org/10.1080/03067319.2024.2408636

El-Saadony MT, Saad AM, Soliman SM et al (2022) Role of nanoparticles in enhancing crop tolerance to abiotic stress: a comprehensive review. Front Plant Sci 13:946717

El-Sayed M, Ayoub HA, Helal M et al (2025) Nanotechnology-enabled soil management for sustainable agriculture: interactions, challenges, and prospective. Environ Sci Nano 12:2128

Fahad S, Bajwa AA, Nazir U et al (2017) Crop production under drought and heat stress: plant responses and management options. Front Plant Sci 8:1147

Faizan M, Sehar S, Rajput VD et al (2021) Modulation of cellular redox status and antioxidant defense system after synergistic application of zinc oxide nanoparticles and salicylic acid in rice (Oryza sativa) plant under arsenic stress. Plan Theory 10:2254

Faizan M, Alam P, Rajput VD et al (2023) Nanoparticle mediated plant tolerance to heavy metal stress: what we know? Sustain For 15:1446

Fang C, Hamilton JP, Vaillancourt B et al (2023) Cold stress induces differential gene expression of retained homeologs in Camelina sativa cv Suneson. Front Plant Sci 14:1271625

Ghaffar S, Abbas A, Naeem-ul-Hassan M et al (2023) Improved photocatalytic and antioxidant activity of olive fruit extract-mediated ZnO nanoparticles. Antioxidants 12:1201

Ghaffari M, Gholizadeh A, Rauf S, Shariati F (2023) Drought-stress induced changes of fatty acid composition affecting sunflower grain yield and oil quality. Food Sci Nutr 11:7718–7731. https://doi.org/10.1002/fsn3.3690

Gholami R, Fahadi Hoveizeh N, Zahedi SM et al (2024) Nanostructure-assisted drought tolerance in olive trees (Olea europaea L.): the role of Fe2O3-graphitic carbon. Frontiers. Plant Sci 15: 1454619

Ghorbani A, Emamverdian A, Pehlivan N et al (2024) Nano-enabled agrochemicals: mitigating heavy metal toxicity and enhancing crop adaptability for sustainable crop production. J Nanobiotechnol 22:91. https://doi.org/10.1186/s12951-024-02371-1

Ghorbanpour M, Mohammadi H, Kariman K (2020) Nanosilicon-based recovery of barley (Hordeum vulgare) plants subjected to drought stress. Environ Sci Nano 7:443–461

Guo S, Hu X, Yu F, Mu L (2024) Heat waves coupled with nanoparticles induce yield and nutritional losses in Rice by regulating stomatal closure. ACS Nano 18:14276–14289. https://doi.org/10.1021/acsnano.3c13165

Hashmi SS, Ibrahim M, Adnan M et al (2024) Green synthesis of silver nanoparticles from *Olea europaea* L. extracted polysaccharides, characterization, and its assessment as an antimicrobial agent against multiple pathogenic microbes. Open. Chemistry 22:20240016. https://doi.org/10.1515/chem-2024-0016

Henry W, Hak WH (2012) Effects of salinity on fresh fruit bunch (FFB) production and oil-to-bunch ratio of oil palm (Elaeis guineensis) planted in reclaimed mangrove swamp areas in Sabah. Oil Palm Bull 65:12–20

Hernández-Hernández H, González-Morales S, Benavides-Mendoza A et al (2018) Effects of chitosan–PVA and Cu nanoparticles on the growth and antioxidant capacity of tomato under saline stress. Molecules 23:178

Hladni N, Jan CC, Jockovic M, et al (2022) Genomic designing for abiotic stress resistant oilseed crops. Springer, Cham

Hofmann T, Lowry GV, Ghoshal S et al (2020) Technology readiness and overcoming barriers to sustainably implement nanotechnology-enabled plant agriculture. Nature Food 1:416–425

Hussain SB, Zahra S, Hussain S et al (2025) Characterization and comparative antibacterial activities of gold nanoparticles synthesized by *Tagetes patula* L. (Asteraceae) Flower Extract. Microsc Res Tech jemt.24829:1869. https://doi.org/10.1002/jemt.24829

Ilyas M, Maqsood MF, Shahbaz M et al (2024) Alleviating salinity stress in canola (Brassica napus L.) through exogenous application of salicylic acid. BMC Plant Biol 24:611. https://doi.org/10.1186/s12870-024-05314-y

Jabbar A, Abbas A, Assad N et al (2023) A highly selective Hg 2+ colorimetric sensor and antimicrobial agent based on green synthesized silver nanoparticles using Equisetum diffusum extract. RSC Adv 13:28666–28675

Jaleel CA, Changxing Z, Jayakumar K, Iqbal M (2009) Low concentration of cobalt increases growth, biochemical constituents, mineral status and yield in Zea mays. J Sci Res 1:128–137

Jian H, Xie L, Wang Y et al (2020) Characterization of cold stress responses in different rapeseed ecotypes based on metabolomics and transcriptomics analyses. PeerJ 8:e8704

Junedi MA, Mukhopadhyay R, Manjari KS (2023) Alleviating salinity stress in crop plants using new engineered nanoparticles (ENPs). Plant Stress 9:100184

Jurkonienė S, Mockevičiūtė R, Gavelienė V et al (2023) Proline enhances resistance and recovery of oilseed rape after a simulated prolonged drought. Plan Theory 12:2718

Khalid MF, Iqbal Khan R, Jawaid MZ et al (2022) Nanoparticles: the plant saviour under abiotic stresses. Nanomaterials 12:3915

Khalofah A, Kilany M, Migdadi H (2021) Phytostimulatory influence of Comamonas testosteroni and silver nanoparticles on Linum usitatissimum L. under salinity stress. Plan Theory 10:790

Khan ZUR, Assad N, Naeem-ul-Hassan M et al (2023) Aconitum lycoctonum L. (Ranunculaceae) mediated biogenic synthesis of silver nanoparticles as potential antioxidant, anti-inflammatory, antimicrobial and antidiabetic agents. BMC Chem 17:128. https://doi.org/10.1186/s13065-023-01047-5

Khan MN, Kaplan A, Iqbal M et al (2025) Cadmium dynamics: beneficial elements and chemical reactions in soil. In: Beneficial elements for remediation of heavy metals in polluted soil. Elsevier, pp 93–139

Kiany T, Pishkar L, Sartipnia N et al (2022) Effects of silicon and titanium dioxide nanoparticles on arsenic accumulation, phytochelatin metabolism, and antioxidant system by rice under arsenic toxicity. Environ Sci Pollut Res 29:34725–34737

Kumar P, Alamri SA, Alrumman SA et al (2022) Foliar use of TiO2-nanoparticles for okra (Abelmoschus esculentus L. Moench) cultivation on sewage sludge–amended soils: biochemical response and heavy metal accumulation. Environ Sci Pollut Res 29:66507–66518

Kumari A, Gupta AK, Sharma S et al (2024) Nanoparticles as a tool for alleviating plant stress: mechanisms, implications, and challenges. Plan Theory 13:1528

Lei Y, He H, Raza A et al (2022) Exogenous melatonin confers cold tolerance in rapeseed (*Brassica napus* L.) seedlings by improving antioxidants and genes expression. Plant Signal Behav 17:2129289. https://doi.org/10.1080/15592324.2022.2129289

Linh TM, Mai NC, Hoe PT et al (2020) Metal-based nanoparticles enhance drought tolerance in soybean. J Nanomater 2020:1–13. https://doi.org/10.1155/2020/4056563

Liu J, Li G, Chen L et al (2021) Cerium oxide nanoparticles improve cotton salt tolerance by enabling better ability to maintain cytosolic K+/Na+ ratio. J Nanobiotechnol 19:153. https://doi.org/10.1186/s12951-021-00892-7
Liu C, Jiang X, Yuan Z (2024) Plant responses and adaptations to salt stress: a review. Horticulturae 10:1221
Magno L, Avice J-C, Morvan-Bertrand A et al (2022) Impacts of heat stress on yield and seed quality in oilseed rape: analysis of the dynamic development of seed storage compounds. In: XXXI international horticultural congress (IHC2022): international symposium on integrative approaches to product quality in 1353, pp 203–210
Majhi S, Sikdar M (2023) How heavy metal stress affects the growth and development of pulse crops: insights into germination and physiological processes. 3 Biotech 13:155
Mangin B, Casadebaig P, Cadic E et al (2017) Genetic control of plasticity of oil yield for combined abiotic stresses using a joint approach of crop modelling and genome-wide association. Plant Cell Environ 40:2276–2291
Mao T, Bao L, Zhang H et al (2024) Mn3O4 nanoenzyme seed soaking enhanced salt tolerance in soybean through modulating homeostasis of reactive oxygen species and ATPase activities. Plan Theory 13:3011
Marinho FJL, Ferreira Neto M, Gheyi HR et al (2021) Use of saline water in irrigation of coconut palm (Cocus nucifera L.). Rev Bras Eng Agríc Ambient 9:359–364
Maryum Z, Luqman T, Nadeem S et al (2022) An overview of salinity stress, mechanism of salinity tolerance and strategies for its management in cotton. Front Plant Sci 13:907937
Nabloussi A, Bahri H, Lakbir M et al (2019) Assessment of a set of rapeseed (Brassica napus L.) varieties under waterlogging stress at different plant growth stages. OCL 26:36
Nadakuduti SS, Laforest LC, Tachev M et al (2023) Heat stress during seed development leads to impaired physiological function and plasticity in seed oil accumulation in Camelina sativa. Front Plant Sci 14:1284573
Nasircilar AG, Ulukapi K, Topcuoglu B et al (2024) Salt and heavy metal stress responses and metal uptake potentials of some leafy vegetables. Agrosystems Geosci Environ 7:e20487. https://doi.org/10.1002/agg2.20487
Naveed M (2012) Potential of rhizobia for sustainable production of non-legumes. In: Crop production for Agriculture improvement
Nawaz A, Rehman HU, Usman M et al (2023) Nanobiotechnology in crop stress management: an overview of novel applications. Discover Nano 18:74. https://doi.org/10.1186/s11671-023-03845-1
Nazari F, Alizadeh AM, Alikord M, et al (2023) Impacts of heavy metals in seed crops and oil seed on human health: a treat to food safety-review. Carpathian J Food Sci Technol 15(2):106–124
Nowroz F, Hasanuzzaman M, Siddika A et al (2024) Elevated tropospheric ozone and crop production: potential negative effects and plant defense mechanisms. Front Plant Sci 14: 1244515
Oliveira HC, Gomes BC, Pelegrino MT, Seabra AB (2016) Nitric oxide-releasing chitosan nanoparticles alleviate the effects of salt stress in maize plants. Nitric Oxide 61:10–19
Omar S, Salim H, Eldenary M et al (2023) Ameliorating effect of nanoparticles and seeds' heat pre-treatment on soybean plants exposed to sea water salinity. Heliyon 9:e21446
Pagano MC, Miransari M (2016) The importance of soybean production worldwide. In: Abiotic and biotic stresses in soybean production. Elsevier, pp 1–26
Paliani S, Ratiani L (2025) Nanotechnology as a transformer of the agro-sector. Georgian Scientists 7:251–256
Ploschuk RA, Miralles DJ, Colmer TD et al (2018) Waterlogging of winter crops at early and late stages: impacts on leaf physiology, growth and yield. Front Plant Sci 9:1863
Poudel S, Vennam RR, Shrestha A et al (2023) Resilience of soybean cultivars to drought stress during flowering and early-seed setting stages. Sci Rep 13:1277

Prasad TNVKV, Swethasree M, Satisha GC et al (2023) Nanoparticulate silica internalization and its effect on the growth and yield of groundnut (*Arachis hypogaea* L.). Environ Sci Technol 57: 5881–5890. https://doi.org/10.1021/acs.est.3c00327

Rajput VD, Minkina T, Kumari A et al (2021) Coping with the challenges of abiotic stress in plants: new dimensions in the field application of nanoparticles. Plan Theory 10:1221

Raza MAS, Shahid AM, Saleem MF, et al (2017) Effects and management strategies to mitigate drought stress in oilseed rape (Brassica napus L.): a review. Zemdirbyste-Agriculture 104(1): 85–94

Raza A, Su W, Jia Z et al (2022) Mechanistic insights into trehalose-mediated cold stress tolerance in rapeseed (Brassica napus L.) seedlings. Front Plant Sci 13:857980

Rui M, Ma C, White JC et al (2018) Metal oxide nanoparticles alter peanut (Arachis hypogaea L.) physiological response and reduce nutritional quality: a life cycle study. Environ Sci Nano 5: 2088–2102

Safdar ME, Qamar R, Javed A et al (2023) Combined application of boron and zinc improves seed and oil yields and oil quality of oilseed rape (Brassica napus L.). Agronomy 13:2020

Salaheldin TA, Mahdi AA, Hendawey MH, Hassan NS (2024) The copper metal and magnetite nanoparticles conjugated with salicylic acid composite stimulated wheat defense mechanism and affected cellular components under heat stress. CNM 9:224–238. https://doi.org/10.2174/2405461508666230606123103

Salem W, El-Deen FN, Ebnalwaleed K, Badry M, Abdel Latef AA (2022) UV-induced antibacterial activity of green-synthesized TiO_2 nanoparticles for the potential reuse of raw surface and underground water. J Plant Growth Regul 41(3):1344–1358

Samanta S, Roychoudhury A (2021) Recent trend in nanoparticle research in regulating arsenic bioaccumulation and mitigating arsenic toxicity in plant species. J Plant Biochem Biotechnol 30:793–812

Santos-Espinoza AM, González-Mendoza D, Ruiz-Valdiviezo VM et al (2020) Changes in the physiological and biochemical state of peanut plants (*Arachis hypogaea* L.) induced by exposure to green metallic nanoparticles. Int J Phytoremediation:1–8. https://doi.org/10.1080/15226514.2020.1856037

Sehgal A, Sita K, Siddique KH et al (2018) Drought or/and heat-stress effects on seed filling in food crops: impacts on functional biochemistry, seed yields, and nutritional quality. Front Plant Sci 9: 1705

Shaffique S, Hussain S, Kang S-M et al (2023) Recent progress on the microbial mitigation of heavy metal stress in soybean: overview and implications. Front Plant Sci 14:1188856

Sharifi P, Bidabadi SS, Zaid A, Abdel Latef AA (2021) Efficacy of multi-walled carbon nanotubes in regulating growth performance, total glutathione and redox state of *Calendula officinalis* L. cultivated on Pb and Cd polluted soil. Ecotoxicol Environ Saf 213:112051

Shelar PV, Mankar GD, Sontakke OP et al (2024) A review of the physio-biochemical and molecular mechanisms of salt tolerance in crop. Curr Agric Res J 12:545

Shukla K, Mishra V, Singh J et al (2024) Nanotechnology in sustainable agriculture: a double-edged sword. J Sci Food Agric 104:5675–5688. https://doi.org/10.1002/jsfa.13342

Siddiqui MH, Al-Whaibi MH, Faisal M, Al Sahli AA (2014) Nano-silicon dioxide mitigates the adverse effects of salt stress on Cucurbita pepo L. Environ Toxicol Chem 33:2429–2437

Singh A, Sengar RS, Rajput VD et al (2022) Zinc oxide nanoparticles improve salt tolerance in rice seedlings by improving physiological and biochemical indices. Agriculture 12:1014

Singh A, Sharma A, Singh O et al (2024) In-depth exploration of nanoparticles for enhanced nutrient use efficiency and abiotic stresses management: present insights and future horizons. Plant Stress 100576:100576

Thabet SG, Alqudah AM (2024) Unraveling the role of nanoparticles in improving plant resilience under environmental stress condition. Plant Soil 503:313–330. https://doi.org/10.1007/s11104-024-06581-2

Thakur N, Yadav AN (2025) Nanotechnology in agriculture: a review on precision farming and sustainable crop production. BioNanoScience 15:243

Tripathi DK, Singh S, Gaur S et al (2018) Acquisition and homeostasis of iron in higher plants and their probable role in abiotic stress tolerance. Front Environ Sci 5:86

Tripathi D, Singh M, Pandey-Rai S (2022) Crosstalk of nanoparticles and phytohormones regulate plant growth and metabolism under abiotic and biotic stress. Plant Stress 6:100107

Ul Hassan Z, Ali S, Ahmad R et al (2017) Biochemical and molecular responses of oilseed crops to heavy metal stress. In: Ahmad P (ed) Oilseed crops, 1st edn. Wiley, pp 236–248

Ullah R, Sher S, Muhammad Z et al (2022) Modulating response of sunflower (*Hellianthus annuus*) to induced salinity stress through application of engineered urea functionalized hydroxyapatite nanoparticles. Microscopy Resand Technique 85:244–252. https://doi.org/10.1002/jemt.23900

Valliyodan B, Ye H, Song L et al (2017) Genetic diversity and genomic strategies for improving drought and waterlogging tolerance in soybeans. J Exp Bot 68:1835–1849

Waraich EA, Ahmad R, Ashraf MY (2011) Role of mineral nutrition in alleviation of drought stress in plants. Aust J Crop Sci 5:764–777

Wu X, Hu J, Wu F et al (2021) Application of TiO2 nanoparticles to reduce bioaccumulation of arsenic in rice seedlings (Oryza sativa L.): a mechanistic study. J Hazard Mater 405:124047

Wu H, Du P, Miao X et al (2024) O-Carboxymethyl chitosan nanoparticles: a novel approach to enhance water stress tolerance in maize seedlings. Int J Biol Macromol 277:134459

Yashveer S, Redhu NS, Singh V, et al (2022) Nanoparticles in agriculture: characterization, uptake and role in mitigating heat stress. Human Health 2(2):160–181

You T, Liu D, Chen J et al (2018) Effects of metal oxide nanoparticles on soil enzyme activities and bacterial communities in two different soil types. J Soils Sediments 18:211–221

Yuan H, Liu Q, Fu J et al (2023) Co-exposure of sulfur nanoparticles and Cu alleviate Cu stress and toxicity to oilseed rape Brassica napus L. J Environ Sci 124:319–329

Zeeshan M, Wang X, Salam A et al (2024) Selenium nanoparticles boost the drought stress response of soybean by enhancing pigment accumulation, oxidative stress management and ultrastructural integrity. Agronomy 14:1372

Zulfiqar HF, Afroze B, Shakoor S, et al (2024) Nanoparticles in agriculture: enhancing crop resilience and productivity against abiotic stresses. IntechOpen, London

The manufacturer's authorised representative in the EU is Springer Nature Customer Service Centre GmbH, Europaplatz 3, 69115 Heidelberg, Germany. If you have any concerns regarding our products, please contact ProductSafety@springernature.com

Printed and bound by CPI Group (UK) Ltd, Croydon, CR0 4YY
01/10/2026
02225923-0019